Lecture Notes in Computer Science 5495

Commenced Publication in 1973
Founding and Former Series Editors:
Gerhard Goos, Juris Hartmanis, and Jan van Leeuwen

Editorial Board

David Hutchison
   Lancaster University, UK

Takeo Kanade
   Carnegie Mellon University, Pittsburgh, PA, USA
Josef Kittler
   University of Surrey, Guildford, UK
Jon M. Kleinberg
   Cornell University, Ithaca, NY, USA
Alfred Kobsa
   University of California, Irvine, CA, USA
Friedemann Mattern
   ETH Zurich, Switzerland
John C. Mitchell
   Stanford University, CA, USA
Moni Naor
   Weizmann Institute of Science, Rehovot, Israel
Oscar Nierstrasz
   University of Bern, Switzerland
C. Pandu Rangan
   Indian Institute of Technology, Madras, India
Bernhard Steffen
   University of Dortmund, Germany
Madhu Sudan
   Microsoft Research, Cambridge, MA, USA
Demetri Terzopoulos
   University of California, Los Angeles, CA, USA
Doug Tygar
   University of California, Berkeley, CA, USA
Gerhard Weikum
   Max-Planck Institute of Computer Science, Saarbruecken, Germany

Mikko Kolehmainen   Pekka Toivanen
Bartlomiej Beliczynski (Eds.)

Adaptive and Natural Computing Algorithms

9th International Conference, ICANNGA 2009
Kuopio, Finland, April 23-25, 2009
Revised Selected Papers

 Springer

Volume Editors

Mikko Kolehmainen
Department of Environmental Sciences, University of Kuopio
PO Box 1627, 70211 Kuopio, Finland
E-mail: mikko.kolehmainen@uku.fi

Pekka Toivanen
Department of Computer Science, University of Kuopio
PO Box 1627, 70211 Kuopio, Finland
E-mail: pekka.toivanen@uku.fi

Bartlomiej Beliczynski
Institute of Control and Industrial Electronics, Warsaw University of Technology
ul. Koszykowa 75, 00-662, Warszawa, Poland
E-mail: B.Beliczynski@ee.pw.edu.pl

Library of Congress Control Number: 2009935814

CR Subject Classification (1998): F.1, F.2, J.3, I.2, D.2.2

LNCS Sublibrary: SL 1 – Theoretical Computer Science and General Issues

ISSN   0302-9743
ISBN-10   3-642-04920-6 Springer Berlin Heidelberg New York
ISBN-13   978-3-642-04920-0 Springer Berlin Heidelberg New York

This work is subject to copyright. All rights are reserved, whether the whole or part of the material is
concerned, specifically the rights of translation, reprinting, re-use of illustrations, recitation, broadcasting,
reproduction on microfilms or in any other way, and storage in data banks. Duplication of this publication
or parts thereof is permitted only under the provisions of the German Copyright Law of September 9, 1965,
in its current version, and permission for use must always be obtained from Springer. Violations are liable
to prosecution under the German Copyright Law.

springer.com

© Springer-Verlag Berlin Heidelberg 2009
Printed in Germany

Typesetting: Camera-ready by author, data conversion by Scientific Publishing Services, Chennai, India
Printed on acid-free paper   SPIN: 12729647   06/3180   5 4 3 2 1 0

Preface

The ICANNGA series of conferences has been organized since 1993 and has a long history of promoting the principles and understanding of computational intelligence paradigms within the scientific community. Originally ICANNGA stood for "International Conference on Artificial Neural Networks and Genetic Algorithms," but in 2005 the conference was renamed to "International Conference on Adaptive and Natural Computing Algorithms," while keeping the acronym ICANNGA. The first ICANNGA conference was held in Innsbruck Austria (1993), then Alés in France (1995), Norwich in the UK (1997), Portoroz in Slovenia (1999), Prague in the Czech Republic (2001), Roanne in France (2003), Coimbra in Portugal (2005) and Warsaw in Poland (2007). Continuing this European tradition, the 9th ICANNGA was held in Kuopio, Finland (2009). The vast majority of ICANNGA conferences is organized by and based at a university.

Drawing on the experience of previous events and following the same general model, ICANNGA 2009 combined plenary lectures and technical sessions. Apart from being a widely recognized conference, it enhanced the possibility to exchange opinions through lectures and discussions, provided a great opportunity to meet new colleagues, as well as to renew old friendships and to facilitate the possibilities for international collaborations. As previously, the conference proceedings are published in the Springer LNCS series.

The size of the ICANNGA conference has varied as has the popularity of different topics over the years. This year we had a compact conference. There were 112 papers submitted, which went through a peer-review process by at least two reviewers. Out of these papers 63 were presented. We are confident that it is an optimal number of papers sustaining the high quality of the ICANNGA conference and supporting the publication standard of Springers LNCS. The most popular topics were: neural networks (18), evolutionary computation (15) and learning (13). From these papers about 50% could be classified as applications, and that actually also holds for the submitted papers.

The ICANNGA 2009 plenary lectures were focused on biological and human phenomena. They were presented by distinguished scientists: adaptive modeling of linguistics and social phenomena (Timo Honkela), artificial and cultured neural networks (Kevin Warwick), digital media (Lars Kai Hansen) and systems biology (David Broomhead).

Besides the scientific program, we had the pleasure to organize social events that mirrored the roots of ICANNGA and the local culture. The organizing team was especially happy with the way the conference participants took part in the traditional Finnish events.

All those people who made this conference possible deserve our gratitude. The members of the Organizing Committee worked one and a half years to create this conference edition: Mikko Kolehmainen, Pekka Toivanen, Yrjö Hiltunen, Erkki

Pesonen, Niina Päivinen and Mauno Rönkkö. Especially, we would like to thank our Conference Secretary Karin Koivisto for her tremendous work. Our technical team kept the wheels rolling during the conference and it consisted of the following persons: Teemu Räsänen, Harri Niska, Jukka-Pekka Skön, Jarkko Tiirikainen, Juha Parviainen, Taneli Laavola, Okko Kauhanen, and Marko Jäntti.

We would like to thank the Advisory Committee for giving us this unique opportunity to organize the ICANNGA conference and their continuous support. We are grateful to the Program Committee and reviewers of the conference for their substantial work in refereeing the papers.

We also wish to thank our publisher, especially Alfred Hofmann, Editor-in Chief of LNCS, and Anna Kramer for their support and collaboration.

On behalf of the Organizing Committee we would like to thank all of you who participated in ICANNGA 2009 and contributed to its success.

May 2009   Mikko Kolehmainen
                                                              Pekka Toivanen
                                                          Bartlomiej Beliczynski

Organization

Organizers

ICANNGA 2009 was organized by the Department of Environmental Sciences, Environmental Informatics, and Department of Computer Science, University of Kuopio, Finland.

Advisory Committee

Rudolf Albrecht	University of Innsbruck, Austria
Bartlomiej Beliczynski	Warsaw University of Technology, Poland
Andrej Dobnikar	University of Ljubljana, Slovenia
Mikko Kolehmainen	University of Kuopio, Finland
Věra Kůrková	Academy of Sciences of the Czech Republic
David W. Pearson	Jean Monnet University of Saint-Etienne, France
Bernardete Ribeiro	University of Coimbra, Portugal
Nigel Steele	Coventry University, UK

Program Committee

Mikko Kolehmainen, Finland (Chair)
Jarmo Alander, Finland
Rudolf Albrecht, Austria
Bartlomiej Beliczynski, Poland
Andrej Dobnikar, Slovenia
Marco Dorigo, Belgium
Antonio Dourado, Portugal
Andrzej Dzielinski, Poland
Yrjö Hiltunen, Finland
Jaakko Hollmén, Finland
Osamu Hoshino, Japan
Marcin Iwanowski, Poland
Sirkka-Liisa Jämsä-Jounela, Finland
Martti Juhola, Finland
Esko Juuso, Finland
Janusz Kacprzyk, Poland
Tadeusz Kaczorek, Poland
Paul C. Kainen, USA
Helen Karatza, Greece

Kostas Karatzas, Greece
Ali Karci, Turkey
Tommi Kärkkäinen, Finland
Mario Koeppen, Germany
Jozef Korbicz, Poland
Věra Kůrková, Czech Republic
Pedro Larranaga, Spain
Kauko Leiviskä, Finland
Amaury Lendasse, Finland
Joachim Marques de Sá, Portugal
Francesco Masulli, Italy
Roman Neruda, Czech Republic
Stanislaw Osowski, Poland
Seppo Ovaska, Finland
Jussi Parkkinen, Finland
Nikola Pavesic, Slovenia
David Pearson, France
Bernardete Ribeiro, Portugal
Henrik Saxén, Finland

Catarina Silva, Portugal
Adam Slowik, Poland
Nigel Steele, UK
Miroslaw Swiercz, Poland
Ryszard Tadeusiewicz, Poland

Tatiana Tambouratzis, Greece
Ari Visa, Finland
Kevin Warwick, UK
Garry Wong, Finland

Organizing Committee

Mikko Kolehmainen (Chair)
Bartlomiej Beliczynski (Past Chair)
Pekka Toivanen (Scientific Program)
Yrjö Hiltunen (Scientific Program)
Erkki Pesonen (Reviewing Process)
Mauno Rönkkö (Website)
Niina Päivinen (Printed Material)
Karin Koivisto (Conference Secretariat)

Reviewers

Jarmo Alander
Rudolf Albrecht
Bartlomiej Beliczynski
Andrej Dobnikar
Antonio Dourado
Andrzej Dzielinski
Nuno Fonseca
Jorge Henriques
Yrjö Hiltunen
Jaakko Hollmén
Osamu Hoshino
Marcin Iwanowski
Sirkka-Liisa Jämsä-Jounela
Martti Juhola
Esko Juuso
Janusz Kacprzyk
Tadeusz Kaczorek
Helen Karatza
Kostas Karatzas
Ali Karci
Tommi Kärkkäinen
Mario Koeppen
Mikko Kolehmainen
Jozef Korbicz
Mikko Korpela
Marek Kowal

Věra Kůrková
Pedro Larranaga
Kauko Leiviskä
Amaury Lendasse
Frank Liu
Noel Lopes
Andrzej Marciniak
Joachim Marques de Sá
Francesco Masulli
Roman Neruda
Harri Niska
Stanislaw Osowski
Seppo Ovaska
Niina Päivinen
Jussi Parkkinen
Nikola Pavesic
David Pearson
Erkki Pesonen
Przemyslaw Pretki
Bernardete Ribeiro
Mauno Rönkkö
Henrik Saxén
Catarina Silva
Krzysztof Siwek
Adam Slowik
Nigel Steele

Mika Sulkava
Miroslaw Swiercz
Ryszard Tadeusiewicz
Tatiana Tambouratzis
Pekka Toivanen

Janne Toivola
Armando Vieira
Ari Visa
Kevin Warwick
Garry Wong

Table of Contents

Neural Networks

Evolutionary Computation

Learning

Soft Computing

Bioinformatics

Applications

Automatic Discriminative Lossy Binary Conversion of Redundant Real Training Data Inputs for Simplifying an Input Data Space and Data Representation

Adrian Horzyk

AGH University of Science and Technology, Department of Automatics
Mickiewicza Av. 30, 30-059 Cracow, Poland
horzyk@agh.edu.pl
http://home.agh.edu.pl/~horzyk

Abstract. Many times we come across the need to simplify or reduce an input data space in order to achieve a better model or better performance of an artificial intelligence solution. The well known PCA, ICA and rough sets can simplify and reduce input data space but they cannot transform real input data vectors into binary ones. Binary training vectors can simplify a training process of neural networks and let them to construct more compact topologies. This paper introduces a new algorithm that reduces input data space and simultaneously automatically lossy transforms real input training data vectors into binary vectors so that they do not lose their discrimination properties. The problem is how to effectively transform real input training data vectors into binary vectors so that an input data space could be simplified and the transformed binary vectors would be enough representative to be able to discriminate all training samples of all classes correctly? The described lossy conversion makes possible to achieve better generalization results for various soft-computing algorithms, can be widely used and avoids the curse of dimensionality problem. This paper introduces a new Automatic Discriminative Lossy Binary Conversion Algorithm (ADLBCA) that is able to solve all these tasks. Generally, no other method can simultaneously and so fast do all these tasks.

1 Introduction

The adaptation processes of various soft-computing algorithms, especially the training process of neural networks, are usually preceded by certain transformations or conversions of training data (TD). It is convenient to simplify an input data space by reducing dispensable inputs before a soft-computing model is created and trained. Almost all soft-computing algorithms have some limitations on input or output values that should be binarised, normalized or narrowed to certain specific ranges of values [1], [2], [3], [4], [6]. The quality of soft-computing training results is also strongly influenced by a preprocessing method of TD. Sometimes, real input vectors have to be converted to binary vectors in order

M. Kolehmainen et al. (Eds.): ICANNGA 2009, LNCS 5495, pp. 1–10, 2009.
© Springer-Verlag Berlin Heidelberg 2009

to make data suitable for a soft-computing training method or to achieve better generalization results. Moreover, an input data space should be reduced in order to construct more compact neural network topologies faster or to achieve better performance and generalization properties.

An input data space can be simplified using the well known PCA, ICA or rough sets [5], [1], [7], [9]. These algorithms can simplify an input data space but cannot perform any binary transformations on TD. Moreover, these methods have the higher computational cost than the introduced ADLBCA. The real input data vectors can be simply converted to binary vectors dividing input data into certain smooth ranges of their real values using various density of ranges as shown in figure 1. In this way, real inputs of each range are transformed into a single binary feature. The problem is how to set up these ranges? What size and density of ranges should be chosen (fig. 1)? The simple smooth ranges can be established experimentally or using various soft-computing algorithms to find out a more suitable set of ranges for some given input data. This solution is very laborious and does not warrant discrimination of converted input data. Another solution is to transform each real input value into a single binary input, but this algorithm produces a huge number of binary inputs that will neither provide a satisfactory generalization nor enable a NN construction algorithm to create a compact NN topology. Furthermore, the unsuitable conversion can spoil generalization properties and results of a later constructed neural network or another soft-computing solution.

The described method has been inspired by the way human eyes process luminous input data into various frequencies of binary spikes that can be afterwards processed by our central nervous system. The eye neuron dendrite projections merge one another, so the same luminous information is usually transformed and processed by more than a single neuron. Many eye neurons convert a wide range of luminous input signals [2], [4], [8]. The similar transformation has been used in the described method in this paper. Ranges of real inputs can merge one another, i.e. they do not need to be disjunctive. A single lossy binarising neuron can gather information from a wide range of input data values (fig. 3). The lossy binarising ranges are selected in such a way that training samples of all classes can be univocally discriminated only if TD are not contradictory. Moreover, these ranges should be computed in such a way that their number can be minimized, an input data space can be simplified and a generalization quality can be maximized. Many neural networks compete for better generalization using various training strategies, neuron functions, numbers of layers, interconnections and initiations of weight parameters. The generalization results depend also on an input data space representation, e.g. the ranges of real input values that neurons can cover and cost-effectively represent. The described algorithm can make many known training processes much easier when the real training input vectors are transformed into binary vectors that represent wide ranges of real input ranges ([1], [2], [3], [6]).

This paper describes an Automatic Discriminative Lossy Binary Conversion Algorithm (ADLBCA) dedicated to various classification tasks and

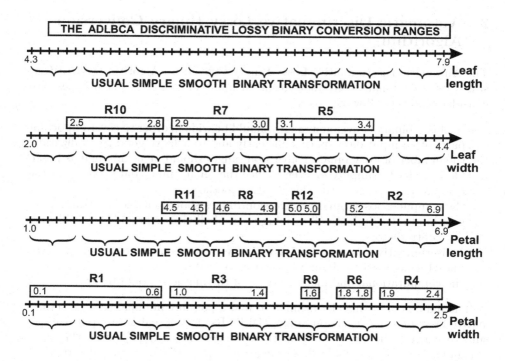

Fig. 1. The comparison of the simple smooth and described ADLBCA transformations of real input values into binary ones for the Iris data from the ML Repository

soft-computing methods which require or prefer binary input vectors. The mentioned thesis is illustrated by means of the Iris, Wine and Heart data from the ML Repository. The Iris data set has been used to demonstrate (figs. 1, 4, 6) how this method transforms and evaluates input data using a special algorithm for a selection of input data ranges.

The presented algorithm cannot be easy compared with other algorithms because there is no other algorithm that is be able to automatically transform real vectors into binary ones, simultaneously simplifying an input data space and warrant discrimination of converted input data. A real input data dimension is hardly comparable with a binary data dimension because these two data spaces have a very different influence on a training process, a topology construction and network elements, e.g. neuron functions. The binary input features have been computed in order to simplify next computations that can be processed by various soft-computing algorithms (tab. 1). On the other hand, simple smooth binary transformations (fig. 1) usually produce many redundant input binary features and cannot warrant discrimination after conversion. The usefulness and performance of the ADLBCA are demonstrated and compared in conjunction with the other soft-computing methods (tab. 1) that use binary input vectors constructed by the ADLBCA instead of original ones.

2   Automatic Discriminative Lossy Binary Conversion Algorithm

The Automatic Discriminative Lossy Binary Conversion Algorithm (ADLBCA) starts its lossy binary conversion from an input data analysis that takes into consideration the following goals:

- the lossy binary conversion ranges should be wide in order to cover important parts of an input data space sufficiently and to achieve good generalization,
- the number of lossy binary conversion ranges should be as minimal as possible in order to simplify or even reduce the binary input data space and the computational cost of the classification method,
- the computed lossy binary conversion ranges should enable the training algorithm to discriminate all training cases of all classes provided that TD are not contradictory,
- a discriminative property of lossy binary conversion ranges should be estimated using statistical analysis of training data,
- the computational and memory costs of the method should be low.

The main goal of this algorithm is to find a possibly minimal set of discriminative lossy binary conversion ranges (DLBCRs) for all real data input features and to convert the values v_t from these ranges $R = [L_m^t; P_m^t]$ into the value $+1$ and other values outside these ranges $R = [L_m^t; P_m^t]$ into the value 0 or -1 depending on the prefered coding: unipolar 1 or bipolar (2). Moreover, an appropriate selection of these ranges can predifferentiate and prediscriminate some training cases and help a following soft-computing algorithm ultimately to discriminate them. Real input values from these ranges R can be transformed into binary values using Lossy Binarizing Neurons (LBNs) (fig. 3) which compute their outputs using equation (1) or (2). First, real data inputs have to be sorted and indexed separately for each input feature. Figure 1 illustrates the sorted Iris data after all input features. The heapsort algorithm should be used because its computational cost is always $O(nlogn)$. The stability of the sorting algorithm does not matter for this method.

After all input data features are sorted, the algorithm starts to search for a minimal set of DLBCRs taking into account the following criteria:

1. The selected range should contain as many cases of the same class and as few cases of different classes as possible,
2. The ranges containing cases of a smaller number of classes are preferred. The best discriminative ranges contain cases from a single class.
3. The ranges can contain training cases from other classes only if they are discriminated by other ranges.

These criteria are important to satisfy the requirements mentioned at the beginning of this paper, especially in view of good generalization properties. This algorithm sorts TD after each feature value and indexes training data for all input features separately and then looks for optimal ranges R in the following way:

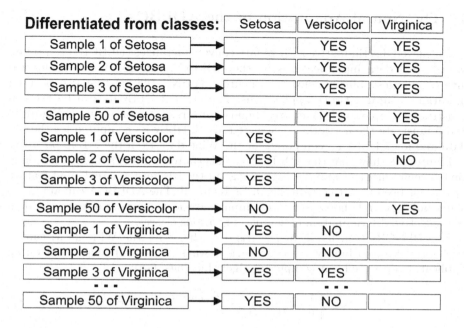

Differentiated from classes:	Setosa	Versicolor	Virginica
Sample 1 of Setosa		YES	YES
Sample 2 of Setosa		YES	YES
Sample 3 of Setosa		YES	YES
...		...	
Sample 50 of Setosa		YES	YES
Sample 1 of Versicolor	YES		YES
Sample 2 of Versicolor	YES		NO
Sample 3 of Versicolor	YES		
...		...	
Sample 50 of Versicolor	NO		YES
Sample 1 of Virginica	YES	NO	
Sample 2 of Virginica	NO	NO	
Sample 3 of Virginica	YES	YES	
...	...	...	
Sample 50 of Virginica	YES	NO	

Fig. 2. The exemplar tables of the discriminated classes for the Iris samples

1. First, all training cases are marked as indiscriminated for all classes except the classes they represent.
2. Next, all yet indiscriminated data cases for all input features are looked through in the sorted order and a range containing a maximal number of training cases and the minimum number of classes are sought. Each range is described by an input feature and its range of values.
3. All yet indiscriminated cases for which the range was chosen are marked as discriminated for all classes which this range does not contain.
4. Next, all fully discriminated training cases for all input features are looked through in order to remove their indexes from the sorted index tables (fig. 2).
5. Steps 2, 3 and 4 are repeated until all TD cases are discriminated from all other classes (fig. 2) or there is no more range (that includes the indiscriminated training samples) to consider.
6. If not all training cases are discriminated and no more ranges can be used to carry out their discrimination, all the atomic ranges are chosen for all input features that contain indiscriminated training cases and are added to the previously selected ranges in steps 2, 3 and 4. The atomic ranges always contain a sequence of cases of a single class or they may represent a few classes but the range is narrowed to a single value (fig. 1). If step 6 occurs it means that some training cases are contradictory or they can be differentiated only by a combination of these ranges.

3   The Neural Adaptation of the ADLBCA

The achieved discriminative lossy binary conversion ranges (DLBCRs) computed for the real data input features by the ADLBCA can be easily transformed into specific losy binary conversion neurons (LBNs) (fig. 3) that can be used in various neural network applications. The created DLBCRs can be used to construct a simple single layer lossy binary conversion neural network. Figures 4 and 5 illustrate the networks constructed for 12 DLBCRs computed for the Iris data and for 7 DLBCRs computed for the Wine data from the ML Repository.

Figures 6-8 show that the ADLBCA can also automatically simplify and sometimes even reduce an input data space, e.g. for the Iris data, the real input feature v_1 has not been used to create the DLBCRs at all. The ADLBCA can automatically find out and choose these real input data features ranges that provide a cost-effective discrimination of samples of various classes.

The DLBCRs (e.g. fig. 4 for the Iris data, fig. 5 for the Wine data) can be transformed into a binary values in various ways ((1) or (2)) dependent on an application or a used soft-computing algorithm (figs. 6-8):

$$f_k^{LB}(v_r, R) = \begin{cases} 1 & L_m^k \le v_r \le P_m^k \ where \ R = [L_m^k; P_m^k] \\ 0 & otherwise \end{cases} \qquad (1)$$

$$f_k^{LB}(v_r, R) = \begin{cases} 1 & L_m^k \le v_r \le P_m^k \ where \ R = [L_m^k; P_m^k] \\ -1 & otherwise \end{cases} \qquad (2)$$

Functions (1) and (2) together with the described ranges simplify the representation of input data vectors but all training samples can be still correctly discriminated. That is one of the most important feature of this algorithm. The ADLBCA is partially similar to PCA, ICM, rough sets, fuzzy sets and even lossy compression algorithms, e.g. jpeg. The acceptable level for ADLBCA is established by the ability to univocally and correctly discriminate all training samples. The lossy compression and simplification reduce minor and unimportant data values and input features transforming the important ones to binary representation of them (figs. 4-5).

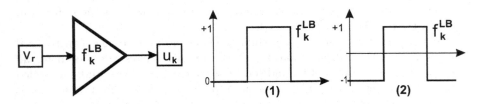

Fig. 3. The lossy binary conversion neuron (LBN) created for the real input feature v_r and for the binary conversion discrimination range R producing the binary output feature u_k. The unipolar (1) and bipolar (2) binary conversion functions.

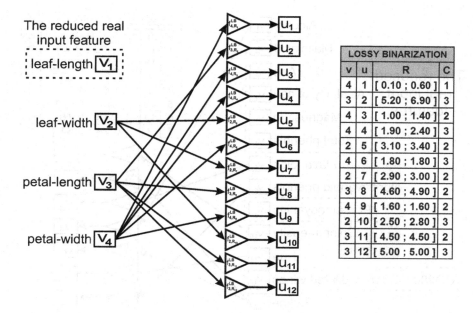

Fig. 4. The lossy binary conversion neural network and the base DLBCRs for the Iris data where u_r - the binary features, v_k - the original real features

4 The ADLBCA Applications

The ADLBCA has been used to convert the Iris, Wine and Heart data input vectors into the binary vectors (figs. 1, 4 and 5) and used for classification using the selected soft-computing algorithms (tab. 1) [1], [3], [4], [6]. The ADLBCA has been also included as a part of the methodology for a construction of the Self-Optimizing Neural Networks 3 (SONN-3) and used to develop better optimized neural network topologies [3]. Figure 6 shows the SONN-3 classifier for the Iris data using binary vectors computed for the DLBCRs (shown in fig. 4). Figures 7 and 8 show the SONN-3 topologies for the Wine and Heart data.

The Wine data have been randomly divided into 78 training and 100 validating data in order to show the generalization abilities on the original real data and on binary transformed data using the ADLBCA. Table 1 describes the parameters of the selected soft-computing algorithms: MLP, RBF, GRNN, SONN-2, SONN-3, SSV Tree, FSM, SVM, IncNet, k-NN [1], [2], [3], [4], [6] which have been used to compare generalization properties of the best solutions found for these data using the built-in Statistica Neural Networks and Ghost Miner 3.0 solvers, automatic designers and cross-validation. The comparisons presented in table 1 demonstrate that when the ADLBCA has been used for the TD preprocessing the considered soft-computing algorithms have always achieved the better training and generalization results, the more reduced input data spaces and the more compact NN topologies or rules etc.

8   A. Horzyk

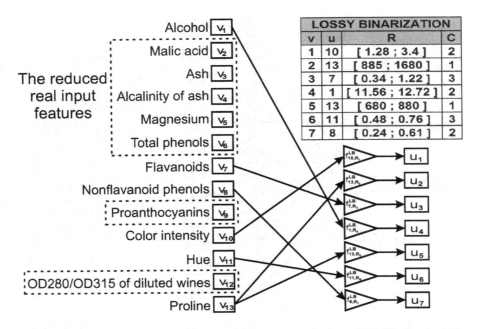

Fig. 5. The lossy binary conversion neural network and the base DLBCRs for the Wine data where u_r - the binary features, v_k - the original real features

5 Conclusions

The new Automatic Discriminative Lossy Binary Conversion Algorithm (ADL-BCA) has been described in this paper. The ADLBCA is able to fully automatically lossy convert real input data vectors into binary vectors. This feature can be used by various soft-computing methods to make further computations more easy or to achieve better generalization results. The presented algorithm lossy compresses real values from the specially computed ranges and transforms them into the binary values. The ADLBCA lossy compression warrants that all important differences that make possible to discriminate training samples from various classes are not lost. Only insignificantly differing real values are grouped together into the described discriminative lossy binary conversion ranges (DLBCRs). The introduced ADLBCA can very fast ($O(nlogn)$) process the described conversion and provide the better discrimination of all training samples of all classes only if they are not contradictory. The main goal of this algorithm is not to reduce an input data space but to simplify it for other soft-computing methods. It can emphasize important data differences and use them to achieve better generalization results. Moreover, the ADLBCA computes the DLBCRs in such a way that a small number of binary inputs and a compact soft-computing solution can be achieved (tab. 1). It also covers an input data space with the most important and discriminative data features. The ADLBCA also automatically reduces some

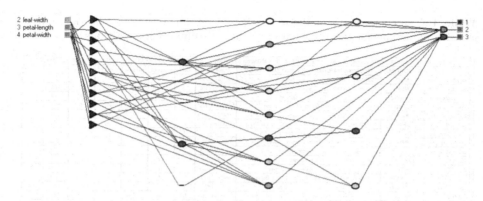

Fig. 6. The SONN-3 classifier effectively using 3 of 4 original input features and 11 of 12 converted lossy binary features created for the Iris data by the ADLBCA

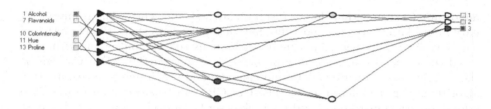

Fig. 7. The SONN-3 classifier effectively using 5 of 13 original input features and 6 of 7 converted lossy binary features created for the Wine data by the ADLBCA

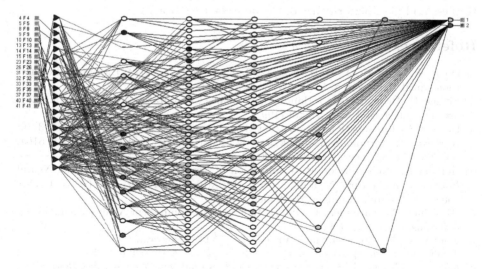

Fig. 8. The SONN-3 classifier effectively using 17 of 44 original input features and 20 converted lossy binary features created for the Heart data by the ADLBCA

Table 1. The comparisons of the best achieved performance, generalization and dimension reductions for the Wine data and the soft-computing algorithms

WINE DATA COMPARISONS (devided into 78 training and 100 validation samples)				USED	TRAIN RESULTS		TEST RESULTS		TOTAL	
TOOL	ALGORITHM	TRAINING PARAMETERS AND TOPOLOGY	input data	DIM	correct	errors	correct	errors	correct	errors
Stat.NN designer	MLP	1000 iter., 21+9+1 neur., 3 layers, steps 3, unit penalty 0	ADLBCA transf.	7	78	0	99	1	177	1
Stat.NN designer	MLP	1000 iter., 36+4+1 neur., 3 layers, steps 3, unit penalty 0	original data	13	78	0	97	3	175	3
Own implement.	SONN-3	16 neur., 4 layers, 32 connections	ADLBCA transf.	5	78	0	96	4	174	4
Own implement.	SONN-2	10 neurons, 3 layers, 23 connections	ADLBCA transf.	6	78	0	96	4	174	4
Stat.NN designer	RBF	1000 iter., 21+23+1 neurons, steps 3, unit penalty 0	ADLBCA transf.	7	78	0	96	4	174	4
Stat.NN designer	GRNN	1000 iter., 21+75+2 neurons, steps 3, unit penalty 0	ADLBCA transf.	7	78	0	96	4	174	4
Ghost Miner 3.0	SSV Tree	Beam search, opt. leaves count, 5 rules	original data	4	76	2	97	3	173	5
Stat.NN designer	RBF	1000 iter., 36+24+1 neurons, steps 3, unit penalty 0	original data	13	78	0	95	5	173	5
Ghost Miner 3.0	SSV Tree	Beam search, opt. pruning degree, 6 rules	original data	4	78	0	94	6	172	6
Ghost Miner 3.0	SSV Tree	BFS, opt. pruning degree, 6 rules	original data	4	77	1	94	6	171	7
Ghost Miner 3.0	FSM	Gaussian fun., tree clusterization	original data	13	77	1	94	6	171	7
Ghost Miner 3.0	Nu SVM	Polynomial kernel, SVM Solver, Nu=0.1, Bias=0.1	original data	13	75	3	94	6	169	9
Ghost Miner 3.0	SSV Tree	BFS, opt. leaves count, 5 rules	original data	3	75	3	93	7	168	10
Ghost Miner 3.0	IncNet	1000 iter., Gauss neur., Bias 1, Max neur. 100	original data	13	78	0	83	17	161	17
Stat.NN designer	GRNN	1000 iter., 36+75+2 neurons, steps 3, unit penalty 0	original data	13	78	0	82	18	160	18
Ghost Miner 3.0	k-NN	Euclidean distance, Fixed k = 1	original data	13	78	0	71	29	149	29
Ghost Miner 3.0	k-NN	Euclidean distance, Auto k, Max k 20	original data	13	78	22	77	23	155	45

minor input features that do not discriminate TD sufficiently. The resultant binary input data space can sometimes have a smaller data space than an original one (figs. 7-8). This algorithm can also help many soft-computing algorithms to avoid the curse of dimensionality problem. The constructive ADLBCA can be successfully used to various data and can cooperate with many soft-computing algorithms (e.g. figs. 6-8, tab. 1). The DLBCRs can be easily transformed into a lossy binary conversion neural network described in this paper. The lossy binary conversion neural network consists of the lossy binary conversion neurons (LBNs) that can be interconnected to other neurons that are trained or adapted by other soft-computing algorithms. The performed experiments confirm effectiveness and high performance of the described ADLBCA.

References

1. Duch, W., Korbicz, J., Rutkowski, L., Tadeusiewicz, R. (eds.): Biocybernetics and Biomedical Engineering, EXIT, Warszawa (2000)
2. Fiesler, E., Beale, R. (eds.): Handbook of Neural Computation. IOP Publishing Ltd. Oxford University Press, Bristol (1997)
3. Horzyk, A.: Introduction to Constructive and Optimization Aspects of SONN-3. In: Kurkova, V., Neruda, R., Koutnik, J. (eds.) ICANN 2008, Part II. LNCS, vol. 5164, pp. 763–772. Springer, Heidelberg (2008)
4. Horzyk, A., Tadeusiewicz, R.: Comparison of Plasticity of Self-Optimizing Neural Networks and Natural Neural Networks. In: Mira, J., Alvarez, J.R. (eds.) ICANN 2005, pp. 156–165. Springer, Heidelberg (2005)
5. Hyvarinen, A., Karhunen, J., Oja, E.: Independent Component Analysis. John Wiley and Sons, Chichester (2001)
6. Jankowski, N.: Ontogenic neural networks, EXIT, Warszawa (2003)
7. Jolliffe, I.T.: Principal Component Analysis. Springer, Heidelberg (2002)
8. Kalat, J.: Biological Psychology. Thomson Learning Inc., Wadsworth (2004)
9. Pawlak, Z.: Rough sets. Theoretical Aspects of Reasoning about Data. Kluwer Academic Publisher, Dordrecht (1991)

On Tractability of Neural-Network Approximation

Paul C. Kainen[1], Věra Kůrková[2], and Marcello Sanguineti[3]

[1] Department of Mathematics, Georgetown University,
Washington, D.C. 20057-1233, USA
kainen@georgetown.edu
[2] Institute of Computer Science, Academy of Sciences of the Czech Republic
Pod Vodárenskou věží 2, Prague 8, Czech Republic
vera@cs.cas.cz
[3] Department of Communications, Computer, and System Sciences (DIST)
University of Genoa, Via Opera Pia 13, 16145 Genova, Italy
marcello@dist.unige.it

Abstract. The tractability of neural-network approximation is investigated. The dependence of worst-case errors on the number of variables is studied. Estimates for Gaussian radial-basis-function and perceptron networks are derived.

1 Introduction

Widely-used connectionistic models take on the form of linear combinations of all n-tuples of functions computable by computational units of various kinds (e.g., perceptrons, radial or kernel units) with trigonometric, Heaviside, Gaussian, or spline activation functions. In contrast to approximation schemes where approximating functions belong to a linear subspace generated by the first n elements of a set of functions with a fixed linear ordering, *fixed-basis approximation*, the approximation scheme used in connectionistic models is sometimes called *variable-basis approximation* or *approximation from a dictionary*.

Bounds on worst-case approximation errors of these connectionistic models typically include several factors, one of which involves the number n of terms in the linear combinations (i.e., the number of computational units) - called *model complexity* - while another involves (often implicitly) the number d of variables (i.e., the number of network inputs). Dependence on the dimension d is often cryptic; i.e., estimates depend on "constants", but the "constants" may grow with the dimension even exponentially. Moreover, the families of functions for which the bounds are valid may become negligibly small for large d.

An emphasis on the model complexity n is certainly reasonable when the number d of variables is fixed, but in modern research (where technology allows ever-increasing amounts of data to be collected) it is natural to also take into account the role of d. Indeed, this is considered in information-based complexity (see [1,2,3]) and more recently this situation has been studied in the context of

M. Kolehmainen et al. (Eds.): ICANNGA 2009, LNCS 5495, pp. 11–21, 2009.
© Springer-Verlag Berlin Heidelberg 2009

functional approximation and neural networks [4,5]. The upper bounds on the approximation error are called *tractable* (with respect to d) when for every fixed n they take on the form of a polynomial in d. Various authors derived upper bounds in the factorized form $\xi(d)\kappa(n)$, for a nonincreasing function $\kappa(n)$ (often of the form $n^{-\alpha}$ for $\alpha > 0$) and a typically unspecified increasing function $\xi(d)$ of the number d of variables. In these cases, tractability is guaranteed when $\xi(d)$ is a polynomial. In some literature (e.g., [6]) the term $\xi(d)$ is referred to as "a constant"; however, it is such only for a fixed value of d. Often, the function $\xi(d)$ is "hidden" in the "big o" notation [7], which, despite its frequent use, may be quite misleading as it focuses only on the dependence on n.

The rate of growth of $\xi(d)$ is not crucial for small values of d. However, for large values of d, the approximation error can even grow exponentially with d as a consequence of exponential growth in $\xi(d)$ (e.g., [6, item 9, p. 940]), when the "curse of dimensionality" [8] strikes. As remarked in [2], in general estimating the dependence of the approximation error on d is much harder than estimating its dependence on n and only few estimates are available.

The purpose of this paper is to use tractability to look at available results from a unifying perspective, and to analyze when neural network approximation, by and for parameterized sets of functions, can be applied. We give dimensional analysis of formulas in terms of the parameters involved.

We compare and contrast various results in which tractability in dimension is achieved for neural-network approximation of "reasonable" families of functions large enough to include such members as the Gaussian function in each dimension. We also discuss cases in which the function $\xi(d)$ decreases - even exponentially fast - with dimension. For neural-network approximation, upper bounds that depend polynomially on d, so guaranteeing tractability, were derived in, e.g., [4,6,9,10]. However, polynomials may not provide much control unless the degrees are quite small. For large d, even quadratic approximation is not going to be sufficient. For this reason, in sections 4 and 5 we focus on the case where dependence on d is *linear* (or better).

The paper is organized as follows. In Section 2 the concept of tractability with respect to the dimension d of the worst-case approximation by functions from a dictionary is introduced. In Section 3, results from approximation theory are used to describe sets of functions for which variable-basis approximation is tractable. These results are applied to approximation by Gaussian radial-basis function networks in Section 4 and to perceptron networks in Section 5. Section 6 is a brief discussion.

2   Worst-Case Tractability in Neural-Network Approximation

Let $(\mathcal{X}_d, \|.\|_{\mathcal{X}_d})$ be a normed linear space of functions or equivalence classes of functions of d variables and let A_d, G_d be two nonempty subsets of $\mathcal{X}_d$. Functions in A_d are to be approximated by linear combinations of elements of G_d.

We use the following terminology: For a subset G of a linear space, span_n means the set of all n-fold linear combinations, i.e.,

$$\text{span}_n\, G = \left\{ \sum_{i=1}^{n} w_i g_i \,\middle|\, w_i \in \mathbb{R}, g_i \in G \right\}.$$

With suitable choices, these sets consist of functions computable by one-hidden layer neural networks, radial-basis functions, kernel models, splines with free nodes, trigonometric polynomials with free frequencies, etc. Other choices for G include Bessel potential functions [5] and Hermite functions [11].

The set G is sometimes called a *dictionary* and the approximation scheme $\text{span}_n\, G$ is called *variable-basis approximation scheme*. The integer n can be interpreted as the *model complexity* measured by the number of computational units.

If A and T are nonempty subsets of a normed linear space $(\mathcal{X}, \|.\|_{\mathcal{X}})$, the *deviation* $\delta(A, T)$ of A from T is the worst-case error in $\|.\|_{\mathcal{X}}$ in approximating functions in A by functions from T

$$\delta(A, T)_{\mathcal{X}} := \sup_{f \in A} \inf_{g \in T} \|f - g\|_{\mathcal{X}}.$$

Clearly, $\delta(A', T') \leq \delta(A, T)$ if $A' \subseteq A$ and $T' \supseteq T$. Also, $\delta(c\,A, c\,T) = c\,\delta(A, T)$, where $c\,A = \{c\,a \mid a \in A\}$.

For variable-basis models, many upper bounds on rates of approximation take on the factorized form

$$\delta(A_d, \text{span}_n\, G_d)_{\mathcal{X}_d} \leq \xi(d)\,\kappa(n), \tag{1}$$

where $\xi : \mathbb{N} \to \mathbb{R}_+$ is a function of the number d of variables of functions in $\mathcal{X}_d$ and $\kappa : \mathbb{N} \to \mathbb{R}_+$ is a nonincreasing function of the model complexity n (often $\kappa(n) = n^{-1/2}$); see, e.g., [6,9,12,13,14,15].

If the upper bound (1) holds with a polynomial $\xi(d)$ in the number d of variables, then the problem of approximating A_d by elements of $\text{span}_n\, G_d$ is called *tractable with respect to d in the worst case* [1], or simply *tractable*.

3   Tractability for Balls in Variational Norms

Sets of functions which can be tractably approximated by $\text{span}_n G$ are given by a theorem on approximation by convex combinations of n elements of a bounded subset of a Hilbert space, derived by Maurey [16], Jones [15], and Barron [6]. The Maurey-Jones-Barron theorem was extended to $(\mathcal{L}^p(\Omega_d), \|.\|_{\mathcal{L}^p(\Omega_d)})$, $p \in (1, \infty)$, in [14], and to the space $(\mathcal{M}(\Omega_d), \|.\|_{\infty})$ of bounded measurable functions on $\Omega_d \subseteq \mathbb{R}^d$ with the essential supremum norm in [17]. These results can be reformulated in terms of a norm, introduced in [18] as an extension of "variation

---

[1] Tractability with respect to other quantities and in different settings has been defined; see, e.g., [1].

with respect to half-spaces" from [12]. This norm, called G-variation and denoted $\|\cdot\|_G$, is defined for any bounded subset of a normed linear space $(\mathcal{X}, \|\cdot\|_{\mathcal{X}})$ as the Minkowski functional (see, e.g., [19, p. 131]) of the $\mathcal{X}$-closure of the symmetric convex hull of G, i.e.,

$$\|f\|_{G,\mathcal{X}} = \|f\|_G := \inf \left\{ c > 0 \,|\, c^{-1}f \in \mathrm{cl}_{\mathcal{X}} \, \mathrm{conv}\,(G \cup -G)\right\}. \tag{2}$$

Note that G-variation can be infinite and that it depends on the ambient space norm; when this norm is clear from the context we merely write $\|f\|_G$.

For a nonempty set Ω and $S \subseteq \Omega$, we denote by χ_S the *indicator function* of S, i.e., $\chi_S(x) = 1$ if $x \in S$, otherwise $\chi_S(x) = 0$. Let $\mathcal{F}$ be any family of indicator functions of subsets of Ω, $\mathcal{S} = \{S \subseteq \Omega \,|\, \chi_S \in \mathcal{F}\}$ be the family of the corresponding subsets of Ω, and A be a subset of Ω. Then A is said to be *shattered* by $\mathcal{F}$ if $\{S \cap A \,|\, S \in \mathcal{S}\}$ is the whole power set of A. The *VC-dimension* of $\mathcal{F}$ is the largest cardinality of any subset A which is shattered by $\mathcal{F}$. The *coVC-dimension* of $\mathcal{F}$ is the VC-dimension of the set $\mathcal{F}' := \{ev_x \,|\, x \in \Omega\}$, where the *evaluation* $ev_x : \mathcal{F} \to \{0, 1\}$ is defined for every $\chi_S \in \mathcal{F}$ as $ev_x(\chi_S) = \chi_S(x)$.

For a normed linear space $(\mathcal{X}, \|\cdot\|)$ and $r > 0$ we denote by $B_r(\|\cdot\|)$ the closed ball of radius r centered at zero, i.e.,

$$B_r(\|\cdot\|) = \{f \in \mathcal{X} \,|\, \|f\| \le r\}.$$

The following theorem gives upper bounds of the form $\xi(d)\kappa(n)$ on the worst-case errors in approximation of functions from balls in G_d-variations by $\mathrm{span}_n \, G_d$.

Theorem 1. *Let d be a positive integer, $(\mathcal{X}_d, \|\cdot\|_{\mathcal{X}_d})$ be a Banach space of d-variable functions, G_d its bounded subset with $s_d = \sup_{f \in G_d} \|f\|_{\mathcal{X}_d}$, and $r_d > 0$. Then for every positive integer n, the following hold.*
(i) If $(\mathcal{X}_d, \|\cdot\|_{\mathcal{X}_d})$ is a Hilbert space, then

$$\delta(B_{r_d}(\|\cdot\|_{G_d}), \mathrm{span}_n \, G_d)_{\mathcal{X}_d} \le s_d \, r_d \, n^{-1/2};$$

(ii) If $\Omega_d \subseteq \mathbb{R}^d$, $p \in (1, \infty)$, and $(\mathcal{X}_d, \|\cdot\|_{\mathcal{X}_d}) = (\mathcal{L}^p(\Omega_d), \|\cdot\|_{\mathcal{L}^p})$, then

$$\delta(B_{r_d}(\|\cdot\|_{G_d}), \mathrm{span}_n \, G_d)_{\mathcal{L}^p(\Omega_d)} \le 2^{1+1/\bar{p}} \, s_d \, r_d \, n^{-1/\bar{q}},$$

where $q = p/(p-1)$, $\bar{p} = \min(p, q)$, and $\bar{q} = \max(p, q)$.
(iii) If $\Omega_d \subseteq \mathbb{R}^d$, $(\mathcal{X}_d, \|\cdot\|_{\mathcal{X}_d}) = (\mathcal{M}(\Omega_d), \|\cdot\|_\infty)$ and $G_d = \Theta_{\Omega_d}$ is a set of some indicator functions on Ω_d such that the co-VC-dimension $h^*_{d,\Theta}$ of Θ_{Ω_d} is finite, then

$$\delta(B_{r_d}(\|\cdot\|_{\Theta_{\Omega_d}}), \mathrm{span}_n \, \Theta_{\Omega_d})_{\mathcal{M}(\Omega_d)} \le 6\sqrt{3} \left(h^*_{d,\Theta}\right)^{1/2} s_d \, r_d \, (\log n)^{1/2} \, n^{-1/2}.$$

Proof. (i) By Maurey-Jones-Barron's estimate restated in terms of G-variation [18], for every bounded subset G_d of a Hilbert space $(\mathcal{X}_d, \|\cdot\|_{\mathcal{X}_d})$, every $f \in \mathcal{X}_d$, and every positive integer n, we get $\|f - \mathrm{span}_n \, G_d\|_{\mathcal{X}_d} \le (s_G^2 \|f\|_{G_d}^2 - \|f\|_{\mathcal{X}_d}^2) \, n^{-1/2}$.

(ii) By [20, Theorem A.4] (which is a slight reformulation of [14, Theorem 5]), for every $f \in \mathcal{L}^p(\Omega_d)$ and every positive integer n, $\|f - \mathrm{span}_n G_d\|_{\mathcal{L}^p(\Omega_d)} \leq 2^{1+1/\bar{p}} s_d r_d n^{-1/\bar{q}}$. So the statement follows again by the definition of deviation.

(iii) This follows by [17, Theorem 3].   □

In the upper bounds in Theorem 1 (i)-(iii), the functions $\xi(d)$ are of the forms $\xi(d) = s_d r_d$, $\xi(d) = h_d^* s_d r_d$, and $\xi(d) = d^{1/2} s_d r_d$, resp. These estimates imply tractability when $s_d r_d$, $h_{d,\Theta}^* s_d r_d$, and $d^{1/2} s_d r_d$, resp., grow polynomially with d increasing. Note that s_d and $h_{d,\Theta}^*$ are determined by the choice of G_d, but r_d can be chosen in such a way that $\xi(d)$ is a polynomial.

4 Tractability for Gaussian Radial-Basis-Function Networks

In this section, we investigate tractability of approximation by Gaussian radial-basis function (RBF) networks.

Let $\gamma_{d,b} : \mathbb{R}^d \to \mathbb{R}$ denote the d-dimensional *Gaussian function of width b*, given by

$$\gamma_{d,b}(x) = e^{-b\|x\|^2}.$$

For every $y \in \mathbb{R}^d$, let τ_y be the translation operator defined for any $g : \mathbb{R}^d \to \mathbb{R}$ by the equation

$$\tau_y(g)(x) = g(x - y).$$

When $b = 1$, we merely write γ_d instead of $\gamma_{d,1}$. The set of *Gaussian radial-basis d-variable functions with varying widths and varying centriods* is denoted by

$$G_d^\gamma = \left\{ \tau_y(\gamma_{d,b}) \,|\, y \in \mathbb{R}^d, b > 0 \right\},$$

while the set of *Gaussian radial-basis d-variable functions with a fixed width b > 0 and varying centroids* is denoted by

$$G_d^{\gamma,b} = \left\{ \tau_y(\gamma_{d,b}) \,|\, y \in \mathbb{R}^d \right\}.$$

Our first estimate holds for Gaussian RBF networks with units of fixed width. A simple calculation shows that for $b > 0$, $\|\gamma_{d,b}\|_{\mathcal{L}^2(\mathbb{R}^d)} = (\pi/2b)^{d/4}$. Thus $s_d = (\pi/2b)^{d/4}$. By Theorem 1 (i), we get the following corollary.

Corollary 1. *Let d be a positive integer, $b > 0$, and $r_d > 0$. Then for every positive integer n*

$$\delta(B_{r_d}(\|.\|_{G_d^{\gamma,b}}), \mathrm{span}_n G_d^{\gamma,b})_{\mathcal{L}^2(\mathbb{R}^d)} \leq \left(\frac{\pi}{2b} \right)^{d/4} r_d \, n^{-1/2}.$$

So in the upper bound from Corollary 1, $\xi(d) = \xi(d,b) = (\pi/2b)^{d/4} r_d$. Thus for $b = \pi/2$, this corollary implies tractability for r_d growing with d polynomially, for $b > \pi/2$, it implies tractability even if r_d is increasing exponentially fast, while for $b < \pi/2$, it merely implies tractability for r_d decreasing exponentially fast. Hence, the width b of Gaussians has a strong impact on the size of radii r_d of balls in $G_d^{\gamma,b}$-variation for which $\xi(d)$ is a polynomial. The narrower the Gaussians, the larger the balls for which Corollary 1 implies tractability.

In [5], upper bounds of the form $\xi(d)\kappa(n)$ on approximation by Gaussian RBF networks were derived for smooth functions defined by smoothing operators in the form of convolutions with certain kernels called Bessel potentials. These potentials are defined by means of their Fourier transforms: for $s > 0$, the *Bessel potential* of order s, denoted by $\beta_{d,s}$, is the function on $\mathbb{R}^d$ with Fourier transform

$$\hat{\beta}_{d,s}(\omega) = (1 + \|\omega\|^2)^{-s/2}.$$

For $q \in [1, \infty)$, d a positive integer, and $s > d/q$, the *Bessel potential space* (with respect to $\mathbb{R}^d$) [21, pp. 134-136], denoted by $(L^{q,s}, \|.\|_{L^{q,s}})$, is defined as

$$L^{q,s}(\mathbb{R}^d) := \{f \mid f = w * \beta_{d,s}, \ w \in \mathcal{L}^q(\mathbb{R}^d)\}$$

$$\|f\|_{L^{q,s}(\mathbb{R}^d)} := \|w\|_{\mathcal{L}^q(\mathbb{R}^d)} \quad \text{for} \quad f = w * \beta_{d,s}.$$

Since the Fourier transform of a convolution is $(2\pi)^{d/2}$ times the product of the transforms, we have $\hat{w} = (2\pi)^{-d/2}\hat{f}/\hat{\beta}_{d,s}$. Thus w is uniquely determined by f and so the Bessel potential norm is well-defined.

The next theorem gives upper bounds on worst-case errors of two sets of smooth functions in approximation by Gaussian RBF networks.

(i) The first set is the ball of radius r_d in the Bessel potential space $L^{1,s}(\mathbb{R}^d)$.

(ii) The second set is the intersection of the ball of radius r_d in $L^{2,s}(\mathbb{R}^d)$ with those functions $f = w * \beta_{d,s}$ such that w has support $\operatorname{supp} w = \operatorname{cl}\{x \in \mathbb{R}^d \mid w(x) \neq 0\}$ of Lebesgue measure bounded by ν_d, i.e.,

$$A_{s,r_d,\nu_d}^{(1)} = \{f \in \mathcal{L}^2(\mathbb{R}^d) \mid f = w * \beta_{d,s}, \ \|w\|_{\mathcal{L}^2(\mathbb{R}^d)} \leq r_d, \ \lambda(\operatorname{supp} w) \leq \nu_d\}.$$

The following upper bounds on worse-case $\mathcal{L}^2$-errors in approximation of these three sets of smooth functions by Gaussian networks were proven in [5, Theorems 4.4 and 5.1] under the additional condition that w is continuous a.e. However, with a more careful argument, only measurability is required. Recall that for complex z with $\operatorname{Re}[z] > 0$, $\Gamma(z) = \int_0^\infty t^{z-1}e^{-t}\,dt$ and $\Gamma(n) = (n-1)!$.

Theorem 2. *Let d, n be positive integers, $s > d/2$, and $r_d > 0$. Then*

$$(i) \ \delta(B_{r_d}(\|.\|_{L^{1,s}(\mathbb{R}^d)}), \operatorname{span}_n G_d^\gamma)_{\mathcal{L}^2(\mathbb{R}^d)} \leq \left(\frac{\pi}{2}\right)^{d/4} \frac{\Gamma(s/2 - d/4)}{\Gamma(s/2)} r_d\, n^{-1/2};$$

$$(ii) \ \delta_n(A_{s,r_d,\nu_d}^{(1)}, \operatorname{span}_n G_d^\gamma)_{\mathcal{L}^2(\mathbb{R}^d)} \leq \left(\frac{\pi}{2}\right)^{d/4} \frac{\Gamma(s/2 - d/4)}{\Gamma(s/2)} \nu_d^{1/2}\, r_d\, n^{-1/2}.$$

Note that $\frac{\Gamma(s/2-d/4)}{\Gamma(s/2)}$ as a function of s is decreasing on the interval $(d/2, \infty)$. The bounds from Theorem 2 are of the form $\xi(d)\kappa(n)$ where $\kappa(n) = n^{-1/2}$, while $\xi(d)$ are multiples of r_d and $\nu_d^{1/2}$ with coefficients going to zero exponentially fast with d increasing.

The shape of the support of the function w determines ν_d for which $f = w * \beta_{d,s}$ belongs to $A^{(1)}_{s,r_d,\nu_d}$, while the $\mathcal{L}^2$-norm of the function w determines r_d. When the support of w is the Euclidean unit ball in $\mathbb{R}^d$, then $f = w * \beta_{d,s}$ belongs to $A^{(1)}_{s,r_d,\nu_d}$ with ν_d equal to $\pi^{d/2}/\Gamma((d+2)/2)$ [22, p. 304]. So in this case, ν_d goes to zero exponentially fast with the dimension d. However, when the support of w is a cube, then growth or decrease of ν_d depends on the size of its side (see also the remarks in [23, Section 18.2]).

5   Tractability for Perceptron Networks

In this section we investigate tractability of worst-case errors in approximation by perceptron networks.

A measurable function $\sigma : \mathbb{R} \to \mathbb{R}$ is called a *sigmoid* when $\lim_{t \to -\infty} \sigma(t) = 0$, and $\lim_{t \to \infty} \sigma(t) = 1$. A special case is the *Heaviside function* $\vartheta : \mathbb{R} \to \mathbb{R}$ defined as $\vartheta(t) = 0$ for $t < 0$ and $\vartheta(t) = 1$ for $t \geq 0$. For $\Omega_d \subseteq \mathbb{R}^d$, let

$$H_d(\Omega_d) = \{\vartheta(e \cdot x + b) : \Omega_d \to \mathbb{R} \,|\, e \in S^{d-1}, \, b \in \mathbb{R}\}$$

denote the *set of functions on Ω_d computable by Heaviside perceptrons* with input weights e in the unit sphere S^{d-1} in $\mathbb{R}^d$ and biases $b \in \mathbb{R}$. For a sigmoid σ, we denote by

$$H^\sigma_d(\Omega_d) = \{\sigma(v \cdot x + b) : \Omega_d \to \mathbb{R} \,|\, v \in \mathbb{R}^d, \, b \in \mathbb{R}\}$$

the *set of functions on Ω_d computable by sigmoidal perceptrons*. When $\Omega_d = \mathbb{R}^d$, we merely write H_d and H^σ_d. For Ω_d bounded, the set $H_d(\Omega_d)$ is equal to the set of all indicator functions of subsets of Ω_d. Thus H_d-variation is sometimes called *variation with respect to half-spaces*.

The next corollary estimates worst-case $\|.\|_\infty$-errors in approximation of balls in H_d-variation.

Corollary 2. *Let d be a positive integer and $r_d > 0$, then for every positive integer n*

$$\delta(B_{r_d}(\|.\|_{H_d}), \mathrm{span}_n H_d)_{\mathcal{M}(\mathbb{R}^d)} \leq 6\sqrt{3}\, d^{1/2}\, r_d\, (\log n)^{1/2}\, n^{-1/2}\,.$$

Proof. The statement follows by Theorem 1 (iii) and the fact that the co-VC dimension of the set H_d of closed half-space indicator functions on $\mathbb{R}^d$ is equal to d [17, p. 162]. □

In the upper bound in Corollary 2, we have $\xi(d) = d^{1/2}\, r_d$. This implies tractability for every r_d growing polynomially with d.

We now consider upper estimates known only for the odd-dimensional case. A real-valued function f on $\mathbb{R}^d$, d odd, is of *weakly controlled decay* [9] if it is d-times continuously differentiable and satisfies for all multi-indices $\alpha \in \mathbb{N}^d$, with $|\alpha| = \sum_{i=1}^d \alpha_i$ and $D^\alpha = \partial^{\alpha_1} \cdot \ldots \cdot \partial^{\alpha_d}$,

(i) $|\alpha| < d \implies \lim_{\|x\|\to\infty} D^\alpha f(x) = 0$, and

(ii) $|\alpha| = d \implies \exists \varepsilon > 0$ such that $\lim_{\|x\|\to\infty} D^\alpha f(x) \|x\|^{d+1+\varepsilon} = 0$.

We denote by $\mathcal{V}(\mathbb{R}^d)$ the set of all functions of weakly controlled decay on $\mathbb{R}^d$. This set includes the Schwartz class of functions rapidly decreasing at infinity as well as the class of d-times continuously differentiable functions with compact supports. In particular, it includes the Gaussian function. Functions in $\mathcal{V}(\mathbb{R}^d)$ have finite Sobolev seminorms defined as

$$\|f\|_{d,1,\infty} := \max_{|\alpha|=d} \|D^\alpha f\|_{\mathcal{L}^1(\mathbb{R}^d)}.$$

Let $A_{r_d}^{(3)}$ denote the intersection of $\mathcal{V}(\mathbb{R}^d)$ with the ball $B_{r_d}(\|\cdot\|_{d,1,\infty})$ of radius r_d in the Sobolev seminorm $\|.\|_{d,1,\infty}$, i.e.,

$$A_{r_d}^{(3)} = \mathcal{V}(\mathbb{R}^d) \cap B_{r_d}(\|\cdot\|_{d,1,\infty}).$$

So, smoothness of functions in $A_{r_d}^{(3)}$ is expressed in terms of a condition on the maxima of $\mathcal{L}^1$-norms of iterated partial derivatives of order $|\alpha| = d$.

Theorem 3. *Let d be an odd integer, $\Omega_d \subset \mathbb{R}^d$ be a set of finite Lebesgue measure, $r_d > 0$, $k_d = 2^{1-d}\pi^{1-d/2}d^{d/2}/\Gamma(d/2) \sim (\pi d)^{1/2}(e/2\pi)^{d/2}$, and σ be any continuous nondecreasing sigmoid. Then for all positive integers n*

(i) $\delta(A_{r_d}^{(3)}|_{\Omega_d}, \operatorname{span}_n H_d(\Omega_d))_{\mathcal{L}^2(\Omega_d)} \le k_d \lambda(\Omega_d)^{1/2} r_d\, n^{-1/2}$

(ii) $\delta(A_{r_d}^{(3)}|_{\Omega_d}, \operatorname{span}_n H_d^\sigma(\Omega_d))_{\mathcal{L}^2(\Omega_d)} \le k_d \lambda(\Omega_d)^{1/2} r_d\, n^{-1/2}$.

Proof. (i) It was shown in [9, Corollary 4.3] that for d odd and every $f \in \mathcal{V}(\mathbb{R}^d)$ one has $\|f\|_{H_d, \mathcal{M}(\mathbb{R}^d)} \le k_d \|f\|_{d,1,\infty}$. It is easy to check that for every Ω_d of finite Lebesgue measure $\|f|_{\Omega_d}\|_{H_d(\Omega_d), \mathcal{L}^2(\Omega_d)} \le \|f|_{\Omega_d}\|_{H_d(\Omega_d), \mathcal{M}(\Omega_d)} \le \|f\|_{H_d, \mathcal{M}(\mathbb{R}^d)}$. As $\sup_{g \in H_d(\Omega_d)} \|g\|_{\mathcal{L}^2(\Omega_d)} = \lambda(\Omega_d)^{1/2}$, the item (i) follows by Theorem 1 (i).

(ii) It was shown in [24] that if $\Omega_d \subset \mathbb{R}^d$ is a set of a finite Lebesgue measure, then in $\mathcal{L}^p(\Omega_d)$ for any continuous nondecreasing sigmoid σ, $H_d^\sigma(\Omega_d)$-variation is equal to $H_d(\Omega_d)$-variation. So the statement follows by (i). $\square$

In the upper bounds from Theorem 3, $\xi(d) = k_d\, \lambda(\Omega_d)^{1/2} r_d$. Since $k_d \sim (\pi d)^{1/2} \left(\frac{e}{2\pi}\right)^{d/2}$ tends to zero exponentially fast, even if the maxima of $\mathcal{L}_1$-norms of partial derivatives grow rather quickly with the dimension d, approximation is still tractable.

The rate of growth of r_d for which Theorem 3 guarantees tractability depends on the shape of the domain Ω_d, too. The same remark as the one after Theorem 2 on the dependence of Lebesgue measures of balls and cubes on d applies here.

The next corollary estimates the worst-case $\mathcal{L}^2$-errors in approximation by perceptron networks of the set $G_d^{\gamma,1} = \{\tau_y(\gamma_d) \,|\, y \in \mathbb{R}^d\}$ of d-variable Gaussians with widths equal to 1 and varying centroids.

Corollary 3. *Let d be an odd integer and $\Omega_d \subset \mathbb{R}^d$ be a set of a finite Lebesgue measure. Then for every positive integer n*

$$\delta(G_d^{\gamma,1}|_{\Omega_d}, \operatorname{span}_n H_d(\Omega_d)_{\mathcal{L}^2(\Omega_d)}) \le (2\pi d)^{3/4} \lambda(\Omega)^{1/2} n^{-1/2}.$$

Proof. It was shown in [9, Corollary 6.2] that $\|\gamma_d\|_{H_d, \mathcal{M}(\mathbb{R}^d)} \leq (2\pi d)^{3/4}$. It is easy to see that for every $G \subset (\mathcal{X}, \|.\|_{\mathcal{X}})$ closed under translation, every f, and every $y \in \mathbb{R}^d$, one has $\|\tau_y(f)\|_{G, \mathcal{X}} = \|f\|_{G, \mathcal{X}}$. Hence $\|\tau_y(\gamma_d)\|_{H_d, \mathcal{M}(\mathbb{R}^d)} \leq (2\pi d)^{3/4}$. Thus, for every Ω_d of a finite Lebesgue measure we get

$$\|\tau_y(\gamma_d)|_{\Omega_d}\|_{H_d(\Omega_d), \mathcal{L}^2(\Omega_d)} \leq \|\tau_y(\gamma_d)|_{\Omega_d}\|_{H_d(\Omega_d), \mathcal{M}(\Omega_d)} \leq \|\tau_y(\gamma_d)\|_{H_d, \mathcal{M}(\mathbb{R}^d)}.$$

As $\sup_{g \in H_d(\Omega_d)} \|g\|_{\mathcal{L}^2(\Omega_d)} = \lambda(\Omega)^{1/2}$, by Theorem 1 (i) for every $y \in \mathbb{R}^d$ we get $\|\tau_y(\gamma_d)|_{\Omega_d} - \mathrm{span}_n H_d(\Omega_d)\|_{\mathcal{L}^2(\Omega_d)} \leq (2\pi d)^{3/4} \lambda(\Omega)^{1/2} n^{-1/2}$, which implies the statement. $\qquad\qquad\square$

In the upper bound from Corollary 3, we have $\xi(d) = d^{3/4} \lambda(\Omega_d)^{1/2}$. This implies that approximation of d-variable Gaussians on domains Ω_d by perceptron networks is tractable when the Lebesgue measures $\lambda(\Omega_d)$ grow polynomially with d. So, Gaussian-basis functions can be replaced by Heaviside perceptron networks with only a polynomial increase in the number of computational units. In particular for d odd and sets $\Omega_d \subset \mathbb{R}^d$ of unit Lebesgue d-dimensional measures, the $\mathcal{L}^2(\Omega_d)$-errors in approximating the Gaussians by linear combinations of n Heaviside perceptrons are at most $(2\pi d)^{3/4} n^{-1/2}$.

6 Discussion

We have described some sets of functions for which approximation by neural networks with n hidden units is tractable in the number d of variables. For such sets, estimates of worst-case errors in the form $\xi(d)\kappa(n)$ with $\xi(d)$ polynomial follow from the Maurey-Jones-Barron theorem and its extensions to $\mathcal{L}^p$-spaces, for $p \in (1, \infty]$. However, other proof techniques for estimation of rates of neural-network approximation (such as those based on Hermite polynomials used in [4] and [25]) may yield other tractability results.

Acknowledgement

V. K. was partially supported by the Project 1ET100300517 of the program "Information Society" of the National Research Program of the Czech Republic and the Institutional Research Plan AV0Z10300504. M.S. was partially supported by a PRIN Grant from the Italian Ministry for University and Research, project "Models and Algorithms for Robust Network Optimization". Collaboration of V. K. and M. S. was partially supported by the 2007–2009 Scientific Agreement among University of Genoa, National Research Council of Italy, and Academy of Sciences of the Czech Republic. P. C. K. was partially funded by Georgetown University.

References

1. Traub, J.F., Werschulz, A.G.: Complexity and Information. Cambridge University Press, Cambridge (1999)
2. Wasilkowski, G.W., Woźniakowski, H.: Complexity of weighted approximation over $\mathbb{R}^d$. J. of Complexity 17, 722–740 (2001)
3. Woźniakowski, H.: Tractability and strong tractability of linear multivariate problems. J. of Complexity 10, 96–128 (1994)
4. Mhaskar, H.N.: On the tractability of multivariate integration and approximation by neural networks. J. of Complexity 20, 561–590 (2004)
5. Kainen, P.C., Kůrková, V., Sanguineti, M.: Complexity of Gaussian radial basis networks approximating smooth functions. J. of Complexity 25, 63–74 (2009)
6. Barron, A.R.: Universal approximation bounds for superpositions of a sigmoidal function. IEEE Transactions on Information Theory 39, 930–945 (1993)
7. Knuth, D.E.: Big omicron and big omega and big theta. SIGACT News 8(2), 18–24 (1976)
8. Bellman, R.: Dynamic Programming. Princeton University Press, Princeton (1957)
9. Kainen, P.C., Kůrková, V., Vogt, A.: A Sobolev-type upper bound for rates of approximation by linear combinations of Heaviside plane waves. J. of Approximation Theory 147, 1–10 (2007)
10. Kůrková, V., Savický, P., Hlaváčková, K.: Representations and rates of approximation of real–valued Boolean functions by neural networks. Neural Networks 11, 651–659 (1998)
11. Beliczynski, B., Ribeiro, B.: Several enhancements to hermite-based approximation of one-variable functions. In: Kůrková, V., Neruda, R., Koutník, J. (eds.) ICANN 2008, Part I. LNCS, vol. 5163, pp. 11–20. Springer, Heidelberg (2008)
12. Barron, A.R.: Neural net approximation. In: Narendra, K. (ed.) Proc. 7th Yale Workshop on Adaptive and Learning Systems, pp. 69–72. Yale University Press (1992)
13. Breiman, L.: Hinging hyperplanes for regression, classification and function approximation. IEEE Transactions on Information Theory 39, 999–1013 (1993)
14. Darken, C., Donahue, M., Gurvits, L., Sontag, E.: Rate of approximation results motivated by robust neural network learning. In: Proceedings of the Sixth Annual ACM Conference on Computational Learning Theory, pp. 303–309. The Association for Computing Machinery, New York (1993)
15. Jones, L.K.: A simple lemma on greedy approximation in Hilbert space and convergence rates for projection pursuit regression and neural network training. Annals of Statistics 20, 608–613 (1992)
16. Pisier, G.: Remarques sur un résultat non publié de B. Maurey. In: Séminaire d'Analyse Fonctionnelle 1980-1981, École Polytechnique, Centre de Mathématiques, Palaiseau, France, vol. I(12)
17. Gurvits, L., Koiran, P.: Approximation and learning of convex superpositions. J. of Computer and System Sciences 55, 161–170 (1997)
18. Kůrková, V.: Dimension-independent rates of approximation by neural networks. In: Warwick, K., Kárný, M. (eds.) Computer-Intensive Methods in Control and Signal Processing. The Curse of Dimensionality, pp. 261–270. Birkhäuser, Basel (1997)
19. Kolmogorov, A.N., Fomin, S.V.: Introductory Real Analysis. Dover Publications Inc. (1970)

20. Kůrková, V., Sanguineti, M.: Error estimates for approximate optimization by the extended Ritz method. SIAM J. on Optimization 15, 461–487 (2005)
21. Stein, E.M.: Singular Integrals and Differentiability Properties of Functions. Princeton University Press, Princeton (1970)
22. Courant, R.: Differential and Integral Calculus, vol. II. Wiley-Interscience, Hoboken (1988)
23. Kainen, P.C.: Utilizing geometric anomalies of high dimension: When complexity makes computation easier. In: Warwick, K., Kárný, M. (eds.) Computer-Intensive Methods in Control and Signal Processing. The Curse of Dimensionality, pp. 283–294. Birkhäuser, Basel (1997)
24. Kůrková, V., Kainen, P.C., Kreinovich, V.: Estimates of the number of hidden units and variation with respect to half-spaces. Neural Networks 10, 1061–1068 (1997)
25. Narcowich, F.J., Ward, J.D., Wendland, H.: Sobolev error estimates and a Bernstein inequality for scattered data interpolation via radial basis functions. Constructive Approximation 24, 175–186 (2006)

Handling Incomplete Data Using Evolution of Imputation Methods

Pawel Zawistowski[1] and Maciej Grzenda[2]

[1] Warsaw University of Technology,
Faculty of Electronics and Information Technologies, Institute of Electronic Systems
ul. Nowowiejska 15/19, 00-665 Warsaw, Poland
pzawist2@elka.pw.edu.pl
[2] Warsaw University of Technology, Faculty of Mathematics and Information Science,
Pl. Politechniki 1, 00-661 Warsaw, Poland
grzendam@mini.pw.edu.pl

Abstract. In this paper new approach to treat incomplete data has been proposed. It has been based on the evolution of imputation strategies built using both non-parametric and parametric imputation methods. Genetic algorithms and multilayer perceptrons have been applied to develop a framework for constructing the imputation strategies addressing multiple incomplete attributes. Furthermore we evaluate imputation methods in the context of not only the data they are applied to, but also the model using the data. The accuracy of classification on data sets completed using obtained imputation strategies has been described. The results outperform the corresponding results calculated for the same data sets completed using standard strategies.

Keywords: imputation methods, incomplete data, genetic algorithm, multilayer perceptron.

1 Introduction

Data quality is always a matter of concern. Whether the task is data mining, machine learning, statistical analysis or any other, the results strongly depend on the quality of the given data. Most data sets are imperfect in some way. Over the years there have been quite a few attempts to define all the kinds of imperfections found in various data sets. Parsons [12] makes a summary of some of those attempts, and presents five imperfection types: *uncertainty, imprecision, incompleteness, inconsistency and ignorance*. In this paper we focus on incompleteness which occurs when some values in a data set are missing. Little and Rubin[11] define three ways in which data may be missing : MCAR (*missing completely at random*), MAR (*missing at random*) and NMAR (*not missing at random or non-ignorable*). MCAR means that whether some value is missing does not depend on any other values present in the data set. MAR is a situation in which a missing value of some attribute may depend on the other attributes, but not on the value itself. NMR is the most difficult situation in which the fact

M. Kolehmainen et al. (Eds.): ICANNGA 2009, LNCS 5495, pp. 22–31, 2009.
© Springer-Verlag Berlin Heidelberg 2009

that a value is missing may depend on the value itself. The method proposed in this paper does not directly require the data to fit to one specific type. However, as method vectors use imputation methods, such requirements may be imposed by the latter.

There are three general approaches in face of incompleteness. First of all, an attempt can be made to acquire and fill in the missing data. This however can be time consuming, difficult, expensive or sometimes not even possible. Another possibility is to revise the data set and delete all the impaired instances or even attributes (this approach is sometimes called *complete case analysis*). When the data are valuable, this is unfortunately not feasible. The last solution is to impute the missing values using a proper method. The imputation approach makes it possible to avoid deleting possibly useful information on one side, but poses a threat of introducing errors into the data set on the other. Yet in many cases imputation is the best solution to incompleteness.

Filling in missing data requires choosing from all the available methods the one that gives the best possible results. One of the most famous is probably the *EM-algorithm* presented, among others, by Dempster et al. [4] and Schafer [13]. The procedure assumes a distribution of the missing values and then fills them in using a two step iterative procedure : the *E-step* estimates the expected values of the missing data and the *M-step* changes the parameters of the distribution to maximize the likelihood of the data. Another interesting iterative method called non invasive imputation has been proposed by Gediga and Düntsch [5]. Abdella and Marwala propose to use neural networks as an imputation method [1]. A whole framework for dealing with incomplete data for use in data mining is described by Wei and Tang [15]. One-class classifiers are proposed as a method by Juszczak and Duin [10]. The kNN algorithm is used as an imputation method by Acũna [2], Batista and Monard [3], Cohen et al. [8] and Jönsson [9]. Finally Hu et al. [8] gives an overview of some popular imputation methods.

When there is a need to fill in missing values to more than one attribute in a data set, the choice of a single method to perform all the imputations becomes difficult, or not even possible. In particular, the latter problem occurs when no method suitable for all the attribute types exists. Furthermore, using a single method to perform all the imputations does not guarantee the possibility of achieving the best results.

In addition, the notion of correct imputation may depend on the problem the data set is used for. Treating incompleteness of a data set is usually a preprocessing step, which is meant to prepare the set to be used by some model. In such situations, the same model can be used when choosing the right imputation methods, i.e. in their evaluation.

In this paper a framework aiming to combine different imputation methods has been proposed. The way evolutionary algorithm can be used to find imputation strategies for data sets with multiple incomplete attributes has been proposed. Moreover, the strategies are evaluated in view of the problem the data set is used for. In other words, the suitability of imputation strategy is evaluated in terms of its impact on the prediction or classification process that is performed using the data set.

The framework combines genetic algorithms and multilayer perceptrons. Genetic algorithms have been used to evolve method vectors representing imputation strategies. Multilayer perceptrons (MLP) have been applied to represent classification models used in test cases. The error rate of classification performed on the data sets imputed using a strategy, provides basis for the fitness function. In other words, the higher the accuracy of the classification performed on the imputed data set, the better the imputation strategy used to fill in missing values is.

The work follows our research on time series prediction using combination of evolutionary algorithms and neural networks [6,7]. The main assumption is to develop frameworks capable of autonomous model development and data processing that do not require expert intervention.

The rest of the paper is organized as follows. Section 2 describes the concept of model-based evaluation. In section 3 precise definition of method vectors is given and an evolutionary algorithm for finding them is proposed. Section 4 contains the description of the framework used to test our propositions. Tests and results are described in section 5. Finally a summary is presented and further work topics are proposed.

2   Model Based Evaluation

When evaluating an imputation method, the distances between the original and the filled in values are often used as the performance measure. This applies to the test cases, in which values are artificially removed and then imputed.

However when the data set is noisy, and this is often the case, such an approach may lead to efforts in recreating noise. Consider a situation in which there is a character recognizing model given and it is meant to be used on a data sample from Fig. 1. The only problem is that the model cannot handle incomplete data and therefore all missing parts have to be filled in. Which imputation methods could be considered as "good" in such a situation? Bearing in mind the goal (which is character recognition) the answer is: all the methods that impute the missing data in a way, which will not distort the character recognition model (i.e. the model should recognize the "K" letter on the sample). There is no need to try to find a method which could recreate the missing part of the sample precisely with all the noise that was originally there.

We want to check the feasibility of measuring performance of imputation methods (in our case method vectors) in context of the model that will be using the filled in data. In real life situations the models, the incomplete data sets are meant for, are often known before performing the imputations. In such situations, knowing which methods perform well for the given model would be beneficial.

Fig. 1. A noisy letter sample with some of the data missing

3   Method Vectors

For a given set of imputation methods Γ, and a given data set D, in which attributes $a_1, a_2, \ldots a_n$ suffer from incompleteness, $V = [m_1, m_2, \ldots, m_n]$ is a method vector, where $m_i \in \Gamma$ is an imputation method meant for filling in missing values for attribute a_i. Thus method vectors are vectors in an n-dimensional *imputation method space* defined by a specific data set D and a specific set of imputation methods Γ. Such vectors are used to fill in the incomplete data sets. The goal is of course to achieve the best results possible, but in order to do that we have to define what exactly is the *best method vector*. To evaluate the method vectors we use a model-based approach and present a definition of the best method vector below.

With all the previous assumptions, let M denote a given model to be used. The role of the model is to address the problem the data set is used for e.g. to perform prediction or classification. Thus, in general $M : \mathbb{R}^N \rightarrow \mathbb{R}^C$, while $D \subset \mathbb{R}^N$ and $n \leq N$. In particular all attributes can miss some values. In the latter case $n = N$. In case of classification the number of output signals produced by model M is $C = 1$. In our test case the model is implemented using multilayer perceptron.

Moreover, $e_M(D)$ denotes the mean absolute error of model M on data set D. Furthermore D_V is the data set created from D by filling in all the missing values using methods from vector V.

Using the above notation, the best method vector $V^* = [m_1, m_2, \ldots, m_n]$ is defined as follows:

$$e_M(D_{V^*}) = \min_V e_M(D_V) \tag{1}$$

Equation 1 holds when all the methods in Γ are non-parametric, however some imputation methods do have parameters. In such situations not only the method vectors have to be found, but also their parameters.

Let

$$P(V) = \{p : p \in P_{m'_1} \times P_{m'_2} \times \ldots \times P_{m'_K}\}$$

denote the parameter space for vector V, where $K \leq n$ and $P_{m'_i}$ is the parameter space for the i-th parametric method. D_V^p denotes the data set created from D

by filling in all the missing values using methods from vector V with parameters $p \in P(V)$.

The objective of finding the best imputation method vector with parameters, means actually finding the best pair $[V^*, p^*]$, where $p^* \in P(V^*)$ and it holds that

$$e_M\left(D_{V^*}^{p^*}\right) = \min_{V, p \in P(V)} e_M\left(D_V^p\right) \tag{2}$$

Now a procedure of performing an imputation using method vectors can be proposed. For a given data model M and a given incomplete data set D, the imputation procedure proposed in Alg. 1 has been used.

Input: Γ - a set of imputation methods to be used, D - an incomplete data set,
      M - a model to be used for data set D
Result: $D_{V^*}^{p^*}$, which is a complete data set to be used by model M
begin
    $[V^*; p^*]$ = find a pair for which Eq. 2 holds;
    $D_{V^*}^{p^*}$ = fill in the incomplete data set D using methods defined in vector V^*
    with parameters p^*;
    return $D_{V^*}^{p^*}$;
end

Algorithm 1. Method vector based imputation procedure

This formulates quite a difficult optimization task, which can be solved using a genetic algorithm. In our approach, the genetic algorithm works on populations of method vectors. Each individual consists of n genes representing imputation methods used for n individual incomplete attributes. To simplify the notation we can assume that we have two types of genes: non-parametric genes - used for non-parametric methods and parametric genes applied for parametric methods. This means that parametric genes are actually pairs $[m, p_m]$, where m is an imputation method, and p_m is the vector of parameter values for this method.

The proposed algorithm uses two types of genetic operators to diversify the population, namely crossover and mutation. Both operators have two versions. One version is used for non-parametric and one for parametric genes.

For a given maximum range R, mutation changes an individual by randomly changing a random number $r \in [0, R]$ of genes. The crossover operator is based on one-point crossover. Both the parametric mutation and parametric crossover depend on specific parametric genes i.e. imputation method each parametric gene represents. This is the case, because the parameter space may and usually is different for each imputation method. For example a kNN based method requires one parameter, while a SOM based method would require at least two. The mutation is applied according to Alg. 2.

In general, in the first stage of the algorithm changes occur both to the set of methods and parameters of methods comprising on imputation strategies can occur. In the final stage of the algorithm, changes are applied to parameters only.

Data: $P = [i_1, \ldots, i_P]$ - current population, $i_j = [g_1^j, \ldots, g_N^j]$ - j-th individual
with genes $g_1^j, \ldots, g_N^j$, T - the threshold turning off non parametric
operations, num - current generation number, MAX - maximal number
of generations, par - the parametric mutations threshold

begin
    $i_j =$ randomly select an individual from P;
    $l = |\{g^j : g^j \text{is a parametric method}\}|$;
    if $l = 0 \wedge num > T \times MAX$ **then**
      | **return** no mutation;
    else
      $p =$ a random number within $[0; 1]$;
      if $l > 0 \wedge ((num > T \times MAX) \vee (p \leq par))$ **then**
        | **return** parametric mutation of individual i;
      else
        | **return** non-parametric mutation of individual i;
      end
    end
end

Algorithm 2. The selection of mutation operator

4 Tests

The goal of the tests was to compare the results that can be achieved using
the proposed algorithm with the standard approach i.e. using single imputation
method to fill in all the missing data. The comparison has been based on the performance of classification models working on the imputed data sets. Multilayer
perceptrons have been used to implement the models.

4.1 Method Set

The method set defining the search space used during our tests consists of the
following methods: *random/median/mode/mean imputation, non-invasive imputation*, the *kNN algorithm* and the *SOM-based imputation*. The latter two
methods have been used in three versions, namely *mean/median* and *mode*.

Random imputation fills in values at random with a distribution based on the
complete parts of the data set. Median, mode and mean imputation fill in the
missing values with the median, mode and mean values, respectively. The kNN
method fills the incomplete data using the mean, median or mode from k nearest
neighbours of the given data row. The SOM-based approach is similar to kNN,
except for the fact that the neighbourhood is found using a self organizing map
to cluster the rows together. Non-invasive imputation is an iterative algorithm
presented in [5].

4.2 Test Preparation

The tests were done on 3 data sets from the *UCI machine learning repository*:
Iris, Congressional Voting Records and Breast Cancer Wisconsin (Diagnostic).

Table 1. Genetic algorithm parameters

Value Description
20 population size
100 number of generations
0.1 mutated population fraction
0.2 crossed over population fraction
0.8 fraction of generation with non-parametric genetic operations

The data sets were split into training and testing parts and they were used to train and test neural networks for each data set. The incomplete rows of the Votes set were put aside for this part of the tests.

MLP has been used to build a model using each data set. Java Neural Network Simulator (JNNS) was used to train a neural network for each of our test data sets. Having all the neural networks, data sets for the genetic algorithm were prepared. Values from 25% and 50% of the rows from the testing parts of Iris and Wisconsin data sets were removed. The Votes set is already incomplete. The obtained incomplete sets were used as the working sets for the rest of our tests. The objective of the algorithm was to minimize the neural network mean square errors achieved on the imputed working set.

4.3   Test Procedure

For each working set - 75% (search part) was used to find the best method vectors and single imputation methods, and 25% (evaluation part) to evaluate the found solutions. Iris set was an exception - because it contains only 150 rows, 10 fold cross-validation was used in that case.

The algorithm settings have been chosen so that to allow numerous runs of the algorithm and ensure adequate data for comparison purposes. In real applications, the number of generations and population members can be increased. The genetic algorithm was run 50 times for each data set with the parameters given in Tab.1. The performance of method vectors was compared with performance of single methods, which were found using another evolutionary algorithm. Algorithm 1 depicts the test procedure.

5   Results

The results of the tests are summarised in Tab. 2. All the values have been obtained on the testing set. The set was used neither to drive the evolution process, nor to train the MLP model. e_V stands for average MSE of the neural network on the data set filled in with method vectors. e_S denotes average MSE of the neural network on the data set filled in with single method used to impute all the missing values in all incomplete attributes. What should be emphasised, the set of individual methods that are investigated in the latter case, contains all the methods that can be used in method vectors.

Data:
$D_1 = [\text{Iris25, Iris50}], D_2 = [\text{Wisconsin25, Wisconsin50, Votes}]$ - test data sets

```
begin
    foreach D ∈ 𝔻₁ do
        tab[1...10] = split D into 10 parts;
        eᵥ = eₛ = 0;
        for i = 1...10 do
            V = solution from the genetic algorithm run on set D − tab[i];
            S = single best method found using set D − tab[i];
            eᵥ+ = 0.1 ∗ ( evaluate solution V on set tab[i]);
            eₛ+ = 0.1 ∗ ( evaluate solution S on set tab[i]);
        end
        return eᵥ, eₛ
    end
    foreach D ∈ 𝔻₂ do
        tab[1...2] = split D into 2 parts;
        V = solution from the genetic algorithm run on set tab[1];
        S = single best method found using set tab[1];
        eᵥ = evaluate solution V on set tab[2];
        eₛ = evaluate solution S on set tab[2];
        return eᵥ, eₛ
    end
end
```

Algorithm 3. The test procedure

Table 2. Test results. N - number of rows, N_{Inc} - number of incomplete rows, σ_{e_V}, σ_{e_S} - standard deviations.

Set	N	N_{Inc}	e_V	σ_{e_V}	e_S	σ_{e_S}
Iris 25%	75	19	0.0730	0.0189	0.2494	0.0430
Iris 50%	75	38	0.0873	0.0347	0.4191	0.0565
Wisconsin 25%	285	72	0.1246	0.0274	0.2208	0.0360
Wisconsin 50%	285	143	0.1061	0.0491	0.3834	0.0594
Votes	280	203	0.0903	0.0060	0.1868	0.0168

The results show that for all data sets a suitable method vector representing an imputation strategy has been found. The error rate of classification models M, denoted by e_V has been significantly lower than the error rate e_S achieved when completing the data with the best individual strategy. Moreover, standard deviation σ_{e_V} shows that the proposed method provides results with highly similar quality in different algorithm runs.

An interesting phenomenon can be seen when comparing the results of *Iris 25%* and *Iris 50%*, and also *Wisconsin 25%* and *Wisconsin 50%*. The performance obtained on the sets with more data missing (50%) does not decrease (in comparison with the 25% versions) as significantly as it could be expected. Moreover, in the case of Wisconsin data sets, the performance even improves.

This is an interesting observation which may have different causes. First of all a greater number of missing values means that the objective function becomes more sensitive to the filled in values, which may make it easier for the genetic algorithm to find the best solutions. Another possibility is that using the model for calculating the objective function clears some errors originally present in the data set, or maybe distorts the data. The observed phenomenon of achieving better performance on potentially worse data deserves further investigation. This will require performing tests on a number of different models and data sets with a larger gradation of incompleteness levels.

To sum up, the results show that the imputation strategy found by the algorithm outperforms individual strategies used for the same data sets. Not only, the improvement of on average 73% in MSE rates has been obtained, but also the minimal improvement was 45%. Therefore, the algorithm helps to address the need for model-driven data imputation.

6 Summary

Method vectors may lead to much better results in data imputation. However finding the best method vectors is not an easy task, as the search space is complicated and has many different potential solutions achieving the same or similar results. The proposed algorithm successfully finds imputation strategies addressing the need of completing incomplete data sets. The error rate achieved when using these imputation strategies is significantly lower than the error rate obtained when completing the data set by the best involved individual imputation method.

The proposed framework allows to adopt multiple imputation methods used to deal with incompleteness of individual attributes. Not only the combination of them, but also their parameters can be set by evolution. Moreover, the framework construction allows to use different decision models. Any model, based on neural networks or other AI techniques that is applied to solve the classification or prediction problem can be included in the framework.

An interesting concept for further studies is to introduce an order in which the imputations take place and allow the methods to use previously filled in values for further imputation steps.

References

1. Abdella, M., Marwala, T.: The use of genetic algorithms and neural networks to approximate missing data in database. In: IEEE 3rd International Conference on Computational Cybernetics (2005)
2. Acuña, E., Rodriguez, C.: The treatment of missing values and its effect in the classifier accuracy. In: Classification, Clustering and Data Mining Applications. Springer, Heidelberg (2004)
3. Batista, G.E.A.P.A., Monard, M.C.: A Study of K-Nearest Neighbour as a Model-Based Method to Treat Missing Data. In: Argentine Symposium on Artificial Intelligence (2001)

4. Dempster, A.P., Laird, N.M., Rubin, D.B.: Maximum Likelihood from Incomplete Data via the EM Algorithm (1977)
5. Gediga, G., Düntsch, I.: Maximum consistency of incomplete data via non–invasive imputation. Artificial Intelligence Review 19 (2003)
6. Grzenda, M.: Load Prediction Using Combination of Neural Networks and Simple Strategies. Frontiers in Artificial Intelligence and Applications 173, 106–113 (2008)
7. Grzenda, M., Macukow, B.: Demand Prediction with Multi-Stage Neural Processing. In: Advances in Natural Computation and Data Mining, pp. 131–141. Xidian University Press, China (2006)
8. Hu, M., Salvucci, S.M., Cohen, M.P.: Evaluation of some popular imputation algorithms. In: Proceedings of the Survey Research Methods Section. American Statistical Association (1998)
9. Jönsson, P., Wohlin, C.: Benchmarking k-nearest neighbour imputation with homogeneous Likert data. Empirical Software Engineering 11(3) (2006)
10. Juszczak, P., Duin, R.P.W.: Combining One-Class Classifiers to Classify Missing Data. Multiple Classifier Systems (2004)
11. Little, R.J., Rubin, D.B.: Statistical Analysis with Missing Data, 2nd edn. John Wiley and Sons, Chichester (2002)
12. Parsons, S.: Current approaches to handling imperfect information in data and knowledge bases. IEEE Transactions on Knowledge and Data Engineering 8(3) (1996)
13. Schafer, J.L.: Analysis of Incomplete Multivariate Data. Chapman & Hall/CRC, Boca Raton (1997)
14. Strike, K., El Emam, K., Madhavji, N.: Software cost estimation with incomplete data. IEEE Transactions on Software Engineering 27(10) (2001)
15. Wei, W., Tang, Y.: A generic neural network approach for filling missing data in data mining. In: IEEE International Conference on Systems, Man and Cybernetics (2003)

Ideas about a Regularized MLP Classifier by Means of Weight Decay Stepping

Paavo Nieminen and Tommi Kärkkäinen

Department of Mathematical Information Technology,
University of Jyväskylä, Finland
{nieminen,tka}@jyu.fi
http://www.mit.jyu.fi/
http://users.jyu.fi/~nieminen/

Abstract. The generalization capability of a multilayer perceptron can be adjusted by adding a penalty (weight decay) term to the cost function used in the training process. In this paper we present a possible heuristic method for finding a good coefficient for this regularization term while, at the same time, looking for a well-regularized MLP model. The simple heuristic is based on validation error, but not strictly in the sense of early stopping; instead, we compare different coefficients using a subdivision of the training data for quality evaluation, and in this way we try to find a coefficient that yields good generalization even after a training run that ends up in full convergence to a cost minimum, given a certain accuracy goal. At the time of writing, we are still working on benchmarking and improving the heuristic, published here for the first time.

Keywords: classification, neural networks, MLP, regularization, heuristic.

1 Introduction

This paper deals with the task of pattern classification. We begin by going through the necessary definitions and equations for those who want to implement a similar system. Later on, we shall turn to less formalized, heuristic ideas.

Let $\{\mathbf{x}_i\}_{i=1}^N, \mathbf{x}_i \in \mathbb{R}^D$ be a set of N vectors of dimension D. Each vector $\mathbf{x}_i$ is known to belong to a class $c_i \in \{1, \ldots, C\}$, coded as a binary vector $\mathbf{y}_i \in \mathbb{R}^C$ such that the components of the i:th vector are $(\mathbf{y}_i)_k = 1$ for $k = c_i$ and $(\mathbf{y}_i)_k = -1$ for $k \neq c_i$. Using the known pairs $(\mathbf{x}_i, \mathbf{y}_i)$ as examples, we aim to train a computer system to associate vectors with a corresponding class. For our current purposes, any vector $\mathbf{y} \in \mathbb{R}^C$ is converted to an exact class representative by choosing c equal to smallest k for which $(\mathbf{y})_k \geq (\mathbf{y})_j$ for all $j \in \{1, \ldots, C\}$, i.e., selection of the index of the largest component, or, in the case of equality, the one with the smallest index. This conversion works fine for vectors that are already approximately close to the encoded class prototype.

We especially want the machine to be able to generalize what it has learned by example, and apply the knowledge to new vectors $\mathbf{x}_q \notin \{\mathbf{x}_i\}_{i=1}^N$ for which nobody

M. Kolehmainen et al. (Eds.): ICANNGA 2009, LNCS 5495, pp. 32–41, 2009.
© Springer-Verlag Berlin Heidelberg 2009

knows the proper class before the machine makes its advisory guess. In this way, we can automatically identify images, sounds, industrial measurements, and other useful things that can be represented as vectors. Artificial neural networks (ANN) are a widely used system for such automatic classification tasks. For more knowledge about ANNs, we refer to the textbook [1].

Of special interest in this study is the multilayer perceptron (MLP), a feed-forward ANN with sigmoidal activation. Such an ANN can be easily (and efficiently) implemented directly from the matrix representation described in prior works [2,3,4]. For any $\mathbf{x} \in \mathbb{R}^D$ we set

$$\mathbf{o}^0 = \mathbf{x}, \quad \mathbf{o}^l = \mathcal{F}^l(\mathbf{W}^l \hat{\mathbf{o}}^{(l-1)}) \quad \text{for} \quad l = 1, \ldots, L \, . \tag{1}$$

By the notation $\mathbf{o}^0$ we mean the vector on the "zeroth" layer, which is considered to be the input vector $\mathbf{x}$ presented to the network. Iterative computation yields the output vectors $\mathbf{o}^l$ for all the other L layers of the network, numbered $l = 1, \ldots, L$. The values depend on the selection of layer-wise neural weight matrices $\mathbf{W}^l$. The hat notation $\hat{\mathbf{o}}^{(l-1)}$ means that the output vector of the previous layer $\mathbf{o}^{(l-1)}$ is extended by an initial coordinate of value one, facilitating the bias mechanism so that the first column of $\mathbf{W}^l$ contains the biases of neurons on layer l. Finally, the notation $\mathcal{F}^l(\cdot)$ denotes the application of a function matrix to a vector. For this study, the function matrices are diagonal, and they consist solely of the traditional hyperbolic tangent activation function on hidden layers and the identity mapping on the output layer. The reasoning behind this choice is presented in [3]. The final output of the MLP resides in the output vector $\mathbf{o}^L$. We use the notation $\mathcal{N}(\{\mathbf{W}^l\})(\mathbf{x}) = \mathbf{o}^L$ to denote the output of the above iterative computation.

A model such as (1) is to be trained somehow. A usual way is to minimize an error function, or *cost function*, computed over the available training data set. The present work is based on the following cost function formulation:

$$J(\{\mathbf{W}^l\}) = \frac{1}{2N} \sum_{i=1}^{N} \left\| \mathcal{N}(\{\mathbf{W}^l\})(\mathbf{x}_i) - \mathbf{y}_i \right\|^2 + \beta \sum_{l=1}^{L} \sum_{(i,j) \in I_l} \frac{1}{2S_l} |\mathbf{W}_{i,j}^l|^2 \tag{2}$$

for $\beta \geq 0$. Here, the index set I_l contains all other indices of the weight matrices except the ones corresponding to the bias-vector (i.e., first column) of $\mathbf{W}^L$ as suggested by the test results in [3] (see also [5], Chapter 9). S_l is the number of elements in the index set I_l. This averaging divisor was not present in the earlier works [3,4].

If we were to use only the first mean-squared-error term, and find a model near the global optimum of that cost function, it is quite likely that the model would *overfit* individual quirks of the training examples, and not be able to generalize into unforeseen vectors. As seen from the formula, we have decided to make experiments with the well-known method of adding a *regularization term*, with a weight coefficient β called the *regularization parameter* [5]. The choice of the weight decay formulation with a single coefficient is more thoroughly explained and contrasted with early stopping in [3].

We are aware of other alternative heuristic methods to overcome the overfitting issue, such as network growing and pruning (or other ways of evolving the architecture), and early stopping by validation [1]. In the future, we will look into incorporating ideas from them, as well as from evolutionary and multiobjective approaches such as those in [6,7,8]. For this paper though, we use a non-evolving, fully connected MLP, and use the more traditional gradient-based training at the core. Early stopping will be involved, in a way.

In Sect. 2 we motivate and describe a heuristic training algorithm that operates using the ideas of cross-validation, brute-force parameter search, regularization, and early stopping. To our knowledge, the method as such has not been published before. In the crowded field of machine learning we can be mistaken, in which case we hope this paper is still useful as a utility assessment and computational experiment on that specific heuristic. In Sect. 3 we show how the heuristic operates in practice, using a benchmark classification problem, and in Sect. 4 we recap, and make remarks of the current and future study of the method.

2 Tentative Proposal of a Heuristic for Finding a Good Regularized MLP

We begin by describing the track of thoughts leading to the algorithm which is then stated in pseudocode.

2.1 Underpinnings of the Heuristic

In essence, our basic idea is to shoot some random starting points into the weight space, and try, by means of some extra control, to hit at least one location that feels promising. Then, in the end, we take a more determined look in the neighborhood of that location.

First of all, because we use local gradient-based optimization while the cost function is known to be non-convex, we cannot avoid the fact that the result depends on the starting point selection. In fact, the cost surface in MLP training is known to be quite tricky for optimization (as a side note, in [9] the cost function is visualized approximately, and several factors affecting the shape, including the effect of weight decay, are pointed out and visualized). Because of this reasoning, we introduce *multiple starting points* as the basis of the algorithm: as an outer loop, we start many times from different random weights.

Another heuristic belief is that, because the cost function is tricky, perhaps we should not be obliged to use the most robust optimization with guaranteed convergence properties with "non-tricky" assumptions. In a way, we always "stop early" before true convergence no matter how we decide when to stop. So, heuristically thinking, we could maybe use a quick-and-dirty optimization that will go towards the closest local optimum if it is easy, and more or less give up if it is not easy – as long as it happens so fast that we can have many restarts. As a first attempt, we imagine a *short spirt of conjugate gradient optimization* with loose accuracy criterion or hard-limited iteration counts will create a desired kind of local step forward.

We need to be able to tell which encountered network is likely to be the best. So, like in early stopping, we train the network using a training set for some epochs further from its current state, and then pause for a moment to do *quality evaluation using a validation set*. But we do a bit more at each of these checkpoints: We change the value of β each time. In the beginning, we set β to a relatively high value, which is known to pull weights towards zero, inhibiting the saturation of neurons, among other virtues. Strong regularization also smoothens the cost landscape, thus maybe accelerating the optimization in the beginning.

Because the first steps are done with a high amount of regularization, the MLP might automatically seek into a friendly position for the later steps, during which the regularization level is diminished. On the other hand, there is also a danger of underfitting; it was definitely not hard to experience this on the first day of trial runs. This looks like a question to think about, and gain some more insight.

As the inner loop, we *decrease β towards zero* with a few logarithmic steps. As a result, we expect the network to become a more aggressive classifier due to some greater weights. Eventually it could start overfitting, if it is complex enough. We refuse to make a guess about when overfitting occurs; instead of ever stopping, we *keep a record* of (i) the best validation error so far, (ii) the exact neural model, $\{\mathbf{W}^l\}_{l=1}^{L}$, that yielded the best result, and (iii) the β that was used while arriving at the best result. Early stopping is unnecessary, because we can always come back to the so-far best result. Early stopping would definitely save time, but now we care more about generalization than training time.

It is assumed that after sufficiently many restarts, we end up with a reasonably good network, and a reasonably good regularization coefficient. In the very end, we use the whole original training set, and strict demands in accuracy to make a *final training session*. We start the final optimization from the so-far best network, using the so-far best regularization coefficient. This way we eventually use the whole available data for training, but we have already incorporated validation measures.

If we were to step the β's with greater density, and with shorter iteration limits, we would be approaching a multi-starting, "early-stopping" search with a gradually decreasing penalty included in the cost function. But we see ours as a different kind of algorithm, since the β stepping is so sparse. And instead of guessing when to stop, we come back later to the sweetest looking spot. It is more like "random multistart with a bit of guidance for each restart". Unfortunately, at this point we cannot report to have found a definitive sequence of β's that would work well regardless of what data set is used, so the stepping strategy remains an open question. Other issues are addressed in Sect. 4.

2.2 Pseudocode

After verbosely describing the ideas and mainstays behind our algorithm, we present our current, tentative proposal as the following pseudocode:

1. Split the original training set into pre-train and validation (half-and-half). Decide a plan for successive betas to try; a basic choice is:

```
beta_plan = [1e-0, 1e-0.5, 1e-1, 1e-1.5, ..., 1e-6]
```

Initialize tracking of best combination:

```
beta_best = undefined.
valE_best = Inf
W_best    = undefined.
```

2. Execute the main loop:

```
For N_restarts times:
    W = random weights

    For each beta in beta_plan:
        W = network trained further from current values,
            using pre-train set, and current beta, loose accuracy.

        valE_this = sum-squared-error on pre-validation set

        If (valE_this < valE_best):
            beta_best = beta
            valE_best = valE_this
            W_best    = W
```

3. After all the N_restarts runs, reload the so-far best combination:

```
beta = beta_best
W    = W_best
```

4. Train the final network using the full training data, starting from W_best,
   using beta_best, tight accuracy demand, many iterations.

2.3   Gradient-Based Local Minimization

To train the MLP classifier further on each step of the algorithm we use a
fast gradient-based local minimization scheme based on the conjugate gradient
method. We recall from [3] the layer-wise matrix representation of the gradient
of Eq. (2):

$$\nabla_{\mathbf{W}^l} J(\{\mathbf{W}^l\}) = \frac{1}{N} \sum_{i=1}^{N} \boldsymbol{\xi}_i^l \, [\hat{\mathbf{o}}_i^{(l-1)}]^T + \frac{\beta}{S_l} \, \widetilde{\mathbf{W}}^l \, ,$$

where

$$\boldsymbol{\xi}_i^L = \mathcal{N}(\{\mathbf{W}^l\})(\mathbf{x}_i) - \mathbf{y}_i, \tag{3}$$

$$\boldsymbol{\xi}_i^l = \mathrm{Diag}\{(\mathcal{F}^l)'(\mathbf{W}^l \, \hat{\mathbf{o}}_i^{(l-1)})\} \, (\mathbf{W}_1^{(l+1)})^T \, \boldsymbol{\xi}_i^{(l+1)} \, . \tag{4}$$

In the formula, $\mathbf{W}_1^{(l+1)}$ denotes a matrix otherwise identical to $\mathbf{W}^{(l+1)}$ but
with the first column, containing the bias values, removed. Furthermore, $\widetilde{\mathbf{W}}^l = [\mathbf{0} \; \mathbf{W}_1^L]$ for $l = L$, and coincides with the whole matrix $\mathbf{W}^l$ for $1 \leq l < L$. This

last detail is necessary because we exclude the output layer's bias weights from the regularization term in the formulation (2).

Assessments of optimization methodology is beyond the scope of this paper. A comprehensive comparison to other training algorithms is being prepared and planned to be published as a technical report; we refer to the author's website for the current research status.

3   Computational Experiments

We have found benchmark problems available in the UCI Machine Learning Repository [10] to be useful. So far, we have run preliminary tests on the following data set titles:

- Iris: classifying flowers; this is a small case that takes little time to compute with, but on the other hand it could be too small for practical assessments. Interesting cases could include for example a surplus 4-100-3 net, like the one illustrated in [9].
- Wine: classifying wine cultivors; we have briefly tried 13-5-4-3 and 13-100-3 nets on this task; without artificial tampering this looks a bit too "easy" for benchmarking.
- PenDigits: the pen digit task with different net architectures has been tried; it is the only one to be present in this paper.
- Thyroid (ANN): the hypothyroid diagnosis task is being used, also with different net architectures.

More data sets will be tried later. Results of the benchmarking are still indecisive, and we postpone their presentation to a follow-up article. Nevertheless, we are already able to show some graphics about how the heuristic is supposed to work. Similar behaviour can be seen on most of the tried architectures and datasets, but so far they need manual tweaking of the parameters.

One major goal of our heuristic is to make the obtained classification accuracy independent of the chosen ANN architecture (i.e., number of layers and neurons), as long as the architecture is above some threshold level of complexity to prevent underfitting. The regularization (found via our heuristic) is responsible for the prevention of overfitting. For the demonstration presented in what follows, we chose a network structure of 16-30-10, which means 16 input variables, 30 neurons on one hidden layer, and 10 outputs for binary encoding of the base-10 digits to be recognized. Based on trials, even a smaller hidden layer would work for this dataset. And, as said, anything reasonably more complex should be usable with our heuristic.

3.1   Progression of the Heuristic

In Fig. 1 we can see how some observables change while the algorithm proceeds over the betas. In the illustration, the succession of β's is actually the 16 values $[\frac{1}{2}, 10^{-1}, 10^{-2}, \ldots, 10^{-7}, 10^{-6}, \ldots, 10]$. Each line in the drawing (replicated once

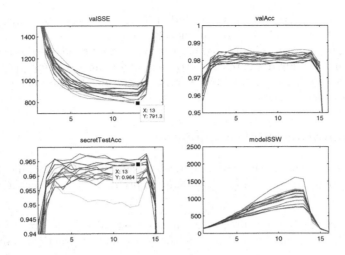

Fig. 1. Progression of the heuristic on the PenDigits 16-30-10 case: As β gradually decreases, and then increases again, the MLP passes a point where we believe it has a good generalization capability. In the end, the final training continues from the best location so-far, using the full training set and greater accuracy goal.

in each quadrant box) corresponds to one launch of a random initial MLP. The horizontal axis contains the checkpoints after which the validation error is measured and β is updated. For Fig. 1 this was after each 100 epochs, summing up to 1600 epochs for each restart. The vertical axes show how the measured value changes from checkpoint to checkpoint.

In the top left box, the vertical axis shows the key measurable, the validation error. From among all the runs, the final decision is made by remembering the minimal value hit of the sum-of-squared errors (SSE) measure on the validation set. It is hoped, and visually confirmed at least in the case of Fig. 1, that the validation SSE reflects quite well the classification accuracy of the test set (which is labeled "secret" in the image because the training algorithm may not use it for any decisions). The secret true accuracy is shown in the lower left box.

On the upper right box, we show the validation accuracy; it is basically the true functionality which may be different from the SSE measure. This is no secret to the algorithm, if we should choose to use it as a measure.

Lastly, on the lower right, we show the sum of squares of the MLP's neuronal connection weights. The effect of regularization is evident, because in the test run illustrated in Fig. 1 we let the β increase again after it touched the value 10^{-7}. In fact, such a there-and-back plan for β's could be a justifiable variation of the proposed algorithm.

The data marker in the lower-left box shows that the test set accuracy was 96.4 % for the 13th selection ($\beta = 0.01$), for which the validation SSE was minimal. For this individual, the model $\{\mathbf{W}\}$ and β got stored, and it can be seen that the heuristic did a rather good job here.

3.2 Comparison to Early Stopping with a Constant Regularization Coefficient

It is clear that the implementation of the algorithm is quite easy to turn into a traditional early-stopping mechanism by imposing a low maximum iteration limit for local optimization and setting the plan of β's to a long series of constant, or very little changing, values. For comparison, we ran the exact same benchmark case as in Fig. 1 through such a stripped-down training with $\beta = 0$. The progression of such algorithm is shown in Fig. 2. The validation error seems to be getting smaller all the time, even though quite slowly. It is hard to tell at which point of this slowing down is a proper time to stop the training. The real-world test accuracy, which should be our secret goal, starts to deteriorate quite soon for some restarts!

The scalings in Figs. 1 and 2 are the same, and we can observe that even though slightly higher test accuracies are momentarily achieved in the simulated early stopping scheme (Fig. 2), there is no clear indication of when to stop training. The heuristic, shown in Fig. 1, on the other hand, provides a controlled way of making such a decision automatically.

At this point, we have made too few benchmark runs to be able to make conclusive statements based on evidence, and the depicted PenDigits case may not be the most illustrative one. But here we have presented, with the preliminary example that we have, one of our current research efforts. By visualizations and experiments such as those demonstrated here, we hope to gain an understanding of the mechanisms and potential heuristics to obtain better generalization for the easy-to-build, efficient-to-use MLPs.

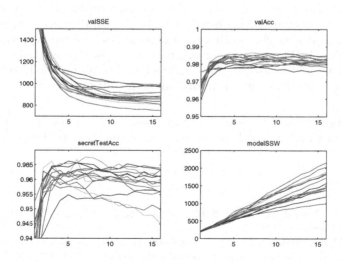

Fig. 2. Progression with constant value of $\beta = 0$ for the PenDigits 16-30-10 case

4 Discussion

We presented a tentative heuristic algorithm for training an MLP to perform pattern classification. Regularization using a weight decay penalty is applied as an attempt to provide generalization capability. The coefficient of the penalty term is changed in a step-wise manner during the training process, and the best combination is recorded to be used as the starting point of the final training phase.

The proposed heuristic is easy to implement, provided that an implementation of an MLP with weight decay cost formulation and a fast local gradient-based optimization algorithm are available. At its current status, the heuristic is a bit rough, and it has not yet been subjected to sufficient benchmark comparisons with other methods. Benchmarking, and probably also improvements are due in the future. Scientific interest is found in looking for justifiable evidence to support or object the underlying ideas. So far, in our computational experiments, we have had to adjust the parameters depending on the test case, which is not good for a system in real use. More insight into the proposed method's functioning is therefore needed.

This is the first time we publish this algorithm that surfaces from our overall research goal of non-linear classification of industrial data. The ideas are based on previous research, documented in [3,4], and computational experiments performed afterwards. A similar system is already working for a specific industrial task, which gives us hope of generalizing the idea and explaining also its theoretical underpinnings in a useful way.

Looking into the future, the heuristic has to be compared more rigorously against other methods, in terms of quality and computational performance. There are many other ways to achieve generalization, briefly noted in the introduction of this paper. It could also be meaningful to introduce some elements from evolutionary algorithms; after all, the MLP already "matures" in our heuristic under varying environmental pressures, and we keep track of an "elite" individual. Why not extend this to a small population, and experiment with mutations and crossover. Also, interesting prospects could be found by using a proper multiobjective optimization methodology.

Acknowledgments. This research was funded by the European Union, European Regional Development Fund, in relation to the project RISC-PROS.

References

1. Haykin, S.: Neural Networks: A Comprehensive Foundation, 2nd edn. Prentice Hall, New Jersey (1999)
2. Hagan, M.T., Menhaj, M.B.: Training feedforward networks with the Marquardt algorithm. IEEE Transactions on Neural Networks 5(6), 989–993 (1994)
3. Kärkkäinen, T.: MLP in layer-wise form with applications to weight decay. Neural Computation 14(6), 1451–1480 (2002)

4. Kärkkäinen, T., Heikkola, E.: Robust formulations for training multilayer perceptrons. Neural Computation 16(4), 837–862 (2004)
5. Bishop, C.M.: Neural Networks for Pattern Recognition. Oxford University Press, Oxford (1995)
6. Stanley, K.O., Miikkulainen, R.: Evolving neural networks through augmenting topologies. Evolutionary Computation 10(2), 99–127 (2002)
7. Abbass, H.A.: Speeding up backpropagation using multiobjective evolutionary algorithms. Neural Computation 15, 2705–2726 (2003)
8. Naval Jr., P.C., Yusiong, J.P.T.: An evolutionary multi-objective neural network optimizer with bias-based pruning heuristic. In: Liu, D., Fei, S., Hou, Z., Zhang, H., Sun, C. (eds.) ISNN 2007. LNCS, vol. 4493, pp. 174–183. Springer, Heidelberg (2007)
9. Kordos, M., Duch, W.: A survey of factors influencing MLP error surface. Control and Cybernetics 33(4), 611–631 (2004)
10. Asuncion, A., Newman, D.: UCI machine learning repository (2007), http://www.ics.uci.edu/~mlearn/MLRepository.html

Connection Strategies in Associative Memory Models with Spiking and Non-spiking Neurons

Weiliang Chen, Reinoud Maex, Rod Adams, Volker Steuber,
Lee Calcraft, and Neil Davey

University of Hertfordshire, College Lane, Hatfield, Hertfordshire AL10 9AB
{W.3.Chen,R.Maex,R.G.Adams,V.Steuber,L.Calcraft,
N.Davey}@herts.ac.uk

Abstract. The problem we address in this paper is that of finding effective and parsimonious patterns of connectivity in sparse associative memories. This problem must be addressed in real neuronal systems, so results in artificial systems could throw light on real systems. We show that there are efficient patterns of connectivity and that these patterns are effective in models with either spiking or non-spiking neurons. This suggests that there may be some underlying general principles governing good connectivity in such networks.

Keywords: Associative Memory, Spiking Neural Network, Small World Network, Connectivity.

1 Introduction

In earlier work [1-3] we have shown how the pattern of connectivity in sparsely connected, associative memories influences the functionality of the networks. The nodes in our networks are given a position, either in a 1D or 2D space. It is then meaningful to talk about issues such as path length, clustering and other concepts familiar from the study of non-random graphs. We have found that networks with only local connectivity do not perform well, as global computation is difficult, whereas random connectivity gives good performance, albeit with a much greater amount of connection fiber. We, and others [3-5], have shown that small world patterns of connectivity can give good performance, with more economical use of resources. However our most efficient networks have been those with *almost* completely local connections [4].

In these experiments we have used large networks (up to 50,000 units) of simple threshold units with no signal delay between nodes. The dynamics is therefore akin to a standard, sparse Hopfield network, although not identical, as we make no requirement for symmetry in connections. In the work presented here we take steps towards much more biologically plausible networks. Firstly we use artificial integrate and fire, spiking neurons and secondly we model signal propagation times according to the geometry of the model. Of course the dynamical behavior of the resulting network is much richer than that of the non-spiking network, but we are now able to investigate the generality of our previous results. Our main finding is that the relation between performance and connectivity in the spiking neural network is surprisingly similar to

M. Kolehmainen et al. (Eds.): ICANNGA 2009, LNCS 5495, pp. 42–51, 2009.
© Springer-Verlag Berlin Heidelberg 2009

that of the more abstract model. This in turn suggests that there may be some general principles at work, which could be of relevance to the analysis of real neuronal networks.

2 Models Examined

Our basic model has a collection of artificial neurons arranged in a ring. The distance between any pair of neurons is taken as the minimum number of steps, on the ring, to get between them. All our networks share two important features. Firstly the networks are regular, so that each neuron has k incoming connections. Secondly the networks are sparse, so that with a network of N units, $k << N$.

With this configuration there are two extremes of connectivity. In a local network, or lattice, each node is connected to those nodes that are closest to it; such networks are known as cellular networks in the context of neural computation, where they are normally 2D lattices. Alternatively the network can have random connectivity, where the probability of any two nodes being connected is k/N, independently of their position. It has been established that whilst local networks have minimum wiring length, they perform poorly as associative memories: pattern correction is a global computation and local connectivity does not allow easy passage of information across the whole network [4]. Randomly connected networks, have very short characteristic path lengths (scaling with log N) and consequently pattern correction is much better, and in fact cannot be improved with any other architecture [4]. However, random networks use a lot of connecting fibre and this has encouraged the investigation of other types of connectivity: it is desirable to find patterns of connectivity that give performance comparable to random networks, but with more economical wiring. It has been established that there are indeed such patterns of connectivity; in particular several researchers have shown that so-called *small world* [6] connectivity can give good performance. We have also shown, that in non-spiking networks, fairly tight Gaussian distributions of connectivity can give very parsimonious networks [2]. In this paper we extend our analysis of how the connectivity affects performance to the more complex dynamics exhibited by networks of integrate and fire spiking neurons.

2.1 Connectivity

N artificial neurons are arranged in a 1-D space with periodic boundary conditions – they can be thought of as occupying a ring, see Figure 1. Each neuron has k incoming connections, and so the network is regular. The reason for this restriction is given in the next section, when discussing the learning rule. The local network has each node connected to its k nearest neighbours, excluding itself (none of our networks has direct self connectivity). Small world networks are constructed using the standard method introduced by Watts and Strogatz [6]. The local network is made progressively more random by rewiring a fraction (p) of the connections to random locations. When $p = 1$ the local network is transformed into a random network.

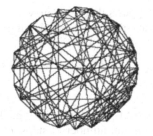
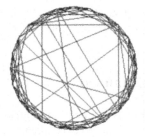

Fig. 1. Units arranged in a simple one-dimensional ring. On the left the units have random connectivity and on the right they have local connectivity and some distal connections – a *small world* model.

We also investigate networks with a Gaussian pattern of connectivity. Here the probability that any two nodes are connected falls as a Gaussian function of distance between the two nodes, see Figure 2. The shape of the Gaussian is parameterised by its standard deviation, σ. Such distributions are particularly interesting as connectivity between individual neurons in the mammalian cortex is thought to be similar [7], see Figure 2.

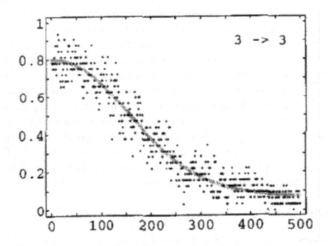

Fig. 2. The probability of a connection between any pair of neurons in layer 3 of the rat visual cortex against cell separation. The horizontal axis is marked in μm. Taken from [7].

2.2 Learning

Before the effect of connectivity can be empirically evaluated the networks must be trained. The simplest approach would be to use the covariance weights of the standard Hopfield network (with or without clipping). This, however, is not a particularly good approach when the networks are sparse and non-symmetric [3]. A more effective method, in this case, is to use standard perceptron learning. In this case, for a given level of connectivity, optimal capacity and performance is obtained when the connectivity is regular, and hence our restriction to regular networks.

The sets of training patterns used consist of random, bipolar or binary vectors, where the probability of any bit being on (+1) is 0.5. The learning process is:

```
Begin with zero weights
Repeat until all units are correct
    Set state of network to one of the ξ^p
    For each unit, i, in turn:
      Calculate its net input h_i^p .
    If (ξ_i^p = on and h_i^p < T) or (ξ_i^p = off and h_i^p > -T)
      then change all the weights to unit i
      according to:
```

$$w_{ij} = w_{ij} + \frac{\xi_j^p}{k} \text{ when } \left(\xi_i^p = on \text{ and } h_i^p < T\right)$$

$$w_{ij} = w_{ij} - \frac{\xi_j^p}{k} \text{ when } \left(\xi_i^p = off \text{ and } h_i^p > -T\right)$$

The value $\xi_i^p = on$ denotes the ith bit of pattern p being $+1$

and the value $\xi_i^p = off$ denotes the value -1 or 0 according to the type of network

T is the learning threshold and here we set $T = 10$.

For the non-spiking network we use the standard bipolar +1/-1 representation. However for the spiking network we use 0/1 binary patterns, as these can then be easily mapped onto the presence or absence of spikes.

2.3   Network Dynamics

2.3.1   Non-spiking Network
These networks use the standard asynchronous dynamics of the Hopfield network: units output +1 if their net input is positive and -1 if negative. As the connectivity is not symmetrical there is no guarantee that the network will converge to a fixed point, but, in practice these networks normally exhibit straightforward dynamics [8]. However, if the network does not converge within 5000 epochs we take the network state at this point as the final state.

2.3.2   Integrate and Fire Spiking Network
The model uses a leaky integrate-and-fire spiking neuron which includes synaptic integration, conduction delays and external current charges. The membrane potential (in *volts*), V, of each neuron in the network is set to 0 if no stimulation is presented, and is referred to as the membrane resting potential. The neuron can be stimulated and change its potential by either receiving spikes from other connected neurons, or

by receiving an external current. If the membrane potential of a neuron reaches a fixed firing threshold, V_{FIRE}, the neuron emits a spike and the potential is reset to resting state ($0mV$) for a certain period (the refractory period). During this period the neuron cannot fire another spike even if it receives very high stimulation. Here the refractory period is set to a reasonable value of $3ms$ [9].

A spike that arrive at a synapse triggers a current, the density of this current (in Amperes per Farad), $I_{ij}(t)$ (where i is the postsynaptic neuron and j is the presynaptic neuron), is given by:

$$I_{ij}(t) = \frac{(t - t_{arrive})}{\tau} \exp\left(1 - \frac{(t - t_{arrive})}{\tau}\right)$$

where t_{arrive} is the time that a spike arrives at node i from node j

so that $t_{arrive} = t_{spike} + delay_{ij}$

The value of $I_{ij}(t)$ will reach a peak τ seconds (the synaptic time constant) after a spike arrives. We set τ to be 2ms.

Two delay modes were used in the model. The fixed delay mode gives each connection a fixed 1ms delay. In the second mode, the delay of spikes (in ms) over a connection is defined by: $delay_{ij} = \sqrt[3]{d_{ij}}$ where d_{ij} is the distance between the two nodes. This gives a rough mapping from a one dimensional ring structure to a more realistic three dimensional system. For a network with 5000 units, the delay will vary between $1ms$ and about $14ms$.

The rate of change of membrane potential is defined by: $\dfrac{dV}{dt} = -\dfrac{V}{\tau_m} + I_{TOTAL}$. Here the first term represents the leak of current density and consequently a decrease in voltage in the neuron. The second term is the total current density entering the cell. It is calculated as the weighted sum of synaptic inputs and any external stimulation:

$$I_{TOTAL} = \sum_j w_{ij} I_j + I_{EXTERNAL}$$

The Injection of External Currents

The network requires an initial stimulation from external currents in order to trigger the first spikes. A simple current injection, which transforms a static binary pattern to a set of current densities is used. Given an input pattern, unit i receives an external current if it is on in that pattern, otherwise the unit receives no external current. Each external current has a density of $3A/F$ and is continually applied to the unit for the first $50ms$ of simulation. This mechanism guarantees that the first spiking pattern triggered in the network is identical to the input pattern. After the first spikes (about 7 ~ 8ms from the start of a simulation), both internal currents caused by spikes, and the external currents, affect the network dynamics. Spike activity continues after the removal of external currents, as the internal currents caused by spike chains become the driving force. The network is then allowed to run for $500ms$, before its final state is evaluated, as will be described in the next section.

3   Performance Measures

The Effective Capacity (*EC*) [10] of a network is a measure of the maximum number of patterns that can be stored in the network with *reasonable* pattern correction still taking place. In other words, it is a capacity measure that takes into account the dynamic ability of the network to perform pattern correction. We take a fairly arbitrary definition of *reasonable* as the ability to correct the addition of 60% noise to within an overlap of 95% with the original fundamental memory. Varying these two percentage figures gives differing values for *EC* but the values with these settings are robust for comparison purposes. For large fully connected networks the *EC* value is about 0.1 of the conventional capacity of the network, but for networks with sparse, structured connectivity *EC* is dependent upon the actual connectivity pattern.

The Effective Capacity of a particular network is determined as follows:

> *Initialise the number of patterns, P, to 0*
> *Repeat*
> > *Increment P*
> > *Create a training set of P random patterns*
> > *Train the network*
> > *For each pattern in the training set*
> > > *Degrade the pattern randomly by adding 60% of noise*
> > > *With this noisy pattern as start state, allow the network to converge*
> > > *Calculate the overlap of the final network state with the*
> > > *original pattern*
> > *EndFor*
> > *Calculate the mean pattern overlap over all final states*
> *Until the mean pattern overlap is less than 95%*
> *The Effective Capacity is then P-1.*

The Effective Capacity of the network is therefore the highest pattern loading for which a 60% corrupted pattern has, after convergence, a mean overlap of 95% or greater with its original value.

Of course this measure is simple to calculate for the network of non-spiking neurons, but its implementation in the spiking network is not as straightforward, as we need to define exactly what is meant by overlap of the network state, a collection of spike events, with a stored pattern. To this end we follow the method of Anishenko [4]. The state of any unit in the network is assumed to be encoded in its firing rate, $r_i(t)$, as measured over a short time window (in our case $20ms$). The overlap of the network state and a binary pattern vector is then defined as the cosine of the angle between the pattern and the vector of firing rates: $O_\xi(t) = \dfrac{\sum_i \xi_i r_i(t)}{\sqrt{\sum_i \xi_i^2 \sum_i r_i^2(t)}}$.

4 Results

We use two patterns of connectivity, small world and Gaussian in networks of 5000 units, with each unit having 100 incoming connections. In the non-spiking network this implies a theoretical maximum loading of up to 200 unbiased random patterns, although in practice the capacity is around 140 patterns. For each type of network results are means over 10 runs. Error bars are not shown, as they are so small as to be virtually invisible.

4.1 Small World Networks

We begin by giving the results of the small world networks, as these include the two extremes of local and random connectivity. Here a local network was progressively rewired, in increments of $p = 0.1$, until a random network with $p = 1$ was reached.

In Figure 3 the results for the non-spiking network, the spiking network with fixed signal propagation delay and the spiking network with cube root delay are given. At the left side of the graph the Effective Capacity of the networks with local connectivity only is shown. All three networks show an EC value of about 20 patterns. At the right side of the graph can be seen the performance of completely rewired networks, a random graph. The performance in this case is much improved, ranging from 44 to 56 patterns. The best performing network is the spiking network with fixed delays. To reiterate the implication of this: a local pattern of connectivity does not support good

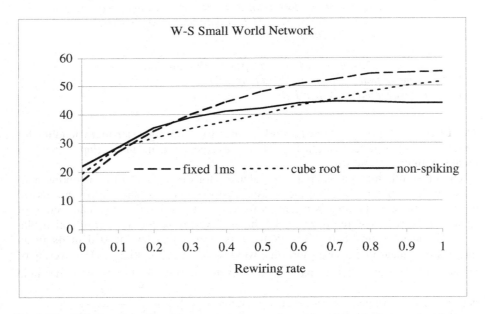

Fig. 3. The Effective Capacity of three types of network: one learning rule, but varying dynamics. Locally connected networks are transformed into random networks by progressive rewiring. The networks are 5000 units with $k = 100$.

integration of information across the whole network, whereas random connectivity provides good global computation in these networks. As our earlier work has already indicated a rewiring rate of about 0.6 gives optimal performance in the non-spiking network. Interestingly the spiking networks continue to improve past this point. It is worth pointing out that none of the well performing networks can be properly described as being small world networks, in the Watts and Strogatz [6] sense. They identified the small world regime at a rewiring level of only about 0.01, when path lengths have dropped, but clustering remains high. At $p = 0.6$ clustering has dropped to a level similar to a random network.

There are two intriguing features of these results. Firstly it is apparent that the very simple non-spiking network acts as a reasonable predictor of the much more complicated integrate and fire spiking network. Secondly the spiking networks, in some circumstances, perform better than their non-spiking cousins. It is not obvious to us why this should be the case.

4.2 Gaussian Networks

In this pattern of connectivity the probability of any two nodes being connected falls with a Gaussian function of their spatial separation. The specific distribution is controlled by σ. In this experiment σ varies from $0.4k$ (40) and then in increments of $0.2k$ (20) to k (100) and thereafter in multiples of k. Remembering that with the size of the networks being 5000 units, the maximum separation between any two nodes is 2500, so that a distribution with $\sigma = 200$, say, is very tight, relative to the size of the complete network.

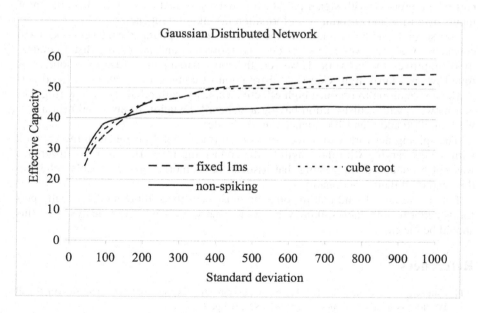

Fig. 4. The Effective Capacity of networks with connection probability following a Gaussian distribution of varying s. The networks are 5000 units with $k = 100$.

The results are shown in Figure 4. At the left hand side of the graph the initial networks have an Effective Capacity of 25-27 patterns. These networks have very tight connectivity distributions, with most connections (~ 95%) made with the 80 units on either side. This has given the network a small improvement on the local network, with connections made to all 50 units on each side. All three types of network then show rapidly improving performance to about $\sigma = 2k$ (200) – here the performance of the three networks is similar with an EC of about 42 patterns. Further widening of the connectivity does not bring much benefit to the non-spiking network; this is not surprising as it is already almost at the performance level of a random network. However both spiking networks continue to improve, passing an EC of 50 at a σ of $4k$.

5 Discussion

In the work presented here we have endeavored to examine the performance of associative memory networks of spiking neurons, in relation to the connectivity in the network, and to compare this performance to the simpler Hopfield type associative memories. Our first finding is that the non-spiking networks provide a reasonably good prediction of the performance of spiking networks with the same connectivity and weights. Moreover this prediction is both qualitative and quantitative. To the best of our knowledge this is the first study to make this direct comparison of these neural models.

In one sense the similarity of the two models could be expected: both types of neuron integrate their input and respond when this net input exceeds a firing threshold. However, in another sense it would not be anticipated. In the non-spiking network continuous time is not modeled. In the spiking model, however, time is an integral part of the process, with signal propagation delays, refractory periods, integration of inputs over time and encoding of information in spiking frequencies.

Our second finding is related to the first result. In spiking neural networks local connectivity alone gives relatively poor performance, and increasing distal connectivity improves the network. However, the most parsimonious use of resources is found when a fairly tight Gaussian distribution of connections is used. A good network configuration to produce high effective capacity with relatively low wiring cost is a network with a distribution having a standard deviation of about 400 (in a network of 5000 nodes and 100 connections per node).

The spiking network with fixed delays performed slightly better than the network with delays varying with the length of the connecting fiber. However the difference was not pronounced, suggesting that associative memories are reasonably robust to this feature of their functionality.

Finally we have found that in some circumstances the spiking model actually performs better than the non-spiking version. Further work is needed to analyse why this should be the case.

References

[1] Calcraft, L., Adams, R., Davey, N.: Gaussian and Exponential Architectures in Small World Associative Memories. In: ESANN, Bruge (2006)
[2] Calcraft, L., Adams, R., Davey, N.: Efficient architectures for sparsely-connected high capacity associative memory models. Connection Science 19, 163–175 (2007)

 [3] Davey, N., Calcraft, L., Adams, R.: High capacity, small world associative memory models. Connection Science 18, 247 (2006)
 [4] Anishchenko, A., Bienenstock, E., Treves, A.: Autoassociative Memory Retrieval and Spontaneous Activity Bumps in Small-World Networks of Integrate-and-Fire Neurons. Submitted to Neural Computation (2005)
 [5] Bohland, J., Minai, A.: Efficient Associative Memory Using Small-World Architecture. Neurocomputing 38-40, 489–496 (2001)
 [6] Watts, D., Strogatz, S.: Collective Dynamics of 'small-world' networks. Nature 393, 440–442 (1998)
 [7] Hellwig, B.: A quantitative analysis of the local connectivity between pyramidal neurons in layers 2/3 of the rat visual cortex. Biological Cybernetics 82, 111 (2000)
 [8] Davey, N., Hunt, S.P., Adams, R.G.: High capacity recurrent associative memories. Neurocomputing 62, 459–491 (2004)
 [9] Kandel, E., Schwartz, J., Jessel, T.: Principles of Neural Science, 4th edn (2000)
[10] Calcraft, L.: Measuring the Performance of Associative Memories, University of Hertfordshire, Technical Report 420 (2005)

Some Enhancements to Orthonormal Approximation of 2D Functions

Bartlomiej Beliczynski

Warsaw University of Technology,
Institute of Control and Industrial Electronics,
ul. Koszykowa 75, 00-662 Warszawa, Poland
B.Beliczynski@ee.pw.edu.pl

Abstract. Some enhancements and comments to approximation of 2D functions in orthogonal basis are presented. This is a direct extension of the results obtained in [2]. First of all we prove that a constant bias extracted from the function contributes to the error decrease. We demonstrate how to choose that bias prooving an appropriate theorem. Secondly we discuss how to select a 2D basis among orthonormal functions to achieve minimum error for a fixed dimension of an approximation space. Thirdly we prove that loss of orthonormality due to truncation of the arguments range of the basis functions does not effect the overall error of approximation and the formula for calculating of the expansion coefficients remains the same. As an illustrative example, we show how these enhanencements can be used.

1 Introduction

In this paper we consider several issues of 2D function approximation in Hilbert space. For a fixed basis functions in a Hilbert space, there always exists the best approximation. Contrary to many neural network schemes contemporarily used, the approximation basis here, is chosen to be orthonormal. This gives several implementation advantages [3]. First of all, the best approximation property is achieved instantly and approximation learning phase is reduced to simple calculation of the expansion coefficients. Any improvement contributing to the decrease of error is done incrementally. Secondly approximation error is calculated and controlled in every approximation step. Thirdly, because calculations are very simple, no searching methods, no matrix inversion, no even simple division is necessary, the method is numerically efficient and robust. Quite easily one can go for thousands of basis functions. As the result, the expansion coefficients may represent well the original function. Such a representation usually requires less space than the original data points. The original function could be recalled by using again basis functions either in continuous or sampled form with any desired sampling rate.

In [2] several enhancements to 1D orthogonal approximation were described. Instead of approximating function f, it was suggested to approximate $f - f_0$, where f_0 was a fixed chosen function. After approximation was done, f_0 was

M. Kolehmainen et al. (Eds.): ICANNGA 2009, LNCS 5495, pp. 52–61, 2009.
© Springer-Verlag Berlin Heidelberg 2009

added to the approximated $f - f_0$. From approximation and data compression point of view, this procedure makes sense if additional efforts put into representation of f_0 are compensated by reduction of the approximation error. The second suggestion made there, was to calculate and collect excessive number of basis functions and its expansion coefficients and select those only which contribute the most to the approximation error reduction. The third problem which was stated and discussed there, was the problem of loosing orthonormality property by basis functions if the set $\mathbb{R}$ of their arguments range was replaced by its subset. When approximating basis is orthonormal, the expansion coefficients are calculated easily. Otherwise these calculations are more complicated. However it was proved there that when truncating the basis functions, despite of loss of orthonormality, we might determined the expansion coefficients as before.

In this paper a generalization of the results of [2] for 2D space is presented and discussed. First of all we demonstrate how to choose f_0 function. Secondly we discuss how to select a 2D basis among orthonormal functions to achieve minimum error for a fixed dimension of approximation space. We consider it, but some practical doubts of efficiency in 2D case occur. The third issue concerning loss of basis orthonormality is also generalized for 2D. The 2D approximation is demonstrated by using color picture approximation. We employ practically very useful 2D Hermite basis.

This paper is organized as follows. In Section 2 basic facts about 2D approximation needed for later use are recalled. In Section 3 we present our main results concerning bias extraction, basis functions selection and prove of correctness for expansion coefficients calculation despite the lack of basis orthonormality. In Section 4 some comments on relationship between neural networks, orthonormal basis and Hermite functions are presented. In Section 5 certain practicalities and an application of our improvements to the image approximation are demonstrated and discussed. In Section 6 conclusions are drawn.

2   Approximation Framework for 2D Functions

In this Section selected facts on 2D function approximation useful for this paper will be recalled for the subsequent use. Let us consider the following function

$$f_{n+1,m+1} = \sum_{i=0}^{n} \sum_{j=0}^{m} w_{ij} g_{ij}, \qquad (1)$$

where $g_{ij} \in \mathcal{G} \subset \mathcal{H}$, and $\mathcal{H}$ is a Hilbert space $\mathcal{H} = (\mathcal{H}, ||.||)$, $w_{ij} \in \mathbb{R}$, $i = 0, \ldots, n$, $.j = 0, \ldots, m$.

For any function f from a Hilbert space $\mathcal{H}$ and a closed (finite dimensional) subspace $\mathcal{G} \subset \mathcal{H}$ with basis $\{g_{00}, \ldots, g_{ij}, \ldots, g_{nm}\}$ there exists a unique best approximation of f by elements of $\mathcal{G}$ ([4]). Let us denote it by g_b. Because the error of the best approximation is orthogonal to all elements of the approximation space $f - g_b \perp \mathcal{G}$, the coefficients w_i may be calculated from the set of linear equations

$$\langle g_{ij}, f - g_b \rangle = 0 \text{ for } i = 0, ..., n, \ j = 0, ..., m \tag{2}$$

where $\langle .,. \rangle$ denotes inner product.

Formula (2) can also be written as

$$\left\langle g_{ij}, f - \sum_{k=0}^{n} \sum_{l=0}^{m} w_{kl} g_{kl} \right\rangle = \langle g_{ij}, f \rangle - \sum_{k=0}^{n} \sum_{l=0}^{m} w_{kl} \langle g_{ij}, g_{kl} \rangle = 0 \tag{3}$$

$$\text{for } i = 0, ..., n, .j = 0, ..., m.$$

Now if 2D basis $\{e_{00}, ..., e_{ij}, ..., e_{nm}\}$ is orthonormal i.e.

$$\langle e_{ij}, e_{kl} \rangle = \left\{ \begin{array}{cc} 1 & \text{if } i = k \text{ and } j = l \\ 0 & \text{elsewhere} \end{array} \right\} \tag{4}$$

then

$$w_{ij} = \langle e_{ij}, f \rangle \tag{5}$$

Finally (1) will take the form

$$f_{n+1,m+1} = \sum_{i=0}^{n} \sum_{j=0}^{m} \langle e_{ij}, f \rangle \ e_{ij}, \tag{6}$$

The squared error, $error_{n+1,m+1} = <f - f_{n+1,m+1}, f - f_{n+1,m+1}>$ of the best approximation of a function f in the basis $\{e_{00}, ..., e_{ij}, ..., e_{nm}\}$ is thus expressible by

$$||error_{n+1,m+1}||^2 = ||f||^2 - \sum_{i=0}^{n} \sum_{j=0}^{m} w_{ij}^2. \tag{7}$$

3   Main Results

3.1   Extracting of Bias

In this section our first enhancement is introduced. Let f be any function from a Hilbert space $\mathcal{H}$. Instead of approximating function f, we suggest to approximate the function $f - f_0$, where $f_0 \subset \mathcal{H}$ is a known function. Later f_0 is added to the approximant of $f - f_0$. Now a modification of (6) will be the following

$$f_{n+1}^{f_0} = f_0 + \sum_{i=0}^{n} \sum_{j=0}^{m} <f - f_0, e_{ij} > e_{ij}, \tag{8}$$

Then approximation error will be expressed as

$$e_n^{f_0} = f - f_{n+1}^{f_0} = f - f_0 - \sum_{i=0}^{n} \sum_{j=0}^{m} <f - f_0, e_{ij} > e_{ij},$$

and similarly to (7) its squared norm

$$||e_{n+1}^{f_0}||^2 = ||f - f_0||^2 - \sum_{i=0}^{n}\sum_{j=0}^{m} < f - f_0, e_{ij} >^2 \qquad (9)$$

Theorem 1. *Let $\mathcal{H}$ be a Hilbert space of functions on a subset of $\mathbb{R}^2$ containing the rectangular region $D = [a_1, a_2] \times [b_1, b_2]$, let f be a function from $\mathcal{H}$, $f \in \mathcal{H}$, $\{e_{00}, ..., e_{ij}, .. ., e_{nm}\}$ be an orthonormal set in $\mathcal{H}$, c be a constant $c \in \mathbb{R}$. Let $f_0 = c1_D$ where 1_D denotes a function of value 1 in the region D and 0 elsewhere, and the approximation formula be the following*

$$f_{n+1}^{f_0} = f_0 + \sum_{i=0}^{n}\sum_{j=0}^{m} < f - f_0, e_{ij} > e_{ij}$$

then the norm of the approximation error is minimized for $c = c_0$ and

$$c_0 = \frac{< f, 1_D > - \sum_{i=0}^{n}\sum_{j=0}^{m} < f, e_{ij} >< e_{ij}, 1_D >}{(a_2 - a_1)(b_2 - b_1) - \sum_{i=0}^{n}\sum_{j=0}^{m} < e_{ij}, 1_D >^2} \qquad (10)$$

Proof. The squared error formula (9) could be expressed as follows $||e_{n+1}^{f_0}||^2 = \left|\left| f - f_{n+1}^{f_0} \right|\right|^2 = ||f||^2 + ||f_0||^2 - 2 < f, f_0 > - \sum_{i=0}^{n}\sum_{j=0}^{m}(\langle f, e_{ij}\rangle - \langle e_{ij}, f_0\rangle)^2 = ||f||^2 + c^2(a_2 - a_1)(b_2 - b_1) - 2c < f, 1_D > - \sum_{i=0}^{n}\sum_{j=0}^{m}(\langle f, e_{ij}\rangle^2 + c^2 \langle e_{ij}, 1_D\rangle^2 - 2c \langle f, e_{ij}\rangle \langle e_{ij}, 1_D\rangle)$. Now differentiating the squared error formula in respect of c and equating it to zero one obtains (10).

Along the Theorem 1 we are suggesting a two step approximation. First f_0 should be calculated and then the function $f - f_0$ will be approximated in a usual way.

Remark 1. One may notice that in many applications c_0 of (10) could well be approximated by

$$c_0 \simeq \frac{< f, 1_D >}{(a_2 - a_1)(b_2 - b_1)} \qquad (11)$$

The right hand side of (11) expresses the mean value of the approximated function f in the region D. A usual choice of D is such as an actual function f arguments range.

3.2   Basis Selection

In a typically stated approximation problem there is a function to be approximated f and a basis $\{e_{00}, ..., e_{ij}, ..., e_{nm}\}$ of approximation. We are looking for the function expansion coefficients. Because orthonormality of the basis, those coefficients contribute directly to the approximation error decrease as shown in (7). So one can calculate an excessive number of expansion coefficients and order them in a descending sequence. Then using as many coefficients as needed for the approximation, one may ensure the fastest decrease of error with respect to the number of basis functions. A theoretical question might be stated. How many coefficients should initially be calculated as to achieve, as the result of ordering, N the most significant ones? This is similar problem as for 1D functions. However for 2D case, the number of 1D components and then 2D approximating functions is usually very large. Thus all selection methods are time and space consuming and this enhancement for 2D system it is less attractive.

3.3   Truncated Basis

According to equation (3) and (5), the best approximation coefficients can easily be calculated if the approximation basis is orthonormal. While implementing, basis functions are represented via their discrete arguments values being truncated into a limited range of its arguments. Thus the basis represented in computer, as a rule, looses its orthonormality property. However formula (5) for calculating weights is still valid. More precisely this is stated in Proposition 1.

Proposition 1. *Let $\mathcal{H}$ be a Hilbert space of functions on a subset of $\mathbb{R}^2$ containing the rectangular region $D = [a_1, a_2] \times [b_1, b_2]$, $\{e_{00}^t, ..., e_{ij}^t, .. ., e_{nm}^t\}$ be an orthonormal set of functions in $\mathcal{H}$, truncated to the region D, let f be a function from $\mathcal{H}$ and f be zero outside of D then*

$$w_{ij} = \langle e_{ij}^t, f \rangle, \ for \ i = 0, ..., n, .j = 0, ..., m.$$

Proof. If basis is orthonormal then the best approximation coefficients are calculated via $w_{ij} = \langle e_{ij}, f \rangle$, for $i = 0, ..., n, .j = 0, ..., m$, but because f is being nonzero only in the region D, thus $\langle e_{ij}, f \rangle = \langle e_{ij}^t, f \rangle$.

Remark 2. If a function to be approximated is defined over a limited range of its argument, the orthonormal basis of approximation is truncated to that range as well, then despite the loss of orthonormality, the best approximation is calculated in the same easy way as in the non truncated case.

4   Neural Networks and Orthonormal Basis Approximation

The great advantage of neural network based function approximation is the universal approximation architecture allowing any $\mathbb{R}^d$ function be approximated with any desired accuracy. Even one-hidden-layer architecture with sufficient number of nodes and an activation function selected from a wide class is appropriate for this task [5]. Neural networks are the approximators usually acting in spaces with non-orthogonal, adaptively selected basis. Neural networks nodes activation functions are the same for all the nodes.

Recently in several publications (see for instance [6], [7]) it was suggested to use Hermite functions as activation functions in neural schemes. Novelty of this approach is that contrary to the traditional neural architecture, every node in the hidden layer has different activation function. It gains several advantages of the Hermite functions and prove to be successful in real data classification tasks

[6]. However, in such approach orthogonality of Hermite functions is not really exploited.

Another opportunity is to look at Hermite functions as RBF functions and adopt RBFN methods to the Hermite learning tasks. However among Hermite functions properties what is really important, is that the Hermite functions are orthonormal. In such case, the approximation is much simpler than in the case of RBF algorithms. No searching methods are necessary.

Various orthonormal basis might be considered: generalized Fourier series, Legendre, Laguerre and Hermite polynomials based, orthonormal wavelets etc. Their particular features determine their usefulness and applications. Among various types of orthonormal basis which might be used for approximation, the Hermite functions draw special attention. A set of Hermite functions is naturally attractive for various approximation, classification and data compression tasks. The Hermite functions are well localized in time and frequency. They are defined on the real numbers set $\mathbb{R}$. They can be recursively calculated. Approximating function coefficients are determined relatively easily to achieve the best approximation property. Since Hermite functions are eigenfunctions of the Fourier transform, the time and frequency spectra are simultaneously approximated. Each subsequent function extends frequency bandwidth within a limited range of well concentrated energy; see for instance [1]. By introducing scaling parameter we may control that bandwidth influencing at the same time the dynamic range of the input argument, till we strike a desirable balance. If Hermite functions are generalized into two variables functions, they retain the same useful properties.

5   Practicalities and Illustrative Example

5.1   2D Hermite Functions

In this Section we will describe some practicalities useful for implementation and an illustrative example. The orthonormal basis chosen for implementation is the Hermite basis. The approximated function expansion coefficients are easily and independently calculated from (5). We will focus our attention to a space of a great practical interest $\mathcal{L}^2(-\infty, +\infty) \times \mathcal{L}^2(-\infty, +\infty)$ with the inner product defined $< f(x,y), g(x,y) >= \int\limits_{-\infty}^{+\infty} \int\limits_{-\infty}^{+\infty} f(x,y)g(x,y)dxdy$. In such space a sequence of linearly independent and bounded functions could be defined as follows

$$h_{ij}(x,y) = h_i(x)h_j(y), \; i = 0, ..., n, \, .j = 0, ..., m$$

where

$$h_i(t) = c_i e^{-\frac{t^2}{2}} H_i(t); \; H_i(t) = (-1)^i e^{t^2} \frac{d^i}{dt^i}(e^{-t^2}); \; c_i = \frac{1}{(2^i i! \sqrt{\pi})^{1/2}}. \quad (12)$$

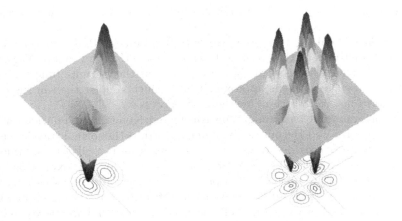

Fig. 1. 2D Hermite functions $h_{01}(x,y)$ (left) and $h_{22}(x,y)$ (right)

The polynomials $H_i(t)$ are Hermite polynomials and the functions $h_i(t)$ are Hermite functions [3]. According to (12) the first several Hermite functions could be calculated

$$h_0(t) = \frac{1}{\pi^{1/4}}e^{-\frac{t^2}{2}}; \qquad h_1(t) = \frac{1}{\sqrt{2}\pi^{1/4}}e^{-\frac{t^2}{2}}2t;$$

$$h_2(t) = \frac{1}{2\sqrt{2}\pi^{1/4}}e^{-\frac{t^2}{2}}(4t^2 - 2); \quad h_3(t) = \frac{1}{4\sqrt{3}\pi^{1/4}}e^{-\frac{t^2}{2}}(8t^3 - 12t)$$

Selected plots of two 2D functions of the Hermite basis are shown in Fig.1.

Described here 2D Hermite functions are direct products of 1D Hermite functions. The last could be evaluated from the direct formula (12). However for large n, Hermite polynomials reach very large values and those calculations are error prone. Another possibility for determining Hermite functions, which is numerically better conditioned, is to use recursive formula (see for instance [1]).

If a function to be approximated is represented by a set of pairs $\{(t_i, f_i)\}_{i=-N}^{N}$, in the range $[-t_{\max}, t_{\max}]$ then starting from $h_0(t)$ and $h_1(t)$ one obtains the Hermite basis and the expansion coefficients from (5).

If sampling is regular with sampling time denoted as T, then the inner product in (5) could be approximated as follows

$$< f, e_{ij} > = \int\limits_{-\infty}^{+\infty}\int\limits_{-\infty}^{+\infty} f(x,y)e_{ij}(x,y)dxdy \approx \sum_{k=-N}^{N}\sum_{l=-M}^{M} f(x_k, y_l)e_{ij}(x_k, y_l)T^2$$

$$(13)$$

From engineering point of view, while approximating a function it is important to meet the time and frequency requirements. These two domains interrelated via Fourier transform could be controlled by scaling parameter $\sigma \in (0, \infty)$. So if one substitutes $t := \frac{t}{\sigma}$ into (12) and modifies c_n to ensure orthonormality, then

$$h_n(t, \sigma) = c_{n,\sigma} e^{-\frac{t^2}{2\sigma^2}} H_n(\frac{t}{\sigma}) \quad \text{where} \quad c_{n,\sigma} = \frac{1}{(\sigma 2^n n! \sqrt{\pi})^{1/2}} \tag{14}$$

and

$$h_n(t, \sigma) = \frac{1}{\sqrt{\sigma}} h_n(\frac{t}{\sigma}) \quad \text{and} \quad \widetilde{h}_n(\omega, \sigma) = \sqrt{\sigma} \widetilde{h}_n(\sigma\omega) \tag{15}$$

Note that h_n as defined by (14) has two arguments whereas h_i as defined by (12) has one argument. These functions are related by (15). Described in this paper 2D approximation uses products of 1D Hermite functions, so those properties are transferred.

5.2   An Illustrative Example of Approximation

In this section we will describe a 2D function approximation example. It will be a simultaneous approximation of three functions in the form of an color image. The color image was particularly chosen, because it is well defined, non-trivial example of 2D function. Quality of approximation may well be visually asses, apart from the rigorous error calculation.

The function is represented by 2272x1704x3≈3.8Mx3 data points and is shown in Fig.2. For each color, the parameter c_0 of (10) was determined $c_0 = [133 \ 137 \ 140]^T$ and f_0 extracted from the function as stated in Theorem 1. Then the number of Hermite functions in each direction were arbitrary selected

Fig. 2. The original image consisting of 2272x1704x3=3.8Mx3 data points

Fig. 3. The approximating image recalled from 2D basis with 151 Hermite functions in x direction and 114 functions in y direction and totally 17kx3 coefficients

as number data points divided by 15. This resulted with 151 Hermite functions in x direction, 114 Hermite functions in y direction and total $17kx3$ expansion coefficients. The parameters σ in both directions were the following $\sigma_x = 81.6$ and $\sigma_y = 70.4$. The function was represented by its expansion coefficients. The approximation basis could have been treated as continuous or sampled with any sampling rates. In Fig.3 the approximating function is shown where the continuous basis functions have been sampled with sampling time the same as for the original image. Of course better approximation could have been achieved in the expense of increasing number of basis functions.

Modification of the approximation equation as in (8) contributes to the error decrease. Without modification, the average squared error per data point was on the level 0.00707,but when the modification was applied as in Fig.3, the error decreased to the level 0.00587, what means about 17% of improvement. The differences were clearly seen on the picture. Further improvement could have be achieved by selecting such Hermite functions which contribute the most to the error decrease. If we consider this process from data compression point of view, one also has to take into account the sequence of indices of Hermite functions used in the basis. This additional data should be stored or transmitted as well and such modification become less attractive.

6   Conclusions

We described some enhancements and one prove of correctness for the 2D orthonormal function based approximation. The results were proved in an abstract way and illustrated by an example of image approximation. These results were immediate generalization of 1D case presented in [2]. Contrary to many neural schemes, this 2D function approximation method is very efficient and numerically robust. We have run approximation with several dozen thousand coefficients. Good numerical properties are due to orthogonal basis, recursive formula for basis calculation and dispensing with searching methods.

References

1. Beliczynski, B.: Properties of the Hermite activation functions in a neural approximation scheme. In: Beliczynski, B., Dzielinski, A., Iwanowski, M., Ribeiro, B. (eds.) ICANNGA 2007. LNCS, vol. 4432, pp. 46–54. Springer, Heidelberg (2007)
2. Beliczynski, B., Ribeiro, B.: Several enhancements to Hermite-based approximation of one-variable functions. In: Kůrková, V., Neruda, R., Koutník, J. (eds.) ICANN 2008, Part I. LNCS, vol. 5163, pp. 11–20. Springer, Heidelberg (2008)
3. Kreyszig, E.: Introductory functional analysis with applications. J.Wiley, Chichester (1978)
4. Kwok, T., Yeung, D.: Objective functions for training new hidden units in constructive neural networks. IEEE Trans. Neural Networks 8(5), 1131–1148 (1997)
5. Leshno, T., Lin, V., Pinkus, A., Schocken, S.: Multilayer feedforward networks with a nonpolynomial activation function can approximate any function. Neural Networks 13, 350–373 (1993)
6. Linh, T.H.: Modern Generations of Artificial Neural Networks and their Applications in Selected Classification Problems. Publishing Hause of the Warsaw University of Technology (2004)
7. Ma, L., Khorasani, K.: Constructive feedforward neural networks using Hermite polynomial activation functions. IEEE Transactions on Neural Networks 16, 821–833 (2005)

Shortest Common Superstring Problem with Discrete Neural Networks

D. López-Rodríguez and E. Mérida-Casermeiro

Department of Applied Mathematics, University of Málaga, Málaga, Spain
{dlopez,merida}@ctima.uma.es

Abstract. In this paper, we investigate the use of artificial neural networks in order to solve the Shortest Common Superstring Problem. Concretely, the neural network used in this work is based on a multivalued model, MREM, very suitable for solving combinatorial optimization problems. We describe the foundations of this neural model, and how it can be implemented in the context of this problem, by taking advantage of a better representation than in other models, which, in turn, contributes to ease the computational dynamics of the model. Experimental results prove that our model outperforms other heuristic approaches known from the specialized literature.

1 Introduction

Many problems in computational biology, such as DNA sequencing [1,2,3,4], and in data compression [5,6], can be formulated as instances of the Shortest Common Superstring Problem (SCSS).

For example, DNA sequencing consists in determining the correct sequence of nucleotides in a DNA molecule. Nucleotides (adenine, cytosine, guanine and thymine) are represented by the alphabet {a,c,g,t}. Currently, the nucleotides of a DNA fragment can be directly determined in laboratories. Once the nucleotides of all fragments have been determined, the sequence assembly problem aims at reconstructing the original molecule from overlapping fragments. SCSS can be viewed as an abstract representation of this particular task.

This problem is defined as follows: Given a set of strings $P = \{s_1, \ldots, s_N\}$, the objective is to find the string S^* of minimum length, such that, for all $i \in \{1, \ldots, N\}$, $s_i \in P$ is a substring of S^*.

Finding such a superstring is known to be a NP-hard problem [7,8]. Furthermore, it is MAX-SNP-hard [9]. Arora, in [10] proved that problems in MAX-SNP-hard do not admit polynomial time approximation schemes unless P=NP.

Among the approximation algorithms used to compute the shortest superstring, we can find a greedy algorithm [9], which consists in merging pairs of strings with maximum overlap, until a unique string is obtained, which is the approximated solution.

This greedy approach is conjectured to be a 2-approximation algorithm [11], meaning that, in the worst case, the solution provided by this method is twice

M. Kolehmainen et al. (Eds.): ICANNGA 2009, LNCS 5495, pp. 62–71, 2009.
© Springer-Verlag Berlin Heidelberg 2009

as long as the optimal solution. However, Blum et al. [9] proved a factor of 4 for this problem.

This greedy algorithm was further improved by Jiang [12] (within a factor of $2\frac{2}{3}$) and later by Sweedyk [13] (obtaining a constant factor of 2.5), the best result up-to-date. However, these algorithms are not easily implementable and, in practice, the original greedy algorithm (reliable and fast), proposed in [9], is preferred, as well as some of its variants.

In this work, we propose the use of a discrete neural network to solve SCSS problem. In the recent years, a Hopfield-like discrete neural network [14] has been presented to solve this problem. However, its use implies the correct fine-tuning of some parameters. Another drawback of that model is that it needed N^2 neurons to represent the solution of the problem when the number of strings to be merged is $|P| = N$.

The neural model proposed in this work is a generalization of Hopfield's discrete model [15], allowing the neurons to take any value in a discrete set. With the help of a simple computational dynamics, this model is able to represent solutions to this problem better than the previous neural approach, just by using N neurons.

The multivalued MREM model has obtained very good results when applied to other combinatorial optimization problems [16,17,18,19], guaranteeing the convergence to local minima of the energy function.

The rest of this paper is structured as follows: in Sec. 2, we present a detailed formulation of the SCSS problem. Later, in Sec. 3, the neural model MREM is described, as well as the implementation of the computational dynamics to solve the problem at hand. In Sec. 4, experimental results of applying our neural model are shown, whereas in Sec. 5 some conclusions and remarks to this work are presented.

2   Description of the Problem

Given an alphabet A, and a set of strings over the alphabet, $P = \{s_1, \ldots, s_n\}$, the Shortest Common Superstring Problem consists in finding a string S^* containing all strings in P as substrings and with minimum length.

Let us define the overlap $s_{i,j}$ between strings s_i and s_j (in this order), as the string of maximum length (denoted by $|s_{i,j}|$) such that it is a suffix for s_i and a prefix for s_j.

The solution to SCSS can be represented as a permutation Π of numbers $\{1, \ldots, n\}$, meaning the order in which strings in S must be arranged to get the solution string $S^* = S_\Pi$.

Thus, the objective function to be minimized is:

$$|S_\Pi| = F(\Pi) = \sum_{i=1}^{n} |s_i| - \sum_{i=1}^{n-1} |s_{\Pi(i),\Pi(i+1)}| \tag{1}$$

where $|s|$ denotes the length of string s. Note that $s_{\Pi(i),\Pi(i+1)}$ is the overlap between 2 consecutive strings in S_Π, corresponding to strings at positions $\Pi(i)$ and $\Pi(i+1)$.

The minimization of the total length of S_Π is here achieved by maximizing the sum of the lengths of the respective overlaps in the corresponding order given by permutation Π.

Note that the solution may not be unique:

Example. Let us consider the set of strings $P = \{$agcct, acgcgt, cgtacg, tgatc, gtgag$\}$ over the alphabet $A = \{$a, c, g, t$\}$. Then, $S_1 =$ cgtacgcgtgagcctgatc and $S_2 =$ tgatcgtacgcgtgagcct are superstrings containing all strings in P, of equal length.

3   The MREM Model

In this section, the fundamentals of the Multivalued REcurrent Model (MREM) [20] are described. This discrete neural network is a generalization of Hopfield's model [21,15] and other binary and multivalued models, such as SOAR [22] and MAREN [23].

3.1   Description of the Neural Network

Let us consider a recurrent neural network formed by N neurons, where the state of each neuron $i \in \mathcal{I} = \{1, \ldots, N\}$ is defined by its output v_i taking values in any finite set $\mathcal{M} = \{m_1, m_2, \ldots, m_L\}$. This set does not need to be numerical. For example, $\mathcal{M} = \{$red, green, blue$\}$ or $\mathcal{M} = \{$Sunday, Monday, \ldots, Saturday$\}$.

The vector V whose components are the corresponding neuron outputs, $V = (v_1, v_2, \ldots, v_N)$, is called state vector. Associated to each state vector, an energy function, similar to Hopfield's, can be defined:

$$E(V) = -\frac{1}{2} \sum_{i=1}^{N} \sum_{j=1}^{N} w_{i,j} f(v_i, v_j) + \sum_{i=1}^{N} \theta_i(v_i) \qquad (2)$$

where

- $W = (w_{i,j})$ is the synaptic weight matrix, expressing the connection strength between neurons.
- $f: \mathcal{M} \times \mathcal{M} \to \mathbb{R}$ is the so-called similarity function, since $f(v_i, v_j)$ measures the similarity between the outputs of neurons i and j.
- $\theta_i: \mathcal{M} \to \mathbb{R}$ is the generalization of the biases $\theta_i \in \mathbb{R}$, present in Hopfield's model.

The aim of the network is to minimize the energy function given by Eq. (2), i.e., to achieve a stable state corresponding to a local (global, when possible) minimum of the energy function, which is usually identified with the objective function of the problem to solve.

The introduction of the similarity function f makes the network very versatile and usually causes a better representation of the problem at hand, see, for example, [24,18,25,26]. It leads to a better representation of problems than other multivalued models, as SOAR and MAREN [23,22], since in those models most of the information enclosed in the multivalued representation is lost by the use of the signum function that only produces values in $\{-1, 0, 1\}$.

Many computational dynamics can be defined for this model, that is, several neuron updating schemes are available provided the versatility of the network.

Usually, neuron updates are made by taking into consideration the input to the network, called synaptic potential. This potential is computed as $U = -\Delta E$, that is, the opposite of the energy increase produced by the studied neuron update. Thus, if E is the current energy value, and E' is the energy value associated to the proposed update, then $U = E - E'$.

If several possible updates $\{V_1, \ldots, V_K\}$ are studied, consider $U_j = E - E'_j$, where E'_j is the energy value associated to the possible new state V_j. In this case, the update is given by the new state achieving the maximum potential $u = U_j = \max\{U_1, \ldots, U_K\}$.

If $u > 0$, then the update reduces the value of the energy function. Otherwise, since no improvement is obtained by that update, the network does not perform the action. In this situation, the network is said to have converged to a stable state.

Stable states correspond to local minima of the energy function, in the sense that, by using the given dynamics, it is not possible to achieve a further improvement of the solution.

3.2   MREM Applied to SCSS

Note that a solution to SCSS problem can be represented as a permutation of the strings, meaning the order in which strings have to be merged to obtain that solution.

Then, we define feasible state vectors as those representing permutations of $\{1, \ldots, n\}$. Thus, any feasible state vector V will represent an ordering of the strings in S. $v_i = k$ means that s_k is placed in the i-th place in the solution string s_V.

It can be observed that the objective function in Eq. (1) for SCSS consists of two terms. The first one, $\sum_i |s_i|$ is fixed, and therefore it is not important at the optimization stage.

The other term, $-\sum_{i=1}^{n-1} |s_{\Pi(i),\Pi(i+1)}|$, can be expressed as the energy function of the MREM model.

By comparing the objective function in Eq. (1) and the energy function of the neural model, in Eq. (2), we can define:

$$w_{i,j} = \begin{cases} 2, \text{ if } j = i+1, i = 1, \ldots, n-1 \\ 0, \qquad\qquad \text{otherwise} \end{cases}$$

$$f(x, y) = |s_{x,y}|$$

$$\theta_i(x) = 0$$

to obtain the desired identification between both functions.

The computational dynamics of this model is based on that the network must remain in a feasible state along iterations. This is the reason for not needing the fine-tuning of parameters, usually present in Hopfield's energy function, as penalty terms for unsatisfied constraints. Furthermore, it is an easily implementable dynamics, and it can be described as follows:

1. Select a random initial feasible state for the network.
2. The net sequentially selects 2 neurons m and p such that $1 \leq m < p \leq N$. Then, the current solution V can be expressed as the concatenation of 3 subsequences, represented by 3 vectors $a = (v_1, \ldots, v_m)$, $b = (v_{m+1}, \ldots, v_p)$ and $c = (v_{p+1}, \ldots, v_N)$.
3. The network studies the updates to different configurations: acb, bac, bca, cab, cba, where

$$acb = (v_1, \ldots, v_m, v_{p+1}, \ldots, v_N, v_{m+1}, \ldots, v_p)$$

$$bac = (v_{m+1}, \ldots, v_p, v_1, \ldots, v_m, v_{p+1}, \ldots, v_N)$$

$$bca = (v_{m+1}, \ldots, v_p, v_{p+1}, \ldots, v_N, v_1, \ldots, v_m)$$

$$cab = (v_{p+1}, \ldots, v_N, v_1, \ldots, v_m, v_{m+1}, \ldots, v_p)$$

$$cba = (v_{p+1}, \ldots, v_N, v_{m+1}, \ldots, v_p, v_1, \ldots, v_m)$$

by computing the corresponding synaptic potentials:

$$U_{abc} = 0 \quad \text{(since there is no change in state vector)}$$

$$U_{acb} = |s_{v_m, v_{p+1}}| + |s_{v_N, v_{m+1}}| - |s_{v_m, v_{m+1}}| - |s_{v_p, v_{p+1}}|$$

$$U_{bac} = |s_{v_p, v_1}| + |s_{v_m, v_{p+1}}| - |s_{v_m, v_{m+1}}| - |s_{v_p, v_{p+1}}|$$

$$U_{bca} = |s_{v_p, v_{p+1}}| + |s_{v_N, v_1}| - |s_{v_m, v_{m+1}}| - |s_{v_p, v_{p+1}}| = |s_{v_N, v_1}| - |s_{v_m, v_{m+1}}|$$

$$U_{cab} = |s_{v_N, v_1}| + |s_{v_m, v_{m+1}}| - |s_{v_m, v_{m+1}}| - |s_{v_p, v_{p+1}}| = |s_{v_N, v_1}| - |s_{v_p, v_{p+1}}|$$

$$U_{cba} = |s_{v_N, v_{m+1}}| + |s_{v_p, v_1}| - |s_{v_m, v_{m+1}}| - |s_{v_p, v_{p+1}}|$$

These expressions are derived from $U = E - E'$, being E the energy associated to the current network state, and E' the energy associated to the corresponding update.

4. The next network configuration is the one decreasing most the energy function value (equivalently, achieving the greatest potential): if U_{ijk} is the maximum in $\{U_{abc}, U_{acb}, U_{bac}, U_{bca}, U_{cab}, U_{cba}\}$, then the next state is $ijk \in \{abc, acb, bac, bca, cab, cba\}$. If $ijk = abc$, there is no change in the state vector.
5. Repeat steps 2 - 4 until convergence is detected, that is, all pairs of neurons have been studied, and no change is done in the configuration of the network (state vector).

Once the network converges, the stable state represents a minimum of the energy function which, in our case, is equivalent to a maximum of the aggregate overlap length in the resulting string, given by S_V.

Algorithm 1. Greedy Heuristic
___
Data: Set $P = \{s_1, \ldots, s_N\}$ of strings.
Result: A string s such that every s_i is a substring of s (intended to have
      minimal length).
begin
    while $|P| > 1$ **do**
        Select two strings $a, b \in P$ with maximal overlap
        Merge a and b into a new string c
        $P \longleftarrow (P \setminus \{a,b\}) \cup \{c\}$
    return *the unique string* $s \in P$.
end

4 Experimental Results

In this section, we compare the efficiency of our neural model MREM to the greedy heuristic presented in [9], which was conjectured to be a 2-approximation algorithm. This greedy heuristic is shown in Algorithm 1.

Two experiments have been performed with these algorithms. The first one consisted on find the SCSS of a set of strings of fixed length, whereas the second allowed to use strings of variable length.

Fixed length string datasets were randomly built according to 3 parameters: string length ($\{6,8,10\}$), number of words in P ($|P| \in \{25, 50, 100\}$) and number of symbols in the alphabet ($|A| \in \{2, 4, 6, 8\}$). For each combination of these parameters, 10 instances were built (that is, 10 sets P), and the algorithms were independently run 100 times to obtain the superstring length results given in Table 1. Note that the greedy algorithm always selected the same solution, whereas MREM achieves different results depending on its random initial state, what helps avoiding local optima of the energy function.

In the last two columns of the tables, we present the improvement made by MREM with respect to the greedy algorithm (in %). Positive values indicate that MREM performed better than the greedy. Note that our neural approach outperformed the greedy algorithm in most cases on average, and always on best result.

For variable length string datasets, the definition of $|P|$ and $|A|$ remain the same, but string length was randomly selected in $\{2, \ldots, 10\}$. Thus, for each value of $|A|$, $|P|$ strings of length between 2 and 10, formed the set P. As before, for each combination of the parameters, 10 sets P were built and 100 independent executions were performed with each one. Superstring length results are given in Table 2. Note that, in all cases, MREM outperformed the greedy algorithm, obtaining shorter superstrings, not only on minimal length, but also on average length.

There are 2 behaviors that can be seen on these tables:

- As the number $|A|$ of symbols in the alphabet increases (for a fixed number of strings, $|P|$), MREM and the greedy algorithm tend to obtain more similar results, reducing the improvement made by MREM over the latter.

Table 1. Average and minimum superstring length comparison between our neural proposal and the greedy algorithm, for fixed length strings

Length	$|P|$	$|A|$	MREM			Greedy		Improvement	
			Average	Best	Time (sec.)	Best	Time (sec.)	Average	Best
6	25	2	67.169	63.1	0.0216	91.5	0.0256	26.59	31.04
		4	106.065	103.1	0.0211	113	0.0234	6.14	8.76
		6	121.516	118.3	0.0217	120.1	0.0239	-1.18	1.5
		8	126.033	123.7	0.0209	124.1	0.0237	-1.56	0.32
	50	2	100.276	92.9	0.1442	168.7	0.091	40.56	44.93
		4	195.324	189	0.1342	209.5	0.0917	6.77	9.79
		6	225.096	219.8	0.1336	236	0.0912	4.62	6.86
		8	238.884	234	0.1447	249.2	0.092	4.14	6.1
	100	2	146.332	136.3	0.9982	326.7	0.4456	55.21	58.28
		4	362.473	352.2	0.7874	383.5	0.4063	5.48	8.16
		6	418.712	409.1	0.8418	424.1	0.4063	1.27	3.54
		8	451.846	443.8	0.8881	458	0.4118	1.34	3.1
8	25	2	112.69	108.1	0.0251	127.9	0.0272	11.89	15.48
		4	158.908	155.7	0.0204	158.9	0.257	-0.01	2.01
		6	170.554	167.8	0.0221	174.3	0.025	2.15	3.73
		8	178.423	176.1	0.0206	179.6	0.0258	0.66	1.95
	50	2	180.117	171.6	0.149	237.6	0.0954	24.19	27.78
		4	291.753	284.8	0.1399	305.1	0.0961	4.37	6.65
		6	323.997	318.6	0.1391	332.1	0.0965	2.44	4.07
		8	339.505	334.9	0.1338	346.1	0.0944	1.91	3.24
	100	2	287.94	275	0.9774	439.7	0.4076	34.51	37.46
		4	547.072	536.8	0.8547	567.1	0.414	3.53	5.34
		6	617.981	609.9	0.8469	633.1	0.4253	2.39	3.66
		8	650.429	642.8	0.9	656.7	0.4273	0.95	2.12
10	25	2	155.939	151	0.0234	176.7	0.0258	11.75	14.54
		4	207.935	204.6	0.0219	215.3	0.0259	3.42	4.97
		6	220.99	218.2	0.0207	220	0.0261	-0.45	0.82
		8	226.615	224.4	0.0206	229.6	0.0262	1.3	2.26
	50	2	284.153	274	0.1614	318.6	0.0984	10.81	14
		4	396.893	390.5	0.1396	399	0.0986	0.53	2.13
		6	425.848	420.9	0.134	432.4	0.0995	1.52	2.66
		8	439.814	434.8	0.1449	445.6	0.0999	1.3	2.42
	100	2	465.955	450.5	0.955	581.9	0.4389	19.93	22.58
		4	740.486	729.7	0.8943	770.4	0.428	3.88	5.28
		6	818.364	809.9	0.8752	834.6	0.4318	1.95	2.96
		8	853.481	845.4	0.8965	867	0.4371	1.56	2.49

- As the number of strings in P increases (for fixed number $|A|$), MREM improves its relative performance with respect to the greedy algorithm. Thus, in real-world problems, MREM may achieve better results than the greedy algorithm.

Table 2. Average and minimum superstring length comparison between our neural proposal and the greedy algorithm, for variable length strings

$|P|$	$|A|$	MREM			Greedy		Improvement	
		Average	Best	Time (sec.)	Best	Time (sec.)	Average	Best
25	2	66.004	62.2	0.0189	77.3	0.0293	14.61	19.53
25	4	101.112	98.5	0.0186	105.7	0.024	4.34	6.81
25	6	109.799	107.3	0.0204	110.4	0.0238	0.54	2.81
25	8	109.829	107.5	0.0216	111.5	0.0238	1.5	3.59
50	2	117.959	110.4	0.1303	142.9	0.0929	17.45	22.74
50	4	181.495	176.1	0.1302	189.2	0.0935	4.07	6.92
50	6	198.456	192.9	0.1382	210.1	0.0972	5.54	8.19
50	8	225.345	220.8	0.1448	230.7	0.0954	2.32	4.29
100	2	201.824	190.8	0.9567	253.8	0.436	20.48	24.82
100	4	351.522	342	0.8342	382	0.4214	7.98	10.47
100	6	388.724	380.6	0.8386	405.2	0.4205	4.07	6.07
100	8	421.852	414.4	0.8033	435.6	0.4248	3.16	4.87

5   Conclusions and Future Work

In this work, a neural model, MREM, is presented to solve the Shortest Common Superstring problem. This problem arises in real-world applications coming from molecular genetics (DNA sequencing) and data compression.

The neural model MREM is a generalization of Hopfield's model. Its main feature is that neuron states can be selected from a discrete set $\mathcal{M} = \{m_1, \ldots, m_L\}$, instead of taking value in $\{-1,1\}$ or $\{0,1\}$. This fact makes the network represent combinatorial optimization problems more easily.

A neural dynamics has been developed and implemented to solve the problem at hand, taking advantage of the representation of a solution as a permutation of the indices of the strings to be merged.

We have tested our approach by comparing it to a greedy algorithm, well-known from the specialized literature. In our results, MREM proved to outperform the greedy algorithm in most cases. It may be of great help in tackling real-world SCSS instances.

As a future work, we plan to:

- Develop a parallel version of the computational dynamics presented in this paper, in order to reduce the computational time used to achieve the solution.
- Introduce some mechanism to avoid local optima of the objective function. The hybridization of MREM with other stochastic techniques (Genetic Algorithms, Simulated Annealing) may be helpful.
- Make a theoretical study on the behavior of this new neural algorithm, in order to confirm the improvement over the greedy algorithm.

Acknowledgements

This work is partially supported by Junta de Andalucía (Spain) under contract TIC-01615, project name Intelligent Remote Sensing Systems.

Authors also wish to thank Prof. Gabriela Andrejková, from University of Kosice, Slovakia, for introducing the Shortest Common Superstring problem to them.

References

1. Ilie, L., Popescu, C.: The shortest common superstring problem and viral genome compression. Fundamenta Informaticae 73(1,2), 153–164 (2006)
2. Lesk, A.: Computational Molecular Biology, Sources and Methods for Sequence Analysis. Oxford University Press, Oxford (1988)
3. Li, M.: Towards a dna sequencing theory (learning a string). In: Proc. 31st Annual Symposium on Foundations of Computer Science, pp. 125–134 (1990)
4. Peltola, H., Soderlund, H., Tarhio, J., Ukkonen, E.: Algorithms for some string matching problems arising in molecular genetics. In: Proc. IFIP Congress, pp. 53–64 (1983)
5. Daley, M., McQuillan, I.: Viral gene compression: complexity and verification. In: Domaratzki, M., Okhotin, A., Salomaa, K., Yu, S. (eds.) CIAA 2004, vol. 3317, pp. 102–112. Springer, Heidelberg (2005)
6. Storer, J.: Data Compression: Methods and Theory. Computer Science Press, Rockville (1988)
7. Garey, M.R., Johnson, D.S.: Computers and Intractability. In: Garey, M.R., Johnson, D.S. (eds.) A guide to the theory of NP-Completeness, W. H. Freeman and Company, New York (1979)
8. Maier, D., Storer, J.: A note on the complexity of the superstring problem. In: Proceedings of the 12th Annual Conference on Information Science and Systems, pp. 52–56 (1978)
9. Blum, A., Jiang, T., Li, M., Tromp, J., Yannakakis, M.: Linear approximation of shortest superstring. Journal of the ACM 41(4), 630–647 (1994)
10. Arora, S., Lund, C., Motwani, R., Sudan, M., Szegedy, M.: Proof verification and hardness of approximation problems. In: 33rd Annual Symposium on Foundations of Computer Science, pp. 14–23 (1992)
11. Turner, J.: Approximation algorithms for the sortest common superstring problem. Information and Computation 83(1), 1–20 (1989)
12. Jiang, T., Jiang, Z., Breslauer, D.: Rotation of periodic strings and short superstrings. In: Proc. 3rd South American Conference on String Processing (1996)
13. Sweedyk, Z.: A $2\frac{1}{2}$-approximation algorithm for shortest superstring. SIAM Journal of Computing 29, 954–986 (1999)
14. Andrejkov, G., Levick, M., Oravec, J.: Approximation of shortest common superstring using neural networks. In: Proc. of 7th International Conference on Electronic Computers and Informatics, pp. 90–95 (2006)
15. Hopfield, J., Tank, D.: Neural computation of decisions in optimization problems. Biological Cybernetics 52, 141–152 (1985)
16. Mérida-Casermeiro, E., Galán-Marín, G., Muñoz-Pérez, J.: An efficient multivalued hopfield network for the travelling salesman problem. Neural Processing Letters 14, 203–216 (2001)

17. Mérida-Casermeiro, E., Muñoz-Pérez, J., Domínguez-Merino, E.: An n-parallel multivalued network: Applications to the travelling salesman problem. In: Mira, J., Álvarez, J.R. (eds.) IWANN 2003. LNCS, vol. 2686, pp. 406–413. Springer, Heidelberg (2003)

18. Mérida-Casermeiro, E., López-Rodríguez, D.: Graph partitioning via recurrent multivalued neural networks. In: Cabestany, J., Prieto, A.G., Sandoval, F. (eds.) IWANN 2005. LNCS, vol. 3512, pp. 1149–1156. Springer, Heidelberg (2005)

19. López-Rodríguez, D., Mérida-Casermeiro, E., Ortiz-de-Lazcano-Lobato, J.M., López-Rubio, E.: Image compression by vector quantization with recurrent discrete networks. In: Kollias, S.D., Stafylopatis, A., Duch, W., Oja, E. (eds.) ICANN 2006. LNCS, vol. 4132, pp. 595–605. Springer, Heidelberg (2006)

20. Mérida-Casermeiro, E.: Red Neuronal recurrente multivaluada para el reconocimiento de patrones y la optimización combinatoria. Ph. D thesis, Universidad de Málaga (2000)

21. Hopfield, J.: Neural networks and physical systems with emergent collective computational abilities, vol. 79, pp. 2254–2558 (1982)

22. Ozturk, Y., Abut, H.: System of associative relationships (soar) (1997)

23. Erdem, M.H., Ozturk, Y.: A new family of multivalued networks. Neural Networks 9(6), 979–989 (1996)

24. Mérida, E., Muñoz, J., Benítez, R.: A recurrent multivalued neural network for the N-queens problem. In: Mira, J., Prieto, A.G. (eds.) IWANN 2001, vol. 2084, pp. 522–529. Springer, Heidelberg (2001)

25. López-Rodríguez, D., Mérida-Casermeiro, E., Ortiz-de-Lazcano-Lobato, J.M., Galán-Marín, G.: k-pages graph drawing with multivalued neural networks. In: de Sá, J.M., Alexandre, L.A., Duch, W., Mandic, D.P. (eds.) ICANN 2007. LNCS, vol. 4669, pp. 816–825. Springer, Heidelberg (2007)

26. Galán-Marín, G., Mérida-Casermeiro, E., López-Rodríguez, D.: Improving neural networks for mechanism kinematic chain isomorphism identification. Neural Processing Letters 26, 133–143 (2007)

A Methodology for Developing Nonlinear Models by Feedforward Neural Networks

Henrik Saxén and Frank Pettersson

Heat Engineering Lab., Åbo Akademi University, Biskopsg. 8, 20500 Åbo, Finland
{henrik.saxen,frank.pettersson}@abo.fi

Abstract. Feedforward neural networks have been established as versatile tools for nonlinear black-box modeling, but in many data-mining tasks the choice of relevant inputs and network complexity is still a major challenge. Statistical tests for detecting relations between inputs and outputs are largely based on linear theory, and laborious retraining combined with the risk of getting stuck in local minima make the application of exhaustive search through all possible network configurations impossible but for toy problems. This paper proposes a systematic method to tackle the problem where an output shall be estimated on the basis of a (large) set of potential inputs. Feedforward neural networks of multi-layer perceptron type are used in the three-stage modeling approach: First, starting from sufficiently large networks an efficient pruning method is applied to detect a pool of potential model candidates. Next, the Akaike weights are used as to select the actual Kullback-Leibler best models in the pool. Third, the hidden nodes of these networks are available for the final network, where mixed-integer linear programming is applied to find the optimal combination of M hidden nodes, and the corresponding upper-layer weights. The procedure outlined is demonstrated to yield parsimonious models for a nonlinear benchmark problem, and to detect the relevant inputs.

Keywords: Non-linear black-box modeling, Neural networks, Information criterion, Structural and parametric optimization.

1 Introduction

In many complex modeling problems encountered in engineering and business, there is a large number of variables that potentially affect the "dependent" variable(s), and the underlying relations between the variables are often poorly known. Neural network have become popular data-driven modeling tools due to their ability to express arbitrary nonlinear relations, but the choices of inputs and network structure still constitute major challenges when real-world problems are tackled. This is due to the requirement to restrict the number of network parameters (weights) to avoid over-fitting, which is a serious problem when data with noise is studied. A simple trial-and-error approach, where different inputs are tested with networks of different complexity, is clearly a tedious and inefficient procedure that does not work on large problems: Many practical data-mining problems are also of a size that does not allow

M. Kolehmainen et al. (Eds.): ICANNGA 2009, LNCS 5495, pp. 72–78, 2009.
© Springer-Verlag Berlin Heidelberg 2009

for an exhaustive search among all possible models, keeping in mind the convergence problems that may be encountered in estimating the parameters of neural networks.

The problem has been tackled mainly by two techniques that can be termed constructive or destructive. In constructive methods (e.g., [1-2]) a small network is allowed to grow according to some criteria until it provides a reasonable solution to the problem at hand. By contrast, in destructive or pruning approaches (e.g., [3-5]), the connections or hidden nodes of a large network are gradually removed until the point is reached when further pruning proves detrimental for the network performance. Methods have also been proposed where both approaches have been used [6]. Even though these methods may lead to parsimonious models, they are often very sensitive to the noise-to-signal ratio and, as a rule, also require laborious retraining, which may still give rise to results sensitive to the successful convergence of the training, since only one network at a time is manipulated. In more recently reported efforts, multiple networks have been entertained, e.g., populations in genetic algorithms, in a simultaneous optimization of weights and network structure [7-10]. Still, the stochastic nature of these search methods may leave interesting model candidates unexplored and the fine-tuning of the weights can also be a problem.

The present paper outlines a method that tackles the problem of finding a (nonlinear) model on the basis of a large number of potential inputs. Feedforward neural networks are used as models, and special attention is paid to the detection of hidden nodes appropriately connected to relevant inputs. Starting from sufficiently large networks a large number of runs with an efficient pruning algorithm is applied to produce promising models capturing the underlying input-output relationships. The best of these models are selected on the basis of the Akaike weights, and the hidden nodes of the selected models, together with their (sparse) input connections, are made available for the final model candidates. Mixed-integer linear programming (MILP) is applied to optimize the upper-layer weights and the selection of the most important hidden nodes for building a final model, gradually increasing its complexity (expressed in terms of numbers of hidden nodes). The next section presents the steps of the method, while the performance of it on a test problem is illustrated in Section 3. The last section concludes the paper with a brief discussion.

2 The Method

2.1 Pruning Method

The pruning algorithm is based on feedforward neural networks of multi-layer perceptron type with a single layer of hidden nonlinear units connected to N inputs $\mathbf{x}$ (weight matrix $\mathbf{W}$) and to a single linear output node y (weight vector $\mathbf{v}$). If the outputs of the hidden nodes are known, $\mathbf{v}$ can be determined by matrix inversion. The algorithm starts with a random $\mathbf{W}$ and equates each element in this matrix, in turn, to zero, determines the optimal upper-layer weights and the corresponding value of the objective function, $\varepsilon = \left\| \mathbf{y} - \hat{\mathbf{y}} \right\|_2^2 / n$, where n is the number of observations (=dim($\mathbf{y}$)).

It detects the connection which at deletion yields the smallest objective function, permanently equates the corresponding weight in $\mathbf{W}$ to zero, and repeats the procedure. For more details, the readers are referred to refs. [11-13].

In order to determine the point where enough connections have been removed, we apply the Akaike information criterion (*AIC*) [14], modified for a small-sample bias adjustment [15]

$$AIC_c = AIC + \frac{2K(K+1)}{n-K-1} \quad \text{with} \quad AIC = -n\ln\varepsilon + 2K. \tag{1a,b}$$

The index is calculated during the pruning and the network corresponding to the minimum AIC_c value is selected to a pool of promising networks.

2.2 Model Selection within the Pool

Since similar networks may appear in the pool, and also some rather poor models (due to the effect of the random initialization of the lower-layer weights prior to the pruning step), an evaluation of the networks in the pool should be made, retaining only the best models. Therefore, the minimum value of the Akaike criterion among the R models in the pool, $AIC_{c,\min}$, was determined. The likelihood of a model g_i on the given data X is [16]

$$L(g_i|X) \propto \exp(-\Delta_i/2) \tag{2}$$

with

$$\Delta_i = AIC_{c,i} - AIC_{c,\min}. \tag{3}$$

A relative likelihood can be defined by normalization

$$\omega_i = \frac{\exp(-\Delta_i/2)}{\sum\limits_{r=1}^{R} \exp(-\Delta_r/2)}, \tag{4}$$

where ω_i is called the Akaike weight, and considered an evidence ratio for the fact that model i is the actual Kullback-Leibler best model amongst the ones available in the set. [16] The networks with an Akaike weight value exceeding a threshold can be retained.

2.3 Selection of Hidden Nodes and Upper-Layer Weights in the Final Model

As the last step of the algorithm, the hidden nodes of the remaining networks, with their input connections and weights, are considered central building blocks to be optimally assembled in the final model. The determination of the best combinations of these hidden nodes, where the total number of included hidden nodes, M, is varied from one to a sufficiently large value, $M_{\max}$, is done by solving mixed-integer linear programming problems (MILP). The MILP problems, which are solved with respect to which nodes to select and the upper-layer weights, are based on a linearization of a problem minimizing ε. Since the problem is quadratic it can easily be represented with a suitable set of linear under-estimators, while the presence of each hidden node in the final optimal network is identified with a binary variable.

3 Illustrative Example

A test function [11] with a three-dimensional input (N = 3) is used to illustrate the performance of the method

$$y = a(x_1^2 + 0.5x_1x_2) + (1-a)(0.5x_2x_3 + x_3^2) + b\eta \qquad (5)$$

This function was designed to yield a varying and nonlinear dependence between the inputs and the output for different values of the parameter $a \in (0,1)$: The output is independent of x_3 for $a = 1$, independent of x_1 for $a = 0$, while for $0 < a < 1$ it depends on all three input variables. The parameter b controls the level of noise in the model. All inputs as well as the noise term were generated as normally distributed random variables with zero mean and unit variance, i.e., η, $x_i = \mathcal{N}(0,1)$. Even though it is usually advantageous to normalize the inputs and outputs before a data-driven modeling effort, it was decided not to do so in the present example, because of the problems associated with interpreting the results in terms of final residual levels.

The method was evaluated on 400 data points for different degrees of complexity of the input-output relation by using $a = 0.1$, 0.5, or 0.9 with either lower ($b = 0.05$) or higher ($b = 0.20$) noise level. The pruning method was executed from different random lower-layer weight matrixes with 15 hidden nodes, yielding 40 smaller sparsely connected networks with 12-15 remaining hidden nodes where AIC_c corresponded to a minimum. These networks were moved to the pool, and the hidden nodes with their lower-layer connections of those of them for which $\omega_i > 0.1$ (cf. Eq. (4)) were collected into the final network. As the last stage of the algorithm, a final network model was determined with a step-wise increase of complexity for $M = 1\ldots15$ hidden nodes, where for every case the most advantageous combination of M hidden nodes was determined by solving a MILP problem.

Figure 1 presents the performance of the final model candidates with growing number of hidden nodes for the low-noise ($b = 0.05$, upper panel) and high-noise ($b = 0.20$, lower panel) cases. The following conclusions can be drawn:

The cases with $a = 0.1$ and $a = 0.9$ show similar behavior in both figures (excluding the excursion in Fig. 1b for $a = 0.9$ and 3-5 hidden nodes): This is expected due to the symmetry of the data with respect to the inputs x_1 and x_3 (cf. Eq. (5)). The case with $a = 0.5$, in turn, is seen to require more hidden nodes for achieving a model with a given error, ε; this agrees with the higher complexity of this model, where all three inputs are important. This case also shows the highest residuals for models of high complexity. A general observation is that models with 6-8 hidden nodes are appropriate for capturing the underlying dependence between the inputs and output.

Another interesting way of analyzing the result is to study the occurrence of the three inputs in the optimal models of different complexity. Figure 2 shows for $a = 0.10$, 0.50 and 0.90, with $b = 0.05$ (Fig. 2a) and $b = 0.20$ (Fig. 2a) the ratio of connections from the hidden nodes to given inputs and the total number of connection from the hidden nodes to the inputs, using the symbols $*$ for x_1, $*$ for x_2 and O for x_3. The top panel of Fig. 2a shows that the only input is x_3 for networks with one or two hidden nodes, and that x_1 is not included among the inputs until the eight hidden node is incorporated. A similar behavior is seen in the bottom panel of the figure, but now with interchanged x_1 (*) and x_3 (O). The occurrence of the second input, x_2, follows a somewhat more complicated pattern due to its role in both parts of Eq. (5), and this variable is present in all models with three or more hidden nodes. In summary, the method has successfully detected the relevant input variables.

As for the effect of noise, it may be concluded that the results are comparatively insensitive to the noise levels, which indicates that the method has been successful in capturing the underlying input-output mapping of the function. However, tests on data with smaller sample size are still needed to verify this statement, as well as an evaluation on real-world problems.

a)

b)

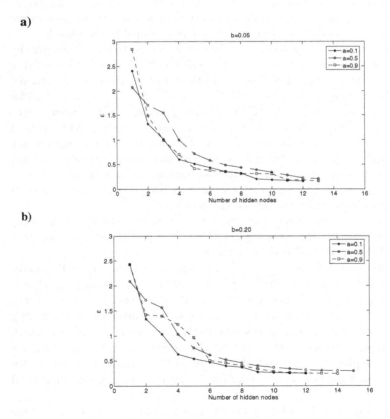

Fig. 1. Evolution of the error of the final model as functions of the number of hidden nodes. a) Low noise level. b) High noise level. The values of a is given in the legends and b above the panels (cf. Eq. (5)).

a)

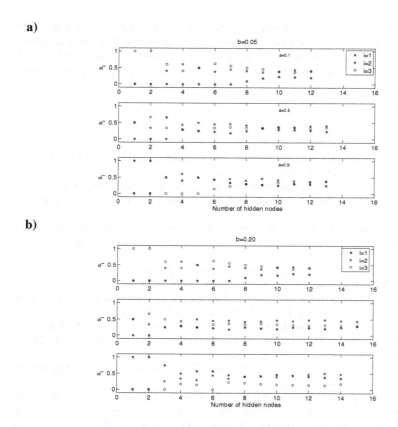

b)

Fig. 2. Share s_i of connections to input x_i in final models of growing number of hidden nodes for $a = 0.10, 0.50$ and 0.90. a) Low noise level, $b = 0.05$, b) High noise level, $b = 0.20$.

4 Conclusions

An algorithm for automatic data-driven generation of nonlinear models has been proposed. In a pruning step, several neural models with a single layer of sigmoid units are evolved by discarding unnecessary lower-layer connections using a statistical criterion for determining the stopping point, where parsimonious models remain. These networks are collected in a pool, and a statistical criterion is applied to select the best ones. Finally, the hidden nodes of the selected networks are used as building blocks of the final model, and MILP is applied to simultaneously select the hidden nodes and train their upper-layer weights for gradually increasing network size. The method has been illustrated on a test problem where the number of relevant inputs, the nonlinearity as well as the signal-to-noise level can be varied, demonstrating the method's ability to detect relevant inputs and to tackle the bias-variance dilemma [17].

In forthcoming work, the method will be applied to real-world problems, where the relevant inputs and their influence on the output variable(s) will be detected on the basis of large data sets from the fields of engineering and business. Future efforts will also be geared towards developing the criteria for how the final model complexity shall be determined.

References

1. Frean, M.: The Upstart Algorithm. A Method for Constructing and Training Feed-forward Neural Networks. Neural Computation 2, 198–209 (1991)
2. Fahlman, S.E., Lebiere, C.: The Cascade-Correlation Learning Architecture. In: Touretzky, D.S. (ed.) Adv. Neural Inf. Proc. Syst., vol. 2, pp. 524–532. Morgan Kaufmann, San Francisco (1990)
3. Le Chun, Y., Denker, J.S., Solla, S.A.: Optimal Brain Damage. In: Touretzky, D.S. (ed.) Adv. Neural Inf. Proc. Syst., vol. 2, pp. 598–605. Morgan Kaufmann, San Francisco (1990)
4. Engelbrecht, A.P.: A new pruning heuristic based on variance analysis of sensitivity information. IEEE Trans. Neural Networks 2, 1386–1399 (2001)
5. Jorgensen, T.D., Haynes, B.P., Norlund, C.C.F.: Pruning artificial neural networks using neural complexity measures. Int. J. Neural Systems 18, 389–403 (2008)
6. Narasimha, P.L., Delashmit, W.H., Manry, M.T., Li, J., Maldonado, F.: An integrated growing-pruning method for feedforward network training. Neurocomputing 71, 2831–2847 (2008)
7. Fogel, D.B.: An Information Criterion for Optimal Neural Network Selection. IEEE Trans. Neural Networks 2, 490–497 (1991)
8. Gao, F., Li, M., Wang, F., Wang, B., Yue, P.: Genetic Algorithms and Evolutionary Programming Hybrid Strategy for Structure and Weight Learning for Multilayer Feedforward Neural Networks. Ind. Eng. Chem. Res. 38, 4330–4336 (1999)
9. Hinnelä, J., Saxén, H., Pettersson, F.: Modeling of the blast furnace burden distribution by evolving neural networks. Ind. Eng. Chem. Res. 42, 2314–2323 (2003)
10. Pettersson, F., Chakraborti, N., Saxén, H.: A Genetic Algorithms Based Multiobjective Neural Net Applied to Noisy Blast Furnace Data. Applied Soft Computing 7, 387–397 (2007)
11. Saxén, H., Pettersson, F.: A simple method for selection of inputs and structure of feedforward neural networks. In: Ribeiro, R., et al. (eds.) Adaptive and Natural Computing Algorithms. Springer Computer Science, Heidelberg (2005)
12. Saxén, H., Pettersson, F.: Method for the selection of inputs and structure of feedforward neural networks. Comput. Chem. Engng. 30, 1038–1045 (2006)
13. Saxén, H., Pettersson, F.: Nonlinear Prediction of the hot Metal Silicon Content in the Blast Furnace. ISIJ Int. 47, 1732–1737 (2007)
14. Akaike, H.: A new look at the statistical model identification. IEEE Trans. Automatic Control 19, 716–723 (1974)
15. Hurvich, C.M., Tsai, C.-L.: Regression and time series model selection in small samples. Biometrika 76, 297–307 (1989)
16. Kullback, S., Leibler, R.A.: On Information and Sufficiency. Annals of Mathematical Statistics 22, 79–86 (1951)
17. Haykin, S.: Neural Networks. A Comprehensive Foundation, 2nd edn. Prentice-Hall Inc., New Jersey (1999)

A Predictive Control Economic Optimiser and Constraint Governor Based on Neural Models

Maciej Ławryńczuk

Institute of Control and Computation Engineering, Warsaw University of Technology
ul. Nowowiejska 15/19, 00-665 Warsaw, Poland
Tel.: +48 22 234-76-73
M.Lawrynczuk@ia.pw.edu.pl

Abstract. This paper discusses a Model Predictive Control (MPC) structure for economic optimisation of nonlinear technological processes. It contains two parts: an MPC economic optimiser/constraint governor and an unconstrained MPC algorithm. Two neural models are used: a dynamic one for control and a steady-state one for economic optimisation. Both models are linearised on-line. As a result, an easy to solve on-line one quadratic programming problem is formulated. Unlike the classical multilayer control system structure, the necessity of repeating two nonlinear optimisation problems at each sampling instant is avoided.

1 Introduction

Model Predictive Control (MPC) algorithms based on linear models have been successfully used for years in advanced large-scale industrial applications [11,13,15]. It is mainly because MPC algorithms have a unique ability to take into account constraints imposed on process inputs (manipulated variables) and outputs (controlled variables) which decide on quality, economic efficiency and safety. Moreover, MPC is very efficient in multivariable process control. To maximise economic gains MPC cooperates with economic optimisation [1,2,4,7,8,9,14,15].

In case of nonlinear processes it is justified to use neural models [5] for economic optimisation and control [6,8,9,10,12,15,16] rather than fundamental models. Fundamental models are usually not suitable for on-line control and optimisation as they are very complicated and may lead to numerical problems (ill-conditioning, stiffness, etc.). Neural models can be efficiently used on-line in MPC and economic optimisation because they have excellent approximation abilities, a small number of parameters and a simple structure. Moreover, neural models directly describe input-output relations of process variables, complicated systems of algebraic and differential equations do not have to be solved on-line.

The way MPC cooperates with economic optimisation is very important because it determines not only possible economic profits but also computational efficiency. In the classical multilayer control system structure [4,15] the control layer keeps the process at given operating points and the optimisation layer calculates these set-points. Typically, because of big computational burden, the

M. Kolehmainen et al. (Eds.): ICANNGA 2009, LNCS 5495, pp. 79–88, 2009.
© Springer-Verlag Berlin Heidelberg 2009

nonlinear economic optimisation task is solved significantly less frequently than the MPC controller executes. Such an approach can result in low economic effectiveness [15]. As an alternative to the classical structure, the MPC algorithm can be supplemented with an additional steady-state target optimisation task which recalculates the operating point determined by the nonlinear economic optimisation layer as frequently as MPC executes [7,9,13,15]. Unfortunately, three optimisation problems have to be solved on-line. Another idea is to integrate economic optimisation and MPC optimisation into one task [8,15].

This paper presents an efficient MPC scheme based on neural models for economic optimisation and control of nonlinear technological processes. The main part of the structure is an MPC economic optimiser/constraint governor which calculates on-line the operating point in such a way that economic gains are maximised and constraints are satisfied. The operating point is next used in an unconstrained MPC algorithm with Nonlinear Prediction and Linearisation (MPC-NPL) [10,15,16]. Two neural models are used: a dynamic model in the control subproblem, a steady-state model in the economic optimisation/constraint governor. Both models are linearised on-line. As a result, the described control system structure requires solving on-line only one quadratic programming problem. Unlike the classical multilayer structure, the necessity of repeating two nonlinear optimisation problems at each sampling instant is avoided.

2   The Classical Multilayer Control System Structure

The structure of the standard multilayer control system is depicted in Fig. 1. The objective of economic optimisation (named local steady-state optimisation) is to maximise the production profit and to satisfy constraints, which determine safety and quality of production. Typically, the economic optimisation layer solves the following problem (for simplicity of presentation Single-Input Single-Output (SISO) process is assumed)

$$\min_{u^s} \left\{ J_E(k) = c_u u^s - c_y y^s \right\}$$

$$u^{\min} \leq u^s \leq u^{\max}$$
$$y^{\min} \leq y^s \leq y^{\max} \tag{1}$$
$$y^s = f^s(u^s, h^s)$$

where u is the input of the process (manipulated variable), y is the output (controlled variable) and h is the measured (or estimated) disturbance, the superscript 's' refers to the steady-state. The function $f^s : \Re^2 \longrightarrow \Re \in C^1$ denotes a steady-state model of the process. Quantities c_u, c_y represent economic prices, $u^{\min}$, $u^{\max}$, $y^{\min}$, $y^{\max}$ denote constraints.

In MPC at each sampling instant k future control increments are calculated

$$\Delta \boldsymbol{u}(k) = [\Delta u(k|k)\ \Delta u(k+1|k)\dots\Delta u(k+N_u-1|k)]^T \tag{2}$$

Only the first element of the determined sequence (2) is applied to the process i.e. $u(k) = \Delta u(k|k) + u(k-1)$. At the next sampling instant, $k+1$, the prediction is shifted one step forward and the whole procedure is repeated.

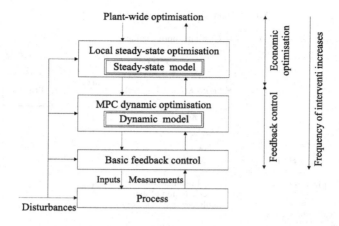

Fig. 1. The classical multilayer control system structure

Let u_{eo}^s be the solution to economic optimisation (1). Using the nonlinear steady-state model, the value y_{eo}^s corresponding to u_{eo}^s is calculated. It is next passed as the desired set-point $(y^s(k) = y_{eo}^s)$ to the MPC optimisation problem

$$\min_{\Delta\mathbf{u}(k)} \left\{ J_{MPC}(k) = \sum_{p=1}^{N} \mu_p (y^s(k) - \hat{y}(k+p|k))^2 + \sum_{p=0}^{N_u-1} \lambda_p (\Delta u(k+p|k))^2 \right\}$$

$$u^{\min} \le u(k+p|k) \le u^{\max}, \quad p = 0,\dots,N_u - 1 \qquad (3)$$

$$-\Delta u^{\max} \le \Delta u(k+p|k) \le \Delta u^{\max}, \quad p = 0,\dots,N_u - 1$$

$$y^{\min} \le \hat{y}(k+p|k) \le y^{\max}, \quad p = 1,\dots,N$$

N and N_u are prediction and control horizons, respectively, $\mu_p \ge 0$, $\lambda_p > 0$.

If the process is nonlinear it is reasonable to use a nonlinear steady-state model in economic optimisation (1) and a nonlinear dynamic model in MPC optimisation (3). As a results, in the standard multilayer control system structure two nonlinear optimisation problems have to be solved on-line.

Typically, the economic optimisation problem is solved less frequently than the MPC controller executes, while the MPC optimisation task has to be solved at each sampling instant. In the case of slowly varying disturbances such an approach gives satisfactory results. On the other hand, in practice disturbances (e.g. flow rates, properties of feed and energy streams) vary significantly and not much slower than the dynamics of the process. In such cases the classical multilayer structure with low frequency of economic optimisation can be economically inefficient [15]. As increasing the frequency of nonlinear economic optimisation is usually not possible in practice, MPC can be supplemented with additional steady-state target optimisation which recalculates the optimal operating point as frequently as MPC executes [7,9,13,15]. Alternatively, economic and MPC tasks can be integrated into one optimisation problem [8,15].

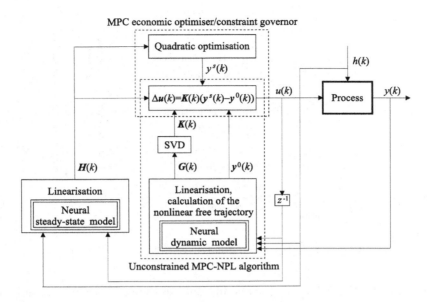

Fig. 2. The configuration of the MPC economic optimiser/constraint governor cooperating with the unconstrained MPC-NPL algorithm

3   A Predictive Control Economic Optimiser and Constraint Governor Based on Neural Models

The configuration of the discussed structure is shown in Fig. 2. An MPC economic optimiser/constraint governor calculates on-line the operating point $y^s(k)$ in such a way that economic profits are maximised and constraints are satisfied. The operating point is next used in an unconstrained MPC algorithm with Nonlinear Prediction and Linearisation (MPC-NPL) [10,15,16] which calculates future control increments $\Delta u(k)$. The economic optimiser/constraint governor directly uses the MPC-NPL control law (dashed boxes in Fig. 2 overlap).

Two nonlinear neural models are used. A dynamic neural model is used in the unconstrained MPC-NPL algorithm, a steady-state neural model is used in the economic optimiser/constraint governor. Both models are linearised on-line taking into account the current state of the process. As a result, only one quadratic programming problem must be solved on-line. Unlike the classical multilayer control system structure, the necessity of repeating two nonlinear optimisation problems at each sampling instant is avoided.

3.1   Neural Models of the Process and On-Line Linearisation

Let the dynamic model of the process under consideration be described by

$$y(k) = f(\boldsymbol{x}(k)) = f(u(k - \tau), \ldots, u(k - n_B), y(k - 1), \ldots, y(k - n_A), \tag{4}$$
$$h(k - \tau_h), \ldots, h(k - n_C))$$

where $f : \Re^{n_A+n_B+n_C-\tau-\tau_h+2} \longrightarrow \Re \in C^1$, $\tau \le n_B$, $\tau_h \le n_C$. A MultiLayer Perceptron (MLP) neural network with one hidden layer and a linear output [5] is used as the function f in (4). Output of the model can be expressed as

$$y(k) = f(\boldsymbol{x}(k)) = w_0^2 + \sum_{i=1}^{K} w_i^2 \varphi(z_i(k)) \tag{5}$$

where $z_i(k)$ are sums of inputs of the ith hidden node, $\varphi : \Re \longrightarrow \Re$ is the nonlinear transfer function, K is the number of hidden nodes. From (4) one has

$$z_i(k) = w_{i,0}^1 + \sum_{j=1}^{I_u} w_{i,j}^1 u(k - \tau + 1 - j) + \sum_{j=1}^{n_A} w_{i,I_u+j}^1 y(k - j) \tag{6}$$

$$+ \sum_{j=1}^{I_h} w_{i,I_u+n_A+j}^1 h(k - \tau_h + 1 - j)$$

Weights of the network are denoted by $w_{i,j}^1$, $i = 1, \ldots, K$, $j = 0, \ldots, n_A + n_B + n_C - \tau - \tau_h + 2$, and w_i^2, $i = 0, \ldots, K$, for the first and the second layer, respectively, $I_u = n_B - \tau + 1$, $I_h = n_C - \tau_h + 1$.

The dynamic neural model is linearised on-line taking into account the current state $\bar{\boldsymbol{x}}(k)$ of the process determined by past input and output signal values corresponding to the arguments of the nonlinear model (4). Using Taylor series expansion, the linear approximation of the model is

$$y(k) = f(\bar{\boldsymbol{x}}(k)) + \sum_{l=1}^{n_B} b_l(\bar{\boldsymbol{x}}(k))(u(k - l) - \bar{u}(k - l)) \tag{7}$$

$$- \sum_{l=1}^{n_A} a_l(\bar{\boldsymbol{x}}(k))(y(k - l) - \bar{y}(k - l))$$

where $a_l(\bar{\boldsymbol{x}}(k)) = -\frac{\partial f(\bar{\boldsymbol{x}}(k))}{\partial y(k-l)}$, $b_l(\bar{\boldsymbol{x}}(k)) = \frac{\partial f(\bar{\boldsymbol{x}}(k))}{\partial u(k-l)}$ are coefficients of the linearised model [8,9,10].

The second MLP network is used as the steady-state model $y^s = f^s(u^s, h^s)$

$$y^s = f^s(u^s, h^s) = w_0^{2s} + \sum_{i=1}^{K^s} w_i^{2s} v_i^s = w_0^{2s} + \sum_{i=1}^{K^s} w_i^{2s} \varphi(z_i^s) \tag{8}$$

where $z_i^s = w_{i,0}^{1s} + w_{i,1}^{1s} u^s + w_{i,2}^{1s} h^s$. Weights of the second network are denoted by $w_{i,j}^{1s}$, $i = 1, \ldots, K^s$, $j = 0, 1, 2$, and w_i^{2s}, $i = 0, \ldots, K^s$.

The steady-state neural model is linearised on-line taking into account the current state of the process determined by $u(k-1)$ and $h(k)$. The linear approximation of the model is

$$y^s = f^s(u^s, h^s)|_{u^s=u(k-1),\ h^s=h(k)} + \boldsymbol{H}(k)(u^s - u(k - 1)) \tag{9}$$

where

$$\boldsymbol{H}(k) = \left. \frac{df^s(u^s, h^s)}{du^s} \right|_{u^s=u(k-1),\ h^s=h(k)} \tag{10}$$

3.2 Economic Optimiser and Constraint Governor Optimisation

As shown in Fig. 2, in the unconstrained MPC-NPL algorithm at each sampling instant k the nonlinear dynamic neural model (4) is used on-line twice: to find a local linearisation (7) and a nonlinear free trajectory. Thanks to the linearisation, the output prediction can be expressed as the sum of a forced trajectory, which depends only on the future (on future input moves $\Delta u(k)$) and a free trajectory $y^0(k)$, which depends only on the past

$$\hat{y}(k) = G(k)\Delta u(k) + y^0(k) \tag{11}$$

where the matrix $G(k)$ of dimensionality $N \times N_u$ contains step-response coefficients of the linearised dynamic model (7), $\hat{y}(k) = [\hat{y}(k+1|k)\ldots\hat{y}(k+N|k)]^T$, $y^0(k) = [y^0(k+1|k)\ldots y^0(k+N|k)]^T$.

The idea of the discussed control system is to remove constraints originally taken into account in the general MPC optimisation problem (3) and use the unconstrained MPC-NPL algorithm. In order to maximise economic profits and satisfy constraints originally present in the original economic optimisation problem (1) and in the MPC optimisation problem, the economic optimiser/constraint governor is used. Thanks to the prediction equation (11), the MPC performance function $J_{MPC}(k)$ in (3) becomes a quadratic function. Optimal future control increments can be calculated analytically, without any optimisation as

$$\Delta u(k) = P^+(k) \begin{bmatrix} S_M(y^s(k) - y^0(k)) \\ 0_{N_u \times 1} \end{bmatrix} \tag{12}$$

where $S_M^T S_M = M$, $S_\Lambda^T S_\Lambda = \Lambda$, $P(k) = \begin{bmatrix} S_M G(k) \\ -S_\Lambda \end{bmatrix}$, $y^s(k) = [y^s(k)\ldots y^s(k)]^T$ is a vector of length N and '+' denotes Moore-Penrose pseudo-inverse which can be effectively calculated by means of Singular Value Decomposition (SVD) of the matrix $P(k) = U(k)\Sigma(k)V(k)^T$

$$P^+(k) = V(k) \begin{bmatrix} diag\left(\frac{1}{\sigma_1(k)}, \ldots, \frac{1}{\sigma_{N_u}(k)}\right) \\ 0_{N \times N_u} \end{bmatrix}^T U^T(k) \tag{13}$$

where $\sigma_1(k), \ldots, \sigma_{N_u}(k)$ are singular values of the matrix $P(k)$. Optimal control increments in the unconstrained MPC-NPL algorithms can be expressed as

$$\Delta u(k) = K(k)(y^s(k) - y^0(k)) \tag{14}$$

where $K(k) = P^+(k) [I_{N_u \times N_u} 0_{N_u \times N}]^T$.

The objective of the economic optimiser/constraint governor is to maximise the production profit and to satisfy constraints taken into account originally in the economic optimisation task (1) and in the rudimentary MPC optimisation task (3). Steady-state and dynamic models are linearised, the suboptimal prediction (11) is used. The quadratic programming problem is

$$\min_{u^s(k),\ \varepsilon^{\min},\ \varepsilon^{\max}} \left\{ c_u u^s(k) - c_y y^s(k) + \rho^{\min} \left\| \varepsilon^{\min} \right\|^2 + \rho^{\max} \left\| \varepsilon^{\max} \right\|^2 \right\}$$

$$u^{\min} \leq \boldsymbol{J} \Delta \boldsymbol{u}(k) + \boldsymbol{u}^{k-1}(k) \leq \boldsymbol{u}^{\max}$$

$$-\Delta \boldsymbol{u}^{\max} \leq \Delta \boldsymbol{u}(k) \leq \Delta \boldsymbol{u}^{\max}$$

$$\boldsymbol{y}^{\min} - \varepsilon^{\min} \leq \boldsymbol{G}(k) \Delta \boldsymbol{u}(k) + \boldsymbol{y}^0(k) \leq \boldsymbol{y}^{\max} + \varepsilon^{\max}$$

$$\varepsilon^{\min} \geq 0, \quad \varepsilon^{\max} \geq 0 \tag{15}$$

$$u^{\min} \leq u^s(k) \leq u^{\max}$$

$$y^{\min} \leq y^s(k) \leq y^{\max}$$

$$y^s(k) = f^s(u^s, h^s)\big|_{u^s = u(k-1),\ h^s = h(k)} + \boldsymbol{H}(k)(u^s(k) - u(k-1))$$

$$\Delta \boldsymbol{u}(k) = \boldsymbol{K}(k)(\boldsymbol{y}^s(k) - \boldsymbol{y}^0(k))$$

where $\boldsymbol{y}^{\min} = \left[y^{\min} \dots y^{\min} \right]^T$, $\boldsymbol{y}^{\max} = \left[y^{\max} \dots y^{\max} \right]^T$ are vectors of length N, $\boldsymbol{u}^{\min} = \left[u^{\min} \dots u^{\min} \right]^T$, $\boldsymbol{u}^{\max} = \left[u^{\max} \dots u^{\max} \right]^T$, $\Delta \boldsymbol{u}^{\max} = \left[\Delta u^{\max} \dots \Delta u^{\max} \right]^T$, $\boldsymbol{u}^{k-1}(k) = \left[u(k-1) \dots u(k-1) \right]^T$ are vectors of length N_u, $\boldsymbol{J}$ is the all ones lower triangular matrix of dimensionality $N_u \times N_u$, $\boldsymbol{M} = diag(\mu_1, \dots, \mu_N)$, $\boldsymbol{\Lambda} = diag(\lambda_0, \dots, \lambda_{N_u-1})$. To cope with infeasibility, output constraints are softened by slack variables (vectors $\varepsilon^{\min}$, $\varepsilon^{\max} > 0$ of length N) [11,15].

Considering the system structure depicted in Fig. 2, at each sampling instant k the following steps are repeated:

1. Linearisation of the steady-state neural model: obtain $\boldsymbol{H}(k)$.
2. Linearisation of the dynamic neural model: calculate coefficients $a_l(\bar{\boldsymbol{x}}(k))$, $b_l(\bar{\boldsymbol{x}}(k))$ of the linearised model and step response coefficients comprising the dynamic matrix $\boldsymbol{G}(k)$.
3. Find the nonlinear free trajectory $\boldsymbol{y}^0(k)$ using the dynamic neural model.
4. Using the SVD decomposition find the matrix $\boldsymbol{K}(k)$ which defines the unconstrained MPC-NPL control law $\Delta \boldsymbol{u}(k) = \boldsymbol{K}(k)(\boldsymbol{y}^s(k) - \boldsymbol{y}^0(k))$.
5. Solve the quadratic programming problem (15).
6. Calculate $\Delta \boldsymbol{u}(k)$ using the unconstrained MPC-NPL control law.
7. Apply the first element of the sequence $\Delta \boldsymbol{u}(k)$, i.e. $u(k) = \Delta u(k|k) + u(k-1)$.
8. Set $k := k + 1$, go to step 1.

Detailed description of on-line linearisation of neural models, calculation of step-response coefficients and of the free trajectory is given in [8,9,10].

4   Simulation Results

The studied process is a neutralisation chemical reactor [3] in which acid (HNO$_3$), base (NaOH) and buffer (NaHCO$_3$) are continuously mixed. The output pH is controlled by manipulating the base flow rate q_3, the buffer flow rate q_2 is the measured disturbance ($u = q_3$, $h = q_2$, $y = pH$). As steady-state and dynamic properties of the process are nonlinear, nonlinear neural models are used.

Three models of the process are used. The fundamental model [3] is used as the real process during simulations. An identification procedure is carried out, two neural models are obtained: a dynamic one ($K = 6$) and a steady-state one ($K^s = 4$) which are next used for MPC and economic optimisation, respectively. Parameters of MPC are: $N = 10$, $N_u = 2$, $\mu_p = 1$, $\lambda_p = 0.01$.

To maximise the production rate the economic performance function is

$$J_E(k) = -q_3^s \qquad (16)$$

The following constraints are imposed on manipulated and controlled variables

$$q_3^{\min} \leq q_3, q_3^s \leq q_3^{\max}, \qquad pH, pH^s \leq pH^{\max} \qquad (17)$$

where $q_3^{\min} = 1 \ ml/s$, $q_3^{\max} = 31.2 \ ml/s$. For the first part of the simulation ($k = 1, \ldots, 119$) $pH^{\max} = 10.5$, for the second part ($k = 120, \ldots, 199$) $pH^{\max} = 10.6$ and for the third part ($k = 200, \ldots, 350$) $pH^{\max} = 10.7$. The scenario of disturbance changes is

$$q_2(k) = 2 - 1.6(\sin(0.008k) - \sin(0.08)) \qquad (18)$$

In the multilayer structure the MPC algorithm with on-line Nonlinear Optimisation (MPC-NO) is used. Three versions of this structure are compared:

a) nonlinear economic optimisation is repeated 70 times less frequently than the MPC-NO algorithm executes,
b) nonlinear economic optimisation is repeated 45 times less frequently than the MPC-NO algorithm executes,
c) the "ideal" multilayer structure in which nonlinear economic optimisation is repeated as frequently as the MPC-NO algorithm executes.

In the first two cases nonlinear MPC optimisation is repeated at each sampling instant on-line whereas nonlinear economic optimisation every 70th or 45th sampling instant, respectively. In the third case two nonlinear optimisation problems are solved at each sampling instant.

Simulation results are depicted in Fig. 3. In the first two cases the frequency of nonlinear economic optimisation is low. It means that changes in the disturbance and in the output constraint are taken into account infrequently. As a result the operating point is constant for long periods. For the whole simulation horizon the economic performance index $J_E = \sum_{k=1}^{350} J_E(k) = \sum_{k=1}^{350}(-q_3(k))$ is calculated. In the first case $J_E = -10211.52$, in the second case $J_E = -10230.63$ (the bigger the negative value the better). In the "ideal" multilayer structure changes in the disturbance q_2 and in the output constraint are taken into account at each sampling instant. The economic performance index improves to $J_E = -10279.68$.

When the discussed structure with the economic optimiser/constraint governor cooperating with the unconstrained MPC-NPL algorithm is used, the trajectory of the system is very close to that obtained in the "ideal" but unrealistic case when two nonlinear optimisation problems are solved at each sampling instant on-line (at economic optimisation and MPC layers). In the discussed structure $J_E = -10272.94$ (only 0.065% worse in comparison with the "ideal" structure). At the same time, the discussed structure is computationally efficient because only one quadratic programming problem is solved on-line.

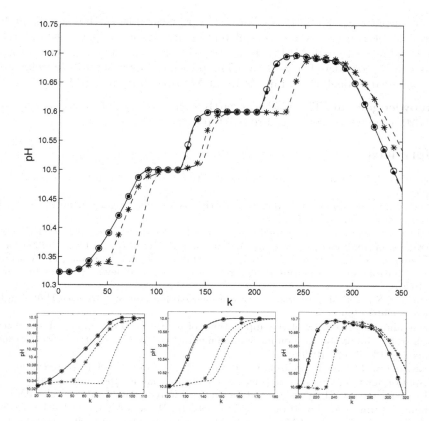

Fig. 3. Simulation results: the multilayer structure with nonlinear economic optimisation repeated 70 (*dashed line*) and 45 (*dashed line with asterisks*) times less frequently than the nonlinear MPC controller, the "ideal" multilayer structure with nonlinear economic optimisation repeated as frequently as the nonlinear MPC controller (*solid line with circles*) and the MPC economic optimiser/constraint governor cooperating with the unconstrained MPC-NPL algorithm (*dashed line with dots*)

5   Conclusions

The constraint governor/economic optimiser maximises economic profits and satisfies constraints. It cooperates with the unconstrained MPC algorithm with Nonlinear Prediction and Linearisation. Neural models are linearised on-line, only one quadratic programming task is solved. Economic results are very close to those obtained in the "ideal" classical multilayer structure in which two nonlinear optimisation problems are solved at each sampling instant. The approach is also computationally efficient in comparison with the steady-state target optimisation structure in which three optimisation problems are solved on-line [9,15].

Neural models are used because they have excellent approximation abilities, a small number of parameters and a simple structure. Moreover, they directly

88   M. Ławryńczuk

describe input-output relations of process variables while in the case of fundamental models complex systems of differential and algebraic equations have to be solved on-line. Although MLP neural networks are used, different types of networks can be considered (e.g. RBF). The presented approach can be relatively easy extended to deal with Multiple-Input Multiple-Output (MIMO) processes.

Acknowledgement. This work was partly supported by Polish national budget funds 2007-2009 for science as a research project.

References

1. Blevins, T.L., Mcmillan, G.K., Wojsznis, M.W.: Advanced control unleashed, ISA (2003)
2. Brdys, M., Tatjewski, P.: Iterative algorithms for multilayer optimizing control. Imperial College Press, London (2005)
3. Chen, J., Huang, T.C.: Applying neural networks to on-line updated PID controllers for nonlinear process control. Journal of Process Control 14, 211–230 (2004)
4. Findeisen, W.M., Bailey, F.N., Brdyś, M., Malinowski, K., Tatjewski, P., Woźniak, A.: Control and coordination in hierarchical systems. J. Wiley and Sons, Chichester (1980)
5. Haykin, S.: Neural networks – a comprehensive foundation. Prentice Hall, Englewood Cliffs (1999)
6. Hussain, M.A.: Review of the applications of neural networks in chemical process control – simulation and online implmementation. Artificial Intelligence in Engineering 13, 55–68 (1999)
7. Kassmann, D.E., Badgwell, T.A., Hawkins, R.B.: Robust steady-state target calculation for model predictive control. AIChE Journal 46, 1007–1024 (2000)
8. Ławryńczuk, M., Tatjewski, P.: Efficient predictive control integrated with economic optimisation based on neural models. In: Rutkowski, L., Tadeusiewicz, R., Zadeh, L.A., Zurada, J.M. (eds.) ICAISC 2008. LNCS, vol. 5097, pp. 111–122. Springer, Heidelberg (2008)
9. Ławryńczuk, M.: Neural models in computationally efficient predictive control cooperating with economic optimisation. In: de Sá, J.M., Alexandre, L.A., Duch, W., Mandic, D.P. (eds.) ICANN 2007. LNCS, vol. 4669, pp. 650–659. Springer, Heidelberg (2007)
10. Ławryńczuk, M.: A family of model predictive control algorithms with artificial neural networks. Int. Journal of Applied Mathematics and Computer Science 17, 217–232 (2007)
11. Maciejowski, J.M.: Predictive control with constraints. Prentice Hall, Englewood Cliffs (2002)
12. Nørgaard, M., Ravn, O., Poulsen, N.K., Hansen, L.K.: Neural networks for modelling and control of dynamic systems. Springer, London (2000)
13. Qin, S.J., Badgwell, T.A.: A survey of industrial model predictive control technology. Control Engineering Practice 11, 733–764 (2003)
14. Saez, D., Cipriano, A., Ordys, A.W.: Optimisation of industrial processes at supervisory level: application to control of thermal power plants. Springer, Heidelberg (2002)
15. Tatjewski, P.: Advanced control of industrial processes, Structures and algorithms. Springer, London (2007)
16. Tatjewski, P., Ławryńczuk, M.: Soft computing in model-based predictive control. Int. Journal of Applied Mathematics and Computer Science 16, 101–120 (2006)

Computationally Efficient Nonlinear Predictive Control Based on RBF Neural Multi-models

Maciej Ławryńczuk

Institute of Control and Computation Engineering, Warsaw University of Technology
ul. Nowowiejska 15/19, 00-665 Warsaw, Poland
Tel. +48 22 234-76-73
M.Lawrynczuk@ia.pw.edu.pl

Abstract. This paper is concerned with RBF neural multi-models and a computationally efficient nonlinear Model Predictive Control (MPC) algorithm based on such models. The multi-model has an ability to calculate predictions over the whole prediction horizon without using previous predictions. Unlike the classical Nonlinear Auto Regressive with eXternal input (NARX) model, the multi-model is not used recursively in MPC, the prediction error is not propagated. The presented MPC algorithm needs solving on-line only a quadratic programming problem but in practice it gives closed-loop control performance similar to that obtained in nonlinear MPC, which hinges on on-line non-convex optimisation.

1 Introduction

Model Predictive Control (MPC) algorithms based on linear models have been successfully used for years in advanced industrial applications [7,12,14]. It is largely because MPC algorithms can take into account constraints imposed on both process inputs (manipulated variables) and outputs (controlled variables), which usually decide on quality, economic efficiency and safety. Moreover, MPC algorithms are very efficient in multivariable process control.

Because properties of many technological processes are nonlinear, different nonlinear MPC techniques have been developed [8,14]. In particular, MPC algorithms based on neural models [1] of processes can be effectively used on-line [2,4,5,6,9,10,11,14]. It is because neural models have excellent approximation abilities, relatively a small number of parameters and a simple structure. Neural models directly describe input-output relations of process variables, complicated systems of algebraic and differential equations (fundamental models) do not have to be solved on-line which may lead to numerical problems (ill-conditioning, stiffness, etc.).

MPC algorithms are very model-based. The role of the model cannot be ignored during model structure selection and identification. The model has to be able to precisely predict future behaviour of the process over the whole prediction horizon. In practice, neural network models are usually trained using the rudimentary backpropagation algorithm which yields one-step ahead predictors. Recurrent neural network training is much more complicated. Naturally, one-step ahead predictors are not suited to be used recursively in MPC for long

M. Kolehmainen et al. (Eds.): ICANNGA 2009, LNCS 5495, pp. 89–98, 2009.
© Springer-Verlag Berlin Heidelberg 2009

range prediction since the prediction error is propagated. It is because of noise, model inaccuracies and underparameterisation, i.e. the order of the model used in MPC is usually significantly lower than the order of the real process or even the proper model order is unknown. To solve the problem resulting from the inaccuracy of one-step ahead predictors in MPC a multi-model approach has been proposed in [3,13] for linear processes. Alternatively, the model of a specialised structure can be used which does not ignore its specific role in MPC [6].

Contribution of this paper is twofold. It describes the RBF neural multi-model and derives a computationally efficient (suboptimal) MPC algorithm with Nonlinear Prediction and Linearisation (MPC-NPL) [5,6,14] based on such models. The algorithm needs solving on-line a numerically reliable quadratic programming approach. In practice, the algorithm gives closed-loop performance comparable to that obtained in fully-fledged nonlinear MPC, in which nonlinear optimisation is repeated at each sampling instant.

2   Model Predictive Control Algorithms

In MPC algorithms [7,14] at each consecutive sampling instant k a set of future control increments is calculated

$$\Delta \boldsymbol{u}(k) = [\Delta u(k|k)\ \Delta u(k+1|k) \ldots \Delta u(k+N_u-1|k)]^T \qquad (1)$$

It is assumed that $\Delta u(k+p|k) = 0$ for $p \geq N_u$, where N_u is the control horizon. The objective is to minimise the differences between the reference trajectory $y^{ref}(k+p|k)$ and predicted values of the output $\hat{y}(k+p|k)$ over the prediction horizon $N \geq N_u$. The following quadratic cost function is usually used

$$J(k) = \sum_{p=1}^{N} \mu_p (y^{ref}(k+p|k) - \hat{y}(k+p|k))^2 + \sum_{p=0}^{N_u-1} \lambda_p (\Delta u(k+p|k))^2 \qquad (2)$$

where $\mu_p \geq 0$, $\lambda_p > 0$ are weighting factors. Only the first element of the determined sequence (1) is applied to the process

$$u(k) = \Delta u(k|k) + u(k-1) \qquad (3)$$

At the next sampling instant, $k+1$, the prediction is shifted one step forward and the whole procedure is repeated.

Since constraints have to be usually taken into account, future control increments are determined from the following optimisation problem (for simplicity of presentation hard output constraints [7,14] are used)

$$\min_{\Delta u(k|k)\ldots \Delta u(k+N_u-1|k)} J(k)$$

subject to

$$u^{min} \leq u(k+p|k) \leq u^{max}, \quad p = 0, \ldots, N_u - 1 \qquad (4)$$
$$-\Delta u^{max} \leq \Delta u(k+p|k) \leq \Delta u^{max}, \quad p = 0, \ldots, N_u - 1$$
$$y^{min} \leq \hat{y}(k+p|k) \leq y^{max}, \quad p = 1, \ldots, N$$

2.1   Classical Modelling and Prediction

Let the Single-Input Single-Output (SISO) process be described by the following discrete-time Nonlinear Auto Regressive with eXternal input (NARX) model

$$y(k) = f(\boldsymbol{x}(k)) = f(u(k - \tau), \dots, u(k - n_B), y(k - 1), \dots, y(k - n_A)) \qquad (5)$$

where $f : \Re^{n_A + n_B - \tau + 1} \longrightarrow \Re$, $\tau \leq n_B$. In MPC predictions are obtained from

$$\hat{y}(k + p|k) = y(k + p|k) + d(k) \qquad (6)$$

where quantities $y(k + p|k)$ are calculated from the nonlinear model (5) used for $p = 1, \dots, N$. In the "DMC type" disturbance model the unmeasured disturbance $d(k)$ is assumed to be constant over the prediction horizon [14].

Using (5) and (6), output predictions for $p = 1, \dots, N$ are calculated from

$$\hat{y}(k + p|k) = f(\underbrace{u(k - \tau + p|k), \dots, u(k|k)}_{I_{uf}(p)}, \underbrace{u(k - 1), \dots, u(k - n_B + p)}_{I_u - I_{uf}(p)}, \qquad (7)$$

$$\underbrace{\hat{y}(k - 1 + p|k), \dots, \hat{y}(k + 1|k)}_{I_{yp}(p)}, \underbrace{y(k), \dots, y(k - n_A + p)}_{n_A - I_{yp}(p)}) + d(k)$$

where $I_{uf}(p) = \max(\min(p - \tau + 1, I_u), 0)$, $I_u = n_B - \tau + 1$, $I_{yp}(p) = \max(p - 1, n_A)$. In MPC the model must be used recursively as predictions depend on predictions calculated for previous sampling instants within the prediction horizon. Although a one-step ahead predictor is given as the result of backpropagation training, it is used for N-steps ahead prediction. Since model inaccuracies, underparameterisation and noise are unavoidable, the prediction error is propagated.

3   MPC-NPL Algorithm Based on Neural Multi-models

3.1   Neural Multi-modelling and Prediction

In the multi-model approach one independent neural model is used for each sampling instant within the prediction horizon. For the sampling instant $k + 1$ (i.e. the first instant of the prediction horizon) the following model is used

$$y(k + 1) = f_1(u(k - \tau + 1), \dots, u(k - n_B), \qquad (8)$$
$$y(k), \dots, y(k - n_A))$$

where $f_1 : \Re^{\min(n_B - \tau + 2, n_B) + \max(2 - \tau, 0) + n_A + 1} \longrightarrow \Re$ is a nonlinear function which describes the first submodel. For the sampling instant $k + 2$ the model is

$$y(k + 2) = f_2(u(k - \tau + 2), \dots, u(k - n_B), \qquad (9)$$
$$y(k), \dots, y(k - n_A))$$

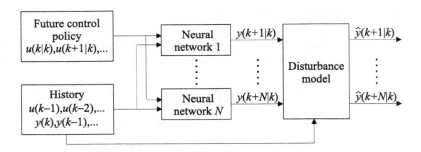

Fig. 1. The neural multi-model used for prediction in MPC

where $f_2 : \Re^{\min(n_B - \tau + 3, n_B) + \max(3 - \tau, 0) + n_A + 1} \longrightarrow \Re$ is a nonlinear function which describes the second submodel. In general, for $p = 1, \ldots, N$, all submodels can be expressed in a compact form as

$$y(k + p) = f_p(\boldsymbol{x}(k + p|k)) = f_p(u(k - \tau + p), \ldots, u(k - n_B), \qquad (10)$$
$$y(k), \ldots, y(k - n_A))$$

where $f_p : \Re^{\min(n_B + p - \tau + 1, n_B) + \max(p - \tau + 1, 0) + n_A + 1} \longrightarrow \Re$ is a nonlinear function which describes the pth submodel.

Fig. 1 shows the neural multi-model used for prediction in MPC. The multi-model consists of N neural networks. One independent neural network is used for each sampling instant within the prediction horizon. These neural networks realise functions $f_1, \ldots, f_N$, inputs of neural networks correspond to the arguments of the multi-model (10), all networks have one output. Because independent submodels are used for consecutive sampling instant of the prediction horizon, the "DMC type" disturbance model cannot be used. Instead of (6), one has to use

$$\hat{y}(k + p|k) = y(k + p|k) + d(k + p|k) \qquad (11)$$

for $p = 1, \ldots, N$. Independent disturbance estimations are

$$d(k + p|k) = y(k) - f_p(k|k - 1) \qquad (12)$$

where $y(k)$ is measured while $f_p(k|k-1)$ is calculated from the multi-model used for the sampling instant k

$$f_p(k|k - 1) = f_p(u(k - \tau), \ldots, u(k - n_B - p), \qquad (13)$$
$$y(k - p), \ldots, y(k - n_A - p))$$

Using (10) and (11), output predictions for $p = 1, \ldots, N$ are calculated from

$$\hat{y}(k + p|k) = f_p(\underbrace{u(k - \tau + p|k), \ldots, u(k|k)}_{I_{uf}(p)}, \underbrace{u(k - \max(\tau - p, 1)), \ldots, u(k - n_B)}_{I_{up}(p)},$$
$$\underbrace{y(k), \ldots, y(k - n_A))}_{n_A + 1} + d(k + p|k) \qquad (14)$$

where $I_{uf}(p) = \max(p - \tau + 1, 0)$, $I_{up}(p) = n_B - \max(\tau - p, 1) + 1$. Unlike predictions (7) calculated from the NARX model (5), they do not depend on predictions for previous sampling instants within the prediction horizon. The multi-model is not used recursively, the prediction error is not propagated.

Although the multi-model is used for long-range prediction in MPC, neural networks are trained as one-step ahead predictors. It is possible because for prediction one independent neural submodel is used for each sampling instant within the prediction horizon. Thanks to the structure of the multi-model it is not used recursively, predictions do not depend on previous predictions.

The multi-model is comprised of N feedforward RBF type neural networks with one hidden layer containing Gaussian functions and linear outputs [1]. Output of the multi-model for the sampling instant $k + p$, $p = 1, \ldots, N$ is

$$
y(k+p|k) = f_p(\boldsymbol{x}(k+p|k)) = w_0^p + \sum_{i=1}^{K^p} w_i^p \exp(-\|\boldsymbol{x}(k+p|k) - \boldsymbol{c}_i^p\|_{\boldsymbol{Q}_i^p})
$$

$$
= w_0^p + \sum_{i=1}^{K^p} w_i^p \exp(-z_i^p(k+p|k)) \tag{15}
$$

where K^p is the number of hidden nodes of the pth network. Vectors $\boldsymbol{c}_i^p$ and diagonal weighting matrices $\boldsymbol{Q}_i^p = diag(q_{i,1}^p, \ldots, q_{i,\max(p-\tau+1,0)-\max(\tau-p,1)+n_A+n_B+2}^p)$ describe centres and widths of the nodes, respectively, weights are denoted by w_i^p, $i = 1, \ldots, K^p$, $p = 1, \ldots, N$. The model (15) is sometimes named Hyper Radial Basis Function (HRBF) neural network in contrast to the ordinary RBF neural networks in which widths of the nodes are constant. Let $z_i^p(k+p|k)$ be sums of inputs of the ith hidden node. Using (14) one has

$$
z_i^p(k+p|k) = \sum_{j=1}^{I_{uf}(p)} q_{i,j}^p (u(k-\tau+1-j+p|k) - c_{i,j}^p)^2 \tag{16}
$$

$$
+ \sum_{j=1}^{I_{up}(p)} q_{i,I_{uf}(p)+j}^p (u(k-\max(\tau-p,1)+1-j) - c_{i,I_{uf}(p)+j}^p)^2
$$

$$
+ \sum_{j=1}^{n_A+1} q_{i,I_{uf}(p)+I_{up}(p)+j}^p (y(k-j+1-p) - c_{i,I_{uf}(p)+I_{up}(p)+j}^p)^2
$$

3.2   MPC-NPL Optimisation Problem

If for prediction the nonlinear multi-model is used without any simplifications, the nonlinear optimisation problem (4) must be solved on-line at each sampling instant. To reduce computational complexity, the MPC algorithm with Nonlinear Prediction and Linearisation (MPC-NPL) [5,6,14] is adopted here. At each sampling instant k the neural multi-model is used on-line twice: to find a local

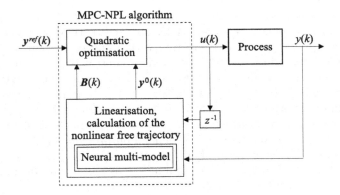

Fig. 2. The structure of the MPC-NPL algorithm

linearisation and a nonlinear free trajectory as shown in Fig. 2. The output prediction is the sum of a forced trajectory, which depends only on future control signals and a free trajectory $\boldsymbol{y}^0(k)$, which depends only on the past

$$\hat{\boldsymbol{y}}(k) = \boldsymbol{B}(k)\boldsymbol{u}_N(k) + \boldsymbol{y}^0(k) \tag{17}$$

where $\hat{\boldsymbol{y}}(k) = [\hat{y}(k+1|k) \dots \hat{y}(k+N|k)]^T$, $\boldsymbol{u}_N(k) = [u(k|k) \dots u(k+N-1|k)]^T$, $\boldsymbol{y}^0(k) = \left[y^0(k+1|k) \dots y^0(k+N|k)\right]^T$ are vectors of length N. The matrix $\boldsymbol{B}(k)$ is calculated from the on-line linearisation of the neural multi-model

$$\boldsymbol{B}(k) = \begin{bmatrix} b_{1,0}(k) & b_{1,1}(k) & \dots & b_{1,N-1}(k) \\ b_{2,0}(k) & b_{2,1}(k) & \dots & b_{2,N-1}(k) \\ \vdots & \vdots & \ddots & \vdots \\ b_{N,0}(k) & b_{N,1}(k) & \dots & b_{N,N-1}(k) \end{bmatrix} \tag{18}$$

Quantities $b_{p,l}(k) = \frac{\partial f_p(\bar{\boldsymbol{x}}(k+p|k))}{\partial u(k+l|k)}$ are calculated analytically from (15) and (16). Linearisation points $\bar{\boldsymbol{x}}(k+p|k)$ are comprised of past input and output signal values corresponding to arguments of the neural multi-model (10)

$$\bar{\boldsymbol{x}}(k+p|k) = [\underbrace{\bar{u}(k-1) \dots \bar{u}(k-1)}_{I_{uf}(p)} \underbrace{\bar{u}(k-\max(\tau-p,1)) \dots \bar{u}(k-n_B)}_{I_{up}(p)} \tag{19}$$

$$\underbrace{\bar{y}(k) \dots \bar{y}(k-n_A)]^T}_{n_A+1}$$

Future control signals are not known in advance, $\bar{u}(k+p|k) = \bar{u}(k-1)$ for $p \geq 0$. In MPC only $N_u \leq N$ future control moves $\Delta \boldsymbol{u}(k)$ are found. One has

$$\boldsymbol{u}_N(k) = \boldsymbol{J}\Delta \boldsymbol{u}(k) + \boldsymbol{u}_N^{k-1}(k) \tag{20}$$

where $\boldsymbol{J} = \begin{bmatrix} \boldsymbol{J}_1 \\ \boldsymbol{J}_2 \end{bmatrix}$ is a matrix of dimensionality $N \times N_u$, $\boldsymbol{J}_1$ is the all ones lower triangular matrix of dimensionality $N_u \times N_u$, $\boldsymbol{J}_2$ is the all ones matrix of dimensionality $(N-N_u) \times N_u$ and $\boldsymbol{u}_N^{k-1}(k) = [u(k-1) \dots u(k-1)]^T$ is a vector of length N. Using (20), the prediction equation (17) is

$$\hat{\boldsymbol{y}}(k) = \boldsymbol{B}(k)\boldsymbol{J}\Delta\boldsymbol{u}(k) + \boldsymbol{B}(k)\boldsymbol{u}_N^{k-1}(k) + \boldsymbol{y}^0(k) \qquad (21)$$

Thanks to using the suboptimal prediction equation (17), the optimisation problem (4) becomes the following quadratic programming task

$$\min_{\Delta\boldsymbol{u}(k)} \left\| \boldsymbol{y}^{ref}(k) - \boldsymbol{B}(k)\boldsymbol{J}\Delta\boldsymbol{u}(k) - \boldsymbol{B}(k)\boldsymbol{u}_N^{k-1}(k) - \boldsymbol{y}^0(k) \right\|_{\boldsymbol{M}}^2 + \|\Delta\boldsymbol{u}(k)\|_{\boldsymbol{\Lambda}}^2$$

subject to

$$\boldsymbol{u}^{\min} \leq \boldsymbol{J}_1\Delta\boldsymbol{u}(k) + \boldsymbol{u}^{k-1}(k) \leq \boldsymbol{u}^{\max} \qquad (22)$$

$$-\Delta\boldsymbol{u}^{\max} \leq \Delta\boldsymbol{u}(k) \leq \Delta\boldsymbol{u}^{\max}$$

$$\boldsymbol{y}^{\min} \leq \boldsymbol{B}(k)\boldsymbol{J}\Delta\boldsymbol{u}(k) + \boldsymbol{B}(k)\boldsymbol{u}_N^{k-1}(k) + \boldsymbol{y}^0(k) \leq \boldsymbol{y}^{\max}$$

where $\boldsymbol{y}^{ref}(k) = \left[y^{ref}(k+1|k)\ldots y^{ref}(k+N|k)\right]^T$, $\boldsymbol{y}^{\min}(k) = \left[y^{\min}\ldots y^{\min}\right]^T$, $\boldsymbol{y}^{\max}(k) = \left[y^{\max}\ldots y^{\max}\right]^T$ are vectors of length N, $\boldsymbol{u}^{\min} = \left[u^{\min}\ldots u^{\min}\right]^T$, $\boldsymbol{u}^{\max} = \left[u^{\max}\ldots u^{\max}\right]^T$, $\boldsymbol{u}^{k-1}(k) = \left[u(k-1)\ldots u(k-1)\right]^T$, $\Delta\boldsymbol{u}^{\max} = \left[\Delta u^{\max}\ldots\Delta u^{\max}\right]^T$ are vectors of length N_u, $\boldsymbol{M} = diag(\mu_1,\ldots,\mu_N)$, $\boldsymbol{\Lambda} = diag(\lambda_0,\ldots,\lambda_{N_u-1})$.

In the MPC-NPL algorithm (Fig. 2) at each sampling instant k the following steps are repeated:

1. Linearisation of the neural multi-model: obtain the matrix $\boldsymbol{B}(k)$.
2. Find the nonlinear free trajectory $\boldsymbol{y}^0(k)$ using the neural multi-model assuming no changes in the control signal from the sampling instant k onwards.
3. Solve the quadratic programming problem (22) to determine $\Delta\boldsymbol{u}(k)$.
4. Apply $u(k) = \Delta u(k|k) + u(k-1)$.
5. Set $k := k+1$, go to step 1.

4   Simulation Results

For simplicity of presentation neural multi-models and the MPC-NPL algorithm are described for SISO processes with one input and one output. On the other hand, as emphasised in the Introduction, MPC algorithms are very efficient when applied to Multiple-Input Multiple-Output (MIMO) processes. Hence, during simulations a methanol-water distillation column shown in Fig. 3 is considered. From the perspective of supervisory MPC algorithms, the plant has two manipulated variables: R – the reflux stream flow rate, V – the vapour stream flow rate and two controlled variables: x_d – the composition of the top product, x_b – the composition of the bottom product. The process exhibits significantly nonlinear behaviour, both steady-state and dynamic properties are nonlinear. As a result, MPC algorithms based on linear models are inefficient [5].

The fundamental model is used as the real process during simulations. It is simulated open-loop to obtain data for training. Two classes of neural RBF models are trained, namely the classical NARX model and the multi-model. Second-order dynamics is assumed while the real order of the process (the fundamental model) is high (neural models are deliberately underparameterised).

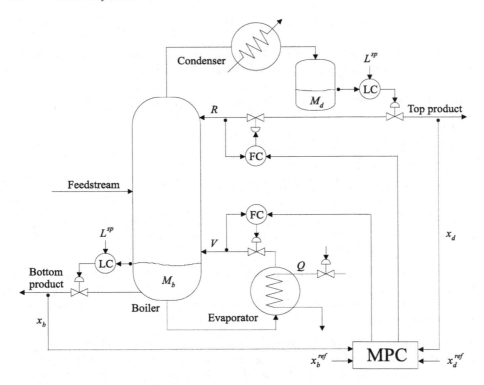

Fig. 3. The distillation column control system structure

The classical NARX model is comprised of two RBF networks with 30 Gaussian functions in the hidden layer and one output. The first network calculates the composition of the top product, the second one – the composition of the bottom product. The multi-model is comprised of RBF neural networks with 10 Gaussian functions in the hidden layer and one output.

When trained (the backpropagation algorithm), both the classical NARX model and the multi-model have similar Mean Squared Errors. In the first case $MSE = 7.8185 \cdot 10^{-2}$, in the second case $MSE = 7.9354 \cdot 10^{-2}$ (for the training data set). Fig. 4 shows step responses of the process and predictions for $N = 10$. Initially $x_d = 0.95$, $x_b = 0.05$. The reflux flow rate R decreases at the sampling instant 0 by 10 $kmol/h$ while the vapour flow rate V is constant, the sampling time is 1 min. The classical NARX neural model correctly calculates the first predictions (for $p = 1$) while for next sampling instants the prediction error is propagated and predictions differ from the real process. Conversely, the neural multi-model correctly predicts behaviour of the process over the whole prediction horizon.

For control two models are used. The fundamental model is used as the real process during simulations. Parameters of MPC are: $N = 10$, $N_u = 3$, $M_p = diag(5, 0.5)$, $\Lambda_p = diag(1.5, 1.5)$. Manipulated variables are constrained: $R^{\min} = R_0 - 20 \, kmol/h$, $R^{\max} = R_0 + 20 \, kmol/h$, $V^{\min} = V_0 - 20 \, kmol/h$, $V^{\max} = V_0 + 20 \, kmol/h$ where $R_0 = 33.3634 \, kmol/h$, $V_0 = 83.3636 \, kmol/h$.

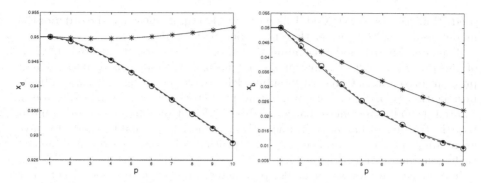

Fig. 4. Step responses (long-range predictions) calculated by the classical NARX neural model (*solid line with asterisks*) and by the neural multi-model (*dashed line with circles*) vs. the real process (*solid line with dots*)

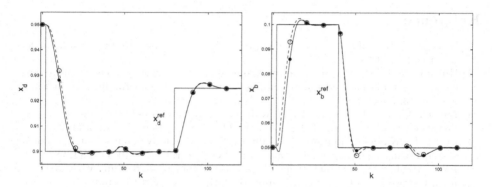

Fig. 5. Simulation results: the MPC-NO algorithm (*solid line with dots*) and the MPC-NPL algorithm (*dashed line with circles*) with the same neural multi-model

Two MPC schemes are compared with the same neural multi-model: the MPC algorithm with Nonlinear Optimisation (MPC-NO) and the described MPC-NPL algorithm. Simulation results are depicted in Fig. 5 for given reference trajectories (x_d^{ref}, x_b^{ref}). The closed-loop performance obtained in the suboptimal MPC-NPL algorithm with quadratic programming is practically the same as in the computationally prohibitive MPC-NO approach in which at each sampling instant a nonlinear optimisation problem has to be solved on-line.

5   Conclusions

The model used in MPC has to be able to make accurate predictions of the process over the whole prediction horizon. The rudimentary backpropagation algorithm gives neural models which are in fact one-step ahead predictors. They are not suited to be used recursively in MPC for long range prediction since the prediction error is propagated. It is particularly important when the order of the model used in MPC is lower than the order of the real process.

The neural multi-model recommended in this paper predicts future behaviour of the process over the whole prediction horizon without taking into account previous predictions. Consecutive submodels are trained using the classical back-propagation algorithm as one-step ahead predictors. The paper also describes a computationally efficient and accurate MPC-NPL algorithm based on the multi-model. The algorithm uses on-line only a numerically reliable quadratic pro-gramming procedure, the necessity of repeating full nonlinear optimisation at each sampling instant is avoided.

To reduce the number of model parameters one can prune neural networks and take into account in the MPC cost function (2) only selected predictions.

Acknowledgement. This work was partly supported by Polish national budget funds 2007-2009 for science as a research project.

References

1. Haykin, S.: Neural networks – a comprehensive foundation. Prentice Hall, Engle-wood Cliffs (1999)
2. Hussain, M.A.: Review of the applications of neural networks in chemical process control – simulation and online implmementation. Artificial Intelligence in Engi-neering 13, 55–68 (1999)
3. Liu, D., Shah, S.L., Fisher, D.G.: Multiple prediction models for long range pre-dictive control. In: Proc. of the IFAC World Congress, Beijing, China (1999)
4. Liu, G.P., Kadirkamanathan, V., Billings, S.A.: Predictive control for non-linear systems using neural networks. Int. Journal of Control 71, 1119–1132 (1998)
5. Ławryńczuk, M.: A family of model predictive control algorithms with artificial neural networks. Int. Journal of Applied Mathematics and Computer Science 17, 217–232 (2007)
6. Ławryńczuk, M.: Suboptimal nonlinear predictive control with structured neural models. In: de Sá, J.M., Alexandre, L.A., Duch, W., Mandic, D.P. (eds.) ICANN 2007. LNCS, vol. 4669, pp. 630–639. Springer, Heidelberg (2007)
7. Maciejowski, J.M.: Predictive control with constraints. Prentice Hall, Englewood Cliffs (2002)
8. Morari, M., Lee, J.H.: Model predictive control: past, present and future. Comput-ers and Chemical Engineering 23, 667–682 (1999)
9. Nørgaard, M., Ravn, O., Poulsen, N.K., Hansen, L.K.: Neural networks for mod-elling and control of dynamic systems. Springer, London (2000)
10. Piche, S., Sayyar-Rodsari, B., Johnson, D., Gerules, M.: Nonlinear model predictive control using neural networks. IEEE Control Systems Magazine 20, 56–62 (2000)
11. Pottmann, M., Seborg, D.E.: A nonlinear predictive control strategy based on radial basis function models. Comp. and Chem. Eng. 21, 965–980 (1997)
12. Qin, S.J., Badgwell, T.A.: A survey of industrial model predictive control technol-ogy. Control Engineering Practice 11, 733–764 (2003)
13. Rossiter, J.A., Kouvaritakis, B.: Modelling and implicit modelling for predictive control. Int. Journal of Control 74, 1085–1095 (2001)
14. Tatjewski, P.: Advanced control of industrial processes, Structures and algorithms. Springer, London (2007)

Parallel Implementations of Recurrent Neural Network Learning

Uroš Lotrič and Andrej Dobnikar

Faculty of Computer and Information Science,
University of Ljubljana, Slovenia
{uros.lotric,andrej.dobnikar}@fri.uni-lj.si

Abstract. Neural networks have proved to be effective in solving a wide range of problems. As problems become more and more demanding, they require larger neural networks, and the time used for learning is consequently greater. Parallel implementations of learning algorithms are therefore vital for a useful application. Implementation, however, strongly depends on the features of the learning algorithm and the underlying hardware architecture. For this experimental work a dynamic problem was chosen which implicates the use of recurrent neural networks and a learning algorithm based on the paradigm of learning automata. Two parallel implementations of the algorithm were applied - one on a computing cluster using MPI and OpenMP libraries and one on a graphics processing unit using the CUDA library. The performance of both parallel implementations justifies the development of parallel algorithms.

1 Introduction

In recent years the commercial computer industry has been undergoing a massive shift towards parallel and distributed computing. This shift was mainly initiated by the current limitations of semiconductor manufacturing. New developments are also reflected in the areas of intensive computing applications by fully exploiting the capabilities of the underlying hardware architecture, and impressive enhancements in algorithm performance can be achieved with a low to moderate investment of time and money.

Today, clusters of loosely coupled desktop computers represent an extremely popular infrastructure for implementation of parallel algorithms. Processes running on computing nodes in the cluster communicate with each other through messages. The message passing interface (MPI) is a standardized and portable implementation of this concept, providing several abstractions that simplify the use of parallel computers with distributed memory [1].

Recently, the development of powerful graphics processing units (GPUs) has made high-performance parallel computing possible by using commercial discrete general-purpose graphics cards [2]. There are many technologies, among which Nvidia's compute unified device architecture (CUDA) is the most popular [3]. It includes C/C++ software development tools, function libraries and a hardware abstraction mechanism that hides GPU hardware architecture from developers.

M. Kolehmainen et al. (Eds.): ICANNGA 2009, LNCS 5495, pp. 99–108, 2009.
© Springer-Verlag Berlin Heidelberg 2009

Algorithms for implementation of various neural networks can take advantage of parallel hardware architectures [4,5,6]. This comes from the concurrency, inherently present in the neural network models themselves. Training of neural network models is computationally expensive and time consuming, especially in cases when large neural networks and/or large data sets of input-output samples are considered. However, each particular neural network model has its own characteristics, and even the same model with different parameters and training data sets may lead to different behaviors on the same parallel hardware. From that point of view, gathering universal solutions is impossible, although the applicable parallelization concepts become very similar.

In this paper the parallel implementations of training algorithm for large recurrent neural networks are considered, taking into account the possibilities of available technologies. In the next section a fully recurrent neural network is presented together with a training algorithm based on the linear reward penalty correction scheme. Furthermore, the possibilities for parallelization are outlined in section three, taking into account the capabilities and limitations of the technologies. Hardware architectures used in the experiments are presented in section four, followed in section five by the experimental setup and the results in terms of processing times and speedups. Finally, the main findings are summarized.

2   Fully Connected Recurrent Neural Network

The recurrent neural network is one of the most general types of neural networks [7]. Feed-back connections enable the recurrent neural network to memorize. A fully connected recurrent neural network with outputs from each neuron connected to all neurons is presented in Fig. 1. Assume a recurrent neural network with m inputs and n neurons, the first l of them being connected to the outputs. At time t, input sample $\mathbf{x}(t)$ together with the current outputs of the

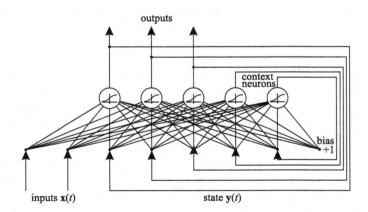

Fig. 1. Fully connected recurrent neural network with $m = 2$ inputs, $n = 5$ neurons and $l = 3$ outputs

neurons $\mathbf{y}(t)$ is presented to the neural network. For easier notation, input to the neurons can be written as the vector $\mathbf{z}(t) = (\mathbf{x}(t), \mathbf{y}(t), 1)$ with $m + n + 1$ elements, the last element representing the bias. During the learning process, knowledge is stored in the weights on connections w_{ij}, where index i runs over the neurons and index j over the elements of the vector $\mathbf{z}$. The output value of i-th neuron is given by the non-linear sigmoid function of the weighted sum

$$y_i(t+1) = \frac{1}{1 + e^{-u_i(t)}} \quad , \quad u_i(t) = \sum_{j=1}^{m+n+1} w_{ij} z_j(t) \quad . \tag{1}$$

The objective of a neural network learning algorithm is to find a set of weights that minimizes the error function on the given data set of input-output samples $(\mathbf{x}(t), \mathbf{d}(t))$, $t = 0, \ldots, T$,

$$\mathcal{E} = \sum_{t=1}^{T} E(t) \quad , \quad E(t) = \frac{1}{2} \sum_{i=1}^{l} e_i(t)^2 \quad , \tag{2}$$

with $e_i(t) = d_i(t) - y_i(t)$ being the difference between the desired and calculated value of the i-th neuron. The recurrent neural networks attempt to acquire the dynamics of the system, and therefore input-output pairs should be presented in causative order.

Many algorithms for recurrent neural network learning are known [7]. The most standard approaches apply gradient-based techniques such as back propagation through time or real time recurrent learning. The problem of both is expensive computation of gradients and slow convergence when large recurrent neural networks are applied. One of alternatives is learning with heuristic approaches that mimic computation of gradients but with much smaller computation requirements. Such an algorithm is the linear reward penalty algorithm, or LRP correction scheme, known from the field of learning automata [8].

The basic idea of the LRP correction scheme is to change the probabilities of possible changes in individual weights (actions), based on a given response from the environment. When an action is rewarded, its probability is increased. Contrarily, when an action is penalized, its probability is decreased. To preserve the total probability of all actions, the probabilities of non-selected actions are proportionally reduced in the first case and increased in the second case.

In neural network learning, an action represents a change in a single weight for a given value Δw [9]. There are two actions associated with each weight: one increases the weight and the other decreases it. In the presented fully connected recurrent neural network there are $N_w = n(m + n + 1)$ weights leading to $N_a = 2N_w$ possible actions, while the response of the environment from the LRP correction scheme is simply represented by the error function given in (2).

At the beginning of the learning process all actions have equal probabilities, $p_k(0) = 1/N_a$, $k = 1, \ldots, N_a$. During the learning process the probabilities and weights are updated according to the following scheme. Suppose that in learning step s the action a is rewarded. In this case the probabilities of actions are updated as

$$p_k(s+1) = p_k(s) + \begin{cases} +\lambda[1 - p_a(s)] , & k = a \\ -\lambda p_k(s) & , & k \neq a \end{cases} , \tag{3}$$

with λ, $0 < \lambda < 1$, being the correction constant, and the weight change is accepted. Conversely, when in learning step s the action a is penalized, the corresponding weight is returned to the previous value and the probabilities of actions become

$$p_k(s+1) = p_k(s) + \begin{cases} -\lambda p_a(s) & , & k = a \\ +\lambda\left[\frac{1}{N_a - 1} - p_k(s)\right] , & k \neq a \end{cases} . \tag{4}$$

3   Exploiting Concurrency in Training Algorithm

Algorithms can be efficiently parallelized by following the methodology proposed by Ian Foster [1]. It consists of four design steps: partitioning, communication, agglomeration and mapping. The focus of the first two is to find as much concurrency as possible, while the latter two consider the requirements of the underlying hardware architecture. In the partitioning step, the data and/or computations are divided into small tasks that can be computed in parallel. In the communication step, data that has to be passed between tasks is identified. Communication represents the overhead of parallel designs and should be kept as low as possible. In the agglomeration step, small tasks are grouped in the agglomerated tasks to improve performance, mainly by reducing communication. In the mapping step the agglomerated tasks are assigned to the processing units. Usually there are as many agglomerated tasks as there are independent processing units.

The pseudo-code of the learning algorithm for the recurrent neural network based on the LRP correction scheme is given in Fig. 2. The most obvious portion of code, suitable for parallelization, is the updating of probabilities, identified by *1* in Fig. 2. The **for** k loop can be partitioned into N_a small tasks, each of them responsible for updating one probability, either by (3) or (4). Small tasks only need to send their results to the task that chooses a new action. It is also straightforward to parallelize the propagation of signals through the neural network. The **for** i loop, identified by *2* in Fig. 2, can also be split into n small tasks, each calculating the output of one neuron following (1). However, the result $y_i(t+1)$ obtained for each task must be broadcasted to all other tasks in order to make the calculation of neuron outputs in the next time step possible. In both identified cases, the computation is not very time demanding; therefore, fast communication is the key issue for successful parallelization.

While using intra-processor communication can still be profitable in the specified situations, inter-processor communication is certainly too slow. In cases of slow communication between processors the computation time of each task must be large compared to the time needed for communication. Unfortunately, the given algorithm does not exhibit such concurrency.

In cases where the number of concurrent processes is small compared to the number of input-output samples T, slight modification of the learning algorithm leads to an efficient parallelization, also for systems with slow inter-processor

```
randomly initialize neural network weights w_ij
initialize probabilities for actions p_k(0)
for s ← 1 to S do
    randomly choose action a and adequately change corresponding weight
    calculate the response of the environment
        initialize variables: E_old ← E, E ← 0, y(0) ← 0
        for t ← 1 to T do          // over all input-output samples *3*
            for i ← 1 to n do      // over all neurons *2*
                calculate y_i(t + 1)
                update error, E ← E + E(t)
        end for t
    update probabilities, if E < E_old use (3) else (4)
        for k ← 1 to N_a do        // over all actions *1*
            update probability p_k(s + 1)
        end for k
end for s
```

Fig. 2. Pseudo-code of the LRP correction scheme for a recurrent neural network. Parts of the code suitable for parallelization are indicated by a number surrounded by two asterisks.

communication. More precisely, instead of parallelizing the for k and for i loops, one can decide to parallelize the for t loop, marked *3* in Fig. 2. The causality between consecutive input-output samples in the for t loop prevents one from directly parallelizing it. Parallelization is only possible if the data set of T input-output samples is split into P parts of approximately T/P samples, on which the response of the environment can be calculated separately and afterwards brought together. Instead of a single initialization of the vector $\mathbf{y}$ at the beginning of the response calculation in Fig. 2, additional initializations are needed for each part separately, which causes transitional phenomena on the outputs of the neurons. The modified portion of the code is presented in Fig. 3.

```
calculate the response of the environment
    initialize variables: E_old ← E, E ← 0
    for r ← 1 to P do          // over all P parts *4*
        initialize variables: y(⌊(r − 1)T/P⌋) ← 0
        for t ← ⌊(r − 1)T/P⌋ + 1 to ⌊rT/P⌋ do
            for i ← 1 to n do
                calculate y_i(t + 1)
            end for i
            update error, E ← E + E(t)
        end for t
    end for r
```

Fig. 3. Modified pseudo-code of the LRP correction scheme for recurrent neural network. The for r loop, suitable for parallelization is indicated by *4*.

In this case suitable partitioning involves splitting the `for` r loop into P processes. Communication is basically needed only to get the cumulative error $\mathcal{E}$. When the processes do not share memory, each process needs its own copy of the weights. Therefore, it is additionally necessary to update the weights after each iteration of the `for` s loop.

4   Parallel Hardware Architectures

Parallelization of the recurrent network learning algorithm based on the LRP correction scheme was examined on two distributed hardware platforms: on a computing cluster and on graphics processing units.

4.1   Commodity Computing Cluster

Currently, the most popular and affordable parallel computers are clusters of commodity desktop computers. Processes running on the computing nodes in a cluster communicate with each other through messages. The Message Passing Interface (MPI), the standardized and portable implementation of communication through messages, is most commonly used to make parallelization on such systems feasible. Unfortunately, commodity clusters are typically not balanced between computation speed and communication speed - the communication network is usually quite slow compared to the speed of the processors. Therefore, in the process of parallel algorithm design it is important to be aware of slow communication.

In the present work, a commodity cluster composed of four nodes, each having an Intel Core Duo 6700 processor running at 2.66 GHz and 2 GB of RAM, is used. The nodes are connected over a 1 Gb Ethernet switch as shown in Fig 4a. DeinoMPI [10] implementation of the MPI standard [1] is used on the Windows XP operating system. The application utilizes MPI and OpenMP [1] function calls.

Many modern commodity clusters are made of dual-core or even quad-core multiprocessors. The MPI standard supports communication between processing cores inside the same multiprocessor in the same way as between processors belonging to distinct computers. In this case the interaction between MPI processes running on the same multiprocessor will happen via message passing. Some additional time can be gained by using only one MPI process per multiprocessor and within this process by forking threads to occupy unused cores. OpenMP is a standardized software library that supports such thread creation and interaction among cores via the concept of shared variables. Because of the lower communication overhead, forking threads with OpenMP function calls inside multiprocessors is preferable to pure MPI implementation because it usually leads to faster programs.

4.2   Graphics Processing Units

Graphics processing units (GPUs) are nowadays extending their initial role as specialized 2D and 3D graphics accelerators to high performance computing

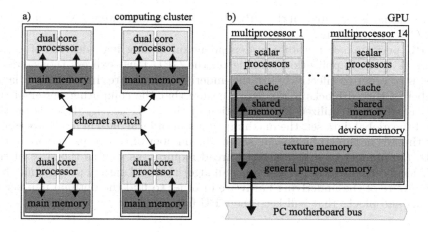

Fig. 4. Hardware architectures of a) a computing cluster and b) a graphics processing unit

devices, specialized for computing-intensive and memory-intensive highly parallel computation. Nvidia's compute unified device architecture (CUDA) is the most popular library, also featuring development tools.

CUDA represents GPUs as computing devices capable of executing a very large number of threads in parallel. For example, the Nvidia GeForce 8800GT GPU, used in our further experiments, consists of 14 multiprocessors, which can use 1 GB of device memory. Each multiprocessor consists of 8 scalar processors with 16 kB of shared memory and 8192 32-bit registers allowing computation in single-precision floating point. Its architecture is schematically represented in Fig. 4b.

GPUs feature memory access bandwidth an order of magnitude higher than ordinary CPUs. For example, when there is no conflict, the shared memory inside the multiprocessor can be accessed as quickly as reading a register. Despite very high bandwidth to the device memory, access to the device memory is faced with very high latency, measured in hundreds of GPU cycles. Besides, CUDA features additional texture memory. Although the shared memory and the device memory are not cached, the texture memory is. Reading data from the texture memory instead of the device memory can thus result in performance benefits.

In our work a desktop computer with an Intel Core Duo 8400 Processor and 4 GB of RAM with 64-bit Windows XP installed hosts two Nvidia GeForce 8800 GT graphics processing units.

According to the CUDA programming model the computation is organized into grids, which are executed sequentially. Each grid is organized as a set of thread blocks, in which threads are executed concurrently and can cooperate together by efficiently sharing data inside a multiprocessor. A maximum of 512 threads can run in parallel in each thread block. Unfortunately, threads in different blocks of the same grid cannot communicate and synchronize with each other. Moreover, thread blocks of the same grid have the same size and their threads execute the same kernel. A kernel is a portion of an application, a function, that is executed on the GPU. It is coded in annotated C/C++ language with CUDA extensions.

GPU performance is radically dependent on finding high degrees of parallelism. A typical application running on the GPU must express thousands of threads in order to effectively use the underlying hardware. In practice, a large number of thread blocks is needed to ensure that the computing power of the GPU is efficiently utilized [3]. Utilization of the GPU heavily depends on the size of the global data set, the maximum amount of local data in multiprocessors that threads in thread blocks can share, the number of thread processors in the GPU, the number of registers each thread requires, as well as the sizes of the GPU local memories. When analyzing an algorithm and data, a programmer has to be aware of the underlying hardware in order to find the optimal number of threads and blocks that will keep the GPU fully utilized.

5   Experimental Work

In this section the performance of the proposed hardware architectures on original and modified learning algorithms is assessed. In all cases the fully connected recurrent neural network was trained to identify an unknown discrete dynamic system, in our case a finite state machine which performs the time-delayed exclusive or xor(3) function, $y(t) = x(t - 2) \oplus x(t - 3)$. There are 1000 binary input-output samples in the training data set.

5.1   Original LRP Correction Scheme

In this case only the for loops indicated by *1* and *2* in Fig. 2 were parallelized. In the case of the computing cluster, the results are given only for the setup in which all four nodes were utilized. Communication between nodes was performed using the MPI library, while parallelization inside the node was done using pragma directives of the OpenMP standard. Source code was compiled using a Microsoft C/C++ compiler. On the other hand, only one GPU was used to parallelize the original algorithm.

Processing times, normalized to 1000 iterations, and speedups of both architectures are given in Fig. 5 for a range of neural network sizes. For comparison, the processing times of the standalone application, exploiting only one core of the Intel Core Duo 6700 processor, are presented.

It is obvious that the cluster is not appropriate for parallelization of the original LRP correction scheme, since communication overwhelms computation by a large margin. A linear relationship between processing time and the number of neurons is expected since the length of the messages increases linearly with the number of neurons. On the GPU, the computation of (1) is performed concurrently for all neurons, and therefore an approximately linear increase in computation time with an increasing number of neurons is observed. The local peeks at 400 and 800 neurons on the speedup curve are the consequence of the GPU hardware architecture.

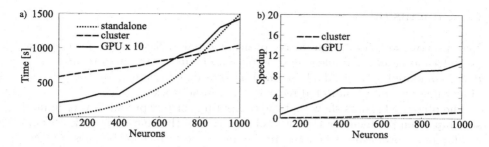

Fig. 5. Performance of the original LRP correction scheme as a function of the number of neurons on different architectures: a) processing time and b) speedup

Although some speedup, defined as the ratio between standalone and parallel computation time, is observed in the case of cluster computing, the usage of nodes is far from efficient, with less than 20% utilization of nodes. In addition the linear dependence of the speedup on the number of neurons shows that the GPU is not fully utilized when the learning algorithm is running on small neural networks.

5.2   Modified LRP Correction Scheme

Due to the unpromising parallelization of the original LRP correction scheme, only the for r loop marked by *4* in Fig. 3 was parallelized on the computing cluster in this case. On the other hand, the second GPU was utilized for parallelization of the for r loop in Fig. 3, while inside each GPU the parallelization scheme from the original algorithm was further used. Processing times for both parallel architectures and the standalone application are given in Fig. 6a.

When parallelizing the for r loop, far more time is spent in processing than in communication. Therefore, the processing on the cluster of four dual core nodes is sped up by approximately a factor of eight. On the GPUs a similar dependence is observed as in the case of the original LRP correction scheme, except that the processing times are approximately halved whilst the speedups are doubled.

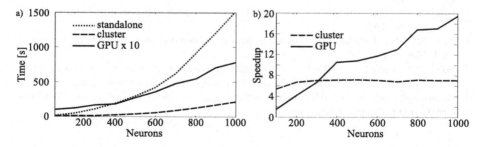

Fig. 6. Performance of the modified LRP correction scheme as a function of the number of neurons on different architectures: a) processing time and b) speedup

6   Conclusion

Neural networks offer a high degree of internal parallelism, which makes them perfect candidates for implementation on parallel hardware architectures. This paper compared two affordable parallel hardware architectures on the problem of learning in a fully connected recurrent neural network.

The presented results show that the computing clusters provide a very limited speedup when parallelizing the internal structure of the neural network. Results are far more promising when the processing is performed in batches and not online. The development of graphics processing units now offers highly parallel hardware platforms to users. The performance of graphics processing units is improving with an increasing number of concurrent operations, and therefore they represent a perfect target platform for neural network computation. Their main drawbacks are computation in single-precision floating point and the development tools, which somehow require that the user understands the particularities of the hardware in order to benefit from it.

References

1. Quinn, M.: Parallel Programming in C with MPI and OpenMP. McGraw Hill, Boston (2003)
2. Halfhill, T.R.: Parallel Processing With CUDA. Microprocessor report (2008), http://www.MPRonline.com
3. Nvidia: Nvidia CUDA Compute Unified Device Architecture, Programming Guide, Version 1.1 (2007), http://nvidia.com/cuda
4. Seiffert, U.: Artifical neural networks on massively parallel computer hardware. In: ESANN 2002 proceedings, Bruges, Belgium, pp. 319–330 (2002)
5. Lotrič, U., Dobnikar, A.: Parallel implementations of feed-forward neural network using MPI and C# on.NET platform. In: Ribeiro, B., et al. (eds.) Adaptive and natural computing algorithms: proceedings of the International Conference in Coimbra, Portugal, pp. 534–537 (2005)
6. Catanzaro, B., Sundaram, N., Keutzer, K.: Fast support vector machine training and classification on graphics processors. In: McCallum, A., Roweis, S. (eds.) Proceedings of the 25th International Conference on Machine Learning, Helsinki, Finland, pp. 104–111 (2008)
7. Haykin, S.: Neural networks: a comprehensive foundation, 2nd edn. Prentice-Hall, New Jersey (1999)
8. Narendra, K., Thathachar, M.A.L.: Learning automata: an introduction. Prentice-Hall, New Jersey (1989)
9. Šter, B., Gabrijel, I., Dobnikar, A.: Impact of learning on the structural properties of neural networks, part 2. LNCS, pp. 63–70. Springer, Heidelberg (2007)
10. Deino Software: DeinoMPI - High Performance Parallel Computing for Windows (2008), http://mpi.deino.net

Growing Competitive Network for Tracking Objects in Video Sequences

J.M. Ortiz-de-Lazcano-Lobato[1], R. M. Luque[1],
D. López-Rodríguez[2], and E.J. Palomo[1]

[1] Department of Computer Science, University of Málaga, Málaga, Spain
{jmortiz,rmluque,ejpalomo}@lcc.uma.es
[2] Department of Applied Mathematics, University of Málaga, Málaga, Spain
dlopez@ctima.uma.es

Abstract. The aim of this paper is to present the use of Growing Competitive Neural Networks as a precise method to track moving objects for video-surveillance. The number of neurons in this neural model can be automatically increased or decreased in order to get a one-to-one association between objects currently in the scene and neurons. This association is kept in each frame, what constitutes the foundations of this tracking system. Experiments show that our method is capable to accurately track objects in real-world video sequences.

1 Introduction

In video surveillance systems, accurate and real-time multiple objects tracking will greatly improve the performance of objects recognition, activity analysis and high level event understanding [1,2,3,4].

Segmentation and tracking of multiple objects is important not only for visual surveillance, but also for other video analysis applications such as video compression, indexing, video archival and retrieval systems [5,6], as well as in robotics [7], human-machine interfaces [8], ambient intelligent systems [9] and augmented reality applications [10].

Most of the work on tracking for visual surveillance is based on change detection [5,11,12,13] or frame differencing [14] if the camera is stationary.

The most popular approach for visual tracking is the adaptive tracking of coloured regions, with techniques such as the particle filtering of coloured regions [10,15] and the Kalman/mean-shift [16], which uses the well known mean-shift algorithm [17] to determine the search region, and the Kalman filter to predict the position of the target object in the next frame.

In this paper, the use of growing competitive neural networks (GCNNs) [18] to perform object tracking is proposed. These networks are derived from the usual competitive neural networks (CNNs) [19]. Their main particularity consists in that this kind of network is able to generate new process units (neurons) when needed, in order to get a better representation of the input space.

In general, CNNs are suitable for data clustering, since each neuron in a CNN is specifically designed to represent a single cluster. In the field of object tracking

M. Kolehmainen et al. (Eds.): ICANNGA 2009, LNCS 5495, pp. 109–118, 2009.
© Springer-Verlag Berlin Heidelberg 2009

in video sequences, such clusters correspond to moving objects. Thus, it seems reasonable to use CNNs as trackers.

However, due to the dynamic nature of a video sequence, objects are constantly appearing and disappearing from the scene, and the method used to track objects should take care of this situation. Consequently, since classical CNNs are not able to manage a variable number of clusters (that is, objects in the scene), the use of GCNNs become a good approach for tracking.

Another advantage of using a neural model to tackle this problem is that neural networks are intrinsically parallel, which allows the processing of all objects in the scene at the same time.

The rest of this paper is structured as follows: in section 2 standard competitive neural networks are briefly described. Section 3 is devoted to the segmentation algorithm, while section 4 explains the tracking system. Finally some experimental results and conclusions are presented in sections 5 and 6 respectively.

2   Standard Competitive Neural Network

The standard competitive neural network [19] forms the kernel of the presented segmentation and tracking modules. Some modifications, which are explained in next sections, have been added to the network in order to obtain a good performance when it is used for these two processes.

The standard competitive neural network has a unique layer of N process units or neurons. This number of neurons is fixed a priori by the user. In each time instant t, an input pattern $\boldsymbol{x}(t)$ is presented to the network and a competition process among the neurons starts. The neuron whose weight vector $\boldsymbol{w}_j$ is closest to the input pattern in the input space is declared the winner. Therefore, the winner neuron is the one which best represents that input pattern:

$$c(t) = \arg \min_{1 \leq j \leq N} \{\|\boldsymbol{x}(t) - \boldsymbol{w}_j(t)\|^2\} \tag{1}$$

Once the winner neuron has been determined, the weight vector $\boldsymbol{w}_{c(t)}$ must be updated in order to incorporate some knowledge from the pattern to the network. Only the winner neuron has gained the right to learn something from the pattern and, thus, it is the only neuron which will be updated in the instant time t. Thus, the standard competitive updated rule is

$$\boldsymbol{w}_i(t) = \begin{cases} \boldsymbol{w}_i(t-1) + \alpha \left(\boldsymbol{x}(t) - \boldsymbol{w}_j(t-1)\right) \text{ if } i = c(t) \\ \boldsymbol{w}_i(t-1) \qquad\qquad\qquad\qquad\quad \text{otherwise} \end{cases} \tag{2}$$

where $\alpha \in [0, 1]$ is named the *learning rate* and determines how important is the information extracted from the current input sample with respect to the information already known from previous training steps.

3   Object Segmentation

Almost every visual surveillance system starts with motion detection. Motion detection aims at segmenting regions corresponding to moving objects from the

The estimated pattern $\hat{\boldsymbol{x}}_j(t)$ is obtained by summing the current object pattern, stored in the weight vector of the neuron, and the averaged change observed in that pattern and computed for the K previous frames.

$$\hat{\boldsymbol{x}}_j(t) = \boldsymbol{w}_j(t-1) + \frac{1}{K-1} \sum_{i=1}^{K-1} (H_j(i+1) - H_j(i)) \tag{4}$$

with $H_j(i)$ the entry which was written down, in the log of the j-th neuron, $t - K - 1 - i$ frames ago.

Finally, the competition rule has also been slightly modified. This time a mask vector $\boldsymbol{m} \in \{0,1\}^D$, with D dimension of the input space, has been added in order to let the user choose which object components should be considered when calculating the quantisation error during the competition process. Then,

$$c(t) = \arg \min_{1 \leq j \leq N} \{\|\boldsymbol{m} \cdot (\boldsymbol{x}(t) - \boldsymbol{w}_j(t))\|^2\} \tag{5}$$

where $\cdot$ means the componentwise product.

4.2   Neurons Birth and Death

The size of the neural network layer should not be fixed a priori because the number of objects which are present in the scene varies from one frame to another. Hence, the proposed network is formed by $n(t)$ neurons in a time instant t, and a mechanism to add new neurons to the network and to eliminate the useless neurons is needed.

When an unknown object appears in the scene, none of the existing neurons is able to represent it accurately and the quantisation error is expected to reach a high value, compared with the error obtained for correctly identified objects. Thus, a new neuron should be created in order to track that new object. A user-defined parameter $\delta \in [0,1]$ has been utilised. It is related to the maximum relative error permitted in the quantisation process. The parameter δ manages the neurons birth by means of the check

$$\forall j \in \{1 \ldots n(t)\} \quad \frac{\|\boldsymbol{x}(t) - \boldsymbol{w}_j(t)\|}{\|\boldsymbol{x}(t)\|} > \delta \tag{6}$$

Notice that if δ is assigned a low value, then (6) ensures that the neurons are updated only if the input pattern $\boldsymbol{x}(t)$ is very close to $\boldsymbol{w}_j(t)$ in the input space. That is, when $\boldsymbol{x}(t)$ and $\boldsymbol{w}_j(t)$ are patterns representing the same object in different frames.

Once the neuron is created, its memory structures are initialised. The input pattern responsible for the birth of the neuron is assigned to the weight vector of the neuron and to the first entry in the neuron log.

$$\boldsymbol{w}_j(t) = \boldsymbol{x}(t) \; ; \; H_j(1) = \boldsymbol{x}(t) \tag{7}$$

On the other hand, if an object leaves the scene then the neuron which represents it should be destroyed. For this purpose, each neuron has a counter C_{die} which

---

Algorithm 1. Main steps of the tracking algorithm

Input: Time instant t and the features of the segmented objects $x_i(t)$
Output: Labelling of the segmented objects
foreach *Segmented object $x_i(t)$* **do**
   | Compute winner neuron by means of Eq. (5);
   | **if** *Eq. (6) is satisfied* **then**
   | | Create a new neuron. Initialize it;
   | **else**
   | | Update the network using equation Eq. (3);
   | **end**
end
Refresh the counter values belonging to the neurons which win a competition;
Decrement all neurons counter values by one;
Check out neuron counters and destroy neurons whose counter value is zero;

---

means the *lifetime* of the neuron, measured in number of training steps, i.e., frames. Each training step, the counter value is decreased by one and, if the value reaches zero then the corresponding neuron is removed. Every time a neuron wins a competition its counter value is changed to the initial value. Therefore, only neurons associated to objects which are not longer in the scene are destroyed, since it is very unlikely for these neurons to win a competition.

5   Results

In this section the results of our tracking approach to detect and track rigid objects are showed. Some traffic sequences are used to prove the effectiveness of our method. In these sequences some common problems appear, such as occlusions, stopped car in the scene or errors happened in the segmentation phase, which must be satisfactorily solved by a robust tracking algorithm. They are provided by a video surveillance online repository [21] and the Federal Highway Administration (FHWA) under the Next Generation Simulation (NGSIM) program.[1]

Each object pattern has nine components: the x-coordinate and y-coordinate of the object centroid, the 2-d coordinates of the upper left corner of the box which bounds the object, the length and width of that bounding box; and the RGB color components. In our experiments the mask vector is set to hide all object components except for the centroid.

In figure 1, the objects of the sequence are detected and tracked along several frames. In figure 2(b) errors in segmentation phase are observed. Two objects are overlapped at the bottom of the image and one object is divided in two blobs at the left. These problems are solved in fig. 2(c) as it can be observed in the objects identified by the numbers 15, 17 and 13.

Other two different sequences are showed in fig. 3 and 4. Figure 3 contains occlusions of the objects in motion caused by background objects, such as trees.

---

[1] Datasets of NGSIM are available at http://ngsim.fhwa.dot.gov/

Fig. 1. Two frames (209, 218) of a traffic sequence are viewed in which the objects are identified and tracked

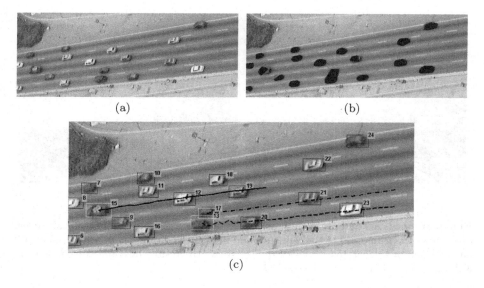

Fig. 2. Problems in the segmentation phase, which are observed in 2(b), are solved in the tracking phase. The trajectory of some objects are plotted in 2(c).

Objects identified in 3(a) by 142 and 154, are robustly obtained in 3(d) despite the segmentation results showed in 3(c).

6   Conclusions

In this work we have presented a new algorithm for moving object detection and tracking in video sequences. This is an important part of video surveillance systems, since these systems need a good starting point to analyse object behaviour. With a reliable tracking algorithm, objects can be easily identified in the video sequence and, using other analysis tools, the behaviour of these objects can be studied, and the system can determine whether there are suspicious/dangerous objects or not.

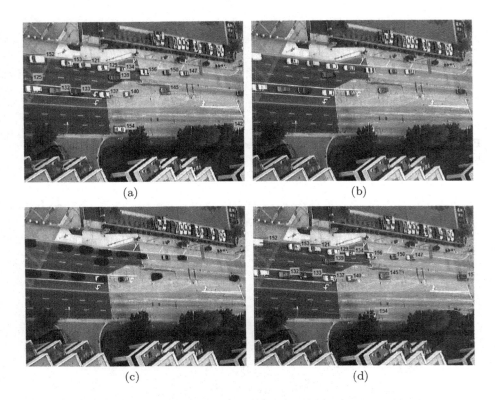

Fig. 3. Stopped cars and car occlusions caused by some trees are observed in this scene. 3(a) shows the frame 740 of the sequence and its identified objects. In 3(b) (frame 747) an occlusion of the car 154 is observed. 3(c) shows the results of segmentation step. In 3(d) the object 154 is correctly tracked.

Fig. 4. Another sequence obtained in [21]. Results of our tracking approach can be observed in 4(a) and 4(b) (frames 245 and 255).

The algorithm proposed in this paper is based on the use of a subtype of the well-known competitive neural networks: the growing competitive neural network (GCNN), which allows the creation and removing of neurons, which are identified to the objects in the scene. Since the number of objects in a video sequence can change from frame to frame, it seems reasonable to permit a change in the number of process units of the network. Thus, a better representation of the foreground objects is obtained.

This new neural model is able to predict the features of each object (location...), by using a log which stores all information known for every object in the last few frames. This allows to deal with several problems produced at the segmentation phase, such as object occlusion or fusion.

Experimental results show that our approach is a reliable and accurate method to detect objects in video sequences publicly available in Internet. In addition, segmentation derived problems can be robustly tackled by this method.

Our future work covers aspects of behavioural analysis, as it is the next logical step in a surveillance system.

Acknowledgements

This work is partially supported by Junta de Andalucía (Spain) under contract TIC-01615, project name Intelligent Remote Sensing Systems.

References

1. Amer, A., Dubois, E., Mitiche, A.: Real-time system for high-level video representation: Application to video surveillance. In: Proceedings of the SPIE International Symposium on Electronic Imaging, pp. 530–541 (2003)
2. Haritaoglu, I., Harwood, D., Davis, L.: w^4: Real-time surveillance of people and their activities. IEEE Trans. Pattern Anal. Mach. Intell. 22(8), 809–830 (2000)
3. Lv, F., Kang, J., Nevatia, R., Cohen, I., Medioni, G.: Automatic tracking and labeling of human activities in a video sequence. In: Proceedings of the 6th IEEE International Workshop on Performance Evaluation of Tracking and Surveillance (2004)
4. Stauffer, C., Grimson, W.: Learning patterns of activity using real time tracking. IEEE Trans. Pattern Anal. Mach. Intell. 22(8), 747–767 (2000)
5. Elgammal, A., Davis, L.: Probabilistic framework for segmenting people under occlusion. In: Proceedings of the 8th IEEE International Conference on Computer Vision, pp. 145–152. IEEE Computer Society, Los Alamitos (2001)
6. Menser, B., Brunig, M.: Face detection and tracking for video coding applications. In: Conference Record of the Thirty-Fourth Asilomar Conference on Signals, Systems and Computers, pp. 49–53 (2000)
7. Böhme, H., Wilhelm, T., Key, J., Schauer, C., Schröter, C., Gross, H., Hempel, T.: An approach to multi-modal human-machine interaction for intelligent service robots. Robot. Auton. Syst. 44, 83–96 (2003)
8. Darrell, T., Gordon, G., Harville, M., Woodfill, J.: Integrated person tracking using stereo, color, and pattern detection. Internat. J. Comput. Vision 37, 175–185 (2000)

9. Hayashi, K., Hashimoto, M., Sumi, K., Sasakawa, K.: Multiple-person tracker with a fixed slanting stereo camera. In: 6th IEEE International Conference on Automatic Face and Gesture Recognition, pp. 681–686 (2004)
10. Grest, D., Koch, R.: Realtime multi-camera person tracking for immersive environments. In: IEEE 6th Workshop on Multimedia Signal Processing, pp. 387–390 (2004)
11. Wren, C., Azarbayejani, A., Darrell, T., Pentland, A.: Pfinder: Real-time tracking of the human body. IEEE Trans. Pattern Analysis and Machine Intelligence 19(7) (1997)
12. Krahnstover, N., Yeasin, M., Sharma, R.: Towards a unified framework for tracking and analysis of human motion. In: Proc. IEEE Workshop Detection and Recognition of Events in Video (2001)
13. Siebel, N., Maybank, S.: Fusion of multiple tracking algorithm for robust people tracking. In: Proc. European Conf. Computer Vision, pp. 373–387 (2001)
14. Lipton, A., Fujiyoshi, H., Patil, R.: Moving target classification and tracking from real-time video. In: Proc. DARPAIU Workshop, pp. 129–136 (1998)
15. Nummiaro, K., Koller-Meier, E., Van Gool, L.: An adaptive color-based particle filter. Image Vision Comput. 21, 99–110 (2003)
16. Comaniciu, D., Ramesh, V.: Mean shift and optimal prediction for efficient object tracking. In: IEEE Int. Conf. Image Processing (ICIP 2000), pp. 70–73 (2000)
17. Comaniciu, D., Ramesh, V., Meer, P.: Real-time tracking of non-rigid objects using mean shift. In: IEEE Conference on Computer Vision and Pattern Recognition, pp. 142–149 (2000)
18. Alahakoon, D., Halgamuge, S.K., Srinivasan, B.: Dynamic self-organizing maps with controlled growth for knowledge discovery. IEEE Trans. Neural Networks 11(3), 601–614 (2000)
19. Ahalt, S., Krishnamurthy, A., Chen, P., Melton, D.E.: Competitive learning algorithms for vector quantization. Neural Networks 3, 277–290 (1990)
20. Luque, R., Domínguez, E., Palomo, E., Muñoz, J.: A neural network approach for video object segmentation in traffic surveillance. In: Campilho, A., Kamel, M.S. (eds.) ICIAR 2008. LNCS, vol. 5112, pp. 151–158. Springer, Heidelberg (2008)
21. Vezzani, R., Cucchiara, R.: Visor: Video surveillance online repository. In: BMVA symposium on Security and surveillance: performance evaluation (2007)

Emission Analysis of a Fluidized Bed Boiler by Using Self-Organizing Maps

Mika Liukkonen[1,*], Mikko Heikkinen[1], Eero Hälikkä[2],
Reijo Kuivalainen[2], and Yrjö Hiltunen[1]

[1] Department of Environmental Science, University of Kuopio, P.O. Box 1627,
70211 Kuopio, Finland
{Mika.Liukkonen,Mikko.Heikkinen,Yrjo.Hiltunen}@uku.fi
[2] Foster Wheeler Power Group, P.O. Box 201, 78201 Varkaus, Finland
{Eero.Halikka,Reijo.Kuivalainen}@fwfin.fwc.com

Abstract. In this study, a self-organizing map (SOM) -based process analysis and parameter approximation method was used to the emission analysis of a circulating fluidized bed process. The aim was to obtain the optimal process parameters in respect to the flue gas nitrogen oxide (NO_x) content in different predefined states of process. The data processing procedure in the research went as follows. First, the process data were processed by using a self-organizing map and k-means clustering to generate subsets representing the separate process states in the boiler. These process states represent the higher level process conditions in the combustion, and can include for example start-ups, shutdowns, and idle times in addition to the normal process flow. Next, optimal areas were discovered from the map within each process state, and the reference vectors of the optimal neurons were used to approximate the values of desired process parameters. In addition, a subtraction analysis of reference vectors was performed to analyze the optimal situations. In conclusion, the method showed potential considering its wider use in the field of energy production.

Keywords: Self-organizing map, Fluidized bed, Optimization, Emission modeling, Parameter estimation, Neural networks.

1 Introduction

The world-wide targets for reducing harmful process emissions are having an increasing effect on the modern-day production of energy. At the meantime, higher demands are set for the efficiency of combustion processes. Therefore it is essential to develop such data analysis and modeling methods that can respond to these challenges. Nevertheless, efficient combustion of fuels with lower emissions is a difficult task in power plants because usually it is not possible to reduce the emissions by cutting the production of power. For example in industry a certain amount of energy is needed to maintain the current level of production. This means that other ways of action must be

---

* Corresponding author.

M. Kolehmainen et al. (Eds.): ICANNGA 2009, LNCS 5495, pp. 119–129, 2009.
© Springer-Verlag Berlin Heidelberg 2009

discovered to reduce the emissions. Fortunately, process data can involve important information on the behavior of the process and on different phe$_x$mena that affect the emissions and the energy efficiency of a combustion process. This information can be extremely valuable when optimizing the process.

Artificial neural networks (ANN) have proved their power and usability in the modeling of different industrial processes [1–4]. ANNs have offered functional applications in miscellaneous fields of industry, for instance in the production of steel, pulp and energy, in chemical and electronics industry, and even in waste water treatment [5–10]. The benefits of neural networks include flexibility, nonlinearity, applicability, adaptivity, a high tolerance of faults and a high computing power [2, 11–12], which has led to a large variety of applications. These strong advantages make ANNs a valuable alternative for modeling method in industrial processes.

1.1 Process States and Their Sub-models

The use of a self-organizing map (SOM) [11], developed by Kohonen in the early 1980s, in the analysis of process states has produced a variety of applications in the past years. Kasslin *et al* [13] have first introduced the concept of process states in 1992 by using a SOM to monitor the state of a power transformer. Later on, Alhoniemi *et al.* [5] have broadened the field of SOM-based applications by using the method in the monitoring and modeling of several industrial processes.

Several of our recent studies have shown [8, 14–15] that different states of the fluidized bed combustion process can be discovered in process data by using the SOM. These states can include for instance start-ups, shut-downs, idle times and different states of normal process flow. There can be major differences in the performance of the process between these conditions, for example the quantities of different emission components may vary greatly. In addition, we have proposed that these upper level process states involve secondary process states, where e.g. the steam flow is lower than normally or the bed temperature is unstable [14]. Sometimes it is crucial to identify these secondary states, because the performance of the process can fluctuate also in a smaller scale but regardless by affecting critically e.g. the formation of an emission component. In this study, however, we concentrate on the upper level states of process, because in this case they seem to provide the best sub-models and the most illustrative examples.

Constructing sub-models in parallel with higher level models opens an interesting perspective to present-day process modeling. This is due to the fact that process states and their related sub-models can offer valuable information on the performance of the process, which is indicated in our earlier studies concerning the wave soldering and the activated sludge treatment process [6, 10]. The sub-model -based approach is realistic for example when it is evident that less detectable but still important condition-related phenomena remain concealed when using generic modeling. Despite being more difficult to distinguish, these events can have substantial effects on the combustion process [14]. Nonetheless, these latent phenomena can be analyzed by identifying different process states and creating sub-models, advancing from a generic model to more detailed models [14].

1.2 Optimization

The theoretical optimum of a variable can be determined by using classic optimization methods, such as extreme-value calculation or linear optimization. These methods are relatively simple to use, needing only one cycle to process. However, the classic methods also make strict demands on the search space, setting limits that are sometimes difficult to achieve [16]. Additionally, a large number of variables can make these methods complicated to use because it is usually necessary to perform a selection of variables or a sensitivity analysis first. More precise representation of reality can be attained by using simulation-supported optimization due to the use of real data as solution suggestions. On the other hand, simulation-supported optimization methods generally necessitate the optimization cycle to be completed several times [16], so they can be lengthy procedures in complicated optimization cases.

The analysis and optimization approach presented here offers a valuable option for real-world applications because it takes generally only a few seconds' processing time for a computer to create the self-organizing feature map, illustrate the results and propose the optimal parameters. In consequence, this enables comparatively fast responding to the fluctuations of the process, and would enable even dynamic optimization and control of process parameters. Evidently the preconditions for swift responses are the regular and frequent updating of modeling data and the appropriate definition of the possible cost function. Additionally, it is important that the data samples represent the whole range of probable machine operation.

2 Process and Data

Fluidized bed combustion is a common technology used in power plants and designed chiefly for solid fuels such as coal. Start-ups are generally the only occasions that involve the use of supporting fuels such as oil or natural gas. An archetypal circulating fluidized bed (CFB) boiler comprises a combustion chamber, a separator and a return leg for the recirculation of the bed particles, whereas the fluidized bed is composed of sand, fuel ash and sulfur capturing material. This granular mixture is fluidized by the combustion air brought in from the bottom of the chamber. Due to high fluidizing velocities, the bed particles are in consistent movement with the flue gases. The particles pass through the main combustion chamber into a separator, where the larger particles are extracted and directed back to the combustion chamber. Meanwhile, the finer particles are separated from the cycle and gathered by using a bag-house filter or an electrostatic precipitator located downstream from the convection section of the boiler. The characteristic combustion temperature in CFB boilers is between 850 and 900 °C.

The raw process data were compressed for the analysis by using a moving average to comprise 10 000 data samples. After averaging the resolution of the data used in modeling was five minutes, the number of variables being 36.

3 Methodology

The self-organizing map (SOM) [11] algorithm was used as a modeling method, whereas the k-means [17] clustering algorithm was used to create sub-categories representing the different states of the process on the map. At the last stage, the reference vectors of the SOM were used to approximate the optimal parameters.

3.1 Self-Organizing Maps (SOM)

The applications based on the self-organizing map [11] algorithm form one of the most diverse application areas of artificial neural networks. Conventionally the SOM has been exploited in many different practical applications, such as exploratory data analysis, pattern recognition, speech analysis, industrial and medical diagnostics, robotics and instrumentation, and even control [11]. Kasslin *et al* [13] and Alhoniemi *et al* [5] have presented SOM-based applications to process monitoring and modeling in several industrial processes. In recent times, the gamut of industrial applications based on the algorithm has become especially diverse [6, 8, 10, 14–15].

The ordinary use of SOM is to make data analysis easier by mapping *n*-dimensional input vectors to classes, or *neurons*, for instance in a two-dimensional lattice (map). The map of neurons reflects dissimilarities in the statistics of the data, and chooses such features that approximate to the distribution of the data samples. The topological organization of the input data is preserved on the SOM by associating the input vectors with common features with the same or neighboring neurons. In addition, the generalized properties of a neuron can be represented with an *n*-dimensional, neuron-specific prototype vector, or *reference vector*. The size of the map can be varied depending on the purpose by changing the number of neurons; the bigger the map, the more details appear.

The analysis with SOM involves an unsupervised learning process. Firstly, the preliminary reference vectors are initialized randomly by picking their values from an even distribution whose limits are determined by the input data. During the learning process the inputs are classified one by one into best matching units (BMU) on the map. Generally the BMU is defined as the neuron whose reference vector has the smallest *n*-dimensional Euclidean distance to the input vector. Even as the BMU is discovered for the input, the nearest neighbors of the BMU become activated as well, according to a predefined neighborhood function (e.g. Gaussian distribution) that is dependent on the topology of the network. Eventually the reference vectors of all activated neurons are updated.

3.2 K-Means Clustering

The k-means algorithm [17] is an extensively used non-hierarchical data clustering method. The basic way to execute the algorithm is to determine k cluster centers at random, and thereafter to direct each data point to the cluster whose mean value is the closest equivalent in the Euclidean-distances-sense. Subsequently, the mean vectors of the data points included to each cluster are computed and used iteratively as new cluster centers. The optimal number of clusters can be discovered e.g. by using the Davies-Bouldin -index [18]. Small values of DB-index correspond to compact clusters whose centers are far from each other. Thus, the optimal number of clusters is the number where the DB-index reaches its minimum. This way the need for knowing the clusters *a priori* can be avoided.

3.3 Determination of Process States

The states of the combustion process were defined by combining the SOM analysis, k-means clustering, and process expertise. Firstly, all the input values were variance scaled. At the next stage, the scaled data were coded into inputs for a self-organizing network, whereby a SOM having 384 neurons in a 24x16 hexagonal arrangement was created. The parameters of the final SOM were defined by experimental testing. The linear initialization and batch training algorithm were used in the training of the map, and the neighborhood function was Gaussian. The map was taught with 10 epochs, and the initial neighborhood had the value of 6. The SOM Toolbox (http://www.cis.hut.fi/projects/somtoolbox/) was used in the analysis under a Matlab (version 7.6) software (Mathworks Inc., Natick, MA, USA, 2008) platform. The k-means method was then used to cluster the reference vectors of the trained map, whereby the optimal number of clusters was defined by using the DB-index. Next, the information gained by clustering was united to expert process knowledge to identify the different states of process outlined by the clusters.

3.4 Subtraction Analysis

Subtraction of reference vectors can identify reasons for events, as the result describes the difference between the two vectors concerned. This form of analysis is called the subtraction analysis here. To perform subtraction analysis within each process state (cluster), the following formula was used:

$$s = C_\gamma - r_{opt} , \qquad\qquad (1)$$

where s is the result vector of subtraction analysis, C_γ is the center vector of cluster γ, and r_{opt} signifies the reference vector of the optimal neuron in cluster γ. This way the variables contributing to low emission rates within each process state are revealed. There are two important aspects to bear in mind when interpreting the results: (*i*) a great absolute value of the variable component in vector s means that the variable has largely affected the reduction of the emission within the process state concerned, and (*ii*) if the value of s is negative, the value of the variable has been higher in the optimal situation than in the cluster in general; in contrast, if the value of s is positive, the value of the variable has been lower in the optimal situation than in the cluster in general.

3.5 Visualization and Optimization

There are a number of ways to visualize a SOM. Perhaps the most common of them is to use 2-dimensional mapping of neurons, and to use color coding to visualize the special features on the map. Nonetheless, one of the most descriptive methods is to use 3-dimensional visualization of the component map. In this approach, the first two dimensions are defined by the arrangement of neurons, while the third dimension illustrates the desired output feature, or vector component. This kind of map is more informative and more decipherable than the conventional 2-dimensional map.

After training the SOM, each neuron is defined by its location on the map grid and by its reference vector, which represents an average description of the data rows assimilated to the neuron concerned. The reference vector has the same

124   M. Liukkonen et al.

dimensionality as the input vectors. Therefore it is possible to regard a neuron as an n-dimensional reference vector, which can be represented as follows:

$$r_m = (r_{m1}, r_{m2}, ..., r_{mn}), (m = 1, ..., M) ,\qquad(2)$$

where n is the number of variables and M refers to the total number of map neurons. If the criteria for optimization are known the vectors can also be used as parameter estimators for optimal situations (e.g. process conditions with low emissions). Thus the reference vector of the optimal neuron is determined by using the following equation:

$$r_{opt} = r_m[\min(E)] ,\qquad(3)$$

where E signifies the computational value for the vector component of the emission compound in the reference vector r_m. Alternatively, E can be a derivative variable, e.g. the result of a specified cost function. Realistic values for different vector components are achieved by undoing data normalization.

4 Results

The clustering of data followed roughly the intensity of the steam flow, as can be seen in Fig. 1 a) and b), where the SOM component plane of the main steam flow and the clusters with the identified process states are represented, respectively. In addition,

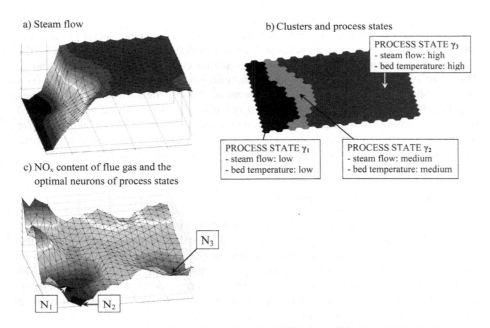

Fig. 1. a) SOM component plane for steam flow, b) SOM with clusters and identified process states, c) SOM component plane for NO$_x$ content of flue gas, where also the optimal neurons (N$_1$, N$_2$ and N$_3$) for respective process states are indicated.

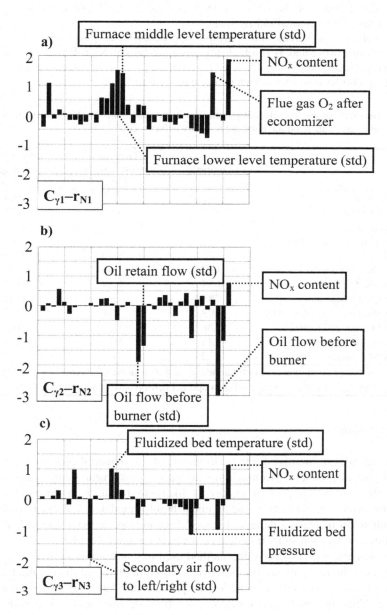

Fig. 2. The results of the subtraction analysis. a) Process state 1 (low steam flow), b) process state 2 (medium steam flow), and c) process state 3 (high steam flow). $C_{\gamma 1}$, $C_{\gamma 2}$ and $C_{\gamma 3}$ indicate the center vectors of clusters γ_1, γ_2 and γ_3, respectively; r_{N1}, r_{N2} and r_{N3} are the reference vectors of neurons N_1, N_2 and N_3 presented in Fig. 1 c). 'std' refers to the standard deviation of the averaged variable.

the component plane of the nitric oxide (NO_x) content with process state -specific optimal neurons is presented in Fig. 1 c).

The results of the subtraction analysis for process states γ_1 (low steam flow), γ_2 (medium steam flow) and γ_3 (high steam flow) are presented in Fig. 2 a), b) and c), respectively. Finally, the optimal values for selected process parameters in the process state γ_3 (high boiler load) are shown in a scaled form in Table 1.

Table 1. Optimal values for selected parameters in process state γ3 (high steam flow). The original values are scaled and the units are not shown due to confidentiality.

	Optimal	Optimal + nearest neighbors ([1]ave)	Optimal + nearest neighbors ([2]std)
Total limestone flow	1,33	1,34	0,04
Main steam flow	65,2	65,1	0,35
Primary air flow	13,1	13,0	0,37
Primary air flow to fuel feed	2,24	2,23	0,05
Secondary air flow	5,31	5,30	0,09
Fluidized bed pressure	1,07	1,06	0,03
Coal conveyor speed	13,1	12,9	0,42
Fluidized bed temperature	283	283	0,52
Furnace middle level temperature	109	109	0,22
Flue gas O_2 after economizer	1,56	1,56	0,01
Oil flow before burner	0,19	0,19	0,02
Flue gas NO_x content	47,2	48,0	0,67

[1]Average of vector components in optimal neuron and its nearest neighbors.
[2]Standard deviation of vector components in optimal neuron and its nearest neighbors.

5 Discussion

The general goal in improving the performance of an industrial process is to reduce the total cost of production to maximize the profit. In the near future, an increasing part of the production cost of energy is likely to result from different process emissions, which are already at the moment highly regulated and even sanctioned. This leads to an increasing demand for modeling and optimization methods that can be exploited in decision support systems or even in integrated applications.

The schematic presentation of the method used in this study is shown in Fig. 3. At the first stage the raw data are pre-processed, which involves all the required actions, such as the normalization of variables, to process the data to a form suitable for modeling. Then the SOM and k-means clustering are used and linked with expert process knowledge to discover the different states of the process. Finally, the subtraction analysis of reference vectors and the optimization are performed within the process states, and the optimal parameters are approximated in the separate states.

The results showed the potential of the method in the modeling of the circulating fluidized bed process. The identification of process states produced clearly defined

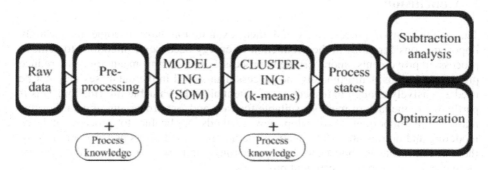

Fig. 3. Schematic presentation of the method used

states, cluster borders following mostly the changes in the intensity of steam flow, in other words the degree of the boiler load. It is obvious that the optimal situation concerning the NO_x formation on the high boiler load is different from the optimal on the low load. This is because it is not possible to obtain the lower level of emissions if a certain amount of steam must be produced. In addition, it must be borne in mind that sometimes the definition of process states may not be as clear as in this case. Thus it is reasonable to identify the process states before optimization is performed.

The subtraction analysis of vectors offers deeper information on factors affecting the reduction of emissions, or NO_x in this case. For instance in respect to high boiler load, 3–6 factors seem to contribute to the lower concentration of nitrogen oxides. In contrast, the corresponding factors in the other states are different. In the process state of medium steam flow (including start-up times), for example, the sufficient flow of oil seems to ensure a lower concentration of NO_x. These observations support the fact that the optimization should be performed within the process states.

The optimization of an industrial process is often a time-consuming and laborious task. By using the method presented the process parameters can be estimated fast and reliably if the modeling data cover the search space widely enough. The noticeable benefits of the presented method are its flexibility, nonlinearity and a strong computing power. The method used is also very illustrative, informative and quite simple to use, and therefore provides a useful, efficient and fruitful way to define the optimal parameters for a real manufacturing process.

The method presented offers several opportunities in the near future. For example, currently only a few emission components are included to international emission trading. However, the regulations restricting different emissions are tightening world-wide due to environmental issues such as global warming. As new emission components are probably included to the trading of emissions, the method can be used for emission cost modeling after the definition of a cost function to be used in the optimization. Optionally, the method can also be used to more general process optimization, for instance to optimize the total profit of producing steam in fluidized bed boilers.

5 Conclusion

The acquisition of process data and their exploitation have become increasingly important considerations in the production of energy because of intentions to achieve process improvements, more efficient processes and, at the mean time, to reduce process emissions. Unfortunately the process data stored in the databases may not be used extensively enough in respect to the potential of the valuable information. This is partly due to the lack of methods suitable for easy and fast processing of data. For this reason it is important to develop new methods applicable to process analysis, modeling and optimization. The analysis method presented offers new possibilities for the development of industrial data processing, and also empowers the process engineers to develop more efficient processes.

References

1. Harding, J.A., Shahbaz, M., Srinivas, K.A.: Data Mining in Manufacturing: A Review. Journal of Manufacturing Science and Engineering 128(4), 969–976 (2006)
2. Haykin, S.: Neural Networks: A Comprehensive Foundation. Prentice Hall, Upper Saddle River, NJ (1999)
3. Meireles, M.R.G., Almeida, P.E.M., Simões, M.G.: A Comprehensive Review for Industrial Applicability of Artificial Neural Networks. IEEE Trans. Industrial Electronics 50(3), 585–601 (2003)
4. Mujtaba, I.M., Hussain, M.A. (eds.): Application of Neural Network and Other Learning Technologies in Process Engineering. Imperial College Press, London (2001)
5. Alhoniemi, E., Hollmén, J., Simula, O., Vesanto, J.: Process Monitoring and Modeling Using the Self-Organizing Map. Integrated Computer Aided Engineering 6(1), 3–14 (1999)
6. Heikkinen, M., Heikkinen, T., Hiltunen, Y.: Process States and Their Submodels Using Self-Organizing Maps in an Activated Sludge Treatment Plant. In: Proc. 48th Scandinavian Conference on Simulation and Modeling (SIMS 2008) [CD-ROM], Oslo University College (2008)
7. Heikkinen, M., Nurminen, V., Hiltunen, T., Hiltunen, Y.: A Modeling and Optimization Tool for the Expandable Polystyrene Batch Process. Chemical Product and Process Modeling 3(1), Article 3 (2008)
8. Heikkinen, M., Kettunen, A., Niemitalo, E., Kuivalainen, R., Hiltunen, Y.: SOM-based method for process state monitoring and optimization in fluidized bed energy plant. In: Duch, W., Kacprzyk, J., Oja, E., Zadrożny, S. (eds.) ICANN 2005. LNCS, vol. 3696, pp. 409–414. Springer, Heidelberg (2005)
9. Liukkonen, M., Hiltunen, T., Havia, E., Leinonen, H., Hiltunen, Y.: Modeling of Soldering Quality by Using Artificial Neural Networks. IEEE Trans. Electronics Packaging Manufacturing 32(2), 89–96 (2009)
10. Liukkonen, M., Havia, E., Leinonen, H., Hiltunen, Y.: Application of Self-Organizing Maps in Analysis of Wave Soldering Process. Expert Systems with Applications 36, 4604–4609 (2009)
11. Kohonen, T.: Self-Organizing Maps. Springer, Heidelberg (2001)
12. Reed, R.D., Marks II, R.J.: Neural Smithing: Supervised Learning in Feedforward Artificial Neural Networks. The MIT Press, Cambridge (1999)

13. Kasslin, M., Kangas, J., Simula, O.: Process State Monitoring Using Self-Organizing Maps. In: Aleksander, I., Taylor, J. (eds.) Artificial Neural Networks 2, vol. I, pp. 1532–1534. North-Holland, Amsterdam (1992)
14. Liukkonen, M., Hälikkä, E., Kuivalainen, R., Hiltunen, Y.: Process State Identification and Modeling in a Fluidized Bed Energy Plant by Using Artificial Neural Networks. In: Proc. Finnish-Swedish Flame Days, The Finnish and Swedish National Committees of the International Flame Research Foundation (IFRF) (2009)
15. Liukkonen, M., Heikkinen, M., Hiltunen, T., Hälikkä, E., Kuivalainen, R., Hiltunen, Y.: Modeling of Process States by Using Artificial Neural Networks in a Fluidized Bed Energy Plant. In: Troch, I., Breitenecker, F. (eds.) Proc. MATHMOD 2009 VIENNA, Full Papers Volume [CD]. ARGESIM - Publishing House, Vienna (2009)
16. Sauer, W., Oppermann, M., Weigert, G., Werner, S., Wohlrabe, H., Wolter, K.-J., Zerna, T.: Electronics Process Technology: Production Modelling, Simulation and Optimisation. Springer, London (2006)
17. MacQueen, J.: Some methods for classification and analysis of multivariate observations. In: Proc. Fifth Berkeley Symposium on Mathematical Statistics and Probability. Statistics, vol. I, pp. 281–297. University of California Press, Berkeley (1967)
18. Davies, D.L., Bouldin, D.W.: A cluster separation measure. IEEE Trans. Pattern Recognition and Machine Intelligence 1(2), 224–227 (1979)

Network Security Using Growing Hierarchical Self-Organizing Maps

E.J. Palomo, E. Domínguez, R.M. Luque, and J. Muñoz

Department of Computer Science
E.T.S.I.Informatica, University of Malaga
Campus Teatinos s/n, 29071 – Malaga, Spain
{ejpalomo,enriqued,rmluque,munozp}@lcc.uma.es

Abstract. This paper presents a hierarchical self-organizing neural network for intrusion detection. The proposed neural model consists of a hierarchical architecture composed of independent growing self-organizing maps (SOMs). The SOMs have shown to be successful for the analysis of high-dimensional input data as in data mining applications such as network security. An intrusion detection system (IDS) monitors the IP packets flowing over the network to capture intrusions or anomalies. One of the techniques used for anomaly detection is building statistical models using metrics derived from observation of the user's actions. The proposed growing hierarchical SOM (GHSOM) address the limitations of the SOM related to their static architecture. Experimental results are provided by applying the well-known KDD Cup 1999 benchmark data set, which contains a great variety of simulated networks attacks. Randomly selected subsets that contain both attacks and normal records from this benchmark are used for training the GHSOM. Before training, a transformation for qualitative features present in the benchmark data set is proposed in order to compute distance among qualitative values. Comparative results with other related works are also provided.

Keywords: Self-organization, network security, intrusion detection system.

1 Introduction

Data clustering is an unsupervised learning method to find most similar groups from input data. According to a similarity measure, data belonging to one group are most similar than data belonging to different groups. The unsupervised learning methods are especially useful when we have no information about the input data and we have to discover the existing groups in data. The input data are usually represented as feature vectors in a multidimensional space.

The self-organizing map (SOM) is being widely used as a tool for knowledge discovery, data mining, detection of inherent structures in high-dimensional data and mapping these data into a two-dimensional representation space [1]. This neural network has been applied successfully in multiple areas since the mapping retains the relationship among input data and preserves their topology, so that

M. Kolehmainen et al. (Eds.): ICANNGA 2009, LNCS 5495, pp. 130–139, 2009.
© Springer-Verlag Berlin Heidelberg 2009

an understanding of the structure of the data is provided. On the other hand, SOMs have some difficulties. First, the network architecture has to be established in advance. Due to the high dimensionality of the input data, it is not easy determine the number and arrangement of neurons that obtains the best results. Second, hierarchical relations among input data are difficult to detect, so that understanding of the data is limited.

The growing hierarchical SOM (GHSOM) intends to solve these limitations. This model has a hierarchical architecture structured in layers, where each layer is composed of different SOMs with an adaptative architecture. The architecture of each SOM is determined during the unsupervised learning process, in such a way that the architecture mirrors the structure of the data. The proposed GHSOM by Rauber et al. [2], uses the metric based on Euclidean distance to compare two input data. However, although this metric is useful for vectors with quantitative values, it is not appropriate for vectors where qualitative values are present due to the fact that qualitative values do not have an order associated. Therefore, the use of a distance measure is not appropriate for qualitative values. Most of related works have mapped these qualitative values into consecutive quantitative values [3,4,5]. Although using this mapping we can apply the Euclidean distance, it assigns different distances among distinct qualitative values.

The GHSOM model is used to implement an Intrusion Detection System (IDS) in a network environment. A pre-processing of the qualitative values to correctly represent quantitative values is proposed in this paper. There are two different approaches commonly used in detecting intrusions [6]. The first approach is known as misuse detection, which detect attacks by storing the signatures of previously known attacks. This approach fails detecting unknown attacks or variants of known attacks and the signature database has to be manually updated. The second approach is known as the anomaly detection approach, where a normal profile is first established. Then, deviants from the normal profile are considered intrusions. Therefore, our IDS is an anomaly detection system, since detects new attack types in addition to normal connections. There is a wide variety of anomaly detection systems proposed using data mining techniques such as clustering, support vector machines (SVM) and neural network systems [7,8,9]. Also, various IDS based on self-organization have been used, however the false positive rates is usually very high [10].

In this work, the KDD Cup 1999 benchmark data set [11,12] has been used for training and testing. This data set has served as the first and only reliable benchmark data set that has been used for most of the research work on intrusion detection algorithms [4]. The wide variety of attacks and the presence of both quantitative and qualitative values have done this data set very appropriate for our experiments.

The remainder of this paper is organized as follows. In Section 2, a description of the GHSOM model and the pre-processing of the qualitative data features are provided. In Section 3, some experimental results after training and testing the IDS with the KDD Cup 1999 benchmark data set are presented. Also, a result comparison with other related works is provided. Section 4 concludes this paper.

2   GHSOM Model

The GHSOM has a hierarchical architecture composed of layers, which consist of several growing SOMs [13]. Initially, the GHSOM consists of a single SOM of 2x2 neurons. During the training, the GHSOM architecture is automatically adapted depending on the input patterns. For each SOM more neurons can be added until reach a certain level of detail in the representation of the data mapped onto the SOM. After the growing of the map, the heterogeneity of each neuron of the map is checked. If the neuron has a bad representation of the data, that is, if the cluster has heterogeneous data, the neuron is expanded in a new map in the next layer of the hierarchy in order to provide a more detailed representation. When the training has finished, the GHSOM mirrors the inherent structure of the input patterns, improving the representation achieved with a single SOM. Therefore, each neuron represents a data cluster, where data belonging to one cluster are more similar than data belonging to different clusters.

An example of the GHSOM architecture is given in Fig. 1. This architecture consists of just two layers; the first layer has just one map with 2x2 neurons, whereas the second layer has two maps expanded from two first-layer neurons. The layer 0 represents all the input data and is used to control the hierarchical growth process.

The metric used in the GHSOM to compare two vectors is based on the Euclidean distance. In many real life problems, there exist qualitative data in addition to quantitative data. For example, in the case of network security, some of the features to analyze contain just qualitative data, such as the protocol type, whose values are TCP, UDP or ICMP. In order to feed the neural network with these data, most of the related works map these data into consecutive

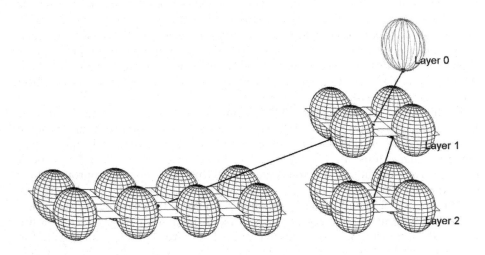

Fig. 1. Sample architecture of a GHSOM

quantitative values. However, since qualitative data have no an order associated, the use of the Euclidean distance is not appropriate for these types of data. For example, if the TCP, UDP and ICMP values are assigned 1, 2 and 3, respectively, and the Euclidean distance is used to compare these values, why is the distance greater between TCP and ICMP than between TCP and UDP? Therefore, we need qualitative values to have the same distance, since qualitative values must just indicate whether they are present or not.

In this paper, this problem has been solved replacing each qualitative feature with a binary vector composed by as many binary features as different possible values that feature can take, as shown in Fig. 2. These binary features are known as dummy variables or dummy features. Let $Q = \{q_1, q_2, ..., q_n\}$ be the set of a n-valued qualitative feature, where q_i is the i-th value of the feature Q. The replacing is defined by the following function

$$f : Q \longrightarrow \{0,1\}^n$$

$$f(q_i) = (\underbrace{0, .., 0}_{i-1}, 1, \underbrace{0, .., 0}_{n-i})$$

This way, the distance among qualitative values is always the same allowing the use of the Euclidean distance or other standard metric, whereas the binary values represent the values of the qualitative feature. Note that in this mapping each qualitative value represents a unit vector of the vector space.

The adaptative growth process of a GHSOM, is controlled by two parameters $\tau 1$ and $\tau 2$, which are used to control the growth of a map and to control the hierarchical growth of the GHSOM, that is, the neural expansion in new maps, respectively. These parameters are the only ones that have to be established in advance. But the automatic adaptation of the GHSOM depends mainly on the error in the representation of each neuron, also called quantization error of the neuron (qe). The qe is a measure of the similarity of data mapped onto each

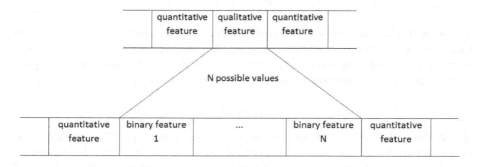

Fig. 2. Replacing qualitative features with dummy features

neuron, where the higher is the qe, the higher is the heterogeneity of the data cluster. The quantization error of the unit i is defined as follows

$$qe_i = \sum_{x_j \in C_i} \|w_i - x_j\| \tag{1}$$

where C_i is the set of patterns mapped onto the neuron i, x_j is the jth input pattern from C_i, and w_i is the weight vector of the neuron i.

Initially, the quantization error at layer 0 must be computed as given in (2), where w_0 is the mean of the all input data I.

$$qe_0 = \sum_{x_j \in I} \|w_0 - x_j\| \tag{2}$$

The initial quantization error qe_0, measures the dissimilarity of all input data and it is used for the hierarchical growth process of the GHSOM together with the $\tau 2$ parameter, following the condition given in (3). That is, the quantization error of a neuron i (qe_i) must be smaller than a fraction (τ_2) of the initial quantization error (qe_0). Otherwise, the neuron is expanded in a new map in the next level of the hierarchy.

$$qe_i < \tau_2 \cdot qe_0 \tag{3}$$

When a new map is created, a coherent initialization of the weight vectors of the neurons of the new map is used as proposed in [14]. This initialization provides a global orientation of the individual maps in the various layers of the hierarchy. Thus, the weight vectors of neurons mirror the orientation of the weight vectors of the neighbor neurons of its parent. The initialization proposed computes the mean of the parent and its neighbors in the respective directions.

The new map created from the expansion of a neuron in the previous level of the hierarchy, it is trained as a single SOM. During the training, the set of input patterns are those that were mapped onto the upper expanded unit. In each iteration t, an input pattern is randomly selected from this data subset. The winning neuron of the map is the neuron with the smallest Euclidean distance to the input pattern, whose index r is defined in (4).

$$r(t) = \arg\min_i \{\|x(t) - w_i(t)\|\} \tag{4}$$

The winner's weight vector is updated guided by a learning rate α, decreasing in time (5). In addition to the winner, the neighbors of the winner are updated depending on a Gaussian neighborhood function h_i and its distance to the winner. This neighborhood function reduces its neighborhood kernel in each iteration.

$$w_i(t + 1) = w_i(t) + \alpha(t)h_i(t)[x(t) - w_i(t)] \tag{5}$$

When the training of the map m is finished, the growing of the map has to be checked. For that, the quantization error of each neuron (qe_i) must be computed in order to compute the mean of the quantization error of the map (MQE_m). If the MQE_m of the map m is smaller than a certain fraction $\tau 1$ of the quantization error of the corresponding unit u in the upper layer, the map stops growing. This stopping for the growth of a map is defined in (6). Otherwise, the map grows to achieve a better level of representation of the data mapped onto the map.

$$MQE_m < \tau_1 \cdot qe_u \tag{6}$$

The growing of a map is done by inserting a row or a column of neurons between two neurons, the neuron with the highest quantization error e and its most dissimilar neighbor d. The neuron d is computed according to the expression (7), where Λ_e is the set of neighbor neurons of e.

$$d = \arg\max_i(\|w_e - w_i\|), \qquad w_i \in \Lambda_e \tag{7}$$

3   Experimental Results

The proposed GHSOM model has been used to implement an IDS. It was trained and tested with the pre-processed KDD Cup 1999 benchmark data set created by MIT Lincoln Laboratory. The purpose of this benchmark was to build a network intrusion detector capable of distinguishing between intrusions or attacks, and normal connections. The 10% KDD Cup 1999 benchmark training data set has been used for training the GHSOM. It contains 494,021 connection records, each of them with 41 features pre-processed from monitoring TCP packets in a military network environment. In the training data set exists 22 attack types in addition to normal records, which fall into four main categories [15]: DoS (denial of service), Probe, R2L (Remote-to-Local) and U2R (User-to-Root). The testing was done with the 10% testing data set, which is composed of 311,029 connection records and 15 additional attack types.

In both data sets three qualitative features are found: protocol type, service and status of the connection flag. As was mentioned in Section 2, qualitative features have to be mapped to numerical values in order to compare two vectors with the Euclidean distance. However, it is not enough to map qualitative values into quantitative values since it assigns an order among qualitative values of a feature. Therefore, each qualitative feature has been replaced with new binary features (dummy features), according to the number of possible values of each feature. In the training data set the number of features has been increased from 41 to 118, since the protocol type feature has 3 different values, the service feature 66 and the status of the connection flag feature 11. The new binary feature has assigned the value 1 if the replaced symbolic feature had assigned the value that represents that new feature or 0 otherwise. This way, among the input data there are just quantitative features but the qualitative values are still represented.

In order to train our IDS, we have selected two data subsets from the total of 494,021 connection records, DS1 with 100,000 connection records and DS2 with 169,000. Both DS1 and DS2 contain the 22 attack types in addition to normal records. The distribution of the selected data mirrors the distribution of the data in the 10% KDD Cup data set, which has an irregular distribution.

The distribution of training and testing data subsets used is given in Table 1, where the 'Unknown' connection category represents the connection records of the new attack types present in the testing data set. Both DS1 and DS2 were trained with 0.1 and 0.07 as value for parameters τ_1 and τ_2, respectively. Thus, the training with DS1 and DS2 data subsets generated a very simple architecture of just two layers in both cases, with just 20 and 17 neurons, respectively.

Table 1. Data distribution of different data subsets

Connection Category	10% Training	DS1	DS2	10% Test
Normal	97278	30416	53416	60593
DoS	391458	64299	110299	223298
Probe	4107	4107	4107	2377
R2L	1126	1126	1126	5993
U2R	52	52	52	39
Unknown	0	0	0	18729

Related works are usually interested in distinguish between attacks or normal records. However, we are also interested in classify an anomaly into its attack type using 23 groups or clusters (22 attack types in addition to normal records) instead 2 groups (normal or attack records). Thus, the detected rate are the attacks that were detected as attacks, the false positive rate are the normal connection records that were detected as attacks, and the identified rate are the connection records that were identified as their correct connection type. The training results using both subsets are given in Table 2. DS1 achieved 100% detection rate and 1.78% false positive rate during the training. DS2 also achieved 100% detection rate, but with a lower false positive rate (0.71%). The identification rate was around 94% and 95% in both cases, respectively.

Table 2. Training results for DS1 and DS2

Training Set	Detected(%)	False Positive(%)	Identified(%)
DS1	100	1.78	94.03
DS2	100	0.71	95.158

The two trained GHSOMs where testing with the entire 311,029 connection records from the 10% KDD Cup 1999 testing data set, where 15 new attack types are found in addition to the rest of attack types and normal records. The testing results are shown in Table 3, where we can notice that both GHSOMs achieved similar results. DS1 and DS2 achieved a detection rate of 99.99%, whereas DS2 false positive rate and identification rate are slightly lower than those for DS1.

Most of the related works have used self-organizing maps to implement an IDS. In [4], a hierarchical Kohonen net (K-Map) was proposed. It was composed of

Table 3. Testing results with 311,029 records for DS1 and DS2

Training Set	Detected(%)	False Positive(%)	Identified(%)
DS1	99.99	3.94	90.73
DS2	99.99	3.72	90.42

three pre-specified levels, where each level is a single K-Map or SOM. Their best result was 99.63% detection rate after testing, but taking into account several limitations. It was trained with a selected 169,000 connection records data set with 22 attack types from the KDD Cup 1999 data set as we used. However, just three attack types were used during the testing, while we used 38 attack types, where 15 attack types were unknown. They used a combination of 20 features that had to be established in advance. Furthermore, 48 neurons were used in each level. From [15], where a self-organizing map was applied to network security, we chose the only one SOM trained on all the 41 features in order to compare results. This neural network achieved a detection rate of 97.31%, but using 400 neurons. An emergent SOM (ESOM) for the Intrusion Detection process was proposed in [16]. They achieved detection rates between 98.3% and 99.81% and false positives between 2.9% and 0.1%. However, the intrusion detection was just limited to DoS attacks and they used a pre-selected subset of 9 features. Also, a large number of neurons were used in order to achieve emergency, between 160x180 and 180x200 neurons. In addition, their best result (99.81% detection rate and 0.1% false positive rate) was just trained and tested with one DoS attack type, the smurf attack. These different results are summarized in Table 4. Note that our false positive rates are higher than the rest of the proposed IDSs due to the lower number of neurons. Nevertheless, we achieved higher detection rates with a more simple architecture, where the data hierarchical structure is easier to understand than other approaches using more complex architectures.

Table 4. Testing results for differents IDSs based on self-organization

	Detected(%)	False Positive(%)	Neurons
GHSOM	99.99	3.72	17
K-Map	99.63	0.34	144
SOM	97.31	0.042	400
SOM (DoS)	99.81	0.1	28800

4   Conclusions

In this paper, a growing hierarchical self-organizing map (GHSOM) is proposed for network security applications. The proposed neural model performs an anomaly detector using the GHSOM. This GHSOM can deal with the limitation of SOMs related to their static architecture. The architecture of the proposed model is composed of several SOMs arranged in layers, where the whole architecture (number of layers, maps and neurons) is established during the training process depending on the input data and mirroring their inherent structure.

The proposed GHSOM has been used to implement an Intrusion Detection System (IDS). Our model was trained and tested with the standard KDD Cup 1999 benchmark data set, which is composed of 494,021 connection records for training and 311,029 for testing. These connection records consist of 41 features, where 3 of them are qualitative. A transformation of these qualitative features into binary features has been proposed to be able to compute the Euclidean distance among qualitative features.

Two different subsets have been selected for the experimental results, DS1 and DS2 with 100,000 and 169,000 connection records, respectively. During the training, we achieved 100% detection rate and 0.71% false positive rate. After testing, a detection rate of 99.99% was achieved with a false positive rate of 3.72%. These results have been compared with those achieved in [4,15,16], which also used the same benchmark data set. For our experiments, we have taken into account all the features and the attack types, both during the training and testing, using a reduced number of neurons, mirroring the inherent structure of the data and without get worried about establish the architecture in advance. Moreover, we are interested in the number of connection records identified as their correct connection record types (identification rate), achieving around the 95% during the training and around 90% after testing.

Acknowledgements. This work is partially supported by the Spanish Ministry of Science and Innovation under contract TIN-07362.

References

1. Kohonen, T.: Self-organized formation of topologically correct feature maps. Biological cybernetics 43(1), 59–69 (1982)
2. Rauber, A., Merkl, D., Dittenbach, M.: The growing hierarchical self-organizing map: Exploratory analysis of high-dimensional data. IEEE Transactions on Neural Networks 13(6), 1331–1341 (2002)
3. Lei, J., Ghorbani, A.: Network intrusion detection using an improved competitive learning neural network. In: 2nd Annual Conference on Communication Networks and Services Research, pp. 190–197 (2004)
4. Sarasamma, S., Zhu, Q., Huff, J.: Hierarchical kohonen net for anomaly detection in network security. IEEE Transactions on Systems Man and Cybernetics Part B-Cybernetics 35(2), 302–312 (2005)
5. Depren, O., Topallar, M., Anarim, E., Ciliz, M.: An intelligent intrusion detection system (ids) for anomaly and misuse detection in computer networks. Expert Systems with Applications 29(4), 713–722 (2005)
6. Denning, D.: An intrusion-detection model. IEEE Transactions on Software Engineering SE-13(2), 222–232 (1987)
7. Lee, W., Stolfo, S., Chan, P., Eskin, E., Fan, W., Miller, M., Hershkop, S., Zhang, J.: Real time data mining-based intrusion detection. In: DARPA Information Survivability Conference & Exposition II, vol. 1, pp. 89–100 (2001)

8. Maxion, R., Tan, K.: Anomaly detection in embedded systems. IEEE Transactions on Computers 51(2), 108–120 (2002)
9. Tan, K., Maxion, R.: Determining the operational limits of an anomaly-based intrusion detector. IEEE Journal on Selected Areas in Communications 21(1), 96–110 (2003)
10. Ying, H., Feng, T.J., Cao, J.K., Ding, X.Q., Zhou, Y.H.: Research on some problems in the kohonen som algorithm. In: International Conference on Machine Learning and Cybernetics, vol. 3, pp. 1279–1282 (2002)
11. Lee, W., Stolfo, S., Mok, K.: A data mining framework for building intrusion detection models. In: IEEE Symposium on Security and Privacy, pp. 120–132 (1999)
12. Stolfo, S., Fan, W., Lee, W., Prodromidis, A., Chan, P.: Cost-based modeling for fraud and intrusion detection: results from the jam project. In: Proceedings of DARPA Information Survivability Conference and Exposition, 2000. DISCEX 2000, vol. 2, pp. 130–144 (2000)
13. Alahakoon, D., Halgamuge, S., Srinivasan, B.: Dynamic self-organizing maps with controlled growth for knowledge discovery. IEEE Transactions on Neural Networks 11, 601–614 (2000)
14. Dittenbach, M., Rauber, A., Merkl, D.: Recent advances with the growing hierarchical self-organizing map. In: 3rd Workshop on Self-Organising Maps (WSOM), pp. 140–145 (2001)
15. DeLooze, L., AF DeLooze, L.: Attack characterization and intrusion detection using an ensemble of self-organizing maps. In: 7th Annual IEEE Information Assurance Workshop, pp. 108–115 (2006)
16. Mitrokotsa, A., Douligeris, C.: Detecting denial of service attacks using emergent self-organizing maps. In: 5th IEEE International Symposium on Signal Processing and Information Technology, pp. 375–380 (2005)

On Document Classification with Self-Organising Maps

Jyri Saarikoski[1], Kalervo Järvelin[2], Jorma Laurikkala[1], and Martti Juhola[1]

[1] Department of Computer Sciences,
33014 University of Tampere, Finland
[2] Department of Information Studies,
33014 University of Tampere, Finland
{Jyri.Saarikoski,Kalervo.Jarvelin,Jorma.Laurikkala,
Martti.Juhola}@uta.fi

Abstract. This research deals with the use of self-organising maps for the classification of text documents. The aim was to classify documents to separate classes according to their topics. We therefore constructed self-organising maps that were effective for this task and tested them with German newspaper documents. We compared the results gained to those of k nearest neighbour searching and k-means clustering. For five and ten classes, the self-organising maps were better yielding as high average classification accuracies as 88-89%, whereas nearest neighbour searching gave 74-83% and k-means clustering 72-79% as their highest accuracies.

1 Introduction

The growth of digital documents and information stored as text in the Internet has been rapid in the recent years. Searching and grouping such documents in various ways have become an important and challenging function. A myriad of documents are daily accessed in the Internet to find interesting and applicable information. Distinguishing in some way interesting documents from the uninteresting ones is, even if a self-evident goal, crucial. For this purpose, computational methods are of paramount importance. We are interested in researching the classification of text documents, also those written in languages other than English.

There are known methods for constructing groups, clusters or models of documents, see for instance [4], [12] and [13]. These machine learning methods have included k nearest neighbour searching, probabilistic methods such as Naïve Bayesian classifiers [5] and evolutionary learning with genetic algorithms [13]. The methods were of the supervised category. We investigated the use of unsupervised Kohonen self-organising maps [8] that seemed to be seldom used in this field. They have been, however, applied to constructing visual maps of text document clusters, in which documents were clustered based on the features they contain. WEBSOM [7] [9] was employed to organize large document collections, but it did not include document classification in the sense to be compared with the current research. Chowdhury and Saha [2] classified 400, 500 and 600 sport articles, whereas Guerro-Bote et al. [6] employed 202 documents from a bibliographic database. Moya-Anegón et al. [10] made domain analysis of documents with self-organising maps, clustering and multidimensional scaling. Instead of

M. Kolehmainen et al. (Eds.): ICANNGA 2009, LNCS 5495, pp. 140–149, 2009.
© Springer-Verlag Berlin Heidelberg 2009

document clustering, we were interested in investigating how accurately and reliably self-organising maps are able to classify documents. Therefore, we constructed self-organising maps on document sets belonging to different known classes and used them to classify new documents. We employed ten-fold crossvalidation runs on our test document collection to assess classification accuracy in the document collection. We also performed comparable tests with k nearest neighbour searching and k-means clustering which employ supervised learning to find a baseline level for the classification of the document data used. In principle, the use of self-organising maps is reasonable, because outside laboratory tests there is not necessarily a reliably classified learning set available.

In the present research, we extend our previous research of using self-organising maps for information retrieval in the same German document collection as in [11]. In the prior work, we studied retrieval from the document collection, the topics of which were associated with some of its documents, and we used both relevant and non-relevant documents in the document sample extracted from the collection. In the present work, our interest was in the classification, in other words separation between document classes. We therefore used only documents relevant to the classes examined.

2 The Data and Its Preprocessing

We used a German document set which was taken from an original collection of 294809 documents [1] from CLEF 2003 of the years 1994 and 1995 (http://www.clef-campaign.org/). The articles were from newspapers such as Frankfurter Allgemeine and Der Spiegel. There were 60 test topics associated with the collection. In every topic there was a relatively small subset of relevant documents. Relevant topics were included in our tests. At first, 20 topics were taken from the 60 topics otherwise randomly. From those 20 selected the smallest classes (topics) were still left out which included 6-25 relevant documents in the collection. Such small document classes would not have been quite reasonable for 10-fold crossvalidation tests, because their average numbers of test documents in test sets would only have been from 0.6 to 2.5, which might have resulted in considerable random influence. Thus, we attained 10 topics (classes) and 425 relevant documents (observations) so that the numbers of the relevant documents of the topics were 27, 28, 29, 29, 34, 39, 44, 53, 55 and 87.

The concept of relevance means here that the association of the documents to the topics had been manually ensured in advance by independent evaluators who had nothing to do with the present research.

To transform pertinent document data into the input variable form for a self-organising map, some preprocessing was required. At first, the German stemmer called SNOWBALL was run to detect word stems like 'gegenteil' from 'Gegenteil' or 'gegenteilig' from all documents and topics chosen. In addition, a list of 1320 German stopwords was used to sieve semantically useless words from them. Stopwords are prepositions like 'ab', articles like 'ein' and 'eine' or pronouns like 'alle', adverbs or other uninteresting "small words", which are mostly uninflected words. They were removed from the documents and topic texts. Thereafter, short words, shorter than four letters, were removed, because they are typically, after stemming, as word prefices rather useless as term words. The last preprocessing phase included the computation of the frequencies of remaining word stems.

The documents and topics were of SGML format. In the following, the first example presents an (abbreviated) SGML document and the second example depicts a topic connected to some other documents. Classification variables were formed on the basis of words occurring in the actual text parts of the documents and topics.

A document:
<DOC>
<TITLE>Ahornblatt nach 33 Jahren vergoldet</TITLE>
<TITLE>Zum 20. Mal Eishockey-Weltmeister</TITLE>
<TITLE>Sieg im Penaltyschießen</TITLE>
<TITLE>Finnland </TITLE>
<TEXT>Das Eishockey-Mutterland Kanada ist nach 33 Jahren wieder die Nummer eins in der Welt. Durch einen 3:2-Erfolg im Penaltyschießen gegen Finnland lösten die Ahornblätter im WM-Finale in Mailand den einst übermächtigen Rivalen und Titelverteidiger Rußland ab, der bereits im Viertelfinale gegen die USA (1:3) ausgeschieden war. Nach regulärer Spielzeit und Verlängerung hatte es 1:1 (0:0, 0:0, 1:1, 0:0) gestanden. Zuvor hatte Brind'Amour (56.) die Führung der Finnen durch Keskinen (47.) ausgeglichen. Im Penaltyschießen zeigten die Kanadier die besseren Nerven, die Finnen verschossen viermal in sechs Versuchen. Robitaille verwandelte den sechsten Penalty für Kanada. Die Kanadier, zuletzt 1961 bei der WM in Genf und Lausanne auf dem Thron, feierten bei den 58. Titelkämpfen ihren 20. WM-Titel und machten damit...
</TEXT>
</DOC>

A topic:
<DE-title> Rechte des Kindes </DE-title>
<DE-desc> Finde Informationen über die UN-Kinderrechtskonvention. </DE-desc>

We computed document vectors for all documents by applying the common vector space model with *tf·idf* weighting for all remaining word stems. Thus, a document is presented in the following form

$$D_i = (w_{i1}, w_{i2}, w_{i3}, ..., w_{it}) \qquad (1)$$

where w_{ik} is the weight of word k in document D_i, $1 \leq i \leq n$, $1 \leq k \leq t$, where n is the number of the documents and t is the number of the remaining word stems in all documents. Weights are given in *tf·idf* form as the product of term frequency (*tf*) and inverse document frequency (*idf*). The former for word k in document D_i is computed with

$$tf_{ik} = \frac{freq_{ik}}{\max_l \{freq_{il}\}} \qquad (2)$$

where $freq_{ik}$ equals the number of the occurrences of word k in document D_i and l is for all words of D_i, $l=1,2,3,..., t-1, t$. The latter is computed for word k in the document set with

$$idf_k = \log \frac{N}{n_k} \qquad (3)$$

where N is equal to the number of the documents in the set and n_k is the number of the documents, which contain word k at least once. Combining equations (2) and (3) we obtain a weight for word k in document D_i

$$w_{ik} = tf_{ik} \cdot idf_k \tag{4}$$

Based on this computation all 425 documents were mapped as document vectors weighted with the *tf·idf* form.

Finally, the length of each document vector was shortened only to include 500 or alternatively 1000 middle (around median) word stems from the total word frequency distribution increasingly sorted. Very often the most and least frequent words are pruned in information retrieval applications, because their capacity to distinguish relevant and non-relevant documents (to a topic) is known to be poor. We chose either 500 or 1000 words, since from several values we found these as good choices for this data in our earlier research [11].

It is worth noticing that document vectors were only computed from a learning set in crossvalidation. Information about its corresponding test set was not used in order to create as a realistic situation as possible, where the system knows an existing learning set and its words in advance, but not those of a test set. Thus, each learning set included its own word set, somewhat different from those of the other learning sets, and the document vectors of its corresponding test set were prepared according to the words of the learning set.

3   Classification with Self-Organising Maps

Kohonen self-organising maps are neural networks that apply unsupervised learning and they have been exploited for numerous visualisation and categorisation tasks [5]. We employed them to study their applicability to divide the test documents into different classes on the basis of document vectors computed. We used the SOM_PAK program written in C (http://www.cis.hut.fi/projects/somtoolbox/) in Helsinki University of Technology, Finland.

In our previous research on the same German document collection [11], we observed that random initialisation, bubble neighbourhood and up to 17×17 nodes were good choices. Different numbers of learning epochs were tested. Finally, as few as 3 coarse and 15 tuning epochs were applied.

The following procedure was implemented.

1. Create a self-organising map using a learning data set.
2. Form the model vector of a node during the learning process of the network. Its dimension is equal to that of the input vectors.
3. Determine a class for a node of the map according the numbers of documents of different classes in the current node. The most frequent document class determines the class of the node. If there are more than one class with the same maximum, label the node according to the class of the document (from the maximum classes) closest to the model vector (learnt during the process) of the node. Consider all nodes in this manner.

After this procedure each node corresponded to some document class. Some node could also remain empty, which would be bypassed during the later process.

Next the classification of a test document set was performed where a test document was compared to the model vector of each node to find which node was the closest (the best fit), on the basis of Euclidean distance, to the test document.

After computing all document vectors of a test set, classification accuracy was computed by checking for every document of a test set j whether it was classified into its correct class.

$$a_j = \frac{c_j}{n_j} 100\% \tag{5}$$

Here c_j ($j=1,..,10$) is equal to the number of the correctly classified documents in test set j and n_j is the number of all documents in that test set. Accuracy a_j was obtained for each test set. Since a random element is involved in the initialisations of neural networks, we repeated 10 tests for every learning and test set pair. For each such crossvalidation pair about 90% of documents were put to a learning set and the rest 10% to its corresponding test set. Documents were selected into learning sets and test sets so that the relative proportions of various kinds of documents were similar in both sets. Thus, 10-fold crossvalidation was applied, which produced 10 times 10 test runs for a test document set. Average classification accuracies were finally calculated from those 100 runs.

4 Nearest Neighbour Searching and *K*-Means Clustering

In order to compare results obtained by self-organising maps, we tested with nearest neighbour searching and k-means clustering by using exactly the same crossvalidation document selections as above for the documents.

Classification with nearest neighbour searching was performed with the following procedure.

1. Search for k nearest neighbours of a test document from a learning set.
2. Compute the majority class from those k documents, i.e. the most frequent document class among the neighbours.
3. Determine the class of the text document on the basis of the preceding step. If there are two or more classes including the same maximum number of documents, select the class randomly from those majority classes.
4. Repeat the former steps for all documents of a test set.

After the nearest neighbour searching, the classification results were assessed for correctness. Values of k were 1, 3, 5, 7 and 9. The Euclidean distance measure was applied. The procedure was run for all 10 pairs of the learning and test sets, for which average classification accuracies were calculated. We employed the Matlab program. Nearest neighbour searching included no such an initialisation property of random character as self-organising maps and clustering. Consequently, the nearest neighbour searching was run only once for every learning and test set pair.

Clustering was accomplished with the Matlab program according to the test protocol similar to that of nearest neighbour searching. The documents of a learning set were clustered into k clusters in the Euclidean space of the document vector variables, when k was equal to 2, 5, 10 and 20. The class of each cluster was determined similarly to the above "voting" principle of nearest neighbour searching. A test set was then dealt with and results computed. This was done 10 times for all 10 learning and test sets to obtain the average results.

5 Results

We tested with the two input vector lengths, 500 and 1000 word stems, either 2, 5 or 10 classes (topics), which respectively included 142, 278 or 425 relevant documents in total. Less than 10 classes (5 or 2 largest classes) were tested in order to see what may happen when we merely restricted ourselves to the largest document classes, i.e. discarded the classes smaller than with 39 or 55 documents. In the following, we present the means and standard deviations of 100 crossvalidation test runs of the self-organising maps and k-means clustering and those of 10 crossvalidation runs of nearest neighbour searching. The crossvalidation division into test and learning sets was identical between all three machine learning methods used.

Table 1 shows the results computed with the self-organising maps. The highest result at each row is written in bold in Tables 1-3. The best 2-class and 5-class situations in Table 1 were with the smallest network of the 25 nodes. Instead, the 10-class condition gave its best results with the networks of 7×7 nodes. The vector lengths used did not yield so unambiguous an outcome. For the self-organising maps, 4.8% of all nodes as minimum were empty with the size of 5×5 nodes and 5 classes. As maximum 66.9% were empty with the size of 13×13 and 2 classes. These empty nodes obtained hits (incorrect classifications) from 0.8% (10 classes) to 5.0% (2 classes) both with the size of 5×5.

Table 2 presents the results of nearest neighbour searching. Its results of all 2-class test alternatives were exceptionally high. This was at least partly due to very different topics of the two classes one being 'children theme' and the other 'nuclear power theme'. The 5-class and 10-class situations were at their best with nearest neighbour searching of k equal to 1. For the 2-class alternatives the longer vector length of 1000 word stems produced better results than the shorter length of 500, but for the 5-class and 10-class alternatives it was vice versa.

The numbers of 2, 5, 10, 20, 40, 60, 80, 100 and 120 clusters were tested for clustering. Table 3 describes most clustering results excluding those of 40, 60, 100 and 120 clusters since these were poorer than the results of 80 clusters. The best results were gained by using the cluster number of 80, except for the 2-class condition. The shorter vectors were better than the longer ones.

Running times of individual learning and test pairs were moderate while using a computer with a 1.6 GHz processor and 1 GB memory. They varied from 1.6 s to 13 s for the self-organising maps. The Matlab implementation of nearest neighbour searching took from 0.4 s to 1.1 s and that of k-means clustering from 1.8 s to 34 s. These do not contain the short time of the preprocessing common to all three.

Table 1. Means and standard deviations of classification accuracies (%) of self-organising maps for 100 test runs

Number of classes	Vector length	Number of nodes				
		5×5	7×7	9×9	11×11	13×13
2	500	**93.2±8.2**	88.4±10.2	77.4±11.1	68.8±11.5	60.6±13.8
	1000	**90.5±8.0**	86.4±9.5	75.7±11.5	67.2±13.1	62.5±14.3
5	500	**87.8±6.2**	86.0±6.9	84.3±7.1	77.5±6.6	73.4±7.9
	1000	**89.0±6.8**	87.0±5.9	83.2±7.3	78.1±7.5	72.2±8.2
10	500	79.2±7.3	**88.1±5.6**	86.7±5.8	82.6±6.6	79.6±6.3
	1000	76.5±5.1	**89.2±5.4**	88.0±5.4	84.2±4.8	80.8±5.6

Table 2. Means and standard deviations of classification accuracies (%) of nearest neighbour searching for 10 test runs

Number of classes	Vector length	Number k of nearest neighbours				
		1	3	5	7	9
2	500	95.1±4.6	97.1±3.7	95.8±3.6	97.9±3.4	**99.2±2.4**
	1000	**99.3±2.1**	**99.3±2.1**	**99.3±2.1**	98.7±4.2	98.7±2.8
5	500	**83.4±6.2**	76.3±7.1	69.0±6.7	70.5±5.4	69.7±8.4
	1000	**74.4±5.5**	60.8±7.7	56.5±9.9	59.0±7.7	59.4±10.9
10	500	**83.3±7.0**	81.8±6.1	80.2±6.8	78.9±5.7	78.5±5.7
	1000	**80.7±6.0**	72.6±6.2	69.7±5.7	67.9±6.9	71.1±5.7

Table 3. Means and standard deviations of classification accuracies (%) of k-means clustering for 100 test runs

Number of classes	Vector length	Number k of clusters				
		2	5	10	20	80
2	500	62.1±5.7	73.7±17.7	**92.4±14.7**	**97.9±7.0**	95.9±5.9
	1000	61.3±1.6	65.0±11.2	76.9±18.3	83.6±17.7	**91.9±9.7**
5	500		52.0±6.9	59.2±7.4	65.5±6.0	**78.5±7.3**
	1000		44.6±9.8	54.4±7.1	58.8±6.6	**72.3±6.9**
10	500			48.1±5.7	56.9±7.3	**73.6±8.3**
	1000			42.7±5.7	52.1±5.7	**71.9±6.2**

Fig. 1 shows an example of the self-organising maps. It includes 10 classes with 383 documents of a learning set when the size of the map was 7×7, the input vector length was 1000 and a random test run was chosen. Its average classification accuracy was 88.8%.

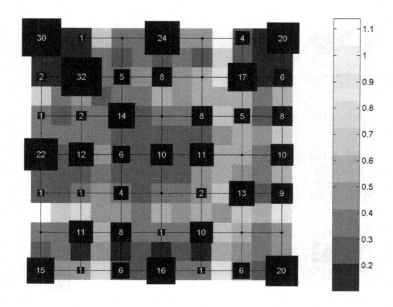

Fig. 1. The numbers of relevant documents of a learning set hit each node are counted in the map. The darker the node, the more compact the concentration of the document group is. The larger the node, the greater the number of documents. The other 42 documents of all 425 documents were not here, but allocated to the test set.

Fig. 2 depicts the same map as Fig. 1, but the nodes are marked with the class identifiers computed. The following list gives the class identifiers, numbers of documents and class titles occurring in Fig 2.

#186 : 24 : Holländische Regierungskoalition
#156 : 25 : Gewerkschaften in Europa
#147 : 26 : Ölunfälle und Vögel
#195 : 26 : Streik italienischer Flugbegleiter
#193 : 31 : EU und baltische Länder
#184 : 35 : Mutterschaftsurlaub in Europa
#150 : 40 : AI gegen Todesstrafe
#152 : 48 : Rechte des Kindes
#190 : 50 : Kinderarbeit in Asien
#187 : 78 : Atomtransporte in Deutschland

To statistically compare the results, the Friedman test [3] was conducted. Since nearest neighbour searching included 10, but the others 100 test runs, the means of the 10 crossvalidations of the latter two methods were first calculated. For the 2-class condition nearest neighbour searching and clustering obtained significantly ($p = 0.004$) better results than the self-organising maps for the vector length of 500. For the length of 1000, nearest neighbour searching was significantly ($p = 0.00005$) better. For the 5-class and 10-class conditions, the self-organising maps outperformed significantly ($p < 0.001$) the other methods with both vector lengths.

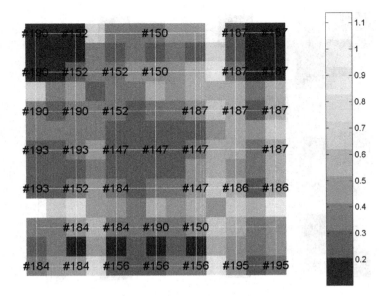

Fig. 2. The class identifiers are attached to the nodes where they beat voting as "majority" classes. Notice that we cannot sum up the numbers of documents from this figure and the preceding list and to compare them directly to those of Fig.1, because the nodes also include some probably incorrect (non-relevant) classifications from "minority" classes.

6   Conclusions

We tested self-organising maps, nearest neighbour searching and k-means clustering with documents from a German newspaper article collection. Except the 2-class alternative which favoured nearest neighbour searching, self-organising maps gave the best results. Table 1 suggests that if more classes are involved, the number of the nodes in a network should increase. On the other hand, for nearest neighbour searching the dispersion of documents to several classes supports the idea to keep to the number k of neighbours equal to 1. Table 3 (k-means) suggests that the number of the cluster is best to set high. Differences caused by the vector lengths were not consistent, but the self-organising maps were mostly somewhat better with the length of 1000 word stems, meanwhile nearest neighbour searching and k-means clustering favoured the length of 500. Doubtless the self-organising maps were effective classifiers for the current data. Excluding the 2-class condition, they outperformed the other two methods, when the self-organising maps gave the average classification accuracies of 88-89%, nearest neighbour searching reached 74-83% and clustering 72-79%. A 2-class condition is an extreme situation. A more realistic alternative contains a greater number of classes. Nearest neighbour searching was the fastest method.

We can continue our research with the current document data and larger document sets. We are going to perform an extensive analysis with additional learning methods.

Acknowledgements

The research was partially funded by Alfred Kordelin Fund and Academy of Finland, projects 120264, 115609 and 124131. SNOWBALL stemmer was by Martin Porter.

References

1. Airio, E.: Word normalization and decompounding in mono- and bilingual IR. Information Retrieval 9(3), 249–271 (2006)
2. Chowdhury, N., Saha, D.: Unsupervised text classification using Kohonen's self organizing network. In: Gelbukh, A. (ed.) CICLing 2005. LNCS, vol. 3406, pp. 715–718. Springer, Heidelberg (2005)
3. Conover, W.J.: Practical Nonparametric Statistics. John Wiley & Sons, New York (1999)
4. Doan, A., Domingos, P., Halevy, A.: Learning to match the schemas of data sources: a multistrategy approach. Machine Learning 50, 279–301 (2003)
5. Duda, R.O., Hart, P.E., Stork, D.G.: Pattern Classification, 2nd edn. John Wiley & Sons, New York (2001)
6. Guerro-Bote, V.P., Moya-Anegón, F., Herrero-Solana, V.: Document organization using Kohonen's algorithm. Information Processing and Management 38, 79–89 (2002)
7. Honkela, T.: Self-Organizing Maps in Natural Language Processing, Academic Dissertation. Helsinki University of Technology, Finland (1997)
8. Kohonen, T.: Self-Organizing Maps. Springer, Berlin (1995)
9. Lagus, K., Kaski, S., Kohonen, T.: Mining massive document collections by the WEBSOM method. Information Sciences 163(1-3), 135–156 (2004)
10. Moya-Anegón, F., Herrero-Solana, V., Jiménez-Contreras, E.: A connectionist and multivariate approach to science maps: the SOM, clustering and MDS applied to library and information science research. Journal of Information Science 32(1), 63–77 (2006)
11. Saarikoski, J., Laurikkala, J., Järvelin, K., Juhola, M.: A study on the use of self-organising maps in information retrieval. To appear in Journal of Documentation (2008)
12. Sebastiani, F.: Machine learning in automated text categorization. ACM Computing Surveys 34(1), 1–47 (2002)
13. Serrano, J.I., del Castillo, M.D.: Evolutionary learning of document categories. Information Retrieval 10, 69–83 (2007)

A Heuristic Procedure with Guided Reproduction for Constructing Cocyclic Hadamard Matrices

V. Álvarez, M.D. Frau, and A. Osuna*

Dpto. Matemática Aplicada I, Universidad de Sevilla, Avda. Reina Mercedes s/n
41012 Sevilla, Spain
{valvarez,mdfrau,aosuna}@us.es

Abstract. A genetic algorithm for constructing cocyclic Hadamard matrices over a given group is described. The novelty of this algorithm is the guided heuristic procedure for reproduction, instead of the classical crossover and mutation operators. We include some runs of the algorithm for dihedral groups, which are known to give rise to a large amount of cocyclic Hadamard matrices.

1 Introduction

A Hadamard matrix is a $n \times n$ square $(-1, 1)$ matrix H_n so that $H_n \cdot H_n^T = nI$. Equivalently, a Hadamard matrix is a square matrix over $\{1, -1\}$ so that its rows are pairwise orthogonal.

The knowledge of Hadamard matrices is a major question for applications in a wide range of different disciplines, as in the design of good (even optimal) error-correcting codes meeting the Plotkin bounds (see [15] for details). A classical reference on Hadamard matrices and their uses is [9].

It may be easily proved that the size n of a Hadamard matrix H_n must be 1, 2 or a multiple of 4. It is conjectured that such a H_n exists for all n divisible by 4. However, the proof of this conjecture remains an important problem in Coding Theory, since there is no evidence of this fact until now.

In fact, there are infinitely many orders multiple of four for which uncertainty about the existence of these matrices has not been removed at all. Furthermore, even in the case that a Hadamard matrix is known to exist for a given order $n = 4t$, there is no algorithm available which outputs a Hadamard matrix of this order $4t$ in reasonable time, as it is pointed out in [14].

The *cocyclic* framework concerning Hadamard matrices was introduced in the 90s [12,13] as a promising context to solve the questions above.

A cocyclic matrix M_f over a finite group $G = \{g_1, \ldots, g_{4t}\}$ of order $|G| = 4t$ consists in a matrix $M = (f(g_i, g_j))$, $f : G \times G \to \{1, -1\}$ being a 2-cocycle over G with coefficients in $\{1, -1\}$, so that

$$f(g_i, g_j)f(g_i g_j, g_k) = f(g_j, g_k)f(g_i, g_j g_k), \qquad \forall\, g_i, g_j, g_k \in G$$

* All authors are partially supported by the research projects FQM–296 and P07–FQM–02980 from JJAA and MTM2008-06578 from MICINN (Spain).

M. Kolehmainen et al. (Eds.): ICANNGA 2009, LNCS 5495, pp. 150–160, 2009.
© Springer-Verlag Berlin Heidelberg 2009

The link between cocyclic and Hadamard matrices was first noticed in [12]. A more recent reference is [11], in which many of the classical and more recently discovered constructions of Hadamard matrices are shown to be cocyclic. This support the idea that cocyclic construction is the most uniform construction technique for Hadamard matrices yet known. Consequently, the *cocyclic Hadamard Conjecture* arises in turn.

The main advantages of working with cocyclic Hadamard matrices may be resumed in the following facts:

- The cocyclic Hadamard test (which claims that it suffices to check whether the summation of every row but the first is zero, see [13] for details) runs in $O(t^2)$ time, better than the $O(t^3)$ algorithm for usual (not necessarily cocyclic) Hadamard matrices.
- The search space is reduced to the set of cocyclic matrices over a given group (that is, 2^s matrices, provided that a basis for 2-cocycles over G consists of s generators), instead of the whole set of $\left(\begin{array}{c} 4t \\ 2t \\ 4t-1 \end{array} \right)$ matrices with entries in $\{-1, 1\}$ consisting of the row $(1, \overset{t}{.}., 1)$ and $4t - 3$ vectors of length $4t$ orthogonal to $(1, \overset{t}{.}., 1)$.

In particular, the work in [5] suggest that the cocyclic framework (c.f. in the table below) may reduce significantly the size of the search space in the general framework (g.f. for brevity) case, as the table below indicates:

t	1	2	3	4	5	6	7	8
c.f.	$O(10^0)$	$O(10^1)$	$O(10^2)$	$O(10^3)$	$O(10^5)$	$O(10^6)$	$O(10^7)$	$O(10^8)$
g.f.	$O(10^1)$	$O(10^9)$	$O(10^{24})$	$O(10^{49})$	$O(10^{82})$	$O(10^{125})$	$O(10^{177})$	$O(10^{238})$

Considerable effort has been devoted to the design of efficient algorithms for constructing cocyclic Hadamard matrices. Exhaustive search is not feasible for orders $4t$ greater than 20 (the search space grows exponentially on t, see [5] for instance). Consequently, alternative methods are required. As far as we know, two different heuristic methods have been proposed until now, in terms of image restorations [6] and genetic algorithms [2].

We present here a new genetic algorithm for constructing cocyclic Hadamard matrices. The main difference with respect to that of [2] is a novel heuristic for reproduction: instead of the usual crossover and mutation operators we shall better use a guided reproduction procedure. Calculations in Section 5 suggest that this new feature improves the original algorithm. This heuristic involves the notions of *i-paths* and *intersections* introduced in [5], to be described further in Section 2.

As it is shown in [5], dihedral groups seems to be the most prolific familiy of groups giving rise to cocyclic Hadamard matrices. We particularize the algorithm to the case of these groups. We also include some runs of the algorithm, which have been worked out in MATHEMATICA 4.0, running on a *Pentium IV 2.400 Mhz DIMM DDR266 512 MB*.

A deeper study on the way in which 2-coboundaries over G have to be combined in order to give rise to cocyclic Hadamard matrices (attending to i-paths and intersections, as described in [5]) would lead to an improvement of the performance of the guided genetic algorithm in a straightforward manner.

We organize the paper as follows. Section 2 collects some general notions and results about cocyclic Hadamard matrices. The algorithm looking for cocyclic Hadamard matrices equipped with the new heuristic for reproduction is described in Section 3. Section 4 is devoted to particularize the algorithm to the case of dihedral group.

2   Generalities about Cocyclic Hadamard Matrices

Consider a multiplicative group $G = \{g_1 = 1, g_2, \ldots, g_{4t}\}$, not necessarily abelian. A cocyclic matrix M_f over G consists in a binary matrix $M_f = (f(g_i, g_j))$ coming from a 2-cocycle f over G, that is, a map $f : G \times G \to \{1, -1\}$ such that

$$f(g_i, g_j)f(g_ig_j, g_k) = f(g_j, g_k)f(g_i, g_jg_k), \qquad \forall\, g_i, g_j, g_k \in G.$$

We will only use normalized cocycles f (and hence normalized cocyclic matrices M_f), so that $f(1, g_j) = f(g_i, 1) = 1$ for all $g_i, g_j \in G$ (and correspondingly $M_f = (f(g_i, g_j))$ consists of a first row and column all of 1s).

Effective methods for constructing a basis $\mathcal{B}$ for 2-cocycles over a given group G are known ([12,13],[7],[4]). Such a basis consists of some representative 2-cocycles (coming from inflation and transgression) and some elementary 2-coboundaries ∂_i, so that every cocyclic matrix admits a unique representation as a Hadamard (pointwise) product $M = M_{\partial_{i_1}} \ldots M_{\partial_{i_w}} \cdot R$, in terms of some coboundary matrices $M_{\partial_{i_j}}$ and a matrix R formed from representative cocycles.

Recall that every *elementary coboundary* ∂_d is constructed from the characteristic set map $\delta_d : G \to \{\pm 1\}$ associated to an element $g_d \in G$, so that

$$\partial_d(g_i, g_j) = \delta_d(g_i)\delta_d(g_j)\delta_d(g_ig_j) \qquad \text{for} \qquad \delta_d(g_i) = \begin{cases} -1 & g_d = g_i \\ 1 & g_d \neq g_i \end{cases} \qquad (1)$$

Although the elementary coboundaries generate the set of all coboundaries, they might not be linearly independent (see [4] for instance). Moreover, since the elementary coboundary ∂_{g_1} related to the identity element in G is not normalized, we may assume that $\partial_{g_1} \notin \mathcal{B}$.

The cocyclic Hadamard test asserts that a cocyclic matrix is Hadamard if and only if the summation of each row (but the first) is zero [13]. In what follows, the rows whose summation is zero are termed *Hadamard rows*.

We now reproduce the notions of *generalized coboundary matrix*, *i-walk* and *intersection* introduced in Definition 2 of [5].

The *generalized coboundary matrix* $\bar{M}_{\partial_j}$ related to a elementary coboundary ∂_j consists in negating the jth-row of the matrix M_{∂_j}. Note that negating a row of a matrix does not change its Hadamard character. As it is pointed out in [5], every generalized coboundary matrix $\bar{M}_{\partial_j}$ contains exactly two negative entries in each row $s \neq 1$, which are located at positions (s, i) and (s, e), for $g_e = g_s^{-1} g_i$. We will work with generalized coboundary matrices from now on.

A set $\{\bar{M}_{\partial_{i_j}} : 1 \leq j \leq w\}$ of generalized coboundary matrices defines an *i-walk* if these matrices may be ordered in a sequence $(\bar{M}_{l_1}, \ldots, \bar{M}_{l_w})$ so that consecutive matrices share exactly one negative entry at the ith-row. Such a walk is called an *i-path* if the initial and final matrices do not share a common -1, and an *i-cycle* otherwise. As it is pointed out in [5], every set of generalized coboundary matrices may be uniquely partitioned into disjoint maximal *i-walks*.

A characterization of Hadamard rows may be easily described attending to *i*-paths.

Proposition 1. *[5] The ith row of a cocyclic matrix $M = M_{\partial_{i_1}} \ldots M_{\partial_{i_w}} \cdot R$ is a Hadamard row if and only if*

$$2c_i - 2I_i = 2t - r_i \tag{2}$$

where c_i denotes the number of maximal i-paths in $\{\bar{M}_{\partial_{i_1}}, \ldots, \bar{M}_{\partial_{i_w}}\}$, r_i counts the number of -1s in the ith-row of R and I_i indicates the number of positions in which R and $\bar{M}_{\partial_{i_1}} \ldots \bar{M}_{\partial_{i_w}}$ share a common -1 in their i^{th}-row.

From now on, we will refer to the positions in which R and $\bar{M}_{\partial_{i_1}} \ldots \bar{M}_{\partial_{i_w}}$ share a common -1 in a given row simply as *intersections*, for brevity.

Equation (2) is the heart of the guided heuristic procedure for reproduction which is applied in the genetic algorithm described in this paper.

3   The Algorithm

The genetic algorithm described in [2] and implemented in [3] is based upon the natural evolution principles of Holland's [10]:

- The population consists of a subset of $4t$ cocyclic matrices M_f over G, $M_f = (f(g_i, g_j))$, which are identified to a binary tuple, the coordinates $(f_1, \ldots, f_s)$ of the 2-cocycle f with regards to the basis $\mathcal{B}$. Accordingly, the coordinates f_i are the genes of the individual f.
- The evaluation function counts the number of Hadamard rows in M_f: the more Hadamard rows M_f posses, the fittest M_f is. In particular, an individual *ind* gives rise to a cocyclic Hadamard matrix if and only if evaluate(*ind*)$= 4t - 1$.
- Crossover combines the features of two parent chromosomes to form two similar offspring by swapping corresponding segments of the parents.
- Mutation arbitrarily alters just one gene of a selected individual (the mutation rate is fixed in 1%).

In the reproduction process, the individuals of the population are paired at random, so that the application of the crossover operator gives rise to another $4t$ individuals, which are added to the population. The generation $i + 1$ is formed from generation i by choosing the $4t$ fittest individuals after the reproduction process.

We now propose a different approach. Instead of the usual crossover and mutation operators described above, we shall better use another heuristic for reproduction. With probability p_r[1], an individual M_f randomly selected from the population gives rise to $4t - 1$ children, so that the $(i + 1)^{th}$-row of the ith-child is Hadamard. Otherwise the usual crossover operator is used, applied over two individuals randomly selected. Generation P_{w+1} is obtained from generation P_w keeping the fittest individuals and replacing a set of less fit individuals with the children just constructed, so that a population of $8t$ individuals is formed. In this process duplicate copies of the same individual are not permitted.

Consequently, the blinded processes of crossover and mutation are now substituted by a completely oriented procedure for reproduction: this way it is guaranteed that anytime an individual exists such that its ith-row is Hadamard.

In order to generate these children, the genes of M_f have to be modified so that equation (2) is satisfied. It is remarkable that the magnitudes c_i and I_i depends heavily on the subset of 2-coboundaries which gives rise to M_f. On the contrary, the magnitude r_i depends only on the representative 2-cocycles implicated in the generation of M_f.

Attending to these facts, a heuristic procedure for reproduction may be straightforwardly defined in the following way. The key idea is to modify the genes of M_f corresponding to 2-coboundaries in such a manner that the magnitudes c_i and I_i are also modified in turn, so that the difference $2c_i - 2I_i$ is closer to the constant value $2t - r_i$.

Depending on whether $2c_i - 2I_i > 2t - r_i$ or $2c_i - 2I_i < 2t - r_i$, we need to increase or decrease I_i (resp. decrease or increase c_i) so that the equality may hold. More concretely:

1. If $2c_i - 2I_i > 2t - r_i$, the algorithm randomly chooses one of the following possibilities:

   - Collapses two different i-paths into just one i-path, so that c_i decreases 1 unit.
   - Introduces a new negative sharing position between R and the product of M_{∂_j}, so that I_i increases 1 unit.

2. If $2c_i - 2I_i < 2t - r_i$, the algorithm randomly chooses one of the following possibilities:

   - Splits one i-path into two different i-paths, so that c_i increases 1 unit.
   - Adds a new i-path, introducing a new 2-coboundary generator, so that c_i increases 1 unit.
   - Eliminates a negative sharing position between R and the product of M_{∂_j}, so that I_i decreases 1 unit.

The way in which these procedures have to be implemented depends on the group G over which 2-cocycles are considered. In the following section we will

---

[1] Experimental results show that a good value for the parameter p_r is 0.8.

explicitly show a pseudo-code of the particular heuristic procedure for reproduction in the case of dihedral groups.

The population is expected to evolve generation through generation until an optimum individual (i.e. a cocyclic Hadamard matrix) is located. This has been the case in the examples showed in the last section.

We include now a pseudo-code of the algorithm.

```
Input: a group (G, ·) of order |G| = 4t
Output: some (eventually one) cocyclic Hadamard matrices over G


\\ the initial population is created
pob ← ∅
fit ← ∅
for i from 1 to 8t{
    ind ← create_new()
    pob ← pob ∪ {ind}
    fit ← fit ∪ {evaluate(ind)}
}
p_r ← 0.8
while (max(fit) < 4t − 1){
\\ reproduction starts
    if random(0, 1) ≤ p_r then{
        j ←random(1, 8t)
        ind_j ← the j^{th}-individual of pob
        list ← guidedreproduction(ind_j)
    else
        i ←random(1, 8t)
        j ←random(1, 8t) ≠ i
        (ind_i, ind_j) ← the (i^{th}, j^{th})-individuals of pob
        list ← usualreproduction(ind_i, ind_j)
    }
    remove in (pob, fit) those entries corresponding to the less
    size(list) fit individuals
    for i from 1 to size(list){
        pob ← pob ∪ {list(i)}
        fit ← fit ∪ {evaluate(list(i))}
    }
}
List the individuals in pob meeting the optimal fitness, 4t − 1
```

Some auxiliar functions have been used, which we describe now:

– create_new() outputs a binary tuple of length s (s being the dimension of the basis $\mathcal{B}$ of 2-cocycles over G), each bit randomly generated as 0 or 1 with the same probability. A deeper knowledge about the properties of the group G might lead to improved versions of this procedure. As a matter of fact, in the case of dihedral groups, the number of 1s should be forced to $2t$, as the tables in [5] suggest, since the density of cocyclic Hadamard matrices seems to be maximum with this rate of 1s.

- evaluate(ind) measures the fitness of the individual ind, that is, counts the number of the Hadamard rows (i.e. those whose summation is zero) in the cocyclic matrix generated by the pointwise product of the matrices related to the 2-cocycles of $\mathcal{B}$ corresponding to the 1s in ind. In particular, an individual ind gives rise to a cocyclic Hadamard matrix if and only if evaluate(ind)= $4t-1$.
- random(min, max) outputs a integer in the range $[min, max]$ randomly generated.
- guidedreproduction(ind) applies the heuristic procedure for reproduction on the individual ind. The output consists in $4t-1$ new individuals, the $(i+1)^{th}$-row of the ith-individual being Hadamard.
- usualreproduction(ind_i, ind_j) applies the usual crossover operator for reproduction on the individuals ind_i and ind_j. The output consists in 2 new individuals.

4   Guided Reproduction on Dihedral Groups

Denote by D_{4t} the dihedral group $\mathbb{Z}_{2t} \times_\chi \mathbb{Z}_2$ of order $4t$, $t \geq 1$, given by the presentation

$$< a, b | a^{2t} = b^2 = (ab)^2 = 1 >$$

and ordering

$$\{1 = (0,0), a = (1,0), \ldots, a^{2t-1} = (2t-1,0), b = (0,1), \ldots, a^{2t-1}b = (2t-1,1)\}$$

In [8] a representative 2-cocycle f of $[f] \in H^2(D_{4t}, \mathbb{Z}_2) \cong \mathbb{Z}_2^3$ is written interchangeably as a triple (A, B, K), where A and B are the inflation variables and K is the transgression variable. All variables take values ± 1. Explicitly,

$$f(a^i, a^j b^k) = \begin{cases} A^{ij}, & i+j < 2t, \\ A^{ij}K, & i+j \geq 2t, \end{cases} \qquad f(a^i b, a^j b^k) = \begin{cases} A^{ij}B^k, & i \geq j, \\ A^{ij}B^k K, & i < j, \end{cases}$$

Let β_1, β_2 and γ denote the representative 2-cocycles related to $(A, B, K) = (-1,1,1), (1,-1,1), (1,1,-1)$ respectively.

A basis for 2-coboundaries is described in [5], and consists of the elementary coboundaries $\{\partial_a, \ldots, \partial_{a^{2t-3}b}\}$. This way, a basis for 2-cocycles over D_{4t} is given by $\mathcal{B} = \{\partial_a, \ldots, \partial_{a^{2t-3}b}, \beta_1, \beta_2, \gamma\}$.

We focus in the case $(A, B, K) = (1, -1, -1)$ (that is, $R = \beta_2\gamma$), since computational results in [8,5] suggest that this case contains a large density of cocyclic Hadamard matrices.

Furthermore, as it is pointed out in Theorem 2 of [5], cocyclic matrices over D_{4t} using R are Hadamard matrices if and only if rows from 2 to t are Hadamard. We have updated the genetic algorithm in turn, so that only rows from 2 to t

are used in order to check whether their summations are zero. Accordingly, the fitness of an individual runs through the range $[0, t-1]$.

In order to define the heuristic procedure for reproduction we need to know how the 2-coboundaries in $\mathcal{B}$ have to be combined to form i-paths, $2 \leq i \leq t$. This information is given in Proposition 7 of [5].

Proposition 2. *[5] For $1 \leq i \leq 2t$, a maximal i-walk consists of a maximal subset in*

$$(M_{\partial_1}, \ldots, M_{\partial_{2t}}) \quad or \quad (M_{\partial_{2t+1}}, \ldots, M_{\partial_{4t}})$$

formed from matrices $(\ldots, M_j, M_k, \ldots)$ which are cyclically separated in $i-1$ positions (that is $j \pm (i-1) \equiv k \bmod 2t$).

We now have enough information about how to combine 2-coboundaries in $\mathcal{B}$ in order to modify the value of $2c_i - 2I_i$, so that $2c_i - 2I_i = 2t - r_i$, that is, the ith-row of our individual being Hadamard.

Notice that since $r_i = 2(i-1)$ for $2 \leq i \leq t$, the cocyclic Hadamard test reduces to $c_i - I_i = t - i + 1$, for $2 \leq i \leq t$.

We include below a pseudo-code of the guidedreproduction procedure described in the section before, particularized to the case of dihedral groups.

```
Input: an individual ind of the population
Output: a list newpob of 4t − 1 individuals, the (i + 1)th-row of the
ith-individual being Hadamard
```

$newpob \leftarrow \emptyset$
for i from 2 to $t\{$
    $ipaths \leftarrow$ list with the maximal i-paths naturally related to ind
    $c \leftarrow$ size of $ipaths$
    $intersec \leftarrow$ intersecting positions of -1s in the ith-row of ind
    $I \leftarrow$ size of $intersec$
    while $c - I \neq t - i + 1\{$
       if $c - I > t - i + 1\{$
          $ind \leftarrow$ decrease$(ipaths, intersec, i - 1, random(1, 2))$
       else$\{$
          $ind \leftarrow$ increase$(ipaths, intersec, i - 1, random(1, 3))$
       $\}$
       recompute the values $ipaths$, c, $intersec$ and I related to ind
    $\}$
    $newpob \leftarrow newpob \cup \{ind\}$
$\}$
$newpob$

Some auxiliar functions have been used, which we describe now:

- decrease($ipaths, intersec, i - 1, j$) tries to decrease the value $c - I$, that is, size($ipaths$)−size($intersec$). This function acts in a different way, depending on the value of $1 \leq j \leq 2$:

  - decrease($ipaths, intersec, i-1, 1$) outputs an individual ind with exactly size($ipaths$) − 1 i-paths. More concretely, it extends one of the i-paths (say p_1, randomly selected) in $ipath$ to the left, until this i-path is connected to a previously existent i-path, say p_2. There are two possibilities now: if $p_1 \neq p_2$, then p_1 and p_2 have been merged into a solely path. On the contrary, if $p_1 = p_2$, then p_1 has been extended to form a i-cycle. In both cases, we have effectively generated a new individual consisting of size($ipaths$) − 1 i-paths.
  - decrease($ipaths, intersec, i - 1, 2$) outputs an individual ind with exactly size($intersec$) + 1 intersections. It suffices to randomly choose a 2-coboundary sharing a negative entry with R in the ith-row, in case that it exists. Otherwise the function
    $$decrease(ipaths, intersec, i - 1, 1)$$
    should be called.

- increase($ipaths, intersec, i - 1, j$) tries to increase the value $c - I$, that is, size($ipaths$)−size($intersec$). This function acts in a different way, depending on the value of $1 \leq j \leq 3$:

  - increase($ipaths, intersec, i - 1, 1$) tries to increase the number of the i-paths in $ipaths$, by splitting an existent i-path into two different i-paths. This is only possible for i-paths consisting of at least three 2-coboundaries. If it is the case, it suffices to delete any 2-coboundary different from the extremes of the i-path. If not, the function
    $$increase(ipaths, intersec, i - 1, 1 + random(1, 2))$$
    is called.
  - increase($ipaths, intersec, i - 1, 2$) tries to increase the number of the i-paths in $ipaths$, by adding a new i-path in $ipaths$ which does not extend any of the previously existent i-paths. This is only possible if a 2-coboundary exists such that it is not adjacent to any of the i-paths in $ipaths$. If it is not the case, the function
    $$increase(ipaths, intersec, i - 1, 2 + (-1)^{random(1,2)})$$
    is called.
  - increase($ipaths, intersec, i - 1, 3$) tries to create an individual ind with size($intersec$)−1 intersections. It suffices to randomly delete a 2-coboundary sharing a negative entry with R in the ith-row, in case that it exists. Otherwise the function
    $$increase(ipaths, intersec, i - 1, random(1, 2))$$
    is called.

5   Examples and Further Work

All the calculations of this section have been worked out in MATHEMATICA 4.0, running on a *Pentium IV 2.400 Mhz DIMM DDR266 512 MB*.

The table below shows some cocyclic Hadamard matrices over D_{4t} (understood as the pointwise linear combinations of the corresponding 2-cocycles of the basis $\mathcal{B}$ described in the preceding section), and the number of iterations and time required (in seconds) as well. Notice that the number of generations is not directly related to the size of the matrices, because of the randomness inherent in any genetic algorithm.

t	iter.	time	product of generators of 2-cocycles over D_{4t}
2	0	0''	$(0,1,0,0,0,0,1,1)$
3	0	0''	$(0,1,1,1,0,0,1,0,0,0,1,1)$
4	0	0''	$(0,1,0,1,1,1,0,0,1,1,1,0,0,0,1,1)$
5	1	0.2''	$(1,1,0,0,1,1,1,1,0,0,1,0,0,0,0,1,0,0,1,1)$
6	1	0.4''	$(1,0,1,0,1,1,1,0,0,0,1,0,1,1,0,1,1,0,0,0,0,1,1)$
7	2	2.87''	$(0,1,1,1,0,0,0,0,0,1,0,1,0,0,0,1,0,0,1,0,0,0,1,0,0,0,1,1)$
8	2	4.54''	$(0,1,0,0,1,1,1,0,0,1,0,0,0,0,0,1,1,1,0,1,0,1,1,0,0,0,1,0,1,$ $0,1,1)$
9	11	51.2''	$(1,1,0,1,1,1,1,0,1,1,1,1,0,1,0,0,1,0,0,0,1,1,0,1,0,1,1,1,0,$ $1,0,0,0,0,1,1)$
10	8	65.11''	$(1,0,0,0,1,0,0,1,1,0,1,1,1,1,0,0,1,0,1,0,0,1,0,1,0,1,0,0,0,$ $0,0,0,0,1,1,1,0,0,1,1)$
11	93	21'	$(1,1,0,1,1,0,1,1,0,1,0,1,0,1,0,0,0,0,0,1,1,1,1,1,0,1,1,0,1,$ $1,1,0,0,1,1,1,0,1,1,1,1,1,0,1,1)$
12	44	18'44''	$(0,0,1,1,0,1,1,0,0,1,0,1,1,1,0,0,0,0,0,0,0,0,1,0,0,1,1,1,$ $0,0,1,0,0,1,0,1,0,1,0,1,0,0,1,0,0,1,1)$
13	40	22'12''	$(1,1,1,0,1,1,1,1,1,1,0,0,0,1,1,0,1,1,1,1,0,0,1,0,1,0,0,1,0,$ $0,1,1,1,1,1,0,1,1,0,1,0,1,0,0,0,1,0,1,0,0,1,1)$

We have experimented 20 runs for each odd value of $3 \leq t \leq 13$ and for each of the values of the parameter p_r over the range $p_r = i/10$, $0 \leq i \leq 10$, which we can not reproduce here due to the page constraints. All of them found some cocyclic Hadamard matrices. Experimentally, the average time and the average number of required iterations suggest that the optimum value for p_r is 0.8.

Unfortunately, the algorithm has not been able to find cocyclic Hadamard matrices for $t > 13$ due to memory limitations: the computer breaks as soon as 5 hours (or equivalently 3000 generations) are reached.

The authors are convinced that improved versions of the algorithm are still to be implemented (for instance, as soon as a method for simultaneously generating a group of Hadamard rows is described). The work in [1] supports this idea.

References

1. Álvarez, V., Armario, J.A., Frau, M.D., Gudiel, F., Osuna, A.: Rooted trees searching for cocyclic Hadamard matrices over D_{4t}. In: AAECC18 Proceedings. LNCS. Springer, Heidelberg (to appear, 2009)
2. Álvarez, V., Armario, J.A., Frau, M.D., Real, P.: A genetic algorithm for cocyclic Hadamard matrices. In: Fossorier, M.P.C., Imai, H., Lin, S., Poli, A. (eds.) AAECC 2006. LNCS, vol. 3857, pp. 144–153. Springer, Heidelberg (2006)
3. Álvarez, V., Armario, J.A., Frau, M.D., Real, P.: Calculating cocyclic Hadamard matrices in *Mathematica*: exhaustive and heuristic searches. In: Iglesias, A., Takayama, N. (eds.) ICMS 2006. LNCS, vol. 4151, pp. 419–422. Springer, Heidelberg (2006)
4. Álvarez, V., Armario, J.A., Frau, M.D., Real, P.: The homological reduction method for computing cocyclic Hadamard matrices. J. Symb. Comput. (2008) (available online 30-September-2008), doi:10.1016/j.jsc.2007.06.009
5. Álvarez, V., Armario, J.A., Frau, M.D., Real, P.: A system of equations for describing cocyclic Hadamard matrices. Journal of Comb. Des. 16 (4), 276–290 (2008)
6. Baliga, A., Chua, J.: Self-dual codes using image restoration techniques. In: Bozta, S., Sphparlinski, I. (eds.) AAECC 2001. LNCS, vol. 2227, pp. 46–56. Springer, Heidelberg (2001)
7. Flannery, D.L.: Calculation of cocyclic matrices. J. of Pure and Applied Algebra 112, 181–190 (1996)
8. Flannery, D.L.: Cocyclic Hadamard matrices and Hadamard groups are equivalent. J. Algebra 192, 749–779 (1997)
9. Hedayat, A., Wallis, W.D.: Hadamard Matrices and Their Applications. Ann. Stat. 6, 1184–1238 (1978)
10. Holland, J.H.: Adaptation in natural and artificial systems. University of Michigan Press, Ann Arbor (1975)
11. Horadam, K.J.: Hadamard matrices and their applications. Princeton University Press, Princeton (2006)
12. Horadam, K.J., de Launey, W.: Cocyclic development of designs. J. Algebraic Combin. 2(3), 267–290 (1993); Erratum: J. Algebraic Combin. (1), 129 (1994)
13. Horadam, K.J., de Launey, W.: Generation of cocyclic Hadamard matrices. In: Computational algebra and number theory. Math. Appl., vol. 325, pp. 279–290. Kluwer Acad. Publ., Dordrecht (1995)
14. Kotsireas, I.S., Pinheiro, G.: A meta-software system for the discovery of Hadamard matrices. In: HPCS 2005 Proceedings, IEEE Computer Society P2343, pp. 17–23 (2005)
15. MacWilliams, F.J., Sloane, N.J.A.: The theory of error-correcting codes. North Holland, New York (1977)

Tuning of Large-Scale
Linguistic Equation (LE) Models
with Genetic Algorithms

Esko K. Juuso

Control Engineering Laboratory
Department of Process and Environmental Engineering
P.O.Box 4300, FI-90014 University of Oulu, Finland
esko.juuso@oulu.fi
http://ntsat.oulu.fi

Abstract. Evolutionary computing is widely used to tune intelligent systems which incorporate expert knowledge with data. The linguistic equation (LE) approach is an efficient technique for developing truly adaptive, yet understandable, systems for highly complex applications. Process insight is maintained, while data-driven tuning relates the measurements to the operating areas. Genetic algorithms are well suited for LE models based on nonlinear scaling and linear interactions. New parameter definitions have been developed for the scaling functions to handle efficiently the parameter constraints of the monotonously increasing second order polynomials. While identification approaches are used to define the model structures of the dynamic models. Cascade models, effective delays and working point models are also represented with LE models, i.e. the whole system is configured with a set of parameters. Results show that the efficiency of the systems improves considerably after the implementation of simultaneous tuning of all parameters.

Keywords: intelligent models, genetic algorithms, linguistic equations.

1 Introduction

Engineering simulators are usually constructed with phenomenological modelling which is based on a thorough understanding of the system's nature and behaviour represented by a suitable mathematical treatment. Data-driven methodologies have been introduced to steady-state modelling, and they can be extended to dynamic systems by using special structures. The resulting black box models are basically flexible surfaces defined by several parameters. In the control area, these approaches are known under the term system identification. Statistical, fuzzy, neural, and neurofuzzy methods are applied first separately, and later as combined approaches in which cascaded models are based on decomposition. The increased complexity creates additional challenges for development and tuning: the number of parameters increases, and various constraints must be taken into account. These problems cannot be solved with basic regression methods.

M. Kolehmainen et al. (Eds.): ICANNGA 2009, LNCS 5495, pp. 161–170, 2009.
© Springer-Verlag Berlin Heidelberg 2009

The *response surface methodology (RSM)* is used for multiple regression between several variables: each output variable y_i can be obtained from the input variables $x_1, x_2, \ldots, x_n$ by a quantitative quadratic relationship represented with a *multiple input, single output (MISO)* model. Arbitrary nonlinear models can be developed by using appropriate calculated variables as inputs, e.g. semi-physical models of the inputs are important in linear modelling, see [1]. In data-driven approaches, *principal component analysis (PCA)* compresses the data by reducing the number of dimensions with minor loss of information [2].

Fuzzy logic emerged from approximate reasoning, and the connection between fuzzy rule-based systems and expert systems is obvious [3,4]. The most popular neural network architecture is the *multilayer perceptron (MLP)* which is very closely connected to backpropagation learning [5]. A *function expansion*

$$y_i = \sum_{k=1}^{m_f} w_{ik} F_k(\boldsymbol{x}) = \sum_{k=1}^{m_f} w_{ik} \, f(\boldsymbol{\beta}_k \cdot (\boldsymbol{x} - \boldsymbol{\gamma}_k)), \tag{1}$$

with some basis functions $F_k(\boldsymbol{x})$, $k = 1, \ldots, m_f$, provides a flexible way to present several types of black box models [6]. These data-driven approaches can handle fairly complex systems, but insight into the process is difficult to maintain.

Complex data-based models can be tuned with *genetic algorithms (GAs)*: penalty functions can be used to reduce complexity, i.e. the number of neurons, layers, rules, active coefficients etc. The main challenge is the efficient coding of the alternatives, because formulating the constraints for the parameters is difficult. Good solutions can be found for simple fuzzy models since the labels have a clear sequence [7]. This idea can be extended into the nonlinear scaling approach used in *linguistic equation (LE)* systems [8,9]. Fitness can be evaluated efficiently with compact, parameterised LE models. In earlier applications, monotonous scaling functions were forced by penalty functions [10]. This facilitates the detection of constraint violations, but in large scale systems the approach is rather ineffective.

This paper presents a new method of selecting parameters for large-scale linguistic equation (LE). The method has been especially adapted for models using genetic algorithms for tuning.

2   Nonlinear Scaling in Modelling

Membership definitions provide nonlinear mappings from the operation area of the (sub)system to the linguistic values represented inside a real-valued interval $[-2, 2]$, denoted as the *linguistic range*, see [9]. The concept of a feasible range is defined as a trapezoidal membership function (Fig. 1), which is based on the support and core areas defined in the fuzzy set theory [11]. The support area is defined by the minimum and maximum values of the variable, $\min(x_j)$ and $\max(x_j)$, respectively. The value range of x_j is divided into two parts by the central tendency value c_j, and the core area, $[(c_l)_j, (c_h)_j]$, is limited by the central tendency values of the lower and upper part.

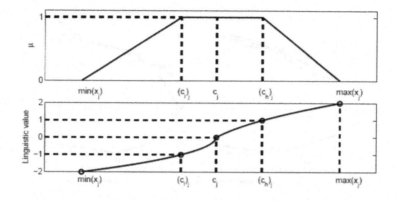

Fig. 1. Feasible range and membership definitions [9]

In current systems, the membership definitions consist of two second order polynomials: one for negative values, $X \in [-2, 0)$, and one for positive values, $X \in [0, 2]$:

$$f_j^- = a_j^- X_j^2 + b_j^- X_j + c_j, \, X_j \in [-2, 0),$$
$$f_j^+ = a_j^+ X_j^2 + b_j^+ X_j + c_j, \, X_j \in [0, 2]. \tag{2}$$

The values X_j are called *linguistic values* because the scaling idea is based on the membership functions of fuzzy set systems. The coefficients of the polynomials are defined by points $\{\min(x_j), -2), ((c_l)_j, -1), (c_j, 0), ((c_h)_j, 1), (\max(x_j), 2\}$. As the membership definitions are used in a continuous form, the functions $f_j^-(X_j)$ and $f_j^+(X_j)$ should be monotonous, increasing functions in order to produce realisable systems. In order to keep the functions monotonous and increasing, the derivatives of the functions f_j^- and f_j^+ should always be positive (Fig. 2).

The inequalities for the core and the support are satisfied with

$$(c_l)_j - \min(x_j) = \alpha_j^- (c_j - (c_l)_j),$$
$$\max(x_j) - (c_h)_j = \alpha_j^+ ((c_h)_j - c_j) \tag{3}$$

if the coefficients α_j^- and α_j^+ are both in the range $\frac{1}{3} \ldots 3$. Corrections are done by changing the borders of the core area, the borders of the support area or the centre point. Additional constraints for derivatives can also be taken into account. The coefficients of the polynomials can be represented by

$$a_j^- = \tfrac{1}{2}(1 - \alpha_j^-) \, \Delta c_j^-,$$
$$b_j^- = \tfrac{1}{2}(3 - \alpha_j^-) \, \Delta c_j^-,$$
$$a_j^+ = \tfrac{1}{2}(\alpha_j^+ - 1) \, \Delta c_j^+,$$
$$b_j^+ = \tfrac{1}{2}(3 - \alpha_j^+) \, \Delta c_j^+, \tag{4}$$

where $\Delta c_j^- = c_j - (c_l)_j$ and $\Delta c_j^+ = (c_h)_j - c_j$. Membership definitions may contain linear parts if some coefficients α_j^- or α_j^+ equals to one (Fig. 2).

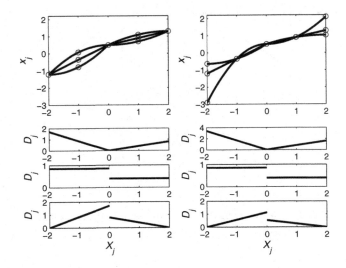

Fig. 2. Feasible shapes of membership definitions f_j and corresponding derivatives D_j: coefficients adjusted with core (left) and support (right). Derivatives are presented in three groups: (1) decreasing and increasing, (2) asymmetric linear, and (3) increasing and decreasing.

The best way to tune the system is to first define the working point and the core $[(c_l)_j, (c_h)_j]$, then the ratios α_j^- and α_j^+ from the range $\frac{1}{3} \dots 3$, and finally to calculate the support $[\min(x_j), \max(x_j)]$. The membership definitions of each variable are configured with five parameters, including the centre point c_j and three consistent sets: the corner points $\{\min(x_j), (c_l)_j, (c_h)_j, \max(x_j)\}$ are good for visualisation; the parameters $\{\alpha_j^-, \Delta c_j^-, \alpha_j^+, \Delta c_j^+\}$ are suitable for tuning; and the coefficients $\{a_j^-, b_j^-, a_j^+, b_j^+\}$ are used in the calculations. The upper and the lower parts of the scaling functions can be convex or concave, independent of each other. Simplified functions can also be used: a linear membership definition only requires two parameters: c_j and $b_j = b_j^+ = b_j^-$ or $\Delta c_j = \Delta c_j^+ = \Delta c_j^-$, since $\alpha_j^+ = \alpha_j^- = 1$ and $a_j^+ = a_j^- = 0$; an asymmetrical linear definition has $\Delta c_j^+ \neq \Delta c_j^-$ and $b_j^+ \neq b_j^-$.

Nonlinear scaling transforms nonlinear models into linear problems [8]. The basic element of a linguistic equation (LE) model is the compact equation

$$\sum_{j=1}^{m} A_{ij} X_j(t - n_j) + B_i = 0, \tag{5}$$

where X_j is a linguistic value of the variable j, $j = 1 \dots m$. The interaction coefficients $A_{ij} \in \Re$ and the bias term $B_i \in \Re$ have a relative meaning, i.e. the equation can be multiplied or divided by any non-zero real number without changing the model. Each variable j has its own time delay n_j compared to the variable with latest time label. Variable time delay is an important factor in many applications, e.g. the time delay in a pipe line depends on the flow velocity. The level of time delay is directly obtained from the level of the variable which affects to the delay. A linguistic equation can be used in any direction.

A LE model with several equations is represented as a matrix equation. In small systems, the directions are usually quite clear. For more complex systems, a set of alternative variable groups is first developed, and all the selected combinations are taken into account in generating the alternatives of the equations. The variable selection and the grouping of variables into 3-5 variable combinations is based on expert knowledge and gathered data [12].

3   Linguistic Equation Models

Nonlinear steady-state models can be constructed with linguistic equations and extended to dynamic systems by dynamic structures. Case-based systems can include both steady state and dynamic models.

3.1   Steady-State LE Models

In LE models, nonlinear scaling is performed twice: first scaling with inverse functions f_j^{-1} from real values to the interval [-2, 2] is performed before applying linguistic equations, and then scaling with functions f_j from the interval [-2, 2] to real values after applying linguistic equations [9]. One variable can be solved from each equation if the other variables with non-zero coefficients are known, i.e.

$$
x_{out} = f_{out} \left(-\frac{1}{A_{i\,out}} \left(\sum_{j=1, j \neq out}^{m} A_{ij}\, f_j^{-1}(x_j) + B_i \right) \right), \tag{6}
$$

where the functions f_j and f_{out} are membership definitions. This is a fairly general expression, e.g. a function expansion (1) is a special case where $\alpha_j = -A_{ij}/A_{i\,out}$, $f_{out} = 1$, and $B_i = 0$. The arguments of variable specific functions f_j^{-1} are defined with location parameters $\gamma_j = c_j$ and dilation parameters β_j. The scaling of LE models is normally asymmetrical in respect to the location c_j.

Each fully nonlinear membership definition requires three parameters more than is needed for normalisation, and each equation is defined by as many parameters as there are variables. LE models with full nonlinear scaling have slightly more parameters than MLP networks if the number of inputs $n_i \geq 3$. General quadratic models have less parameters than these LE models if the number of inputs $n_i \leq 6$. In *LE* models, the number of parameters decreases drastically if linear or asymmetric linear definitions are used for some variables. In cascade models, intermediate variables can be handled as linguistic values, i.e. linguistification is only needed for the original inputs and delinguistification for the final outputs.

3.2 Dynamic LE Models

The basic form of a linguistic equation (LE) model is a static mapping, as in fuzzy set systems and neural networks. Therefore, dynamic models include several inputs and outputs originating from a single variable. Nonlinear scaling reduces the number of input and output signals needed for modelling nonlinear systems. For the default LE model, all the degrees of the polynomials of the *parametric models* become very low. The model

$$Y(t) + a_1 Y(t-1) = b_1 U(t - n_k) + e(t) \tag{7}$$

is a special case of (5) with three variables, $Y(t), Y(t-1)$ and $U(t - n_k)$, in the linguistic range, the interaction matrix $A = [1\ a_1 - b_1]$, and the bias term $B = 0$. The need for higher order models can be tested by applying classical identification with different polynomial degrees to the data after scaling with membership definitions. The membership definitions have captured the real meaning of the variables if higher degrees do not improve the model considerably.

Linear *state-space models* can be used in LE models by including the coefficient matrices A_S, B_S, C_S and D_S in the interaction matrix

$$A = \begin{pmatrix} A_S & B_S & -I_x & 0 \\ C_S & D_S & 0 & -I_y \end{pmatrix} \tag{8}$$

to handle the state variables, $\boldsymbol{X}(t)$, and the inputs, $\boldsymbol{U}(t)$, in the linguistic range. The identity matrices I_X and I_Y introduce variables $\boldsymbol{X}(t+1)$ and $\boldsymbol{Y}(t)$ into the model. The bias term $B = 0$. Equations can be used sequentially to solve one variable per equation. Through nonlinear scaling LE models can combine several linear local models. In addition, gradual changes can be taken into account in membership definitions and interaction coefficients.

The membership definition of the variable y does not depend on time; the bias term $B = 0$. An alternative approach is to also create membership definitions for the change Δy and to obtain the derivative directly from the corresponding LE model. The output, the derivative of the variable y, is integrated with numerical integration methods. In addition, step size control is often needed to adapt the simulation to changing operating conditions, since LE models are developed for wide operating areas. An appropriate handling of *time delays* extends the operating area of the model considerably. Delays should be assessed against process knowledge, especially if normal on-line process data is used [8]. Effective delays depend on working conditions (process case). In LE models, this extension requires two to five additional parameters per modelled delay.

3.3 Case-Based LE Models

An effective approach to the data-based modelling of complex nonlinear systems is to partition the data set into subsets and to approximate each subset with a special LE model. Decomposition can be based on expert knowledge or data-based clustering methods applied on original data, scaled data, or principal

components obtained from the data or the scaled data. Linguistic fuzzy clustering or linguistic neural networks based on nonlinear scaling detect clusters of
different *geometrical shapes*, and thus considerably reduce the number of submodels needed [13]. Takagi-Sugeno fuzzy models [14] can be extended linguistic
TS models, where the local linear models are LE models. The *clustering variables* are not necessarily used as inputs in the case models. Each case may even
have a completely different set of variables.

Case models are normal LE models: each equation has two to five variables,
membership definitions of which can be partly general and partly case specific.
In small dynamic models, a single equation includes all the interactions, i.e. variables affecting the working point of the model are included in the model. The
equation systems of larger models are sets of equations in which each equation
describes an interaction between two to five variables. Case models and fuzzy
reasoning are used for the detection of operating conditions, and hybrid systems can be constructed by introducing fuzzy decisions to selected submodels.
Transitions between the submodels are smooth if the LE models are nonlinear.

3.4 Tuning LE Models

Structural restrictions of the LE models are beneficial for tuning and adaptation.
Manual tuning consists two independent parts: scaling and interactions. Both
can be defined manually but the constraints must be taken into account when
defining the scaling functions. Neural tuning reduces the error between the model
and the training data [8]: membership definitions are tuned for one variable of the
equation at a time with a linear neural network. The centre point c_j is obtained
by mean or median, and the linear network is used to obtain the coefficients a_j^-,
b_j^-, a_j^+ and b_j^+. For matrix equations, only one variable from each equation can
be selected for tuning. Recursive implementations of the regression or the linear
network make on-line adaptation possible.

4 Genetic Tuning

Genetic algorithms can handle the whole system simultaneously by comparing
the effects of the parameters defined by the scaling functions, interactions, time
delays, and the weights of the models. The fitness values of the models are
obtained through simulation. Genetic tuning is controlled by population size,
number of bits, and probabilities of crossover and mutation.

4.1 Parameters

The main part of LE modelling is to tune the parameters of the scaling functions. The centre points c_j are the real values corresponding to the normal (or the
origin); the other parameters of the membership definitions are defined by the
differences $\{c_j - (c_l)_j, (c_l)_j - \min(x_j), (c_h)_j - c_j, \max(x_j) - (c_h)_j\}$. Penalty functions are needed to ensure monotonously increasing functions, i.e. inequalities

Table 1. Ranges for the parameters of the scaling functions

Parameter	Lower limit	Upper limit
c_j	$\frac{1}{4}(3\overline{x_j} + min(x_j))$	$\frac{1}{4}(3\overline{x_j} + max(x_j))$
Δc_j^-	$\frac{1}{4}(\overline{x_j} - min(x_j))$	$\frac{3}{4}(\overline{x_j} - max(x_j))$
Δc_j^+	$\frac{1}{4}(max(x_j) - \overline{x_j})$	$\frac{3}{4}(max(x_j) - \overline{x_j})$

(3) are filled. No penalties are needed for the parameters $\{c_j, \Delta c_j^-, \Delta c_j^+, \alpha_j^-, \alpha_j^-\}$ if the coefficients α_j^- and α_j^+ are restricted to the range $\frac{1}{3}\dots 3$. For asymmetric linear functions, only three parameters are tuned since $\alpha_j^- = \alpha_j^+ = 1$. For linear definitions, the number of parameters is reduced to two because $\Delta c_j^- = \Delta c_j^+$. In dynamic models, the definitions of variables are time invariant.

Interaction coefficients A_{ij} and bias terms B_i can be included in genetic tuning the same way as the parameters of membership definitions. Variable time delays are handled by introducing additional parameters corresponding to the membership definitions of the time delays. Genetic tuning can be extended into multimodel systems by introducing corresponding parameters. For example, working point models contain scaling functions, interactions, and effective time delays. In comparision to single models, cascade models also introduce additional parameters.

4.2   Coding and Population

All the parameters are handled in the same way by genetic algorithms. In current applications, real numbers are represented with binary coding, i.e. the population of the solution is represented with chromosomes described as bits. Different parts of the bit sequences define different parameters. The coded real values are then used in simulation to obtain the fitness of the model. The initial population is constructed by using limited value ranges: the ranges of the parameters used in the scaling functions are shown in Table 1, and all the interaction coefficients and bias terms have the same default range $[-1, 1]$. Outliers need to be removed when defining the minimum and maximum values. Expert knowledge can be used to modify the ranges. It is also possible to concentrate the initial population close to the corner points obtained from data analysis (Fig. 1).

4.3   Fitness

The fitness values are obtained by simulating the alternatives defined by the parameters. The fitness of each equation i, $i = 1,\dots,n$, is evaluated within the linguistic range with the residual. To keep the solutions comparable, the combined coefficient vector $(A_{i1} A_{i2} \dots A_{im} B_i)$ is normalised. Additional selection rules are needed for the case-based systems aimed for the detection of the operating conditions.

The objective function consists of a weighted average of several performance measures and penalty terms. Since the residuals have a linguistic meaning, all the measures can be scaled into the linguistic range [−2, 2]: the mean absolute error (MAE) is already in the range [0, 2], and the mean square error (MSE) can also be scaled. The correlation coefficients and R^2 values provide similar fitness measures when scaled to the range [−2, 2] but the scaling functions can be chosen in a more flexible way. Maximum error and relative error are scaled into the range [0, 2]. Comparison of the calculated values and the measurements in the scaled range provides two additional performance measures: the slope and the base of the linear model. To avoid overfitting, penalties are defined for the complexity of the model, e.g. number of inputs, equations and submodels.

5   Results

The genetic tuning approach has been successfully tested in various applications. The dynamic model of a solar collector field consists of four submodels which all have the same ARX structure: the outlet temperature as a state variable, irradiation and oil flow as input variables, and effective time delay, which depends on the oil flow [15]. A working point model, which relates the difference of the outlet and inlet temperatures to the irradiation and ambient temperature, provides smooth transitions between the submodels: start-up, low, normal, and high operation. Also directions of these interactions are known for all the models. The full nonlinear model with 105 parameters was tuned and tested with process data obtained during four test campaigns. The data of 40 days collected with a sample time of five seconds was divided into train and test data. The dynamic models consists of three interactive submodels in two applications: temperature, humidity, and granular size in fluidised bed granulation [16]; alkali, lignin and dissolved solids in a batch cooking process [17]. The ARX structure with 41 parameters was used in the granulation models, but in the cooking models the changes in outputs were calculated using LE models containing 43 parameters.

Genetic tuning is controlled by population size, number of bits, and probabilities of crossover and mutation. A good solution has high correlation and R^2 values, and small values for errors, base values, and penalties, and a slope value close to one. The number of parameters is large making the real value coded GAs an interesting area of future research.

6   Conclusions

The LE approach provides compact models in which all the parameters have clear constraints, and the whole system can be assessed with expert knowledge. This advantage becomes increasingly important in cascade and interactive models. Genetic algorithms are well suited to LE systems: the tuning results improved considerably when the new parameter definitions were used and all the parameters were handled simultaneously.

References

1. Ljung, L.: System Identification - Theory for the User, 2nd edn. Prentice Hall, Upper Saddle River (1999)
2. Jolliffe, I.T.: Principal Component Analysis, 2nd edn., p. 487. Springer, New York (2002)
3. Dubois, D., Prade, H., Ughetto, L.: Fuzzy logic, control engineering and artificial intelligence. In: Verbruggen, H.B., Zimmermann, H.-J., Babuska, R. (eds.) Fuzzy Algorithms for Control, International Series in Intelligent Technologies, pp. 17–57. Kluwer, Boston (1999)
4. Driankov, D., Hellendoorn, H., Reinfrank, M.: An Introduction to Fuzzy Control. Springer, Berlin (1993)
5. Rummelhart, D.E., Hinton, G.E., Williams, R.J.: Learning internal representations by error propagation. In: Rummelhart, D.E., McClelland, J. (eds.) Parallel Data Processing, pp. 318–362. MIT Press, Cambridge (1986)
6. Ljung, L.: Perspectives on system identification. In: Chung, M.J., Misra, P. (eds.) Plenary papers, milestone reports & selected survey papers, 17th IFAC World Congress, Seoul, Korea, July 6-11, pp. 47–59. IFAC (2008)
7. Lotvonen, S., Kivikunnas, S., Juuso, E.: Tuning of a fuzzy system with genetic algorithms and linguistic equations. In: 5th European Congress on Intelligent Techniques and Soft Computing -EUFIT 1997, vol. 2, pp. 1289–1293. Mainz, Aachen (1997)
8. Juuso, E.K.: Fuzzy control in process industry: The linguistic equation approach. In: Verbruggen, H.B., Zimmermann, H.-J., Babuska, R. (eds.) Fuzzy Algorithms for Control, International Series in Intelligent Technologies, pp. 243–300. Kluwer, Boston (1999)
9. Juuso, E.K.: Integration of intelligent systems in development of smart adaptive systems. International Journal of Approximate Reasoning 35, 307–337 (2004)
10. Juuso, E.K.: Modelling and simulation in development and tuning of intelligent controllers. In: 5th Vienna Symposium on Mathematical Modelling, Vienna, Austria, Argesim Report 30, February 8-10, pp. 6–10. Argesin Verlag, Vienna (2006)
11. Zimmermann, H.-J.: Fuzzy set theory and its applications. Kluwer Academic Publishers, Dordrecht (1992)
12. Ahola, T., Juuso, E., Leiviskä, K.: Variable selection and grouping in a paper machine application. International Journal of Computers, Communications & Control 2(2), 111–120 (2007)
13. Juuso, E.K.: Linguistic equations for data analysis: FuzzEqu toolbox. In: TOOLMET 2000 Symposium - Tool Environments and Development Methods for Intelligent Systems, pp. 212–226. Oulun yliopistopaino, Oulu (2000)
14. Takagi, T., Sugeno, M.: Fuzzy identification of systems and its applications to modeling and control. IEEE Trans. Syst., Man, & Cybern. 15(1), 116–132 (1985)
15. Juuso, E.K.: Intelligent dynamic simulation of a solar collector field. In: Simulation in Industry, 15th European Simulation Symposium ESS 2003, pp. 443–449. SCS, Gruner Druck (2003)
16. Juuso, E.K.: Intelligent modelling of a fluidised bed granulator used in production of pharmaceuticals. In: SIMS 2007 - The 48th Scandinavian Conference on Simulation and Modeling, pp. 101–108. Linköping University Electronic Press, Linköping (2007)
17. Juuso, E.K.: Forecasting batch cooking results with intelligent dynamic simulation. In: 6th EUROSIM Congress on Modelling and Simulation, vol. 2, p. 8. University of Ljubljana, Ljubljana (2007)

Elitistic Evolution: An Efficient Heuristic for Global Optimization

Francisco Viveros Jiménez[1], Efrén Mezura-Montes[2], and Alexander Gelbukh[3]

[1] Universidad del Istmo Campus Ixtepec, Cd. Ixtepec, Oaxaca, México
pacovj@hotmail.com
[2] Laboratorio Nacional de Informática Avanzada, Xalapa, Veracruz, México
emezura@lania.mx
[3] Centro de Investigación en Computación del Instituto Politécnico Nacional,
D.F. México
gelbukh@gelbukh.com

Abstract. A new evolutionary algorithm, Elitistic Evolution (termed EEv), is proposed in this paper. EEv is an evolutionary method for numerical optimization with adaptive behavior. EEv uses small populations (smaller than 10 individuals). It have an adaptive parameter to adjust the balance between global exploration and local exploitation. Elitism have great influence in EEv' proccess and that influence is also controlled by the adaptive parameter. EEv' crossover operator allows a recently generated offspring individual to be parent of other offspring individuals of its generation. It requires the configuration of two user parameters (many state-of-the-art approaches uses at least three). EEv is tested solving a set of 16 benchmark functions and then compared with Differential Evolution and also with some well-known Memetic Algorithms to show its efficiency. Finally, EEv is tested solving a set of 10 benchmark functions with very high dimensionality (50, 100 and 200 dimensions) to show its robustness.

1 Introduction

The global optimization problem is unsolved because of its high level of complexity. Many different scopes were generated in recent years. Evolutionary algorithms is one of the many different approaches that have been used in recent years. Evolutionary algorithms (**EAs**) are stochastic techniques based on the Darwinian principle of the *survival of the fittest*. Most EAs offer important advantages such robustness, reliability, global search capability and low information requirement. Some representative techniques of EAs are: Genetic Algorithms [5], Evolutive Strategies [1] and Differential Evolution [2].

EAs are population-based techniques. Each individual of the population represent a candidate solution (commonly conformed at least by 30 individuals). EAs generate offspring using a crossover operator. Normally, some population individuals are selected to be parents of the offspring individuals. Then, sligth alterations to some individuals are performed with a mutation operator. Finally,

M. Kolehmainen et al. (Eds.): ICANNGA 2009, LNCS 5495, pp. 171–182, 2009.
© Springer-Verlag Berlin Heidelberg 2009

old solutions are replaced with some specific new solutions. Normally, EAs use the number of generations as termination condition. A generation represent the proccess of creating new solutions and replacing the old ones. Evaluations of the objetive function (**FEs**) are performed in each iteration. Most of state-of-the-art technique use at least three user parameters.

This paper describes a new EA called Elitistic Evolution (**EEv**). EEv has the following major differences with a traditional EA:

1. Uses small populations (smaller than 10 individuals).
2. Uses an adaptive parameter to adjust the balance between global exploration and local exploitation in all the processes.
3. Elitism have great influence in EEv' proccess and that influence is also controlled by the adaptive parameter.
4. Its mutation operator generate the offspring.
5. It uses a new crossover operator. EEv' crossover operator allows a recently generated offspring individual to be parent of other offspring individuals of its generation.
6. It uses two replacement politics balanced by the adaptive parameter.
7. It requires the configuration of two user parameters.

The adaptive nature of EEv allows the algorithm to have precision and high convergency speed, even in complex problems with very high dimensionality ($N > 50$). The performance and features of this technique are shown by testing it in a set of 16 well-known functions taken from the literature [4,8]. The paper also presents a performance comparison with DE/rand/1/bin and some well-known Memetic Algorithms. However, it is important to remember the No Free Lunch Theorems for Search [3]. No Free Lunch Theorems state that an algorithm which performs well on some test functions, will not necessarily be competitive in a different set of problems.

The sections are organized as follows. The following section contains a detailed description of EEv. Section 3 contains the experimental design and results. Section 4 concludes this paper.

2   Elitistic Evolution

EEv is a population-based stochastic optimizer. The two main foundations of EEv are:

1. **Elitism.** The best individual have special considerations in each EEv stage the elite individual have special considerations.
2. **Adaptive Behavior.** An adaptive parameter control the behavior of all stages of an EEv iterations. The two main behaviors affected by this parameter are: the effect of elitism and the balance of global vs local exploration. The step size used by the mutation operator on each dimension is also adaptive.

Algorithm 1 describes an EEv iteration. The following differences are observed between EEv and a common EA:

1. Each X^g individual became a parent.
2. Mutation is performed before crossover.
3. Crossover replaces offspring instead of generating new individuals.
4. Mutation, crossover and replacement are affected by an adaptive parameter called Enviromental Pressure (C).

Algorithm 1. Algorithm for any g iteration of EEv

---
1   Recalculation of adaptive parameters;
2   Copy each X^g individual in n^g;
3   Perform $mutation_{rs}(n^g, C, \overrightarrow{b})$ to each n^g individual;
4   Replace each n^g with a new n^g created by $crossover(X^g, n^g, C)$;
5   Evaluate the new n^g individuals;
6   **for** *the first C individuals of X^{g+1}* **do**
7     $\lfloor$ Use the C best individuals of the union of X^g and n^g;
8   **for** *the remaining $P - C$ individuals of X^{g+1}* **do**
9     $\lfloor$ Select $P - C$ random individuals from n^g;

EEv needs the adjustment of two parameters which are:

- Population size (P) which tells how many individuals will form the population. $P > 2$.
- Base mutation (B) used as the initial step size for the mutation operator. $B \in [0.0, 1.0]$.

A large P or B value increase the diversity of possible solutions, promoting the global exploration of the search space, and, consequently, demoting the exploitation capacity. The balance between P and B is crucial for the efficiency of the algorithm.

EEv uses two adaptive parameters which are:

- $\overrightarrow{b}$, the N step sizes for random step mutation operator, where N is the number of decision variables in the problem. $\overrightarrow{b}$ have an initial value of B in each N dimension.
- C. Indicates the number of individuals to be affected by local exploration processes. It adjust the balance between global and local search. $C \in [1, P]$. Lower C values promote global exploration; higher C values promote local exploitation. C value depends on the success of the search of the better solution. C have an initial value of 1. The specific effects of C are explained in the description of each stage below.

2.1   Recalculation of Parameters

C and $\overrightarrow{b}$ values change depending on the search success of the last generation, determined by the comparison of the best fitness values of the current and
previous generations. Algorithm 2 shows the recalculation for C and $\overrightarrow{b}$. We
take $F(X_{best}^0) = F(X_{best}^1)$, where X_{best}^g is the elite individual. If better results
are found, then C is decreased, encouraging global exploration, and $\overrightarrow{b}$ values
are adaptively increased to provide a more precise exploration. Otherwise, C is
increased and $\overrightarrow{b}$ values are decreased to encourage local exploitation.

Algorithm 2. Recalculation of adaptive parameters

---

1 **if** $F(X_{best}^{g-1}) > F(X_{best}^g)$ **then**
2   **if** $C > 1$ **then**
3          $C = C - 1$;
4      $\overrightarrow{b} = |X_{best}^{g-1} - X_{best}^g/(upbound-lowbound)| \times (1.0 + rnd(0.0, 1.0) \times (G-g)/G)$;
5   **if** a $\overrightarrow{b}$ *value is equal to 0* **then**
6   Replace it with $B \times (1.0 - rnd(0.0, 1.0) \times g/G)$;

7 **else**
8   **if** $C < P$ **then**
9          $C = C + 1$;
10     $\overrightarrow{b} = \overrightarrow{b} \times (1.0 - rnd(0.0, 1.0) \times g/G)$;

2.2   Mutation

Mutation operator is very similar to Hill-climbing method. This operator generate the offspring individuals by performing a random number of sligth alterations
in some dimensions of each X^g individual. The steps stored in $\overrightarrow{b}$ represent the
maximum percent of search space that the mutation can perform. Algorithm 3
describes the mutation operator.

EEv mutation operator is based on the mutation technique proposed in [6].
It generates a new individual close to the original point. Due to the adaptive
behavior of EEv we have to consider the following facts:

1. The first C mutations represent local exploitation and the last $P - C$ mu-
   tationts represent global exploraitions. That is because the first C mutations are affected in an sligther way due to the calculus of the M ratio (see
   equation 1 below).
2. The operator tends to perform smaller changes at later stages of the process
   due to decreasing of the values $\overrightarrow{b}$.
3. The number of alterations depends on C value. .

$$M = \begin{cases} \overrightarrow{b_k} & \text{if } i \leq C \\ B \times (1.0 - rnd(0.0, 1.0) \times g/G) & \text{if } i > C \end{cases} \qquad (1)$$

Algorithm 3. Mutation algorithm for an i individual

---

1 $alterations = rnd(\lceil nv \times (C/P) \rceil, N)$;
2 **for** *all alterations* **do**
3 $\quad$ $k = rnd(1, N)$;
4 $\quad$ Calculate M with equation 1 ;
5 $\quad$ **repeat**
6 $\quad\quad$ | $aux = n_{k,i}^{g} + (upbound_k - lowbound_k) \times rnd(-M, M)$;
7 $\quad$ **until** $(aux > lowerbound_k)$ and $(aux < upperbound_k)$;
8 $\quad$ $n_{k,i}^{g} = aux$;

Algorithm 4. Crossover operator algorithm

---

1 **for** *each n_i^{g} offspring individual* **do**
2 $\quad$ $k = rnd(0, P - C) + 1$;
3 $\quad$ $l = rnd(0, P - C) + 1$;
4 $\quad$ $m = rnd(1, P)$;
5 $\quad$ $c_1 = rnd(0.0, 1.0)$;
6 $\quad$ $c_2 = rnd(0.0, 1.0 - c_1)$;
7 $\quad$ $c_3 = 1.0 - c_2 - c_1$;
8 $\quad$ $n_i^{g} = c_1 \times n_k^{g} + c_2 \times X_l^{g} + c_3 \times n_m^{g}$;

2.3 Reproduction

The reproduction stage creates the final offspring individuals. Algorithm 4 illustrates the crossover operator. This crossover operator uses two offspring individuals, k and m, and one population individual l. We have to consider the following facts:

- The crossover operator allows each offspring individual to have a chance of being selected as a parent of the following alterations to offspring individuals.
- Each offspring individual will be always at a point between its three parents.
- Any X^{g} and n^{g} individual have the posibility of remain intact.
- The crossover can take into account only two individuals: two n^{g} individuals or one individual from X^{g} and n^{g}.

C value controls the selection of l and k: when $C = 1$, any individual can be selected; when $C = P$, elitism is ensured, promoting exploration around the best individual. Evaluation of the offspring individuals is performed at the end of this stage.

2.4 Replacement

EEv uses two replacement processes: non-generational replacement, borrowed from $(\mu + \lambda)$-ES [1], and random generational replacement. The first replacement method provides convergence capabilities, while the second one provides exploration capabilities.

As in the previous stages, the balance between these two processes is controlled by C: the first C individuals of the new generation population are selected by non-generational replacement and the remaining ones are select randomly from the offspring individuals. Elitism is ensured by non-generational replacement.

The first C individuals of the new generation are ordered by their fitness values, and EEv replacement ensures that the first population individual always be the elite; no search for the elite is required at any stage.

3   Experimental Setup

We followed the experimental setup similar to [7] with 6 tests:

1. Evaluation of performance;
2. Study of the adaptive behavior;
3. Evaluation of sensitivity to population size;
4. Evaluation of sensitivity to the B parameter;
5. Study of scalability;
6. Comparison with MAs.

The benchmark functions are specified in table 1. The functions f were taken from [4], and functions F, from the test suite for CEC 2005 Special Session on real-parameter optimization [8]. In each test, we conducted 50 trials with each function. All the experiments were performed using a Pentium 4 PC with 512 MB of RAM, in C Linux environment.

3.1   Evaluation of Performance

It was based on the test proposed in [8]. The *function error value* ε for a solution x is $f(x) - f(x^*)$, where x^* is the global optimum of f [7]. The maximum number **MAX** of evaluations of f was $10,000\,N$, where N is the dimension of the problem. The fitness evaluation criteria were as follows:

1. **Error** is a compound value formed by all trials as $AVG(\vec{\varepsilon}) \pm STDDEV(\vec{\varepsilon})$.
2. **Evaluation** is another compound value formed by all trials as $AVG(\vec{\alpha}) \pm STDDEV(\vec{\alpha})(\beta)$, where α is the number of function evaluations (**FEs**) required to reach certain error value before MAX value and β is the number of trials with an α value. For all the functions, this value was 10^{-6} except for F_8, F_9, and F_{10}, where the accuracy was 10^{-2}.

We compared the performance of EEv and DE/rand/1/bin. The tests were conducted on a set of sixteen functions with $N = 30$. Error and evaluation values are reported on Table 2. Best results are marked in boldface. We used $P = 5$ and $B = 0.6$ in this test. The *error* and *evaluation* values for DE/rand/1/bin were calculated using a $P = N$, $CR = 0.9$ and $F = 0.9$ as suggested in [7].

Table 1. Test functions

Unimodal functions
Separable ───────────────────────────────
f_{sph} Sphere model
F_1 Shifted sphere function
Non-separable ───────────────────────────
F_3 Shifted rotated high conditioned elliptic function
F_4 Shifted Schwefel's problem 1.2 w/ noise in fitness
Multimodal functions
Separable ───────────────────────────────
f_{sch} Generalized Schefel's problem 2.26
f_{ras} Generalized Rastrigin's function
F_9 Shifted Rastrigin's function
Non-separable ───────────────────────────
f_{ros} Generalized Rosenbrock's function
f_{ack} Ackley's function
f_{grw} Generalized Griewank's function
f_{sal} Salomon's function
f_{whi} Whitley's function
$f_{pen1,2}$ Generalized penalized functions
F_8 Shifted rotated Ackley's function with global optimum on bounds
F_{10} Shifted Rotated Rastrigin's function

We observe that EEv:

- Has fast convergence: the reported evaluation values are small.
- Is consistent: the standard deviation is relatively small as compared with the mean value on all the test problems.
- Is competitive: it overperforms DE/rand/1/bin on nine out of sixteen functions.
- Is faster than DE/rand/1/bin. It reaches the accuracy value on ten out of sixteen functions using less FEs than DE/rand/1/bin.

EEv' success key features are:

1. Adaptive restart mechanism. EEv' restart mechanism allows partial population restart, accelerating convergence.
2. Mutation operator. The operator is very similar to a Hill-Climbing operator, giving EEv a similar behavior to a memetic algorithm. The adaptive step size vector allows a more efficient exploration for the current search situation.
3. Use of recently crossover' generated offspring individuals.

3.2   Study of the Adaptive Behavior

Experiments with $P = 5$ and $B = 0.6$ were conducted to evaluate the adaptive behavior of EEv. The tests were conducted on a set of sixteen functions with

Table 2. Comparison between EEv and DE/rand/1/bin error and evaluations values on problems with $N = 30$

Error	EEv	DE/Rand/1/Bin
f_{sph}	**4.35E − 22 ± 8.37E − 22**	$5.73E - 17 \pm 2.03E - 16$
F_1	$1.67E - 12 \pm 4.37E - 12$	**3.58E − 81 ± 1.36E − 81**
F_3	**9.21E + 05 ± 9.22E + 04**	$3.63E + 06 \pm 9.22E + 05$
F_4	$6.80E + 04 \pm 1.62E + 04$	**5.54E + 01 ± 6.37E + 01**
f_{ras}	**4.32E − 12 ± 1.34E − 11**	$2.55E + 01 \pm 8.14E + 00$
F_9	**1.59E − 01 ± 2.59E − 01**	$2.43E + 01 \pm 6.22E + 00$
f_{sch}	$1.67E + 03 \pm 6.41E + 01$	**4.90E + 02 ± 2.34E + 02**
f_{ros}	**1.42E + 01 ± 7.28E + 00**	$5.20E + 01 \pm 8.56E + 01$
f_{whit}	**2.33E + 01 ± 1.94E + 01**	$3.10E + 02 \pm 1.07E + 02$
f_{pen1}	**5.82E − 25 ± 4.80E − 25**	$4.56E - 02 \pm 1.31E - 01$
F_8	**2.02E + 01 ± 2.65E − 02**	$2.09E + 01 \pm 1.33E - 01$
f_{pen2}	**9.01E − 24 ± 1.12E − 23**	$1.44E - 01 \pm 7.19E - 01$
f_{grw}	$3.08E - 02 \pm 1.57E - 02$	**2.66E − 03 ± 5.73E − 03**
f_{ack}	$1.04E - 07 \pm 1.50E - 07$	**1.70E − 09 ± 1.32E − 09**
f_{sal}	$1.04E + 00 \pm 6.58E - 01$	**2.52E − 01 ± 8.14E + 00**
F_{10}	$3.34E + 02 \pm 3.24E + 01$	**7.33E + 01 ± 6.66E + 01**

Evaluations		
f_{sph}	**93827.6 ± 1157.87(50)**	$148650.8 \pm 6977.7(50)$
F_1	**97640.80 ± 1356.82(50)**	$153450.1 \pm 5780.4(50)$
f_{ras}	**111151.8 ± 7367.1(50)**	−
F_9	**115550.8 ± 11845.2(43)**	−
f_{pen1}	**67299.5 ± 1279.93(50)**	$160955.2 \pm 63176.3(43)$
f_{pen2}	**78541.6 ± 1529.95(50)**	$156016.9 \pm 31515.8(48)$
f_{ras}	**200743.1 ± 31447.5(8)**	−
f_{ros}	**127230(1)**	−
f_{ack}	**164930.0 ± 15637.1(49)**	$215456.1 \pm 9721.4(\mathbf{50})$
f_{grw}	**104585.31 ± 17771.17(16)**	$190292.5 \pm 63478.8(\mathbf{38})$

$N = 30$. Frequencies and mean values of C parameters were taken from a random test and reported in Table 3. The following observations can be obtained from the experiments:

- Elitism is predominant in EEv's optimization process. A statistic mode of P for C parameter is present on all the problems. Additionally, the mean value of C on the test is equal to 4.94 (very close to P).
- C parameter is sensitive to the kind of search space EEv is working with. Each problem needs a different amount of global and local exploration.

3.3 Sensitivity to Population Size

Experiments with different P values were conducted to evaluate the sensitivity of EEv to variations of population size. The tests were conducted on a set of ten

Table 3. C frequencies for a random test run with $P = 5$, $B = 0.6$ and $N = 30$

	f_{sph}	F_1	F_3	F_4	F_9	f_{ras}	F_{sch}	f_{ros}	f_{ack}	f_{grw}
1	15	11	18	2	4	4	3	11	11	15
$P-3$	84	71	247	4	46	46	27	128	61	61
$P-2$	365	281	1366	4	289	315	161	598	306	301
$P-1$	2258	1457	7094	13	1760	1960	1399	3556	1506	1550
P	57278	58180	51275	59977	57901	57675	58410	55707	58116	58073
μ	4.94	4.96	4.82	4.99	4.95	4.95	4.96	4.91	4.96	4.96

functions with $N = 30$. The P values used were 3, 10, and 30. Table 4 shows the error values obtained in the test with 10 benchmark functions.

Performance of EEv is sensitive to population size in all the cases. EEv works best with very small populations. EEv works better with smaller P values in 7 out of 10 test cases. It is important to observe that:

– A smaller P value increments the probability of elitism.
– Smaller P values amplify the effect of the C parameter: major changes in the proportion of individuals involved in exploitation and exploration occur between generations.

3.4 Sensitivity to the B Parameter

Experiments with different B values were conducted to evaluate the sensitivity of EEv to variations of B parameter. The tests were conducted on a set of ten functions with $N = 30$. The B values used were 0.1, 0.45, and 0.8.

B controls the initial and maximum step size for $\overrightarrow{b}$ – the adaptive parameter that contains the step size to be used by the mutation operator. The mutation operator is very similar to a hill-climbing algorithm. It allows the generation of offspring individuals around a parent with a maximum step size specified by $\overrightarrow{b}$. Table 4 shows the error values obtained by using different B values in the test with sixteen benchmark functions. That table shows that:

– Performance in EEv is more sensitive to the B parameter.
– B value is problem-dependent. However, EEv tends to have similar behavior on problems with similar features.
– Greater B values encourage a more optimal exploration.
– EEv performs competitively without fine adjustment of B.

3.5 Scalability Test

Experiments with different N dimensions values were conducted to evaluate the robustness of EEv. The tests were conducted on a set of ten functions. The N values used were 100, and 200. The error values obtained are reported in Table 5.

Table 4. EEv Error values obtained for different P and B values on problems with $N = 30$

$B = 0.6$	$P = 3$	$P = 10$	$P = 30$
f_{sph}	$\mathbf{6.15E-27 \pm 8.24E-27}$	$4.50E-18 \pm 5.51E-18$	$1.68E-12 \pm 1.69E-12$
F_1	$\mathbf{1.90E-14 \pm 1.02E-14}$	$1.40E-10 \pm 2.56E-10$	$3.23E-07 \pm 1.69E-06$
F_3	$\mathbf{7.37E+05 \pm 8.16E+04}$	$1.30E+06 \pm 1.31E+05$	$3.08E+06 \pm 1.88E+05$
F_4	$\mathbf{3.96E+04 \pm 1.04E+04}$	$7.14E+04 \pm 1.46E+04$	$4.08E+04 \pm 2.69E+03$
f_{ras}	$6.63E-02 \pm 1.69E-01$	$\mathbf{2.91E-08 \pm 8.33E-08}$	$3.32E-02 \pm 1.75E-01$
f_{sch}	$\mathbf{1.61E+03 \pm 9.52E+01}$	$1.68E+03 \pm 1.24E+02$	$1.66E+03 \pm 1.02E+02$
f_{grw}	$4.03E-02 \pm 2.30E-02$	$2.69E-02 \pm 1.79E-02$	$\mathbf{1.00E-02 \pm 5.79E-03}$
f_{pen1}	$\mathbf{1.77E-29 \pm 1.24E-29}$	$2.11E-19 \pm 5.45E-19$	$1.33E-10 \pm 6.18E-10$
f_{ack}	$\mathbf{2.24E-10 \pm 1.60E-10}$	$2.41E-03 \pm 6.97E-03$	$5.13E-01 \pm 2.93E-01$
f_{ros}	$1.52E+01 \pm 9.89E+00$	$1.76E+01 \pm 9.64E+00$	$\mathbf{7.23E+00 \pm 1.10E+01}$
$P = 5$	$B = 0.1$	$B = 0.45$	$B = 0.8$
f_{sph}	$\mathbf{1.53E-23 \pm 7.16E-23}$	$1.11E-22 \pm 9.36E-23$	$2.46E-22 \pm 2.97E-22$
F_1	$\mathbf{1.03E-12 \pm 2.03E-12}$	$2.17E-12 \pm 8.33E-12$	$2.12E-12 \pm 8.33E-12$
F_3	$9.47E+05 \pm 1.52E+05$	$\mathbf{9.00E+05 \pm 7.33E+04}$	$9.27E+05 \pm 1.29E+05$
F_4	$1.57E+05 \pm 5.38E+04$	$8.16E+04 \pm 2.26E+04$	$\mathbf{5.94E+04 \pm 6.83E+03}$
f_{ras}	$3.67E+01 \pm 4.22E+00$	$3.31E-02 \pm 1.75E-01$	$\mathbf{1.70E-12 \pm 2.51E-12}$
f_{sch}	$5.67E+03 \pm 2.42E+02$	$1.83E+03 \pm 1.40E+02$	$\mathbf{1.61E+03 \pm 9.44E+01}$
f_{pen1}	$\mathbf{1.74E-25 \pm 7.85E-25}$	$1.31E-24 \pm 2.71E-24$	$9.81E-25 \pm 4.31E-25$
f_{ack}	$1.13E-07 \pm 2.43E-07$	$1.80E-07 \pm 4.28E-07$	$\mathbf{9.21E-08 \pm 1.13E-07}$
f_{ros}	$2.08E+01 \pm 5.19E+01$	$3.87E+01 \pm 2.29E+01$	$\mathbf{1.46E+01 \pm 2.02E+01}$

This table also show a comparison between EEv, DE/rand/1/bin, and DEahc-SPX. DEahcSPX is a Memetic Algorithm which overcomes DE/rand/1/bin performance and was proposed in [7]. The error values from this technique were taken also from [7]. We obtain the following conclusions:

- EEv is a robust algorithm.
- EEv mantains its performance on problems with very high dimensionality.
- EEv is highly competitive on problems with $N \geq 100$. It overperforms the other two techniques on all test problems.

3.6   Comparison with Memetic Algorithms

A comparison with two well-knowm memetic algorithms was conducted. We select two techniques: minimal generation gap (**MGG**) [9] and generalized generation gap (**G3**) [10]. Two crossover operators were selected for this paper: unimodal normal distribution crossover (**UNDX**) and parent centric crossover (**PCX**). The selected combinations were MGG+UNDX and G3+PCX. The tests were conducted on a set of ten functions with $N = 30$. Error values are reported in Table 6. Best results are marked in boldface. The *error* values for both algorithms were taken from [7]. We reach the following conclusions:

Table 5. Comparison between EEv, DE/rand/1/bin and DEahcSPX error values in ten problems with $N = 100$ and $N = 200$

$N = 100$	EEv	DEahcSPX	DE/Rand/1/Bin
F_{sph}	$\mathbf{1.39E-21 \pm 5.66E-22}$	$5.01E+01 \pm 8.94E+01$	$4.28E+03 \pm 1.27E+03$
F_{ras}	$\mathbf{6.635E-02 \pm 1.69E-01}$	$4.75E+02 \pm 6.55E+01$	$8.30E+02 \pm 6.51E+01$
F_{sch}	$\mathbf{5.79E+03 \pm 2.36E+02}$	$2.48E+04 \pm 2.71E+03$	$2.54E+04 \pm 2.15E+03$
F_{ros}	$\mathbf{3.66E+01 \pm 1.77E+01}$	$1.45E+05 \pm 1.11E+05$	$3.33E+08 \pm 1.67E+08$
F_{ack}	$\mathbf{9.30E-09 \pm 1.12E-08}$	$1.91E+00 \pm 3.44E-01$	$8.81E+00 \pm 8.07E-01$
F_{sal}	$\mathbf{2.39E+00 \pm 2.39E-01}$	$3.11E+00 \pm 5.79E-01$	$1.02E+01 \pm 7.91E-01$
F_{whit}	$\mathbf{5.67E+02 \pm 1.95E+02}$	$4.06E+10 \pm 6.57E+10$	$5.44E+15 \pm 5.07E+15$
F_{pen1}	$\mathbf{1.13E-23 \pm 7.33E-24}$	$4.33E+00 \pm 1.75E+00$	$6.20E+06 \pm 7.38E+05$
$N = 200$	EEv	DEahcSPX	DE/Rand/1/Bin
F_{sph}	$\mathbf{6.58E-21 \pm 1.87E-21}$	$7.01E+03 \pm 1.07E+03$	$1.26E+05 \pm 1.06E+04$
F_{ras}	$\mathbf{3.64E-01 \pm 2.96E-01}$	$1.53E+03 \pm 8.31E+01$	$2.37E+03 \pm 7.24E+01$
F_{sch}	$\mathbf{1.18E+04 \pm 3.71E+02}$	$6.61E+04 \pm 1.44E+03$	$6.66E+04 \pm 1.32E+03$
F_{ros}	$\mathbf{3.61E+01 \pm 2.91E+01}$	$1.11E+08 \pm 2.63E+07$	$2.97E+10 \pm 3.81E+09$
F_{ack}	$\mathbf{3.02E-09 \pm 7.34E-10}$	$8.45E+00 \pm 4.13E-01$	$1.81E+01 \pm 2.26E-01$
F_{sal}	$\mathbf{4.54E+00 \pm 2.09E-01}$	$1.10E+01 \pm 4.38E-01$	$3.69E+01 \pm 1.80E+00$
F_{whit}	$\mathbf{4.09E+03 \pm 1.14E+03}$	$4.21E+13 \pm 1.74E+13$	$3.13E+18 \pm 9.48E+17$
F_{pen1}	$\mathbf{5.98E-23 \pm 1.26E-23}$	$2.27E+01 \pm 5.73E+00$	$3.49E+08 \pm 7.60E+07$

Table 6. Comparison between EEv error values and two MAs in problems with N=30

	EEv	MGG+UNDX	G3+PCX
F_{sph}	$4.35E-22 \pm 8.37E-22$	$1.37E-11 \pm 1.94E-11$	$\mathbf{3.58E-81 \pm 1.36E-81}$
F_{ras}	$\mathbf{4.32E-12 \pm 1.34E-11}$	$1.35E+00 \pm 1.03E+00$	$1.75E+02 \pm 3.37E+01$
F_{sch}	$\mathbf{1.67E+03 \pm 6.41E+01}$	$4.12E+03 \pm 1.72E+03$	$4.04E+03 \pm 1.09E+03$
F_{whit}	$\mathbf{2.33E+01 \pm 1.94E+01}$	$4.28E+02 \pm 3.82E+01$	$3.44E+02 \pm 2.97E+00$
F_{pen1}	$\mathbf{5.82E-25 \pm 4.80E-25}$	$4.93E-02 \pm 3.50E-02$	$4.35E+00 \pm 6.94E+00$
F_{ack}	$\mathbf{1.04E-07 \pm 1.50E-07}$	$8.23E-07 \pm 4.64E-07$	$1.48E+01 \pm 4.17E+00$
F_{pen2}	$\mathbf{9.01E-24 \pm 1.12E-23}$	$4.39E-04 \pm 2.20E-03$	$1.50E+01 \pm 1.58E+01$
F_{sal}	$1.04E+00 \pm 6.58E-01$	$\mathbf{1.50E-01 \pm 4.95E-02}$	$4.64E+00 \pm 4.74E+00$
F_{grw}	$3.08E-02 \pm 1.57E-02$	$\mathbf{2.96E-04 \pm 1.48E-03}$	$1.07E-02 \pm 1.30E-02$
F_{ros}	$1.42E+01 \pm 7.28E+00$	$2.81E+01 \pm 1.23E+01$	$\mathbf{4.18E+00 \pm 9.68E+01}$

1. EEv is a competitive technique. It overperforms the other two techniques on six out of ten functions.
2. Uses one less parameter than the other two techniques: MGG and G3 require the adjustment of P, μ, and λ.
3. The algorithm is simplier than the two memetic algorithms.

4   Conclusions and Future Work

The paper describes a new evolutionary method called EEv. EEv is a population-based technique that uses only two user-defined parameters (one less than the majority of the state-of-the-art techniques). The adaptive parameter C allows EEv to reach an effective balance between local exploitation and global exploration.

The tests show that EEv is a competitive approach: it performs well in most test cases and have great efficiency on problems with high dimensionality. In addition to that, it is very fast and precise in some cases: it reach the target accuracy value faster than DE/rand/1/bin. However, it has difficulty solving unimodal non-separable functions. We conclude that EEv performs better with small populations EEV is sensitive to B parameter value. B value depends on the problem features.

More comparative work and further studies should be carried out to provide a more detailed analysis and refinement. Future work with constrained functions should be performed to observe the behavior of EEv. New mechanisms should be tested to improve EEv's performance on shifted functions.

References

1. Beyer, H.-G., Schwefel, H.-P.: Evolution strategies: a comprehensive introduction. Natural Computing 1(1) (2002)
2. Storn, R., Price, K.: Differential Evolution - a simple and efficient heuristic for global optimization. Journal of Global Optimization 11(4) (1997)
3. Wolpert, D.H., Macready, W.G.: No free lunch theorems for optimization. IEEE Trans. on Evolutionary Computation (1997)
4. Mezura-Montes, E., Coello, C.C.A., Velazquez, R.J.: A comparative study of differential evolution variants for global optimization. In: Proceedings of the 8th annual conference on Genetic and evolutionary computation (2006)
5. Goldberg, D.E.: Genetic Algorithms in Search, Optimization and Machine Learning. Addison Wesley, Reading (1989)
6. Viveros, J.F.: DSE: An Hybrid Evolutionary Algorithm with Mathematical Search Method. Special issue journal Research in Computing Science (2008)
7. Noman, N., Iba, H.: Accelerating Differential Evolution Using an Adaptive Local Search. IEEE Transactions on Evol. Comput. 12(1) (2008)
8. Suganthan, P.N., Hansen, N., Liang, J.J., Deb, K., Chen, Y.-P., Auger, A., Tiwari, S.: Problem Definitions and Evaluation Criteria for the CEC 2005 Special Session on Real-Parameter Optimization. Nanyang Technol. Univ., Singaporem IIT Kanpur, India, KanGal Rep. 2005005 (2005)
9. Satoh, H., Yamamura, M., Kobayashi, S.: Minimal generation gap models for GAs considering both exploration and exploitation. In: Proc. IIZUKA 1996 (1996)
10. Deb, K., Anand, A., Joshi, D.: A computationally efficient evolutionary algorithm for real-parameter optimization. Evol. Comput. 10(4) (2002)

Solving the Multiple Sequence Alignment Problem Using Prototype Optimization with Evolved Improvement Steps

Jiří Kubalík

Department of Cybernetics
Czech Technical University in Prague
Technická 2, 166 27 Prague 6, Czech Republic
kubalik@labe.felk.cvut.cz

Abstract. This paper deals with a Multiple Sequence Alignment problem, for which an implementation of the Prototype Optimization with Evolved Improvement Steps (POEMS) algorithm has been proposed. The key feature of the POEMS is that it takes some initial solution, which is then iteratively improved by means of what we call evolved hypermutations. In this work, the POEMS is seeded with a solution provided by the Clustal X algorithm. Major result of the presented experiments was that the proposed POEMS implementation performs significantly better than the other two compared algorithms, which rely on random hypermutations only. Based on the carried out analyses we proposed two modifications of the POEMS algorithm that might further improve its performance.

Keywords: sequence alignment, optimization, evolutionary algorithms.

1 Introduction

Sequence alignment (SA) plays an important role in molecular sequence analysis. SA is a way of arranging the primary sequences of DNA, RNA, or protein to identify regions of similarity that may be a consequence of functional, structural, or evolutionary relationships between the sequences. All of those 3 types of sequences are referred to as domains. In case of more than two sequences the problem is called Multiple Sequence Alignment (MSA).

The sequence of a DNA molecule can be modelled as a string over a 4-character alphabet, each character representing one of the four nucleotides that make up DNA. RNA can be modelled in a similar way. Proteins are chains of amino acids and can be represented as strings over a 20-character alphabet. Both the amino acids and the nucleotides are commonly referred to as residues. Gaps are inserted between the amino acid residues so that residues with identical or similar characters are aligned in successive columns. Aligned sequences of nucleotide or amino acid residues are typically represented as rows of letters within a matrix. In this paper we focused onto alignment of proteins.

The MSA problem is strongly NP-hard [10] thus there does not exist any exact algorithm that solves larger instances of this problem optimally. Hence,

M. Kolehmainen et al. (Eds.): ICANNGA 2009, LNCS 5495, pp. 183–192, 2009.
© Springer-Verlag Berlin Heidelberg 2009

heuristic algorithms are an option for finding at least approximate solutions to the optimal ones. The widely-used representative of this class of algorithms is Clustal algorithm. It gradually builds an alignment by first estimating the evolutionary distance between all sequences to be aligned and then aligns the sequences in order of decreasing similarity using dynamic programming method. Although the dynamic programming is a globally optimal algorithm, the Clustal itself does not guarantee finding the global optimum as it optimizes the alignment in a pair-wise manner taking just two sequences into account at a time.

Very often metaheuristics such as evolutionary algorithms (EAs) were adapted to this problem to find solutions of high quality in reasonable time. The best known work of evolutionary algorithms used for MSA problem was introduced by Notredame and Higgins in [6] in 1996. Their Sequence Alignment by Genetic Algorithm (SAGA) follows the general principles of a standard genetic algorithm. The main focus of SAGA was put onto evolutionary operators – crossover and mutation. There are in total 22 operators implemented in SAGA. In Horng's et al. implementation of an evolutionary algorithm for protein alignment before dynamic programming is applied onto sequences they are "prepared" using evolutionary algorithm with 1-point crossover and four mutation operators [2]. GASP proposed in [5] uses segment profiles to generate the diversified initial population and tries to prevent the destruction of conserved regions by crossover and mutation operations. Additionally, the highest-scoring individual of each generation is optimized using progressive method. Thomsen et al. proposed a simple evolutionary algorithm called MSAEA in [9]. Instead of evolving the alignments from randomly generated initial alignments, they investigated the effects of seeding an EA with already well-fit solution. The MSAEA starts with a population of identical seed alignments derived from the well-known alignment program Clustal X [7]. MSAEA was then used to further improve the initial solution. Five mutation operators were used to alter the candidate alignment solutions during the evolutionary run, for details see [9]. They also experimented with different crossover operators and concluded that the recombination operators were not able to improve the candidate solutions.

In this paper we propose an implementation of the POEMS algorithm for solving the MSA problem. POEMS is an iterative optimization algorithm that searches in each iteration for such a modification of current solution, called prototype, that improves its quality. If an improving modification is found then the modified prototype is considered a prototype for the subsequent iteration. Modifications are represented as sequences of elementary actions (simple mutations in standard EAs), defined specifically for the problem at hand. An evolutionary algorithm is used to search for the best action sequences. Thus, the transition steps taken between two subsequent prototype states can be viewed as *evolved hypermutations*. First results show, that POEMS performs significantly better than evolutionary algorithms that rely on just randomly generated mutations, like the MSAEA does.

The rest of this paper is structured as follows. The considered MSA problem and alignment quality score are described in section 2. In section 3, the adaptation of POEMS to MSA problem is proposed. Section 4 describes the used test data sets and configuration of the POEMS and other two compared algorithms. Results achieved with our approach are analysed in section 5. Section 6 concludes and suggests directions for further analyses and improvements of the proposed approach.

2   Problem Formulation

In this work we assume a set of n sequences $S = S_1, S_2, \ldots, S_n$ to be aligned. Each sequence consists of characters over the alphabet of 20 amino acids plus symbol '-', which stands for a single gap. Originally, the sequences are of different length $l_1, l_2, \ldots, l_n$. We equalized the length of all sequences to the value of l' calculated as $l' = 1.2 \times l_{max}$, where $l_{max} = max(l_1, l_2, \ldots, l_n)$. So, the sequences were filled in with gaps up to the length l'. Initially, all the gaps added to the alignment are placed to the far right hand side of corresponding sequence. The same length alignment scheme was used in [9].

In order to solve the sequence alignment problem, we need a measure to determine how good an alignment is. The most widely-used scoring scheme employs *sum of pairs* (SOP) scoring function together with a *gap penalty* (GP). The SOP score rewards matches and penalizes mismatches between two amino acids at a particular location calculated over all pairs of amino acids $s_{i,k}$ and $s_{j,k}$ in the alignment according to the following formula

$$SOP(S) = \sum_{i=1}^{n-1} \sum_{j=i+1}^{n} \sum_{k=1}^{l'} M(s_{i,k}, s_{j,k}) \qquad (1)$$

where M is a symmetric matrix containing scores for substituting a residue with another one (or a gap). The symmetric matrix is referred to as a *substitution matrix* and is derived from statistical analysis of residue substitution data from sets of reliable alignments of highly related sequences. In this work we consider widely-used BLOSUM62 matrix [1].

We used an *affine gap penalty* (AGP) calculated for each sequence in the alignment and every gap in the sequence according to the following formula:

$$AGP(S) = \gamma + \delta \times (k - 1), \qquad (2)$$

where γ is the gap opening penalty, δ is the gap extension penalty, and k is the length of the gap. Gaps at the sequences terminal regions are treated with no penalty. In this paper we used $\gamma = 10$ and $\delta = 1$.

The optimization task of MSA is then to achieve the alignment by introducing gaps into sequences such that the SOP score of two aligned sequences is maximized and the GP is minimized at the same time. The two objectives are combined in the following final maximization criterion

$$AlignmentScore = SOP - GPS. \qquad (3)$$

3   POEMS

Prototype Optimization with Evolved iMprovement Steps (POEMS) [3] is an iterative optimization approach that employs an EA for finding the best modification of the current solution, called *prototype*, in each iteration. Modifications are represented as a sequence of fixed length of primitive actions defined specifically for the problem at hand. Action sequences are assessed based on how well/badly they modify the current prototype, which is passed as an input parameter to the EA. After the EA finishes, it is checked to determine whether the best evolved sequence improves the current prototype or not. If an improvement is achieved, then the modified prototype is considered as a new prototype for the next iteration. Otherwise the current prototype remains unchanged. The iterative process stops after a specified number of iterations. In other words, the POEMS can be considered as an iterative algorithm with *evolved hypermutations*. An outline of the POEMS algorithm is shown in Algorithm 1.

Prototype initialization. POEMS is an iterative algorithm and as such its performance strongly depends on the initial prototype from which the iterative optimization starts. Here, the prototype is initialized with a solution obtained by Clustal X algorithm, similarly to the seeding of the first population in MSAEA.

Actions. The EA employed in POEMS evolves linear chromosomes of length *MaxGenes*, where each gene represents an instance of a certain action chosen from a set of elementary actions defined for the given problem. Each gene is represented by a record, with an attribute *action_type* followed by parameters of the action. Besides actions that truly modify the prototype, there is also a special type of action called *nop* (no operation). Any action with *action_type = nop* is interpreted as a void action with no effect on the prototype, regardless of the values of its parameters. Chromosomes can contain one or more instances of the *nop* operation. This way a variable effective length of chromosomes is implemented. However, sequences that do not change the prototype at all (they

---

Algorithm 1. Prototype Optimization with Evolved Improvement Steps

```
1  begin
2  |   Prototype ← InitializePrototype()
3  |   i ← 1
4  |   while not TerminationCondition() do
5  |   |   BestSequence ← RunEA(Prototype)
6  |   |   Candidate ← ApplyTo(BestSequence, Prototype)
7  |   |   if IsBetterThan(Candidate, Prototype) then
8  |   |   |   Prototype ← Candidate
9  |   |   i ← i + 1
10 |   return Prototype
11 end
```

Algorithm 2. Iterative Evolutionary Algorithm Used in POEMS

```
1  begin
2  |   Population⁽⁰⁾ ← InitializeActionSequences()
3  |   Evaluate(Population⁽⁰⁾)
4  |   i ← 1
5  |   while not TerminationCondition() do
6  |   |   Parents ← Select(Population⁽ⁱ⁻¹⁾)
7  |   |   if Rand()< P_cross then
8  |   |   |   Children ← Crossover(Parents)
9  |   |   else
10 |   |   |   Children ← Mutation(Parents)
11 |   |   Evaluate(Children)
12 |   |   Replacement ← FindReplacement(Population⁽ⁱ⁻¹⁾)
13 |   |   Population⁽ⁱ⁾ ← Replace(Population⁽ⁱ⁻¹⁾, Replacement, Children)
14 |   |   i ← i + 1
15 |   BestSequence ← BestOf(Population⁽ⁱ⁻¹⁾)
16 |   return BestSequence
17 end
```

Line 1: **begin**
Line 2: $Population^{(0)} \leftarrow$ InitializeActionSequences()
Line 3: Evaluate($Population^{(0)}$)
Line 4: $i \leftarrow 1$
Line 5: **while not** TerminationCondition() **do**
Line 6: $Parents \leftarrow$ Select($Population^{(i-1)}$)
Line 7: **if** Rand()$< P_{cross}$ **then**
Line 8: $Children \leftarrow$ Crossover($Parents$)
Line 9: **else**
Line 10: $Children \leftarrow$ Mutation($Parents$)
Line 11: Evaluate($Children$)
Line 12: $Replacement \leftarrow$ FindReplacement($Population^{(i-1)}$)
Line 13: $Population^{(i)} \leftarrow$ Replace($Population^{(i-1)}$, $Replacement$, $Children$)
Line 14: $i \leftarrow i + 1$
Line 15: $BestSequence \leftarrow$ BestOf($Population^{(i-1)}$)
Line 16: **return** $BestSequence$
Line 17: **end**

---

are composed entirely of *nop* actions) are fatally penalized in order to avoid a convergence to useless trivial modifications.

There are the following four types of *active actions*[1] used in this work:

- PassGaps(*sequenceId, aminoId*). This action is an analogy of the *PassGaps* operator used in [9]. First, it finds in sequence *sequenceId* an amino acid with the particular identifier *aminoId*. If the amino acid is placed directly next to a gap then the amino acid is moved to the other end of the gap.
- MoveBlock(*sequenceId, aminoId, direction*). First, a contiguous block of amino acids that contains given amino acid *aminoId* is found in sequence *sequenceId*. Then the block is moved one position to the left or right according to the value of *direction*.
- InsertGap(*sequenceId, aminoId*). First, amino acid *aminoId* is found in sequence *sequenceId*. Then a single gap is inserted right before the amino acid, if there are still some terminal gaps left at the right hand side end of that sequence.
- RemoveGap(*sequenceId, aminoId*). First, amino acid *aminoId* is found in sequence *sequenceId*. Then, if there is a gap right in front of the amino acid a single gap is removed from the gap and put to the far right hand side of that sequence.

Operators. At the beginning of each EA, the chromosomes (action sequences) are initialized by random. Then the chromosomes are varied by means of crossover

---

[1] Note, that the actions can only operate with gaps that are present in each sequence. No gaps can be either introduced or removed to/from any sequence, since the length of sequences l' is invariant.

and mutation operators. A generalized uniform crossover introduced in [3] and mutation operator that changes either the *action_type* or action parameters are used in this work.

Niching replacement strategy. In order to prevent EA from converging towards action sequences with a minimal number of active actions a niching replacement strategy is proposed for the used iterative EA, see Algorithm 2. A population of size N is split into niches $\{niche_1, niche_2,\ldots,niche_{MaxGenes}\}$ of equal size $N/MaxGenes$ and each niche $niche_i$ can contain only sequences with the number of active actions greater or equal than i. This is achieved so that each newly generated action sequence is inserted into the population only if an admissible replacement is found in the current population by function **FindReplacement()**. An action sequence AS_{old} can be replaced by the new one AS_{new} with i active actions iff $fitness(AS_{new})$ is better than or equal to $fitness(AS_{old})$ and AS_{old} is in $niche_j$ such that $j \leq i$. This way both the quality and diversity of action sequences in the population are ensured.

4   Experimental Setup

4.1   Used Test Data Sets

Table 1 shows the protein sequence data sets that we used in our experiments. All six data sets were selected from the first reference set from the BAliBASE database (version 2, http://bess.u-strasbg.fr/BioInfo/BAliBASE2/) [8], which is a publicly available suite of alignment benchmarks.

4.2   Configuration of Compared Algorithms

Parameters of POEMS were set the same for all data sets as follows

- $MaxGenes = 5$,
- $NicheSize = 20$,
- $PopulationSize = 100$,
- Number of fitness evaluations in each iteration: 1000,
- Number of iterations: 100 (100.000 fitness evaluations in total),
- $P_{Cross} = 0.5$, $P_{Mutate} = 0.5$,
- Tournament selection with $N = 2$.

The proposed POEMS implementation was compared to two other mutation-based algorithms. The first one denoted EA is an analogy of MSAEA algorithm. The evolutionary model is the same as in MSAEA. From a population of λ individuals μ new individuals are created by applying mutations to parental individuals. $(\lambda+\mu)$-selection is then used to select the best λ number of individuals as parents in the new population. Contrary to MSAEA, individuals are not varied by simple mutations. Instead individuals are subjected to hypermutations of the form of action sequences of length 1 to $MaxGenes$. In this paper, EA with $\lambda = 15$ and $\mu = 30$ that run for 3333 generations was considered.

Table 1. BAliBASE data sets used in MSAEA experiments. NSEQ is the number of sequences. LSEQ are length parameters of original sequences. MAL is the maximal alignment length calculated as $MAL = 1.2 \times LClustal$, where $LClustal$ is a length of the alignment obtained by Clustal X for the original sequences. SEQID is the percent residue identity.

Data set	NSEQ	LSEQ (min,max,avg)	MAL	SEQID
1aboA_ref	5	(49,80,63.6)	96	$< 25\%$
1hfh_ref	5	(116,132,121.2)	164	$20 - 40\%$
1amk_ref	5	(242,254,248.2)	308	$> 35\%$
1gtr_ref	5	(419,436,428.2)	539	$> 35\%$
1gpb_ref	5	(796,828,809.8)	1020	$> 35\%$
1taq_ref	5	(806,928,865.2)	1128	$> 35\%$

Second approach denoted BestImprovementSearch (BIS) is the POEMS variant where just the starting population of action sequences is generated in each EA. BIS run for 100 iterations with 1000 random action sequences generated in each iteration. Hence, no evolution of the action sequences took place in BIS.

5   Results

We run 500 independent runs with each of the compared algorithms. Table 2 provides statistics calculated from 500 results produced by each of the compared algorithms. We can see that POEMS achieves the best average and median value on all data sets. Figure 1 illustrates that the median values obtained with POEMS are significantly higher than medians of the other two algorithms at the 5% significance level since the notches in boxplots do not overlap[2]. The differences between the means are also confirmed by Wilcoxon rank-sum test as it rejects the null hypothesis that the means are equal at the 1% significance level for all data sets. Histograms drawn from the top 100 results also show that the best distribution of high quality results is achieved with the POEMS algorithm; the histograms are not presented in the paper due to limited space. Moreover, POEMS finds the best ever solution on all data sets but the 1gpb_ref data set.

If we take a closer look at boxplots for 1amk_ref data set we observe unusual boxplot parameters obtained for EA and POEMS. In both cases the median matches or is very close to the lower bound of a boxplot. Detailed analyses revealed this is due to a large number of copies of the same result. EA got trapped 284 times at solution(s) of alignment score of 6107 while POEMS got stuck 131 times at solution(s) of final alignment score of 6114. We analyzed the results obtained by POEMS on 1amk_ref data set and found out that there were two modifications of one alignment that have the same score of 6114 and differed in just one amino acid position. In particular, there was one sequence in the alignment with a gap and one amino acid that if placed on either side of the gap yield the same final alignment score. Thus, once the POEMS algorithm

---

[2] Though, they are very close to each other in case of the 1aboA_ref data set.

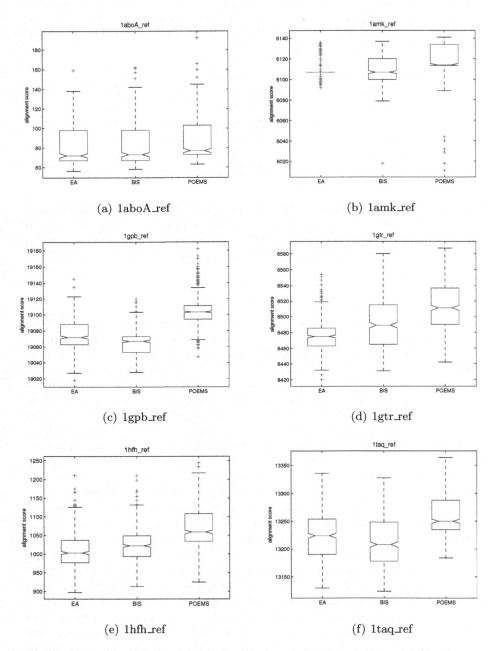

(a) 1aboA_ref

(b) 1amk_ref

(c) 1gpb_ref

(d) 1gtr_ref

(e) 1hfh_ref

(f) 1taq_ref

Fig. 1. Boxplots calculated from the set of 500 results obtained by each algorithm

Table 2. Average, median and standard deviation values calculated from 500 results produced by the compared algorithms

		1aboA_ref	1amk_ref	1gpb_ref	1gtr_ref	1hfh_ref	1taq_ref
Clustal X		-4	5898	18943	8333	688	12981
EA	best	159	6136	19145	8554	1211	13336
	avg	80.0	6108.5	19073.7	8475.1	1011.1	13223.9
	median	72	6107	19072	8475	1003	13224
	stdev	16.7	8.0	18.2	21.0	50.6	41.7
BIS	best	162	6137	19120	8580	1210	13328
	avg	80.1	6108.9	19065.1	8489.9	1025.1	13213.9
	median	73	6107	19067	8489	1022	13208
	stdev	18.1	13.0	14.4	31.1	47.2	41.9
POEMS	best	**192**	**6141**	**19182**	**8587**	**1244**	**13364**
	avg	87.3	6118.8	19105.0	8511.0	1068.3	13261.9
	median	**77**	**6114**	**19103**	**8511**	**1059**	**13250**
	stdev	20.2	15.8	20.7	31.4	150.6	34.3

arrived at one of these two variants of that alignment, it was strongly attracted by the other variant as it causes minimal deterioration in quality of the current prototype among all the hypermutations that effectively change it. This suggests, that POEMS can systematically get trapped under certain circumstances. There are two possible remedies of this problem

- Population based variant of POEMS. A population of fixed size with best-so-far solutions could be maintained such that only unique solutions can be stored there. In each iteration a prototype would be chosen by random. Thus, the possibility of getting stuck between several equally-fit optima would be eliminated.
- Using a tabu list of winning action sequences generated in recent iterations. In the current version of POEMS, action sequences that do not change the prototype are fatally penalized. This modification suggest that also action sequences that appear in the tabu list will be fatally penalized. This way, the chance that solutions would oscillate between several equally-fit optima would be eliminated reduced.

6   Conclusions and Future Work

This paper deals with well-known multiple sequence alignment problem, for which an implementation of the Prototype Optimization with Evolved Improvement Steps algorithm has been proposed. This algorithm is seeded with a solution provided by the Clustal X algorithm and the solution is then further improved by means of what we call *evolved hypermutations*.

We presented promising results that show that the concept of evolved hypermutations is a viable way to utilize EA for problems where it is hard to design an effective recombination operator. We observed that POEMS outperformed two other approaches - an EA with $(\lambda + \mu)$ evolutionary model and the best improvement search - both relying on randomly generated hypermutations.

However, POEMS is an iterative algorithm and as such it can get stuck in a local optimum or can get trapped by a couple of local optima of the same quality. We proposed two modifications of POEMS algorithm that might further improve its performance. These modifications will be subject of our future research.

Acknowledgement

I would like to thank A. Šlachta for helpful discussions. Research described in this paper has been supported by the research program No. MSM 6840770012 "Transdisciplinary Research in Biomedical Engineering II" of the CTU in Prague.

References

1. Henikoff, S., Henikoff, J.G.: Amino acid substitution matrices from protein blocks. Proc. Natl. Acad. Sci. USA 89, 10915–10919 (1992)
2. Horng, J.-T., Wu, L.-C., Lin, C.-M., Yang, B.-H.: A genetic algorithm for multiple sequence alignment. Soft Computing 9(6), 407–420 (2005)
3. Kubalik, J., Faigl, J.: Iterative Prototype Optimisation with Evolved Improvement Steps. In: Collet, P., Tomassini, M., Ebner, M., Gustafson, S., Ekárt, A. (eds.) EuroGP 2006. LNCS, vol. 3905, pp. 154–165. Springer, Heidelberg (2006)
4. Liu, L.-F., Huo, H.-W., Wang, B.-S.: HGA-COFFEE: Aligning Multiple Sequences by Hybrid Genetic Algorithm. In: Li, X., Wang, S., Dong, Z.Y. (eds.) ADMA 2005. LNCS (LNAI), vol. 3584, pp. 464–473. Springer, Heidelberg (2005)
5. Lv, Y., Li, S., Zhou, C., Guo, W., Xu, Z.: Improved genetic algorithm for multiple sequence alignment using segment profiles (GASP). In: Li, X., Zaïane, O.R., Li, Z.-h. (eds.) ADMA 2006. LNCS (LNAI), vol. 4093, pp. 388–395. Springer, Heidelberg (2006)
6. Notredame, C., Higgins, D.G.: SAGA: Sequence alignment by genetic algorithm. Nucleic Acids Research 24, 1515–1524 (1996)
7. Thompson, J.D., Gibson, T.J., Plewniak, F., Jeanmougin, F., Higgins, D.G.: The ClustalX windows interface: flexible strategies for multiple sequence alignment aided by quality analysis tools. Nucleic Acids Research 24, 4876–4882 (1997)
8. Thompson, J.D., Plewniak, F., Poch, O.: BAliBASE: A benchmark alignment database for the evaluation of multiple alignment programs. Bioinformatics 15, 87–88 (1999)
9. Thomsen, R., Fogel, G.B., Krink, T.: Improvement of Clustal-Derived Sequence Alignments with Evolutionary Algorithms. In: Proceedings of the Congress on Evolutionary Computation 2003, vol. 1, pp. 312–319 (2003)
10. Wang, L., Jiang, T.: On the complexity of multiple sequence alignment. Journal of Computational Biology 1(4), 337–348 (1994)

Grid-Oriented Scatter Search Algorithm

Antonio Gómez-Iglesias[1], Miguel A. Vega-Rodríguez[2],
Miguel Cárdenas-Montes[3], Enrique Morales-Ramos[4],
and Francisco Castejón-Magaña[1]

[1] National Fusion Laboratory, CIEMAT, Madrid, Spain
{antonio.gomez,francisco.castejon}@ciemat.es
http://www.ciemat.es
[2] Dep. of Technologies of Computers and Communications,
University of Extremadura, Cáceres, Spain
mavega@unex.es
[3] Dep. of Basic Research, CIEMAT, Madrid, Spain
miguel.cardenas@ciemat.es
[4] ARCO Research Group, University of Extramdura, Cáceres, Spain
enmorales@alumnos.unex.es

Abstract. Scatter search (SS) is an evolutionary algorithm (EA) becoming more important in current researches as the increasing number of publications shows. It is a very promising method for solving combinatorial and nonlinear optimisation problems. This algorithm is being widely implemented for solving problems not taking long, but in case of processes requiring of high execution times likely to be executed using grid computing there is not an implementation for it. Some problems arise when we try to execute this algorithm using the grid, but once they are solved, the obtained results are really promising for many complex and scientific applications like, for example, applications for optimising nuclear fusion devices. Using concurrent programming and distributed techniques associated to the grid, the algorithm works as it could do it in a single computer.

Keywords: Scatter Search, Distributed Programming, Concurrent Programming, Grid Computing.

1 Introduction

Scatter search is a metaheuristic process using formulations based back to the 1960s for combining decision rules and problem constraints. SS works over a set of solutions, combining them to get new ones that improve the original set. As main difference with other evolutionary methods, such as genetic algorithms (GA), SS is not based on large random populations but in strategic selections among small populations. While GAs usually work with populations of hundreds or thousands of individuals, SS uses a set of around 10 different solutions.

The algorithm can be divided into three different phases:

– Generation of an initial set of solution vectors by heuristic processes and selection of a subset of best solutions called *reference solutions*.

M. Kolehmainen et al. (Eds.): ICANNGA 2009, LNCS 5495, pp. 193–202, 2009.
© Springer-Verlag Berlin Heidelberg 2009

– Creation of new individuals based on linear combinations of subsets of the current reference solutions.
– Extraction of the best solutions generated to be used as starting points for a new application of the first step. These steps are repeated until reaching a specified iteration limit.

The scatter search template [9] has served as main reference for many implementations up to date, as long as the dispersion patterns created by these designs have been found useful in several application areas. But this template does not work when the fitness function evaluation takes long to get the value because the delays included and the waiting time make impossible to evaluate a long number of solutions.

In case of optimisation problems related to many scientific areas, the required time to get a fitness value is extremely high, so traditional computation cannot be used to perform a full optimisation process. Modern paradigms like grid offer the computational resources and capabilities to carry out these optimisation problems but they are not easy to use [5][7]. The development of the grid has created a way that could lead to an increase in the performance of this kind of algorithms in terms of execution time and problem size. But a high level of expertise is required to develop and execute grid applications because many problems can arise because of the special behaviour of the grid [11][12]. Up to now, many investigations have been carried out using parallel architectures and GAs [2][6], but grid has not yet been deeply used with complex EA. Here we propose a generic implementation and show an example of SS algorithm using grid capabilities without human supervision. Our goal consists of getting a new SS template which could be used by scientists and developers without problems with the grid because of a transparent interaction with this paradigm.

The rest of the paper is organised as follows: section 2 gives a description of the elements and methods of SS algorithm based on the proposed implementation in [13]. Following this, section 3 is devoted to the distributed fitness calculation using the grid. Section 4 describes an example implementation developed following the previous explanations. Finally, section 5 shows some conclusions and future work.

2   Scatter Search Algorithm

Fig. 1 shows the basic scheme of SS algorithm. All the stages are deeply explained in many related works [9][13]. The design of these stages is very wide and general, with the aim of anyone can implement other techniques for any of them (as shown in this paper).

2.1   Concurrent Implementation

The implementation traditionally proposed for this algorithm [9] lacks of getting solutions for fitness functions when these functions take long. For many problems we can get a new generation after a few seconds, but when each fitness

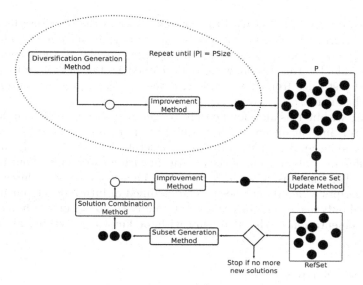

Fig. 1. SS algorithm scheme

value requires of several minutes or even hours working out its evaluation, this template does not offer a good result. Furthermore, we could have a scenario where the researchers involved in a specific problem could not have access to all the computational power required but they could have access to the grid, especially when this access is very easy to obtain.

For these reasons we have developed a new version for this template which can run concurrently as many fitness evaluations as needed. To do this we are using Unix threads within C++ code and child processes [16]. Once a thread has been created, we create a child process which will interact with the grid while the parent process waits for the child to finish. This parent process also looks for the time that the child process is taken to finish its tasks. This is important considering that sometimes grid jobs take longer than foreseen, to finish being better to finish these jobs and resend them again.

3   Grid Based SS Implementation

Grid cannot be compared with supercomputers or special parallel machines which can offer excellent results in term of efficiency. Nevertheless the distributed paradigm of the grid as well as the number of computational resources available makes this a very promising alternative to use SS with complex optimisation problems.

To automate the execution of the SS algorithm using the grid without changing the source code we have developed a method that reads the entire configuration used by the algorithm from a set of XML files which can be easily modified by the user.

But, if we want to interact with the command line in the User Interface (UI), trying to perform complex operations, such as proxy management, calling off any job or resending failed jobs, it is challenging to invoke all these commands from C++ code. For this reason, we have developed a set of python scripts which are invoked just once from C++ code and they manage jobs interacting with the metascheduler.

Our goal within this development is to get a system which can interact with the grid, implementing all the required functionalities to work with this paradigm, ending up a simple template or skeleton that can be easily modified to get a different configuration of the algorithm without worrying about the grid component. For this reason, we have introduced a set of new procedures as well as python scripts which do not need to be modified in future developments. Developers will only need to modify the selection and combination methods, to get a new implementation of the algorithm without knowing anything about grid computing.

3.1 Python Scripts

To get a non-supervised system we have developed a set of python scripts which interact with the metascheduler and the proxy to manage all the required processes in a proper way.

These scripts do not only provide functionalities related to the management of jobs, but also with the whole grid environment: loading the metascheduler, detecting some failures of the grid infrastructure or proxy management.

3.2 Job Submission

Job submission consists of the generation of the JDL (*Job Description Language*) file, and also of the generation of the input information associated to each individual. This generation will depend on the problem we are solving. Thereby, the developer will have to implement a method to create the input information based on the structure used within the template, being a simple task.

Once this input information (consisting of several configuration files, or a folder structure, or any kind of information required for the job to run) has been created, the thread which is managing the individual will create the JDL and submit it.

4 Real Application Implementation

In this section we consider the optimisation of a nuclear fusion device (TJ-II, located in Madrid, Spain) by means of SS and a complex fitness function which tries to improve the equilibrium of this device by minimising the value given by the eq. 1. In a device with a good equilibrium in the confined particles we have fewer particles leaving the desired trajectories so the possibilities to get fusion reactions increase.

Nowadays there is a global concern about the problem of global warming and the use of fossil fuels like our main energy supply. Several possibilities are growing as future energy sources, being the nuclear fusion among the most promising ones.

Nuclear fusion is the process by which multiple atomic nucleus join together to form a heavier nucleus. The fusion of two light nuclei generally releases energy while the fusion of heavy nuclei absorbs energy [14].

$$F_{targetfunction} = \sum_{i=1}^{N} \left\langle \left| \frac{\vec{B} \times \vec{\nabla} \, |B|}{B^3} \right| \right\rangle_i \tag{1}$$

In this equation, i represents the different magnetic surfaces within a magnetic confinement nuclear fusion device and B is the magnetic field. The computational cost of this equation is really high due to the number of operations to perform and the number of iterations i.

There are some modelling tools which can simulate the behaviour of nuclear fusion devices. The computational resources needed for these tools are not only extremely high, but also take long to finish a single simulation. The number of possible configurations for these devices is also high, so the number of different tests to perform can require a long execution time.

The workflow we need to execute to measure the equilibrium in the TJ-II, by solving the expression 1, is widely explained in the related work [10]. Here we introduce the most important components of this workflow. Furthermore, there are many dependencies among these programs related to the kind of information which they need to exchange as well as some checks which must be performed to assure that the final result is correct.

- MGRID: it generates a file with the current of the coils and the magnetic fields generated by the coils.
- VMEC (*Variational Moments Equilibrium Code*), a three-dimensional magnetohydrodynamic (MHD) equilibrium solver [4][8] is used to calculate the configuration of the magnetic surfaces in a stellarator.
- wout2flx: it transforms the original output given by VMEC into a comprehensible format containing only the useful information we need.
- Fitness Calculation: with the output of *wout2flx*, this application solves the eq. 1. This method takes long to work out the fitness value because of the high number of iterations to perform.

This workflow is executed on a Worker Node (WN) of the grid. The way this workflow is managed is via a python script which extracts all the required files from a compressed file, generates the executables, removes unnecessary files and controls the execution order.

The time required by this workflow to get the fitness value varies from a few minutes up to some hours. For this reason it is impossible to optimise this problem by means of a single computer.

This workflow, running on the grid, could be defined as *Parameter Sweep Applications* (PSAs), because it can be executed many times with different input parameters [5]. This kind of problems are structured as sets of computational tasks mostly independent: there are few task-synchronization requirements, or data dependencies, among tasks. For this reason, these applications (workflow in our case) are perfect for becoming the fitness function in our SS algorithm. These PSAs applications are commonly executed on clusters, but they can benefit by means of grid computing due to the number of computational resources and its distributed paradigm, with is suitable for a parallel execution model.

4.1 SS Implementation

Here we summarize the changes introduced to the template to adapt the algorithm to our requirements. As said, the configuration of the algorithm is indicated by means of XML configuration files. Our requirements are related to how we perform the combination and how to measure the diversity among the elements of the population.

Diversification Method. We are using a diversification method based on Greedy Randomized Adaptive Search Procedure (GRASP) constructions [15]. This method uses a local search procedure [1] to improve the generated solutions. The evaluation function given for this GRASP method is the fitness value obtained for the solution when a value is added to one of the elements within the solution. The resulting solution is the best overall solution found with this local search. This process requires of the evaluation of many solutions, so it can be deactivated in order to work with no so good configurations but, at least, the number of iterations which can be calculated in the same time increases.

Combination. To show an implementation of the algorithm we have chosen a mutation-based procedure to mix individuals and obtain a new one. This mutation-based procedure uses the sample standard deviation of each chromosome in the whole population to perform the mutation.

This function assures some level of convergence in the values of the chromosomes even though this convergence is only noticed after a long number of generations. Each selected chromosome is added or subtracted (randomly) with a value between 0 and the standard deviation for that value. The mutation using the standard deviation value could be also used. The quality of solutions becomes better, because of the convergence obtained with this function using elite configurations as base elements to get new solutions.

Diversity Measure. In our case, to measure the diversity among the elements in the population we use the normalised value of each chromosome of any individual. This normalisation is calculated using the eq. 2. In other cases this normalisation is not needed, but for this problem the difference among the possible values of the chromosomes (some of them take values in the range $\{0-1\}$ while others in the range $\{0-10^{11}\}$) makes impossible to use the current value of each chromosome.

$$Norm\_value_i = \frac{value_i - Min_i}{Max_i - Min_i} \tag{2}$$

Finally, with the following expression (eq. 3) we can obtain the distance between two individuals, p and q by:

$$d(p,q) = \sum_{i=1}^{n} |p_i - q_i| \tag{3}$$

Using this expression we can get some levels of diversity within the elements in the population which is also one of the objectives of the SS algorithm. This expression has been also implemented within the template we have developed because it can be used in any kind of problem with a *real* representation of chromosomes. It is also very insightful so any developer could easily understand this function in case of any change would be required.

Improvement Method. The improvement method we have to use cannot perform many evaluations because of the high execution time required for each of thiese evaluations. For this reason, we have developed a method that introduces small variations in the values of the solution and checks is the solution is better than the original one. The result is not as optimal as it could be, in cases of using algorithmgs such as the one proposed by Nelder and Mead [3], but the execution time reduces drastically. Anyway, for problems suitable of using Nelder and Mead method, we allow the possibility of use it just by choosing the appropiate option in the configuration file.

4.2   Managing the Grid

As previously mentioned, there are some problems related to the overload of the UI when the number of waiting processes becomes high. To show the different alternatives we have, the program looks for the number of threads to be launched at any time (this has been also added to the template):

1. If the number of threads is lower than 10, each thread launches a wait command, after sending the job the grid, to wait for its job to finish. Thus, once the job has finished, immediately the master process knows it.
2. If the number of threads is equal or higher than 10, each thread only launches the job. Once all the threads have ended, the master process launches a child process which will look every 30 seconds for the jobs to finish.

4.3   Results

Here we present some results obtained running our example in the grid. During this run, some problems appeared in the grid infrastructure, but the system could recover from these failures. Firstly we show some results related to the time required to evaluate some iterations in the SS. After it, we present some results focused on the evolution of the fitness function within the population along the generations.

Table 1. Configuration of the SS algorithm

Parameter	Value
Number of iterations	40
Number of chromosomes	90
Size of population	100
Size of RefSet	10
Local Search	Deactivated

Table 2. Execution Times for 40 Generations

Test	Scatter Search Algorithm
Total Wall Clock Time	535:36:25
Cumulative CPU Time of Workers	9,095:41:07

The configuration for SS is as shown in Table 1:

With this configuration, the execution time required for these evaluations are as shown in Table 2. The results show the execution time, not the time waiting in queues. This time depends on the number of jobs submitted by other users or the number of resources available, so it can change. This table also shows the aggregated execution time required by of all the processes.

All the times are in the *hh:mm:ss* format. The required time to obtain these results, was 535:36:25 (more than 22 days). Considering the total execution time being more than 9,095 hours, which is almost 379 days, the advantage of using grid computing becomes clear. The required time to get these results depends on the number of computational resources available as well as the number of jobs submitted by other users. In the ideal case of having at least the same number of WN as number of maximum jobs at the same time and no more users, this time will be lower, but this is just an ideal case which is difficult to reach in real environments.

After an iteration, the elements in the population and in the *RefSet* are stored in XML files, so it becomes easier to analyse the results and see how the fitness value evolves. Besides, the execution times of all the jobs are also stored, being this useful to get results as previously shown. With this configuration, the best value found for the expression 1 is $7.056280E + 03$ which represents a great improvement compared to the configurations used up to now in this devices, where the values for this function are over $6.0E + 16$. The resulting design allows getting a better confinement of particles within this device, so the probabilities to obtain fusion reactions increase, improving the efficiency of the device.

The speedup was 16.98, while the efficiency, 0.33. In terms of speedup is difficult, or even impossible, to compare these results with those obtained using supercomputers, because of the special paradigm of the grid, where the number of jobs waiting in the queues produces that the waiting time could be high. But in terms of productivity the results are very encouraging.

4.4 Conclusions about the Implementation of This Example

All the developments needed to implement all the functionalities for this example have been carried out easily by means of the template previously explained. The only difficulties arise from the own functionalities we could think of implementing. The approach shown only uses a set of functions which can be easily understood and implemented once the template works properly within a grid environment.

One of the problems that appeared was related to the use of XML files. UI does not have installed by default the *libxml* library. To solve this we have included the source files of this library in a compressed file which is uncompressed and compiled before compiling the SS algorithm. This library has been added to the template so future developments will not need to worry about this issue. An overload of the UI was found during the development of this example, and the solution has been added to our template in order to avoid this problem in the future.

The results obtained with this algorithm for the problem of optimisation of nuclear fusion devices are really encouraging and aim to perform different tests with different configurations of the algorithm to look for better configurations.

5 Conclusions and Future Work

In this paper we have shown a template to use SS algorithm and the grid. It can be easily modified to accept any kind of combination or selection methods as well as different configurations of the algorithm. This template will allow us to solve different problems, most of them related to nuclear fusion devices optimisation using SS and all the capabilities of the grid. The example shown in this paper will be the starting point for a set of different tests we are willing to carry out in order to get optimal configurations for this kind of devices.

Grid computing appears to be an appropriate environment to make possible bringing results from computationally intensive tasks. In the results obtained in the implementation proposed we can see how the execution time has been reduced thanks to this paradigm.

This new template can be easily modified to work with parallel architectures, just developing some parts of the code but keeping the threads implementation.

Some problems related to the template, specially focused on the UI overload, were found along the implementation of the example, but after solving them for this example, the solutions were added to the template. Thereby future developments can be carried out focusing only on the problem itself but nothing related to the general topics of SS and the grid.

Finally, as future work we think of:

1. Developing a web service oriented version via WSRF services.
2. Developing new replacement and selection methods instead of the proposed in this paper in order to see how this affects the results.

3. Performing a deep study of the results obtained with different configuration of the algorithm for the example proposed to get a set of optimised configurations of the nuclear fusion device.
4. As long as with the VMEC output many functions can be calculated, a multi-objective implementation of the SS using the grid is a promising approach to improve a fusion device considering a longer number of functions.

References

1. Aarts, E., Lenstra, J.K.: Local Search in Combinatorial Optimization. Princeton University Press, Princeton (2003)
2. Alba, E., Tomassini, M.: Parallelism and Evolutionary Algorithms. IEEE Transactions on Evolutionary Computation 6(6) (2002)
3. Avriel, M.: Nonlinear Programming: Analysis and Methods. Dover Publishing, New York (2003)
4. Bellan, P.M.: Fundamentals of Plasma Physics. Cambridge University Press, Cambridge (2003)
5. Berman, F., Hey, A., Fox, G.C.: Grid Computing. In: Making the Global Infrastructure a Reality. John Wiley & Sons, Chichester (2003)
6. Cantú-Paz, E.: A Survey of Parallel Genetic Algorithms. Calculateurs Parallèles, Réseaux et Systèmes Répairs 10(2), 141–171 (1998)
7. Foster, I., Kesselman, C.: The Grid: Blueprint for a New Computing Infrastructure. Morgan Kaufmann, San Francisco (1999)
8. Freidberg, J.: Plasma Physics and Fusion Energy. Cambridge University Press, Cambridge (2007)
9. Glover, F.: A Template for Scatter Search and Path Relinking. In: Hao, J.-K., Lutton, E., Ronald, E., Schoenauer, M., Snyers, D. (eds.) AE 1997. LNCS, vol. 1363, pp. 13–54. Springer, Heidelberg (1998)
10. Gómez-Iglesias, A., et al.: Grid Computing in order to Implement a Three-Dimensional Magnetohydrodynamic Equilibrium Solver for Plasma Confinement. In: PDP 2008 (2008)
11. Joseph, J., Fellenstein, C.: Grid Computing. Prentice Hall, Englewood Cliffs (2003)
12. Juhász, Z., Kacsuk, P., Kranzlmüller, D.: Distributed and Parallel Systems: Cluster and Grid Computing. Springer, Heidelberg (2005)
13. Laguna, M., Martí, R.: Scatter Search Methodology and Implementations. Kluwer Academic Publishers, Dordrecht (2003)
14. Miyamoto, K.: Plasma Physics and Controlled Nuclear Fusion. Springer, Heidelberg (2005)
15. Pitsolulis, L.S., Resende, M.G.C.: Greedy Randomized Adaptive Search Procedures. In: Handbook of Applied Optimization. Oxford University Press, Oxford (2002)
16. Stevens, W.R., Rago, S.A.: Advanced Programming in the Unix Environment. Addison-Wesley, Reading (2005)

Agent-Based Gene Expression Programming for Solving the RCPSP/max Problem

Piotr Jędrzejowicz and Ewa Ratajczak-Ropel

Department of Information Systems
Gdynia Maritime University, Poland
{pj,ewra}@am.gdynia.pl

Abstract. The paper proposes combining a multi-agent system paradigm with the gene expression programming (GEP) to obtain solutions to the resource constrained project scheduling problem with time lags. The idea is to increase efficiency of the GEP algorithm through parallelization and distribution of the computational effort. The paper includes the problem formulation, the description of the proposed GEP algorithm and details of its implementation using the JABAT platform. To validate the approach computational experiment has been carried out. Its results confirm that the agent based gene expression programming can be considered as a promising tool for solving difficult combinatorial optimization problems.

1 Introduction

The paper proposes combining a multi-agent system paradigm with the gene expression programming (GEP) to obtain solutions to the resource constrained project scheduling problem with time lags (RCPSP/max). In recent years the RCPSP/max problem has attracted a lot of attention and many exact and heuristic algorithms have been proposed for solving it [2], [12], [16]. There have been several multi-agent approaches proposed to solve different types of optimization problems. One of them is the concept of an asynchronous team (A-Team), originally introduced in [15]. On the other hand the gene expression programming proposed in [6] as a method for solving different kinds of optimization problems seems to be a population-based approach, which can easily yield to parallelization and decentralization. GEP is a kind of evolutionary algorithm that evolves computer programs, which can take many forms: mathematical expressions, neural networks, decision trees, polynomial constructs, logical expressions, etc. GEP has been successfully used for solving many optimization and especially combinatorial optimization problems ([8], [14]).

In this paper we propose a JABAT-based implementation of GEP algorithm intended for solving instances of the RCPSP/max problem. JABAT is a multi-agent platform which has been designed as a tool enabling design of the A-Team solutions [10], [3]. The algorithm based on GEP was implemented as optimization agent in JABAT. The approach has been validated experimentally. Section 2 of the paper contains the RCPSP/max problem formulation. Section 3 gives

M. Kolehmainen et al. (Eds.): ICANNGA 2009, LNCS 5495, pp. 203–212, 2009.
© Springer-Verlag Berlin Heidelberg 2009

some details on gene expression programing. Section 4 provides details of the GEP implementation for solving the RCPSP/max problem and describes fine-tuning phase of the proposed algorithm. Section 5 provides a brief description of the platform used to implement GEP algorithm as an A-Team. Section 6 describes computational experiment carried-out. Section 7 contains conclusions and suggestions for future research.

2   Problem Formulation

The resource-constrained project scheduling problem with minimal and maximal time lags (RCPSP/max) consists of a set of $n+2$ activities $V=\{0, 1, \ldots, n, n+1\}$, where each activity has to be processed without interruption to complete the project. The dummy activities 0 and $n+1$ represent the beginning and the end of the project. The duration of an activity j, $j = 1, \ldots, n$ is denoted by d_j where $d_0 = d_{n+1} = 0$. There are r renewable resource types. The availability of each resource type k in each time period is r_k units, $k = 1, \ldots, r$. Each activity j requires r_{jk} units of resource k during each period of its duration where $r_{1k} = r_{nk} = 0$, $k = 1, ..., r$. Each activity $j \in V$ has a start time s_j which is a decision variable. There are generalised precedence relations (temporal constraints) of the start-to-start type with time lags $s_j - s_i \geq \delta_{ij}$, $\delta_{ij} \in Z$, defined between the activities.

The structure of a project can be represented by an activity-on-node network $G = (V, A)$, where V is the set of activities and A is the set of precedence relationships. The objective is to find a schedule of activities starting times $S = [s_0, \ldots, s_{n+1}]$, where $s_0 = 0$ (project always begins at time zero) and resource constraints are satisfied, such that the schedule duration $T(S) = s_n + 1$ is minimized.

The RCPSP/max as an extension of the RCPSP belongs to the class of NP-hard optimization problems [1], [5]. The objective is to find a makespan minimal schedule that meets the constraints imposed by the precedence relations and the limited resource availabilities.

3   Gene Expression Programming

Gene expression programming (GEP) as a kind of the genetic programming algorithm was proposed by Candida Ferreira in [6]. The fundamental difference between gene expression programming (GEP), genetic algorithms (GAs) and genetic programming (GP) resides in the nature of the individuals. In GAs the individuals are linear strings of fixed length (chromosomes); in GP the individuals are nonlinear entities of different sizes and shapes (parse trees); and in GEP the individuals are encoded as linear strings of fixed length (the genome or chromosomes) which are afterwords expressed as nonlinear entities of different sizes and shapes (i.e., simple diagram representations or expression trees).

The genes in GEP have a special structure. The genome or chromosome consists of a linear, symbolic string of fixed length composed of one or more genes.

Each gene may be composed of two parts: the head and the tail or one part: the tail only. The head contains symbols that represent both functions and terminals, whereas the tail contains only terminals. GEP chromosomes are usually composed of more than one gene of equal length but different classes (multigenic chromosomes). For each problem, the number of genes, as well as the length of the head, are chosen a priori. It is important, that the genes must be built such that any modification made in them always results in a structurally correct genes. There are several special gene operators proposed for GEP.

Using GEP for solving different combinatorial optimization problems was discussed in [6] and [7], also for scheduling problems: traveling salesperson problem and task assignment problem. The main components of the GEP algorithm used for solving combinatorial optimization problems include:

- gene structure,
- population of individuals,
- fitness function and selection rules,
- reproduction operators.

In the combinatorial optimization problems the multigenic chromosomes are considered where each gene are composed of the tail only. In one chromosome different classes of genes may be used. Genes are not subject to any additional restrictions - a gene may be created as a set of any terminals combination from the set fixed for the class to which the gene belongs.

The chromosomes which are parts of individuals from the initial population are generated randomly. Most of them or even all may not represent correct solutions. The next generation is formed basing on a simple elitism rule - the best individual (or one of the best) from each generation is cloned into the next generation.

In GEP it is important to use suitable fitness function to control the fitness of solutions. To make the evolution efficient some different fitness functions were proposed in [7]. In most cases they are based on the absolute or relative error. For the scheduling problems, where minimal solution is the best, the function $f_x = T_g - t_x + 1$ was proposed, where x is the individual, t_x is the length of the schedule and T_g is the the length of the largest schedule encoded in the chromosomes of the current population. Individuals in GEP are selected according to fitness by the roulette-wheel sampling [9].

To reproduce generations of individuals in GEP except a simple elitism and the roulette-wheel sampling several class of operators are used. They are named as: insertion, transposition, recombination and mutation. For combinatorial optimization problems two additional classes of operators were proposed: deletion/insertion and permutation [7].

4   Using GEP for Solving RCPSP/max

The foundations of gene expression programming have been used to construct the algorithm for solving RCPSP/max. A high level pseudocode of this algorithm, denoted as GEP4RCPSPmax is presented in figure 1.

```
GEP4RCPSPmax(GenerationNumber, PopulationSize)
{
  Create initial population Pop of size PopulationSize
  for(i=1; i<GenerationNumber; i++)
  {
    Pop.mutationConstruct(0.02)
    Pop.mutationLC(0.08)
    Pop.geneDelIns(0.04)
    Pop.inversion(0.04)
    Pop.inversion2g(0.02)
    Pop.restrictedPermutation(0.06)
    Pop.onePointRecombination(0.6)
    Pop.twoPointRecombination(0.8)
    Pop.nPointRecombination(0.8)
    Pop.geneRecombination(0.5)
    Remember the best individual
    Create next generation
  }
}
```

Fig. 1. The pseudocode of GEP4RCPSPmax algorithm

Each chromosome representing the schedule consists of n+2 genes. Genes represent project activities in sequence from 0 to $n+1$. In each gene the information of activity starting time is coded as a non negative integer value. In each gene values from the same interval $c_j^{gep} \in [0, h]$, $j = 0, \ldots, n + 1$ are stored. To calculate the real starting time for each activity the following formula is used: $c_j = c_j^e + c_j^{gep}\%(c_j^l - c_j^e)$, where c_j^e is the earliest, and c_j^l is the latest possible starting time for the activity j.

To calculate the fitness the function described in section 3 is used. The initial population of solution is generated randomly. Each individual is used to produce a solution and for each individual the fitness function value is calculated. In case of failure the fitness function is decreased by the duration of these activities which caused the failure.

Individuals for the next generation are selected by the roulette-wheel sampling with fitness as the criterion. The best individual from each generation is cloned to the next. To evolve population some classic GEP operators are used together with several of their modifications. The considered GEP operators are: gene deletion/insertion (geneDelIns), inversion, restricted permutation, one and two point recombination and gene recombination. The modified ones include: two genes inversion, n point recombination and mutation. All considered operators except mutationConstruct and mutationLC are used only once for one chromosome in one generation. Descriptions of these operators follows:

  – mutationConstruct - the RCPSP/max SGSU (serial generation scheme with unscheduling) [11] algorithm is used for the randomly chosen individual;

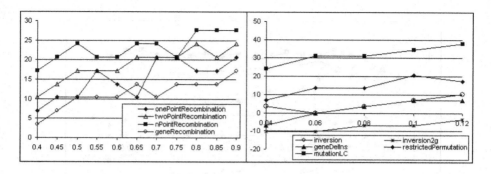

Fig. 2. Percentage of the maximum fitness for different probability values for operators from the crossover class (on the left) and the mutation class (on the right)

- mutationLC - the new gene value is randomly selected from the fixed set of values;
- geneDelIns - one individual is randomly selected from the population, the gene from chromosome is selected randomly and moved in the new randomly chosen place in the chromosome, the genes between the two selected places are moved as well;
- inversion - one individual is randomly selected from the population, two points from the chromosome are selected randomly and the sequence of genes between them is inverted;
- inversion2g - inversion where the sequences of only two genes are used;
- restrictedPermutation - one individual is randomly selected from the population, two genes from randomly selected positions of chromosome are replaced;
- onePointRecombination - in the considered case it is equivalent to the one point crossover operator;
- twoPointRecombination - in the considered case it is equivalent to the two point crossover operator;
- nPointRecombination - two individuals are randomly selected from the population, and the two childes are created in such a way that for each gene position the gene from one randomly selected parent is allowed to the first child and the gene from the other parent is allocated to the second child;
- geneRecombination - two genes from randomly selected position are swapped between two randomly selected individuals from the population.

To evaluate the effectiveness of the approach several fine-tuning computational experiments have been carried out. In the experiments three factors have been investigated:

- effectiveness of the different operators,
- relationships between different operators,
- number of generations needed.

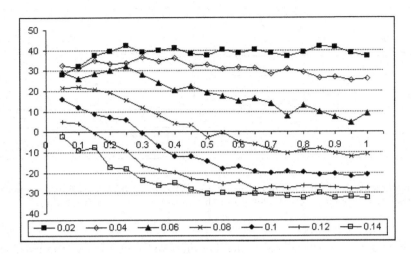

Fig. 3. Percentage of mean fitness for different probability values for all operators from the crossover class

In the fine-tuning experiments the considered operators have been divided into two broad classes: the mutation class and the crossover class. The percentage of maximum fitness obtained for different values of probability of applying various operators are shown in figure 2. The maximum fitness is understood as fitness value achived by the best solution. The experiment results suggest that the most effective operators for RCPSP/max are: nPointRecombination, two-PointRecombination, mutationLC and restrictedPermutation. In the final computational experiments higher probabilities for applying these operators have been used.

The next experiment investigates relationships between operators. Actually, relationships between the two considered classes have been considered, where each operator from the same class has the same probability value. The operators from mutation class have been investigated for probability values from 0.02 to 0.14, and operators from crossover class for probability values from 0.05 to 1. Results as the mean and maximum percentage fitness are presented in figure 3 and 4. As it is shown in these figures, probability values that maximize the mean fitness do not necessarily maximize the maximum fitness and vice versa. In the final experiment probability values have been set at the level in the middle of interval given by values maximizing the maximum fitness and the mean fitness respectively. The values used in the proposed GEP4RCPSPmax algorithm are shown in figure 1.

The last experiment within the fine-tuning stage aimed at investigating the influence of the number of generation evolved. In most cases to produce the best individual it has been enough to evolve 20 generations. However in case of inversion, restrictedPermutation and mutationLC the best individuals have been

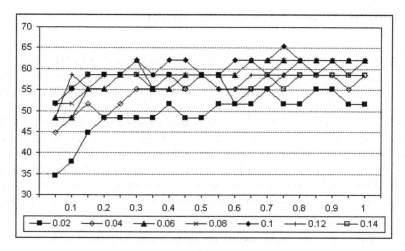

Fig. 4. Percentage of maximal fitness for different probability values for all operators from the crossover class

found after having evolved 80–90 generation. Thus in the final experiment the number of generations has been set to 100.

5   Implementing GEP4RCPSPmax Algorithm as an A-Team

To implement the proposed GEP algorithm as a team of agents the JABAT platform described in details in [10] [3], and [4] has been used. JABAT is a middleware allowing to design and implement A-Team architectures for solving various combinatorial optimization problems. The problem-solving paradigm on which the proposed system is based can be best defined as the population-based approach.

JABAT produces solutions to combinatorial optimization problems using a set of optimization agents, each representing a, so called, improvement algorithm. Each improvement algorithm when supplied with a potential solution to the problem at hand, tries to improve this solution. To escape getting trapped into the local optimum an initial population of solutions (individuals) is generated or constructed. Individuals forming an initial population are, at the following computation stages, improved by independently acting agents, thus increasing chances for reaching the global optimum.

JABAT may be used to implement the GEP4RCPSPmax algorithm in several ways. In case described in this paper the GEP4RCPSPmax algorithm has been implemented as an improvement algorithm embedded within the dedicated optimization agent. During computation several instances of this agent are used in parallel to speed up the process of improving individuals stored in the common memory.

6   Computational Experiment Results

To validate the proposed approach computational experiment has been carried out using benchmark instances of the RCPSP/max from PSPLIB [13]. The experiment involved computation with a fixed number of optimization agents used by the GEP4RCPSPmax algorithm. The respectie results are shown in table 1. The discussed results have been obtained using 5 optimization agents. At the beginning the population of solutions in the common memory of JABAT has been generated randomly. During computation solutions from common memory are drawn by optimization agents which try to improve their fitness through tuning the GEP4RCPSPmax algorithm. Each solution drawn by an optimization agent is incorporated into the population of 49 solutions generated internally by this agent. Such population is then evolved in 100 generations. If the best

Table 1. Experiment results for agent-based GEP4RCPSPmax

Number of activities	Mean RE	% FS	Mean CT
10	0.8 %	97.33 %	0.48 s
20	4.94 %	94.02 %	0.63 s
30	9.57 %	83.91 %	0.89 s

Table 2. Experiment results for ISES [2]

Number of activities	Mean RE	% FS	Mean CT
10	0.99 %	100.00 %	0.71 s
20	4.99 %	100.00 %	4.48 s
30	10.37 %	100.00 %	22.68 s

Table 3. Experiment results for B&B [2]

Number of activities	Mean RE	% FS	Mean CT
10	0.00 %	100.00 %	–
20	4.29 %	100.00 %	–
30	9.56 %	98.92 %	–

Table 4. Experiment results for C-BEST [2]

Number of activities	Mean RE	% FS	Mean CT
10	0.00 %	100.00 %	–
20	3.97 %	100.00 %	–
30	8.91 %	100.00 %	–

individual thus produced is an improvement as compared with the solution originally drown from the common memory then the improved solution replaces the worst individual in the JABAT common memory.

The computation results have been evaluated in terms of the mean relative error (Mean RE) calculated as the deviation from the lower bound, percent of feasible solutions (% FS) and mean computation time (Mean CT). The maximum computational time for one instance of the problem has, in all cases been not longer then 1 second during computations on 2 PC's with 2.0 GHz processors.

The results obtained by the proposed agent-based implementation of the GEP4RCPSPmax algorithm can be compared with these reported in the literature. Such a comparison are presented in tables 2, 3, and 4. It can be observed that the presented algorithm is efficient and the results are comparable. The advantege of the presented approach seems to be computation times, although they are not always easy to identify and there are some difficulties in comparing them precisely. The disadvantege is rather low percent of feasible solutions produced, which should be improved in future.

7   Conclusions

Experiment results show that the proposed agent-based GEP implementation is an effective tool for solving resource-constrained project scheduling problems with time lags. Presented results are comparable with the others known from the literature. It is worth noticing that the computation times are very short, even for more complex problem instances. Implementing such algorithm is quite simple, but some effort is needed to select GEP operators during the fine-tuning stage. The most important disadvantage of this algorithm is possible lack of solutions in some cases. This can be partly eliminated by using more then one instance of the algorithm.

Future research will concentrate on finding and testing more effective operators and eliminating a possibility of not finding a feasible solution at all. The other direction of research would be to using GEP algorithm in cooperation with other algorithms, like for instance local search, tabu search or simulated annealing in one agent-based system.

References

1. Bartusch, M., Mohring, R.H., Radermacher, F.J.: Scheduling project networks with resource constraints and time windows. Annual Operational Research 16, 201–240 (1988)
2. Cesta, A., Oddi, A., Smith, S.F.: A Constraint-Based Metod for Project Scheduling with Time Windows. Journal of Heuristics 8, 108–136 (2002)
3. Barbucha, D., Czarnowski, I., Jedrzejowicz, P., Ratajczak, E., Wierzbowska, I.: JADE-Based A-Team as a Tool for Implementing Population-Based Algorithms. In: Proc. VI Int. Conf. on Intelligent Systems Design and Applications, vol. 3, IEEE Computer Society, Los Alamitos (2006)

4. Barbucha, D., Czarnowski, I., Jedrzejowicz, P., Ratajczak-Ropel, E., Wierzbowska, I.: e-JABAT–An Implementation of the Web-Based A-Team, Intelligent Agents in the Evolution of Web and Applications. Studies in Computational Intelligence, vol. 167, pp. 57–86. Springer, Heidelberg (2009)
5. Blazewicz, J., Lenstra, J., Rinnooy, A.: Scheduling subject to resource constraints: Classification and complexity. Discrete Applied Mathematics 5, 11–24 (1983)
6. Ferreira, C.: Gene Expression Programming: A New Adaptive Algorithm for Solving Problems. Complex Systems 13(2), 87–129 (2001),
   http://www.gene-expression-programming.com/webpapers/Ferreira-CS2001/Introduction.htm
7. Ferreira, C.: Gene Expression Programming: Mathematical Modeling by an Artificial Intelligence, on-line book (2002),
   http://www.gene-expression-programming.com/GepBook/Introduction.htm
8. Ferreira, C.: Gene Expression Programming: Mathematical Modeling by an Artificial Intelligence, 2nd edn. Springer, Heidelberg (2006)
9. Goldberg, D.E.: Genetic Algorithms in Search, Optimization, and Machine Learning. Addison-Wesley, Reading (1989)
10. Jedrzejowicz, P., Wierzbowska, I.: JADE-Based A-Team Environment. In: Alexandrov, V.N., van Albada, G.D., Sloot, P.M.A., Dongarra, J. (eds.) ICCS 2006. LNCS, vol. 3993, pp. 719–726. Springer, Heidelberg (2006)
11. Neumann, K., Schwindt, C., Zimmermann, J.: Project Scheduling with Time Windows and Scarce Resources. Springer, Heidelberg (2002)
12. Neumann, K., Schwindt, C., Zimmermann, J.: Resource-Constrained project Scheduling with Time Windows. In: Recent developments and new applications, Perspectives in Modern Project Scheduling, pp. 375–407. Springer, Heidelberg (2006)
13. PSPLIB, http://129.187.106.231/psplib
14. Wilson, S.W.: Classifier Conditions Using Gene Expression Programming, IlliGAL Report No. 2008001, University of Illinois at Urbana-Champaign, USA (2008)
15. Talukdar, S., Baerentzen, L., Gove, A., de Souza, P.: Asynchronous Teams: Cooperation Schemes for Autonomous, Computer-Based Agents, Technical Report EDRC 18-59-96, Carnegie Mellon University, Pittsburgh (1996)
16. Valls, V., Ballestin, F., Barrios, A.: An evolutionary algorithm for the resource-constrained project scheduling problem subject to temporal constraints. In: Proc. of PMS 2006, Tenth International Workshop on Project Management and Scheduling, Nakom, Poznan, pp. 363–369 (2006)

Feature Selection from Barkhausen Noise Data Using Genetic Algorithms with Cross-Validation

Aki Sorsa and Kauko Leiviskä

University of Oulu, Control Engineering Laboratory
P.O. Box 4300, FIN-90014
{aki.sorsa,kauko.leiviska}@oulu.fi

Abstract. Barkhausen noise is used in non-destructive testing of ferromagnetic materials. It has been shown to be sensitive to material properties but the reported results are more or less qualitative. The quantitative prediction of the material properties from the Barkhausen noise signal is challenging. In order to develop reliable models, the feature selection is critical. The feature selection method applied in this study utilizes genetic algorithms with cross-validation based objective function. Cross-validation is used because the amount of data is limited. The results show that genetic algorithms can be successfully applied to feature selection. The obtained results are reliable and rather consistent with the results obtained earlier.

Keywords: Barkhausen noise, feature selection, genetic algorithms, cross-validation.

1 Introduction

The Barkhausen noise (BN) measurement is an intriguing technique for non-destructive testing of ferromagnetic materials. That is due to the fast response, low costs and rather simple equipment. The origin of the BN is in the movements of magnetic domain walls when the material is placed in a varying external magnetic field. As the walls move, they get trapped behind the pinning sites. The walls break away from the pinning sites when the external field strength exceeds a certain limit leading to a fast change in the magnetization of the material. [1] A typical BN signal is presented in Figure 1.

BN has been studied widely and it has been noticed to be very sensitive to different material properties such as microstructure [2], residual stress [3],[4] and hardness [2]. However, the results from the studies are more or less qualitative and thus there is a need for quantitative prediction of material properties. A typical approach in the studies is that a single feature calculated from the BN signal is compared to the studied material property. Often the calculated feature is either the root mean square value (RMS) [1] or the BN energy [3], which is obtained by integrating the BN signal over one BN envelope (Figure 1). A more recent approach considers the BN signal as a function of the applied magnetic field [2],[4]. The resulting BN profile can be used to calculate features such as peak height, position and width.

M. Kolehmainen et al. (Eds.): ICANNGA 2009, LNCS 5495, pp. 213–222, 2009.
© Springer-Verlag Berlin Heidelberg 2009

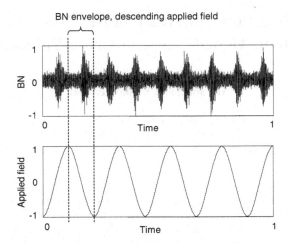

Fig. 1. A typical BN signal in scaled units (above) and the applied magnetic field (below). The BN envelope is illustrated for the descending magnetic field.

The quantitative prediction of the material properties from the BN signal is challenging due to the random nature of the phenomenon [4]. Also, some material properties are not measured which leads to unexplained variations in the calculated features. Furthermore, the reliable feature selection is essential concerning the prediction of material properties from the BN signal. Thus, this study concentrates on the automatic feature selection with genetic algorithms. Cross-validation is used in the objective function because only a limited amount of data is available and thus the probability of chance correlations is high [6]. The probability can be decreased by using cross-validation [6].

The present authors have studied the quantitative prediction of residual stress from the BN signal [5]. The algorithm was separated into four steps: feature generation, grouping and selection and the model parameter identification. Most of the features were obtained from the BN profile. Also, some features were readily obtained from the measuring device and some were calculated directly from the signal. The feature generation resulted in a group of features from which the significant features were selected based on the correlations. The same set of features is used also in this study. [5]

2 Methodology

2.1 Feature Selection

The feature selection requires three components: the search algorithm, the mathematical modelling procedure and the objective function [6]. The mathematical modelling procedure can be for example multivariable linear regression (MLR) [7],[8], partial least squares regression (PLSR) [9], principal component regression (PCR) [10],[11] or artificial neural networks (ANN) [12].

The objective function is critical to obtain solutions with desired properties such as high predictivity [6]. The objective function is also critical in avoiding the overfitting of the model. Traditionally, either correlation or prediction error based objective functions have been used. However, they have been shown to be biased and thus are not to be used in variable selection. A better objective function is based on cross-validation. [8]

The simplest method for cross-validation is data splitting, where the data is split into training and validation sets. The data can be split chronologically or randomly. Data splitting is a good option for cross-validation if there is an excess amount of data available. However, if the amount of data is limited, the data splitting reduces the amount of data greatly for both the training and the validation purposes. In such cases, it is advantageous to use resampling methods for cross-validation. [13]

In resampling methods, all the data is used for training and validation of the model. Leave-one-out cross-validation (LOO CV) extracts one data point for validation and uses the rest of the data for training. The same procedure is repeated until all the data points are used for validation to obtain an estimate of the averaged prediction error. [8] In k-fold cross-validation, the data is separated randomly to K subsets of almost equal sizes. One of the subsets is extracted for validation and the others are used for training. The procedure is repeated K times to obtain an estimate of the average prediction error. The k-fold cross-validation is considered to give better results in variable selection than the LOO CV method. [8] For example, [9] used k-fold cross-validation method for variable selection with good results.

There are many reported search algorithms for the feature selection. Stepwise forward and backward selections are simple methods available for the feature selection. In the forward selection, one variable is added to the model at a time. Similarly, one variable at a time is removed from the model in the backward selection. [12] The procedure is continued until the addition or removal of features does not result in an increase in the prediction accuracy. [12] used the forward selection followed by the backward elimination to select the variables for the ANN model. In [10], they used the backward selection as a part of their algorithm aiming for robust variable selection for the PLSR model. However, the drawback of these simple methods is that the search ends when the algorithm finds a minimum being it local or global [6].

Instead of the simple search methods, some more advanced methods have been proposed, such as genetic algorithms. Genetic algorithms was used by [10] and [11] for selecting the significant components for the PCR model. In [11], they used an internal validation set as a part of the objective function to avoid overfitting. The Durbin-Watson criterion and the penalty constant for the number of components were used by [10] to obtain robust feature sets for the modelling. Cross-validation was used in the objective function by [7] to find the significant variables for an MLR model.

2.2 Multivariable Linear Regression Models

The multivariable linear regression model is used in this study. The equation of the MLR model is [13]

$$y = Xb + e.$$ (1)

Above, y is the output variable matrix ($N \times M$), X is the input variable matrix ($N \times P$), b is the matrix of the regression coefficients (parameters) of the model ($P \times M$) and e is the matrix of the modelling residuals ($N \times M$). N is the number of data points, M is the number of output variables and P is the number of features. The model equation given above is linear but the model may also include nonlinear terms. The parameters are typically defined with the ordinary least squares method if the model is linear concerning the fitting parameters [13].

2.3 Genetic Algorithms

Genetic algorithm is an optimization method inspired by the evolution where genetic operators steer the population towards the global optimum. The solution to the optimization problem is coded into the chromosomes which compose the population. Binary or real-valued coding can be used. In the binary coding, the information is stored as binary digits, while real values are used with the real-valued coding. The information in the chromosomes is decoded and passed through an objective function to obtain the fitness values for each chromosome. The fitness value determines the suitability of the solution to the problem. [14]

Genetic operators, reproduction and mutation, regulate the evolution of the population. In reproduction, selected parents are crossed to create offspring having properties from both parents. The parent selection methods favor the chromosomes with better fitness values and thus the better chromosomes are more likely to reproduce. Such a procedure steers the population towards the better solutions. Typical parent selection methods are tournament and roulette wheel methods. Random changes are generated to the population through mutation, which prevents the population from converging to local optima. To obtain desired characteristics for the evolution of the population, probabilities are defined for crossing and mutation. Many methods are available for crossing and mutation but those are not reported here. Along with the above mentioned mechanisms, also elitism can be used in generating the new population. In elitism, a fraction of the best chromosomes are directly moved to the new population. Thus the best solutions never disappear from the population. [14]

3 Feature Selection Algorithm

3.1 Applied Genetic Algorithm

The applied genetic algorithm uses binary coding with one bit per each variable. The information in the chromosomes is interpreted as follows: if the bit is 0, the feature is not selected and if the bit is 1, the feature is selected. The evolution of the population is regulated by the probabilities defined for crossing (p_c) and mutation (p_m). New populations are generated until the predefined number of generations (n_{gen}) is reached. It is also possible to stop the algorithm when the solution have converged [14].

The parents for the crossing are determined by the tournament selection method with the number of candidates being 2 in each tournament as suggested in [14]. The winner of the tournament (the most suitable chromosome) is selected as a parent. After the parent selection, a random number is generated and if it is higher than the crossing probability, the selected parents are crossed and the generated offspring are placed into the new population. Otherwise the parents are placed into the new population. The crossing utilizes the one-splitting point method where the splitting point is defined randomly and the parents' chromosome segments after the splitting point are switched. [14] The crossing procedure is continued until the predefined population size (n_{pop}) is reached. In the mutation, each bit in the population is subjected to a possible mutation. A random number is generated for each of the bits and if it is higher than the mutation probability the bit is changed. The final step in generating the new population is to apply elitism as the worst chromosome of the new population is replaced by the best chromosome of the previous population. This prevents the best solution from disappearing from the population through the genetic operations. The parameters for the applied genetic algorithm in this study are: $n_{pop} = 200$, $n_{gen} = 20$, $p_c = 0.9$ and $p_m = 0.02$.

Even though a genetic algorithm is likely to find the global optimum, there is a chance that the algorithm finds only a local optimum during the predefined number of generations (n_{gen}). That is because the evolution of the population depends on the initial population which is generated randomly. Thus the optimization with the genetic algorithm is repeated 50 times in this study. Based on the previous study by the present authors, we expect that the suitable number of features is rather small [5]. Thus the initial population is generated so that 90 % of the bits are 0 and only 10 % are 1.

3.2 The Objective Function

The objective function utilizes the MLR model which predicts the output variable based on the calculated features from the BN signal. The MLR model and its least squares solution is presented in Section 2.1. The prediction error of the MLR model is evaluated through the k-fold cross-validation procedure. The subset extracted for validation is denoted by X_k while the remaining training set is denoted by X_{-k}. In the k-fold procedure, each of the subsets is used for the validation resulting in K submodels with their parameter vectors denoted by $b_{-1}, ..., b_{-k}, ... b_{-K}$. The prediction errors for the validation sets are computed from [9]

$$\hat{e}_k = y_k - X_k \hat{b}_{-k} . \tag{2}$$

After all the subsets are used for the validation, the prediction errors are collected in one vector, $\hat{e}$. Then, the sum of the squared error of prediction (*SSEP*) is given by

$$SSEP = \hat{e}^T \hat{e} . \tag{3}$$

As mentioned earlier, equation 3 is not a good objective function for the variable selection. Thus an additional penalty term is added to form the final objective function [8]

$$J = SSEP \times \lambda p . \tag{4}$$

Above, λ is the penalty constant and p is the number of input variables in the model. In this study, 10-fold cross-validation is used ($K = 10$) and the penalty constant is set to one ($\lambda = 1$). Even though the identification/validation procedure is repeated K times with different validation sets, the value of the objective function still depends strongly on the randomly created subsets [15]. To decrease the significance of the subset generation, the k-fold algorithm is repeated 11 times and the median of the *SSEP* values is calculated and used in the objective function.

4 Results and Discussion

4.1 Used Data Sets

The data used in this study is the same as used before by the present authors. Thus the data acquisition and the feature generation are not reported here but can be found from [5]. The data set includes 51 features and 115 data points. Before the feature selection, the features are normalized.

4.2 The Suitable Number of Features

As described in the previous section, the optimizations with the genetic algorithm are repeated 50 times. The number of features in the best solutions varies between 4 and 7. The majority of the best results include 4 variables as shown in Figure 2a. This indicates that the suitable number of features is 4. Figure 2b shows the average value of the objective function and the average *SSEP* as a function of the number of features. The values of the objective function tend to increase and the values of the *SSEP* tend to decrease with increasing number of features. The decrease of the prediction error is expected based on the reported results [11]. The question is that if the additional features are significant or insignificant. The addition of insignificant features to the model decreases the prediction accuracy for the future observations [6]. In other words, the overfitted models predict the data set used for the model identification well but fail in predictions with new data sets.

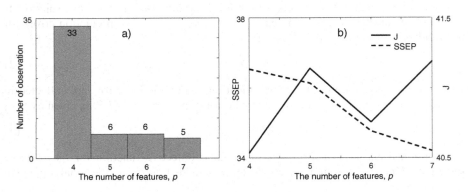

Fig. 2. a) The number of features obtained in 50 optimizations with the genetic algorithm. The majority of the optimizations resulted in a set including 4 features. b) The average *SSEP* and *J* as a function of the number of features in the best solution. The average *SSEP* tends to decrease and the average *J* tends to increase with increasing number of features.

4.3   The Most Significant Features

The examination of the significance of the features is started by finding the best feature sets corresponding to each number of features (p = 4-7). For each set, p models are identified by removing one feature at a time from the feature set. The identified models are then evaluated by calculating the value of the objective function and the relative change in the *SSEP*. The relative changes in the *SSEP* for the identified models are presented in Table 1. Table 1 shows that with the feature set including 4 features, all the features are significant, because removing any of the features leads to at least a 15 % increase in the *SSEP*. Table 1 shows also that the feature sets having more than 4 features all include at least one feature that is not very significant. This indicates that the suitable number of features is 4 as already indicated by the results given in Figure 2.

Table 1. The significance of single variables to the sum of squared error of prediction

Removed feature	x_1	x_{13}	x_{16}	x_{37}			
Change of SSEP (%)	22.5	24.2	32.2	15.4			
Removed feature	x_{15}	x_{22}	x_{33}	x_{37}	x_{45}		
Change of SSEP (%)	28.3	4.3	12.7	20.5	28.3		
Removed feature	x_1	x_{13}	x_{16}	x_{26}	x_{46}	x_{47}	
Change of SSEP(%)	28.8	26.9	40.6	7.4	9.8	11.5	
Removed feature	x_1	x_{13}	x_{16}	x_{22}	x_{33}	x_{37}	x_{45}
Change of SSEP (%)	5.6	8.0	11.7	1.9	3.9	8.9	5.6

The feature set with 4 features given in Table 1 is the set including the most significant features. From the 33 optimizations giving the 4 feature set as the best solution, 32 give that set as the solution. Based on this and the evidence given in Figure 2 and Table 1, it can be concluded that in this case, the features x_1, x_{13}, x_{16} and x_{37} are to be used as the input variables in the modelling of the residual stress.

4.4   Comparison of the Results with the Earlier Results

To our knowledge, there exist no similar studies concerning the data-based modelling of residual stress from the BN measurements. Thus the obtained results are only compared to the results presented in [5] by the present authors. The comparison shows that the selection algorithm discussed in this paper gives more reliable results. The algorithm utilized in the previous study used only correlations to determine the significant features and resulted in the feature set having 6 features. As showed in this study, the set of 6 features includes insignificant features which decreases the prediction accuracy for the future predictions. Furthermore, it has been stated that correlation-based feature selection with no cross-validation gives biased results [8].

The results of the earlier and this study are somewhat consistent. Feature x_1 was also selected in the earlier study and is the RMS value of the BN signal corresponding to the overall Barkhausen activity [5]. Feature x_{13} is the coercivity which can be related to the peak position of the BN profile [4]. The peak position was selected in the earlier study and corresponds to the field strength leading to the maximum

Barkhausen activity [5]. Feature x_{37} is a fitting parameter of the trapezoidal function and was also used in the earlier study [5]. A new feature found by the selection algorithm used in this study is feature x_{13} which is the loop area obtained directly from the measuring device. In this study, no features corresponding to the BN profile are selected, while the earlier study used two such features [5].

4.5 The Suitability of the Objective Function

The objective function used in this study utilizes k-fold cross-validation and the penalty constant for the number of features. Considering the obtained results (Figure 2 and Table 1), it can be said that the used objective function is suitable for the feature selection. Similar results have been reported also earlier with similar objective functions [8]. The use of k-fold cross-validation ensures that the results are reliable and that the features with possible chance correlations are not included in the model. Also, the possible influence of the training/validation data generation [15] is minimized by using the median value from the 11 repetitions of the k-fold algorithm. The need for cross-validation when applying the squared error based objective functions has also been shown by other studies [8],[9]. Furthermore, the penalty added for the number of features regulates the optimization in the way that only significant features are obtained from the feature selection.

4.6 The Applicability of Genetic Algorithms to Variable Selection

This study shows that genetic algorithms can be successfully applied to the feature selection problem. However, Figure 2 shows that the formulation of the objective function is critical considering the results. With no penalty constant, there would have been a much bigger set of suitable features. Probably the variation in the results would have also been greater, because the prediction accuracy depends not only on the significance of the features but also on the number of features as shown in Figure 2. Thus it is possible to obtain the same prediction accuracy with a few significant features or with a greater amount of insignificant features.

The drawback of genetic algorithms is that the calculations are rather time-consuming. However, the results from genetic algorithms have been compared to the results obtained through the exhaustive search [10]. In the exhaustive search procedure, all the possible feature combinations are tested. The results are thus very good but the method is computationally very expensive. It has been shown that the results obtained with genetic algorithms are in good agreement with the results from the exhaustive search [10]. Furthermore, the time-consumption of the calculations can be decreased by, for example, an early stopping criterion, which has been used by [11]. Also, the objective function may utilize the stepwise search algorithms to accelerate the convergence of the results [7]. In such an approach, the solution in the chromosome is considered as an initial guess for the simple search algorithms. Another possible way to decrease the computational burden is to generate the initial population through the stepwise search procedures. In such a case, different data sets must be used in the initialization of population to guarantee the diversity of the population.

5 Conclusions

Barkhausen noise is used in the non-destructive testing of ferromagnetic materials. The quantitative prediction of material properties from the BN signal is challenging and thus the reliable feature selection methods are required to develop robust models. In this study, the binary-coded genetic algorithm with the cross-validation based objective function is used in the feature selection from the data set obtained through the BN measurements. The k-fold cross-validation procedure is used in the objective function to avoid misleading results originating from the chance correlations due to the limited amount of data. The objective function also utilizes a MLR model. The cross-validation procedure is repeated several times to avoid the influence of the training/validation data generation to the results. Even though genetic algorithms are likely to find the global optimum, the optimizations are repeated 50 times. The results from the optimizations clearly showed that the suitable number of features in this study is 4. The most significant features were also found with high confidence. Those features were the RMS value of the BN signal, coercivity, the fitting parameter of the trapezoidal function and the loop area obtained from the measuring device. The results obtained in this study are somewhat consistent with the results obtained earlier by the present authors.

As a conclusion, it can be said that the used objective function performed well. All the selected features were significant according to the validation carried out by identifying the models without each of the selected features. Also, similar models were identified for the feature sets with more features convincing that 4 features are to be used. This study also showed that genetic algorithms can be applied successfully to the feature selection problem. However, the results depend strongly on the used objective function.

Acknowledgements

This study is carried out under the INTELBARK research project funded by Finnish Academy.

References

1. Lindgren, M., Lepistö, T.: Effect of prestraining on Barkhausen noise vs. stress relation. NDT&E International 34, 337–344 (2001)
2. Blaow, M., Evans, J.T., Shaw, B.A.: Effect of hardness and composition gradients on Barkhausen emission in case hardened steel. Journal of Magnetism and Magnetic Materials 303, 153–159 (2006)
3. Stefanita, C.-G., Atherton, D.L., Clapham, L.: Plastic versus elastic deformation effects on magnetic Barkhausen noise in steel. Acta Materialia 48, 3545–3551 (2000)
4. Stewart, D.M., Stevens, K.J., Kaiser, A.B.: Magnetic Barkhausen noise analysis of stress in steel. Current Applied Physics 4, 308–311 (2004)
5. Sorsa, A., Leiviskä, K., Santa-aho, S.: Prediction of residual stress from the Barkhausen noise signal. In: NDT 2008, Cheshire, UK, BiNDT, September 15-18, p. 10 (2008)

6. Baumann, K.: Cross-validation as the objective function for variable-selection techniques. TrAC Trends in Analytical Chemistry 22, 395–406 (2003)
7. Jouan-Rimbaud, D., Massart, D.L., de Noord, O.E.: Random correlation in variable selection for multivariate calibration with a genetic algorithm. Chemometrics and Intelligent Laboratory Systems 35, 213–220 (1996)
8. Wisnovski, J.W., Simpson, J.R., Montgomery, D.C., Runger, G.C.: Resampling methods for variable selection in robust regression. Computational Statistics & Data Analysis 43, 341–355 (2003)
9. Anderssen, E., Dyrstad, K., Westad, F., Martens, H.: Reducing over-optimism in variable selection by cross-model validation. Chemometrics and Intelligent Laboratory Systems 84, 69–74 (2006)
10. Barros, A.S., Rutledge, D.N.: Genetic algorithm applied to the selection of principal components. Chemometrics and Intelligent Laboratory Systems 40, 65–81 (1998)
11. Depczynski, U., Frost, V.J., Molt, K.: Genetic algorithms applied to the selection of factors in principal component regression. Analytica Chimica Acta 420, 217–227 (2000)
12. Benoudjit, N., Cools, E., Meurens, M., Verleysen, M.: Chemometric calibration of infrared spectrometers: selection and validation of variables by non-linear models. Chemometrics and Intelligent Laboratory Systems 70, 47–53 (2004)
13. Harrell Jr., F.E.: Regression Modeling Strategies with Applications to Linear Models, Logistic Regression, and Survival Analysis. Springer, New York (2001)
14. Michalewicz, Z.: Genetic Algorithms + Data Structures = Evolution Programs. Springer, New York (1996)
15. Wessels, L.F.A., Reinders, M.J.T., Hart, A.A.M., Veenman, C.J., Dai, H., He, Y.D., van't Veer, L.J.: A protocol for building and evaluating predictors of disease state based on microarray data. Bioinformatics 21, 3755–3762 (2005)

Time-Dependent Performance Comparison of Evolutionary Algorithms

David Shilane[1], Jarno Martikainen[2], and Seppo J. Ovaska[2]

[1] University of California, Berkeley
[2] Helsinki University of Technology
dshilane@berkeley.edu, martikainen@iki.fi, sovaska@cc.hut.fi

Abstract. We propose a statistical methodology for comparing the performance of evolutionary algorithms that iteratively generate candidate optima over the course of many generations. Performance data are analyzed using multiple hypothesis testing to compare competing algorithms. Such comparisons may be drawn for general performance metrics of any iterative evolutionary algorithm with any data distribution. We also propose a data reduction technique to reduce computational costs.

Keywords: Bootstrap, evolutionary algorithms, multiple hypothesis testing, optimization, performance comparison, statistics, time series.

1 Introduction

Many optimization procedures iteratively estimate a function's global optimum via a stochastic process over the course of many generations. The algorithm's quality may be summarized in terms of a *performance curve* such as the average or median result as a function of generation. [11] establish a general framework for the statistical performance comparison of EAs at a single generation. We seek to extend this methodology to compare EAs' performance curves over a range of generations. The proposed comparison establishes an experimental framework that analyzes sampled data using multiple hypothesis testing. Previous research into the design and comparison of EAs has considered a variety of approaches. Because [13] have shown that no single optimization algorithm can best solve all problems, we typically select among a number of candidate algorithms in particular settings. [2] compares EAs through an analysis of variance (ANOVA) study of computational parameters such as the mutation probability and their effect on performance. Similarly, [10] and [1] apply ANOVA and [7] a generalized linear model to aid in the design of effective EAs in particular contexts. However, these techniques may only be applied to the study of particular measures of performance under specific distributional assumptions. By contrast, [11] establish a general procedure for the performance comparison of evolutionary algorithms without these assumptions. Within this framework, statistical sampling is used to collect performance data for each algorithm, and a multiple hypothesis testing procedure [3, 4] based on bootstrap resampling [5] of the data is used to identify significant performance differences. This approach allows the

M. Kolehmainen et al. (Eds.): ICANNGA 2009, LNCS 5495, pp. 223–232, 2009.
© Springer-Verlag Berlin Heidelberg 2009

user to compare the performance of two algorithms at a single generation for general data generating distributions [9] and performance measures. Here, we seek to adapt the procedure of [11] to compare EA performance over the course of many generations. Additionally, we present a data reduction technique that estimates the test results in a more computationally feasible manner. We also provide an illustrative case study that seeks to compare the mean performance of four candidate EAs.

2   Time Series Data

An iterative optimization algorithm a produces at each generation an estimate of the global optimum for a function $f : \mathbb{R}^D \to \mathbb{R}$ with $D \in \mathbb{Z}^+$. If we allow a to run for $G \in \mathbb{Z}^+$ generations, then at each generation $g \in \{1, \ldots, G\}$, the algorithm produces a *point estimate* $X_{ag} = (X_{ag1}, \ldots, X_{agD})$ of the global optimum and a corresponding *fitness* value $f(X_{ag})$. The point estimate may be considered *cumulatively optimal* at generation g if it better optimizes f than all previous estimates. Many EAs employ an *elitist selection* mechanism [6] that retains the cumulatively optimal estimate until it is improved upon at a later generation. For algorithms that employ an elitist selection mechanism, the algorithm's iterative estimate of the function f's global optimum at generation $g + 1$ fundamentally depends upon that obtained at generation g; indeed, if the optimization procedure cannot improve upon the previous estimate, both quantities are the same. Because each generational result depends upon the previous generation's estimate, the fitness values may be viewed as a highly dependent G-dimensional time series data structure $Y_a = (Y_{a1}, \ldots, Y_{aG}) = (f(X_{a1}), \ldots, f(X_{aG}))$. Suppose we wish to study the performance of EAs in an algorithm set A, each of which will run for G generations. In order to compare these competing procedures, the researcher must designate a performance curve $\mu(Y_a)$ for each algorithm $a \in A$ at each generation $g \in \{1, \ldots, G\}$. This G-dimensional parameter is a function of the a's data generating distribution and may be selected according to the researcher's preferences. While parameters such as the trimmed mean, median, or percentiles are all salient in particular contexts, a typical choice for the performance curve is the G-dimensional vector-wise expected (mean) value of the algorithm's estimate of the global optimum as a function of generation:

$$\mu_a \equiv \mu(Y_a) \equiv E[Y_{ag}]; \qquad a \in A; \qquad g \in \{1, \ldots, G\}. \tag{1}$$

Because EAs follow a complex stochastic process, we seek to estimate the performance curve based on sampled data. Collecting data from $n_a \in \mathbb{Z}^+$ independent, identically distributed trials of algorithm a results in n_a time series observations $Y_{ia} = (Y_{ia1}, \ldots, Y_{iaG})$, $i \in \{1, \ldots, n_a\}$. Using the data collected, we can estimate a's performance according to a statistic $\hat{\mu}(Y_a)$. For the parameter (1), the sample (empirical) mean is used:

$$\hat{\mu}_a \equiv \hat{\mu}(Y_a) = \left(\frac{1}{n_a} \sum_{i=1}^{n_a} Y_{ia1}, \ldots, \frac{1}{n_a} \sum_{i=1}^{n_a} Y_{iaG} \right); a \in A. \tag{2}$$

[11] establish a general framework for performance comparison that may consider either an algorithm's relative improvement of its initial performance given an initial candidate solution or the absolute performance obtained from sampling an initial value on each trial. In the latter setting, a performance comparison of two algorithms at a single generation is determined in the test of a single hypothesis. In this work, we will adopt the convention of testing a single hypothesis per generation.

3   Performance Comparison

The performance curve μ_a (1) can be estimated according to $\hat{\mu}_a$ (2), and competing algorithms may be ranked at each generation. As in [11], a multiple hypothesis testing procedure [3] is appropriate for performance comparison. For each pair of algorithms $a, b \in A$, the researcher must establish *null hypotheses* that define a difference in performance curves at each generation. Though other parameters may be employed, a typical set of hypotheses is the equality of means (1):

$$H : [\mu(Y_{a1}) - \mu(Y_{b1}) = 0; \dots, \mu(Y_{aG}) - \mu(Y_{bG}) = 0] ; a, b \in A; a \neq b. \quad (3)$$

Depending upon the researcher's preferences, asymptotic performance (e.g. the performance in the last 100 generations) may be the most meaningful measure. In this case, the structure of (3) may be altered to include only the salient generation intervals. In order to test the null hypotheses, we need to estimate the standard error of the observed performance difference at each generation, which relies upon the following estimate of the variance vector $\sigma^2(Y_a)$:

$$\hat{\sigma}^2(Y_a) = \left(\frac{1}{n_a} \sum_{i=1}^{n_a} [Y_{ia1} - \hat{\mu}(Y_{a1})]^2, \dots, \frac{1}{n_a} \sum_{i=1}^{n_a} [Y_{iaG} - \hat{\mu}(Y_{aG})]^2 \right) ; a \in A. \quad (4)$$

It should be noted that the bootstrap estimate of variance in (4) divides by the sample size n_a [5], whereas others prefer the unbiased estimate that instead divides by $n_a - 1$. This latter estimate may be substituted at the user's discretion for large sample sizes because these quantities differ by only a small amount. Using the statistics (2) and the estimated variances (4), the hypotheses (3) may be tested using two sample t-statistics:

$$t = \left(\frac{\hat{\mu}(Y_{a1}) - \hat{\mu}(Y_{b1})}{\sqrt{\frac{\hat{\sigma}^2(Y_{a1})}{n_a} + \frac{\hat{\sigma}^2(Y_{b1})}{n_b}}}, \dots, \frac{\hat{\mu}(Y_{aG}) - \hat{\mu}(Y_{bG})}{\sqrt{\frac{\hat{\sigma}^2(Y_{aG})}{n_a} + \frac{\hat{\sigma}^2(Y_{bG})}{n_b}}} \right) ; a, b \in A, a \neq b. \quad (5)$$

If other performance curves are used in place of the expected value, the above test statistics may be modified with only small changes. In particular, μ_a and μ_b may be estimated with empirical estimates $\hat{\mu}_a$ and $\hat{\mu}_b$, which are obtained by applying the specified performance function to the observed data. Similarly, the denominator is given by the standard error of the performance difference

$\hat{\mu}_a - \hat{\mu}_b$, which may be derived analytically or estimated using the bootstrap [5]. The utility of the general framework of [11] is that the comparison methodology is otherwise identical once the test statistics are specified.

The remainder of the hypothesis testing procedure is otherwise identical to that proposed in [11]. A bootstrap method is used to estimate the joint distribution of the test statistics (5), and a multiple testing procedure (MTP) is selected to control a desired Type I Error Rate at level $\alpha \in (0,1)$. Because time series data structures produce a highly dependent null hypothesis structure, a joint MTP is necessary; a marginal MTP that tests each hypothesis independently of all others is not appropriate in this case [3]. Although the choice of the Type I Error Rate is left to the researcher, using the False Discovery Rate (FDR) may provide results that can be easily interpreted within a scientific context. The FDR Type I Error Rate is defined as the mean proportion of false positives among the rejected hypotheses. By controlling an MTP at FDR level α, we can ensure with probability $1 - \alpha$ that the average proportion of false positives is α, which provides the user with a measure of reliability for the results. An MTP that controls the FDR at level α ensures that an average proportion of $1 - \alpha$ of the rejected hypotheses reflect true performance differences. Additionally, if FDR results were collected for a larger number of generations G^*, then we could also expect a proportion of $1 - \alpha$ of the rejected hypotheses in the range $[G + 1, \ldots, G^*]$ to be reliable.

MTPs may be summarized in terms of adjusted p-values and confidence region plots. For each hypothesis, the adjusted p-value is the minimum value of α necessary to reject the hypothesis. Confidence regions depict a plausible range of values for the true performance difference. At each generation, we reject the null hypothesis (3) if and only if the confidence region does not contain zero. Because confidence regions are a function of the data, they either contain or do not contain the true performance difference at each generation; however, if the comparison experiment is repeated a large number of times, a proportion of $1 - \alpha$ of all confidence regions produced would contain the true performance difference curve. Estimated confidence regions are currently available for bootstrap-based MTPs controlling the Family-Wise Error Rate (FWER), which is defined as $P(V > 0)$, and the generalized Family-Wise Error Rate (gFWER) $P(V > k), k \in \mathbb{Z}^+$ [12]. However, deriving FDR confidence regions is currently an active area of research.

Although the bootstrap is an effective tool in hypothesis testing applications, this technique is computationally intensive and may require many resamplings (i.e. $B \geq 10000$) to produce accurate results. Because of the inherent dependence of time series data structures, we propose limiting the testing to data collected at regular generational intervals. Although less comprehensive than testing at every generation, the computational savings will typically outweigh the small loss in accuracy. In this setting, confidence regions would be interpolated from the limited test results. In the case of a performance curve comparison based on interval sampling, the main question of interest is how large to set the interval size h. In practice, the researcher may choose among candidate values of h in terms of the relative improvement in a metric such as the mean squared error.

4   Example: A Performance Curve Comparison

To illustrate the proposed methodology, we analyzed performance curve data from several competing EAs [6] for the following variant of Ackley's function:

$$f = -c_1 \exp\left(-c_2\sqrt{\frac{1}{D}\sum_{d=1}^{D}X_d^2}\right) - \exp\left(\frac{1}{D}\sum_{d=1}^{D}\cos(c_3 X_d)\right) + c_1 + \exp(1) \quad (6)$$

with the following parameters supplied for this example:

$$c_1 = 20;\ c_2 = 0.2;\ c_3 = 2\pi;\ D = 10;\ X_d \in (-20, 20),\ d \in \{1, \ldots, D\}.$$

Minimizing Ackley's function (6) is a canonical optimization problem because it has a large number of local optima and a known solution. An EA described in [11] was applied to this problem in a study to select among four candidate mutation rates. The four corresponding algorithms will be indexed by the set $A = \{2, 4, 6, 8\}$ whose elements respectively denote the gene-wise percentage mutation rate. These EAs were identical in all other aspects. We chose the expected value $\mu(Y_a)$ as our performance curve (1). A total of $n_a = 100$ trials of each algorithm $a \in A$ were conducted to collect time series data and estimate $\mu(Y_a)$ with (2). Each trial was conducted for a total of $G = 10000$ generations with data recorded at each generation. The sample mean performance curve for each algorithm is plotted in Figure 1. On average, it appears that EA 4 best minimizes the Ackley function (6) for approximately the first 2000 generations, and it is thereafter eclipsed by EA 6, which appears to outperform all other procedures for the duration of the trials. In order to substantiate the validity of these claims, we conducted pairwise comparisons of the algorithms using the procedure of Section 3.

We tested each pair of algorithms $a, b \in A$; $a \neq b$ for a difference in mean performance at each generation. The null hypothesis (3) states that there exists no difference in expected performance between EAs a and b at each generation. We tested this null hypothesis using the boostrap based SSMaxT [3] FWER-controlling MTP and also using the FDR Conservative MTP, both of which controlled their respective Type I Error Rates at level $\alpha = 0.05$ with $B = 10000$ bootstrap re-samplings. The results of these tests are contained in Table 1. The **Rejections** column of Table 1 shows the number of rejected hypotheses in the pairwise test. Because the null hypothesis (3) has a two-sided alternative, a rejection may correspond to a significant performance difference in either direction and may be determined by examining the sample mean performance curve plot of Figure 1. The pairwise FDR Conservative tests of $\mu_2 - \mu_4$, $\mu_4 - \mu_8$, and $\mu_6 - \mu_8$ all result in significant performance differences at all $G = 10000$ generations. As shown in Figure 1, EA 4 appears to better optimize (6) compared with EAs 2 and 8, and because all G hypotheses were rejected, we conclude that EA 4 performs significantly better than EAs 2 and 8 for the duration of this experiment. Likewise, we also conclude that EA 6 significantly outperforms EA 8 across all generations studied.

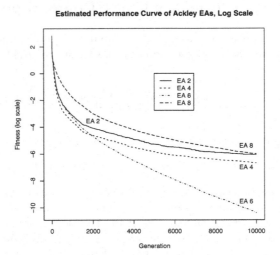

Fig. 1. Sample mean fitness (log scale) as a function of generation in 4 EAs seeking to optimize the Ackley function (6). Each plot is an estimate of the EA's expected value performance curve at each generation based upon $n_a = 100$ trials.

Table 1. Summary results for pairwise performance curve comparisons of four EAs seeking to optimize the Ackley function (6) based upon multiple hypothesis testing of the null hypothesis (3) at each generation $g \in \{1, \ldots, G\}$

Null	TI Error	MTP	Rejections	MIG	Max Adjp	Preferred EA
$\mu_2 - \mu_4 = 0$	FWER	SSMaxT	9952	261	-	4
$\mu_2 - \mu_4 = 0$	FDR	Conserv.	10000	0	0.0256	4
$\mu_2 - \mu_6 = 0$	FWER	SSMaxT	8707	1394	-	6
$\mu_2 - \mu_6 = 0$	FDR	Conserv.	8824	1243	-	6
$\mu_2 - \mu_8 = 0$	FWER	SSMaxT	8159	10000	-	2
$\mu_2 - \mu_8 = 0$	FDR	Conserv.	8272	10000	-	2
$\mu_4 - \mu_6 = 0$	FWER	SSMaxT	8960	2285	-	6
$\mu_4 - \mu_6 = 0$	FDR	Conserv.	9004	2285	-	6
$\mu_4 - \mu_8 = 0$	FWER	SSMaxT	9984	29	-	4
$\mu_4 - \mu_8 = 0$	FDR	Conserv.	10000	0	0.005	4
$\mu_6 - \mu_8 = 0$	FWER	SSMaxT	9989	22	-	6
$\mu_6 - \mu_8 = 0$	FDR	Conserv.	10000	0	0.005	6

For each pairwise comparison, the maximum insignificant generation (**MIG**) column of Table 1 indicates that last generation at which the null hypothesis is not rejected. Using these values and the mean performance plot in Figure 1, we can draw conclusions about the range at which particular algorithms outperform others. For instance, EA 2 outperforms EA 8 over much of the observed spectrum, which results in 8159 rejections for the FWER SSMaxT test. However, the null hypothesis is not rejected at the 10000th generation, meaning that these EAs's performances do not differ significantly at the final generation. Moreover, we see that EA 6 creates significant separation in performance from EA 2 for all generations after the 1394th generation in the FWER SSMaxT test, and likewise EA 6 significantly outperforms EA 4 at all generations after 2285. For the three pairwise performance comparisons that resulted in rejections of all G null hypotheses, we have displayed the maximum adjusted p-value (**Max Adjp**) in Table 1 of the G simultaneous tests. In the case of the FDR Conservative tests of $\mu_4 - \mu_8$ and $\mu_6 - \mu_8$, all hypotheses were rejected with $p \leq 0.005$. Finally, the **Preferred EA** column of Table 1 displays a qualitative overall judgment of the preferred algorithm. As a heuristic standard, we choose to prefer an algorithm if it performs significantly better than another for at least 75 percent of the generations sampled. The comparisons of EAs 2 and 4 to EA 6 are also of interest. In both tests, EA 6 performs significantly worse than the others at early generations but later overtakes both algorithms. In each of these comparisons, EA 6 significantly outperforms the competing algorithm for the duration of the final 7500 generations.

These observations may be further substantiated in the depiction of FWER 0.95 confidence regions for each of the pairwise SSMaxT tests. In each of the comparisons, the null hypothesis is rejected at generation g if and only if the confidence region does not contain the value zero at that generation. When significant performance differences exist, the confidence region will lie below zero if the first algorithm better minimizes (6), and this region will lie above zero if the second algorithm performs significantly better. For a maximization problem, the situation is reversed. As an example, the 95% confidence region for the test of $\mu_4 - \mu_6 = 0$ is depicted in Figure 2 with attention restricted to generations greater than 1000 in order to provide a magnified view. These results indicate that EA 4 significantly outperforms EA 6 in the early generations. Somewhat before the 2000th generation, the upper bound of the confidence region crosses the line $y = 0$, and the performance difference between the two algorithms is insignificant until the lower bound crosses this line somewhat after the 2000th generation. EA 6 significantly outperforms EA 4 for the remainder of the generations considered because the entire confidence region is above the line $y = 0$. Figure 2 also contains estimated confidence regions produced by restricting comparison to data collected at every 100th generation. The full confidence region was estimated using a linear interpolation at the missing generational values. Qualitatively, the estimated confidence regions approximate the full data regions reasonably well. Furthermore, the computational time required to perform all six pairwise tests of the 4 algorithms studied was reduced from approximately two days to 15 minutes using the **multtest** package [8] of the R statistical programming language on a server with a 2.4 GHz processor and approximately 3.4 GB of RAM. In this application, it is reasonable to conclude that the remarkable computational savings justifies the small loss in accuracy of the confidence region

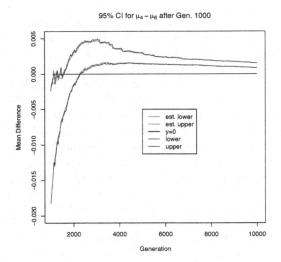

Fig. 2. 0.95 confidence region for the test of $\mu_4 - \mu_6$ with attention restricted to generations after 1000 to provide enhanced magnification

plots and test results. The pairwise comparisons suggest clear performance differences between the EAs over many of the $G = 10000$ generations studied. EA 4 significantly outperforms EAs 2 and 8 at nearly all generations. EA 4 initially outperforms EA 6 by a significant margin but is eventually overtaken. Table 2 displays conclusions drawn from a closer inspection of the FDR comparison between EAs 4 and 6 at a variety of generational intervals. After frequent lead changes in the first 68 generations, EA 4 outperformed EA 6 through generation 1862. However, this performance difference became insignificant at generation 1548. Likewise, EA 6 insignificantly outperformed EA 4 from generation 1863 to generation 2285 and significantly outperformed EA 4 thereafter. We conclude that EA 4 is preferred in most of the first 1547 generations, EA 6 performs best after generation 2285, and the EAs are approximately equal in between.

The proposed methodology offers a convenient and flexible framework for evaluating algorithmic performance and designing adaptive optimization strategies. The researcher may choose any desired performance curve and does not need to rely upon distributional assumptions for the data collected. These techniques may be applied to compare arbitrary sets of stochastic algorithms in any optimization setting. Stochastic algorithms may be compared to deterministic procedures with only small changes to the hypothesis structure (3) and test statistics (5). Furthermore, the proposed data reduction technique provides an avenue for researchers to estimate the results of a full generational analysis in a more computationally tractable manner. Performance comparison is largely a retrospective procedure for validating experimental results. As such, it is not designed to seek out the best candidate optimum for the problem at hand; indeed,

Table 2. A comparison of EAs 4 and 6 at a variety of generational intervals in the Ackley function case study. The preferred EA is selected by mean performance at each generation, and this performance is classified as either significant or insignificant based upon the results of the FDR multiple hypothesis test of equality in means.

Generation Interval	Preferred EA	Difference
$1 - 68$	Either	Insignificant
$69 - 127$	4	Insignificant
$128 - 1128$	4	Significant
$1129 - 1466$	4	Mixed
$1467 - 1547$	4	Significant
$1548 - 1862$	4	Insignificant
$1863 - 2285$	6	Insignificant
$2286 - 10000$	6	Significant

running any one of the four candidates of Section 4 for all the computational time allotted to our comparison would certainly have produced a better result than any obtained in our study. However, statistical performance comparison may be especially helpful in applications that are sufficiently similar to well-studied examples and may aid the design of heuristic procedures in many settings.

Acknowledgments

The authors wish to thank Sandrine Dudoit for her suggestions. The contribution of S. J. Ovaska was funded by the Academy of Finland under Grant 124721.

References

[1] Castillo-Valdivieso, P.A., Merelo, J.J., Prieto, A., Rojas, I., Romero, G.: Statistical analysis of the parameters of a neuro-genetic algorithm. IEEE Transactions on Neural Networks 13(6), 1374–1394 (2002)

[2] Czarn, A., MacNish, C., Vijayan, K., Turlach, B., Gupta, R.: Statistical exploratory analysis of genetic algorithms. IEEE Transactions on Evolutionary Computation 8(4), 405–421 (2004)

[3] Dudoit, S., van der Laan, M.J.: Multiple Testing Procedures and Applications to Genomics. Springer, New York (2008)

[4] Dudoit, S., van der Laan, M.J., Pollard, K.S.: Multiple testing. Part I. Single-step procedures for control of general Type I error rates. Statistical Applications in Genetics and Molecular Biology 3(1), Article 13 (2004)

[5] Efron, B., Tibshirani, R.: An Introduction to the Bootstrap. Chapman and Hall, Boca Raton (1994)

[6] Fogel, D.B.: Evolutionary Computation: Toward a New Philosophy of Machine Intelligence. Wiley-IEEE Press, Hoboken (2000)

[7] François, O., Lavergne, C.: Design of evolutionary algorithms – a statistical perspective. IEEE Transactions on Evolutionary Computation 5(2), 129–148 (2001)

[8] Pollard, K.S., Dudoit, S., van der Laan, M.J.: Bioinformatics and Computational Biology Solutions Using R and Bioconductor. In: Multiple Testing Procedures: the multtest Package and Applications to Genomics. Statistics for Biology and Health, vol. 15, pp. 249–271. Springer, Heidelberg (2005)

[9] Pollard, K.S., van der Laan, M.J.: Choice of a null distribution in resampling-based multiple testing. Journal of Statistical Planning and Inference 125(1-2), 85–100 (2004)

[10] Rojas, I., González, J., Pomares, H., Merelo, J.J., Castillo, P.A., Romero, G.: Statistical analysis of the main parameters involved in the design of a genetic algorithm. IEEE Transactions on Systems, Man, and Cybernetics – Part C 32(1), 31–37 (2002)

[11] Shilane, D., Martikainen, J., Dudoit, S., Ovaska, S.J.: A general framework for statistical performance comparison of evolutionary computation algorithms. Information Sciences 178(14), 2870–2879 (2008)

[12] van der Laan, M.J., Dudoit, S., Pollard, K.S.: Augmentation procedures for control of the generalized family-wise error rate and tail probabilities for the proportion of false positives. Statistical Applications in Genetics and Molecular Biology 3(1), Article 15 (2004)

[13] Wolpert, D.H., Macready, W.G.: No free lunch theorems for optimization. IEEE Transactions on Evolutionary Computation 1(1), 67–82 (1997)

Multiobjective Genetic Programming
for Nonlinear System Identification

Lavinia Ferariu and Alina Patelli

"Gh. Asachi" Technical University of Iasi,
Department of Automatic Control and Applied Informatics,
D. Mangeron 53A, 700050, Iasi, Romania
lferaru@ac.tuiasi.ro, apatelli@ac.tuiasi.ro

Abstract. The paper presents a novel identification method, which makes use of genetic programming for concomitant flexible selection of models structure and parameters. The case of nonlinear models, linear in parameters is addressed. To increase the convergence speed, the proposed algorithm considers customized genetic operators and a local optimization procedure, based on QR decomposition, able to efficiently exploit the linearity of the model subject to its parameters. Both the model accuracy and parsimony are improved via a multiobjective optimization, considering different priority levels for the involved objectives. An enhanced Pareto loop is implemented, by means of a special fitness assignment technique and a migration mechanism, in order to evolve accurate and compact representations of dynamic nonlinear systems. The experimental results reveal the benefits of the proposed methodology within the framework of an industrial system identification.

Keywords: genetic programming, multiobjective optimisation, nonlinear system identification.

1 Introduction

Creating mathematical descriptions for dynamic systems using measurements of plant input - output variables remains a difficult task, especially because, in almost all industrial applications, poor *a priori* information about the model structure is available [1]. Usually, the identification methodologies select optimal models using a set of predefined structures. Though, the multitude of computational models illustrates that no model architecture can be uniformly better than the others. As consequence, advanced approaches have to ensure the adaptation of the model topology subject to the problem to solve [2].

In that context, genetic programming techniques become attractive, as they can efficiently breed a population of possible models working simultaneously on the model parameters and structure [3], [4]. Within the framework of system identification, the efficiency of genetic programming techniques could also be related to the inherent benefits of evolutionary algorithms (the ability to cope with ill - behaved problem domains, multimodality, discontinuity, time - variance, noise) [5].

M. Kolehmainen et al. (Eds.): ICANNGA 2009, LNCS 5495, pp. 233–242, 2009.
© Springer-Verlag Berlin Heidelberg 2009

The present approach addresses nonlinear models, linear in parameters, which may provide the approximation of any dynamic nonlinear bounded function with any desired degree of accuracy [6]. Most of the other identification approaches configure the structure of these models using a large set of regressors and/or deterministic rules for adding or excluding the model terms. On the other hand, the suggested evolutionary design procedure permits a flexible configuration of possible linear combinations of regressors, encoded in tree - based individuals. Each potential model may include any combination of lagged plant variables, subject to the use of $\{+, *\}$ operators.

The quality of the nonlinear potential models is evaluated with respect to their accuracy and complexity order. The multiobjective optimisation favours the selection of accurate and simple tree - based individuals, with expected good generalisation capabilities. The technique is able to reduce the risk of producing over fitted models, frequently encountered within the context of genetic programming. Given the specific context of system identification, the accuracy objective is assigned with a higher priority. To fit that end, special mechanisms are used, such as a customised fitness assignment procedure and a migration – based evolution.

Moreover, the convergence speed of the algorithm is improved by means of hybridisation with QR decomposition, acting as a local optimisation procedure. Also, enhanced genetic operators have been implemented to encourage the production of fitter individuals.

The paper is organised as follows. Section 2 briefly describes the mathematical basis in relation to the nonlinear identification problem to solve. The main steps of the classic multiobjective evolutionary loop are discussed in section 3, whilst section 4 browses through the enhancements implemented to improve the algorithm performances. Section 5 is devoted to experimental results and addresses the identification of an industrial system. Conclusions are outlined in section 6.

2 Nonlinear Model

The suggested approach evolves input - output nonlinear models, linear in parameters, which have been proved to be universal approximators of nonlinear dynamics [6]. For the sake of simplicity, the case of single input - single output systems is addressed:

$$\hat{y}(k) = \sum_{i=1}^{r} c_i F_i(\mathbf{x}(k)), \quad c_i \in R, \, k = 1,..p \,, \text{ where} \tag{1}$$

$$\mathbf{x}(k) = [u(k),...,u(k - n_u), y(k - 1),..., y(k - n_y)] \,. \tag{2}$$

Here, k stands for the current time instant, the system input and output are denoted with u and y respectively, $\hat{y}$ indicates the estimated output provided by the designed model, $\mathbf{x}$ is a vector containing the current and the lagged values of plant input and output, n_u and n_y are the maximum permitted input and output lags.

Nonlinear functions F_i are called regressors and represent atomic combinations (products) of terminals, namely $\mathbf{x}$ vector elements, considered to any exponent. The potential models result as polynomial (linear) combinations of regressors. The model can be rewritten in matrix-based formalism:

$$
\begin{bmatrix}
F_1(\mathbf{x}(1)) & .. & \cdots & F_r(\mathbf{x}(1)) \\
F_1(\mathbf{x}(2)) & .. & \cdots & F_r(\mathbf{x}(2)) \\
\vdots & & \vdots & \vdots \\
F_1(\mathbf{x}(p)) & .. & \cdots & F_r(\mathbf{x}(p))
\end{bmatrix}
\begin{bmatrix}
c_1 \\ c_2 \\ \vdots \\ c_r
\end{bmatrix}
=
\begin{bmatrix}
\hat{y}(1) \\ \hat{y}(2) \\ \vdots \\ \hat{y}(p)
\end{bmatrix}
\tag{3}
$$

or

$$
\mathbf{F} \cdot \mathbf{c} = \hat{y}, \quad \mathbf{F} \in R^{p \times r}, \mathbf{c} \in R^r, \hat{y} \in R^p \, . \, \cdot
\tag{4}
$$

The regressor matrix $\mathbf{F}$ encodes the structure of the model and vector $\mathbf{c}$ contains its parameters. During the evolutionary loop, the algorithm maintains the variety of the regressors F_i by means of genetic operators. QR decomposition has been deployed as a local optimisation procedure, acting in Lamarckian sense, in order to provide convenient numerical computation of vector $\mathbf{c}$.

The performances of the models are evaluated subject to accuracy and parsimony. That means the genetic programming aims to produce a compact model, having few F_i regressors, for which good approximation capabilities are expected to be obtained even on data different than that used for evaluation during the evolutionary loop.

3 Multiobjective Evolutionary Design Procedure

The approach demands the maximum input and output lags, along with a representative training data set $S = \{(u_i, y_i)\}$, $i = 1, .., p$, including output and input plant variables measurements, able to illustrate the behaviour of the dynamic system. Having no other *a priori* information about the model structure, the algorithm starts the exploration of the search space with a random initial population of tree-encrypted potential models.

The individuals are built by means of recursive combinations between the elements of vector $\mathbf{x}$ (terminal nodes) and function/operator nodes [3], [7], [8]. According to (1), any allowed structure could be obtained using the set of operators $O = \{+, *\}$. Examples of tree-encrypted individuals are indicated in Fig. 1.

The terminal and operator sets have to comply with closure and sufficiency requirements in order to be able to produce a valid model [3]. Firstly, the closure propriety is met, because every operator of O accepts as input parameter any value and type returned by any element of O, as well as any value and type of any terminal of set $\mathbf{x}$. Secondly, the sufficiency property states that the terminal and operator sets should be complex enough to properly encode the desired solution. In that case O is minimally sufficient and the absence of extraneous functions/operators may bring important benefits related to the exploration capabilities of the algorithm. As consequence, the sufficiency propriety is satisfied if an appropriate number of lagged values are included in vector $\mathbf{x}$. Adequate maximum lags (n_u and n_y) could be chosen by trial and error, without involving rich expertise and/or extensive tuning effort, as it is not mandatory to configure a minimally sufficient **x** set. The algorithm has the ability to build the model structures selecting only a convenient subset of elements from **x**.

At each generation of the evolutionary loop, the lot of the current batch of trees is sent to the reproduction pool. Here, the genetic operators – crossover and mutation – work for producing new potential models, called offspring. The resulted offspring are afterwards reunited with their parents in an intermediary population out of which the survivors are chosen by means of a dominance analysis, as described in the following.

The performances of the tree-encrypted individuals are evaluated subject to their accuracy and parsimony, via a multiobjective approach. One objective guarantees the achievement of accurate models in terms of S, as it demands the minimisation of the total squared output error (SEF) computed over the whole training data set. Another objective encourages the selection of simple models and it requests the minimisation of the **C**omplexity **F**unction (CF), which indicates the number n of regressors, existing inside the model M:

$$SEF(M) = \frac{1}{2} \sum_{i=1}^{p} (y_i - \hat{y}_i)^2 \; , \; CF(M) = n \; . \tag{5}$$

These objectives are conflicting, meaning that they have different optimum points. As consequence, the problem admits an infinite set of Pareto - optimal solutions. For any Pareto-optimal solution no improvement could be achieved towards an objective direction without performances degradation subject to at least one of the remaining objectives. In that context, the multiobjective optimisation procedure has not only to generate solutions close to the Pareto-optimal front, but it also has to preserve the population diversity, as each Pareto-optimal solution may illustrate another possible trade-off between the considered objectives.

To solve the multiobjective optimisation, Deb's algorithm based on dominance analysis could be used [9]. Within a population P of individuals, a solution is considered nondominated if, compared with any other solution of the population P, it is fitter with respect to at least one objective function. At insertion, the intermediary extended population of offspring and parents is separated in several Pareto sets of different orders. All nondominated individuals describe the first order Pareto front. If the solutions included in first Pareto front are eliminated, the nondominated solutions determined on the rest of the population represent the second order Pareto front, and so on. Within each front, the individuals furthest away from the rest get to keep their raw fitness value, whilst the others receive a slightly diminished fitness, according to the proximity of their neighbours [9]. The fitness values scaled between 0 and 1 represent the selection probabilities during insertion.

4 Genetic Loop Enhancements

The suggested **E**nhanced **M**ulti **O**bjective **O**ptimisation genetic programming procedure (EMOO) implements several special mechanisms, able to provide increased convergence speed and improved exploration capabilities.

Trees are built from randomly selected elements of vector **x** connected by means of operators from set O, therefore they encode terminals, not regressors, as the desired model (1) requires. To remedy the setback, all trees are transformed from their raw terminal based form to a regressive based form, exploiting the equation:

$$a \cdot (b + c) = a \cdot b + a \cdot c \, . \tag{6}$$

All "+" nodes, successors of "*" nodes, are "lifted" from the leaf level towards the root level, so that in the end, no "+" nodes would be situated on levels lower than "*" nodes. This adjustment facilitates the hybridisation with the QR decomposition, which aids the genetic programming in rapidly finding proper model parameters.

In order to support the genetic operators in effectively optimising the model structure without spoiling the effects of QR decomposition, several customisations have been proposed. The crossover operator randomly selects one cut point in each of the two parents and then swaps the resulting sub-trees, thus obtaining the two offspring. If the selected sub-trees encode the same regressor with similar parameter values, the interchange would be pointless as it would bring no improvement relative to the parents' performances. Even if the parameters of two similar sub-trees are not the same, that is most likely a sign that QR decomposition needs "more time" to adjust their values and swapping the sub-trees would delay that effect. In Fig. 1, nodes 5, 6 and 7 of the first parent, as well as nodes 4, 5 and 6 of the second parent store the same regressor. Therefore, the regressor root along with all the other nodes on the path to the tree root will be eliminated from the potential cut-point list. Thus, the potential cut-point list is drastically reduced, saving the time and computational effort of redundant or harmful sub-tree swaps. Even so, not all flaws of the crossover – QR decomposition collaboration have been removed. In the late stages of the algorithm the phenomenon of compensation may emerge, meaning that well adapted individuals may contain several regressors with low coefficients, while the same performances may be achieved by only one regressor with a higher coefficient value or a greater exponent. That is why mutation has been improved to handle not only the alteration of terminal names, but also to change their exponents.

Several enhancements have been implemented to deal with the multiobjective optimisation, during the evolutionary loop. From the application standpoint it is more important to produce accurate models than simple ones. In that context, the suggested design procedure associates a higher priority to *SEF* minimisation. To increase the

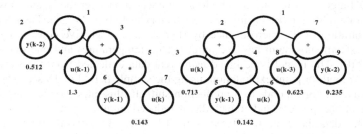

Fig. 1. Parents in the reproduction pool

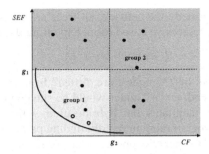

Fig. 2. Population clustering for multiobjective optimisation

selection pressure in favour of *SEF* compliant trees, the initial population has been split into two subpopulations. The first one is evolved via a single objective procedure based solely on *SEF*, while the second subpopulation is evolved subject to a multiobjective optimisation. Once at *No_m* generations the migration is permitted, the two populations exchange their best trees, thus increasing the weight of the *SEF* minimisation objective, without completely ignoring the lowest priority objective.

The subpopulation which evolves subject to multiobjective optimisation undergoes a dominance-based procedure. At each generation, the goal values are updated according to the mean performances of the population, each goal being associated to an objective. All the trees that comply with both goals are assigned with fitness values based only on their *SEF* related performances. The rest of the trees, situated outside the boundaries set by the considered goals, are evaluated according to the classic Pareto procedure. The maximum fitness value receivable by the individuals in the second group is lower than the smallest of fitness values assigned to the first group. In Fig. 2, the solid curve represents the desired optimal front. The assignment of the trees in group 1 with the highest fitness values encourages the individuals to "shift" during the evolutionary loop towards a convenient zone of the Pareto optimal front (see individuals marked with "o" in Fig. 2).

5   Application

The EMOO algorithm was used to build a nonlinear model for the steam subsystem of the **E**vaporation **S**tation (ES) from the sugar factory of Lublin, Poland. The ES increases the concentration of the sucrose juice by running it through five consecutive heating units. The steam subsystem has one input (steam temperature) and one output (steam pressure) [10]. The model is designed using real data collected from the sugar factory during one month of plant exploitation, using the sample period of 10 sec. The selected learning data set corresponds to a production shift. It illustrates the maximum possible excitation of the process and it includes a reduced number of missing or uncertain values. The data have been filtered with 4th order Butterworth filters and, then, decimated using each 10th sampled value. The validation of the model is done with respect to the testing data set, which includes measurements acquired from the previous month of plant exploitation. No analytical model of the nonlinear plant is available.

The experimental comparative study calls in sequel the suggested evolutionary method already described (EMOO) and two other genetic programming procedures. The upgraded elitist **S**imple **O**bjective **O**ptimisation (SOO) procedure evolves the tree-based individuals only subject to *SEF* minimisation. The **M**ulti **O**bjective **O**ptimisation (MOO) procedure considers the same level of priority for the involved objectives and does not make use of the migration mechanism and of the fitness computation algorithm described in Section 4.

Establishing adequate values for the algorithm parameters is a context dependent issue which could be addressed by trial and error. As indicated in Table 1, the experiments were done using different values of training data set length (p), maximum input lag (n_u), maximum output lag (n_y), number of individuals per generation and maximum number of training generations.

Firstly, the performances of SOO have been analysed and compared against EMOO. The experimental conclusions are summarised in Table 2. The SOO algorithm assesses the performances of the current population's individuals only with respect to their accuracy, completely ignoring the complexity criterion. As a direct consequence, higher maximum lags lead to increased number of regressors. That behaviour may tend to become disturbing, as the solution's generalisation capabilities decreases. Models M1 through M3 show poorer *SEF* values over the validation data set, caused by overfitting. Another factor that leads to unsatisfactory generalisation behaviour is the length and content of the training data set. The third important setback of SOO procedures has to do with their elitist nature. Eliminating the offspring that show worse performances than their parents has a good impact on the algorithm's convergence speed, but toward the end of the evolutionary loop, the diversity of the population is severely diminished, as the final generations get saturated with fairly identical trees. At this point, no matter how many training steps are considered, the *SEF* value stops improving (M4$\rightarrow$M6).

The same experiments were carried out using MOO procedure. The MOO algorithm considers the insertion based on a dominance analysis described in Section 3. This strategy reduces the risk of low population diversity towards the end of the evolutionary loop and facilitates a better coverage of the problem search space, with a noticeable improvement in model accuracy. Though, the convergence speed is lower than in the case of the SOO alternative. The first order front generated by the MOO procedure in M1 case (Table 1) is depicted in Fig. 3. The figure also contains the first order front selected from the population of the last generation produced by EMOO.

Table 1. Algoritm parameters sets used for experimental trials

Model ID	p	Lags	individuals number/ generations number
M1	197	$n_u=3, n_y=3$	50/50
M2	197	$n_u=5, n_y=4$	50/50
M3	197	$n_u=7, n_y=6$	50/50
M4	200	$n_u=3, n_y=2$	60/80
M5	297	$n_u=3, n_y=2$	60/100
M6	350	$n_u=3, n_y=2$	80/200

Table 2. SOO vs EMOO algorithm performances

| Model | CF | | MSEF (SEF/p) | | | |
| | | | Training | | Validation | |
	SOO	EMOO	SOO	EMOO	SOO	EMOO
M1	7	4	0.736	0.700	0.740	0.720
M2	9	5	0.700	0.650	0.875	0.670
M3	12	4	0.650	0.630	0.890	0.700
M4	8	5	0.740	0.700	0.710	0.715
M5	9	4	0.735	0.650	0.750	0.780
M6	8	5	0.745	0.635	0.730	0.750

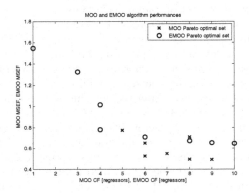

Fig. 3. EMOO advantages over MOO

Supplementary results achieved for EMOO are listed in Table 2 and Table 3. The EMOO yields substantially better results. As revealed in Fig. 3, the EMOO is focused on a convenient zone of the Pareto front, as it considers high priority for *SEF* minimisation. Moreover, EMOO preserves a reduced number of regressors, even for larger maximum lags (M1→M3). Accurate models are obtained relatively fast, without altering the diversity of populations.

The EMOO procedure is more flexible than the standard MOO as it features two extra refined tuning mechanisms. The first one refers to the special fitness assignment mechanism. As a general rule, too simple trees cannot encrypt accurate models. Therefore, if the *CF* goal value is just as restrictive as the *SEF* goal value, the population will rapidly be taken over by rudimentary models, unable to provide good accuracy. Results are better, as the *CF* goal laxity is increased (M7→M9). The second tuning mechanism refers to the migration technique. Once every *No_m* generations the two subpopulations exchange genetic material. If the migration rate is too high, the selection pressure imposed by the individuals with low *SEF* values from the SOO subpopulation will counterbalance the efforts of the MOO complementary procedure (M10), leading to results similar to the standard elitist SOO (accurate, but complex). A rare migration produces results which resemble the classic MOO case (M11). All the results in Table 3 were obtained using the following parameters: 197 training data points (*p*), 3 input lags, 2 output lags, 50 individuals per generation and 50 training

Table 3. EMOO algorithm performances

| Model no | Goals | | No_m | CF | MSEF (*SEF/p*) | |
	SEF	*CF*			Training	Validation
M7	average	average	10	1	7.535	12.35
M8	average	1.5*average	10	4	0.732	0.750
M9	average	2*average	10	5	0.625	0.680
M10	average	2*average	3	8	0.725	0.750
M11	average	2*average	25	3	0.890	0.900

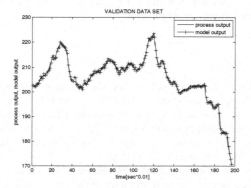

Fig. 4. Model validation

generations. In M9 case, the model with the lowest *SEF* value achieved at the last generation was evaluated over the validation data set (Fig. 4). The selected model is accurate and simple (8 regressors), featuring good generalisation capacity (at validation, the mean squared error value is 0.705 and the absolute relative error is achieved between $[8.198*10^{-5}, 0.0254]$).

6 Conclusions

The enhanced multiobjective optimisation genetic programming algorithm described in this paper is a new instrument in system identification. Its ability to provide flexible yet compact models with good performances both on training and validation sets, without any prerequisites other than training/ testing data, makes it compatible with the special requirements of complex systems featuring dynamic nonlinearities. The search procedure is unsupervised and improves simultaneously the model structure and parameters, excluding the need of any off-line term reduction mechanisms.

Due to the linear in parameters formalism that the generated models are compliant with, a hybridisation with a local optimisation procedure is facilitated. Along with the random distribution of the initial population, the enhanced genetic operators guarantee a speedy convergence and a good variety of the population at any generation. By assessing the generated models with respect to multiple objectives, the genetic search becomes more refined. Isolating the individuals compliant with adjustable objective goals enforces the priority of the accuracy requirement over the complexity one. Moreover, the separation in two subpopulations, evolved simultaneously, one according to a simple objective technique (accuracy) and the other to a multiobjective technique (accuracy and complexity) allows the generated models to embody the advantages of both evolutionary procedures.

The proposed method is time and resource consuming, therefore it is recommended for complex nonlinear systems where no rich *a priori* data is available and high accuracy models are required. Though, the automatic generation of potential solutions could bring important benefits to the total design time, if compared with the case of "trial and error" model configuration.

References

1. Flemming, P.J., Purshouse, R.C.: Evolutionary Algorithms in Control Systems Engieering: A Survey. Control Engineering Practice 10, 1223–1241 (2002)
2. Ferariu, L., Voicu, M.: Nonlinear System Identification Based on Evolutionary Dynamic Neural Networks wih Hybrid Structure. In: 16th IFAC Congress. Elsevier, Prague (2005)
3. Koza, J.R.: Genetic Programming: On the Programming of Computers by Means of Natural Selection. MIT Press, Cambridge (1992)
4. Rodriguez-Vasquez, K., Fonseca, C.M., Flemming, P.J.: Identifying the Structure of Nonlinear Dynamic Systems Using Multiobjective Genetic Programming. IEEE Trans. on Systems Man and Cybernetics, Part A – Systems and Humans 34, 531–534 (2004)
5. Bäck, T., Fogel, D., Michalewicz, Z.: Evolutionary Computation 2. In: Advanced Algorithms and Operators. Institute of Physics Publishing, Bristol (2000)
6. Wey, H., Billings, S.A., Lui, J.: Term and Variable Selection for Nonlinear Models. Int. J. Control 77, 86–110 (2004)
7. Kumar, A.V., Balasubramaniam, P.: Optimal Control for Linear Singular Systems Using Genetic Programming. Applied Mathematics and Computation 192, 78–89 (2007)
8. Rodriguez-Vasquez, K., Flemming, P.J.: A Genetic Programming/NARMAX Approach to Nonlinear System Identification. In: 2nd International Conference on Genetic Algorithms in Engineering Systems: Innovations and Applications, GALESIA, pp. 409–414 (1997)
9. Deb, K.: Multiobjective Optimization using Evolutionary Algorithms. John Wiley and Sons, Chichester (2001)
10. Marcu, T., Mirea, L., Ferariu, L., Frank, P.M.: Miscellaneous Neural Networks Applied to Fault Detection and Isolation of an Evaporation Station. In: 4th IFAC Symposium on Fault Detection, Supervision and Safety for Technical Processes. Elsevier, Budapest (2000)

NEAT in HyperNEAT Substituted with Genetic Programming

Zdeněk Buk, Jan Koutník, and Miroslav Šnorek

Computational Intelligence Group
Department of Computer Science and Engineering
Faculty of Electrical Engineering
Czech Technical University in Prague
{bukz1,koutnij,snorek}@fel.cvut.cz
http://cig.felk.cvut.cz

Abstract. In this paper we present application of genetic programming (GP) [1] to evolution of indirect encoding of neural network weights. We compare usage of original HyperNEAT algorithm with our implementation, in which we replaced the underlying NEAT with genetic programming. The algorithm was named HyperGP. The evolved neural networks were used as controllers of autonomous mobile agents (robots) in simulation. The agents were trained to drive with maximum average speed. This forces them to learn how to drive on roads and avoid collisions. The genetic programming lacking the NEAT complexification property shows better exploration ability and tends to generate more complex solutions in fewer generations. On the other hand, the basic genetic programming generates quite complex functions for weights generation. Both approaches generate neural controllers with similar abilities.

1 Introduction

In training of artificial neural networks using evolutionary algorithms a method of encoding the neural networks into individuals in the evolution is needed. Basically, there are two types of encodings. Direct encoding of either connection weights or a network structure causes individuals and therefore the search space complexity growth. Indirect encoding that utilizes a system, which develops the neural network from information encoded by the individual can overcome such drawback of direct encoding.

One of the most perspective algorithm of neural network weights and structure encoding is the hypercube encoding invented by Ken Stanley [2,3] as HyperNEAT algorithm. The HyperNEAT algorithm consists of a function or a set of functions, which generates weights for neural network. The neural network consist of neurons placed in a rectangular mesh called substrate. The substrate coordinates serve as inputs for the function. The function output is the weight of connection between two neurons. In the original HyperNEAT algorithm, the function is called CPPN (Compositional Pattern Production Network) and is constructed from a set of nodes with scalar product on their inputs and non-linear function at their outputs. The network structure reminds a neural network

M. Kolehmainen et al. (Eds.): ICANNGA 2009, LNCS 5495, pp. 243–252, 2009.
© Springer-Verlag Berlin Heidelberg 2009

(therefore, it is named network rather than a function). The CPPN in the HyperNEAT is generated using the NEAT (NeuroEvolution of Augmenting Topologies) [3] algorithm. NEAT is a type of evolutionary algorithm, which evolves the network from a simple form featuring complexification and niching.

The NEAT algorithm is the component that we decided to replace with a different style of weights encoding and generation in our approach. Rather, we use genetic programming, which generates functions that compute weights for connections among neurons in the substrate.

The application domain is a control of autonomous agents in simulated environment. The agents are equipped with sensors with scalable resolution. Encoding of neurons weights allows the resolution of the sensory input to be changed independently of the size of the individual that contains the weight generating function. The previous experiments presented in [4] show that the NEAT can produce recurrent neural network, which can control the agent to move through the simulated environment with maximum average speed. The fitness of the evolved neural network is the average speed of the controlled agent. Our goal is to replace the NEAT algorithm with genetic programming and compare it to the original HyperNEAT algorithm.

1.1 Related Work

Many techniques for evolution of either weights or structure of neural networks were already developed such as Analog Genetic Encoding [5,6,7], Continual Evolution Algorithm [8], GNARL [9], Evolino [10] and NeuroEvolution of Augmenting Topologies (NEAT) [11]. The NEAT algorithm became a part of the HyperNEAT algorithm as a tool for evolution of CPPNs.

HyperNEAT algorithm was already applied to control artificial agents in food gathering problem [2]. It was shown that HyperNEAT is capable of large scale networks evolution ($> 8 \cdot 10^6$ connections). The simulated agent was equipped with concentric sensors for food in particular directions linked with effector, which drives the agent to the direction. In our approach, the sensors are organized in polar rays with particular angular and distance resolution. The sensors are sensitive to the surface color.

The agents can share portion of one substrate together [12]. The substrate splits to local but linked areas. The agents can exploit cooperative behavior afterwards.

Agents can complete common goals also with a minimum information from the sensors with evolutionary trained feed forward networks as well [13]. In the case, the agents exhibit reactive behavior.

This paper is organized as follows. Section 2 describes the HyperGP algorithm. Section 3 describes the simulation environment and the agent setup. Section 4 describes the experimental results, performance of HyperGP is compared with HyperNEAT. Final section concludes the paper.

2   HyperGP Algorithm

2.1   Genetic Programming

Our implementation of GP uses recombination and mutation operators and direct representation of mathematical expressions. The final expression is a function of up to 4 variables (x_1, y_1, x_1, y_2), which was named CPPF (Compositional Pattern Production Function). During the evolution the expressions are generated from the sets of elementary functions and atoms (variables and constants). The operators are:

- **Random expression generator** generates expressions with defined depth. The functions list (patterns) defines the set of basic function that the final expressions will be composed of. The atoms list contains the variables (input variables for the final expressions/functions) and constants. At the beginning the first function is randomly chosen from the list and then the generator is applied recursively on this expression - in this case not only functions but also the atoms are used in random selection. The maximal depth of the expression can be specified, so this value is decremented whenever the generator is recursively run on some subexpression.
- **Mutation** selects random place/subexpression in the given expression and replaces the subexpression by some randomly generated one (with respect to the maximum depth of the expression). It splits the expression into particular subexpressions, then randomly chooses one of them, and using the random expression generator replaces it by some random expression. The depth parameter is also used, so the final (mutated) expression fulfills the maximal depth condition.
- **Recombination** combines two expressions. Crossover position is selected randomly in both expressions. It generates the positions of all subexpressions in two given parent expressions (individuals). Then it replaces the subexpression in the first individual by the subexpression from the second individual. So it keeps the positions and swaps the subexpressions. The maximal depth condition is still fulfilled.

2.2   Evolution

We are working with the fixed size population of individuals (expressions). The GP algorithm generates in each iteration the set of offsprings (the number of the offsprings does not depend on the size of the population) using the mutation and recombination operators. The parent population and the offspring population is then joined into one set (pool) and using the selection function (based on the fitness value of the individuals) the new population is created (all the unused individuals are deleted).

3   Experimental Setup

3.1   Simulation Environment

Experiments with the agents were performed in simulation environment called ViVAE (Visual Vector Agent Environment) featuring easy design of simulation scenarios in SVG vector format [4]. There are two types of surfaces in the simulation (road and grass) with different frictions. The grass has friction 5 times higher than the road.

ViVAE supports number of different agents equipped with various sensors for surfaces and other objects in the scenario. In the current experiment, scenario with one agent was user, see Figure 1.

3.2   Agent Setup

The agent in the simulation is controlled by neural network controller constructed by the HyperGP or HyperNEAT. The agent is driven by two simulated wheels and is equipped with a number of sensors. The controlling neural network is organized in a single layer of possibly fully interconnected perceptron (global) type neurons (neurons compute biased scalar product, which is transformed by bipolar logistic sigmoidal function). Steering angle is proportional to inverse actual speed of the robot.

The sensors as well as the neural network are spread in a substrate. Neurons and sensors are addressed with polar coordinates, see Figure 2. Two of the neurons in the output substrate are dedicated to control acceleration of the wheels. The neurons are marked with the red color in Figure 5.

Each individual in the evolution contains three different CPPFs, see Figure 2. Function f_i generates weights between input sensors an the neurons, function f_b

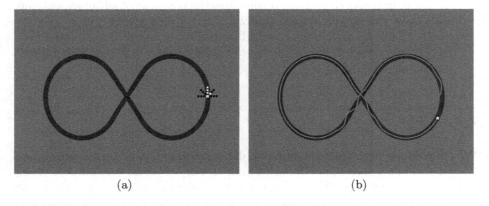

(a)   (b)

Fig. 1. Experimental scenario with an agent placed in starting positions (a). The agent has 3x5 sensors array. The color of the sensor represents a surface friction mapped to 0 and 1 for the neural network input. Track of the trained agent is depicted in (b).

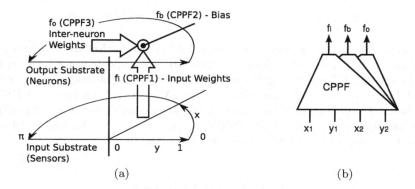

Fig. 2. Organization of the HyperGP substrate. There were two distinct substrates used (a) and three different GP function trees (CPPF) were evolved. Function f_i is weight between input substrate (sensors) and substrate with the neurons. Second function f_b generates bias for neurons in the upper substrate. For bias calculation third and fourth CPPF inputs are set to 0. Last function f_o represents connection weights among neurons in the upper substrate.

computes biases of the neurons and function f_o expresses connection weights among the neurons in the upper layer. Since the bias is a property of a neuron in the output substrate, inputs x_2 and y_2 are setup to 0 for f_b computation.

3.3   HyperGP Setup

The HyperGP algorithm described in Section 2 was executed with the following set of functions, from which the functions are selected.

$$ x+y, \quad x \cdot y, \quad \sin x, \quad \cos x, \quad \tan^{-1} x, \quad \sqrt{|x|}, \quad |x|, \quad e^{-x^2}, \quad e^{-(x-y)^2} \quad (1) $$

The list of atoms used in the functions is the following one:

$$ x_1, \quad x_2, \quad x_3, \quad x_4, \quad -1, \quad \text{RandomReal(-5,5)} \quad (2) $$

Besides the CPPF inputs, there is extra negative multiplier and random constant between -5 and 5. Depth of the expression was set up to 3.

3.4   HyperNEAT Setup

We have used our own implementation of the HyperNEAT algorithm. The NEAT part resembles Stanley's original implementation. The HyperNEAT extension is inspired mainly by the David D'Ambrosio's HyperSharpNEAT[1]. Following function have been used as output function of the NEAT nodes:

---

[1] Both Stanley's original NEAT implementation and D'Ambrosio's HyperSharpNEAT can be found on http://www.cs.ucf.edu/~kstanley

$$\frac{2}{1 + e^{-4.9\,x}} - 1, \quad x, \quad e^{-2.5\,x^2}, \quad |x|, \quad sin(x), \quad cos(x) \tag{3}$$

The parameter settings are summarized in Table 1. Note, that we have extended the original set of constants which determine the genotype distance between two individuals (C_1, C_2 and C_3) by the new constant C_{ACT}. The constant C_{ACT} was added due to the fact that, unlike in classic NEAT, we evolve networks (CPPNs) with heterogeneous nodes. C_{ACT} multiplies the number of not matching output nodes of aligned link genes. The CPPN output nodes were limited to bipolar sigmoidal functions in order to constrain the output.

Table 1. HyperNEAT parameters

Parameter	Value
population size	100
CPPN weights amplitude	3.0
CPPN output amplitude	1.0
controller network weights amplitude	3.0
distance threshold	15.0
distance C_1	2.0
distance C_2	2.0
distance C_3	0.5
distance C_{ACT}	1.0
mating probability	0.75
add link mutation probability	0.3
add node mutation probability	0.1
elitism per species	5%

4   Experimental Results

4.1   HyperGP with Mutation Only

All experimental results are collected from 10 runs of each algorithm. The HyperGP was executed 2×10 times. In the first set, the mutation only was used as the genetic operator. The convergence of the HyperGP algorithm is plotted in Figures 3. Sub-figure (a) contains convergence plots for the 10 experiment runs. Sub-figure (b) contains plot of the whole population in one experiment run (50 iterations). The individuals are sorted according to their fitness (from left to right). We consider the fitness of 0.83 to be enough for the robot to follow the road. HyperGP with mutation only reaches that fitness in 20 generations (median). Mean fitness reached in the 10 runs is 0.875.

Following set of functions is the one generated in one of the experiment run:

$$f_b = e^{-\left(e^{-\frac{1}{2}(x_2 - x_3)^2} - x_2\right)^2} \tag{4}$$

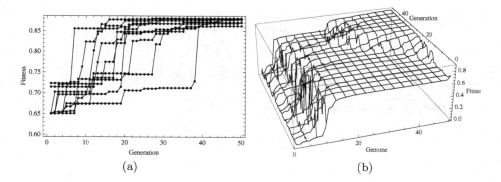

Fig. 3. Convergence of HyperGP algorithm with mutations only. Plot (a) contains linearly interpolated convergence of 10 independent experimental runs. Figure (b) displays how the genomes with the fitness approaching the local optimum are growing in the population. All 50 individuals are sorted by their fitness. Solution with fitness of 0.65 is very common in the initial population. In the populations after 30 generations solution with fitness close to 0.88 spreads in the population.

$$f_o = \sin x_1(x_3 + \sin x_4) \tag{5}$$

$$f_i = \exp\left(\left(\left(e^{-e^{-2\left(x_2 - \tan^{-1}(x_4)\right)^2} + x_3 + x_4}\right)^2 - |x_1|\right)^2\right) \tag{6}$$

4.2   HyperGP with Crossover

In the second set of runs, the crossover operator was added (see Figure 4). Probability of the crossover is 0.75, probability of the mutation is 0.25. We can see that the performance of the algorithm has decreased. The target fitness (0.83) was reached in 7 out of 10 runs only. Algorithm convergence is slowed down. Average fitness reached after 50 evolution generations is of 0.82. Additional increasing of the crossover probability decreased the performance of the algorithm.

4.3   HyperNEAT

The HyperNEAT was setup according to Table 1. The algorithm was executed 10 times. Convergence of the algorithm is plotted in Figure 6. The red line in the figure appears in 50th generation, in which the HyperGP was stopped. The target fitness of 0.83 was reached in 92th generation (median). We can observe that the HyperGP algorithm outperforms the HyperNEAT in the speed of the convergence.

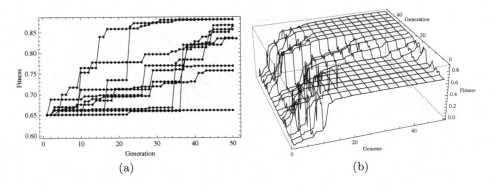

(a)   (b)

Fig. 4. Convergence of HyperGP algorithm with crossover probability of 0.75 and mutation probability of 0.25. Plot (a) contains linearly interpolated convergence of 10 independent experimental runs. Figure (b) displays how the genomes with the fitness approaching the local optimum are growing in the population. All 50 individuals are sorted by their fitness.

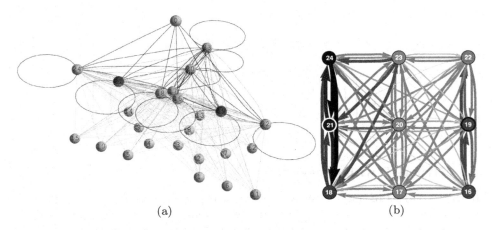

(a)   (b)

Fig. 5. Visualization of the evolved neural network controlling the agents. Figure (a) displays the complete network including inputs. Layer of 3×3 neurons is depicted in figure (b). The network consists of two layers. The bottom layer represents input sensors in a grid of 3x5 inputs. The upper layer contains 9 (3 × 3) neurons. The neurons are mapped into a substrate in polar coordinates to match shape of the input substrate. Red spheres represent neurons (numbers 19 and 21) that steer the agent wheels. The most upper sphere represents bias for the neurons. Connections are displayed with lines. The most visible lines represent connections with stronger synaptic values. The neurons in the neuron layer are emphasized in figure (b). The weights of connections between the neurons are represented by a thickness of arrows. Recurrent weight of a neuron is represented by a color of the disk representing the neuron. Darker neuron has a stronger recurrent self-connection.

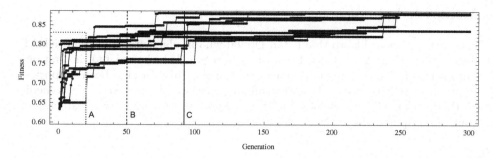

Fig. 6. Convergence of HyperNEAT algorithm in 300 generations performed. The Hy-perGP algorithm was stopped after 50 generations (dashed line - B) and reached target fitness (0.83) in after 20 generations (median, dotted line - A). The HyperNEAT algo-rithm reached the fitness of 0.83 after 92 generations (median, solid line - C).

5   Conclusion and Future Work

In this paper we present results of experiment with generation of recurrent neu-ral network weights using hyper-cubic encoding and multidimensional functions generated by genetic programming. We have replaced the authentic NEAT algo-rithm in the original HyperNEAT by genetic programming, the algorithm was named HyperGP. The neural networks were used as controllers for mobile agents in simulated environment. Fitness function of the particular agents is the average speed of the agent, which forces it to drive on road.

Both HyperNEAT and HyperGP generate suitable solution with agents mov-ing with possibly maximum speed through the scenario, following the roads. HyperNEAT algorithm utilizes it's complexification feature and the resulting CPPNs are simple but the evolution reaches the desired fitness after 92 genera-tions. HyperGP algorithm with same size of population reaches the same fitness in 20 generations.

The HyperGP algorithm has a better explorative property and generates more complex functions in early state of the evolution. The HyperNEAT algorithm sets up the weights in the CPPN in the early evolution phase. More CPPN nodes are added in the latter evolution phase. Both algorithms require more testing and some tuning is required as well.

Both mutation and crossover operators were tested. The best results were reached with mutation operator only. The crossover operation causes major changes in the function. The mutation operator is more tender to the func-tion structure than the crossover. Besides, the problem is sensitive to the CPPF structure and combination of two functions together can completely change the function output within the desired range.

The future work will involve experiments with different presets of both approaches to generation of the weights functions as well testing on different scenarios and goals in the mobile agents.

Acknowledgment

The authors would like to thank Jan Drchal who implemented the HyperNEAT algorithm in Java. As well as CTU student Petr Smejkal who created the first implementation of the agent simulation environment. This work has been supported by the research program "Transdisciplinary Research in the Area of Biomedical Engineering II" (MSM6840770012) sponsored by the Ministry of Education, Youth and Sports of the Czech Republic.

References

1. Koza, J.R.: Genetic Programming: On the Programming of Computers by Means of Natural Selection. The MIT Press, Cambridge (1992)
2. D'Ambrosio, D.B., Stanley, K.O.: A novel generative encoding for exploiting neural network sensor and output geometry. In: GECCO 2007: Proceedings of the 9th annual conference on Genetic and evolutionary computation, pp. 974–981. ACM, New York (2007)
3. Gauci, J., Stanley, K.: Generating large-scale neural networks through discovering geometric regularities. In: GECCO 2007: Proceedings of the 9th annual conference on Genetic and evolutionary computation, pp. 997–1004. ACM, New York (2007)
4. Drchal, J., Koutník, J., Šnorek, M.: Hyperneat controlled robots learn to drive on roads in simulated environment. In: Submitted to IEEE Congress on Evolutionary Computation (CEC 2009) (2009)
5. Mattiussi, C.: Evolutionary synthesis of analog networks. Ph.D thesis, EPFL, Lausanne (2005)
6. Dürr, P., Mattiussi, C., Floreano, D.: Neuroevolution with Analog Genetic Encoding. In: Parallel Problem Solving from Nature - PPSN iX, vol. 9, pp. 671–680. Springer, Heidelberg (2006)
7. Dürr, P., Mattiussi, C., Soltoggio, A., Floreano, D.: Evolvability of Neuromodulated Learning for Robots. In: The 2008 ECSIS Symposium on Learning and Adaptive Behavior in Robotic Systems, pp. 41–46. IEEE Computer Society, Los Alamitos (2008)
8. Buk, Z., Šnorek, M.: Hybrid evolution of heterogeneous neural networks. In: Kůrková, V., Neruda, R., Koutník, J. (eds.) ICANN 2008, Part I. LNCS, vol. 5163, pp. 426–434. Springer, Heidelberg (2008)
9. Angeline, P.J., Saunders, G.M., Pollack, J.B.: An evolutionary algorithm that constructs recurrent neural networks. IEEE Transactions on Neural Networks 5, 54–65 (1993)
10. Schmidhuber, J., Wierstra, D., Gagliolo, M., Gomez, F.: Training recurrent networks by evolino. Neural computation 19(3), 757–779 (2007)
11. Stanley, K.O., Miikkulainen, R.: Evolving neural networks through augmenting topologies. Evolutionary Computation 10, 99–127 (2002)
12. D'Ambrosio, D.B., Stanley, K.O.: Generative encoding for multiagent learning. In: GECCO 2008: Proceedings of the 10th annual conference on Genetic and evolutionary computation, pp. 819–826. ACM, New York (2008)
13. Waibel, M.: Evolution of Cooperation in Artificial Ants. Ph.D thesis, EPFL (2007)

Simulation Studies on a Genetic Algorithm Based Tomographic Reconstruction Using Time-of-Flight Data from Ultrasound Transmission Tomography

Shyam P. Kodali[1,2], Kalyanmoy Deb[1,2],
Sunith Bandaru[1], Prabhat Munshi[1], and N.N. Kishore[1]

[1] Indian Institute of Technology, Dept. of Mechanical Engg., Kanpur, India 208016
{kodalisp,deb,sunithb,pmunshi,nnk}@iitk.ac.in
[2] Dept. of Business Technology, Helsinki School of Economics,
P.O. Box 1210, FI-00101, Helsinki, Finland

Abstract. Results of simulation studies on the application of genetic algorithms (GA) for solving an inverse problem, tomographic reconstruction, using time-of-flight (TOF) data from ultrasound transmission tomography are presented. The TOF data is simulated without taking into consideration the diffraction effects of ultrasound which is reasonably valid when the impedance mismatch in the specimen under consideration is small. The proposed GA based reconstruction algorithm is described and the results for a number of cases are discussed. The sensitivity of the proposed algorithm is studied for various GA parameters viz. the population size, maximum number of generations, crossover probability, and mutation probability. A time complexity analysis of the proposed algorithm shows that the reconstruction times and number of unknowns bears a near quadratic relation enabling the prediction of reconstruction times when dealing with higher resolutions. The performance of proposed algorithm to the reconstruction when TOF data is contaminated with noise is also analyzed and presented. The results obtained are found to be consistent for a wide range of resolutions, type, size, and shape of inclusions.

Keywords: Tomography, Reconstruction, Genetic Algorithms, Inverse Problems.

1 Introduction

Ultrasonic tomography (UT) has been in use for a long period [1] and when the material of test specimen and inclusions are known to have approximately uniform characteristics, UT provides an easy and cost effective way of reproducing the shape, size and location of the inclusion. Acoustic wave attenuation and TOF are two reconstruction parameters which can be used for this purpose. TOF data without considering ray bending is used in the present work.

M. Kolehmainen et al. (Eds.): ICANNGA 2009, LNCS 5495, pp. 253–262, 2009.
© Springer-Verlag Berlin Heidelberg 2009

Popular reconstruction methods for projection data obtained from UT include transform methods like convolution back projection (CBP) [7] and series expansion methods represented by algebraic reconstruction technique (ART) [8]. The tomogram obtained gives a gradation of the physical property being reconstructed over the material-inclusion boundary rather than a clear cut edge. Further transform methods require complete set of projection data for reconstruction which may not be available in a number of practical problems and here lies the motivation for using GA.

GA's are search and optimization techniques based on the dynamics of natural selection and genetics. First proposed by John Holland in 1975, GA's are now being put to use in a wide range of applications [5]. Their versatility is due to the fact that they can handle continuous as well as discrete problems in almost the same way. Also since they work with a population rather than a single initial point global convergence is most certainly ensured. ART on the other hand, suffers from the inherent possibility of getting entrapped in a local optimum.

The principles of various tomographic techniques and their fields of application are described in [6]. The mathematical basis for transmission computed tomographic imaging using straight line reconstruction equations is discussed in [9]. The applicability of GA to ultrasound tomography using simulated TOF is demonstrated in [2] and [10]. Comparative studies on the performance of different reconstruction algorithms applied to non-destructive evaluation with limited data are presented in [3]. The present work is an extension to our earlier one [10].

The process of reconstruction consists of two major steps, the first is acquisition of TOF data and the second is using the acquired data to reconstruct the specimen under consideration. The following sections convey the adopted approach; in section 2 we explain the simulation procedure adopted for acquiring TOF data, in section 3 we discuss the GA based reconstruction algorithm developed, in section 4 we present our results and finally we present our conclusions in section 5.

2   Simulation of Time-of-Flight Data

To simulate the TOF of ultrasound rays, the specimen under consideration is represented as a grid of certain integer values corresponding to different materials. A number of ultrasound sources(S) and detectors(D) are positioned around the specimen in what is termed as modified cross-hole geometry [4]. For any configuration, the sources are actuated in sequence one at a time, and from each of these ultrasound rays travel to each of the detectors, giving us TOF data equal to the product of the number of sources and the number of detectors. For example, considering S=6 sources, D=6 detectors and C=6 configurations we get SxDxC=216 readings [10], which is essentially the input to our reconstruction algorithm. The ray coverage for this case is shown in Fig. 1(a). The TOF is taken to be the arrival time corresponding to the first peak of the signal sensed by the detector as shown in Fig. 1(b). If experimental data is available for these set of sources and detectors, the same can be used in our study. Else, we simulate the TOF by using the following procedure.

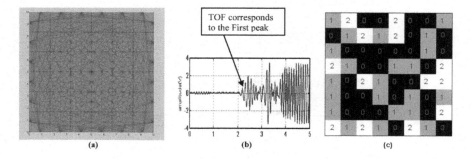

Fig. 1. (a) Ray coverage, (b) Ultrasound signal indicating TOF, (c) Visual display of a member from the initial population with 3 materials

The simulation of TOF assumes that the ultrasound rays follow straight paths from source to the detector. This assumption is reasonable when the impedance mismatch within the specimen is small. The simulated TOF for a ray originating from j-th source and terminating at the k-th detector is estimated using equation (1):

$$tof(r) = \sum_{m}^{M} \frac{l_m^{(r)}(j,k)}{v_m},$$ (1)

where,

M = number of cells intercepted by the ray,
$l_m^{(r)}(j,k)$ = length of ray intercepted by m-th cell along the ray path,
v_m = velocity of propagation of ultrasound through the m-th cell.

3   Proposed GA Based Reconstruction Algorithm

The flowchart of the proposed reconstruction algorithm is shown in Fig. 2. We describe the salient features of this algorithm here. Specifically, we are looking for a particular distribution of the inclusion(s) which best agrees with the simulated data obtained.

To begin with, a population of solutions with random distributions of three materials with different velocity of propagation of ultrasound through them, is created. Fig. 1(c) shows the visual representation of one of the members from the initial population of solutions created when reconstructing a specimen of resolution 8x8. It can be seen that the algorithm starts its search with a population of such solutions having random distributions of three materials considered to be present in the specimen to be reconstructed. The three materials are coded using integers 0, 1, and 2, and are shown with colors black, peach, and white respectively in the visual displays shown in this paper. The initial population members have a coarse grid-size and are refined during the reconstruction process. The

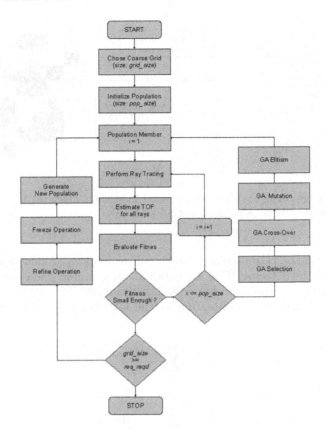

Fig. 2. Flowchart of the GA based reconstruction algorithm

idea is to proceed in steps towards the final solution of required resolution. The best possible solution with a relatively coarser initial grid serves as a seed for the next finer grid. The initial coarse grid is so chosen that its repeated doubling gives a value near, preferably equal to the final resolution required. Each population member is now evaluated for its fitness as defined by equation (2):

$$\phi(i) = \sum_{l=1}^{C} \sum_{j=1}^{S} \sum_{k=1}^{D} (GA\_Pop\_tof^{(i)}(l, j, k) - Specimen\_tof(l, j, k))^2, \qquad (2)$$

where,

$\phi(i)$ = fitness of the i-th member of GA population,
S = number of sources,
D = number of detectors,
C = number of configurations(usually $\binom{4}{2}$=6),
$GA\_Pop\_tof^{(i)}(l, j, k)$ = TOF considering the i-th member of GA population,
$Specimen\_tof(l, j, k)$ = TOF considering specimen to be reconstructed.

Next, selection operation is performed to emphasize good population members in mating pool, from which child population is created. It is observed that tournament selection operator [5] performs better than roulette wheel selection operator [5].

To create new solutions, crossover operation is performed on the population members of the mating pool created by selection operation. A block-crossover operator is employed wherein new solutions (children) are created by swapping corresponding portions of the grid between two mating pool members (parents). The parents from mating pool, the location and the size of portions to be swapped are picked randomly. Fig. 3 illustrates the block-crossover operation.

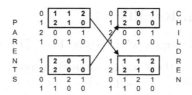

Fig. 3. Illustration of block-crossover operation

Next a few randomly chosen children are subjected to mutation operation, two variations are studied. The first bitwise-mutation, where cells of the member eligible for mutation are randomly assigned values other than the current value in the selected cell. The second termed block-mutation finds the value that appears most number of times in the selected cell itself and its eight surrounding cells and mutates the values in all nine cells to this value. Bitwise and block-mutations are illustrated in Fig. 4.

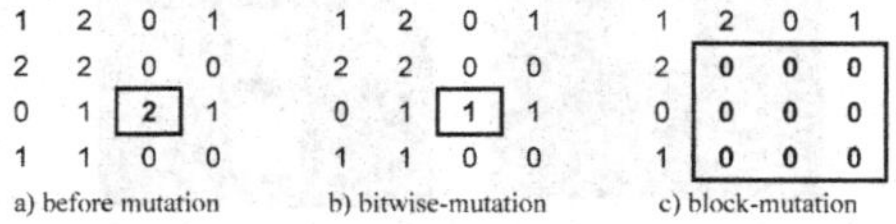

1	2	0	1		1	2	0	1		1	2	0	1
2	2	0	0		2	2	0	0		2	0	0	0
0	1	2	1		0	1	1	1		0	0	0	0
1	1	0	0		1	1	0	0		1	0	0	0

a) before mutation   b) bitwise-mutation   c) block-mutation

Fig. 4. Illustration of mutation operations

To ensure good solutions propagate through subsequent generations, the children population obtained after mutation and the initial population are combined and the best of these equal to the initial population size, are picked as elites. After this the initial population is reset with elites and the GA steps repeated for a certain number of generations. The best member from the elites is reported as the reconstructed solution.

The resulting solution with a coarse grid is now used to generate a new initial population with a higher grid size, say double the current grid-size. For this each cell is divided into four and given the same value as in the parent cell. The next step is to freeze cells which would no more be treated as variables. To identify such cells, each cell is compared with surrounding eight cells. If they all match, the cell under consideration is frozen. After refinement and freezing, the base-population member obtained serves as a seed for creating new initial population of larger grid size. The new initial population is created by randomly assigning values to the unfrozen cells leaving the frozen cell values unaltered. The GA procedure is repeated till the required resolution is obtained. The termination criteria used is the number of generations per step and whether quality of solutions obtained with two successive grid sizes. For larger resolutions estimated time using equation (3) could also be used.

4   Reconstruction Results

4.1   Reconstructed Images with Noise Free Data

To demonstrate the robustness of the algorithm more than 250 different configurations in terms of material, shape, size, and location of the inclusions with required resolutions from 6x6 to 64x64 were reconstructed. Reconstruction results are observed to be fairly consistent for a majority of the cases analyzed. Some representative results of successful reconstructions are illustrated in Fig. 5. In each of the figures shown, the left part is the visual display of the specimen to be reconstructed (used for simulating the TOF data) while the right part is the reconstructed image. Different colors in these correspond to different materials.

From the results obtained, it is observed that a crossover probability of 0.8, mutation probability of 0.2, number of generations per step of 1000, and a population size equal to three times the resolution required yield consistent results for

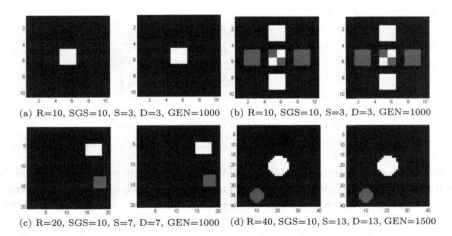

(a) R=10, SGS=10, S=3, D=3, GEN=1000   (b) R=10, SGS=10, S=3, D=3, GEN=1000

(c) R=20, SGS=10, S=7, D=7, GEN=1000   (d) R=40, SGS=10, S=13, D=13, GEN=1500

Fig. 5. Few successful reconstruction results (R: resolution required, SGS: starting grid size, S: number of sources, D: number of detectors, GEN: generations per step)

different configurations. The minimum number of sources and detectors required is such that simulated TOF data is roughly 55–65% of number of unknowns. Although reconstructing coarser grids required much smaller number of generations and smaller population size the goal is to estimate a common set of GA parameters that are robust.

The algorithm failed to capture the exact image in some cases with the parameter settings described above. This is the case when wave-propagation velocities through the materials considered are very close and when the inclusions are present at many locations arbitrarily. For such situations the fitness of different population members is very close and the algorithm cannot make definite decisions in picking up the best population members. However even for these cases the algorithm shows a trend towards correct solution with some parameter tuning. In Fig. 6 we present a few unsuccessful cases and the corresponding improved results with parameter tuning.

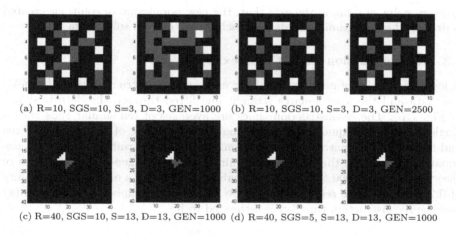

(a) R=10, SGS=10, S=3, D=3, GEN=1000   (b) R=10, SGS=10, S=3, D=3, GEN=2500

(c) R=40, SGS=10, S=13, D=13, GEN=1000   (d) R=40, SGS=5, S=13, D=13, GEN=1000

Fig. 6. Representative unsuccessful and corresponding improved reconstruction results

4.2   Fitness History

Fig. 7(a) shows how the best fitness and mean fitness values of a population vary with generation number for the result shown in Fig. 5(d). The high values of error at the beginning is due to the fact that initially we start with a smaller grid size (10 in this case) and the TOF data against which the error is evaluated is that of the final required grid size (40 in this case). Also, it can be observed that for each of the grid sizes both the best fitness and mean fitness values reach a peak and decrease rapidly followed by an almost asymptotic decrease in these values. This means that if we were to start with SGS=R, it would have taken much larger number of generations to achieve the desired solution or the solution would get stuck at a local optima with fewer generations. Also observe that, it is

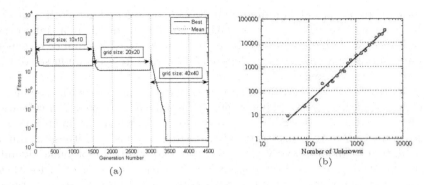

(a)

(b)

Fig. 7. (a) Fitness history for the reconstructed image in Fig. 5 (d), (b) Log-Log plot of number of unknowns versus reconstruction time (noise free TOF data)

at every point of refining the grid that, the new population of solutions created is driving the algorithm towards approaching the desired solution faster

4.3   Reconstruction Time-Complexity

A logarithmic plot of number of unknowns versus reconstruction times for noise free data and the least squares straight line fitted through the data is shown in Fig. 7(b). The reconstruction times are averages of ten similar runs with a single inclusion of square shape covering roughly 9–10% of the specimen area and located in the centre of the specimen. In each of the runs, a population size equal to three times the resolution required, number of generations for each of the steps equal to 1000, a crossover probability of 0.8, and a mutation probability of 0.2 was used. The regression equation between the number of unknowns (x) and the reconstruction time (y) is estimated to be

$$y = 0.009(x)^{1.8057}. \tag{3}$$

The regression equation (3) is nearly quadratic; using which reconstruction times for higher resolutions could be estimated and used as termination criteria.

4.4   Reconstructed Images with Noisy Data

The data from experiments in real world always differs from that simulated under ideal conditions due to presence of noise. To assess the suitability of the proposed algorithm to noisy data, the simulated TOF is modified by adding randomly up to a maximum percentage of noise and is then used as input. Fig. 8(a) shows the variation of average root mean square error (RMS) in reconstruction with respect to the percentage noise level added for two different configurations. The averages are based on ten runs for each of the maximum percentage noise level. In Fig. 8(b) and (c), we show the actual and reconstructed images for the GA run having median performance. It is observed that as the noise level is increased,

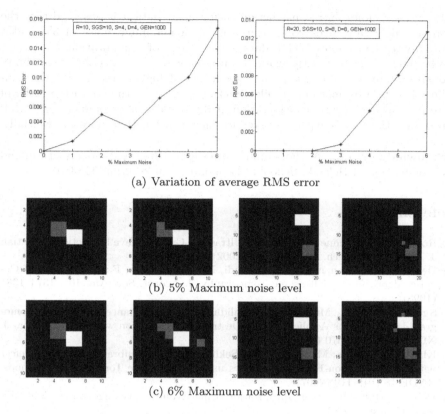

(a) Variation of average RMS error

(b) 5% Maximum noise level

(c) 6% Maximum noise level

Fig. 8. Reconstruction results with noisy data

the number of times an image is reconstructed with an error increases. For the case of 1%, 2% and 3% noise level this is only one out of ten runs, whereas for 4% and more this is five to ten out of ten runs.

5   Conclusions

The proposed algorithm is tested for a wide range of configurations and yields satisfactory results. It is observed that for a majority of cases, a population size three times the final required grid size and 1000 generations are suitable for reconstruction.

A study of the reconstruction times revealed polynomial variation with number of unknowns, enabling one to estimate the reconstruction times for larger grid sizes.

A preliminary study on the effect of crossover and mutation probabilities shows no consistent patterns. However further investigation into the role of these operators is encouraged. One variation of the mutation operator, block-mutation is found to be effective for large grid sizes.

From the studies on reconstructions using noisy data it is observed that the proposed algorithm gives good results up to a maximum noise of 3% added randomly to the simulated TOF data for a number of configurations.

The authors are currently working on enhancing the algorithm's capability and performance in reconstructing specimen's of higher resolutions, thus enabling detection of finer defects like cracks etc. Also for the algorithm to be of more practical relevance, consideration of ray bending is recommended. Parallelization of the code is expected to reduce the reconstruction times substantially.

Acknowledgements. Authors appreciate support by the Academy of Finland and Foundation of Helsinki School of Economics (under grant 118319).

References

1. Rose, J.L.: A Baseline and Vision of Ultrasonic Guided Wave Inspection Potential. J. Press. Vessel Tech. 124, 273–282 (2002)
2. Delsanto, P.P., Romano, A., Scalerandi, M., Moldoveanu, F.: Application of Genetic Algorithms to Ultrasonic Tomography. J. Acoust. Soc. Am. 104, 1374–1381 (1998)
3. Subbarao, P.M.V., Munshi, P., Muralidhar, K.: Performance of Iterative Tomographic Algorithms Applied to Non-Destructive Evaluation with Limited data. J. NDT&E 30, 359–370 (1997)
4. Khare, S., Razdan, M., Munshi, P., Sekhar, B.V., Balasubramaniam, K.: Defect detection in Carbon-Fiber Composites using Lamb-Wave Tomographic Methods. RNDE 18, 101–119 (2007)
5. Deb, K.: Multi-Objective Optimization using Evolutionary Algorithms. John Wiley, Chichester (2001)
6. Kak, A.C., Slaney, M.: Principles of Computerized Tomographic Imaging. IEEE Press, Los Alamitos (1998)
7. Lewitt, R.M.: Reconstruction Algorithms: Transform Methods. Proc. IEEE 71, 390–408 (1983)
8. Censor, Y.: Finite Series-Expansion Reconstruction Methods. Proc. IEEE 71, 409–419 (1983)
9. Greenleaf, J.F.: Computerized Tomography with Ultrasound. Proc. IEEE 71, 330–337 (1983)
10. Kodali, S.P., Bandaru, S., Deb, K., Munshi, P., Kishore, N.N.: Applicability of Genetic algorithm Based Tomographic Reconstruction Using Time-of-Flight Data from Ultrasound Transmission Tomography. In: Proc. GECCO (2008)

Estimation of Sensor Network Topology Using Ant Colony Optimization

Kensuke Takahashi[1], Satoshi Kurihara[2], Toshio Hirotsu[3], and Toshiharu Sugawara[4]

[1] Waseda University, Tokyo, Japan
k.takahashi@isl.cs.waseda.ac.jp
[2] Osaka University, Osaka, Japan
[3] Toyohashi University of Technology, Aichi, Japan
[4] Waseda University, Tokyo, Japan
sugawawra@waseda.jp

Abstract. We propose a method for estimating sensor network topology using only time-series sensor data without prior knowledge of the locations of sensors. Along with the advances in computer equipment and sensor devices, various sensor network applications have been proposed. Topology information is often mandatory for predicting and assisting human activities in these systems. However, it is not easy to configure and maintain this information for applications in which many sensors are used. The proposed method estimates the topology accurately and efficiently using ant colony optimization (ACO). Our basic premise is to integrate ACO with the reliability of acquired sensor data for the adjacency to construct the accurate topology. We evaluated our method using actual sensor data and showed that it is superior to previous methods.

1 Introduction

Recently, computer technology has been miniaturized, and the costs of computing equipment are decreasing. This has enabled the development of sensors with communication capabilities. In line with this progress, various applications have been proposed in the field of sensor networks. Therefore, a huge number of sensors are deployed in different environments to detect events and gather real-world data. Then the sensor data are transmitted via wired and wireless LAN to providers of context-aware services to support human activities.

However, one serious problem in this kind of application is the arrangement of sensors. In particular, the topological relationships between the sensors, or the *sensor network topology*, reflect the physical connectivity of the real-world environment. They are usually mandatory for sensor network applications that assist human activities. However, labor required for manual configuration and maintenance of topological relationship information, or the *adjacencies of sensors*, increases in proportion to the number of sensors, and mis-configurations tend to occur. In addition, sensors may fail to operate, or new ones may need to be added in the environment. As a result, the topology changes. These facts clearly indicate a need for automatic configuration of sensor network topological relationships.

M. Kolehmainen et al. (Eds.): ICANNGA 2009, LNCS 5495, pp. 263–272, 2009.
© Springer-Verlag Berlin Heidelberg 2009

Our objective in this research is to automatically identify the sensor network topology in accordance with human activities. Topology required for our target applications differs from that based on distances between sensors. For example, although the nearest sensor from sensor A is B in Fig. 1, these are separated by a wall, and it is possible that sensor C rather than sensor B will react after sensor A due to a human's movement. *Adjacency* in this paper means the neighboring relationship in terms of human activities. This kind of information reflecting human activities plays an important role in a number of applications such as tracking a person's movement by using coordinated sensors (or sensor agents) [1], assigning missions (tasks) to the appropriate sensor agents [2], and carrying out services for human assistance by foreseeing persons' activities [3].

A number of studies have been done on localization and topological structures of sensors. In the domain of ad hoc networks, methods that can determine the sensor structure by using the received strengths of their radio transmissions have been proposed, such as [4]. The received signal strength can be used to identify directly transmittable local sensors, but the derived topology structure may not be identical to that derived from the actual time-series sensor data that reflect human activities. In addition, we cannot assume that all sensors have wireless transmission capability. [5] proposed an algorithm to estimate the adjacent sensors only from time-series sensor data in their environments. However, that study assumed that the number of people in the environment is known and that they walk at almost the same speed, we believe that such assumptions are implausible in real-world applications. On the contrary, [6] proposed a method that estimates sensor network topology from time-series data using a pheromonal model [7]. This method does not require these implausible assumptions and, in this sense, it is a promising approach for actual sensor network systems. However, its convergence is quite slow; it requires a huge amount of time-series sensor reaction data to acquire the acceptable results. [8] also proposed a method to estimate adjacency relationships between sensors using only time-series sensor reaction data without making implausible assumptions. This is more efficient, and the resulting topology is more accurate than that in [6]. We found that, however, the accuracy of the estimated topology depends on the environments.

In this paper, we propose a new method by improving the method in [8] to more accurately estimate adjacency relationships between sensors from time-series sensor data. The proposed method is also based on the ant colony optimization (ACO), a kind of pheromonal model like that in [6]. A key idea of this method is to use the estimated

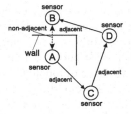

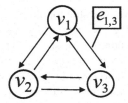

Fig. 1. Example of sensor network topology derived from human activities

Fig. 2. Depiction of sensor network structure

reliability of sensor data for topology estimation; that is, the amount of pheromone is controlled by the reliability. This considerably improves the accuracy of the resulting topology. ACO was inspired by the behavior of ants in finding paths from their colony to a food source. ACO is robust and adaptable to dynamic changes in the environment, and various kinds of optimization problems has been solved by using ACO-based approaches. For example, [9] proposed a routing protocol in wireless sensor networks using ACO. Thus, ACO is very attractive for our needs because the sensor network topology in our application may also dynamically change due to the failure or addition of sensors.

This paper is organized as follows: We first describe the issues addressed in this paper. Next, we explain how the reliability of sensor data is estimated; this estimation method is based on the analysis of actual sensor data in different environments. After that, we explain the details of our proposed method of adjacency estimation. Finally, we experimentally evaluate the proposed method using actual datasets acquired from three environments, and show that our method can generate a more accurate sensor network topology than conventional methods.

2 Problem Description

2.1 Sensor Network Topology

We express the topology of a sensor network as a weighted directed graph $G = (V, E)$, where a vertex $v_i \in V$ represents a sensor in the environment, and an edge $e_{i,j} \in E$ represents a direct path from the position of sensor v_i to that of sensor v_j. The weight of $e_{i,j}$ is expressed as $\tau_{i,j}$, where $\tau_{i,j}$ expresses the adjacency likelihood of v_i being adjacent to v_j. Figure 2 shows an example of a directed graph for three sensors.

In this study, the graph $G = (V, E)$ is initially assumed to be a complete graph in which every pair of vertices is connected by an edge. This graph is called the *main graph*.

2.2 Sensor Data

Our sensors are infrared wired sensors that react when a person walks in front of them. Data from sensors are the set of sensor identifiers that react at time t. Thus, they are expressed as $O_t = \{v_t^1, v_t^2, \ldots, v_t^n\}, v_t^j \in [1, N]$, where $N = |V|$, the number of sensors in the environment, and v_t^j is the sensor identifier. For example, $O_1 = \{1, 3, 5\}$ means that sensor identifiers 1, 3, and 5 react at $t = 1$. Our method analyzes data gathered by dividing every time-interval T, that is, O_t to O_{t+T-1}, at every turn.

2.3 Target Problem and Issues

Suppose that a number of sensors are deployed in an environment, as shown in Fig. 3. When a person walks in the direction of the arrow, v_1, v_2, and v_3 react in turn at time t_1, t_2, and t_3. Therefore, the sensor data are $O = \{O_{t_1} = \{v_1\}, O_{t_2} = \{v_2\}$, and $O_{t_3} = \{v_3\}\}$. We can expect that sensors in an adjacency relationship react in a serial manner according to the human's movement. Our method aims to estimate the

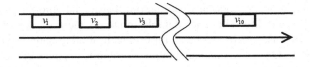

Fig. 3. A route example

local sensor network topologies expressed by the *subgraph* of the main graph derived from the time-series data, and this indicates the local adjacency of sensors. Then the global sensor network topology is built by combining the subgraphs. For example, in Fig. 3, the subgraph, $G' = \{V = \{v_1, v_2, v_3\}$, and $E = \{e_{1,2}, e_{2,3}\}\}$, was created from $O = \{O_{t_1}, O_{t_2}, O_{t_3}\}$. The edges of the main graph, which correspond to the edges of the subgraph, are weighted. However, the following challenging issues arise.

1. When multiple people walk in an environment at almost the same time, sensors that are not adjacent can react in a serial manner.
2. Sensors may incorrectly react by a misreaction of sensors and noise data.

Because issue 1 is more frequently observed proportional to the number of people in the environment, it more strongly affects the sensor network estimation.

Our proposed method to deal with these issues consists of two phases, a *dataset generation phase* and an *adjacency estimation phase*. In the *dataset generation phase*, the reliability of sensor data estimated from actual sensor data is used to calculate the weights of edges, and these weights are used to control the agents' movements. In this paper, *reliability* refers to the degree that sensors that react in a serial manner are really adjacent. The details are described in the following section. The *adjacency estimation phase* estimates the adjacencies of sensors using the data generated in the *dataset generation phase* based on a pheromonal model [7] like [6]. Unlike [6] however, we introduce the weight according to both adjacency and non-adjacency of edges in the sensor network.

3   Reliability of Sensor Data for Adjacency

It is probable that the reliability of sensor data for adjacency estimation depends on the sensor reaction frequency during a certain time interval T: if more people walk in the environment at the same time, the accuracy of the adjacency acquired from the sensor data decreases. Because we cannot determine the number of people in the environment, we assume that if the sensor reaction frequency during T is high, the more people there are moving simultaneously in the environment. Thus, we have to investigate how low the reliability gets according to the sensor reaction frequency.

First, let us define the *reliability* of sensor data, $reliability(n)$, as follows:

$$reliability(n) \stackrel{\text{def}}{=} 100 \times \frac{|E_{sub}(n) \cap E_{correct}|}{|E_{sub}(n)|}, \tag{1}$$

where $E_{sub}(n)$ is the set of edges in the subgraph generated by $O_{T(n-1)}$ to O_{Tn-1}, and $E_{correct}$ is the set of actual adjacent edges. $E_{correct}$ is derived from the actual layout map.

We defined the sensor reaction frequency, $freq(n)$, as follows:

$$freq(n) \overset{\text{def}}{=} 100 \times \frac{\sum_{t=T(n-1)}^{Tn-1} |O_t|}{T \times N}, \tag{2}$$

where N is the number of sensors in the environment.

We investigated the relationships between the sensor reaction frequencies and the reliability by comparing the sensor data and actual sensor location information in three different environments. The results are plotted in Fig. 4. Even though the data were acquired from different environments, this figure shows a similar pattern: the relationship between the sensor reaction frequencies and the accuracy seems to be approximated by a power law. Figure 4 indicates that the reliability of the dataset for adjacency estimation drastically decreases. This feature is used to control the amount of pheromone in the dataset generation phase.

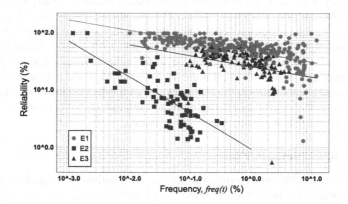

Fig. 4. Relationship between sensor reaction frequencies and adjacencies

4 Estimation of Topology

4.1 Dataset Generation Phase

The purpose of the dataset generation phase is to reduce the impact of incorrect sensor data in accordance with the estimated reliability of the sensor data, where incorrect refers to successive time-series reaction data generated by non-adjacent sensors.

First, in this phase, the subgraphs that correspond to local sensor-network topology are generated from the n-th sensor data $O_{T(n-1)}$ to O_{Tn-1}. Although the method in [8] focused on the sensor reaction time intervals, our method generates subgraphs by investigating whether or not any two sensors reacted in a series. For example, suppose that the sensor data are $O = \{O_{n_1} = \{v_1\}, O_{n_2} = \{v_2\}, O_{n_3} = \{v_3\}, O_{n_3+\alpha+1} = \{v_{10}\}$, and $O_{n_3+\alpha+2} = \{v_{11}\}\}$. Two subgraphs, $G_1 = \{\{v_1, v_2, v_3\}, \{e_{1,2}, e_{2,3}\}\}$, and $G_2 = \{\{v_{10}, v_{11}\}, \{e_{10,11}\}\}$, can be generated. Note that α is an interval time that determines whether or not a new subgraph is generated. In the above example, because the reaction interval time between v_3 and v_{10} is larger than α, two subgraphs were generated. We called nodes such as v_1 and v_{10} *start nodes* and nodes such as v_3 and v_{11} *end nodes*. Even if the reaction time intervals between any two sensors are short, our method identifies that these are not adjacent if one of them reacts right after another sensor.

Next, the weight of each edge of the generated subgraph is calculated. As discussed in Section 3, the relationship between the reaction frequency and the reliability can be characterized by a power law. Therefore, $w(n)$, the weight of each edge of the subgraph at n turns of the simulation, is defined using the reaction frequency $freq(n)$ as follows:

$$w(n) \stackrel{\text{def}}{=} a \times freq(n)^{-b}, \tag{3}$$

This weight can reduce the impact of the sensor data if the reliability of the data is low, and it can increase the impact of the sensor data if the reliability of the data is high.

In equation (3), a and b are constants whose values are determined from the investigation described in the previous sections. These constant values may depend on the environmental structure, but we believe that the structure's basic features (power law) must be invariant. These constants' values are used to calculate the weights of edges. The relative values of weights are important so the value is more effective. Therefore, we use the average values of a and b derived from three different environments. Of course, estimating automatically the optimal value of these constant values from known information (e.g. the size of an environment and the number of sensors) is important, and this is something to be investigated in the future.

4.2   Adjacency Estimation Phase

The purpose of the adjacency estimation phase is to estimate the global sensor network topology by calculating the weight $\tau_{i,j}(n)$, which is also the amount of remaining pheromone on edge $e_{i,j}$ calculated by ACO.

First, N agents are placed on the start node, where N is the number of sensors in the environment. Then, each agent moves toward the end node. Each agent selects the next node stochastically based on the values of $\{\tau_{i,j}(n)\}$. Each agent prefers the edge where a lot of $\tau_{i,j}(n)$ remains. Therefore, the algorithm can focus the search on the edges that have a high adjacency likelihood determined by a previous search. Two types of agents are used in this phase. The first type of agent at node v_i selects the edge $e_{i,j}$ with the probability $p_{i,j}^k$, which is calculated as follows:

$$p_{i,j}^k(n) = \frac{\tau_{i,j}(n)}{\sum_{k,k \neq i} \tau_{i,k}(n)}. \tag{4}$$

The second type of agent selects the next node not based on its dependence on $\tau_{i,j}(n)$, but due to a certain constant probability. The second type of agent can prevent the algorithm from falling into a local optimal solution. The ratio of first to second type agents is seven to three in this paper.

Each edge of the main graph is deposited with pheromone depending on the number of agents that move to each edge of the subgraph. Limiting the amount of pheromone by setting a maximal value prevents the search from diverging. The amount of pheromone on the edges of main graph $\{\tau_{i,j}(n)\}$ evaporates at a constant rate. As a result, the latest search information is reflected relatively strongly. Suppose that the number of agents that move to the edge $e_{i,j}$ is $A_{i,j}(n)$, the rate of decrease is ρ, and the maximal value of the pheromone is τ_{max}. Each edge is deposited with pheromone following equation (5).

$$\tau'_{i,j}(n) = (A_{i,j}(n) \times w(n)) + (1 - \rho)\tau_{i,j}(n)$$

$$\tau_{i,j}(n+1) = \begin{cases} \tau_{max} & \text{if } \tau'_{i,j}(n) > \tau_{max} \\ \tau'_{i,j}(n) & \text{otherwise} \end{cases}, \tag{5}$$

where $\tau_{max} = 500$, and $\rho = 0.3$ were used in this paper.

We also focused on the non-adjacency. The set of nodes of each subgraph is the set of sensors that react in turn. We assumed that the adjacent likelihood was high. On the other hand, the difference in the main graph and a subgraph is assumed to be the set of non-adjacency. Then, suppose that the set of edges in a subgraph is $E_{sub}(n)$, that in the main graph it is E_{main}, and the difference between them is $E_{not}(n) \stackrel{\text{def}}{=} E_{main} - E_{sub}(n)$. Therefore, $E_{not}(n)$ is the set of non-adjacent edges. Then, the amount of pheromone $\tau_{i,j}(t)$ of $e_{i,j} \in E_{not}(n)$ decreases following equation (6). Limiting the amount of pheromone by setting a minimal value of the pheromone τ_{min} prevents the search from diverging.

$$\tau'_{i,j}(n) = (1 - \rho)\tau_{i,j}(n) - w(n)$$

$$\tau_{i,j}(n+1) = \begin{cases} \tau_{min} & \text{if } \tau'_{i,j}(n) < \tau_{min} \\ \tau'_{i,j}(n) & \text{otherwise} \end{cases}, \tag{6}$$

where $\tau_{min} = 0$ was used in this paper.

4.3   Determination of Adjacency Relationships

We determine the sensors' adjacency from the values of $\{\tau_{i,j}(n)\}$, which are determined from the two phases mentioned above. A normal cumulative distribution function generated from the mean $\mu(n)$ and variance $\sigma^2(n)$ of $\{\tau_{i,j}(n)\}$ is used to estimate the adjacencies of sensors. The normal cumulative distribution function $f(x)$ are calculated as follows:

$$f(x) = \frac{1}{2}\left(1 + \frac{2}{\sqrt{\pi}}\int_0^{\frac{(x - \mu(n))}{\sqrt{2\sigma^2(n)}}} \exp(-t^2)dt\right).$$

The adjacency of edges $e_{i,j}$ is determined by the following conditional equation (7) and is expressed by the adjacency matrix $A = \{a_{i,j}\}$, where if $e_{i,j}$ is adjacent, $a_{i,j} = 1$; otherwise, $a_{i,j} = 0$.

$$a_{i,j} = \begin{cases} 0 & \text{if } f(\tau_{i,j}(n) + \tau_{j,i}(n)) < prob \\ 1 & \text{otherwise} \end{cases}, \tag{7}$$

where $prob = 0.6$ was used in this paper.

5   Evaluation

5.1   Experimental Environment

We evaluated our algorithm using actual sensor data and compared it with the results induced using the methods in [6,8]. The sensor data used in this evaluation were collected in three different environments.

(a) Twenty-one days of sensor data were collected in environment E1. Figure 5 shows the correct topological relationships of the sensor network in the E1 environment, which had 14 sensors.

(b) Nine days of sensor data were collected in environment E2. Figure 6 shows the correct topological relationships of the sensor network in the E2 environment, which had 49 sensors.

(c) Six days of sensor data were collected in environment E3, which had 45 sensors.

Note that the sensors we used were infrared reflection sensors.

Our method uses the error rate err, which is defined as follows:

$$err \stackrel{\text{def}}{=} 100 \times \left(\frac{1}{N^2 - N}\right) \sum_{i,j,i \neq j} \left(a_{i,j} - a'_{i,j}\right)^2, \tag{8}$$

where N is the number of sensors in the environment, the matrix $A = \{a_{i,j}\}$ is the estimated adjacency matrix, and $A' = \{a'_{i,j}\}$ is the correct adjacency matrix according to the actual layout map.

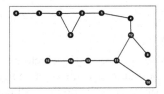

Fig. 5. Correct topological map of sensor network of environment E1

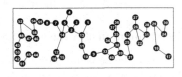

Fig. 6. Correct topological map of sensor network of environment E2

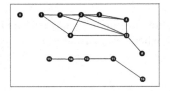

Fig. 7. Estimated topological map of environment E1 using our method

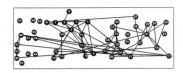

Fig. 8. Estimated topological map of environment E2 using our method

5.2 Experimental Result

Figures 7 and 8 show the estimated topological maps generated using our proposed method. Figure 9 plots the improvement in error rates over time. A comparison of the error rates is in Table 1. It indicates that our method can reduce the error rates to approximately $1/8$ to $1/2$ that of the other methods.

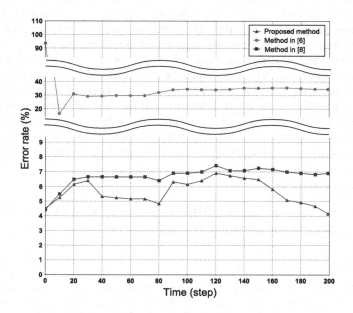

Fig. 9. Changes in error rates over time in environment E2

Table 1. Comparison of error rates

Environment	Proposed method	Method in [6]	Method in [8]
E1	9.18%	31.6%	31.6%
E2	4.16%	34.0%	6.25%
E3	10.9%	21.0%	25.3%

Although our proposed method achieves a lower error rate of estimated adjacency relationships in environment E3, it is not quite as good as that of environments E1 and E2. The possible reasons for this are:

1. A sensor reacts frequently in places where movement of people occurs frequently (e.g. a room entrance), and a reaction in a serial manner with a non-adjacent sensor confuses the algorithm.

2. Because the experimental environment was a laboratory in a university, people often move around simultaneously (e.g. when a lecture starts, they go to lunch).

Thus, we intend to this kind of environment is the future.

6 Conclusion

We proposed a method for estimating sensor network topology using only time-series sensor data. This method enables us to estimate the reliability of sensor data and weight each edge depending on the reliability. Therefore, with this method, we can reduce the impact of incorrect sensor data that are acquired when multiple people move around at nearly the same time, causing a mis-reaction of the sensors. This method also makes it possible to estimate the adjacency using ACO. Therefore, we can use this method to analyze sensor data that reflect the previous results. Then, by searching randomly with constant probability, the method does not fall into a local optimal solution. This method can reflect the latest search information relatively strongly by decreasing the result with the constant rate. As a result, with this method, we can improve the error rate of estimated adjacency compared to conventional methods such as those in ref. [6,8].

Acknowledgements

This work was supported largely by SCOPE program of the Ministry of Internal Affairs and Communications under contract 071607001.

References

1. Horling, B., Mailler, R., Lesser, V.: A Case Study of Organizational Effects in a Distributed Sensor Network. In: Proceedings of the AAAI 2004 Workshop on Agent Organizations: Theory and Practice, San Jose, California, pp. 23–30. AAAI Press, California (2004)
2. Bar-Noy, A., Brown, T., Johnson, M.P., Porta, T.F.L., Liu, O., Rowaihy, H.: Assigning sensors to missions with demands. In: ALGOSENSORS, pp. 114–125 (2007)
3. Weiser, M.: The computer for the 21st century. Scientific American 265(3), 66–75 (1991)
4. Montillet, J.P., Braysy, T., Oppermann, I.: Algorithm for nodes localization in wireless ad-hoc networks based on cost function. In: IWWAN 2005 (2005)
5. Marinakis, D., Dudek, G.: Topological mapping through distributed, passive sensors. In: Proceedings of the International Joint Conference on Artificial Intelligence (IJCAI 2007), Hyderabad, India, January 2007, pp. 2147–2152 (2007)
6. Tamaki, H., Fukui, K., et al.: Automatic acquisition of sensor-network topology based on pheromone communication model. In: Fourth International Conference on Networked Sensing Systems, 2007. INSS 2007, p. 292 (2007)
7. Dorigo, M., Di Caro, G.D.: The Ant Colony Optimization Meta-Heuristic, pp. 11–32. McGraw-Hill, New York (1999)
8. Takahashi, K., Sugawara, T.: Estimation of sensor-network topology from time-series sensor data using ant colony optimization method. In: Swarm Intelligence Symposium, 2008. SIS 2008, September 2008, pp. 1–6. IEEE, Los Alamitos (2008)
9. Okdem, S., Karaboga, D.: Routing in wireless sensor networks using ant colony optimization. In: AHS 2006: Proceedings of the first NASA/ESA conference on Adaptive Hardware and Systems, Washington, DC, USA, pp. 401–404. IEEE Computer Society, Los Alamitos (2006)

Scalability of Learning Impact on Complex Parameters in Recurrent Neural Networks

Branko Šter and Andrej Dobnikar

Faculty of Computer and Information Science, University of Ljubljana
Tržaška 25, 1000 Ljubljana, Slovenia
branko.ster@fri.uni-lj.si

Abstract. The impact of problem extents and network sizes on learning in recurrent neural networks is analysed in terms of structural parameters of related graphs. In previous work the influence of learning on the changes of the typical parameters such as characteristic path length, clustering coefficient, degree distribution and entropy, was investigated. In the present work the focus is enlarged to the scaling problem of the learning paradigm. The results prove the scalability of learning procedures due to the retained dynamics of the parameters during learning with different problem extents and network sizes.

Keywords: recurrent neural networks, structural parameters, scalability, problem extent, network size, learning automata.

1 Introduction

Since the first studies of random graphs by Erdos and Renyi [1], complex networks from the real world became a target of numerous investigations [2,3,4,5,6]. It was recognized that the evolution of real networks and their topologies are governed by certain robust organizing principles [3]. Though systems like the World Wide Web and Internet are indeed changing with their exploitation, neural networks as complex systems normally modify their parameters (weights) through learning procedures. The structural properties of complex networks, or their related graphs, are usually quantified by characteristic path lengths (L), clustering coefficients (C) and degree distributions (P), [3,5]. In our previous work [7] we added entropy (H) as a new structural parameter, based on the degree distribution, because it clarifies the difference between different properties of complex systems. In that work we showed that the clustering coefficient increases with the learning procedure of the recurrent neural network (RNN), the average length is not changing significantly, while the maximal degree and entropy of the related graph are again increasing with the training iterations. According to the standard classification of complex systems we determined that the trained RNNs are neither typical small-world (SW) nor typical scale-free (SF), but possess properties of both. Besides, the degree distributions consistently exhibit increased entropy. It was argued that the learning of neural networks (RNN) is one of many robust organizing principles, observed in nature, that

M. Kolehmainen et al. (Eds.): ICANNGA 2009, LNCS 5495, pp. 273–282, 2009.
© Springer-Verlag Berlin Heidelberg 2009

changes complex structures into a special type of networks, which may indicate a new class of complex and biologically inspired systems with low L, high C and different degree distribution P with increased entropy H.

The temporal complexity of learning algorithms creates serious limitations in terms of practical neural network sizes and/or problem extents. This implies the relevance of scalability of the learning paradigm and explains our focus on learning procedures with different network sizes and different problem extents and their impact on structural parameters. We discovered that a simple, learning automata based learning algorithm works faster than the standard RTRL algorithm when dealing with large network sizes (above 100 neurons), and thus enables experiments with sizes above 100 neurons.

The paper is organized as follows. After s short presentation of the background theory of complex systems and RNNs, the graphs related to RNNs and the transformation of RNNs to graphs is shown. It is followed by a description of entropy as a new feature parameter of complex systems. The experimental work is explained next with emphasis on different problem extents and RNN sizes. The results are given in terms of the dynamics of the structural parameters, provided with the corresponding explanations. In conclusion, some comments and ideas for future work are outlined.

2   Background Theory

2.1   Complex Systems

Complex systems can usually be viewed as networks, with typical examples being the Internet, WWW, neural networks, etc., where the nodes are routers, documents, neurons, and links between them communication links, hyperlinks (URLs) and weights, respectively. Complexity is the study of the behaviour of macroscopic collections of units that are endowed with the potential to evolve over time [5].

To represent complex networks, graphs are usually used. A graph is described by a pair $G = (V, E)$, where V is a set of N nodes or vertices and E is a set of e edges, where each edge connects two nodes. The structural properties of graphs are quantified by the characteristic path length L (also the average shortest path length), the clustering coefficient C and the degree distribution P [3].

Characteristic path length L is defined as:

$$L = \sum_{i=1}^{N} \sum_{j=1, j \neq i}^{N} d(i, j).\tag{1}$$

where $d(i, j)$ is the distance between nodes i and j, defined as the number of edges along the shortest path connecting them, and each edge connects two nodes.

Clustering coefficient C shows the cliquishness of a typical neighbourhood, or average fraction of existing connections between the nearest neighbours of a vertex, which is a local property. For a node i it is defined by the expression:

$$C_i = \frac{2N_i}{k_i(k_i - 1)}, \tag{2}$$

where k_i is the degree of node i or the number of nodes connected to it by edges and N_i is the actual number of neighbours of the node i, and the number of all possible edges between the neighbours is $\binom{k_i}{2} = k_i(k_i - 1)/2$. The clustering coefficient of a graph is the average over all vertices: $C = \frac{1}{N}\sum_{i=1}^{N} C_i$.

$P(k)$ is the probability that a randomly selected node has exactly k edges (degree k). The mean degree is $\bar{k} = \frac{1}{N}\sum_{i=1}^{N} k_i$.

Besides regular networks or graphs, there are three main groups of networks (graphs), differing in their structural properties or topologies: random networks (R), small-world networks (SW) and scale-free networks (SF). There are topological differences between graphs corresponding to regular, partially random or completely random connections. SW networks can be constructed from ordered lattices by random rewiring of edges or by the addition of connections between randomly selected vertices, in both cases with probability p (0.1-0.3).

SW networks have small L, just like R networks, but much greater C. SW networks fall between regular and random graphs. The shape of the degree distribution is similar to that of random graphs. It has a peak at $\bar{k}$ and decays exponentially for large and small k.

SF networks have degree distributions with power-law tails:

$$P(k) \sim k^{-\gamma}. \tag{3}$$

This sort of distribution typically occurs when a random network grows with the preferential attachment. Many large networks from the real world (WWW, Internet, scientific literature, metabolic networks, etc.) exhibit a power-law degree distribution and therefore belong to the SF type of networks.

It has been of great interest recently [8,9,10] to investigate the complex features of biologically inspired models, such as neural networks. In [7], a new structural parameter, entropy (H), was introduced in order to make the difference between recurrent neural networks (RNNs) with random weights and trained RNNs clearer. It was proved that trained RNNs have neither typical SW nor typical SF features. However, they do exhibit increased entropy.

We already mentioned the two important attributes that greatly influence the learning procedures of neural networks, problem extents or loads and network sizes or the number of neurons. The natural question that arises is whether the learning procedure retains the dynamics of structural parameters for different attribute values (fulfills the scalability feature) or not. This will be the main issue dealt with in the rest of the paper.

2.2   Fully Connected Recurrent Neural Networks

Only fully connected RNNs are considered in this paper because they have the most general topology, where each unit (neuron) is connected with every other unit. Because we intended to model dynamical systems, we initially used the

RTRL (real-time recurrent learning) algorithm, which is gradient-based and ensures good learning convergence. Unfortunately, the training of large RNNs with RTRL is very time-consuming, which is the reason we looked for a more efficient algorithm. We found that a learning automata based algorithm (L_{R-P}) works faster on large RNNs than RTRL. It enabled us to extend network size above 100 neurons, i.e. up to 250 neurons.

The outputs of certain units in RNNs represent outputs of the network, while the other units are called context-units because they provide information relevant to sequence-processing problems. Inputs to the network are fed to each unit. Such an RNN can be trained to simulate a dynamic system. The RNN is trained to produce the sequence of desired outputs if fed with the input sequence. The real-time learning gets its input from a source which produces a non-periodic sequence that is long enough.

For the purposes of the experimental work, much larger RNNs are applied than actually needed for the selected tasks. This can be biologically justified by the manner in which living organisms also use a larger number of neurons for simple functions performed at a particular moment.

2.3   The L_{R-P} Algorithm

Learning automata [11] are simple mathematical models, acting in an interaction with a stochastic environment, which can be presented in the form of a stochastic model. Learning automata perform actions and the environment responds with more or less favourable responses. They are capable of modifying their properties or parameters in such a way as to maximize favourable responses from the environment. One of their key properties is the ability to act immediately or on-line, not requiring a time-consuming collecting of statistics of environment responses. Thus they provide an on-line adaptation to the environment.

The correction schemes of learning automata update the probabilities of performing individual actions according to environment responses. In the simplest case the environment responds in a binary manner: reward or punishment. When an action is rewarded (a favourable response from the environment), its probability should consequently increase, while, when the action is punished, its probability should consequently decrease. To preserve the total probabilities of all actions, in the first case the probabilities of other actions should decrease, and in the second case the probabilities of other actions should increase.

Correction schemes of learning automata may also be applied for training neural networks [12]. In this case an action corresponds to a small perturbation of an individual weight: one action corresponds to an increase in the weight for a fixed value Δw, while the other action corresponds to a decrease in the same weight for Δw. The number of actions N_A of the LA equals $2N_W$, where N_W is the total number of weights in the RNN. A decrease in error E_{total} along a specified sequence due to a change in an individual weight represents a favourable response, or reward; an increase, on the other hand, represents an unfavourable response, or punishment.

Let N_W be the number of free parameters of the RNN (weights and biases). The correction scheme L_{R-P} (linear reward-penalty) updates $N_A = 2N_W$ perturbation probabilities after the i-th perturbation in the following manner. If the i-th action is rewarded in the m-th step, its probability increases as

$$p_i(m+1) = p_i(m) + \lambda[1 - p_i(m)], \qquad (4)$$

while probabilities of other $N_A - 1$ actions decrease as

$$p_{j \neq i}(m+1) = (1 - \lambda)p_j(m). \qquad (5)$$

On the other hand, if the i-th action is punished, its probability decreases as

$$p_i(m+1) = (1 - \lambda)p_i(m), \qquad (6)$$

and probabilities of other actions increase as

$$p_{j \neq i}(m+1) = \frac{\lambda}{N_A - 1} + (1 - \lambda)p_j(m). \qquad (7)$$

λ is a correction parameter, with $0 < \lambda < 1$. In this case the change in weight is abolished subsequently.

In the beginning, all action probabilities are equal, $p_j(0) = 1/N_A$, $j = 1, .., N_A$, and later they are continuously updated. In this way the probability vector approximates the gradient in a stochastic manner. While moving in the weight space, this algorithm adaptively updates the direction of the approximate gradient.

2.4   Transformation of an RNN to Graph

In order to comply with the fundamental theory of complex networks, we convert RNNs to undirected graphs. It would also be possible to use directed and/or weighted graphs, with appropriately redefined parameters L and C.

The transformation from the RNN to an undirected graph was performed as follows. Nodes i and j are considered connected when either $|w_{ij}| \geq \Theta$ OR $|w_{ji}| \geq \Theta$, Θ being the threshold value. It is also possible to use the AND operator, instead of OR. OR and AND are binary logical operators, supremal and infimal, respectively. Θ is defined from the trained RNN as follows: when all the weights with absolute values lower than Θ are neglected in the network (replaced with 0), the performance of the RNN on the task should not decrease substantially. In order to better comply with undirected graphs, self-links were disallowed, so $w_{ii} = 0$ for all neurons i. Another modification was the nonapplication of biases in neurons, again to better comply with undirected graphs. Using these restrictions, the training of the RNN was rendered more difficult, but nevertheless, the RNN was able to learn its tasks.

For each RNN, we first calculated the connectivity of the trained network and then set the thresholds for earlier RNNs (during the same training process) such that they have the same connectivity, i.e. the number of edges e must remain constant. The reason for this procedure is that final weights are typically larger than initial ones (between -0.5 and 0.5) and if a constant threshold were applied, graphs corresponding to trained RNNs would have more edges than graphs corresponding to initial RNNs. In this way some of the structural parameters would be changed not because of the structure, but solely because of larger e. We want, however, that the amount of connectivity does not influence the graph parameters. The possible difference in structural parameters should therefore depend only on the inner structure of the graph. In this way the RNNs recorded during training are in fact normalized, and it is easier to compare the RNNs at different stages during training, despite different connectivities, also caused by the final weights being larger than the initial ones.

3   Entropy as a Feature Parameter of Complex Systems

Preliminary experiments showed that during training the binomial degree distribution of the initial random network tends to approach the degree distribution of SF networks. Some hubs regularly occur and the rest of the distribution is moved towards smaller degrees, as in SF networks, except that it stops sooner, so the final distribution does not reach a power-law distribution.

Since this is not a usual distribution, we considered a well known property of probability distributions, Shannon's entropy. We found that entropy of a binomial distribution is increased when random links are prewired to hubs. As this process induces the same type of transformation, it will be explained next.

Shannon's entropy of the binomial distribution gives:

$$H = - \sum_{k=0}^{N-1} P(k) \log \binom{N-1}{k} + (N-1) \log 2 \,, \qquad (8)$$

where we take $p = 0.5$ to simplify the equation. If a random edge is prewired to a node with a large degree l (hub), then the degree of the hub becomes $l+1$, and thus frequencies f for degrees l and $l+1$ change as $\Delta f(l+1) = 1$ and $\Delta f(l) = -1$. At the same time a random node with degree $k+1$ loses a connection, $\Delta f(k+1) = -1$, and the frequency of degree k gets incremented: $\Delta f(k) = 1$. Since $P(k) = f(k)/F$, where F denotes $\sum_{k=0}^{N-1} f(k)$, probabilities change as:

$$\Delta P(k+1) = P'(k+1) - P(k+1) = -1/F \quad \text{with probability } P(k+1) \quad (9)$$
$$\Delta P(k) = P'(k) - P(k) = 1/F \quad \text{with probability } P(k+1) \,, \qquad (10)$$

where P' denotes the new, updated values. It is more convenient to observe what happens to each degree k, i.e. to calculate $\Delta P(k)$ - due to both effects: $-1/F$ and $1/F$. The probability of each degree k is changed by $-1/F$ with probability $P(k)$ and by $1/F$ with $P(k+1)$. The expected change of $\Delta P(k)$ is therefore:

$$E[\Delta P(k)] = \frac{P(k+1) - P(k)}{F} \,. \qquad (11)$$

The term $(N-1)\log 2$ from Eq. 8 cancels out and the expected change in entropy due to disconnecting an edge (change in degrees k and $k+1$) is distributed over all nodes but the hub and can be written as:

$$E[\Delta H_1] = -\sum_{k=0}^{N-2} \log \binom{N-1}{k} E[\Delta P(k)] \qquad (12)$$

$$= -\frac{1}{F} \sum_{k=0}^{N-2} \log \binom{N-1}{k} (P(k+1) - P(k))$$

$$\approx \frac{1}{F} \sum_{k=0}^{N-2} \left[\log \binom{N-1}{k+1} - \log \binom{N-1}{k} \right] P(k+1).$$

On the other hand, the change in entropy due to the hub getting connected is

$$\Delta H_2 = -\log \binom{N-1}{l} (P'(l) - P(l)) - \log \binom{N-1}{l+1} (P'(l+1) - P(l+1))$$

$$= \log \binom{N-1}{l} \frac{1}{F} - \log \binom{N-1}{l+1} \frac{1}{F}. \qquad (13)$$

Since both terms are positive, their sum as the expected change of the total entropy of the network is also positive, which proves the incrementation of the network's entropy with the number of learning iterations.

There is another possible explanation, an intuitive one. As learning usually means acquiring new knowledge or a certain amount of new information, it seems appropriate to consider that entropy as average information is increased with learning.

4   Experimental Work

The influence of the training of RNNs on the structural parameters of the related graphs was studied on the task of dynamic identification of unknown finite state machines, performing the time-delayed XOR(d) function of two subsequent inputs, delayed by d. For the purpose of scalability of learning, network sizes from 50 to 250 (by 50) neurons and problem extents from XOR(2)to XOR(5), $\Delta d = 1$, were tested.

Typical results of the structural parameters during training are depicted in Fig. 1. It is notable that, instead of training iterations, the x-axis contains the mean squared error (MSE) on the training set. The reason is that training times for such large networks vary greatly. It would be difficult to calculate the statistics of runs of different lengths. This is why we decided to measure the progress of training by the MSE. All RNNs were trained until the MSE dropped below 0.001. In this way different training lengths can be normalized.

As mentioned before, the clustering coefficient is increasing with the progress of training and the average length is not changing significantly, while entropy of the related graph is again increasing with the progress of training.

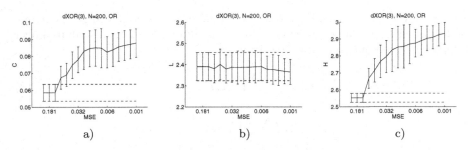

Fig. 1. Delayed XOR(3) with a 200-neuron RNN: (a) characteristic path length L, (b) clustering coefficient C, and (c) entropy H, all versus training iterations. Mean values and standard deviations over 10 runs are indicated.

As the main issue of the paper is the scalability feature of the learning impact on the structural parameters, the experiments of RNN learning (one of which is described in Fig. 1) are repeated with different network sizes and various problem extents, as explained before. The results are depicted in Fig. 2. Only initial and final values of parameters C, L and H are given. C is increased during training in all settings (Fig. 2a), i.e. in all sizes N and task parameters d. It is also obvious that C is smaller in larger RNNs. On the other hand, in larger networks (above 150 neurons) C is more increased relatively regarding its initial value than in smaller ones (below 150 neurons). The task size d has no major influence on the results.

Parameter L does not change significantly during training, which is obvious from Fig. 1b. This is the reason why in Fig. 2b its initial and final values are practically equal. However, L is larger in large networks.

Entropy H is increased during the training process, in accordance with Fig. 1c. This is the reason that the *final* surface is above the *initial* one. Besides, H also increases with the number of neurons N, while the task size seems to have no major impact on H.

5   Conclusion

This paper deals with the special learning algorithm of recurrent neural networks, based on the learning automata theory. The emphasis is on the scaling feature of the learning paradigm. Thus we investigate the influence of problem extent and network size on the structural parameters of the neural network related graphs. It is shown that the main characteristics of the learning process are preserved for all combinations of the attributes (extent, size) under investigation, which means that the average shortest path is practically unchanged with learning, and that the clustering coefficient and entropy increase along with it. In addition, we show that the clustering coefficient is decreased with the number of neurons, and is nearly unchanged with the problem size. The average path length and entropy are increased, also due only to the network size. The problem size plays no important role.

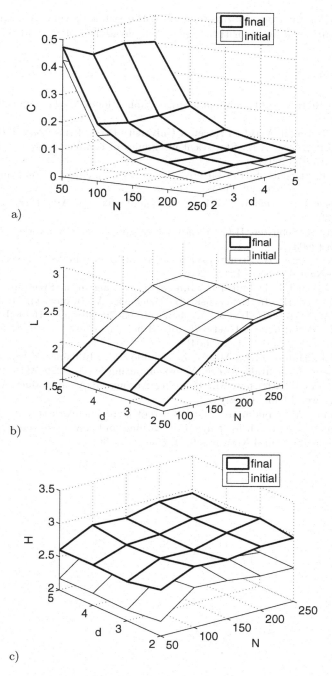

Fig. 2. Initial and final a) clustering coefficient C, b) mean shortest path L, and c) entropy H for different sizes of the RNN (number of neurons N) on different delayed XOR tasks (delay d)

Our plans for future work are related to the study of very large neural networks, efficient learning algorithms and their parallel implementation.

References

1. Erdos, P., Renyi, A.: On random graphs. Publicationes Mathematicae 6, 290–297 (1959)
2. Watts, D.J.: Small Worlds. Princeton University Press, Princeton (1999)
3. Reka, A., Barabasi, A.L.: Statistical mechanics of complex networks. Rev. Mod. Phys. 74, 47–97 (2002)
4. Watts, D.J., Strogatz, S.H.: Collective dynamics of 'small-world' networks. Letters to Nature 393(4), 440–442 (1998)
5. Dorogovtsev, S.N., Mendes, J.F.F.: Evolution of networks. Adv. Phys. 51(4), 1079–1187 (2002)
6. Bornholdt, S., Schuster, H.G.: Handbook of Graphs and Networks. Wiley-VCH, Weinheim (2003)
7. Dobnikar, A., Ster, B.: Structural Properties of Recurrent Neural Networks. Submitted to Neural Process Lett. (2008)
8. Ster, B., Gabrijel, I., Dobnikar, A.: Impact of Learning on the Structural Properties of Neural Networks. In: Beliczynski, B., Dzielinski, A., Iwanowski, M., Ribeiro, B. (eds.) ICANNGA 2007. LNCS, vol. 4432, pp. 63–70. Springer, Heidelberg (2007)
9. Kim, J.B.: Performance of networks of artificial neurons: The role of clustering. Phys. Rev. E 69, 045101/1–4 (2004)
10. Torres, J.J., Munoz, M.A., Marro, J., Garrido, P.L.: Influence of topology on the performance of a neural network. Neurocomputing 58(60), 229–234 (2004)
11. Narendra, K., Thathachar, M.: Learning Automata: An Introduction. Prentice-Hall, Englewood Cliffs (1989)
12. Sundareshan, M., Condarcure, T.: Recurrent neural network training by a learning automaton approach for trajectory learning and control system design. IEEE Transactions on Neural Networks 9(3), 354–368 (1998)

A Hierarchical Classifier with Growing Neural Gas Clustering

Igor T. Podolak and Kamil Bartocha

Institute of Computer Science, Faculty of Mathematics and Computer Science,
Jagiellonian University,
Łojasiewicza 6, Kraków, Poland
uipodola@theta.uoks.uj.edu.pl

Abstract. A novel architecture for a hierarchical classifier (HC) is defined. The objective is to combine several weak classifiers to form a strong one, but a different approach from those known, *e.g.* AdaBoost, is taken: the training set is split on the basis of previous classifier misclassification between output classes. The problem is split into overlapping subproblems, each classifying into a different set of output classes. This allows for a task size reduction as each sub-problem is smaller in the sense of lower number of output classes, and for higher accuracy. The HC proposes a different approach to the boosting approach.

The groups of output classes overlap, thus examples from a single class may end up in several subproblems. It is shown, that this approach ensures that such hierarchical classifier achieves better accuracy. A notion of generalized accuracy is introduced.

The sub-problems generation is simple as it is performed with a clustering algorithm operating on classifier outputs. We propose to use the Growing Neural Gas [1] algorithm, because of its good adaptiveness.

1 Introduction

A *classifier* is a model which assigns an example attribute vector to one of predefined classes [2,3]. In machine learning several methods are known for training model architectures, among them hierarchical. Training a single architecture to achieve a set accuracy, *e.g.* a single neural network, can take a long time. On the other hand, it is possible to combine several weak models to obtain a hierarchical model giving good training and generalization rate, *i.e.* correct classification of examples not used in training, *e.g.* boosting methods like AdaBoost [4,5,6].

The proposed approach of Hierarchical Classifier (HC) defines methodology for building a hierarchical classifier automatically. In short, first a simple weak classifier is built for the whole problem, then sub-problems are built by grouping together examples from classes that were mistaken frequently. This step is done by means of clustering in the results space, *not* the input space. Each of the clusters, which may overlap, forms a new sub-problem for which a new weak classifier is built, and the process is repeated recursively until a set accuracy is reached. HC is especially suited for problems with several output classes, which was the main motivation. HC model was first introduced in [7,8,9].

M. Kolehmainen et al. (Eds.): ICANNGA 2009, LNCS 5495, pp. 283–292, 2009.
© Springer-Verlag Berlin Heidelberg 2009

The *HC* approach differs substantially from an AdaBoost like approaches, where several simple classifiers are built by modifying the probability density function for choosing the examples for training. *HC*, assumes that if a classifier assigns examples from different true classes to some common subset, then these classes are similar, and these classes can form subproblems. *I.e.* the *HC* splits the input space on the base of classes recognized.

One of the problems in *HC* is the choice of a clustering algorithm. The aim of one used in an *HC* is to group frequently mistaken classes into overlapping clusters. This greatly enhances the accuracy of the whole *HC*, as will be shown. The *HC* architecture found reflects the inabilities of individual node classifiers to cope with the problem. The Growing Neural Gas (GNG) [1] algorithm is proposed for this tasks because of its good adaptation ability.

The paper presents the classifier design, then the clustering approach and adaptation of GNG. Finally, experiments are described and a discussion on results is given.

2   Definition of the Problem and Model

Definition 1. *Let* $\mathcal{D} = \{(x_k, \mathcal{C}^k)\} \subsetneq \mathcal{X} \times \mathcal{Q}$, *where* $x_k \in \mathcal{X}$ *is an input attribute vector, and true class* $\mathcal{C}^k \in \mathcal{Q} = \{\mathcal{C}_k, k = 1, \ldots, K\}$, *be a finite set of training pairs. A classifier* $\mathfrak{Cl}$ *is a mapping*

$$\mathfrak{Cl} : \mathcal{X} \longmapsto \mathcal{Q} \ . \tag{1}$$

A multilayer perceptron (MLP) may generalize well, but it may be hard and costly to find the optimum architecture, *i.e.* the number of hidden layers and the number of neurons [10,11,12].

2.1   Hierarchical Classifier Structure

The proposed hierarchical classifier (*HC*) has a tree-like structure with classifiers $\mathfrak{Cl}^i : \mathcal{X}^i \longmapsto \mathcal{Q}^i$ at nodes, which map examples from $\mathcal{X}^i \subset \mathcal{X}$ into a subset of classes $\mathcal{Q}^i \subsetneq \mathcal{Q}$. The set of classes $\mathcal{Q}^0$ for the root classifier is the set of all classes $\mathcal{Q}$, *i.e.* $\mathcal{Q}^0 \equiv \mathcal{Q}$, while the class sets for all the classifiers at other nodes are the proper subsets of the whole class set, *i.e.* $\mathcal{Q}^i \subsetneq \mathcal{Q}$.

The idea is to train all node classifiers only to the point when their accuracy is only a little better than that of a random classifier. These shall be called *weak*:

Definition 2. *Let* $\mathcal{D} = \{(x^{(n)}, \mathcal{C}^{(n)}) | n = 1, \ldots, N\}$ *be a training set of N for a problem of classification of input vectors x into a finite set of K classes* $\{\mathcal{C}_i | i = 1, \ldots, K\}$, *with prior probability* $P(\mathcal{C}_i)$ *such that* $\sum_{k=1}^{K} P(\mathcal{C}_i) = 1$.

We say that classifier $\mathfrak{Cl}$ *is weak if there exists a constant* $0 < \epsilon < 1/2$ *such, that the risk function*

$$R(\mathfrak{Cl}) = \sum_{n=1}^{N} \sum_{k=1}^{K} \ell(x^{(n)}, \mathcal{C}^{(n)}, \mathfrak{Cl}(x^{(n)}) = \mathcal{C}_k) P(\mathcal{C}_k) P(x^{(n)}) < 1/2 - \epsilon \ , \tag{2}$$

where $C^{(n)}$ is the the true class for the input vector of attributes $x^{(n)}$, i.e. $C^{(n)} = C_k$, and $\ell()$ is a $0-1$ loss function, equal to 1 if $\mathfrak{Cl}(x^{(n)}) \neq C_k$.

The whole HC classifier is constructed as follows:

1. Build the root classifier $\mathfrak{Cl}^0$ using the whole set $\mathcal{D}$,
2. find frequently mistaken classes and group them in clusters $\mathcal{Q}^i = \{C_k^i\}_{k=1}^{K_i}$ ($\mathcal{Q}^i$'s may overlap)
3. select $\mathcal{D}^i$ such, that $\forall(\mathcal{Q}^i, \mathcal{D}^i \subsetneq \mathcal{D})\forall x \in \mathcal{D}^i : true\_class(x) \in \mathcal{Q}^i$,
4. for each $\mathcal{D}^i$ build a subclassifier $\mathfrak{Cl}$ repeating steps (1)-(3) recursively until $R[\mathfrak{Cl}] < target\_risk\_value$.

After training, the posterior probability that a class $\mathcal{C}$ is the correct prediction given input vector X is computed using a Bayesian approach according to

$$P(\mathcal{C}|X) = \sum_{\mathcal{Q}^i} P_{\mathcal{Q}^0}(\mathcal{Q}^i|X)P_{\mathcal{Q}^i}(\mathcal{C}|X) \ , \tag{3}$$

where $P_{\mathcal{Q}^0}(\mathcal{Q}^i|X)$ is the (predicted by the root classifier $\mathfrak{Cl}^0$ with output classes from $\mathcal{Q}^0$) probability that the correct class for example X belongs to $\mathcal{Q}^i$, and $P_{\mathcal{Q}^i}(\mathcal{C}|X)$ is the probability of class $\mathcal{C}$ computed by a subclassifier $\mathfrak{Cl}^i$ with training examples from cluster $\mathcal{D}^i$. This can be accomplished by the so called *modified classifier* described below, constructed by an independent clustering algorithm.

2.2 The Modified Classifier

After the clusters are found, $\mathfrak{Cl}$ is transformed into a *modified classifier* $\mathfrak{Cl}_{mod}$

$$\mathfrak{Cl}_{mod}^i : X \longmapsto Q^i \ , \tag{4}$$

whose task is now to predict a cluster from the set $Q^i = \{Q^{ij}, j = 1, \ldots, J_i\}$, $Q^{ij} = \{C_k^{ij}, k = 1, \ldots, K^{ij}\}$, where J_i is the number of clusters, and K^{ij} the number of individual classes in cluster Q^{ij}. An example is shown in Fig. 1. The clusters are found by inspecting the inaccuracies of the parent classifier composing classes that are frequently mistaken together. This follows from the hypothesis, that if the parent classifier $\mathfrak{Cl}$ is able to correctly discern most examples from some two given classes C_i and C_j, then there is no need to dwell on this problem anymore. If, on the other hand, $\mathfrak{Cl}$ frequently mistakes classes C_i and C_j, then a new cluster $\mathcal{Q} = \{C_i, C_j\}$ should be constructed and a new classifier $\mathfrak{Cl}(i,j)$ trained to discriminate these classes.

If the original cluster of classes $\mathcal{Q}$ recognized by classifier $\mathfrak{Cl}$ were divided into non-overlapping clusters $\mathcal{Q}^i, i = 1, \ldots, K, \bigcap \mathcal{Q}^i = \emptyset$, then

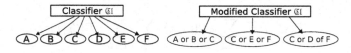

Fig. 1. An example original classifier $\mathfrak{Cl}$ and a modified one $\mathfrak{Cl}_{mod}$

- if an example x with true class C_t was classified by $\mathfrak{Cl}_{mod}$ into $\mathcal{Q}^a, C_t \in \mathcal{Q}^a$, then subclassifier $\mathfrak{Cl}^a$ could enhance the overall accuracy with true class,
- if, on the other hand, $x \in C_t$ was classified into $\mathcal{Q}^b, C_t \notin \mathcal{Q}^b$, then subclassifier $\mathfrak{Cl}^b$ would not have the chance to correct the current classification of $\mathfrak{Cl}$.

A solution is to find overlapping clusters to minimize *generalized risk*:

Definition 3. *The generalized risk* $GR(\mathfrak{Cl}^i)$ *of a classifier* $\mathfrak{Cl}^i$ *is the ratio of the number of examples classified to clusters that do not include their true classes to the total number of examples:*

$$GR(\mathfrak{Cl}^i) = \sum_{n=1}^{N} P(x^{(n)}) \ell(x^{(n)}, C^{(n)}, \mathfrak{Cl}^i_{mod}(x^{(n)}) = \mathcal{Q}^{ij}) \; , \tag{5}$$

where $C^{(n)}$ *is the true class of example* $x^{(n)}$ *and the* $\mathfrak{Cl}_{mod}$ *selects* $\mathcal{Q}^{ij}$, *and* $\ell()$ *is a* $0 - 1$ *loss function*

$$\ell(x^{(n)}, C^{(n)}, \mathfrak{Cl}^i_{mod}(x^{(n)}) = \mathcal{Q}^{ij}) = \begin{cases} 1 \text{ if } C^{(n)} \in \mathfrak{Cl}^i_{mod}(x^{(n)}) = \mathcal{Q}^{ij} \; , \\ 0 \text{ otherwise} \end{cases} \tag{6}$$

$\mathfrak{Cl}^i_{mod}$ is constructed by clustering the output classes from $\mathcal{Q}^i$:

Definition 4. *For a given classifier* $\mathfrak{Cl}^i : \mathcal{X} \longmapsto \mathcal{Q}^i$, *trained with a set of pairs* $\mathcal{D}^i$, *the clustering algorithm for HC*

$$\mathfrak{Clust} : \mathcal{Q}^i \longmapsto Q^i \; , \tag{7}$$

partitions the set of classes into a set of overlapping clusters $Q^i = \{Q^{ij}, j = 1, \dots, J_i\}$. *Clustering is based on* $\mathfrak{Cl}$ *classification.*

The aim of $\mathfrak{Clust}$ is to minimize $GR(\mathfrak{Cl}^i)$. $\mathfrak{Cl}^i_{mod}$ may be called a δ-*strong* classifier

Definition 5. *Let* $\delta > 0$. *A classifier* $\mathfrak{Cl}$ *is* δ-*strong if the risk function*

$$R(\mathfrak{Cl}) < \delta \; . \tag{8}$$

The modified version $\mathfrak{Cl}^i_{mod}$ for a classifier $\mathfrak{Cl}^i$ is constructed as follows

1. Using $\mathcal{D}^i = \{(x^{(n)} : C^{(n)} = C_i)\}$ train a weak $\mathfrak{Cl}^i : \forall x \in \mathcal{D}^i \mathfrak{Cl}^i(x) = [P(C_1), \dots, P(C_K)]$ so that $\forall x \sum_{k=1}^{K_i} P(C_k|x) = 1$,
2. perform clustering $Q^i = \{Q^{ij}, j = 1, \dots, J_i\}$, $Q^{ij} = \{C_k^{ij}, k = 1, \dots, K^{ij}\}$ so that $GR(\mathfrak{Cl}^i) < \delta$ for some δ,
3. now $\mathfrak{Cl}^i_{mod} = [P(\mathcal{Q}^{i1}), P(\mathcal{Q}^{i2}), \dots, P(\mathcal{Q}^{iJ_i})]$, where

$$P(\mathcal{Q}^{ij}) = \sum_{C_k \in \mathcal{Q}^{ij}} P(C_k|x) / \sum_j \sum_{C_k \in \mathcal{Q}^{ij}} P(C_k|x) \; , \tag{9}$$

and $\sum_j P(\mathcal{Q}^{ij}|x) = 1$.

2.3   Clustering Algorithm

Final classification of an example x is computed according to formula

$$P(\mathcal{C}_j|x) = \sum_i P(\mathcal{Q}^i|x)P(\mathcal{C}_j|\mathcal{Q}^i,x) \; , \tag{10}$$

where $P(\mathcal{Q}^i|x)$ is the probability of cluster $\mathcal{Q}^i$ (computed by the parent classifier) and $P(\mathcal{C}_j|\mathcal{Q}^i,x)$ is the $\mathcal{C}_j$ class probability computed by the subclassifier that recognizes examples from classes in $\mathcal{Q}^i$.

Let $\mathfrak{Clust}^i$ select a cluster set $Q^i = \{\mathcal{Q}^{ij}, j = 1,\ldots,J_i\}$ for classifier $\mathfrak{Cl}^i$ recognizing K_i classes. A number of conditions and objectives should be met:

1. $J < K_i$, $\tag{11}$
2. $\forall k \exists j \mathcal{C}_k \in \mathcal{Q}^{ij}$ $\tag{12}$
3. $\neg \exists\; l, m\;\; \mathcal{Q}^{il} \subseteq \mathcal{Q}^{im}$, $\tag{13}$
4. $\forall\; j\; |\mathcal{Q}^{ij}| > 1 \wedge |\mathcal{Q}^{ij}| < K_i$, $\tag{14}$
5. $\neg \exists\; k\; \forall\; j\; C_k \in \mathcal{Q}^{ij}$. $\tag{15}$

The objectives of $\mathfrak{Clust}$ may be contradictory: the $GR(\mathfrak{Cl})$ value has to be minimized, which may result in clusters' heavy overlap while, for computation efficiency reasons, the number of clusters needs to be kept low with small cluster overlap. In the paper HC is implemented with a multilayer perceptron, although classifier type that returns a class probability distribution vector may be used.

2.4   Growing Neural Gas for Clustering

GNG, proposed by Fritzke [1], develops clusters that describe well a given distribution of points. It builds connected groups of neurons starting with 2 neurons connected with an edge. At a given moment there are t neurons $N = \{n_1, n_2, \ldots, n_t\}$ connected with edges from $E = \{\{n_i, n_j\}, n_i, n_j \in N \wedge n_i \neq n_j\}$.

Each neuron is associated with a vector $w \in \mathbb{R}^K$, where K is the number of possible classes. Each edge describes the neighborhood of a neuron and has an age value. For each example x, the GNG finds the nearest (winning) neuron to the classification of x by $\mathfrak{Cl}$, a $[0,1]^K$ vector

$$n_{win} \leftarrow \arg\min_{n\in N}\|\mathfrak{Cl}(x) - w(n)\| \; , \tag{16}$$

and the second nearest neuron, and if those neurons are not connected, an edge is added with zero age. The neuron associated with the winning one is moved towards the point $\mathfrak{Cl}(x)$, and the distance $\|n - n_{win}\|$ is added to the accumulated error of the winning neuron. All edges from the winning neuron have their age incremented and all that exceed a maximum age, are removed.

Every given number of iterations, a neuron n_{err} with the highest accumulated error is found, then another with the highest error n_{serr} from its neighbors (i.e. connected with n_{err}), and a new neuron n_{new} is added in between (connected with new edges) with an associated weights $w_{n_{new}} = (w_{n_{err}} + w_{n_{serr}})/2$. Error of n_{new} is set to the mean of n_{err} and n_{serr} errors. This procedure results in forming groups of connected neurons that represent GNG clusters.

Each $x \in D$ is presented to classifier $\mathfrak{Cl}$ resulting in a class distribution vector, which is presented in turn to GNG. The winning neuron n is found, and the true class of x is added to the cluster that n belongs to. For each GNG cluster G a histogram H_G is defined as $H_G : \mathcal{Q} \longrightarrow \mathbb{N} \cup \{0\}$ such that $H_G(\mathcal{C}_i)$ is the number of times an example from class $\mathcal{C}_i$ activated a neuron from G. Each GNG cluster corresponds to an HC cluster of classes $\mathcal{Q}_G$

$$\mathcal{Q}_G = \{\mathcal{C}_i : H_G(\mathcal{C}_i) > d_G\} \ , \tag{17}$$

where d_G is a threshold. The set of d_G values are adapted so that the clustering constraints (11)-(15) are met, selecting $\forall x \, d_G = \frac{2}{3} \max\limits_{\mathcal{C}} H_G(\mathcal{C})$. If there exists $\mathcal{C}_i : \neg \exists j : \mathcal{C}_i \in \mathcal{Q}^j$, then select $k = \arg \max\limits_{G} H_G(\mathcal{C}_i)$ and $\mathcal{Q}^G = \mathcal{Q}^G \cup \{\mathcal{C}_i\}$.

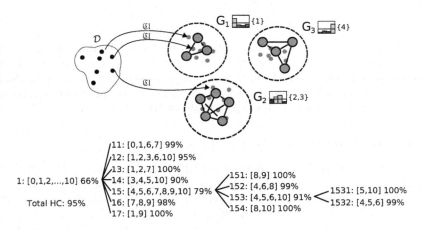

Fig. 2. GNG generated clusters and their associated labels (top) and a final HC for the vowel problem [13]

3 HC Accuracy

For simplicity we study a two–level HC with risk function

$$R(\mathfrak{Cl}) = \sum_{n=1}^{N} \sum_{k=1}^{K} \sum_{i=1}^{|\mathcal{Q}|} \ell(x^{(n)}, \mathcal{Q}^{(n)}, \mathfrak{Cl}_{mod}^0(x^{(n)}))\ell(x^{(n)}, \mathcal{C}^{(n)}, \mathfrak{Cl}^i(x^{(n)})) \ , \tag{18}$$

where $\mathcal{Q}^{(n)} \in Q$ is a cluster of classes where the true class $\mathcal{C}^{(n)}$ of $x^{(n)}$ belongs to. Since the modified root classifier $\mathfrak{Cl}_{mod}^0$ is δ-*strong*, it is possible to show that a two level HC with weak classifiers at nodes performs better than the root $\mathfrak{Cl}^0$.

Proposition 1. *A two level HC, with a weak root classifier $\mathfrak{Cl}^0$ with risk less than $1/2 - \epsilon$ for some $0 < \epsilon \leq 1/2$, which has a generalized risk less than some δ, is at least weak classifier if the children classifiers $\mathfrak{Cl}^i$ have the expected risk*

$$E[R(\mathfrak{Cl}^i)] < 1/2 - \epsilon - \mu \ , \tag{19}$$

where the μ has to be greater than

$$\mu > \frac{\delta(1/2 + \epsilon)}{1 - \delta} \ , \tag{20}$$

Proof. Take the expected value of the risk for a two-level HC, and split it into two terms: first, where the modified classifier $\mathfrak{Cl}^0_{mod}$ classifies an example into a correct cluster, and a second one responsible for incorrect cluster assignments.

$$
\begin{aligned}
E[R(\mathfrak{Cl})] = \sum_{n=1}^{N} \sum_{k=1}^{K} \sum_{\mathcal{Q}^i} & E(\ [\arg\max_i P_{\mathfrak{Cl}^0}(\mathcal{Q}^i|x) == \mathcal{Q}^{(n)} \wedge \mathcal{Q}^{(n)} \ni \mathcal{C}^{(n)}] \\
& \wedge [\arg\max_m P_{\mathfrak{Cl}^i}(\mathcal{C}_m|X) = \mathcal{C}^{(n)}]) P(\mathcal{C}_k) P(x^{(n)}) \\
+ \sum_{n=1}^{N} \sum_{k=1}^{K} \sum_{\mathcal{Q}^i} & E(\ [\arg\max_i P_{\mathfrak{Cl}^0}(\mathcal{Q}^i|x) == \mathcal{Q}^{(n)} \wedge \mathcal{Q}^{(n)} \not\ni \mathcal{C}^{(n)}] \\
& \wedge [\arg\max_m P_{\mathfrak{Cl}^i}(\mathcal{C}_m|x) = \mathcal{C}^{(n)}]) P(\mathcal{C}_k) P(x^{(n)}) \ ,
\end{aligned}
$$

where $\mathcal{Q}^{(n)}$ is a cluster that contains the true class $\mathcal{C}^{(n)}$ of example $x^{(n)}$ and the arg max is used to select the class (or cluster) that maximizes the given value.

This separation is possible, since sums for pairs of examples n and class k are counted once. Therefore each can be classified to more than one cluster, some classifications may be incorrect, but each cluster $\mathcal{Q}^i$ is counted only once.

The first term applies to all correct classifications of $\mathfrak{Cl}^0$ to clusters, which is equivalent to the $\mathfrak{Cl}^0_{mod}$ operation, and which has a risk value less than δ, and a fraction of at least $1 - \delta$ examples are expected to be correctly classified. The second term are the incorrect cluster classifications with risk δ, so

$$
\begin{aligned}
E[R(\mathfrak{Cl})] = (1 - \delta) \sum_{n=1}^{N} \sum_{k=1}^{K} \\
E([\arg\max_m \sum_{i:\mathcal{C}^{(n)} \in \mathcal{Q}^i} P^0_{\mathfrak{Cl}_{mod}}(\mathcal{Q}^i|x^{(n)}) P_{\mathfrak{Cl}^i}(\mathcal{C}_m|x^{(n)}) = \mathcal{C}^{(n)}]) P(\mathcal{C}_k) P(x^{(n)}) \\
+ \delta \sum_{n=1}^{N} \sum_{k=1}^{K} \sum_{i:\mathcal{C}^{(n)} \notin \mathcal{Q}^i} E([\arg\max_m P_{\mathfrak{Cl}^i}(\mathcal{C}_m|x^{(n)}) = \mathcal{C}^{(n)}]) P(\mathcal{C}_k) P(x^{(n)}) \ .
\end{aligned}
$$

Since in the second term the true class does not belong to the selected cluster, therefore the subclassifiers cannot recognize it, and the value for this part of $E[]$ sums up to 1 as all examples are incorrectly classified. Thus

$$E[R(\mathfrak{Cl})] = (1 - \delta) \sum_{n=1}^{N} \sum_{k=1}^{K}$$

$$E([\arg\max_{m} \sum_{i:\mathcal{C}^{(n)} \in \mathcal{Q}^i} P^0_{\mathfrak{Cl}_{mo\vartheta}}(\mathcal{Q}^i|x^{(n)})P_{\mathfrak{Cl}^i}(\mathcal{C}_m|x^{(n)}) = \mathcal{C}U(n)])P(\mathcal{C}_k)P(x^{(n)})$$

$$+\delta \ .$$

As the weak classifiers left are to have expected risk less then $1/2 - \epsilon - \mu$, and all other terms sum up to 1, it follows that

$$E[R(\mathfrak{Cl})] = (1 - \delta)(1/2 - \epsilon - \mu) + \delta \ , \tag{21}$$

Since HC classifier $\mathfrak{Cl}$ is to be at least weak, it must be that

$$E[R(\mathfrak{Cl})] < E[R(\mathfrak{Cl}^0)] < 1/2 - \epsilon \ , \tag{22}$$

Substituting (22) in (21) we get

$$\mu > \frac{\delta(1/2 + \epsilon)}{1 - \delta} \ . \tag{23}$$

This is easily accomplished, because the subclassifiers are to solve tasks with a smaller number of classes, therefore simpler.   □

4   Experiments

Several experiments were done with data sets from the UCI repository [13,14]. The results were compared with results obtained by Setiono with his hidden neuron number optimized networks N2CS2 [15] as can be seen in Tab. 1.

Each network in a node classifier was a simple feed-forward network with 1 hidden layer, where the number of neurons is given columns of Tab. 1. Column "formula hidden" had the number of neurons computed according to $\lfloor \sqrt{in + out} \rfloor$, with in – number of input attributes, and out – number of subproblem's classes.

The impact of the HC architecture on the classification accuracy for the primary-tumor problem can be seen in Tab. 1, with sample accuracies presented for the whole HC and for the root classifier. It is clear that the proposed architecture manages not only to increase the accuracy of the classification process, but also achieves a more stable result on the training set. It can be seen that the small size of single networks makes them very weak. Though, starting from 2 hidden neurons, the hierarchical classifier begins to achieve good results.

The accuracy for the primary-tumour problem is worse, but this is also the case for other networks (e.g the results of Setiono in Tab. 1) as it is a hard problem for machine learning. Since the individual networks are small, it is possible to train them fast obtaining good results. Since each level of the classifiers' tree is built of independent networks, it is easy to parallelize the task. The GNG algorithm proved to be fast and obtaining a stable clusterization with a high generalization rate. Compared to the original algorithm (a simple SAHN clustering approach), it is easier to obtain a correct hierarchical architecture automatically.

Table 1. The impact of HC architecture on accuracy (top). Cross validation training and test accuracy rates for some training sets from the UCI Repository [13,14] (bottom part)

trial no.	3 hidden				7 hidden			
	train		test		train		test	
	$\mathfrak{C}\mathfrak{l}^0$	HC	$\mathfrak{C}\mathfrak{l}^0$	HC	$\mathfrak{C}\mathfrak{l}^0$	HC	$\mathfrak{C}\mathfrak{l}^0$	HC
run no. 1	.41	.64	.29	.31	.59	.79	.34	.41
run no. 2	.39	.67	.24	.34	.58	.78	.36	.39
run no. 3	.40	.64	.34	.41	.60	.79	.36	.39
run no. 4	.47	.66	.39	.41	.60	.76	.43	.48
run no. 5	.45	.67	.31	.46	.61	.82	.43	.43

	1 hidden		2 hidden		3 hidden	
	train	test	train	test	train	test
audiology	$.32 \pm .01$	$.24 \pm .02$	$.89 \pm .00$	$.67 \pm .01$	$.97 \pm .00$	$.78 \pm .01$
primary-tumour	$.25 \pm .00$	$.25 \pm .01$	$.47 \pm .00$	$.36 \pm .01$	$.59 \pm .00$	$.39\ pm.00$
vowel	$.0.9 \pm .00$	$.07 \pm .00$	$.71 \pm .01$	$.62 \pm .01$	$.84 \pm .01$	$.76 \pm .01$
zoo	$.67 \pm .03$	$.66 \pm .05$	$.96 \pm .00$	$.87 \pm .02$	$1.0 \pm .00$	$.92 \pm .01$
soybean	$.18 \pm ..00$	$.09 \pm .00$	$.95 \pm .00$	$.90 \pm .00$	$.97 \pm .00$	$.92 \pm .00$

	4 hidden		5 hidden		6 hidden	
	train	test	train	test	train	test
audiology	$98. \pm .00$	$.73 \pm .02$	$.99 \pm .00$	$.71 \pm .00$	$1.0 \pm .00$	$.72 \pm .01$
primary-tumour	$63. \pm .00$	$.44 \pm .01$	$.69 \pm .00$	$.43 \pm .01$	$.70 \pm .01$	$.42 \pm .02$
vowel	$.96 \pm .00$	$.87 \pm .00$	$.96 \pm .00$	$.89 \pm .00$	$.97 \pm .00$	$.89 \pm .00$
zoo	$1.0 \pm .00$	$.95 \pm .01$	$1.0 \pm .00$	$.95 \pm .00$	$1.0 \pm .00$	$.93 \pm .01$
soybean	$.98 \pm .00$	$.91 \pm .00$	$.98 \pm .00$	$.92 \pm .00$	$.99 \pm .00$	$.93 \pm .00$

	7 hidden		formula hidden		Setiono	
	train	test	train	test	train	test
audiology	$.99 \pm .00$	$.73 \pm .01$	$1.0 \pm .00$	$.78 \pm .00$	$.95 \pm .01$	$.79 \pm .01$
primary-tumour	$.74 \pm .00$	$.42 \pm .01$	$.81 \pm .00$	$.42 \pm .00$	$.62 \pm .01$	$.46 \pm .01$
vowel	$.97 \pm .00$	$.88 \pm .00$	$.99 \pm .00$	$.93 \pm .00$	$.94 \pm .01$	$.89 \pm .01$
zoo	$1.0 \pm .00$	$.94 \pm .01$	$1.0 \pm .00$	$.94 \pm .00$	$1.0 \pm .00$	$.94 \pm .01$
soybean	$.99 \pm .00$	$.93 \pm .00$	$.99 \pm .00$	$.93 \pm .00$	$.96 \pm .01$	$.93 \pm .01$

5   Discussion

This paper presented the HC hierarchical classifier architecture, together with it's enhancement by the use of a Growing Neural Gas (GNG) algorithm. The classifier is composed of several simple classifiers, which are defined to be weak. The composition of such classifiers gives a strong hierarchical classifier. At each node the original problem is split into sub-problems, and a new sub-classifier is built for each of them. A notion of generalized risk was introduced. Previously the authors have used a simple clustering algorithm adapted for extraction of overlapping clusters. The adoption of GNG was spurred by its reported good approximation of input environment together with clear forming of connected

GNG clusters. These clusters were adapted to correspond well with the *HC* cluster structure. The histogram based method was introduced for overlapping clustering. GNG proved to be a good solution.

Experimental results presented compare well with other optimum architecture methods, and *HC* structure is built automatically. Only some rule needs to be used for the selection of node classifiers size, but these can be kept very simple, as can be seen in Tab. 1. Proposed *HC* is rather a scheme since any type of classifier can be used for a node classifier.

Work is continuing to further integrate the GNG clustering with the *HC* scheme, so that clustering process may continue alongside node classifiers training. This would be especially helpful in a changing environment and enable node training parallelizing training a classifier and a clustering scheme.

References

1. Fritzke, B.: A growing neural gas network leatns topologies. In: Tesauro, G., Touretzky, D.S., Leen, T.K. (eds.) Advances in Neural Information Processing Systems, pp. 625–632 (1995)
2. Hand, D., Mannila, H., Smyth, P.: Principles of data mining. MIT Press, Cambridge (2001)
3. Hastie, T., Tibshirani, R., Friedman, J.: The elements of statistical learning. Springer, Heidelberg (2001)
4. Schapire, R.: The strength of weak learnability. Machine Learning 5, 197–227 (1990)
5. Eibl G Pfeiffer, K.P.: Multiclass boosting for weak classifiers. Machine Learning 5, 189–210 (2005)
6. Freund, Y., Schapire, R.E.: A decision-theoretic generalization of on-line learning and an application to boosting. Journal of Computer and System Sciences 55(1), 119–139 (1997)
7. Podolak, I.T.: Hierarchical classifier with overlapping class groups. Expert Systems with Applications 34(1), 673–682 (2008)
8. Podolak, I.T., Biel, S.: Hierarchical classifier. In: Wyrzykowski, R. (ed.) PPAM 2005. LNCS, vol. 3911, pp. 591–598. Springer, Heidelberg (2006)
9. Podolak, I.T.: Hierarchical rules for a hierarchical classifier. In: Beliczynski, B., Dzielinski, A., Iwanowski, M., Ribeiro, B. (eds.) ICANNGA 2007. LNCS, vol. 4431, pp. 749–757. Springer, Heidelberg (2007)
10. Andrews, R., Diederich, J., Tickle, A.B.: Survey and critique of extracting rules from trained artificial neural networks. Knowledge–Based Systems 8, 373–389 (1995)
11. Duch, W., Setiono, R., Żurada, J.M.: Computational Intelligence Methods for Rule-Based Data Understanding. Proceedings of the IEEE 92(5), 771–805 (2004)
12. Setiono, R., Leow, W.K.: FERNN: An Algorithm for Fast Extraction of Rules from Neural Networks. Applied Intelligence 12, 15–25 (2000)
13. Newman, D.J., Hettich, S., Blake, C.L.: UCI repository of machine learning databases (1998)
14. Zwitter, M., Sokolic, M.: Primary tumor data set (1998)
15. Setiono, R.: Feedforward neural network construction using cross validation. NeuralComputation 13, 2865–2877 (2001)

A Generative Model for Self/Non-self Discrimination in Strings

Matti Pöllä

Adaptive Informatics Research Centre,
Helsinki University of Technology,
FI-02015 Espoo, Finland
matti.polla@tkk.fi

Abstract. A statistical model is presented as an alternative to negative selection in anomaly detection of discrete data. We extend the use of probabilistic generative models from fixed-length binary strings into variable-length strings from a finite symbol alphabet using a mixture model of multinomial distributions for the frequency of adjacent symbols in a sliding window over a string. Robust and localized change analysis of text corpora is viewed as an application area.

1 Introduction

Finding anomalies in a collection of data has been one of the most important research areas in the field of artificial immune systems (AIS), i.e., computational methods inspired by the information processing of biological immune systems. The negative selection algorithm (NSA) by Forrest et al. [1] was originally presented as an immunology-inspired method for classifying bit strings into *self* or *non-self* when training samples are available only from the *self* class. This is achieved by producing an initial detector collection which is pruned according to negative selection based on the available self samples–that is: any detector that matches a self sample is rejected. The remaining collection of detectors are then used as an instance-based description of non-self data.

Recently, statistical methods have been shown to have good performance in anomaly detection tasks [2]. Among these, the one-class support vector machine [3] and probabilistic generative models [4] have been used as an alternative to NSA-based learning. In the following sections, a generative model based on multinomial distributions is presented for anomaly detection in variable-length strings from an arbitrary symbol vocabulary.

In Section 2 we review the principle of negative selection based self/non-self discrimination, and in Section 3 we review a generative model for fixed-length binary strings and a related model for character frequencies. In Section 4 we present a generative model using multinomial distributions which can be seen as a hybrid of the models of [4] and [5]. Section 5 discusses the properties of natural language in terms of applying the developed model. Experimental results using a simplified example and a natural language corpus are presented in

M. Kolehmainen et al. (Eds.): ICANNGA 2009, LNCS 5495, pp. 293–302, 2009.
© Springer-Verlag Berlin Heidelberg 2009

Section 6. Topics for further research are outlined discussion in Section 7 with some conclusions in Section 8.

2 Anomaly Detection Using Negative Selection

The negative selection approach to anomaly detection [6] employs an instance-based representation of the unseen data (non-self). The set of all data vectors $\mathcal{U}$ contains the self set $\mathcal{S} \subset \mathcal{U}$ from which a set of self samples $\mathbf{s} \in \mathcal{S}$ are available in the training phase. The self samples are used to prune an initial (often stochastically generated) set $\mathcal{D}_0$ of detector strings such that all detectors $\mathbf{d} \in \mathcal{D}_0$ which have high affinity (similarity) with samples from $\mathcal{S}$ are removed. The affinity function $u(\mathbf{s}, \mathbf{d}) \to \mathbb{R}$ maps the similarity of two vectors into a real value and can be customized to suit the application at hand.

In the pruning phase all self-matching detector candidates are removed from the initial set of detectors according to the discrimination rule

$$m_\tau = \left\{ \begin{array}{ll} u(\mathbf{s}, \mathbf{d}) \geq \tau, & \text{self} \\ u(\mathbf{s}, \mathbf{d}) < \tau, & \text{non-self} \end{array} \right. \quad \forall \mathbf{s} \in \mathcal{S}, \quad \forall \mathbf{d} \in \mathcal{D}_0 \qquad (1)$$

After the censoring phase, any new sample $\mathbf{x}$ can be classified into non-self if a match between $\mathbf{x}$ and a detector $\mathbf{d} \in \mathcal{D}$ is found according to (1).

Compared to simply classifying according to a thresholded similarity with self samples (positive selection), the NSA has the benefit of being able to make the classification decision to non-self based on a single match between a detector and a data sample, whereas positive selection would require matching with each self sample before assigning $\mathbf{x}$ to the non-self class.

Originally, the NSA was used for fixed-length binary strings and affinity was measured using bitwise-similarity metrics such as the Hamming distance or the related r-contiguous and r-chunk matching rules [7]. Since then, the NSA has been extended with various matching rules and data representation schemes from binary data into multidimensional real-valued vector data [8].

However, as an instance-based learning scheme the NSA suffers from the curse of dimensionality problem. Stibor et al. [9,10,11] have shown that in the case of matching bit strings using the r-contiguous bit rule there is no method to generate detectors efficiently as the problem can be reformulated as a k-CNF satisfiability problem. While the unique properties of NSA can be useful in some application domains, the process of searching for non-self matching detectors has limitations in scaling for high-dimensional data. This result motivates the use of statistical affinity measures over negative selection for strings from a non-binary alphabet.

3 Related Work

3.1 Finite Bernoulli Mixture Models

Stibor [4] presented the use of finite multivariate Bernoulli mixtures as a generative model for detecting anomalies in l-dimensional bit strings. In this model, a bit string $\mathbf{x} \in \{0,1\}^l$ is considered to be generated by an l-dimensional Bernoulli distribution. In this discrete distribution the outcome of each bit can be either 1 with probability $P(x = 1) = \Theta$ or 0 with probability $P(x = 0) = 1 - \Theta$. The one dimensional probability distribution $P(x|\Theta) = \Theta^x(1 - \Theta)^{1-x}$ can be extended for l-dimensional bit strings into

$$P(\mathbf{x}|\Theta) = \prod_{i=1}^{l} \Theta_i^{x_i}(1 - \Theta_i)^{1-x_i}, \quad x_i \in \{0,1\} \tag{2}$$

where the parameter vector $\Theta = (\Theta_1, \Theta_2, ..., \Theta_l)$ contains the probabilities for each bit position.

To take into consideration the internal correlations in the data set $\mathcal{X} = \{\mathbf{x_1}, ..., \mathbf{x}_{|\mathcal{X}|}\}$, a linear mixture of M distributions can be used such that the mixture proportions of each component is defined by a parameter $\boldsymbol{\alpha} \in \mathbb{R}^M, \sum_{m=1}^{M} \alpha_m = 1$ and the probability of the mixture model generating the string $\mathbf{x}$ is thus

$$P(\mathbf{x}|\overline{\Theta}, \boldsymbol{\alpha}) = \sum_{m=1}^{M} \alpha_m P(\mathbf{x}|\Theta_m) \tag{3}$$

where the matrix $\overline{\Theta} = (\Theta_1, \Theta_2, ..., \Theta_M)$ contains the parameter vectors of each mixture component.

To find the maximum likelihood estimates for parameters for $\boldsymbol{\alpha}$ and $\overline{\Theta}$ the EM algorithm [12] can be used to iteratively switch between computing the posterior probabilities $P(m|\mathbf{x}, \boldsymbol{\alpha}, \overline{\Theta})$ (E step) and re-estimating $\boldsymbol{\alpha}$ and $\overline{\Theta}$ (M step). In the resulting mixture model, discrimination between self and non-self is done according to the thresholded probability (3) such that any string $\mathbf{x}$ for which $P(\mathbf{x}|\overline{\Theta}, \boldsymbol{\alpha}) \geq \tau$ is classified as self and all other for which $P(\mathbf{x}|\overline{\Theta}, \boldsymbol{\alpha}) < \tau$ are classified as non-self.

3.2   Negative Representation of Character Statistics

In [11] Stibor et al. have shown that the use of the r-chunk matching rule becomes infeasible when the binary alphabet $\Sigma = \{0,1\}$ is changed into a larger symbol vocabulary. In specific, to generate a sufficient amount of detectors the r parameter needs to be close to the string length which results in an infeasible space complexity. Applying the r-chunk matching rule directly to text data where the size of the symbol vocabulary is typically above 20 is thus considered of little use.

Recently, Pöllä and Honkela [5] have used a probabilistic model to generate a negative description of a text document by examining the frequencies of individual characters in a sliding window of w characters. Using a character unigram model, the frequency x_i ($0 \leq x_i \leq w$) of a specific character $i \in \Sigma$ in a multiset of w character has a Binomial distribution

$$P(x_i = k|p) = \binom{w}{k} p^k (1 - p)^{w-k} \tag{4}$$

where p is the unigram probability of the character. This property is then used to produce a description of all character frequencies $x_i = k$ in a window of w adjacent characters which are not observed in the available data. This approach can be considered as a compromise between a negative selection algorithm and a probabilistic self-model since the idea of non-self detectors is used but without the need for inefficient negative selection of detectors since the size of the initial detector collection is limited to $|\mathcal{D}_0| = |\Sigma|(w + 1)$.

4   Multinomial Mixture Model

By combining the ideas of modeling self using a parameterized distribution and the sliding window of characters approach, the two can be combined into a generative model using a multinomial distribution and define non-self as any string for which the probability of being generated by the statistical model does not reach a selected threshold frequency.

Let Σ be a symbol alphabet and let D be a string from Σ. The size (cardinality) of the alphabet is denoted as $|\Sigma|$ and the length of the document as $|D|$. Further, let $\mathbf{x}$ be a $|\Sigma|$-dimensional categorical random variable counting the frequency of each symbol $i \in \Sigma$ in a window of w adjacent symbols in D. Assuming an independent probability Θ_i for each symbol in Σ, the probability of $\mathbf{x}$ has a multinomial distribution

$$P(\mathbf{x}|\Theta) = \frac{w!}{\prod_{i=1}^{|\Sigma|} x_i!} \prod_{i=1}^{|\Sigma|} \Theta_i^{x_i} \tag{5}$$

where $\sum_{i=1}^{|\Sigma|} \Theta_i = 1$ and $\sum_{i=1}^{|\Sigma|} x_i = w$. To fit this model for a specific dataset $\mathcal{X} = \{\mathbf{x}_1, \mathbf{x}_2, ..., \mathbf{x}_{|D|-w+1}\}$ we can find the maximum likelihood estimate for parameters Θ by maximizing the likelihood function

$$\mathcal{L}(\Theta|\mathcal{X}) = \prod_{j=1}^{|\mathcal{X}|} P(\mathbf{x}_j|\Theta) = \prod_{j=1}^{|\mathcal{X}|} \left(\frac{w!}{\prod_{i=1}^{|\Sigma|} x_{ji}!} \prod_{i=1}^{|\Sigma|} \Theta_i^{x_{ji}} \right) \tag{6}$$

where x_{ji} is the frequency of character i in the jth training sample resulting in

$$\Theta_{\mathrm{ML}} = \frac{1}{|\mathcal{X}|} \sum_{j=1}^{|\mathcal{X}|} \mathbf{x}_j \tag{7}$$

where $|\mathcal{X}|$ is the number of available training samples.

Depending on the data set $\mathcal{X}$ at hand, a single multinomial model can be insufficient to capture the internal correlations in the data and a finite mixture model is justified. For a mixture of M multinomials, the probability of $\mathbf{x}$ is

$$P(\mathbf{x}|\alpha, \overline{\Theta}) = \sum_{m=1}^{M} \alpha_m P(\mathbf{x}|\Theta_m) = \sum_{m=1}^{M} \alpha_m \frac{w!}{\prod_{i=1}^{|\Sigma|} x_i!} \prod_{i=1}^{|\Sigma|} \Theta_{mi}^{x_i} \tag{8}$$

where the coefficients α_m define the mixture proportions of the multinomials parameterized by $\overline{\Theta} = (\Theta_1, \Theta_2, ..., \Theta_M)$. However, for the mixture model, the optimal values for α and $\overline{\Theta}$ cannot be solved analytically. As in [4] the EM algorithm can be used to alternate between determining the posterior probabilities and computing new parameter values. The E- and M-steps for a multinomial mixture model (as presented in [13]) are as follows:

- E-step: use the current parameters $\overline{\Theta}$ and α to compute the posterior probability of each sample $\mathbf{x}_j$ being generated by mixture component m

$$
\begin{aligned}
P(m|\mathbf{x}_j, \overline{\Theta}, \alpha) &= \frac{P(\mathbf{x}_j|m, \overline{\Theta}, \alpha)P(m)}{P(\mathbf{x}_j)} \\
&= \frac{\alpha_m \prod_{i=1}^{|\Sigma|} \Theta_{mi}^{x_i}}{\sum_{m'=1}^{M} \alpha_{m'} \prod_{i=1}^{|\Sigma|} \Theta_{m'i}^{x_i}}
\end{aligned}
\tag{9}
$$

- M-step: compute new parameters $\overline{\Theta}^{(t+1)}$ and $\alpha^{(t+1)}$ according to the new posterior probabilities

$$
\alpha_m^{(t+1)} = \frac{1}{|\mathcal{X}|} \sum_{j=1}^{|\mathcal{X}|} P(m|\mathbf{x}_j, \overline{\Theta}^{(t)}, \alpha^{(t)})
\tag{10}
$$

$$
\Theta_{mi}^{(t+1)} = \frac{\sum_{j=1}^{|\mathcal{X}|} x_{ji} P(m|\mathbf{x}_j, \overline{\Theta}^{(t)}, \alpha^{(t)})}{\sum_{r=1}^{|\Sigma|} \sum_{j=1}^{|\mathcal{X}|} x_{jr} P(m|\mathbf{x}_j, \overline{\Theta}^{(t)}, \alpha^{(t)})}
\tag{11}
$$

where x_{ji} is the frequency of character i in the jth training vector and Θ_{mi} is the parameter Θ_i of the mth component of the mixture.

After computing the parameters $\overline{\Theta}$ and α for a dataset $\mathcal{X}$, the discrimination between self and non-self can be made by setting a threshold probability τ such that any $\mathbf{x}$ for which $P(\mathbf{x}|\overline{\Theta}, \alpha) \geq \tau$ is classified as self and $P(\mathbf{x}|\overline{\Theta}, \alpha) < \tau$ as non-self.

To correctly classify all samples $\mathbf{s} \in \mathcal{S}$ into self, the threshold probability should be set to

$$
\tau = \min\{P(\mathbf{s}|\overline{\Theta}, \alpha)\}, \quad \forall \mathbf{s} \in \mathcal{S}
\tag{12}
$$

in order to have the threshold probability as high as possible while still classifying the self samples correctly.

5   Applying Multinomials for Textual Data

Difficulties in statistical modeling of language are often related to the problem of data sparsity (i.e., insufficient amount of available training data in relation to the dimensionality of the data). Analyzing documents using a "bag-of-words" model is a common approach to gain information about the content though losing much information in ignoring word order. A bag-of-characters representation of text extends this tradeoff even further as only a fraction of the entropy in the text is preserved. Thus the character-based analysis is limited to simpler tasks such as anomaly detection, language identification [14] and authorship attribution [15].

Anomaly detection in textual data is closely related to the problem of document classification in information retrieval where document membership in a category is often viewed as a posterior probability using a statistical model. The one-class support vector machine has been applied in document classification [16] tasks with various document representation schemes [17]. A bag-of-words based multinomial model has been used by Novovičová and Malík [13] in document classification with improved results compared to a naïve Bayesian classifier. Blei et al. [18] have also used mixture models based on latent Dirichlet allocation (LDA) where the mixture components represent document topics.

6   Experiments

6.1   Mixture Model for a 4-Symbol Vocabulary

A simple training set consisting of four-character strings from the vocabulary $\Sigma = \{a,b,c,d\}$ is used to fit a multinomial mixture of two components for self/non-self discrimination. The training data set consists of strings where the characters have a strong correlation such that each string in the training data set consists of an equal amount of 'a' and 'b' or alternatively 'c' and 'd' (e.g. 'baab', 'bbaa', 'dccd', or 'cdcd'). The multinomial parameters $\overline{\Theta}$ are initialized randomly in $[0,1]^4$ and the mixture coefficients are initially set to $\alpha = (0.5, 0.5)$. After 30 iteration rounds using EM, the mixture model has learned the parameters

$$\begin{bmatrix} \alpha_1 \\ \alpha_2 \end{bmatrix} = \begin{bmatrix} 0.475 \\ 0.525 \end{bmatrix} \qquad \begin{bmatrix} \Theta_1 \\ \Theta_2 \end{bmatrix} = \begin{bmatrix} 0.07 & 0.07 & 0.43 & 0.43 \\ 0.41 & 0.41 & 0.09 & 0.09 \end{bmatrix}$$

A conditional probability distribution for this mixture model is shown in Figure 1a with various probability contours for selecting the threshold probability τ in Figure 1b. Classification regions are shown for a threshold frequency of $\tau = 0.1129$ in Figure 1c.

Table 1 presents a listing of all $\binom{4+4-1}{4} = 35$ possible 4 character multisets in a descending order of probability according to the mixture model. For example, by setting $\tau = 0.09$ the model classifies each permutation of strings "aabb" and "ccdd" as self and everything else as non-self.

6.2   Anomaly Detection in Written English

Anomaly detection in written natural language was simulated by using short segments from the Reuters corpus[1] and modifying a part of the string to test

---

[1] http://about.reuters.com/researchandstandards/corpus/

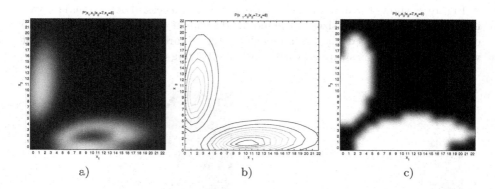

Fig. 1. Conditional probability distributions $P(x_1, x_3 | x_2 = 7, x_4 = 8)$ for strings (a), contour lines for various threshold probabilities τ (b) and decision regions for self (white) and non-self (black) for a given τ (c)

Table 1. List of all 35 possible 4-character multisets in a descending order of probability

$\mathbf{x}$	D example	$P(\mathbf{x}|\alpha, \Theta)$	$\mathbf{x}$	D example	$P(\mathbf{x}|\alpha, \Theta)$
(0 0 2 2)	"cdcd"	**0.098831**	(0 4 0 0)	"bbbb"	0.015445
(2 2 0 0)	"abab"	**0.092671**	(3 0 1 0)	"aaac"	0.013075
(0 0 3 1)	"cccd"	0.065887	(3 0 0 1)	"aaad"	0.013075
(0 0 1 3)	"dddc"	0.065887	(0 3 1 0)	"bbbc"	0.013075
(3 1 0 0)	"aaab"	0.061781	(0 3 0 1)	"bbbd"	0.013075
(1 3 0 0)	"bbba"	0.061781	(2 0 1 1)	"aacd"	0.012973
(2 1 1 0)	"aabc"	0.039224	(1 1 2 0)	"abcc"	0.012973
(2 1 0 1)	"aabd"	0.039224	(1 1 0 2)	"abdd"	0.012973
(1 2 1 0)	"abbc"	0.039224	(0 2 1 1)	"bbcd"	0.012973
(1 2 0 1)	"abbd"	0.039224	(1 0 3 0)	"accc"	0.011022
(1 0 2 1)	"accd"	0.033065	(1 0 0 3)	"addd"	0.011022
(1 0 1 2)	"acdd"	0.033065	(0 1 3 0)	"bccc"	0.011022
(0 1 2 1)	"bccd"	0.033065	(0 1 0 3)	"bddd"	0.011022
(0 1 1 2)	"bcdd"	0.033065	(2 0 2 0)	"aacc"	0.006487
(1 1 1 1)	"abcd"	0.025946	(2 0 0 2)	"aadd"	0.006487
(0 0 4 0)	"cccc"	0.016472	(0 2 2 0)	"bbcc"	0.006487
(0 0 0 4)	"dddd"	0.016472	(0 2 0 2)	"bbdd"	0.006487
(4 0 0 0)	"aaaa"	0.015445			

the sensitivity of detection. As a preprocessing stage, a lowercase conversion was made and all punctuation was removed from the data to limit the symbol vocabulary into 26 characters ('a' to 'z'). A 20-character string form the corpus was selected at random and a single multinomial model was computed from the string by setting $w = 20$. A random segment was then replaced to simulate an edit in the original string and the probability of the model generating the modified string was used to detect the anomaly.

Figure 2a shows the detection rate (proportion of successful detections) for various window lengths and sizes of the modified segment when a substring of 1 to 5 characters was replaced with a random character. Figure 2b shows the same result when a substring was swapped with another substring of the Reuters corpus.

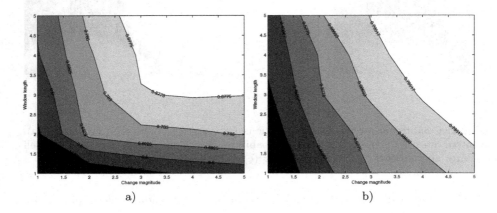

a)   b)

Fig. 2. Detection rate of replacing a substring of 1 to 5 characters with a random string (a) and another segment of the Reuters corpus (b) into the original string D ($|D| = 20$). Window length on the vertical axis. Mean result of 1000 trials.

In Figure 3 the probability (5) of the multiset of characters is shown for each $|D| - w + 1$ window positions for $|D| = 200$ and $w = 100$. An edit in the original string has resulted in probability values which are lower than the threshold $\tau = 4 \cdot 10^{-17}$ and the change is thus detected.

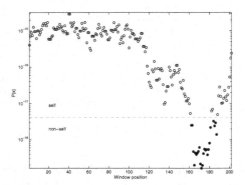

Fig. 3. Probability of each character window in a document. The changed part in the document is detected in the region where the probability is below the threshold τ (black dots).

7   Discussion and Future Work

There are several open questions related to applying the multinomial mixture model for anomaly detection in natural language. Using mixture models involves making a compromise between the model complexity and the approximation accuracy by selecting an appropriate number of mixtures. In the confined example of Section 6.1 the number of components could be easily defined *a priori* using information on the correlation structure of the data. However, the problem of selecting an appropriate model complexity (i.e., using the Akaike information criterion) is a relevant topic for further research. Also, as the method was presented as a general tool for discrete data, the same analysis could be used for word or morpheme level analysis of corpora.

8   Conclusions

Biologically inspired anomaly detection based on negative selection suffers from the curse of dimensionality when extending standard NSA algorithms to non-binary strings. Recent work on statistical models for self/non-self discrimination are thus expected to be more successful for textual data. A generative model for variable-length strings from a general finite symbol alphabet was presented for the application of change detection in textual data. The use of multinomial models on the character and word level was discussed.

Our experiments on artificial data showed that the use of a probability based similarity measure in binary classification is justified especially if there are strong correlations in the data and if information on the symbol order in a set of w adjacent symbols can be omitted for anomaly detection. A large scale experiment on natural language was considered necessary to evaluate the performance of the proposed model in practical settings.

References

1. Forrest, S., Perelson, A.S., Allen, L., Cherukuri, R.: Self-nonself discrimination in a computer. In: Proceedings of the 1994 IEEE Symposium on Research in Security and Privacy, Oakland, CA, pp. 202–212. IEEE Computer Society Press, Los Alamitos (1994)
2. Stibor, T.: An empirical study of self/non-self discrimination in binary data with a kernel estimator. In: Bentley, P.J., Lee, D., Jung, S. (eds.) ICARIS 2008. LNCS, vol. 5132, pp. 352–363. Springer, Heidelberg (2008)
3. Schölkopf, B., Smola, A.J.: Learning with Kernels: Support Vector Machines, Regularization, Optimization, and Beyond. MIT Press, Cambridge (2001)
4. Stibor, T.: Discriminating self from non-self with finite mixtures of multivariate Bernoulli distributions. In: Proceedings of Genetic and Evolutionary Computation Conference – GECCO, pp. 127–134. ACM Press, New York (2008)
5. Pöllä, M., Honkela, T.: Change detection of text documents using negative first-order statistics. In: Proceedings of AKRR 2008, The Second International and Interdisciplinary Conference on Adaptive Knowledge Representation and Reasoning, Porvoo, Finland, September 2008, pp. 48–55 (2008)

6. D'haeseleer, P.: An immunological approach to change detection: theoretical results. In: Proceedings of the 9th Computer Security Foundations Workshop, pp. 18–26. IEEE Computer Society Press, Los Alamitos (1996)
7. de Castro, L.N., Timmis, J. (eds.): Artificial Immune Systems: A New Computational Intelligence Approach. Springer, Heidelberg (2002)
8. González, F.A.: Anomaly detection using real-valued negative selection. Genetic programming and evolvable machines. Journal of Genetic Programming and Evolvable Machines, 4–383 (2003)
9. Stibor, T., Timmis, J., Eckert, C.: The link between r-contiguous detectors and k-CNF satisfiability. In: Congress on Evolutionary Computation – CEC, pp. 491–498. IEEE Press, Los Alamitos (2006); revised and extended version
10. Stibor, T., Mohr, P., Timmis, J., Eckert, C.: Is negative selection appropriate for anomaly detection? In: GECCO 2005: Proceedings of the 2005 conference on Genetic and evolutionary computation, pp. 321–328. ACM, New York (2005)
11. Stibor, T., Bayarou, K.M., Eckert, C.: An investigation of R-chunk detector generation on higher alphabets. In: Deb, K., et al. (eds.) GECCO 2004. LNCS, vol. 3102, pp. 299–307. Springer, Heidelberg (2004)
12. Dempster, A.P., Laird, N.M., Rubin, D.B.: Maximum-likelihood from incomplete data via the EM algorithm. Journal of Royal Statistical Society B 39, 1–38 (1977)
13. Novovičová, J., Malík, A.: Application of multinomial mixture model to text classification. In: Perales, F.J., Campilho, A.C., Pérez, N., Sanfeliu, A. (eds.) IbPRIA 2003. LNCS, vol. 2652, pp. 646–653. Springer, Heidelberg (2003)
14. Cavnar, W.B., Trenkle, J.M.: N-gram-based text categorization, pp. 161–175 (1994)
15. Keselj, V., Peng, F., Cercone, N., Thomas, C.: N-gram-based author profiles for authorship attribution (2003)
16. Manevitz, L.M., Yousef, M.: One-class SVMs for document classification. Journal of Machine Learning Research 2, 139–154 (2001)
17. Srihari, X.W.R., Zheng, Z.: Document representation for one-class SVM. In: Boulicaut, J.-F., Esposito, F., Giannotti, F., Pedreschi, D. (eds.) ECML 2004. LNCS, vol. 3201, pp. 489–500. Springer, Heidelberg (2004)
18. Blei, D.M., Ng, A.Y., Jordan, M.I.: Latent Dirichlet allocation. Journal of Machine Learning Research 3, 993–1022 (2003)

On the Efficiency of Swap-Based Clustering

Pasi Fränti and Olli Virmajoki

Department of Computer Science, University of Joensuu, Finland
{franti,ovirma}@cs.joensuu.fi

Abstract. Random swap-based clustering is very simple to implement and guaranteed to find the correct clustering if iterated long enough. However, its quadratic dependency on the number of clusters can be too slow in case of some data sets. Deterministic selection of the swapped prototype can speed-up the algorithm but only if the swap can be performed fast enough. In this work, we introduce an efficient implementation of the swap-based heuristic and compare its time-distortion efficiency against random and deterministic variants of the swap-based clustering, and repeated k-means.

Keywords: Clustering, prototype swap, probabilistic method, k-means, efficiency.

1 Introduction

The clustering can be found by a sequence of *prototype swaps* and by fine-tuning their exact location by k-means as demonstrated in Fig. 1. A simple but effective approach is to select the swap randomly: select the prototype to be swapped randomly and replace it to the location of a randomly selected data vector. Despite being simple to implement and efficient in practice, the quadratic dependency on the number of clusters can be a limiting factor in cases when there are a large number of clusters.

Several deterministic swap-based methods have been considered in literature by selecting the prototype to be swapped as the one that increases the cost function value least [1], [2], or by merging two existing clusters [3], [4], [5] following the spirit of agglomerative clustering. The new location of the prototype can be chosen either by considering locations of all possible data vectors [3], splitting an existing cluster [3], [6], or by using some heuristic such as selecting the cluster with the largest variance [1]. The swap-based approach has also been used for solving in a related p-median problem [7].

The main drawback of the above mentioned approaches is their computational complexity. Even though the correct clustering can be obtained by much fewer iterations in comparison to random swap, the time spent for the selecting the best prototype can easily overweigh the efficiency of the random swap heuristic. Another drawback is that the deterministic swap may get stuck into a locally optimal solution if randomness is completely eliminated in the process. In terms of time-distortion efficiency, a good compromise uses deterministic prototype selection and random relocation [8].

In this paper, we propose a faster implementation for the deterministic selection of the prototype, and compare it to the existing swap-based variants: random swap,

M. Kolehmainen et al. (Eds.): ICANNGA 2009, LNCS 5495, pp. 303–312, 2009.
© Springer-Verlag Berlin Heidelberg 2009

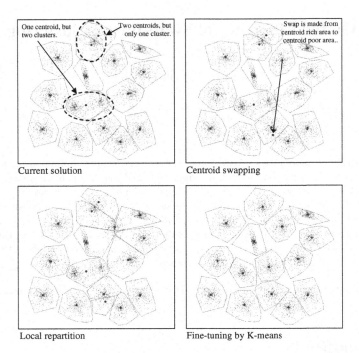

Fig. 1. Demonstration of the swap-based clustering for a sample initial solution for $N=5000$ data points from $M=15$ clusters

deterministic swap, hybrid of these two, and repeated k-means. We give time complexity analysis of the proposed method and show that it is as fast as the random heuristic in a single iteration, and is capable of finding the same clustering result in a fewer iterations. We compare experimentally in which cases the proposed approach is more favorable than the random swap.

2 Swap-Based Clustering

The efficiency of a swap-based clustering depends on how many iterations (swaps) are needed, and how much time each iteration takes. In a recent study, it was shown that for a given probability of success, the efficiency of the random swap [9], [10] has linear dependency on the number of data vectors, quadratic dependency on the number of clusters, and inverse dependency on dimensionality [11]. This is good in most cases but potentially faster method could be implemented if the swap could be found calculated more efficiently.

Another advantage of the random swap method is that it is extremely simple to implement. In comparison to k-means consisting of two steps (centroid step and partition step), it has only one additional step: prototype swap. In most cases, this step is independent on the metric used and it is trivial to implement, which makes it highly useful for practical purposes. Pseudo code of the algorithm is sketched in Fig. 2, which produces for a given input data set (X), a set of prototypes (C), and the partition

Swap-based clustering $(X) \longrightarrow C, P$

$C \leftarrow$ Select random representatives (X);
$P \leftarrow$ Optimal partition(X, C);
REPEAT T times
$\quad (C^{new}, j) \leftarrow$ Swap(X, C);
$\quad P^{new} \leftarrow$ Local repartition(X, C^{new}, P, j);
$\quad C^{new}, P^{new} \leftarrow$ K-means(X, C^{new}, P^{new});
$\quad$ IF $f(C^{new}, P^{new}) < f(C, P)$ THEN
$\quad\quad (C, P) \leftarrow C^{new}, P^{new}$;
RETURN (C, P);

Fig. 2. Structure of the general swap-based clustering
(For details and source code: http://cs.joensuu.fi/sipu/soft)

of the clustering (P) as the output. Here f is the optimization function to be minimized; typically mean square error (MSE).

2.1 Number of Iterations

At first sight, the probability for a successful swap seems to be rather small and a long number of iterations would therefore be needed. However, it is not necessary to find exactly the correct prototype for removal but selecting a neighbor cluster is sufficient. In practice, the number of iterations (T) needed depends on the probability of failure (q), number of clusters (M), and the size of neighborhood (α) as follows [11]:

$$T = \Theta\left(-\ln q \cdot \frac{M^2}{\alpha^2} \right) \tag{1}$$

The clustering, however, can be found by much fewer swaps if the algorithm knows which prototype should be swapped and where it should be re-located. The motivation of this paper is to develop a fast deterministic swapping technique that would be more efficient than the random swap, but at the same time, avoid the problem of getting stuck into a local minimum, which can easily happen when heuristic algorithms are designed.

The task seems to be straightforward but is far from trivial to implement in practice. Methods for prototype selection and replacement can be analytically constructed by considering all possible prototypes for the swap and selecting the locally optimal choice at each step. However, the challenge is to make this without excessive computation. Otherwise the method would be outperformed by the random swap technique since it can performed much faster, and significantly larger number of candidate swaps can be tested in a much shorter time.

2.2 Selecting the Prototype to Be Swapped

Swapping the prototype involves two design questions: which prototype will be swapped and where it should be relocated. The swapping makes one cluster to disappear and creates a new cluster elsewhere in the data space.

Several simple heuristic criteria have been considered for the selection: *cluster with smallest size* or *smallest variance* but these are too naïve to work in practice. Another approach is to *merge* two existing clusters as in agglomerative clustering [2], or applying split-and-merge strategy as in [4], [5]. These are possible but operating the clusters as entity can restrict the clustering too much and is subject to the problem of getting stuck into local minimum.

Considering the clustering as an optimization problem (e.g. minimizing mean square error), the most logical approach is to swap the prototype whose removal from its current location increases the cost function least. This kind of local optimization strategy has been successfully applied in several clustering algorithms [1], [2], [3], [4], [8], and also used in p-median problem [7], which differs from the clustering problems in the sense that the prototypes (facilities) are restricted to a predefined set and not free parameters as in clustering.

2.3 Efficient Implementation of the Selection

To implement the above approach, a so-called *removal cost* must be calculated for each cluster as follows. First, for a given data vector (x_i) associated in a cluster (p_i), its second nearest prototype (q_i) is solved:

$$q_i = \arg \min_{\substack{1 \le j \le m \\ j \ne p_i}} \left\| x_i - c_j \right\|^2 \tag{2}$$

The removal cost for cluster (j) can now be estimated by summing up the differences if the vectors in the cluster are repartitioned to their second nearest one (q_i):

$$D_j = \sum_{p_i = j} \left(d(x_i, c_{q_i}) - d(x_i, c_j) \right) \quad \forall \; j \in [1, M] \tag{3}$$

where c denotes the prototypes, and d the distance function. A more accurate estimation in [6] takes into account that the prototypes will be updated after the repartition, and should therefore be calculated as follows:

$$D_j = \sum_{p_i = j} \left(\frac{\left| n_{q_i} \right|}{\left| n_{q_i} + 1 \right|} d(x_i, c_{q_i}) - d(x_i, c_j) \right) \tag{4}$$

Here n_{qi} refers to the size of the secondary cluster. The drawback of these approaches is that they take N distance calculations for each of the M clusters. Thus, the overall time complexity of the swap step becomes $O(NM)$. This can be significantly larger than that of the random swap, which takes only $O(\alpha N)$ time per swap [11] where α is the number of neighbors.

However, since only a few prototypes will change after the swap, it is more efficient to maintain both data structures (p and q indices) after each swap as in [2, 7]. When a prototype is swapped, all vectors in the old cluster will be repartitioned from the primary partition (p) to their secondary partition (q). The changes in the secondary partition (q) can be updated as described in [2]. New secondary partitions must be resolved for all affected vector by a full search. This causes the following necessary updates:

$$\forall\, p_i = j: \quad p_i \leftarrow q_i \tag{5}$$

$$\forall q_i = j: q_i = \arg\min_{\substack{1 \le j \le m \\ j \ne p_i}} \left\| x_i - c_j \right\|^2 \tag{6}$$

Secondly, the same information must also be updated after the cluster addition. The new prototype attracts vectors from the neighboring clusters. Partition of every vector that is closer to the new prototype (c_j) than the one of its current cluster, must be be updated as follows:

$$p_i \leftarrow j \tag{7}$$

$$q_i \leftarrow p_i \tag{8}$$

The full search involved in the *argmin* operation requires $O(M)$ time. The cluster removed and the cluster added both have α neighbors, each consisting of N/M vectors, on average. The total time complexity of the swap step therefore reduces to $O(2\alpha \cdot M \cdot N/M) = O(\alpha N)$, which is the same complexity as that of the random swap.

2.4 Adding New Prototype

We consider every possible data vector as the potential new location for the swapped prototype. Calculating the cost of each of them is possible [3] but this would be very inefficient taking $O(\alpha N^2)$ time as there are N possible locations, and each of them would require $O(\alpha N)$ time to perform the local repartition. To relax the time requirements, we divide the problem into two sub tasks, which will be considered independently of each other:

  1. Select an existing cluster.
  2. Select the location within this cluster.

It is expected that the choice of the cluster is more important, and the exact location within the cluster is less significant since k-means will be applied to take care of the local fine-tuning of the prototype. Thus, the idea is to first select the cluster, and then add the prototype somewhere inside this cluster. This reduces the number of choices down to M and the corresponding time complexity to $O(\alpha NM)$. However, this is still too much.

Heuristic selection can also be considered. The most natural is to choose the cluster having the highest distortion [1]. This would reduce the time complexity of this step to $O(M+N) = O(N)$, where the first term originates from the selection and the second from the repartition. The cost (actually benefit) of adding the prototype c_j can be calculated as follows:

$$E_j = \sum_{p_i = j} d(x_i, c_j) \tag{9}$$

We consider the following heuristics:

  - Current centroid + ε [1].
  - Furthest vector.

-   Halfway of current centroid and furthest vector.
-   Random.

These heuristics can be performed in O(1)-O(N/M) time plus O(αN) time due to the repartition, which can be performed independently on the selection. The main alternatives for selecting the new location of the prototype are summarized in Table 1. On the basis of this analysis, we select the heuristic variant for the deterministic swap.

Table 1. Main variants for creating a new cluster

Variant	Which cluster	Which location	Time complexity
Full search	Try all	Try all	O(αN^2)
Optimal cluster	Try all	Any heuristic	O(αNM)
Heuristic	Largest distortion	Any heuristic	O(αN)
Random	Random	Random	O(αN)

2.5 Demonstration of the Deterministic Swap

The deterministic swap is demonstrated in Fig. 3 for a random initial solution. The removal and addition cost for each cluster are listed in Table 2. For the removal,

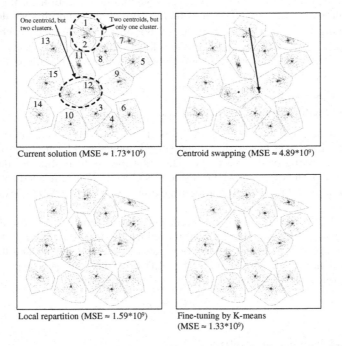

Current solution (MSE ≈ 1.73*10^9)   Centroid swapping (MSE ≈ 4.89*10^9)

Local repartition (MSE ≈ 1.59*10^9)   Fine-tuning by K-means (MSE ≈ 1.33*10^9)

Fig. 3. Demonstration of deterministic swap for a given initial solution using MSE as the cost function

Table 2. Removal and addition cost each cluster presented in Fig. 3

j	Removal cost (D_j)	Addition cost (E_j)	j	Removal cost (D_j)	Addition cost (E_j)
1	0.80	0.39	9	9.90	1.42
2	1.04	0.64	10	11.09	1.26
3	5.48	1.09	11	11.47	0.61
4	5.66	0.92	12	12.17	4.70
5	6.50	0.76	13	14.61	0.94
6	7.67	1.01	14	16.41	0.93
7	8.47	0.45	15	16.68	1.41
8	9.10	0.75			

clusters 1 and 2 are the best choices. Their absence would increase the distortion much less than the removal of any other cluster; the algorithm will choose the cluster 1. For relocation, cluster 12 causes currently the highest distortion by a large margin, and will be chosen. In this example, the *furthest vector* heuristic was applied.

2.6 Combinations of Random and Deterministic Swap

The main problem of the deterministic variant is that it ends up in sub-optimal local optima, and certain amount of randomness is still needed in order to optimize the clustering beyond this limit. The following combinations of random and deterministic techniques are therefore considered:

- RR = random removal, random addition (=random swap).
- RD = random removal, deterministic addition.
- DR = deterministic removal, random addition.
- DD = deterministic removal, deterministic addition.

Time complexities of these variants are summarized in Table 3. From the deterministic removal, the proposed faster implementation of DR and DD are denoted here as D^2R and D^2D, respectively.

The bottleneck of the deterministic swap is the removal, which dominates the processing time. Considering the analysis in [11], the deterministic removal is more efficient than its random counter part if $M < \alpha T$, assuming that it always selects the correct cluster in 1 iteration, and random removal within T iterations. With the data set shown in Fig.3, this is the case ($M=15$, $\alpha \approx 4$, $T=20$). The deterministic removal also stabilizes the solution faster than random removal, which is shown by the decreased time required by the k-means iterations.

The increased (theoretical) time complexity required by the deterministic addition, on the other hand, is insignificant in comparison to its random counter-part. It is therefore possible that the RD variant might be a good compromise between the random swap (RR) and deterministic swap (DD). However, the results even with the

Table 3. Summary of the time complexities of the main variants

	Random removal		Deterministic removal		Deterministic removal with updating data	
	RR	RD	DR	DD	D^2R	D^2D
Removal	$O(1)$	$O(1)$	$O(MN)$	$O(MN)$	$O(\alpha N)$	$O(\alpha N)$
Addition	$O(1)$	$O(N)$	$O(1)$	$O(N)$	$O(1)$	$O(N)$
Local repartition+ K-means	$O(\alpha N)$	$O(\alpha N)$	$O(\alpha N)$	$O(\alpha N)$	$O(\alpha N)$	$O(\alpha N)$
Algorithm in total	$O(\alpha N)$	$O(\alpha N)$	$O(MN)$	$O(MN)$	$O(\alpha N)$	$O(\alpha N)$

slower method in [8] revealed the opposite: the deterministic removal and random addition (DR) provides better time-distortion performance than either DD or RD variants despite its higher time complexity of single iteration. This is also the reason why the D^2R variant is worth considering and is proposed here.

For the k-means component, we use the fast variant [12] that utilizes the information of the prototype activity from previous iteration. Since most of the prototypes stabilize fast, it achieves a remarkable speed-up (both in theory and practice) in comparison to full search with rather simple algorithm. Potentially further speed-up could be potentially obtained by joining it with kd-tree as in [13], or with triangular inequality rule as in [14].

3 Experiments

In the following, we cluster four image data sets [2] by Hewlett-Packard Pavilion with 2.20 GHz ADM Athlon XP 3200+ processor and 512 MB memory. The vectors in *Bridge* are 4×4 non-overlapping blocks, in *Miss America* 4×4 difference blocks of two subsequent frames in a video sequence. The set *House* consists of RGB color values, and *Europe* consists of differential coordinates.

Time-distortion performance of the selected methods is compared in Figure 4 and Table 4. The following variants are compared: *repeated k-means* (RKM), random swap (RS), deterministic swap with random addition (DR) and its time-efficient implementation (D^2R). K-means results are obtained by the fast exact variant as proposed in [12].

The proposed algorithm (D^2R) outperforms its straightforward counterpart (DR) [8] and provides the same clustering quality than the random swap (RS) [9] with a similar or faster speed in case of all data sets except the 2-dimensional set Europe. This is also the only set for which the swap-based clustering does not provide any benefit over the repeated k-means at all. With the higher dimensional data sets, the method works better.

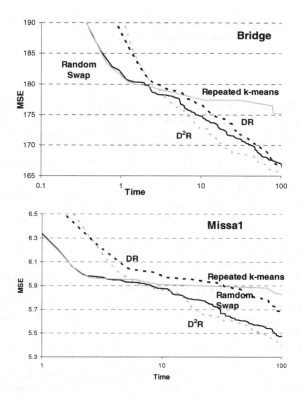

Fig. 4. Time-distortion efficiency of the different swap variants

Table 4. Summary of the clustering quality within a given time constraint

Data Set	Clustering quality (10s)				Clustering quality (100s)			
	RKM	RS	DR	D^{2}R	RKM	RS	DR	D^{2}R
Bridge	179.93	174.88	176.66	172.89	175.18	167.06	166.43	165.45
House	6.66	6.44	6.73	6.44	6.48	6.03	6.16	6.06
Miss America	5.91	5.88	6.01	5.87	5.84	5.47	5.68	5.40
Europe	17.89	20.01	31.63	31.63	6.17	6.27	8.83	13.46

4 Conclusions

Random swap method is very simple to implement and finds the correct clustering when iterated long enough. However, its quadratic dependency on the number of clusters motivated us to develop a faster method for selecting the prototype to be swapped deterministically. We have compared several random and deterministic variants. The methods also generalize for the case of unknown number of clusters by replacing the cost function by a suitable cluster validity index, and by having three operations: cluster removal, cluster addition and random swap.

References

1. Fritzke, B.: The LBG-U method for vector quantization – an improvement over LBG inspired from neural networks. Neural Processing Letters 5(1), 35–45 (1997)
2. Fränti, P., Virmajoki, O.: Iterative shrinking method for clustering problems. Pattern Recognition 39(5), 761–765 (2006)
3. Likas, A., Vlassis, N., Verbeek, J.J.: The global k-means clustering algorithm. Pattern Recognition 36, 451–461 (2003)
4. Kaukoranta, T., Fränti, P., Nevalainen, O.: Iterative split-and-merge algorithm for VQ codebook generation. Optical Engineering 37(10), 2726–2732 (1998)
5. Frigui, H., Krishnapuram, R.: Clustering by competitive agglomeration. Pattern Recognition 30(7), 1109–1119 (1997)
6. Fränti, P., Kaukoranta, T., Nevalainen, O.: On the splitting method for vector quantization codebook generation. Optical Engineering 36(11), 3043–3051 (1997)
7. Resende, M.G.C., Werneck, R.F.: A fast swap-based local search procedure for location problems. Ann. Oper. Res. 150(1), 205–230 (2007)
8. Fränti, P., Tuononen, M., Virmajoki, O.: Deterministic and randomized local search algorithms for clustering. In: IEEE Int. Conf. on Multimedia and Expo. (ICME 2008), pp. 837–840. Hannover, Germany (2008)
9. Fränti, P., Kivijärvi, J.: Randomised local search algorithm for the clustering problem. Pattern Analysis and Applications 3(4), 358–369 (2000)
10. Kanungo, T., Mount, D.M., Netanyahu, N., Piatko, C., Silverman, R., Wu, A.Y.: A local search approximation algorithm for k-means clustering. Computational Geometry 28(1), 89–112 (2004)
11. Fränti, P., Virmajoki, O., Hautamäki, V.: Probabilistic clustering by random swap algorithm. In: IAPR Int. Conf. on Pattern Recognition (ICPR 2008), Tampa, FL, USA (2008)
12. Kaukoranta, T., Fränti, P., Nevalainen, O.: A fast exact GLA based on code vector activity detection. IEEE Trans. on Image Processing 9(8), 1337–1342 (2000)
13. Lai, J.Z.C., Liaw, Y.-C.: Improvement of the k-means clustering filtering algorithm. Pattern Recognition 41(12), 3677–3681 (2008)
14. Elkan, C.: Using the triangle inequality to accelerate k-means. In: Int. Conf. on Machine Leearning (ICML 2003), pp. 147–153 (2003)

Sum-of-Squares Based Cluster Validity Index and Significance Analysis[*]

Qinpei Zhao, Mantao Xu, and Pasi Fränti

Department of Computer Science, University of Joensuu
Box 111, Fin-80101 Joensuu
Finland
{zhao,franti}@cs.joensuu.fi, mantao.xu@carestreamhealth.com

Abstract. Different clustering algorithms achieve different results with certain data sets because most clustering algorithms are sensitive to the input parameters and the structure of data sets. The way of evaluating the result of the clustering algorithms, cluster validity, is one of the problems in cluster analysis. In this paper, we build a framework for cluster validity process, while proposing a sum-of-squares based index for purpose of cluster validity. We use the resampling method in the framework to analyze the stability of the clustering algorithm, and the certainty of the cluster validity index. For homogeneous data based on independent variables, the proposed clustering validity index is effective in comparison to some other commonly used indexes.

1 Introduction

Clustering is an unsupervised process which intends to discover the unknown structure of data sets accurately. There are a number of clustering algorithms [1] based on different strategies and they are developed to satisfy with different needs from the data sets. The common sense is that there is no general algorithm applicable to all kinds of data sets. The problem comes up that how to evaluate the effect of clustering algorithms on different data sets. Cluster validity provides the way of validating the quality of clustering algorithms and the means of discovering the natural structure of the data sets. If cluster analysis is to make a significant contribution, much more attention must be paid to the cluster validity issues. Cluster validity measures are the methods, which can not only compare the results of two different sets of clustering algorithms to determine the better one, but determine the "correct" number of clusters in the data set.

Amounts of cluster validity indexes have been proposed. Milligan and Cooper [2] have presented a comparison study over thirty validity indexes for hierarchical clustering algorithms whereas Dimitriadou et al [3] conducted their comparison study over fifteen validity indexes for the case of binary data. Different indexes under different situations achieve different results. We introduce several indexes mentioned in these two literatures for purpose of comparison.

---

[*] Thanks to Nokia Foundation for financial support.

M. Kolehmainen et al. (Eds.): ICANNGA 2009, LNCS 5495, pp. 313–322, 2009.
© Springer-Verlag Berlin Heidelberg 2009

We separate the indexes in this paper into two types, one is sum-of-squares based type, and the other is classical type. The methods in the first type measure the dispersion of the data points within a cluster and between the clusters respectively. The indexes are:

- Ball and Hall [4], the maximum value of the successive difference is determined as the optimal number of clusters.
- Calinski and Harabasz [5], the minimum value of the successive difference is determined as the optimal number of clusters.
- Hartigan [6], the minimum value of the successive difference is determined as the optimal number of clusters.
- Xu [7], the maximum value can be determined as the optimal number of clusters, the successive difference is applicable but not necessary.

The classical measures are mostly proposed in different area and perform well to some extend. These measures share the advantage of using the maximum or minimum value as the optimal number of clusters.

- Dunn's index [8], the maximum of the index value is determined as the optimal number of clusters.
- Davies-Bouldin index [9], the minimum of the index value is determined as the optimal number of clusters.
- Xie-Beni's separation index [10], the minimum of the index value is determined as the optimal number of clusters.
- Bayesian Information Criterion [11], which is a model selection criteria. The first local maximum is determined as the optimal number of clusters.
- Silhouette Coefficient [12], the maximum of the index value is determined as the optimal number of clusters.

Applications of resampling method, such as bootstrapping, subsampling, or cross validation to cluster validity are not new in the cluster validity. Peck et al. [13] developed a bootstrap-based procedure to obtain approximate confidence bounds on the number of clusters in the "best" clustering. Ben-Hur et al. [14] presented a method that exploited measurements of the stability of clustering solutions obtained by perturbing the data set. Cluster validation by prediction strength [15] considered clustering as a classification problem, which used the way of cross validation technique. Dudoit and Fridlyand [16] introduced a prediction-based sampling method, CLEST, in which, the data was first split into two non-overlapping sets. Then the learning set was clustered and a classifier was built using the obtained labels; the test set was also clustered and the obtained labels were compared using an external index.

We establish a framework of cluster validity process with resampling methods to validate the clustering algorithm and the validity index. Moreover, a sum-of-squares based index is proposed. The rest of the paper is organized as follows. We introduce the framework of the cluster validity in Section 2. The proposed index is formulated in Section 3. Experiments on the proposed method are presented in Section 4, in which the results on both artificial generated and real data sets are also displayed. Two clustering algorithms are applied in the experiment. A further step on variability and certainty analysis is introduced in Section 5. Conclusions and future work are drawn in Section 6.

2 Related Work

Cluster validity relates to the clustering algorithms. The fundamental clustering problem is to partition a given data set into groups, so that the points in the same group are more similar to each other than the points in different groups. Thus, one way of the cluster validity is to analyze within-between group variance.

Let $X = \{x_1, x_2 \ldots x_n\}$ be a set of data with n samples. Suppose the samples in X have hard labels that mark them as representatives of m non-overlapping clusters, says $C = \{C_1, C_2 \ldots C_m\}$. The clustering algorithm is to find the optimal partition $P = \{P_1, P_2 \ldots P_m\}$. The most important parameter among them is the parameter m, the number of clusters, because most of the clustering algorithms require the parameter m as the input and thus the clustering result is also affected by it.

Given the data set X, a specific clustering algorithm, and a fixed range of number of clusters, the basic procedure of the cluster validity involves the following steps:

- Fix the data sets with external information.
- Repeat the clustering algorithm successively for the number of clusters, m from a predefined minimum m_{min}, to a predefined maximum m_{max}.
- Get the clustering results: partitions and codebooks. Calculate the index value of each number of clusters.
- Plot the "number of clusters vs. index metric" graph and select the m at which the partition appears to be the "best" according to how the index is optimized.
- Compare the detected number of clusters (m^*) with the "external information" to prove the effectiveness of the index.

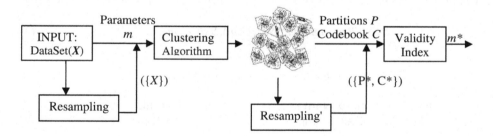

Fig. 1. Scheme diagram of cluster validity process

The clustering algorithm can be any of the existing algorithms. We use the Random Local Search algorithm (RLS) [17] in the validity procedure. The RLS clustering algorithm shares the advantage of both the k-means and the local search. To eliminate the effect on index from the clustering algorithm, K-means clustering, the most typical clustering algorithm is also tested in this paper.

Based on this procedure, we can easily have the scheme diagram of cluster validity in Fig.1. To estimate the stability of the clustering algorithm, we could use resampling method as is shown in the *resampling* part. Furthermore, in order to exclude the effect of data sets and clustering algorithm, another resampling method is employed, as the *resampling'* part shows. This part will be shown in section 4 in detail.

Basically, comparison is essential to prove the effectiveness. The two types' indexes mentioned above are compared to the proposed index in the experiments section.

3 Proposed Method

In cluster analysis, the within group variance and between group variance can be calculated by *sum-of-squares within cluster* (SSW) and *sum-of-squares between clusters* (SSB) respectively. We analysis the existing index based on SSW and SSB, and then propose a sum-of-squares based method, so-called WB-index.

The value of SSW is defined as:

$$SSW(C,m) = \frac{1}{n} \sum_{i=1}^{m} \sum_{j \in C_i} \| x_j - C_{P(j)} \| \tag{1}$$

which is minimized over all *m*-partitions C in the clustering procedure. According to ANOVA, the *total sum-of-squares* (SST) can be decomposed into two parts that are SSW and SSB for any partition C.

$$SSB(C,m) = \frac{1}{n} \sum_{i=1}^{m} n_i \| C_i - \overline{x} \| \tag{2}$$

where n_i is the number of elements in each cluster, and $\overline{x}$ is the mean value of the whole data set, m is the number of clusters. Hence, we can now define a generalized *within-between cluster type* (SSWB) in Eq.3, which is a function of the SSW or SSB:

$$SSWB = function(SSW(C,m), SSB(C,m)) \tag{3}$$

Table 1. Sum-of-squares based indexes

No.	Index Name	Formula
1	Ball & Hall	SSW/m
2	Calinski&Harabasz	$CH = \dfrac{SSB/(m-1)}{SSW/(n-m)}$
3	Hartigan	$H-index = -\log(SSW/SSB)$
4	Xu	$Xu = d\log(\sqrt{SSW/(dn^2)}) + \log(m)$

The sum-of-squares based methods above (table.1) are all based on the property of the SSW and SSB. We study these indexes in Fig.1. As in Fig1.(a) shows, the trends of normalized SSW and SSW/SSB are almost same, indicating that the factor of the SSW has a more important effect in the ratio of SSW/SSB. In other *WB-type* indexes except for Xu's index, we find that they either monotonously increase/decrease or need additional knee point detection method, such as successive difference in order to get the optimal number of clusters. Xu's index has clear minimum knee point; however, our experiments in section 4 will show it doesn't work well on real data sets.

Thus, we propose a simpler sum-of-square method, WB-index as:

$$WB = m \cdot SSW \: / \: SSB \tag{4}$$

We emphasize the effect of SSW with multiplying the number of clusters. The advantages of the proposed method are that it determines the number of clusters by minimal value of it without any knee point detection method, and it is easy to be implemented.

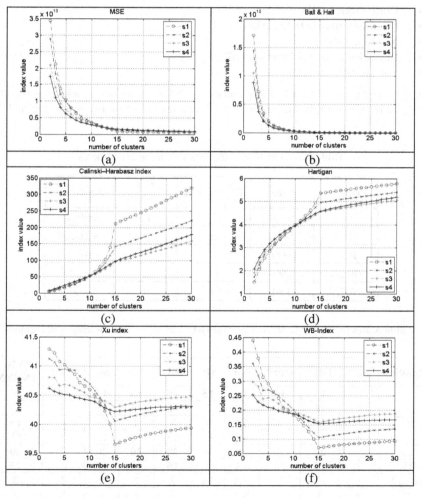

Fig. 2. (a). Comparison of SSW and SSW/SSB; (b)-(f). Comparison of several sum-of-square based indexes with four artificial data sets (s1-s4).

4 Experimental Results

In this paper, we test the methods with the data sets in table.2. The data sets s1 to s4 are generated with varying complexity in terms of spatial data distributions, which have 5000 vectors scattered around 15 predefined clusters with a varying degrees of overlap. The datasets a1 and R15 are generated in 2-dimensional Gaussian distribution. Iris and Breast are the real data sets obtained from the UCI Machine Learning Repository. Iris is a four-dimensional data set, containing three classes of 50 instances each, in which each class refers to a type of iris plant. The second real data set is the Wisconsin breast cancer data set (Wolberg and Mangasarian, 1990).

For purpose of comparison, we test five other classic measures:

- Dunn's index (DI)
- Davies-Bouldin's Index (DBI)
- Xie-Beni (XB)
- Bayes Information Criterion (BIC)
- Silhouette Coefficient (SC)

In the special case of $m=1$, SSW equals to SST. Clustering algorithm is therefore performed by $m=[2,30]$ in the case of S1-S4, and $m=[2,10]$ in the case of the real data sets.

Table 2. Information of the data sets in the experiments

DataSet	Size	Dimension	# of clusters	Generated
s1-s4	5000	2	15	artificial
a1	3000	2	20	artificial
R15	600	2	15	artificial
Breast	699	11	2	real
iris/Iris	150	4	3	real

Table 3. Results using the RLS (with 5000 RLS iterations and 2 K-means iterations)

DataSet	BH*	CH*	Har*	Xu	DI	DBI	XB	SC	BIC*	WB-INDEX
s1-s4	3	15	15	15	15	15	15	15	15	15
	3	15	4	15	7	15	15	15	4	15
	4	15	4	15	16	8	4	15	4	15
	3	15	3	15	25	13	13	15	5	15
a1	3	20	3	20	34	20	20	20	3	20
R15	3	15	15	15	2	15	15	15	8	15
Breast	3	3	3	NA	14	2	2	2	2	2
Iris	3	3	3	NA	2	2	2	9	6	3

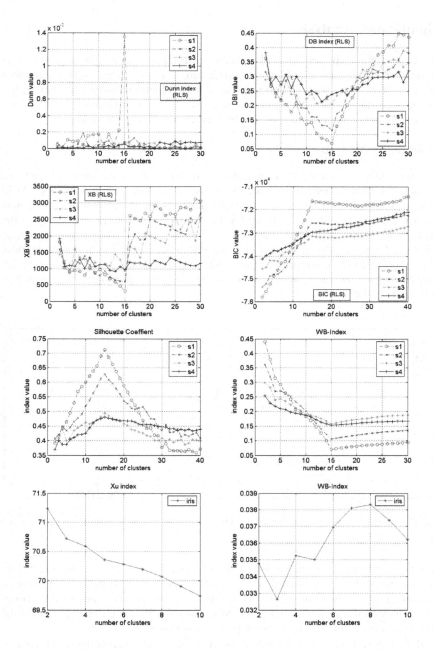

Fig. 3. The results with different validity indexes and data sets. The results of Xu's index and the proposed index on real data set Iris are on the last row. It is unable to find the minimum value of Xu's index as the optimal number of clusters as it is monotonously decreasing.

4 Significance Analysis

The results of the experiments with different clustering algorithms and data sets demonstrate that the proposed index can provide an accurate estimation of the number of clusters, which also shows the effectiveness of the cluster validity. Fundamentally, we can demonstrate the proposed index as it shows in the experiments. Moreover, we want to confirm the results in this section by further significance analysis.

4.1 Variability Analysis

With an uncertain distribution of the results, resampling method can be employed as a natural approach for the variability estimation associated with each index value. As in the process shows (Fig.1), we could resample on the original data set (X), get a new data set (X^*) and apply the new data set for the validation procedure again. Repeat the resampling B times, deal with the B times index values to get the statistical significance. However, the RLS clustering algorithm is designed with randomization, in which there is random swapping of the code vectors. Hence, we keep the data set unchanged and utilize the randomization of the clustering algorithm by running B times to analyze the results.

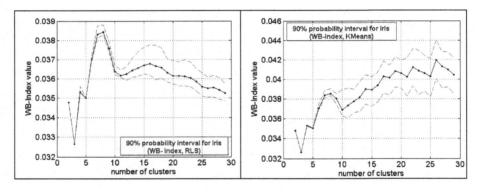

Fig. 4. 90% probability interval of the WB-index with the RLS and KMeans clustering on the data set Iris

Quartile range is one of the measures used to estimate variability. We use it into our scheme to analyze the variability of each index value. With the same setting of input parameters, fix the number of clusters, and run the clustering algorithm B times to get B values on the same number of clusters. Then the 5th and 95th percentiles of the B index values are calculated to get 90% probability range.

Iris data as a real data set is a representative to be tested. According to the results, only the proposed method and the BIC with knee detection get the correct number on Iris. In this case, both of the clustering algorithms with the same data set and index are tested. We run $B = 100$ times of the clustering algorithm with the same input parameters setting. The 90% probability interval with the RLS and K-means is shown respectively in Fig.4, and the dash line is the boundary of the range. It is clear that the range $m = [2, 3]$ strongly indicates the optimal number of clusters with the RLS clustering; and $m = [2, 5]$ with the K-means clustering. The range of m is wider with the K-means clustering than the RLS clustering. Thus we can conclude that RLS clustering is more stable than the K-means and the variability of the K-means on $m = 3$ is convincingly larger than that of the RLS on the Iris.

4.2 Certainty Analysis

We develop another way to prove the certainty of the proposed index, which employs the resampling method. As Fig.1 shows, the effect of the validity index is affected by the data set and the clustering algorithm. In this case, resampling the data set cannot prevent the effect coming from the clustering algorithm. Hence, we process the resampling method on the partitions getting from the clustering to avoid this problem.

In the first run of the validity procedure, a set of partitions (P) is generated. Basically, this set of partitions is the optimal one according to the clustering algorithm. A WB-index value (WBI) is obtained on P. We permute the original partitions (P) by B times, get $\{P^*\}$, and recalculate the index values $\{WBI^*\}$. As the optimal value of the WB-index should be as small as possible, we can estimate the certainty by counting the probability that $WBI^* \leq WBI$.

$$P = \frac{No.(WBI^* \leq WBI)}{TotalNo.(WBI^*)} \tag{5}$$

The smaller the probability P is, the more certainty the method obtains. It is not practical to calculate all possible permutations due to the involved time. Generally, at least $B =1000$ times permutations should be done. In this paper, 1000 random permutations were performed on the partitions (Fig.5). It indicates the certainty of the index, as the observed optimal value is much smaller than any of the values obtained under permutation.

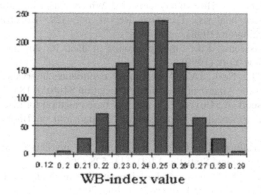

Fig. 5. Distribution of the WB-index on Iris data set (m=3) for 1000 permutations of the partitions with the RLS clustering. The "optimal" value of the WBI is very extreme by reference to this distribution (WBI = 0.032653).

5 Conclusions

We represented a framework with the resampling step for the estimation on the stability of the clustering algorithm and the variability of the validity index in cluster validity process. In addition, we proposed a new sum-of-squares based index which indicates simplicity and good prospect compared to other indexes. Based on the proposed index, we completed the whole process of the cluster validity.

References

1. Xu, R., Wunsch, D.: Survey of clustering algorithms. IEEE Transactions on Neural Networks 16(3), 645–678 (2005)
2. Milligan, G.W., Cooper, M.C.: An examination of procedures for determining the number of clusters in a data set. Psychometrika 50, 159–179 (1985)
3. Dimitriadou, E., Dolnicar, S., Weingassel, A.: An examination of indexes for determining the number of clusters in binary data sets. Psychometrika 67(1), 137–160 (2002)
4. Ball, G.H., Hubert, L.J.: ISODATA, A novel method of data analysis and pattern classification (Tech. Rep. NTIS No. AD 699616). Standford Research Institute, Menlo Park (1965)
5. Calinski, T., Harabasz, J.: A dendrite method for cluster analysis. Communication in statistics 3, 1–27 (1974)
6. Hartigan, J.A.: Clustering algorithms. Wiley, New York (1975)
7. Xu, L.: Bayesian Ying-Yang machine, clustering and number of clusters. Pattern Recognition Letters 18, 1167–1178 (1997)
8. Dunn, J.C.: Well separated clusters and optimal fuzzy partitions. Journal of Cybernetica 4, 95–104 (1974)
9. Davies, D.L., Bouldin, D.W.: Cluster separation measure. IEEE Transactions on Pattern Analysis and Machine Intelligence 1(2), 95–104 (1979)
10. Xie, X.L., Beni, G.: A validity measure for fuzzy clustering. IEEE Trans. on Pattern Analysis and Machine Intelligence 13(8), 841–847 (1991)
11. Frayley, C., Raftery, A.: How many clusters? Which clustering method? answers via model-based cluster analysis. Technical Report no. 329, Department of Statistics, University of Washington (1998)
12. Kaufman, L., Rousseeuw, P.J.: Finding Groups in data. In: An Introduction to cluster analysis. Wiley, New York (1990)
13. Peck, R., Fisher, L., Ness, J.V.: Approximate confidence intervals for the number of clusters. Journal of the American Statistical Association 84(405), 184–191 (1989)
14. Ben-Hur, A., Elisseeff, A., Guyon, I.: A stability based method for discovering structure in clustered data. In: Proceedings of the Pacific Symposium on Biocomputing, pp. 6–17 (2002)
15. Tibshirani, R., Walther, G.: Cluster validation by prediction strength. Journal of Computational & Graphical Statistics 14(3), 511–528 (2005)
16. Dudoit, S., Fridlyand, J.: A prediction-based resampling method for estimating the number of clusters in a dataset. Genome Biology 3(7) (June 2002)
17. Fränti, P., Kivijärvi, J.: Randomized local search algorithm for the clustering problem. Pattern Analysis and Applications 3(4), 358–369 (2000)

Supporting Scalable Bayesian Networks Using Configurable Discretizer Actuators

Isaac Olusegun Osunmakinde and Antoine Bagula

Department of Computer Science, Faculty of Sciences, University of Cape Town,
18 University Avenue, Rhodes Gift, 7707 Rondebosch, Cape Town, South Africa
{segun,bagula}@cs.uct.ac.za

Abstract. We propose a generalized model with configurable discretizer actuators as a solution to the problem of the discretization of massive numerical datasets. Our solution is based on a concurrent distribution of the actuators and uses dynamic memory management schemes to provide a complete scalable basis for the optimization strategy. This prevents the limited memory from halting while minimizing the discretization time and adapting new observations without re-scanning the entire old data. Using different discretization algorithms on publicly available massive datasets, we conducted a number of experiments which showed that using our discretizer actuators with the Hellinger's algorithm results in better performance compared to using conventional discretization algorithms implemented in the Hugin and Weka in terms of memory and computational resources. By showing that massive numerical datasets can be discretized within limited memory and time, these results suggest the integration of our configurable actuators into the learning process to reduce the computational complexity of modeling Bayesian networks to a minimum acceptable level.

Keywords: Intelligent Systems, Massive datasets, Bayesian Networks, Discretization, Scalability.

1 Introduction

Bayesian network models often formulate the core reasoning component of some intelligent systems because of their suitability in handling complex problems [1]. Researchers and practitioners have stressed that learning such models from environments captured as massive datasets is computationally intensive [2] [3]. In practice, it is convenient to say that massive datasets are relatively defined based on the capacity of the machine used for learning and are also dependent on users' execution urgency. For example 50,000 records of dataset may be massive for a machine and moderate or small for another. The intensity on the datasets implies that too much of computational time is expended and limited memory space may crash during these operations. This affects business and research deliveries, and may hinder the growing usage of Bayesian networks in industries that keep massive datasets to build intelligent systems. From our practical knowledge, improving the performance of discretization is obviously a sound basis for optimizing Bayesian networks'

M. Kolehmainen et al. (Eds.): ICANNGA 2009, LNCS 5495, pp. 323–332, 2009.
© Springer-Verlag Berlin Heidelberg 2009

learning. Intelligent system engineers do not want to wait too long to make the reasoning component ready for use.

Most of the existing conventional algorithms load entire massive datasets onto the limited memory for discretizations and a column is processed one at a time. Time is expended for loading, discretizing and probably saving back, which could otherwise have been minimized. Carrying out the discretization process on massive datasets whose size is more than the available allocated memory may currently not be practically feasible. Achieving this requires scalability of discretization methods which is very challenging.

A number of fairly recent studies have developed good conventional algorithms to discretize datasets but they fall short in considering the scalability of their approaches [4] [5]. Among the rationales in this scalability research is studying how massive are the datasets used in the existing discretization approaches. For examples, Li et al.[4] suggested feature selection heuristics for discretizing bio-medical data where they evaluated it on a notable lung-cancer dataset of 10,000 records. Lee's supervised algorithm [6] is similar to Li et al. but he used the entropy of intervals for discretization. Also, Lee used a maximum of 3,163 records of hypothyroid dataset to evaluate his work. Out of the 16 datasets used by Dougherty et al. [5], the maximum size is Australian dataset with 6,650 records.

The computational times and memory usages of the methods described above are not known though these are two important parameters upon which the scalability of discretizing massive datasets for learning networks depends. In an attempt to address scalability of discretizations for the core component of intelligent systems, the available open source network learning applications (e.g. Weka [7] and Hugin [8]) force users to discretize all numeric values of the attributes present in the datasets. However, certain numeric attributes in real life are not necessarily required to be discretized. In this research, we proposed configurable discretizer agent actuators which dynamically scale limited memory and improve computational time efficiency. In a number of comparative evaluations, the actuators outperform the conventional discretization approaches in speed and memory management respectively. Our major contributions are:

- The development of a new generalized configurable discretizer actuators and its system model to optimize the core intelligent system component through discretization processes.
- The evaluation of this configurable actuator on publicly massive datasets using different discretization algorithms implemented in Weka and Hugin systems.

The rest of this paper is arranged as follows: in section 2, we introduce the background of Bayesian networks, discretization algorithms, dynamic memory management scheme and the agent architecture as the theoretical foundations of our configurable discretizer actuators. Section 3 presents the system model and the configuration of the discretizer agent actuators. Section 4 presents the experimental evaluations of the discretization time and memory scalability using publicly available datasets from UCI [9] (University of California Irvine) repository used by intelligent systems researchers. We conclude the paper in section 5.

2 Theoretical Background

2.1 Bayesian Network Models

A Bayesian network model is formally defined as a directed acyclic graph (DAG) represented as G = {X(G), A(G)}, where X(G) = {X_1,...,X_n}, vertices (variables) of the graph G and $A(G) \subseteq X(G) \times X(G)$, set of arcs of G. The network requires discrete random values such that if there exists random variables X_1, . . ., X_n with each having a set of some values x_1, . . ., x_n then, their joint probability density distribution is defined in equation 1;

$$pr(X_1,...,X_n) = \prod_{i=0}^{n} pr(X_i \mid \pi(X_i))$$ (1)

where $\pi(X_i)$ represents a set of probabilistic parent(s) of child X_i [1]. A parent variable otherwise refers to as *cause* has a dependency with a child variable known as *effect*. Every variable X with a combination of parent(s) values on the graph G captures probabilistic knowledge as conditional probability table (CPT). A variable without a parent encodes a marginal probability. Learning the suitable networks from massive datasets is computationally intensive as stated above.

2.2 Discretization Algorithms

Discretization algorithms are techniques which are used as preprocessing key operations in learning Bayesian models [5] [6]. They classify numerical data into their corresponding interval values relatively to the patterns in the data attributes. Weka and the Hugin systems use discretization algorithms which are built around the simple binning and minimum description length (MDL) methods [1]. Simple binning include an equal-width method using an unsupervised discretization approach which divides attribute values into k equal sizes. The seed k is supplied by users while equal-width finds maximum and minimum attribute values and they are used to determine data intervals. The Hellinger-based algorithm uses interval entropy function E(.) as a justification for quality discretization to accommodate any datasets. The entropy of any interval between a and b is shown in equation 2 [6].

$$E([a, b]) \equiv \sqrt{\left| \sum_i \left(\sqrt{pr(x_i)} - \sqrt{pr(x_i \mid \overline{ab})} \right)^2 \right|}$$ (2)

As a basis of the algorithm, the values x_i of the target attribute being discretized are sorted accordingly and they form a column of intervals. The probability distribution of x_i is represented as $pr(x_i)$. The scheme in the next subsection is therefore adopted to prevent the out-of-memory problems in the learning processes.

2.3 Dynamic Memory Management Scheme

The dynamic memory management scheme used in the Loci framework [10] is an economical solution which manages the memory by allocation and de-allocation of data structures based on the lifetime of data structures. Thus, in order to accommodate discretization of large datasets within a limited memory, we extracted and interpreted the scheme from [10] as follows: (i) pre-allocation of memory to data structures, (ii) incorporate relevant memory management operations, (iii) invoke loop scheduling techniques, and (iv) recycle memory from data structures.

Pre-allocation with partitioning of the entire memory alone in scheme i does not benefit space saving until the others in the sequence are involved. Many parallel algorithms exploit scheme i as a trade-off to optimize speed but suffer from peak memory requirement. A possible relevant management operation in ii is the use of remote memory or secondary storage devices, for example. These management operations are generalized concepts of virtual memory. A virtual memory is a multilevel store which gives a large process an impression that it has more primary memory to itself, while it actually uses external disk devices as a supplement [11]. In iii, examples of loop scheduling techniques are multiple nested iterations, recursions, synchronizations, etc. Also in iv, at the end of every schedule or lifetime, memory is recovered from data structures after its execution. Thus, this scheme empowers a system to accommodate massive datasets within a limited memory without a halting problem. The next subsection also describes the basic agent architecture as fundamentals of our configurable actuators.

2.4 A Basic Agent Architecture

Among the classes of agents used in intelligent systems, the software agent as related to this work perceives from the components of *environments* through *sensors* and acting upon the environment through *actuators* [1]. According to Russell [1], a software agent can sense its environment using file content or network packets and also uses writing files or packets as actuators to act on the environment. From Russell's illustration, when environment is perceived, some forms of machine learning algorithms are used to interpret the *percepts*. They consequently generate the instructions required by the actuators to carry out actions on the environment.

The positions of the agent and the environment are often far apart which possess distributed properties. It illustrates that agents can be sent over a network to carry out specific tasks and can also provide services to other components on a given machine. It is deduced from here that agent actuators can be characterized with mobility as they include their required information in their description. Their independence influences the design of components for distributed agents which motivate the development of the configurable discretizer agent actuators in this study. Section 3 now describes the proposed configurable actuators.

3   The Generalized Configurable Discretizer Actuators

3.1   The System Model for the Actuators

Figure 1 depicts the system model that we used to accomplish complete scalable discretization. If either space or time is optimized, it is an incomplete scalability as a trade off is not beneficial to the networks used in intelligent systems. Our strategy combines the memory management scheme in subsection 2.3 and the architecture in subsection 2.4. In this strategy, an actuator is dedicated and sent to discretize values of one or more attributes. For balancing purposes, a number of actuators, rather than all, are heuristically set by users and concurrently distributed at a time. Discretization time is faster as the actuators act on more than one attributes at a time. As the actuators complete discretization of some attributes in a pass, they are returned to the symmetric processors that reschedule them for subsequent attributes. With this, memory is continually and dynamically allocated which then recycles each time there is scheduling of actuators for discretization.

We now define the major components of Figure 1 as follows: discretizer *agent actuators* described in the previous paragraph, discretization *algorithm, massive environment* (or datasets), storage of previous discretized *parameters* and subsequent *observations* made after discretizing the massive datasets. The algorithm which resides on the limited memory of a machine generates tuples of intervals for the actuators to discretize values of the attributes remotely. We adopted the Hellinger-based algorithm in Lee's work [6] as a proof of concept since this research focuses on supporting the optimization of the core reasoning component of intelligent systems through scalable discretizations. The component of the massive dataset (or environment) is kept away from the limited memory and its attributes' values are acted upon concurrently in a secondary storage or across a network. This provides a competitive advantage in developing countries where discretization process can be accidentally suspended probably due to electricity power failure but modeling continues where the process stops.

Also, the previous parameters are used adaptively to discretize subsequent observations instead of re-scanning the entire old massive dataset. The last tuples of intervals if the data patterns remain the same, the data types for all attributes, etc are examples of previous parameters. The configuration used by the discretizer actuators is designed and described in the next subsection.

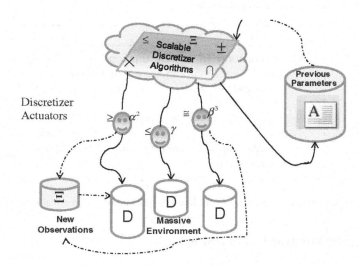

Fig. 1. System Model for Discretizer Agent Actuators

3.2 The Configurable Discretizer Actuators

We designed and configured these actuators as shown in Figure 2 with dynamic packets of information to act upon the environments. The content of the packet consists of the *control information* and the *environments*. The control information provides dynamic set of instructions that the actuators need to use to act upon the environments.

The constituents of the control information depicted in Figure 2 are as follows: *source-address* (e.g. agent-actuator-id), *destination* is any universal resource locator of the data (e.g. secondary storage or network machine address), *node-ids* (or attribute names) and *actions* (e.g. advance discretization scripts using the interval bins) taken by the actuators. The constituents retain their usual meanings as described. The environment acted upon is the schema table (or dataset) at various destinations. The configuration of the actuators can be expanded or modified as new functionalities are provided. Thus, our discretizer actuators are concurrently distributed because they are lightweight, mobile and independent which are suitable on single user machine and distributed architecture. Section Four brings our theory to practice.

```
<Agent-actuator: = id-0    Destination:    =    URL
              Environment: = schema-name>
        <Node Id: = id-1; Bins: = Interval-Tuples>
                                    </Node>
        <Node Id: = id-n; Bins: = Interval-Tuples >
                                    </Node>
        <Action: = Advance-Disc-scripts> </Action>
     </Agent-actuator >
```

```
<Agent-actuator: = id-n    Destination:    =    URL
              Environment: = schema-name>
        <Node Id: = id-1; Bins: = Interval-Tuples >
                                    </Node>
        <Node Id: = id-n; Bins: = Interval-Tuples >
                                    </Node>
        <Action: = Advance-Disc-scripts> </Action>
     </Agent-actuator >
```

Fig. 2. Configuration of the discretizer actuators convert numerical to discrete datasets

4 Experimental Evaluations

One of the objectives of our proposed discretizer actuators is to bring theory to practice with an emphasis on applications and practical work. The algorithms compared are Hellinger's algorithm using our actuators, Weka and Hugin algorithms.

They are experimented on three public [9] massive datasets including (1) El-Nino, (2) Census-Income (KDD) and (3) Pseudo periodic synthetic time series. The El-Nino data set contains oceanographic and surface meteorological readings. The Census-Income (KDD) contains weighted census dataset. Finally, the pseudo dataset is designed for testing indexing schemes in time series.

In practice, the major contributing factors that affect discretizations and modeling performances are the number of instances, columns and number of states (distinct values) in each column of the datasets. The three datasets have varying sizes with over 178,080, 200,000, and 100,000 instances respectively. They include 11, 9 and 10 numeric columns respectively. The pseudo dataset has the worst scenario because its number of instances is equal to the number of its distinct 100,000 states.

4.1 Experiment 1: Comparing Algorithms

The objective here is to find the impact of our configurable actuators on the algorithms. The results depicted by Table 1 are a summary of the average performance of the three algorithms on the three datasets in terms of speed and memory used by the configurable actuators. For each experiment, the speed includes the time to save back into the secondary memory other than leaving the results on the volatile RAM. In all the cases, the results revealed that our configurable actuators using Hellinger's algorithm discretized successfully and was ready to proceed to modelling while the other algorithms (Weka and Hugin) suffered from memory problems by exhibiting an "out of memory" or a "towards memory failure" states. Observe in Table 1 that the Hellinger algorithm performed tremendously better than the other algorithms when we consider the results provided by the highest (or best) number of actuators in each dataset. These results suggest that using our configurable discretizer agent actuators with the Hellinger's algorithm is an economically scalable solution which supports the optimization of Bayesian intelligent modeling.

4.2 Experiment 2: Comparing Execution Speed

From the results in Table 1, we specifically compared the discretization speeds of configurable actuators using Hellinger, Weka and Hugin on the El-Nino dataset stored remotely on a secondary storage. In the same vein with Weka and Hugin which use a processor, we discretized the massive datasets with one symmetric processor (or actuator). This set of experiments was successfully repeated by distributing and concurrently increasing the number of configurable actuators while recording the discretization time as shown in Figure 3. The results show that using the Hellinger's algorithm, an increase in the number of actuators makes the discretization process faster.

In contrast, when looking at the Weka and Hugin discretizations in Figure 3, one can observe that varying the number of actuators did not improved the discretization time. By comparing the discretization time of the highest (best) number of actuators used to the usual one processor of Weka and Hugin, within the allocated limited memory, our configurable actuator using Hellinger's algorithm is faster than Weka and Hugin by 83% in Figure 3. A similar performance pattern is revealed like Figure 3 when we adapted new observations to the previous discretization parameters (intervals used for the old datasets). By cross validation [1], 15% each of the datasets

Table 1. Comparing Configurable Actuators using Hellinger, Weka and Hugin algorithms

Data Sets	Methods	Number of Actuators	Speed (secs)	Mem-usage (MB)	Status
El-Nino (178,080)	Configurable actuators using Hellinger	1 2 3 4 5 11	264 148 105 71 69 33	17.3 17.6 17.8 17.9 18.0 18.2	Ready to Model
	Weka	1	201	59.0	Out of Memory
	Hugin	1	200	66.8	Towards Memory failure
Census-Income-KDD (200,000)	Configurable actuators using Hellinger	1 2 3 4 9	177 96 78 54 29	17.4 18.1 20.2 20.8 22.6	Ready to Model
	Weka	1	173	39.2	Out of Memory
	Hugin	1	176	39.6	Towards Memory failure
Pseudo (100,000)	Configurable actuators using Hellinger	1 2 3 4 5 10	174 104 79 66 64 54	23.4 23.9 25.1 26.5 26.9 29.0	Ready to Model
	Weka	1	169	51.2	Out of Memory
	Hugin	1	165	67.2	Out of Memory

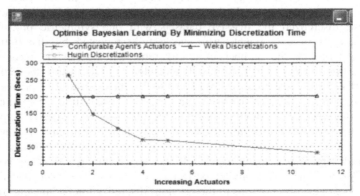

Fig. 3. Increasing number of actuators on El-Nino dataset minimizes (or speeds up) discretization time better than Weka and Hugin discretizations

were selected at random as new observations and were discretized using the previous parameters. Minimization of the discretization time results was also recorded by increasing the actuators similarly to Figure 3.

4.3   Experiment 3: Comparing Memory Usage

The results described in experiments 1 and 2 above show that users who are not opportune to be in a networking environment or who cannot afford a suitable one, can safely discretize massive data on a machine with limited memory by distributing our configurable actuators. One can observe in Figure 4 from the Weka and Hugin discretizations that varying the number of actuators does not improve on memory usage because all the records are loaded onto the memory at a time. The details of occupied megabytes of memory can be seen in Table 1 which reveals halt states.

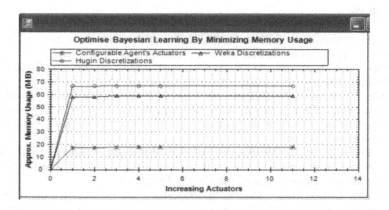

Fig. 4. Concurrent distribution of actuators on El-Nino dataset minimizes memory usage better than Weka and Hugin discretizations

From the results in Figure 4, our configurable actuators using Hellinger's algorithm successfully managed the same limited memory by concurrently exploiting secondary storage resources on remote locations (e.g. hard disk on a machine or on workstations). Though there are slight increases in memory usage as the number of actuators increases, one can observe in Figure 4 that our actuators reduce the memory usage to a minimum acceptable level. For example, this shows that the configurable actuators save 69.2% and 72.8% of the limited memory from crashing as compared to Weka and Hugin discretizations in Figure 4. This once again supports our claim that users cannot afford to trade off between *time* and *space* in real life Bayesian learning via discretization.

5   Concluding Remarks and Future Work

We have proposed in this paper the development of configurable actuators for the discretization of massive datasets as a supportive optimization solution to the computational problems arising in intelligent systems. Experimental results revealed that the use of the configurable actuator is an economically scalable solution to the problem which does not require purchasing expensive hardware. The results support the claim that using our configurable actuators with the Hellinger's algorithm leads to better memory usage and faster discretization of massive datasets compared to conventional algorithms such as Weka and Hugin discretizations.

This study shows that the configurable discretizer actuators can potentially become a more powerful scalable solution that puts an end to the computational problems raised by the learning of network models.

Acknowledgements. Our appreciation goes to the University of Cape Town and the Complex Adaptive Systems (Pty.) Ltd, for their financial supports.

References

1. Russell, S., Norvig, P.: Artificial Intelligence, A Modern Approach, 2nd edn., p. 07458. Prentice Hall Series Inc., New Jersey (2003)
2. Chickering, D., Heckerman, D., Meek, C.: Large-Sample Learning of Bayesian Networks is NP-Hard. The Journal of Machine Learning Research 5, 1287–1330 (2004)
3. William, H., Haipeng, G., Benjamin, P., Julie, S.: A Permutation Genetic Algorithm For Variable Ordering In Learning Bayesian Networks From Data. In: Proceedings of Genetic and Evolutionary Computation Conference, pp. 383–390. Morgan Kaufmann Publishers Inc., San Francisco (2002)
4. Li, J., Liu, H., Wong, L.: Mean-entropy Discretized Features are Effective for Classifying High-dimensional Biomedical data. In: Proceedings of the 3rd ACM SIGKDD Workshop on Data Mining in Bioinformatics, Washington, DC (2003)
5. Dougherty, J., Kohavi, R., Sahami, M.: Supervised and unsupervised discretization of continuous features. In: 12th International Conference on Machine Learning (1995)
6. Lee, C.: A Hellinger-based discretization method for numeric attributes in classification learning. Knowledge-Based Systems 20, 419–425 (2007)
7. Witten, I.H., Eibe, F.: Data Mining Practical Machine Learning Techniques and Tools, University of Waikato - WEKA. Morgan Kaufmann, San Francisco (1999),
   http://www.cs.waikato.ac.nz/~ml/weka/
8. Olesen, K.G., Lauritzen, S.L., Jensen, F.V.: aHugin: A system creating adaptive causal probabilistic networks. In: Dubois, D., Wellman, M.P., D'Ambrosio, B., Smets, P. (eds.) Proceedings of the Eighth Conference on Uncertainty in Artificial Intelligence, pp. 223–229. Morgan Kaufmann, San Mateo (1992),
   http://hugin.sourceforge.net/download/
9. Newman, D., Hettich, S., Blake, C., Merz, C.: UCI Repository of Machine Learning Databases (University of California, Department of Information and Computer Science, Irvine,CA) (1998),
   http://www.ics.uci.edu/~mlearn/MLRepository.html
10. Zhang, Y., Luke, E.A.: Dynamic Memory Management in the Loci Framework. In: Sunderam, V.S., van Albada, G.D., Sloot, P.M.A., Dongarra, J. (eds.) ICCS 2005. LNCS, vol. 3515, pp. 790–797. Springer, Heidelberg (2005)
11. Graham, R.M.: Principles of Systems Programming. John Wiley & sons Inc., New York (1975)

String Distances and Uniformities

David W. Pearson and Jean-Christophe Janodet

University of Saint-Etienne, 18 r. Pr. Lauras, F-42000 St-Etienne
{david.pearson,janodet}@univ-st-etienne.fr

Abstract. The Levenstein or edit distance was developed as a metric for calculating distances between character strings. We are looking at weighting the different edit operations (insertion, deletion, substitution) to obtain different types of classifications of sets of strings. As a more general and less constrained approach we introduce topological notions and in particular uniformities.

Keywords: edit distance, classification, topology, uniformities.

1 Introduction

The Levenstein (or edit) distance was introduced in the paper [1]. It has been used in various applications concerning textual data. One particular application is linked with linguistics and natural language processing where we want to find words that are in some way "close" to a given word or sets of words. This is the initial motivation for our research. We decided to use the Levenstein metric as a starting point for our work, with the understanding that it would not satisfy all of our needs. In particular, we believe that trying to place a metric structure on a natural language may be too strong a condition. This meant that we needed to look for a structure that is less rigid and rigourous than a metric. This paper is the result of our preliminary investigations.

Metric spaces are nice to work with because the idea of distance between objects is well defined, straight forward and in most cases easy to compute. However, sometimes a metric is too strong a condition to require for certain problems. At the other end of the scale we have topological spaces, where the only thing that you can say about points in the space is that they are neighbours of each other. So, a topological space can be too general for certain problems.

A uniform space (or a uniformity) lies somewhere between the two [2,3,4,5]. In a uniformity we have a notion of closeness rather than distance and we can make statements like point a is as close to point b as point c is to point d. We believe that uniformities present potentially interesting properties for text processing and, in particular, we would like to define a uniformity for strings for classification purposes.

For our initial investigations we have used the *edit distance* to define our uniformities. Also called the *Levenshtein distance*, this distance measures the minimum number of deletion, insertion and substitution operations needed to transform one string into another [1,6,7]. This distance, and its variants where each operation has a weight, has been used in many fields including Computational Biology [8,9], Language Modelling [10,11], Pattern Recognition [12,13] and Machine Learning [14,15].

M. Kolehmainen et al. (Eds.): ICANNGA 2009, LNCS 5495, pp. 333–339, 2009.
© Springer-Verlag Berlin Heidelberg 2009

We think of a classification problem of strings as defining a topology (and uniformity) for the strings, *i.e.*, strings in the same class are neighbours. We must add at this point that we refer to classification in a somewhat unrigourous fashion in that the resulting classes may overlap due to the uniform structure. A true classification would result in disjoint classes. Therefore we consider that the weights of the edit distance parameterize the uniformity. When we change the weights the uniformity may or may not change. We are interested in finding the critical parameter values where the uniformity changes.

This paper is composed of three main sections. In the following section we present the relevant theoretical background on uniformities. Then, in the next section we show how we can define uniformities for sets of strings. An example is developed in the last section before finally concluding.

2   Covering Uniformities

There are at least two ways of defining a uniformity: entourage uniformities and covering uniformities. It can be shown that they provide equivalent structures and that the choice of entourage or covering is governed by the application. The entourage approach is very popular nowadays [2,3], but we have found the covering approach to be better adapted to our needs [4,5].

Let X be any fixed space. A *covering* for X is a collection $\mathcal{C}$ of sets $C_i \subseteq X$ such that $\bigcup_i C_i = X$, for $C_i \in \mathcal{C}$. Given two coverings $\mathcal{U}$ and $\mathcal{V}$, $\mathcal{U}$ is said to *refine* $\mathcal{V}$, denoted $\mathcal{U} < \mathcal{V}$, if for all $U_i \in \mathcal{U}$, there exists $V_j \in \mathcal{V}$ such that $U_i \subseteq V_j$.

For a covering $\mathcal{C}$ and a subset $A \subseteq X$, the star of A is defined as follows:

$$*(A, \mathcal{C}) = \bigcup \{C_i \in \mathcal{C} : C_i \cap A \neq \emptyset\}.$$

Given two coverings $\mathcal{U}$ and $\mathcal{V}$, we say that $\mathcal{U}$ *star refines* $\mathcal{V}$, denoted $\mathcal{U} <^* \mathcal{V}$, if for all $U \in \mathcal{U}$, there exists $V \in \mathcal{V}$ such that $*(U, \mathcal{U}) \subseteq V$. In this case, the sets in $\mathcal{V}$ can be thought of as twice as big as those of $\mathcal{U}$ [5].

We now introduce the following definition for a *covering uniformity*. A family μ of coverings is called a *uniformity* if it satisfies the following conditions:

1. if $\mathcal{U}, \mathcal{V} \in \mu$, then there exists $\mathcal{W} \in \mu$ such that $\mathcal{W} < \mathcal{U}$ and $\mathcal{W} < \mathcal{V}$,
2. if $\mathcal{U} \in \mu$ and $\mathcal{U} < \mathcal{V}$ then $\mathcal{V} \in \mu$ and
3. every element of μ has a star refinement in μ.

Some texts refer to this definition as a preuniformity or a non-separating uniformity, we shall simply use the term uniformity. A separation condition can be added and some authors refer to that as a uniformity, but other authors refer to it as a Hausdorff uniformity. The separation condition is not necessary in our case.

The notion of a *normal sequence of coverings* in a uniformity is simply a sequence $\mathcal{U}_n$ such that $\cdots \mathcal{U}_{n+1} <^* \mathcal{U}_n <^* \mathcal{U}_{n-1} \cdots$.

If $\mathcal{U} \in \mu$ and $y \in X$ then a point $x \in X$ is said to be $\mathcal{U}$-*close* to y, denoted $|y - x| < \mathcal{U}$, if there exists $U \in \mathcal{U}$ such that $\{x, y\} \subseteq U$.

Finally, let d be a distance over X. For any $x \in X$, we define the ϵ-sphere around x as $S(x, \epsilon) = \{y \in X : d(x, y) \leq \epsilon\}$. Clearly, if $S(x, \epsilon) \subseteq \mathcal{U} \in \mu$, then every $y \in S(x, \epsilon)$ is $\mathcal{U}$-close to x.

3   Strings and Uniformities

An *alphabet* Σ is a finite nonempty set of symbols called *letters*. For the sake of clarity, we shall use $\Sigma = \{a, b\}$ as a fixed alphabet throughout the rest of this paper. A *string* $w = x_1 \ldots x_n$ is any finite sequence of letters. We write Σ^* for the set of all strings over Σ. Let $|w|$ denote the length of w and λ the empty string.

Following [1], we consider three sorts of edit operations:

- a pair $(x : y) \in \Sigma \times \Sigma$ is called *a substitution of letter x by letter y*,
- a pair $(x : \lambda)$ with $x \in \Sigma$ is called *a deletion of letter x*, and
- a pair $(\lambda : y)$ with $y \in \Sigma$ is called *an insertion of letter y*.

Moreover, we assume that a matrix C assignes a weight to every operation. *E.g.*,

C	λ	a	b
λ	0	1	2
a	1	0	1.5
b	2	1.5	0

The *edit distance* between two strings w_1 and w_2, denoted $d(w_1, w_2)$, is the minimum weight of every sequence of substitutions, deletions and insertions that allows one to transform w_1 into w_2. More formally, d is recursively defined as follows:

$$d(w_1, w_2) = \min \begin{cases} 0 & \text{if } w_1 = w_2 = \lambda \\ C(x : \lambda) + d(w_1', w_2) & \text{if } w_1 = xw_1' \\ C(\lambda : y) + d(w_1, w_2') & \text{if } w_2 = yw_2' \\ C(x : y) + d(w_1', w_2') & \text{if } w_1 = xw_1', w_2 = yw_2' \end{cases}$$

It is well-known [7] that if C defines a metric over $(\Sigma \cup \{\lambda\})$, then $d(w_1, w_2)$ can be efficiently computed in time $\mathcal{O}(|w_1| \cdot |w_2|)$ by means of dynamic programming [6]. Assuming that C defines a metric means that the matrix C has a null diagonal, is positive, symmetric and for all $x, y, z \in \Sigma \cup \{\lambda\}$,

$$C(x : y) \leq C(x : z) + C(z : y). \tag{1}$$

In such a case, the edit distance is determined by only three weights: $C(a : b)$, $C(a : \lambda)$ and $C(b : \lambda)$. We group these values together into a vector $p = \begin{bmatrix} C(a : b) \\ C(a : \lambda) \\ C(b : \lambda) \end{bmatrix}$. Thus, using any fixed p that satifies Eq.(1), we can compute the edit distance between any two strings.

Our main interest is in string classification and so we assume that we have a set of strings, W of cardinality m, and some idea of which strings are together in classes. Let us consider for any $w \in W$, the ϵ-sphere around w: $S(w, \epsilon) = \{w' \in W : d(w, w') \leq \epsilon\}$. Using the edit distance and the ϵ-spheres, we can now calculate uniformities for sets of strings.

To begin with, we compute the distances between all the strings in W using the edit distance with some fixed value of the parameter vector p. If we list all the strings in W in some order horizontally and vertically then the result is simply a symmetric $m \times m$ matrix D with zeros along the diagonal. We take all the elements above the diagonal of this matrix and list them in lexicographical order. Thus if

$$
D = \begin{bmatrix}
0 & d_{12} & d_{13} & \cdots & d_{1m} \\
d_{21} & 0 & d_{23} & \cdots & d_{2m} \\
\vdots & \vdots & \cdots & \ddots & \vdots \\
d_{m1} & d_{m2} & \cdots & \cdots & 0
\end{bmatrix}
$$

where $d_{ij} = d(w_i, w_j)$, then we define the vector

$$
x = \begin{bmatrix}
d_{12} \\
d_{13} \\
\vdots \\
d_{ij} \\
\vdots \\
d_{m-1m}
\end{bmatrix}
$$

We want to adjust the parameter vector p to give the required classification. As the vector x above is dependent on p, we indicate this by $x(p)$ and thus consider x to be a mapping $x : \mathbb{R}^3 \to \mathbb{R}^n$ where $n = \frac{m^2 - m}{2}$ is the number of elements of D above the diagonal. Due to the condition on the values of the three parameters, p is restricted to certain areas of $\mathbb{R}^3$. The standard edit distance uses $p = \begin{bmatrix} 1 & 1 & 1 \end{bmatrix}^T$ and so we will define the admissible values of p based on this point. Define the following vectors $p_0 = \begin{bmatrix} 1 \\ 1 \\ 1 \end{bmatrix}$,

$p_1 = \begin{bmatrix} 0.5 \\ 1 \\ 1 \end{bmatrix}$, $p_2 = \begin{bmatrix} 1 \\ 0.5 \\ 1 \end{bmatrix}$ and $p_3 = \begin{bmatrix} 1 \\ 1 \\ 0.5 \end{bmatrix}$. Then, for our purposes, we can say that the admissible values of p can be defined by the simplex $P = t_0 p_0 + t_1 p_1 + t_2 p_2 + t_3 p_3$ where $t_i \geq 0$ and $\sum_{i=0}^{3} t_i = 1$.

Once a value of p has been chosen, the uniformity is defined by varying x and ϵ in $S(x, \epsilon)$. Clearly different uniformities can be defined for the same value of p, depending on the chosen values of x and ϵ. Another point to mention is that certain choices of x and ϵ will not lead to correct uniformities simply because of the star refinement property that is required. These points are best illustrated by an example, which we present in the following section.

4 Example

Let the set of strings be the following $W = \{aaab, abab, bba, baba, bbaab\}$. Applying the classical edit distance we have the following table:

W	$aaab$	$abab$	bba	$baba$	$bbaab$
$aaab$	0	1	3	3	2
$abab$	1	0	2	2	2
bba	3	2	0	1	2
$baba$	3	2	1	0	3
$bbaab$	2	2	2	3	0

We can find 3 coverings from this. First of all $S(aaab, 1)$, $S(bba, 1)$ and $S(bbaab, 1)$ supply us with

$$\mathcal{U}_2 = \{aaab, abab\}, \{bba, baba\}, \{bbaab\},$$

then $S(aaab, 2)$ and $S(bba, 2)$ give us

$$\mathcal{U}_1 = \{aaab, abab, bbaab\}, \{abab, bba, baba, bbaab\},$$

then finally $S(aaab, 3)$ gives us

$$\mathcal{U}_0 = \{aaab, abab, bba, baba, bbaab\}.$$

It can be verified that $\mathcal{U}_2 <^* \mathcal{U}_1 <^* \mathcal{U}_0$. So, with this choice of p all the strings are $\mathcal{U}_0$-close, $\{aaab, abab, bbaab\}$ and $\{abab, bba, baba, bbaab\}$ are $\mathcal{U}_1$-close and finally $\{aaab, abab\}$, $\{bba, baba\}$ and $\{bbaab\}$ are $\mathcal{U}_2$-close.

To see how the uniformity changes when p changes we carry out the same exercise but with $p = p_1$ as described above. With this value for p the distances are the following

W	$aaab$	$abab$	bba	$baba$	$bbaab$
$aaab$	0	0.5	2	1.5	1.5
$abab$	0.5	0	1.5	2	1.5
bba	2	1.5	0	1	2
$baba$	1.5	2	1	0	2
$bbaab$	1.5	1.5	2	2	0

Using these distances we can now define the following covers:

$$\mathcal{V}_3 = S(aaab, 0.5), S(bba, 0.5), S(baba, 0.5), S(bbaab, 0.5)$$
$$= \{aaab, abab\}, \{bba\}, \{baba\}, \{bbaab\}$$
$$\mathcal{V}_2 = S(aaab, 1), S(bba, 1), S(bbaab, 1)$$
$$= \{aaab, abab\}, \{bba, baba\}, \{bbaab\}$$
$$\mathcal{V}_1 = S(bbaab, 1.5), S(bba, 1.5)$$
$$= \{aaab, abab, bbaab\}, \{abab, bba, baba\}$$
$$\mathcal{V}_0 = S(aaab, 2)$$
$$= \{aaab, abab, bba, baba, bbaab\}$$

and it can be verified that $\mathcal{V}_3 <^* \mathcal{V}_2 <^* \mathcal{V}_1 <^* \mathcal{V}_0$. Here we see that $\mathcal{U}_0 = \mathcal{V}_0$, but the other sets in the different levels are not the same and so the $\mathcal{U}$-uniformity and the $\mathcal{V}$-uniformity are not the same.

Changing p once again and using $p = p_2$ as defined above, leaving the details out we have the following uniformity

$$\mathcal{W}_3 = \{aaab\}, \{abab\}, \{bba, baba\}, \{bbaab\}$$
$$\mathcal{W}_2 = \{aaab, abab\}, \{bba, baba, bbaab\}$$
$$\mathcal{W}_1 = \{aaab, abab, bbaab\}, \{abab, bba, baba, bbaab\}$$
$$\mathcal{W}_0 = \{aaab, abab, bba, baba, bbaab\}$$

with $\mathcal{W}_3 <^* \mathcal{W}_2 <^* \mathcal{W}_1 <^* \mathcal{W}_0$.

These three uniformities are clearly different. To see how the changes occur we traced out the vector x at various points between p_1 and p_2 by setting $p = (1-t)p_1 + tp_2$ with t ranging from 0 to 1 in increments of 0.1. The individual components of the vector x can be seen in figure 1. The uniformity actually changes between the values $t = 0.5$ and $t = 0.6$, i.e., for values of p between $p = \begin{bmatrix} 0.75 \\ 0.75 \\ 1 \end{bmatrix}$ and $p = \begin{bmatrix} 0.8 \\ 0.7 \\ 1 \end{bmatrix}$.

We notice that with the $\mathcal{V}$-uniformity the two strings $aaab$ and $abab$ are always together but bba and $baba$ are separated in the $\mathcal{V}_3$-uniformity. Whilst for the $\mathcal{W}$-uniformity bba and $baba$ remain together but $aaab$ and $abab$ get separated in $\mathcal{W}_3$.

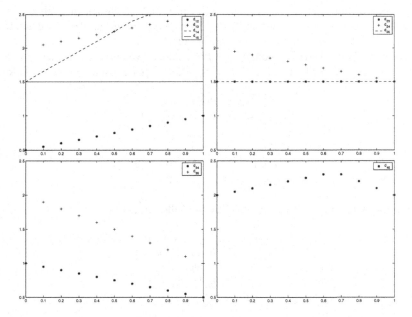

Fig. 1. The distances for the strings $aaab$ (top left), $abab$ (top right), bba (bottom left) and $baba$ (bottom right)

5   Conclusion

We have introduced an approach to string classification based on uniformities. We believe that this approach has potential because it falls between one which is too general based on a topology and one which is too rigorous based on a metric.

We are fully aware of the fact that we need a metric to actually calculate the ϵ-spheres and thus the uniformity, but we wanted to test our ideas in the first instance and so we used the Levenstein distance to advance more quickly. We believe that the results obtained so far are promising and so we are continuing along these lines. Our work is now concentrating on how to define a uniformity and carry out calculations without the need of a metric.

References

1. Levenshtein, V.I.: Binary codes capable of correcting deletions, insertions, and reversals. Doklady Akademii Nauk SSSR 163(4), 845–848 (1965)
2. Kelley, J.L.: General Topology. D. Van Nostrand (1955)
3. James, I.M.: Topologies and Uniformities. Springer, Heidelberg (1999)
4. Howes, N.R.: Modern Analysis and Topology. Springer, Heidelberg (1995)
5. Willard, S.: General Topology. Addison-Wesley, Reading (1970)
6. Wagner, R., Fisher, M.: The string-to-string correction problem. Journal of the ACM 21, 168–178 (1974)
7. Crochemore, M., Hancart, C., Lecroq, T.: Algorithms on Strings. Cambridge University Press, Cambridge (2007)
8. Gusfield, D.: Algorithms on Strings, Trees, and Sequences - Computer Science and Computational Biology. Cambridge University Press, Cambridge (1997)
9. Durbin, R., Eddy, S.R., Krogh, A., Mitchison, G.: Biological Sequence Analysis. Cambridge University Press, Cambridge (1998)
10. Amengual, J.C., Sanchis, A., Vidal, E., Benedí, J.M.: Language simplification through error-correcting and grammatical inference techniques. Machine Learning Journal 44(1-2), 143–159 (2001)
11. Dupont, P.E., Amengual, J.-C.: Smoothing probabilistic automata: An error-correcting approach. In: Oliveira, A.L. (ed.) ICGI 2000. LNCS (LNAI), vol. 1891, pp. 51–64. Springer, Heidelberg (2000)
12. Navarro, G.: A guided tour to approximate string matching. ACM Computing Surveys 33(1), 31–88 (2001)
13. Chávez, E., Navarro, G., Baeza-Yates, R.A., Marroquín, J.L.: Searching in metric spaces. ACM Computing Surveys 33(3), 273–321 (2001)
14. Becerra-Bonache, L., de la Higuera, C., Janodet, J.C., Tantini, F.: Learning balls of strings from edit corrections. Journal of Machine Learning Research 9, 1823–1852 (2008)
15. Delhay, A., Miclet, L.: Analogical equations in sequences: Definition and resolution. In: Paliouras, G., Sakakibara, Y. (eds.) ICGI 2004. LNCS (LNAI), vol. 3264, pp. 127–138. Springer, Heidelberg (2004)

Emergent Future Situation Awareness: A Temporal Probabilistic Reasoning in the Absence of Domain Experts

Isaac Olusegun Osunmakinde and Antoine Bagula

Department of Computer Science, Faculty of Sciences, University of Cape Town,
18 University Avenue, Rhodes Gift, 7707 Rondebosch, Cape Town/South Africa
{segun,bagula}@cs.uct.ac.za

Abstract. Dynamic Bayesian networks (DBNs) are temporal probabilistic models for reasoning over time which are rapidly gaining popularity in modern Artificial Intelligence (AI) for planning. A number of Hidden Markov Model (HMM) representations of dynamic Bayesian networks with different characteristics have been developed. However, the varieties of DBNs have obviously opened up challenging problems of how to choose the most suitable model for specific real life applications especially by non-expert practitioners. Problem of convergence over wider time steps is also challenging. Finding solutions to these challenges is difficult. In this paper, we propose a new probabilistic modeling called Emergent Future Situation Awareness (EFSA) which predicts trends over future time steps to mitigate the worries of choosing a DBN model type and avoid convergence problems when predicting over wider time steps. Its prediction strategy is based on the automatic emergence of temporal models over two dimensional (2D) time steps from historical Multivariate Time Series (MTS). Using real life publicly available MTS data on a number of comparative evaluations, our experimental results show that EFSA outperforms popular HMM and logistic regression models. This excellent performance suggests its wider application in research and industries.

Keywords: Dynamic Bayesian Networks, Situation Awareness, Prediction, Multivariate Time Series.

1 Introduction

Industrial practitioners and researchers observe multivariate time series (MTS) from their daily business activities or dynamical systems (e.g. medical systems, retail, sensor networks, etc). Complex hidden relationships (or patterns) are often embedded among the variables that describe such activities within and across the time steps. Some classical methods such as neural networks and statistical logistic regression models have been applied to predict such hidden patterns but they fall short of proving their prediction capabilities [1]. Using more sophisticated approaches, these hidden patterns can be revealed from the historical MTS to predict risks, or guide actions to be taken at particular future times. Any retail business may for instance intend to know which products require declaration of discounts among selected

M. Kolehmainen et al. (Eds.): ICANNGA 2009, LNCS 5495, pp. 340–349, 2009.
© Springer-Verlag Berlin Heidelberg 2009

outlets in specific months for next year. This is an example of prediction over time within a multitude of complex situations and dynamic Bayesian networks are well suited for reasoning over time in complex environments [2] [3].

The Hidden Markov Model (HMM) is a common and most simple form of DBNs which has gained its wide applicability in speech recognition [3] [4]. Recently, researchers have developed many HMM representations of dynamic Bayesian networks with different characteristics. Murphy [2] proposed some variants of HMM as explicit representations of DBN such as: hierarchical HMM, coupled HMM and factorial HMM. Shenoy [5] presented another DBN model for Brain-Computer Interfaces. As experts, they explicitly modeled the hidden network structure and dependencies between different brain states.

However, the varieties of DBNs have obviously opened up challenging problems of how to choose the most suitable model for various real life applications. Some prediction models also suffer from convergence or exponential problems over wider time steps [3]. That is, the prediction steps get stuck towards zero or tend towards infinity. To complicate the situation further, the challenges have made technologies such as the DBNs too complex for non-experts including practitioners [6] (seasoned software programmers, managers, etc).

In an attempt to address some of these challenges, Deviren [4] presented Structural learning of DBN and also applied it to speech recognition. Their DBN was learnt under a number of assumptions from experts. For instance, they observed stationary assumption which made their DBN leads to a repeated network structure at each time step. In reality, situations in some time steps may change. This is evident that most of these existing DBNs approximate their models. That is, they do not truly (or completely) emerge the network structures and probability distributions but the basis of DBN requires modeling both [2] [3]. Murphy [2] confirms that the HMM, which is the basis of most of these representations of the existing DBNs, are limited in their expressive power. Finding solutions to these challenges is difficult.

In this paper, we propose a new probabilistic modeling called Emergent future Situation Awareness (EFSA) technology which predicts trends over finite future time steps. The EFSA eliminates the worries of choosing a good DBN model and avoids convergence problems. Its automatic and complete emergence of temporal models (network structure and probability distributions) over time from historical MTS is the strategy of the prediction. Our major contributions are as follows:

- The derivation of a temporal probabilistic theory for the new EFSA technology which predicts trends over future time steps in the absence of domain experts.
- The development of the EFSA algorithm which facilitates the mitigation of the worries of practitioners and researchers for choosing a DBN model from the multitude of varieties for specific applications.
- Using a 2D strategy to avoid convergence problems when predicting longer time steps, our EFSA model supports wider applicability for all users: experts and non experts.

The rest of the paper is organized as follows. In section 2, we present the theoretical backgrounds of the dynamic Bayesian networks and our previous ESA (emergent situation awareness) technology. The details of the proposed EFSA technology are presented in section 3. In section 4, we evaluate the performance of the EFSA's consistency and accuracy, and benchmark it with the HMM and the logistic regression models. This paper concludes in section 5.

2 Theoretical Background

2.1 Dynamic Bayesian Networks (DBNs)

Dynamic Bayesian networks are temporal probabilistic models which are often referred to as an extension of the Bayesian network (BN) models in artificial intelligence [2] [3]. A Bayesian belief network is formally defined as a directed acyclic graph (DAG) represented as $G = \{X(G), A(G)\}$, where $X(G) = \{X_1,...,X_n\}$, vertices (variables) of the graph G and $A(G) \subseteq X(G) \times X(G)$, set of arcs of G. The network requires discrete random values such that if there exists random variables X_1, ..., X_n with each having a set of some values x_1, ..., x_n then, their joint probability density distribution is defined in equation 1;

$$pr(X_1,...,X_n) = \prod_{i=0}^{n} pr(X_i \mid \pi(X_i))$$
(1)

where $\pi(X_i)$ represents a set of probabilistic parent(s) of child X_i [3]. A parent variable otherwise referred to as *cause* has a dependency with a child variable known as *effect*. Every variable X with a combination of parent(s) values on the graph G captures probabilistic knowledge (distribution) as a conditional probability table (CPT). A variable without a parent encodes a marginal probability.

However, the inability of the BNs to capture time as temporal dependencies facilitated the developments of various ways of modelling the dynamic Bayesian networks presented at the introduction. The variables and the CPTs of the BNs are similar to the states and the probabilities used in the temporal dependencies of the DBNs. According to [3], a DBN is suitable for modelling environment that emerges (changes) over time and has the capability to predict future behaviour of the environment. Any DBN observes the first-order of Markov model which states that, future event V_{t+1} is independent of the past given the present V_t [3]. Since DBN handles complex situations of multiple dependent events of Markov model over time, researchers [2] [3] present the following three parameters required to construct a DBN model: prior matrix, $Pr(V_0)$; transition matrix, $Pr(V_t \mid V_{t-1})$; and sensor matrix, $Pr(E_t \mid V_t)$. The prior matrix defines the initial probability distribution of states V_0 at the start of emergence of DBNs. The transition matrix describes time dependencies for the transitions of DBN states V_t. Also, the sensor matrix captures the probabilistic distributions from the relationships of observation variables E_t at any time step.

In conjunction with the DBN matrices, equation 2 shows the combined joint probability distribution for any temporal model up to a finite time t.

$$Pr(V_0, V_1,...,V_t, E_1,...,E_t) = Pr(V_0) \prod_{i=1}^{t} Pr(V_i \mid V_{i-1}) Pr(E_i \mid V_i)$$
(2)

The emergence of our DBN technology is based on the theoretical principles underpinning situation awareness [9] in order to make anticipatory planning.

2.2 The ESA Technology

The ESA [11] is an innovative technology, which completely emerges temporal models and reveals the hidden behavior of what is currently happening over time in any domain of interest. Formally, let $\{V^{t}, E^{t}\}$ represent the set of state and observed DBN variables in ESA at time t. The DBNs are emerged over all the non-negative current time steps $t \in T$, such that $T = \{t_1, t_2 \dots t_n\}$ and the interlinked probabilistic relationships at each time step t is represented in equation 3. Equation 3 represents the interconnections of changing networks (or frames) and probability distributions over the time steps.

$$pr(V_1^{t_1}, V_2^{t_1} \dots, V_m^{t_1}) \underset{\equiv}{\overset{\Delta}{}} pr(V_1^{t_2}, V_2^{t_2} \dots, V_m^{t_2}) \underset{\equiv}{\overset{\Delta}{}} pr(V_1^{t_n}, V_2^{t_n} \dots, V_m^{t_n}) \tag{3}$$

where $\underset{\equiv}{\overset{\Delta}{}}$ implies equivalence is *not* true generally. The attractive performance of the ESA encourages its successful applications in many areas, most notably in project management [11]. On the other hand, the ESA falls short of predicting into the future. We therefore conjecture that a variant of the ESA called the EFSA is required to efficiently predict into the future based on the historical time steps.

3 The Proposed EFSA Technology

3.1 Theoretical Derivations of the EFSA

Researchers [3] assert that prediction too far (wider time lag) into the future converges to a fixed point (i.e. remains constant for all time). In order to minimize the convergence problem, the EFSA predicts future trends using the strategy of a two dimensional (2D) time steps. The first dimensional space of time steps $\{t_1, t_2 \dots t_n\}$ monitors the behavioural current patterns as used in the ESA. The second dimensional time steps $\{T_1, \dots, T_m\}$ observes the historical patterns for each of the time steps $\{t_1, t_2 \dots t_n\}$. This is an extension of any period T in the ESA. Therefore in practice, at the end of the tn of Tm, the EFSA updates further future trends to ensure accuracy. Let the DBN variables V^{t} span the space of 2D time steps represented in the system of equations 4.

$$T_1 = \{ v_i^{t_1}, v_j^{t_2} \dots, v_\alpha^{t_n} \}; \ T_2 = \{ v_i^{t_1}, v_j^{t_2} \dots, v_\alpha^{t_n} \}; \dots ; T_m = \{ v_i^{t_1}, v_j^{t_2} \dots, v_\alpha^{t_n} \}$$
$$\text{for each } i, j, \alpha = 1, 2, \dots \ell \tag{4}$$

ℓ is the length of the DBN variables and m is the length of the history. All the changing parameterizations (the DAGs and the probability distributions) of the DBN in the EFSA are now carried out across the historical time steps $T_1 \dots T_m$. That is, the emergence (or learning of the temporal models) takes place across the links:

$$\{ v_i^{t_1}, v_i^{t_1} \dots, v_i^{t_1} \}, \dots, \{ v_\alpha^{t_n}, v_\alpha^{t_n} \dots, v_\alpha^{t_n} \}.$$

Once the temporal probabilistic model emerges, prediction with reasoning now acts on the model. From Markovian principle [3] which states that next states of a system depend on the finite history of the previous states, we can now have multiple n predictions from the space of 2D time steps in equation 4 into the future time steps λ as follows in equation 5.

$$T_{m+}\lambda = \{ v_i^{t_{1}+\lambda}, v_j^{t_{2}+\lambda} ..., v_\alpha^{t_{n}+\lambda} \}, \tag{5}$$

$$\Rightarrow \Pr(v_i^{t_{1}+\lambda}) = \Pr(v_i^{t_{1}+\lambda} \mid E_i^{1:t_{1}}); \Pr(v_j^{t_{2}+\lambda}) = \Pr(v_j^{t_{2}+\lambda} \mid E_j^{1:t_{2}}); \cdots ;$$

$$\Pr(v_\alpha^{t_{n}+\lambda}) = \Pr(v_\alpha^{t_{n}+\lambda} \mid E_\alpha^{1:t_{n}}), \quad \text{for some } \lambda_{>0}$$

In equation 5, E_i, E_j...E_α are the set of evidences or observations of V_i, V_j,...V_α respectively made so far within the space of time steps. Equation 5 is therefore the set of predictions that can be computed by the Bayesian inference algorithms such as Variable elimination, etc [3]. The variable elimination implemented in [10] was integrated as the inference engine in the EFSA due to its efficiency. Therefore, the EFSA performs a multiple future predictions from the space of time.

3.2 The Description of the EFSA

The emergence of DBN or temporal probabilistic model of the EFSA is often a task of BN learning algorithms provided it can learn across the time steps. Therefore, the outlines of learning DBNs automatically from EFSA algorithm are now refined from [11] as follows:

INPUT (D_s : Multivariate Time Series - MTS)

---

1. While D_s = MTS,
    [i] Set t, the frame count, to 1.
    [ii] Set T, the historical time step, to 1, 2, ..., m
    [iii] Let $d_t \in D_s$, $\forall\ t = 1, 2, . . ., n$.
    [iv] For each $t <= n$,
    [v]   For each T $<= m$,
        • Select frame d_t into $\{ d_t \}$
    [vi]   Increment T by 1.
        • Invoke Learning_Algorithms ($\{d_t\}$).
        • Store the emerged frame in n by m matrix B.
    [vii] Increment t by 1.
2. Return the DBN in B
3. Predict the next n time steps using inference engine, then exit.

Fig. 1. Emergent Future Situation Awareness Algorithm

In Figure 1, all parameters retain their usual meanings and d_t is a frame dataset at time t. It is selected into set $\{d_t\}$ over T for learning frames of the DBN. Any Bayesian learning algorithms such as [7] [8] can be used as a subroutine. The algorithms carry out the intra-slice and inter-slice learning over time. Each variable in step T must have parents in step T-1. We integrated the genetic algorithm [8] to learn the DBNs due to its efficiency.

4 Performance Evaluation

One of the objectives of our proposed EFSA technology is to bring theory to practice (implementation) with an emphasis on applications and practical work. The HMM and logistic regression models have been used in our experiments as a baseline of comparison with our EFSA model. The logistic regression model is a function of dependent variable over the independent variables [1].

4.1 Experiment 1: Comparing EFSA Consistency with Other Popular Models

Our intention here was to determine whether the EFSA can predict multiple n-time steps consistently. As a proof of concept, we carried out the evaluations of the three models on three MTS datasets - DIABETES and SENSOR datasets from UCI repository [12], and a real life RAINFALL dataset obtained from a Southern African country (Botswana). The treatment records from the behavior of a diabetes patient were captured electronically as MTS. It contains several treatment measurement codes such as 33 (regular insulin dose), 48 (unspecified blood glucose), etc. It is expected here to predict treatment measurements required in future for the patient based on the historical behavioral patterns. For instance, equation 6 below is a situation which predicts how much of the minimum (about 7 units) measurement of regular insulin dose will be required for the next 12 months in the year 1991.

$$Pr(Measurement^{t+\lambda} <= 7 \ units \mid Code^{\,t} = 33) \tag{6}$$

for all t $\subset$ T, where in diabetes MTS, t = $\{Jan...Dec\}$ and T = $\{1988, 1989, 1990\}$.

A common empirical technique to evaluate the performance of Bayesian network technologies is to use a basic cross validation [3]. We adopted the cross validation approach by setting 1991 time step as actual test data and learnt (or trained) the DBN model across 1988 to 1990 time steps. The EFSA reasons with the temporal model and acts by predicting over time as described in equation 6. This experiment was repeated using basic HMM constructed dynamically on the fly using GeNle Bayesian software [13]. Similarly, equation 6 was also repeated using the logistic regression model implemented in R statistical software [14]. The actual and the predicted results were recorded in each experiment and are shown in Table 1.

The sensor dataset captures the traffic of people flowing in and out of main door of a CalIt2 building at UCI. Our objective here is to be able to predict counts of people for every half an hour over future weeks based on the historical behavioural patterns of traffic. For instance in equation 7, we want to predict the possibility of counting average number (between 8 and 17) of people that flows out of the building for the next 5 weeks.

$$Pr(Count^{t+\lambda} = `8<=17` \ people \mid Flow^{t} = `outflow`) \tag{7}$$

for all $t \subset T$, where in Sensor MTS, $t = \{Week\text{-}1...Week\text{-}5\}$ and $T = \{July, Aug, Sept\}$. We also adopted the cross validation approach by setting October time step as actual test data and learnt (or trained) the DBN model across 15 weeks of July to Sept time steps. The EFSA reasons with the temporal model and acts by predicting over time as described in equation 7. This experiment was also repeated using other models. The actual and the predicted results were recorded in each experiment and are shown in Table 1.

The real life rainfall dataset was obtained from a Southern African country (Botswana) to access onsets of rainfall for Farmers to understand their varying planting dates. Our objective here is to be able to predict normal onset at any station over future months in every coming year. For instance, equation 8 predicts the normal

Table 1. Performance Comparison of Future Predictions on Three Situations Among EFSA, HMM and Logistic Regression Model

Data Sets	Time Steps	Actual (%)	EFSA (%)	HMM (%)	Logistic(%) Regression
Diabetes	Jan	40.15	73.26	70.43	87.30
	Feb	50.79	58.7	72.45	87.31
	Mar	74.64	90.99	79.45	87.41
	Apr	65.42	80.44	89.69	87.42
	May	72.51	69.08	89.07	87.46
	Jun	60.14	59.69	90.16	87.48
	July	69.43	76.54	93.28	87.51
	Aug	61.09	75.69	95.31	87.53
	Sept	55.17	85.48	95.36	87.56
Sensor	Wk-1	19.55	12.46	24.13	25.29
	Wk-2	20.88	13.39	23.07	26.18
	Wk-3	24.44	18.69	33.17	37.07
	Wk-4	27.11	18.81	33.41	37.95
	Wk-5	3.11	2.76	21.01	38.01
Rainfall	Jan	70.22	53.57	58.24	45.88
	Feb	72.75	56.67	51.12	45.62
	Mar	72.39	55.48	44.09	45.33
	Apr	61.95	55.29	41.20	45.04
	May	62.34	56.67	40.02	44.72
	Jun	78.21	62.67	39.59	44.38
	Jul	85.73	59.81	38.90	44.03
	Aug	78.85	58.38	38.89	43.65
	Sept	83.45	58.00	36.87	43.25
	Oct	80.30	60.05	25.65	42.82
	Nov	89.21	62.31	25.14	42.38
	Dec	91.26	65.05	24.69	41.90

onset of rainfall over future months for a given station number 2. This may include more complex conditions, like considering how sea anomalies affect the onset and wind, as shown in equation 9 which other methods such as regression model struggle to handle [2]. For the purpose of comparison, we keep it simpler as equation 8.

$$Pr(Onset^{t+\lambda} = \text{'normal'} \mid Station^t = 2) \qquad (8)$$
$$Pr(Onset^{t+\lambda} = \text{'normal'} \mid Station^t = 2, Sea\_Anom > 0.5, wind < 7.7units) \qquad (9)$$

for all $t \subset T$, where in Rainfall MTS, $t = \{Jan...Dec\}$ and $T = \{1971,...2000\}$.

We also adopted the cross validation approach by setting year 2001 time step as actual test data and learnt (or trained) the DBN model across 1971 to 2000 time steps. The EFSA also reasons with the temporal model and acts by predicting over time as described in equation 8. This experiment was also repeated using HMM and logistic regression models. The actual and the predicted results were recorded in each experiment and are shown in Table 1.

The Table reveals the consistencies or how each model captures the direction of predictive patterns. That is, increase or decrease in predictions from one time step to the next when compared with the actual results. For instance, one can see sensor results in Table 1 as the EFSA prediction increases from 12.46 in week-1 to 13.39 in week-2 and this corresponds to a rise in the actual results. In view of this, one can see generally in Table 1 that the EFSA has the best prediction directions (consistency) of 50%, 100% and 70% for the Diabetes, Sensor and the Rainfall datasets respectively. This results from the fact that the EFSA truly evolves its network and probability distribution with the aid of its 2D strategy of the predictions.

4.2 Experiment 2: Comparing EFSA Accuracy with Other Popular Models

We conducted another set of experiments to measure how accurate are the predictions of the models compared to the actual results. The accuracy is simply computed as the difference between a 100% and the percentage error deviation, where the error is the absolute difference between actual value from the test data and the predicted value from the models, divided by the actual value [3]. The prediction accuracies on all datasets are computed from Table 1 and for instance only the accuracy results of diabetes are specifically recorded in Table 2 accordingly. Figure 2 therefore compares the performance accuracies of the EFSA with other models on Diabetes results in Table 2. The objective here is to improve the prediction accuracy. Observe the predictions of the regression model on diabetes and rainfall in Table 1 as it tends towards a convergence problem.

Over the time steps, the average accuracies of the EFSA, HMM and the regression models on sensor predictions are 72.49%, 62.38% and 50.74% respectively. Also, the average accuracies of the three techniques on rainfall predictions are 76.76%, 51.79% and 57.87% respectively. This shows that the overall accuracy of the EFSA as improved than others is *74.3%*. Thus, the EFSA is more consistent and performs better with future predictions within the multivariate time series.

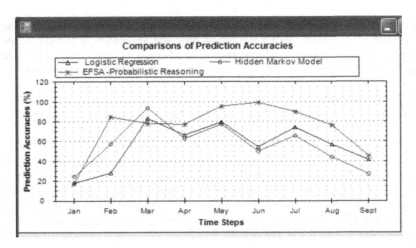

Fig. 2. Temporal probabilistic reasoning of the EFSA improves performance of prediction accuracies on Diabetes better than the HMM and the logistic regression model

Table 2. Evaluation of Accuracy for Future Predictions on Diabetes Situations Among EFSA, HMM and Logistic Regression Models

Dataset	Time Steps	EFSA (%)	HMM (%)	Logistic(%) Regression
	Jan	17.53	24.58	17.90
	Feb	84.43	57.35	28.11
	Mar	78.09	93.56	82.89
	Apr	77.04	62.90	66.37
	May	95.27	77.16	79.38
Diabetes	Jun	99.25	50.08	54.54
	July	89.76	65.65	73.96
	Aug	76.10	43.98	56.72
	Sept	45.45	27.15	41.29
Average Accuracy		**73.65**	**55.82**	**55.68**

5   Concluding Remarks and Future Work

In this paper, we developed and presented the EFSA technology as a new temporal probabilistic reasoning for consistent multiple predictions into the future in the absence of domain experts. This shows that non experts now have fewer worries in choosing from the multitude of DBN types for real life applications.

This study shows that the EFSA can potentially become a powerful temporal probabilistic model used by both experts and non-experts to predict future trends in anticipatory planning. This technology simply emerges from environments and predicts in any domain of interest. The improved overall 74.3% accuracy of the EFSA over the 56.66% of HMM and 54.76% of logistic regression model when evaluated on the domains of the three datasets guarantees the reliability of the EFSA in many diverse areas. The relative efficiency of the EFSA suggests its wide application to make DBNs much simpler for use by researchers and in industries. We are currently developing an economic scalable model for handling massive MTS for the EFSA.

Acknowledgements. Our appreciation goes to the University of Cape Town and Complex Adaptive Systems (Pty.) Ltd. for their financial supports.

References

1. Wong, M.L., Leung, K.S.: An efficient data mining method for learning Bayesian networks using an evolutionary algorithm-based hybrid approach. IEEE Transactions on Evolutionary Computation 8, 378–404 (2004)
2. Murphy, K.: Dynamic Bayesian networks representation, inference and learning, Ph.D thesis, UC Berkeley, Computer Science Division (2002)
3. Russell, S., Norvig, P.: Artificial Intelligence (A Modern Approach), 2nd edn., p. 07458. Prentice Hall Series Inc., New Jersey (2003)
4. Deviren, M., Daoudi, K.: Structural Learning of Dynamic Bayesian Networks in Speech Recognition. In: Proceedings of Eurospeech, Aalborg, Denmark (2001)
5. Shenoy, P., Rao, R.P.N.: Dynamic Bayesian Networks for Brain-Computer Interfaces. In: Advances in NIPS, vol. 17. MIT Press, Cambridge (2005)
6. Silva, E., Plazaola, L., Ekstedt, M.: Strategic Business and IT Alignment: A Prioritized Theory Diagram. In: Proceedings of PICMET, Turkey (2006)
7. Larranaga, P., Kuijpers, C., Murga, R., Yurramendi, Y.: Learning Bayesian Network Structures by Searching for the Best Ordering with Genetic Algorithms. IEEE Transactions on Systems, Man, and Cybernetics, 487–493 (1996)
8. Osunmakinde, I.O., Potgieter, A.: Emergence of Optimal Bayesian Networks from Datasets without Backtracking using an Evolutionary Algorithm. In: Proceedings of the Third IASTED International Conference on Computational Intelligence, Banff, Alberta, Canada, pp. 46–51. ACTA Press (2007) ISBN: 978-0-88986-672-0
9. Endsley, M.R.: Theoretical underpinnings of situation awareness: a critical review. In: Situation Awareness Analysis and measurement, pp. 3–32. Lawrence Erlbaum Associates, Mahwah (2000)
10. Cozman, F.: JavaBayes, Bayesian Networks in Java, University of Sao Paulo (2001), http://www.cs.cmu.edu/~javabayes/Home/
11. Balikuddembe, J.K., Osunmakinde, I.O., Potgieter, A.E.: Software Project Profitability Analysis Using Temporal Probabilistic Reasoning. In: IEEE CS proceedings of the International Conference on Advanced Software Engineering & Its Applications, Washington, pp. 99–102 (2008) ISBN:978-0-7695-3432-9
12. Newman, D., Hettich, S., Blake, C., Merz, C.: UCI Repository of Machine Learning Databases (University of California, Department of Information and Computer Science, Irvine, CA, (1998), http://www.ics.uci.edu/~mlearn/MLRepository.html
13. GeNle 2.0, Decision Systems Laboratory, University of Pittsburgh (2006), http://genie.sis.pitt.edu
14. R Development Core Team: A Language and Environment for Statistical Computing, R Foundation for Statistical Computing, Vienna, Austria (2008) ISBN 3-900051-07-0, http://www.R-project.org

Efficient Hold-Out for Subset of Regressors

Tapio Pahikkala, Hanna Suominen, Jorma Boberg, and Tapio Salakoski

Turku Centre for Computer Science (TUCS)
University of Turku, Department of Information Technology
Joukahaisenkatu 3-5 B, FIN-20520 Turku, Finland
`firstname.lastname@utu.fi`

Abstract. Hold-out and cross-validation are among the most useful methods for model selection and performance assessment of machine learning algorithms. In this paper, we present a computationally efficient algorithm for calculating the hold-out performance for sparse regularized least-squares (RLS) in case the method is already trained with the whole training set. The computational complexity of performing the hold-out is $O(|H|^3 + |H|^2 n)$, where $|H|$ is the size of the hold-out set and n is the number of basis vectors. The algorithm can thus be used to calculate various types of cross-validation estimates effectively. For example, when m is the number of training examples, the complexities of N-fold and leave-one-out cross-validations are $O(m^3/N^2 + (m^2 n)/N)$ and $O(mn)$, respectively. Further, since sparse RLS can be trained in $O(mn^2)$ time for several regularization parameter values in parallel, the fast hold-out algorithm enables efficient selection of the optimal parameter value.

1 Introduction

In this paper, we consider the regularized least-squares (RLS) algorithm (see, e.g., [1]), a kernel-based learning method that is also known as the kernel ridge regression [2], least-squares support vector machine [3], and Gaussian process regression (see, e.g., [4]). RLS has been shown to have a state-of-the-art performance regression and classification and it has been applied in various practical tasks in which kernel based learning algorithms are needed (see e.g. [5]). It has also been modified for other problems such as ranking [6,7].

This popular machine learning method has, however, an inefficiency limitation in large-scale problems; the computational complexity of training an RLS learner together with a nonlinear kernel function is $O(m^3)$, where m is the number of training examples. This may be too tedious when the number of training examples is large.

To make the RLS algorithm more efficient, sparse versions have been considered. In these only a subset of training examples, often called the basis vectors, are used as regressors while the whole set is still used in the training process. This decreases the training complexity to $O(mn^2)$, where $n << m$ is the number of basis vectors (see, e.g., [1,8]). Here, we use the term sparse RLS when referring to RLS with the subset of regressors approach.

M. Kolehmainen et al. (Eds.): ICANNGA 2009, LNCS 5495, pp. 350–359, 2009.
© Springer-Verlag Berlin Heidelberg 2009

In addition to deriving faster machine learning algorithms, it is essential to develop computationally efficient methods for the related performance estimation and model selection. In particular, fast cross-validation (CV) algorithms and their approximations have been proposed (see, e.g., [7,9,10,11,12]) for the RLS-based learning algorithms. In CV, a part of the training data is held out from training to be used for testing and this hold-out procedure is repeated several times. For more thorough discussion about the CV methods and their critical points, we refer to [13].

But how to perform CV efficiently with sparse RLS? Firstly, sparse RLS regression can also be considered as a standard RLS regression using a certain type of modified kernel function [14]. This allows the use of the efficient hold-out algorithms for standard RLS proposed by [10,11]. The computational complexity of these are $O(|H|^2m)$, where $|H|$ is the size of the hold-out set. However, the presence of the coefficient m may make the algorithms too expensive in practice, especially if it is used to calculate CV estimates. For example, the computational complexity of leave-one-out CV (LOOCV) using this approach would be $O(m^2)$, which is more expensive than the training process of sparse RLS if $m > n^2$.

The second approach is to design CV algorithms especially for sparse RLS. Recently, [9] proposed this kind of LOOCV algorithm. Its computational complexity is $O(mn^2)$ which makes it much more practical than the LOOCV algorithm of standard RLS used together with the modified kernel function.

In this paper, we propose an even faster algorithm for computing hold-out estimates for sparse RLS. Its computational complexity is $O(|H|^3+|H|^2n)$. Consequently, our algorithm can be used to calculate various types of CV estimates efficiently. For example, when the sizes of the hold-out sets are sufficiently small, the computational complexity of N-fold CV is no larger than that of training sparse RLS. This is the case especially for LOOCV, whose computational complexity is only $O(mn)$. Further, our hold-out algorithm can be used to efficiently select the optimal regularization parameter value, because it can be combined with the fast method for training sparse RLS with several parameter values in parallel.

2   Sparse Regularized Least-Squares

We first formalize the methods of RLS and the subset of regressors method. We start by considering the hypothesis space $\mathcal{H}$. For this purpose we define so-called kernel functions. Let $\mathcal{X}$ denote the input space, which can be any set, and $\mathcal{F}$ denote an inner product space we call the feature space. For any mapping $\Phi : \mathcal{X} \to \mathcal{F}$, the inner product $k(x,x') = \langle \Phi(x), \Phi(x') \rangle$ of the images of the data points $x, x' \in \mathcal{X}$ is called a kernel function.

Using k, we define for a set $X = \{x_1, \ldots, x_m\}$ of data points a symmetric kernel matrix $K \in \mathbb{R}^{m \times m}$, whose entries are given by $K_{i,j} = k(x_i, x_j)$ and $\mathbb{R}^{m \times m}$ denotes the set of real $m \times m$ -matrices. For simplicity, we assume that K is strictly positive definite, that is, $A^\mathrm{T}KA > 0$ for all $A \in \mathbb{R}^m, A \neq 0$. The strict positive definiteness of K can be ensured, for example, by adding ϵI to K, where $I \in \mathbb{R}^{m \times m}$ is the identity matrix and ϵ is a small positive real number.

We consider the RLS algorithm as a variational problem (for a more comprehensive introduction, see, e.g., [1])

$$h = \operatorname*{argmin}_{h \in \mathcal{H}} \left(\sum_{i=1}^{m} (h(x_i) - y_i)^2 + \lambda \|h\|_k^2 \right), \tag{1}$$

where y_i are the output labels corresponding to the training inputs x_i, $\| \cdot \|_k$ is the norm in the reproducing kernel Hilbert space (RKHS) $\mathcal{H}$ determined by the kernel function k (see, e.g., [15]). The first and the second term of the right hand side of (1) are called the cost function and the regularizer, respectively, and $\lambda \in \mathbb{R}_+$ is a regularization parameter.

The solution of (1) has, by the representer theorem (see, e.g., [15]), the form

$$h(x) = \sum_{i=1}^{m} a_i k(x, x_i), \tag{2}$$

where coefficients $a_i \in \mathbb{R}$. Accordingly, we only need to solve a regularization problem with respect to a finite number of $a_i, 1 \leq i \leq m$. Let $A = (a_1, \ldots, a_m)^{\mathrm{T}} \in \mathbb{R}^m$ be a vector determining the minimizer of h.

To express (1) in a matrix form, we overload our notation and write $h(X) = KA \in \mathbb{R}^m$. This column vector contains the label predictions of the training data points obtained with the function h. Further, according to the properties of the RKHS determined by k, the regularizer can be written as $\lambda \|h\|_k^2 = \lambda A^{\mathrm{T}} K A$. Now, we can rewrite the algorithm (1) as

$$A = \operatorname*{argmin}_{A \in \mathbb{R}^m} \left((Y - KA)^{\mathrm{T}} (Y - KA) + \lambda A^{\mathrm{T}} K A \right).$$

Its solution

$$A = (KK + \lambda K)^{-1} KY = (K + \lambda I)^{-1} Y \tag{3}$$

is found by first differentiating $(Y - KA)^{\mathrm{T}}(Y - KA) + \lambda A^{\mathrm{T}} K A$ with respect to A, setting the derivative to be zero, and solving it with respect to A.

The computation complexity of calculating the coefficient vector A from (3) is dominated by the inversion of a $m \times m$ -matrix, and hence it can be performed in $O(m^3)$ time. This may be too tedious when the number of training examples is large. However, several authors have considered sparse versions of RLS, where only a part of the training examples, called basis vectors, have a nonzero coefficient in (2). This means that when the training is complete, only the basis vectors are needed when predicting the outputs of the new data points, which makes the prediction more efficient than it is with the standard kernel RLS regression. Another advantage of sparse RLS is that its training complexity is only $O(mn^2)$, where n is the number of basis vectors. Further, as we will show below, there are efficient CV and regularization parameter selection algorithms for sparse RLS that are analogous to the ones for standard RLS.

In this paper, we do not pay attention to the approach that is used to select the set of basis vectors. The only detail we point out is related to the computation of the hold-out or CV performance, since this is our main topic. Namely, the selection of the basis vectors should not change if a part of the training examples is held out and sparse RLS is trained with the rest of the examples. One suitable selection method is, for example, a random sub-sampling when the hold-out sets are also selected randomly. In fact, it was found in [1] that simply selecting the basis vectors randomly has no worse learning performance than more sophisticated methods.

Before continuing to the definition of sparse RLS, we introduce some notation. Let $\mathcal{M}_{\Xi \times \Psi}$ denote the set of matrices whose rows and columns are indexed by the index sets Ξ and Ψ, respectively. Below, with any matrix $M \in \mathcal{M}_{\Xi \times \Psi}$ and index set $\Upsilon \subseteq \Xi$, we use the subscript Υ so that a matrix $M_\Upsilon \in \mathcal{M}_{\Upsilon \times \Psi}$ contains only the rows that are indexed by Υ. For $M \in \mathcal{M}_{\Xi \times \Psi}$, we also use $M_{\Upsilon \Omega} \in \mathcal{M}_{\Upsilon \times \Omega}$ to denote a matrix that contains only the rows and the columns that are indexed by any index sets $\Upsilon \subseteq \Xi$ and $\Omega \subseteq \Psi$, respectively.

We now follow [8,1,14] and define the sparse RLS algorithm using the above defined notation. Let $F = \{1, \ldots, m\}$. Instead of allowing functions like in (2), we only allow

$$h(x) = \sum_{i \in B} a_i k(x, x_i),$$

where the set indexing the n basis vectors $B \subset F$ is selected in advance. In this case, the coefficient vector $A \in \mathbb{R}^n$ is a vector whose entries are indexed by B. The label predictions for the training data points can be obtained from

$$h(X) = (K_B)^{\mathrm{T}} A \tag{4}$$

and the regularizer can be rewritten as $\lambda A^{\mathrm{T}} K_{BB} A$. Therefore, the coefficient vector A is the minimizer of

$$(Y - (K_B)^{\mathrm{T}} A)^{\mathrm{T}} (Y - (K_B)^{\mathrm{T}} A) + \lambda A^{\mathrm{T}} K_{BB} A. \tag{5}$$

The minimizer of (5) is found by setting its derivative with respect to A to zero. It is

$$A = P^{-1} K_B Y, \tag{6}$$

where

$$P = K_B (K_B)^{\mathrm{T}} + \lambda K_{BB} \in \mathcal{M}_{B \times B}. \tag{7}$$

The matrices K_{BB} and $K_B (K_B)^{\mathrm{T}} = (KK)_{BB}$ are principal sub-matrices of the positive definite matrices K and KK, respectively, and hence the matrix P is also positive definite and invertible (see e.g. [16, p. 397]). In contrast to (3), the matrix inversion involved in (6) can be performed in $O(n^3)$ time. Since $n << m$, the overall computational complexity of (6) is dominated by the complexity of calculating $K_B (K_B)^{\mathrm{T}}$ which is $O(mn^2)$.

We now reformulate sparse RLS so that its coefficient matrix can be efficiently calculated for different values of the regularization parameter. Note that this reformulation is already known in the machine learning community. Let

$$K_{BB} = CC^{\mathrm{T}} \tag{8}$$

be the Cholesky factorization of K_{BB}, where $C \in \mathcal{M}_{B \times B}$ is a lower triangular matrix with strictly positive diagonal entries. Moreover, let

$$C^{-1}K_B(K_B)^{\mathrm{T}}(C^{-1})^{\mathrm{T}} = V\Lambda V^{\mathrm{T}} \tag{9}$$

be the eigen decomposition of $C^{-1}K_B(K_B)^{\mathrm{T}}(C^{-1})^{\mathrm{T}}$, where $V \in \mathcal{M}_{B \times B}$ is the matrix containing the eigenvectors, and $\Lambda \in \mathcal{M}_{B \times B}$ is a diagonal matrix containing the eigenvalues of the decomposition. Further, let

$$\widetilde{\Lambda} = (\Lambda + \lambda I)^{-1}$$

and

$$Q = (C^{-1})^{\mathrm{T}}V \in \mathcal{M}_{B \times B}. \tag{10}$$

Then, the matrix P^{-1} can be expressed as

$$
\begin{aligned}
P^{-1} &= (K_B(K_B)^{\mathrm{T}} + \lambda K_{BB})^{-1} \\
&= (K_B(K_B)^{\mathrm{T}} + \lambda CC^{\mathrm{T}})^{-1} \\
&= (C^{-1})^{\mathrm{T}}(C^{-1}K_B(K_B)^{\mathrm{T}}(C^{-1})^{\mathrm{T}} + \lambda I)^{-1}C^{-1} \\
&= (C^{-1})^{\mathrm{T}}(V\Lambda V^{\mathrm{T}} + \lambda I)^{-1}C^{-1} \\
&= (C^{-1})^{\mathrm{T}}V\widetilde{\Lambda}V^{\mathrm{T}}C^{-1} \\
&= Q\widetilde{\Lambda}Q^{\mathrm{T}}.
\end{aligned} \tag{11}
$$

The computational complexities of calculating (8), (9) and (10) are $O(n^3)$.

Now, let us first calculate $Q^{\mathrm{T}}K_B Y$ (in $O(nm)$ time) and store it in memory. Then, the solution (6) can be computed for different values of the regularization parameter from

$$A = Q\widetilde{\Lambda}Q^{\mathrm{T}}K_B Y,$$

with a complexity $O(n^2)$. This is because the multiplication of the shifted and inverted eigenvalues $\widetilde{\Lambda}$ with $Q^{\mathrm{T}}K_B Y$ can be performed in $O(n)$ time and the multiplication of the resulting matrix from left by Q can be performed in $O(n^2)$ time.

3   Fast Computation of Hold-Out Error Estimates for Sparse RLS

We note (see, e.g., [14]) that the sparse approach can also be considered as performing a standard RLS regression using the following type of modified kernel function

$$\tilde{k}(x, x') = k(x, X)(K_{BB})^{-1}k(X, x), \tag{12}$$

Therefore, a straightforward way to construct hold-out estimates for sparse RLS would be to use the hold-out algorithms proposed by [10,11] for the standard RLS regression using the modified kernel function (12). The computational complexity of this approach is $O(|H|^2 m)$, where m is the number of training examples and $|H|$ is the size of the hold-out set. The presence of the coefficient m may make it computationally too expensive in practice, especially if it is used to calculate CV estimates. Further, it can be shown that if a data point in the hold-out set is a basis vector, its effect is not completely removed from the training process, because the kernel (12) depends on it.

We now introduce a faster algorithm for calculating a hold-out performance estimates, whose computational complexity is $O(|H|^2(|H| + n))$, where $n << m$ is the number of basis vectors. Consequently, our algorithm can be used to calculate various types of CV estimates efficiently. For example, when the sizes of the hold-out sets are sufficiently small, the computational complexity of N-fold CV is no larger than that of training sparse RLS. This is the case especially for LOOCV.

As previously, $F = \{1, \ldots, m\}$ and $B \subset F$ are the index set for the whole training data set and the set indexing the basis vectors, respectively. Let $H \subset F$ denote the set of indices of the hold-out data points, and let $\overline{H} = F \setminus H$, $E = H \cap B$, and $L = \overline{H} \cap B$. Further, let $h_{\overline{H}}$ be the function obtained by training the sparse RLS algorithm without using the training examples indexed by H. Then, $h_{\overline{H}}(X_H)$ consists of the output values for the hold-out data points X_H that are predicted by $h_{\overline{H}}$. According to (6), the coefficient vector corresponding to $h_{\overline{H}}$ is

$$G^{-1} K_{L\overline{H}} Y_{\overline{H}},$$

where

$$G = K_{L\overline{H}} K_{\overline{H}L} + \lambda K_{LL}.$$

The entries of this coefficient vector are indexed by L. Therefore, according to (4), the output values corresponding to the hold-out set H can be obtained from

$$h_{\overline{H}}(X_H) = K_{HL} G^{-1} K_{L\overline{H}} Y_{\overline{H}}. \tag{13}$$

This is, of course, computationally too expensive to be used in CV, but fortunately, it is possible to calculate the outputs more efficiently when we have trained in advance a sparse RLS learner with the whole data set.

Proposition 1. *Suppose that we have trained sparse RLS by calculating (8), (9), and (10), and we have the following matrices stored in memory:*

$$K_B \in \mathcal{M}_{B \times F} \tag{14}$$

$$\Lambda \in \mathcal{M}_{B \times B} \tag{15}$$

$$Q \in \mathcal{M}_{B \times B} \tag{16}$$

$$(K_B)^T Q \in \mathcal{M}_{F \times B} \tag{17}$$

$$K_B Y \in \mathcal{M}_{B \times 1} \tag{18}$$

$$Q^T K_B Y \in \mathcal{M}_{B \times 1}. \tag{19}$$

Then, the hold-out predictions for a set H can be calculated from

$$h_{\overline{H}}(X_H) = -(J - I)^{-1}R, \tag{20}$$

where

$$J = U\widetilde{\Lambda}U^T - U\widetilde{\Lambda}(Q_E)^T(Q_E\widetilde{\Lambda}(Q_E)^T)^{-1}Q_E\widetilde{\Lambda}U^T, \tag{21}$$

$$R = U\widetilde{\Lambda}Z - U\widetilde{\Lambda}(Q_E)^T(Q_E\widetilde{\Lambda}(Q_E)^T)^{-1}Q_E\widetilde{\Lambda}Z, \tag{22}$$

$$U = ((K_B)^TQ)_H - K_{HE}Q_E, \text{ and} \tag{23}$$

$$Z = Q^TK_BY - (Q_E)^T(K_BY)_E - (Q^TK_B)_{BH}Y_H + (Q_E)^TK_{EH}Y_H. \tag{24}$$

The computational complexity of this calculation is $O(|H|^2(|H| + n))$.

Proof. We start by showing the tenability of (20) and continue by considering the computational complexities. Recall from (13) that the output matrix for the hold-out set can be obtained from

$$h_{\overline{H}}(X_H) = K_{HL}G^{-1}K_{L\overline{H}}Y_{\overline{H}}, \tag{25}$$

where

$$\begin{aligned} G &= K_{L\overline{H}}K_{\overline{H}L} + \lambda K_{LL} \\ &= K_L(K_L)^T - K_{LH}K_{HL} + \lambda K_{LL} \\ &= P_{LL} - K_{LH}K_{HL} \end{aligned}$$

and P is defined in (7). Now, due to the positive definiteness of K, both G and P_{LL} are always invertible (see e.g. [16]). Let

$$W = (P_{LL})^{-1}.$$

Using the block inverse formula (see, e.g., [16, p. 18–19]), we get

$$W = (P^{-1})_{LL} - (P^{-1})_{LE}((P^{-1})_{EE})^{-1}(P^{-1})_{EL}. \tag{26}$$

Now, we observe that

$$G^{-1} = (W^{-1} - K_{LH}K_{HL})^{-1},$$

and using the Sherman-Morrison-Woodbury formula (see, e.g., [16, p. 18–19]), we obtain

$$G^{-1} = W - WK_{LH}(-I + K_{HL}WK_{LH})^{-1}K_{HL}W. \tag{27}$$

The invertibility of the matrix $-I + K_{HL}WK_{LH}$ can be shown by considering the matrix

$$\begin{bmatrix} I & -K_{HL} \\ -K_{LH} & W^{-1} \end{bmatrix}. \tag{28}$$

Since $G = W^{-1} - K_{LH}K_{HL}$, and I are invertible, the invertibility of the matrix (28) follows from the Schur's determinantal formula (see, e.g., [16, p. 21]).

Therefore, also the matrix $I - K_{HL}WK_{LH}$ is invertible, again due to the Schur's determinantal formula.

We continue by observing that

$$
\begin{aligned}
U &= ((K_B)^T Q)_H - K_{HE}Q_E \\
&= K_{HL}Q_L \text{ and} \\
Z &= Q^T K_B Y - (Q_E)^T (K_B Y)_E - (Q^T K_B)_{BH} Y_H + (Q_E)^T K_{EH} Y_H \\
&= (Q_L)^T K_L Y - (Q_L)^T K_{LH} Y_H \\
&= (Q_L)^T K_{L\overline{H}} Y_{\overline{H}}.
\end{aligned}
$$

$$(29)$$

$$(30)$$

According to (26), (11), and (29), we get

$$
\begin{aligned}
K_{HL}WK_{LH} &= K_{HL}(P^{-1})_{LL}K_{LH} - K_{HL}(P^{-1})_{LE}((P^{-1})_{EE})^{-1}(P^{-1})_{EL}K_{LH} \\
&= K_{HL}(Q\tilde{\Lambda}Q^T)_{LL}K_{LH} \\
&\quad - K_{HL}(Q\tilde{\Lambda}Q^T)_{LE}((Q\tilde{\Lambda}Q^T)_{EE})^{-1}(Q\tilde{\Lambda}Q^T)_{EL}K_{LH} \\
&= U\tilde{\Lambda}U^T - U\tilde{\Lambda}(Q_E)^T(Q_E\tilde{\Lambda}(Q_E)^T)^{-1}Q_E\tilde{\Lambda}U^T \\
&= J.
\end{aligned}
$$

Analogously, according to (26), (11), and (30), we get $K_{HL}WK_{L\overline{H}}Y_{\overline{H}} = R$. Finally, by substituting (27) into (25), we get

$$
\begin{aligned}
h_{\overline{H}}(X_H) &= K_{HL}(W - WK_{LH}(J - I)^{-1}K_{HL}W)K_{L\overline{H}}Y_{\overline{H}} \\
&= (I - J(J - I)^{-1})R \\
&= ((J - I)(J - I)^{-1} - J(J - I)^{-1})R \\
&= -(J - I)^{-1}R.
\end{aligned}
$$

We now consider the computational complexity of using (20). The matrix U can be calculated from (14), (16), and (17) using (23) in $O(|H|^2 n)$ time. Moreover, the matrix Z can be calculated from (14), (16), (17), (18), and (19) using (24) in $O(|H|n)$ time. The computational complexity of calculating J and R using (21) and (22) is $O(|H|^2(|H| + n))$. This is because multiplication of an $|H| \times n$ -matrix with a diagonal matrix $\tilde{\Lambda}$ can be computed in $O(|H|n)$ time, the matrix inversion involved in the calculations needs $O(|H|^3)$ time, and all the other matrix products need at most in $O(|H|^2 n)$ time when performed in the optimal order. Finally, we substitute these matrices in (20) from which the solution is obtained by inverting a matrix in $O(|H|^3)$ time. $\quad\square$

The calculation of the matrices (14) – (19) needs $O(mn^2)$ time, and hence the computational complexity of training sparse RLS as in Proposition 1 is the same as that of training sparse RLS in the ordinary way. When we have trained sparse RLS as in Proposition 1, we can use the efficient hold-out method for calculating various types of CV estimates. We can, for example, to perform N-fold CV by partitioning the training set into N approximately equally sized folds and average the results of the individual hold-out estimates. The number of training examples in each fold is then $|H| \approx m/N$. According to Proposition 1, the computational complexity of each hold-out calculation is $O(|H|^2(|H|+n))$, and hence the overall computational complexity of N-fold CV is

$$O\left(m|H|^2 + mn|H|\right) = O\left(\frac{m^3}{N^2} + \frac{m^2n}{N}\right).$$

We observe that when $|H| \approx n$, the computational complexity of CV is equal to that of the sparse RLS training. If we consider this as the largest tolerable computational complexity, the proposed hold-out method is too expensive for larger hold-out sets.

On the other hand, the method is less complex for smaller hold-out sets, like $O(mn)$ for the extreme case of LOOCV. Therefore, it can be used, for example, to select the value of the regularization parameter λ efficiently from a set of candidate values. The computational complexity of the regularization parameter selection is equal to the complexity of CV times the size of the set of candidate values, since the initialization phase of Proposition 1 does not have to be repeated for different candidate values.

4   Conclusion

Hold-out and CV are among the most important methods for model selection and performance evaluation of machine learning algorithms, and therefore their computationally efficient implementations are sought after. In this paper, we presented a computationally efficient algorithm for calculating hold-out performance estimates for sparse RLS when it has been trained in advance with the whole data set.

The computational complexity of training sparse RLS is $O(mn^2)$, where m is the size of the training set and n is the number of basis vectors. We showed that the hold-out estimates for trained sparse RLS can be computed in $O(|H|^2(|H|+n))$ time, where $|H|$ is the size of the hold-out set. Consequently, the algorithm can be used to calculate various types of CV estimates effectively. For example, the complexities of N-fold CV and LOOCV for m training examples are $O(m^3/N^2 + (m^2n)/N)$ and $O(mn)$, respectively.

Sparse RLS can be trained in $O(mn^2)$ time for several regularization parameter values in parallel, and this property can also be combined with the fast hold-out calculation. Therefore, cross-validation can be used to efficiently select the optimal value of the regularization parameter.

Acknowledgments

This work has been supported by Tekes, the Finnish Funding Agency for Technology and Innovation (grant 40020/07) and by Academy of Finland (grant 128061).

References

1. Rifkin, R.: Everything Old Is New Again: A Fresh Look at Historical Approaches in Machine Learning. Ph.D thesis, Massachusetts Institute of Technology (2002)
2. Saunders, C., Gammerman, A., Vovk, V.: Ridge regression learning algorithm in dual variables. In: Proceedings of the Fifteenth International Conference on Machine Learning, pp. 515–521. Morgan Kaufmann Publishers Inc., San Francisco (1998)
3. Suykens, J.A.K., Vandewalle, J.: Least squares support vector machine classifiers. Neural Processing Letters 9(3), 293–300 (1999)
4. Rasmussen, C.E., Williams, C.K.I.: Gaussian Processes for Machine Learning (Adaptive Computation and Machine Learning). The MIT Press, Cambridge (2005)
5. Pahikkala, T., Pyysalo, S., Boberg, J., Järvinen, J., Salakoski, T.: Matrix representations, linear transformations, and kernels for disambiguation in natural language. Machine Learning 74(2), 133–158 (2009)
6. Pahikkala, T., Tsivtsivadze, E., Airola, A., Boberg, J., Salakoski, T.: Learning to rank with pairwise regularized least-squares. In: Joachims, T., Li, H., Liu, T.Y., Zhai, C. (eds.) SIGIR 2007 Workshop on Learning to Rank for Information Retrieval, pp. 27–33 (2007)
7. Pahikkala, T., Tsivtsivadze, E., Airola, A., Järvinen, J., Boberg, J.: An efficient algorithm for learning to rank from preference graphs. Machine Learning 75(1), 129–165 (2009)
8. Smola, A.J., Schölkopf, B.: Sparse greedy matrix approximation for machine learning. In: Langley, P. (ed.) Proceedings of the Seventeenth International Conference on Machine Learning, pp. 911–918. Morgan Kaufmann, San Francisco (2000)
9. Cawley, G.C., Talbot, N.L.C.: Fast exact leave-one-out cross-validation of sparse least-squares support vector machines. Neural Networks 17(10), 1467–1475 (2004)
10. Pahikkala, T., Boberg, J., Salakoski, T.: Fast n-fold cross-validation for regularized least-squares. In: Honkela, T., Raiko, T., Kortela, J., Valpola, H. (eds.) Proceedings of the Ninth Scandinavian Conference on Artificial Intelligence (SCAI 2006), Espoo, Finland, Otamedia, pp. 83–90 (2006)
11. An, S., Liu, W., Venkatesh, S.: Fast cross-validation algorithms for least squares support vector machine and kernel ridge regression. Pattern Recognition 40(8), 2154–2162 (2007)
12. Rifkin, R., Lippert, R.: Notes on regularized least squares. Technical Report MIT-CSAIL-TR-2007-025, Massachusetts Institute of Technology (2007)
13. Suominen, H., Pahikkala, T., Salakoski, T.: Critical points in assessing learning performance via cross-validation. In: Honkela, T., Pöllä, M., Paukkeri, M.S., Simula, O. (eds.) Proceedings of the 2nd International and Interdisciplinary Conference on Adaptive Knowledge Representation and Reasoning (AKRR 2008), Helsinki University of Technology, pp. 9–22 (2008)
14. Quiñonero-Candela, J., Rasmussen, C.E.: A unifying view of sparse approximate gaussian process regression. Journal of Machine Learning Research 6, 1939–1959 (2005)
15. Schölkopf, B., Herbrich, R., Smola, A.J.: A generalized representer theorem. In: Helmbold, D., Williamson, R. (eds.) COLT 2001 and EuroCOLT 2001. LNCS, vol. 2111, pp. 416–426. Springer, Heidelberg (2001)
16. Horn, R., Johnson, C.R.: Matrix Analysis. Cambridge University Press, Cambridge (1985)

Improving Optimistic Exploration in Model-Free Reinforcement Learning

Marek Grześ and Daniel Kudenko

Department of Computer Science, University of York
Heslington, York, YO10 5DD, United Kingdom
{grzes,kudenko}@cs.york.ac.uk

Abstract. The key problem in reinforcement learning is the exploration-exploitation tradeoff. An optimistic initialisation of the value function is a popular RL strategy. The problem of this approach is that the algorithm may have relatively low performance after many episodes of learning. In this paper, two extensions to standard optimistic exploration are proposed. The first one is based on different initialisation of the value function of goal states. The second one which builds on the previous idea explicitly separates propagation of low and high values in the state space. Proposed extensions show improvement in empirical comparisons with basic optimistic initialisation. Additionally, they improve anytime performance and help on domains where learning takes place on the subspace of the large state space, that is, where the standard optimistic approach faces more difficulties.

1 Introduction

The main feature of reinforcement learning is that it can deal with stochastic control when the system is hard to model but easy to simulate. The process of building an explicit, mathematical model of the environment may be as difficult as the control problem itself. If, however, the system can be observed either in real time or through a software simulator, the reinforcement learning approach can be used to approximate the value function, or the optimal policy to control such a system [1]. The key problem in reinforcement learning (RL), either when learning takes place in a real (situated) or simulated system, is the exploration-exploitation tradeoff. It is a problem of action selection which leads to a constant dilemma of two contradicting objectives. The first one is exploitation, that is, maximisation of the reward based on the current policy. And, the second one concerns exploration of the environment in order to improve approximation of the policy in order to perform better in the future [2]. There exist solutions for the small class of bandit problems for which formal correctness was proved. They do not apply, however, to the multi-state case. For this reason heuristic approaches are necessary to deal with exploration-exploitation tradeoff in general RL [3]. Thrun [4] has surveyed and proposed commonly used categorisation of such techniques.

M. Kolehmainen et al. (Eds.): ICANNGA 2009, LNCS 5495, pp. 360–369, 2009.
© Springer-Verlag Berlin Heidelberg 2009

One of the simplest, albeit effective, approaches to the problem of exploration in RL is a straightforward, optimistic initialisation of the value function [2]. The value function of all states is in this case initialised to the highest possible value. The learning algorithm is in this way encouraged to go to unexplored parts of the state space whereas the value function is modified according to received payoff. This can be applied in the most straightforward form as an undirected [4], that is based only on the current content of the Q-table, exploration, but this is also the key element of some other exploration strategies for both model-free, IEQL+ [5], and model-based, R-max [6,7], reinforcement learning. The formal justification for optimism under uncertainty is, for example, due to Brafman and Tannenholtz [6].

An essential advantage of optimistic initialisation is that it provides broad exploration, and as a result of this, it is difficult to miss highly rewarded final states. The current policy which is being learned with optimistic initialisation is either optimal or leads to efficient learning. The disadvantage of this approach is that it may take a lot of time for the algorithm to get rid of optimism, propagate actual costs of actions, and converge to a final policy. In this case exploration can be seen rather as a process of constant reduction of optimism based on a real payoff from the simulation. It leads to situations when the learning agent can go through relatively long trajectories even after many episodes of learning because it is constantly driven by its optimism. In this paper we investigate how to truncate this optimistic wandering in order to converge faster to optimal solution, and also to be able to obtain a reasonable policy as soon as possible, that is, to improve anytime properties [8] which are in fact weak in case of basic optimistic exploration. The first straightforward solution is to simply initialise the value of the final state to a higher value than all other states. In this way positive information can be backpropagated from the goal state, and the algorithm starts exploiting earlier. Another solution which is build on the previous observation leads to a new algorithm in which the propagation of the value function is divided into two stages: learning with standard optimistic initialisation, and backpropagation of the higher positive value of the final state with an explicit border line between low and high values.

Specifically, in this paper we investigate how to improve learning of SARSA [2], model-free reinforcement learning algorithm, with optimistic initialisation of the value function. The contribution of this paper can be summarised as follows:

- we show and empirically evaluate how small changes to the initialisation of the value function can improve the obtained exploration,
- following reasoning from the first contribution a new algorithm to improve optimistic exploration in model-free RL is proposed and evaluated,
- additional applicability of our algorithm to special requirements is also explained or evaluated: learning on the unknown limited area of the state space, and anytime requirements [8].

It is worth noting that optimistic initialisation can be seen as a high level paradigm which can be used with more specific exploration strategies which determine how actions are selected. In our analysis ϵ-greedy action selection is used [2]. It allows to control the ratio of greedy actions (i.e., with highest Q-values) chosen during learning. Generally optimistic initialisation and ϵ-greedy can be seen as orthogonal, complementary techniques.

Optimistic exploration together with our improvements are discussed in Sections 2 and 3. Experimental evaluation is contained in Sections 4, 5 and 6. Concluding discussion is in Section 7.

2   Optimistic Exploration

For the ease of presentation our discussion throughout the paper is based on the family of stochastic shortest path problems where there is a cost (negative reward) given for actions executed in the system, and the objective is to learn how to reach the goal state with a minimum cost (maximal reward) [1]. The generalisation beyond this assumption is discussed in the final part of this section. When learning in this setting, the optimistic exploration is achieved in a natural way. When:

$$\forall_s \forall_a Q(s, a) = 0$$

and the action cost is a negative scalar, R, this represents the optimism under uncertainly principle [2]. The key observation, which according to our best knowledge has been never explicitly discussed in the literature, is that different initialisation of the value function of the goal state changes the character of obtained exploration, when it is based on the current content of the Q-table (so called undirected exploration [4]), and, for example, ϵ-greedy action selection. Specifically, when:

$$\forall_{s \notin G} \forall_a Q(s, a) = -I, \; \forall_{g \in G} \forall_a Q(g, a) = 0, \tag{1}$$

or

$$\forall_{s \notin G} \forall_a Q(s, a) = 0, \; \forall_{g \in G} \forall_a Q(g, a) = I, \tag{2}$$

where I is a positive scalar and G the set of goal states, this is not any more a truly optimistic initialisation. In this case the learning process is unchanged only until the higher value from the final state (in Equations 1 and 2 the value of goal states is always higher) is backpropagated to the given areas of the state space. Once this higher value reaches given entries in the Q-table, it has impact on the exploration, and in particular the agent is encouraged to follow directions from which higher values are propagated.

The question remains of what the ratio I/R should be. The first heuristic choice is to set I to the value approximately equal (in terms of the order of magnitude) to the absolute value of the cost of the longest path to the goal. According to our empirical tests this approach works well in practice. Additionally, a more detailed analysis of the impact of the value of I is in Section 6.

It is worth highlighting that the reinitialisation proposed in this section can be done without any additional knowledge about the domain. Even the knowledge about the goal state is not required. When the goal state is reached for the first time, its value function can be simply initialised to a new value.

Another important observation is that with the type of reward which is considered here, the same effect can be achieved with a straightforward modification of the reward which is given upon entering the goal state. An artificial high positive value can be used instead of the negative action cost. This relation does not hold, however, with another common type of reward where all rewards are zero except the positive reward upon entering the goal state. In this case optimistic initialisation is achieved by assigning to all Q-entries the value equal or higher than the final reward. The property which we discuss at the beginning of this section can be in this case achieved by assigning a higher value to the goal state than to all remaining states, that is, in the same way as with rewards based on the action cost.

3   Improving Optimistic Exploration

Here, a novel algorithm is proposed which builds on the observation described in the previous section. Our principal heuristic guess is that propagation of the high value of the value function from goal states can improve convergence as the agent will be encouraged to exploit towards higher values. In the basic form, described in Section 2, those higher values, when backpropagated, interact with lower values of non-goal states through standard temporal-difference learning of SARSA. The idea of the algorithm proposed in this section is to treat in a different way transitions from lower to higher values. The concept presented here can be easily explained through Algorithm 1. which is named IOE which stands for improved optimistic exploration. The first property of our algorithm is in lines $7 - 9$ where the Q-values of the goal state are reinitialised to the value of zero. The same properties of the algorithm can be obtained by replacing $-I$ with 0 in line 1, and 0 with I in line 8, and modifying inequalities appropriately (see Equations 1 and 2). This shows how the idea which is discussed in Section 2 can be implemented in the SARSA algorithm. A more challenging situation is considered here where the goal state is not known. If knowledge about the goal state is available, lines $7 - 9$ can be removed, and $Q(g, \cdot)$, where $g \in G$, initialised to the value of 0 directly in line 1. The extension to this framework, which represents the essence of this section, is in lines $10 - 18$ of Algorithm 1.. In this part, the algorithm deals with transitions from lower (initialised to $-I$) to higher (backpropagated from $Q(g, \cdot) = 0$, where $g \in G$) values in the Q-table. In line 11, $Q(s, a)$ is shifted to the domain of high values by a direct assignment of the sum of the step penalty, and the value function of the next state.

For better presentation of the concept which is discussed here, an exemplary content of the Q-table when learning with Algorithm 1. is placed in Figure 1. It is a random walk domain in which there is a chain of ten states. There are two stochastic actions (90% of success) left and right, and the goal is to navigate from state 0 (the leftmost state) to state 9 (the rightmost state). The cost of each action is -1. The figure shows the content of the Q-table at the beginning of the fourth learning episode. It can be easily observed that high and low values are propagated. States $6 - 9$ in Figure 1 were reached by higher values which

Algorithm 1. The IOE algorithm

1: Initialise $Q(s,a) = -I$ for all s, a
2: **repeat** {for each episode}
3:     Initialise s, choose a from s using policy derived form Q
4:     **repeat** {for each step of episode}
5:         Take action a, observe reward r, s'
6:         Choose a' from s' using policy derived form Q
7:         **if** $s' \in G$ **then**
8:             $Q(s',a') = 0$
9:         **end if**
10:        **if** $Q(s',a') > -I$ and $Q(s,a) \le -I$ **then** {from lower to higher values}
11:            $Q(s,a) = Q(s',a') + r$
12:            **if** $s' \in G$ **then**
13:                **break** {end of the episode}
14:            **else**
15:                $s \leftarrow s'$, $a \leftarrow a'$
16:                **continue** {go to the next step}
17:            **end if**
18:        **end if**
19:        $\delta = r + \gamma Q(s',a') - Q(s,a)$, $Q(s,a) = Q(s,a) + \alpha\delta$
20:        $s \leftarrow s'$, $a \leftarrow a'$
21:    **until** state s is terminal
22: **until**

	State	0	1	2	3	4	5	6	7	8	9
Action	Left	-101.74	-101.69	-101.52	-101.37	-100.98	-100.69	-100.4	-100.2	-100.1	-100
	Right	-101.75	-101.67	-101.56	-101.31	-100.94	-100.74	-3.02	-2.18	-1	0

Fig. 1. The content of the Q-table of a 10-state random walk at the beginning of the fourth learning episode

encourage the agent to move towards the goal. The remaining states have lower values, but these values are optimistic with a given model of reward, and the agent is encouraged to explore following standard optimistic exploration strategy.

4   Experimental Design

The second part of the paper contains empirical study of the proposed algorithm. Specifically, three RL algorithms are compared: standard SARSA with optimistic initialisation (Optimistic), SARSA with modified initialisation of the value function for the goal state (Semi-optimistic), and Algorithm 1. (IOE). The following values of relevant parameters were applied. The learning rate α is 0.1 in the first episode and is linearly decreased to the value of 0.01 in the last episode (property required by the convergence proofs of the SARSA algorithm). The ϵ-greedy action selection is used. For brevity we selected the value of ϵ which yielded the best performance of the basic Optimistic version. On all problems the value of ϵ was either not significant, or values 0 and 0.01 were the best. For this reason the value of 0 is used with the Optimistic algorithm, that is, the algorithm greedily always selects the best action. Both Semi-optimistic and IOE require some randomness in the selection of actions. After initial experimentation, $\epsilon = 0.05$ was selected and used with Semi-optimistic and IOE in experiments reported in this paper. The order of magnitude of the value of I in the IOE algorithm was 6.

Our analysis is based on stochastic shortest path problems [1]. Detailed description of used domains is in Section 5. In all these domains the value -1 was used as an action cost. In this way initialisation of all Q-entries with the value of zero naturally yields an optimistic initialisation. Since domains are episodic and actions have non-zero cost, the discount factor, $\gamma = 1$, can be used.

Results presented in all figures in Section 6 represent the average of 10 runs. Statistical analysis with two sample t test, and the level of significance at $P <$ 0.05, was used to evaluate differences in performance [9].

5   Domains

For better understanding of the problem discussed in this paper and more extensive evaluation of the algorithm which is proposed, four RL domains are used.

Random walk. The chain of states constitutes the state space in this environment which has been often used in the RL literature [10,11]. In our design there are 128 states and the task is to learn how to get to the rightmost state when starting from the leftmost state. There are two actions: left and right. The environment is stochastic. Each action may fail with probability 0.1, and in such case another action is chosen. The value of I was 128 in the Semi-optimistic configuration.

S-maze. This is an instance of the navigation maze problems. This particular design comes from [12]. In our case a scaled larger version was used. Each grid position from the original task (see, e.g., [2, pp. 235]) was uniformly divided into 64 squares which yields 72×48 states. There are eight actions which lead to an adjacent cell if it is not the border nor an obstacle. In such situations actions do not have any effect. Actions are stochastic. With probability 0.1 an action can fail in which case one of the remaining actions is chosen with a uniform probability. The value of I was 200 in the Semi-optimistic configuration.

Mountain Car. The Mountain Car problem has been commonly used in the literature to test new reinforcement learning algorithms. It is a two-dimensional world consisting of a U-shaped valley and a car placed at the bottom of the valley. The car must move back and forth to gain enough momentum to escape the valley [2]. An experiment was terminated and the agent placed in a random position after reaching the goal state. Two different discretisations were used with 11×21 (these smallest values were taken from [13]), and 100×100 intervals on correspondingly the position and velocity of the car. The value of I was 100 in the Semi-optimistic configuration.

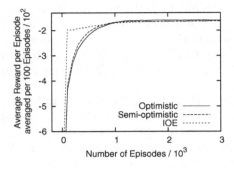
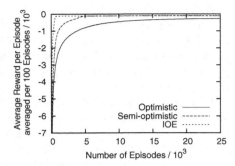

Fig. 2. Results for the random walk task

Fig. 3. Results for the S-maze task

Maze. The last domain is a variant of the maze used by Wiering [14]. It is in our case a 100×100 maze where the start state is located in the position $(0, 0)$. Every non-start and non-goal state can be an obstacle with probability 0.2. For each run a new set of obstacles was drawn according to this rule. The goal is to reach the terminal state. To gain better insight into compared algorithms we test two different configurations which differ in the location of the goal state. The following positions of the goal state were used: $(99, 99)$, and $(49, 49)$. The value of I was 200 in the Semi-optimistic configuration.

6   Results

The first experiment was performed on the random walk task (RW). Results are shown in Figure 2. It is easy to notice that the IOE algorithm worked much better in the initial phase of learning than two remaining algorithms. The difference between IOE and Semi-optimistic is statistically significant from approximately episode 35 to 600. The difference between Optimistic and Semi-optimistic is, however, not statistically significant even though some improvement of Semi-optimistic can be observed in the figure. This task is easier than remaining domains and Semi-optimistic approach was not able to yield significant improvement to the basic optimistic exploration.

The S-maze is two dimensional, and in this case it is more difficult to balance exploration/exploitation tradeoff than in RW. Results reported in Figure 3 show that extensions to basic optimistic exploration improve convergence of the learning algorithm. IOE is better than Semi-optimistic with statistical significance between episodes 120 and 4.5×10^3. Semi-optimistic is in turn better than Optimistic after approximately 600 episodes. The advantage of simple modifications to the algorithm which we propose in this paper are evident in this experiment.

In the next experiment the Mountain Car (MC) domain is evaluated. Two versions are reported. Results for discretisaiton 11×21 are in Figure 4, and discretisation 100×100 in Figure 5. In the first case the state space is significantly smaller, and the Optimistic version performs relatively well compared to

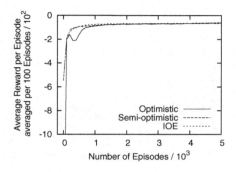 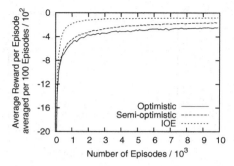

Fig. 4. Results for the Mountain Car task with discretisation 11×21

Fig. 5. Results for the Mountain Car task with discretisation 100×100

other algorithms (approximately between episodes 350 and 500 Semi-optimistic is better than Optimistic with statistical significance). There is no statistically significant difference between Semi-optimistic and IOE on this smaller state space. The influence of different learning strategies is however more evident on the bigger state space (see Figure 5). After 500 episodes the difference between IOE and Semi-optimistic becomes statistically significant. There is no statistical significance between Semi-optimistic and Optimistic. This result shows that IOE is a successful algorithm and can yield better improvement than straightforward reinitialisation implemented in Semi-optimistic.

The main contribution of this paper is the improvement of standard optimistic exploration. In results discussed in previous paragraphs it was shown that the proposed extension can yield significant improvement to this learning strategy. The aim of the next study was to check our extensions to optimistic exploration in situations when basic optimistic exploration may encounter problems. For this experiment the last domain, Maze, was used. By placing the goal state in different locations we want to change the ratio between the likely (the environment is stochastic) distance to the goal and the overall number of states. When the goal state is placed closer to the start state, the agent can focus its learning on the limited area of the state space. This poses problems to optimistic exploration since it leads to exhaustive exploration of the state space, and many states are unnecessary visited many times [15]. Following this way of reasoning, in Figure 6 results with the goal state in the most distant position, (99, 99), are reported. In the second version, the goal state is placed in position (49,49) (the centre of the state space). Here, there are more opportunities for unnecessary exploration behind the goal state. Results of this run are in Figure 7. Illuminating conclusions can be drawn from the comparison of Figures 6 and 7. To compare these two runs we compute the ratio of the episode number when Semi-optimistic reaches the horizontal line in the graph and the corresponding value for Optimistic. For results in Figure 6 this ratio for Semi-optimistic/Optimistc is 0.21. For results in Figure 7 this value is 0.13. When comparing these two ratios, it can be observed that Semi-optimistic yielded better improvement when the goal

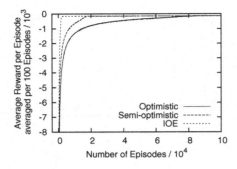
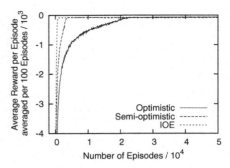

Fig. 6. Results for the 100 × 100 maze with the goal state at position 99 × 99

Fig. 7. Results for the 100 × 100 maze with the goal state at position 49 × 49

state was in position (49,49). It shows that Semi-optimistic can better deal with the growing significance of the domain property studied in this paragraph. IOE has approximately the same low ratio in both cases.

In the IOE algorithm, there is an explicit border between low and high values, and the algorithm was not sensitive to the ratio I/R in our experiments. The Semi-optimistic algorithm propagates both types of values using standard updates, and this ratio is of higher importance. In our additional experiments to check sensitivity of Semi-optimistic to the value of I, it was observed that values with similar order of magnitude, as those mentioned in Section 5, lead to the same results. Differences are observed when the order of magnitude of I is significantly higher (e.g., $I = 10^6$). In this case the first phase of learning is similar as with low values of I and Semi-optimistic is better than Optimistic. The difference is in the latter periods of learning, when the learning curve shows unstable performance.

7   Conclusion

There is an interest in optimism under uncertainty in both model-free [5] and model-based [6,7] RL. In this paper methods to improve optimistic exploration in the model-free SARSA algorithm are studied. The first contribution of this paper is an explicit analysis of the impact of different initialisation of the goal state, or equivalently modification of the final reward (in domains with a non-zero cost of actions) on the optimistic exploration. The improvement to standard optimistic initialisation was shown. According to our best knowledge this problem has never been explicitly studied with this regard. The next contribution constitutes the IOE algorithm in which there is an explicit border between high and low values. Additionally, both extensions are shown to yield an improvement when the optimistic agent learns on the limited area of the state space. Anytime properties of our solutions are also very good, since the reasonable performance is achieved very early. This is the main problem of basic optimistic initialisation. When high values reach the goal state in Semi-optimistic or IOE, the policy has already good anytime properties (according to our empirical results even very close to optimal) and the learning process can be stopped if a reasonable solution is required quickly.

Our results are based on the tabular representation of the state space with a distinct value for each state-action pair. The idea of different initialisation of the value function of goal states can be easily implemented with function approximation with local basis functions (e.g., tile coding [16]). This would be an interesting topic for future research.

Acknowledgment

This research was sponsored by the United Kingdom Ministry of Defence Research Programme.

References

1. Bertsekas, D.P., Tsitsiklis, J.N.: Neuro-Dynamic Programming. Athena Scientific (1996)
2. Sutton, R.S., Barto, A.G.: Reinforcement Learning: An Introduction. MIT Press, Cambridge (1998)
3. Kaelbling, L.P., Littman, M.L., Moore, A.P.: Reinforcement learning: A survey. Journal of Artificial Intelligence Research 4, 237–285 (1996)
4. Thrun, S.: Efficient exploration in reinforcement learning. Technical Report CMU-CS-92-102, Carnegie Mellon University, Computer Science Department (1992)
5. Meuleau, N., Bourgine, P.: Exploration of multi-state environments: Local measures and back-propagation of uncertainty. Machine Learning 35(2), 117–154 (1999)
6. Brafman, R.I., Tennenholtz, M.: R-max - a general polynomial time algorithm for near-optimal reinforcement learning. Journal of Machine Learning Research (2002)
7. Szita, I., Lőrincz, A.: The many faces of optimism: a unifying approach. In: Proceedings of the International Conference on Machine Learning (2008)
8. Anderson, M.L., Oates, T.: A review of recent research on metareasoning and metalearning. AI Magazine 28, 12–16 (2007)
9. Cohen, P.R.: Empirical methods for artificial intelligence. MIT Press, Cambridge (1995)
10. Sutton, R.: Learning to predict by the methods of temporal differences. Machine Learning 3, 9–44 (1988)
11. Singh, S., Dayan, P.: Analytical mean squared error curves for temporal difference learning. Machine Learning 32, 5–40 (1998)
12. Sutton, R.S.: Integrated architectures for learning, planning, and reacting based on approximating dynamic programming. In: Proceedings of the 7th International Conference on Machine Learning, pp. 216–224 (1990)
13. Epshteyn, A., DeJong, G.: Qualitative reinforcement learning. In: Proceedings of the 23rd International Conference on Machine Learning, pp. 305–312 (2006)
14. Wiering, M., Schmidhuber, J.: Efficient model-based exploration. In: Proceedings of the 5th international conference on simulation of adaptive behavior: From animals to animats, pp. 223–228 (1998)
15. Russell, S.J., Norvig, P.: Artificial Intelligence: A Modern Approach, 2nd edn. Prentice Hall, Englewood Cliffs (2002)
16. Lin, C.S., Kim, H.: Cmac-based adaptive critic self-learning control. IEEE Transactions on Neural Networks 2, 530–533 (1991)

Improving Visualization, Scalability and Performance of Multiclass Problems with SVM Manifold Learning

Catarina Silva[1,2] and Bernardete Ribeiro[2]

[1] School of Technology and Management, Polytechnic Institute of Leiria, Portugal
[2] Dep. Informatics Eng., Center Informatics and Systems, Univ. of Coimbra, Portugal
catarina@dei.uc.pt, bribeiro@dei.uc.pt

Abstract. We propose a learning framework to address multiclass challenges, namely visualization, scalability and performance. We focus on supervised problems by presenting an approach that uses prior information about training labels, manifold learning and support vector machines (SVMs).

We employ manifold learning as a feature reduction step, nonlinearly embedding data in a low dimensional space using Isomap (Isometric Mapping), enhancing geometric characteristics and preserving the geodesic distance within the manifold. Structured SVMs are used in a multiclass setting with benefits for final multiclass classification in this reduced space. Results on a text classification toy example and on ISOLET, an isolated letter speech recognition problem, demonstrate the remarkable visualization capabilities of the method for multiclass problems in the severely reduced space, whilst improving SVMs baseline performance.

1 Introduction

Multiclass learning is the problem of assigning labels to instances where the labels are drawn from a finite set of elements and is being increasingly required by modern applications, such as text classification, protein function classification, speech recognition, music categorization and semantic scene classification. The most common approach to such problems is to build upon classification learning algorithms for binary problems, i.e. problems in which the set of possible labels is of size two. Among these algorithms, support vector machines (SVMs) are accepted as one of the best performing methods in many domains [1,2]. When applied to multiclass classification, SVMs are mostly used in their binary version, by reducing a single multiclass problems into multiple binary problems. For instance, a common method is to build a set of binary classifiers where each classifier distinguishes between one of the labels to the rest [3].

The alternative explored in this work is to make use of structured SVMs [4] and cast them to solve multiclass classification problems. The rationale is that having a tool that handles structured outputs, such as graphs or trees, it is possible to build a multiclass classifier [5].

In multiclass classification the challenges are numerous. Feature selection and dimensionality reduction methods must take into account the relevance of features not only to a particular class, as in the binary setting, but to their impact

M. Kolehmainen et al. (Eds.): ICANNGA 2009, LNCS 5495, pp. 370–379, 2009.
© Springer-Verlag Berlin Heidelberg 2009

on all classes. Initial feature selection and dimensionality reduction are usually carried out in the feature space as a pre-processing step. Several supervised and unsupervised techniques can be applied. Manifold learning strategies, like Isomap (Isometric Mapping) [6], are effective for extracting nonlinear structures from high-dimensional data in pattern recognition [7]. Finding the structure behind the data may be important for a number of reasons, such as data visualization and performance improvement. Graphical depiction of the training and testing sets can potentially be crucial in multiclass applications, since it makes possible to quickly give large amounts of information to a human operator [8]. To this purpose it is appropriately assumed that the data lies on a statistical manifold, or a manifold of probabilistic generative models [9]. It can be regarded as a supervised learning method, where the training labels play a central role. In such a scenario, manifold learning can be used not only with the traditionally associated algorithms, such as K-Nearest Neighbors (K-NN), but also with state-of-the-art kernel-based machines like support vector machines (SVMs) [1].

In this contribution we extend previous work by the authors [10], generalizing its application to multiclass problems. Specifically, we propose the use of manifold learning, with a Isomap based nonlinear algorithm that uses training label information in the dimensionality reduction step, combined with structured multiclass SVMs based on structured SVMs.

The rest of the paper is organized as follows. In the next section, we set the foundations and background for multiclass problems and for the multiclass support vector machines (SVMs) approach. In Section 3, we introduce manifold learning as a supervised dimensionality reduction method. In Section 4, we introduce our approach for the use of manifold learning in multiclass problems, with an Isomap-based nonlinear dimensionality reduction algorithm combined with multiclass SVMs. Experiments and results are described and analyzed in Section 5. Finally, Section 6 addresses conclusions and future work.

2   SVM Multiclass Classification

SVMs are inherently two-class classifiers. The most common technique to implement SVM multiclass classification with $|\mathcal{C}|$ classes in practice has been to build $|\mathcal{C}|$ *one-versus-rest* classifiers (commonly referred to as *one-versus-all*), and to choose the class that classifies the test datum with greatest margin. Another strategy is to build a set of *one-versus-one* classifiers, and to choose the class that is selected by the most classifiers. Although this involves building $|\mathcal{C}|(|\mathcal{C}| - 1)/2$ classifiers, the time for training classifiers may actually decrease, because the training data set for each classifier is much smaller.

However, these are not very elegant approaches to solving multiclass problems. A better alternative is provided by the construction of multiclass SVMs, where we build a two-class classifier over a feature vector $\Phi(\mathbf{x}, y)$, derived from the pair consisting of the input features ($\mathbf{x}$) and the class of the datum (y). At test time, the classifier chooses the class

$$y = \arg \max_{y'} \mathbf{w}^T \Phi(\mathbf{x}, y). \tag{1}$$

where $\mathbf{w}$ represents the set of weights that defines the learning machine. The margin during training is the gap between this value for the correct class and for the nearest other class, and so the quadratic program formulation will require that

$$\forall_i \forall_{y \neq y_i} \mathbf{w} \Phi(\mathbf{x}_i, y_i) - \mathbf{w} \Phi(\mathbf{x}_i, y) \geq 1 - \xi_i, \qquad (2)$$

This general method can be extended to give a multiclass formulation of various kinds of linear classifiers. It is also a simple instance of a generalization of classification where the classes are not just a set of independent, categorical labels, but may be arbitrary structured objects with relationships defined between them, usually referred to as structured SVMs.

The algorithm used in this contribution, described in [4], is based on Structured SVMs [5]. It can implement the conventional winner-takes-all (WTA) multiclass classification described in [3]. It learns mappings involving complex structures in polynomial time. A possible application, pertinent to our work is multiclass classification. The multiclass task is tackled by generalizing large margin methods to the broader problem of learning structured responses. The naive approach of treating each structure as a separate class is often intractable, since it leads to a multiclass problem with a very large number of classes. This problem is surpassed specifying discriminant functions that exploit the structure and dependencies within the set of classes $\mathcal{C}$. SVM multiclass uses an algorithm that is different from the one in [3]. It follows the work of Collins [11] on perceptron learning with a similar class of discriminant functions.

Let $\mathcal{C} = \{y_1, \ldots, y_K\}$ be the set of classes and $\mathbf{w} = (\mathbf{v}_1', \ldots, \mathbf{v}_k')'$ be a stack of vectors, where v_k is a weight vector associated with the kth class y_k. Following Crammer and Singer [3], one can then define $F(\mathbf{x}, y_k; \mathbf{w}) = <\mathbf{v}_k, \Phi(\mathbf{x})>$, where $\Phi(\mathbf{x})$ denotes an arbitrary input representation. These discriminant functions can be equivalently represented by defining a joint feature map as follows $\Psi(\mathbf{x}, \mathbf{y}) \equiv \Phi(\mathbf{x}) \otimes \Lambda^c(\mathbf{y})$. Here Λ^c refers to the orthogonal (binary) encoding of the label y and $\otimes$ is the tensor product which forms all products between coefficients of the two argument vectors.

3   Manifold Learning

Many approaches have been proposed for dimensionality reduction, such as the well-known methods of principal component analysis (PCA) [12], independent component analysis (ICA) [13] and multidimensional scaling (MDS) [14]. All these methods are well understood and efficient and have thus been widely used in visualization and classification. Unfortunately, they share a common inherent limitation: they are all linear methods while the distributions of most real-world multiclass data problems are nonlinear.

An emerging nonlinear dimensionality reduction technique is manifold learning, which is the process of estimating a low-dimensional structure which underlies a collection of high-dimensional data. Manifold learning can be viewed as implicitly inverting a generative model for a given set of observations [15]. Let Y be a d dimensional domain contained in a Euclidean space $\mathbb{R}^d$. Let $f : Y \rightarrow \mathbb{R}^D$ be a smooth embedding for some $D > d$. The goal of manifold learning is to recover Y and f given N points in $\mathbb{R}^D$. Isomap [6] provides an implicit description of the mapping f (or f^{-1}). Given $X = \{\mathbf{x}_i \in \mathbb{R}^D | i = 1 \ldots N\}$ find

$Y = \{\mathbf{y}_i \in \mathbb{R}^d | i = 1 \dots N\}$ such that $\{\mathbf{x}_i = f(\mathbf{y}_i) | i = 1 \dots N\}$. Without imposing any restrictions of f, the problem is ill-posed. The simplest case is a linear isometry, i.e. f is a linear mapping from $\mathbb{R}^d \to \mathbb{R}^D$, where $D > d$.

In Isomap [6] the local neighborhood of each example is preserved, while trying to obtain highly nonlinear embeddings with manifold learning. For data lying on a nonlinear manifold, the *true distance* between two data points is the geodesic distance on the manifold, i.e. the distance along the surface of the manifold, rather than the straight-line Euclidean distance. The main purpose of Isomap is to find the intrinsic geometry of the data, as captured in the geodesic manifold distances between all pairs of data points. The approximation of geodesic distance is divided into two cases. In the case of neighboring points, Euclidean distance in the input space provides a good approximation to geodesic distance. In the case of faraway points, geodesic distance can be approximated by adding up a sequence of *short hops* between neighboring points. Isomap shares some advantages with PCA and MDS, such as computational efficiency and asymptotic convergence guarantees, but with more flexibility to learn a broad class of nonlinear manifolds. The Isomap algorithm takes as input the distances $d(\mathbf{x}_i, \mathbf{x}_j)$ between all pairs $\mathbf{x}_i$ and $\mathbf{x}_j$ from N data points in the high-dimensional input space. The algorithm outputs coordinate vectors $\mathbf{y}_i$ in a d-dimensional Euclidean space that best represent the intrinsic geometry of the data. Isomap is accomplished following these steps:

Step 1. Construct neighborhood graph: Define the graph G over all data points by connecting points $\mathbf{x}_i$ and $\mathbf{x}_j$ if they are closer than a certain distance ε, or if $\mathbf{x}_i$ is one of the K nearest neighbors of $\mathbf{x}_j$. Set edge lengths equal to $d(\mathbf{x}_i, \mathbf{x}_j)$.

Step 2. Compute shortest paths: Initialize $d_G(\mathbf{x}_i, \mathbf{x}_j) = d(\mathbf{x}_i, \mathbf{x}_j)$ if $\mathbf{x}_i$ and $\mathbf{x}_j$ are linked by an edge; $d_G(\mathbf{x}_i, \mathbf{x}_j) = +\infty$ otherwise. Then for each value of $k = 1, 2, \dots, N$ in turn, replace all entries $d_G(\mathbf{x}_i, \mathbf{x}_j)$ by $min\{d_G(\mathbf{x}_i, \mathbf{x}_j), d_G(\mathbf{x}_i, \mathbf{x}_k) + d_G(\mathbf{x}_k, \mathbf{x}_j)\}$. The matrix of final values $\mathbf{D}_G = \{d_G(\mathbf{x}_i, \mathbf{x}_j)\}$ will contain the shortest path distances between all pairs of points in G.

Step 3. Apply MDS to the resulting geodesic distance matrix to find a d-dimensional embedding.

This is an unsupervised procedure and constitutes a preprocessing step for classification. Basically it performs a transformation from a high dimensional input data space into a lower dimensional feature space. Then a classifier, for instance, K-NN can be applied to the resulting data. However, the mapping function given by Isomap is only implicitly defined. Therefore, it should be learned by nonlinear interpolation techniques, such as generalized regression neural networks, which can then transform the new test data into the reduced feature space before prediction.

3.1   Supervised Isomap

In the supervised version of Isomap [16], the information provided by the training class labels is used to guide the procedure of dimensionality reduction. The training labels are used to refine the distances between inputs. The rationale is that both classification and visualization can benefit when the inter-class dissimilarity is larger than the intra-class dissimilarity. However, this can also make the algorithm overfit the training set and can often make the neighborhood graph of the input data disconnected. To achieve this purpose, the Euclidean distance $d(\mathbf{x}_i, \mathbf{x}_j)$ between two given observations $\mathbf{x}_i$ and $\mathbf{x}_j$, labeled y_i and y_j respectively, is replaced by a dissimilarity measure [16]:

$$
D(\mathbf{x}_i, \mathbf{x}_j) = \begin{cases} \sqrt{1 - e^{\frac{-d^2(\mathbf{x}_i, \mathbf{x}_j)}{\gamma}}} & y_i = y_j, \\ \sqrt{e^{\frac{-d^2(\mathbf{x}_i, \mathbf{x}_j)}{\gamma}} - \alpha} & y_i \neq y_j. \end{cases} \tag{3}
$$

Note that the Euclidean distance $d(\mathbf{x}_i, \mathbf{x}_j)$ is in the exponent and the parameter γ is used to avoid that $D(\mathbf{x}_i, \mathbf{x}_j)$ increases too rapidly when $d(\mathbf{x}_i, \mathbf{x}_j)$ is relatively large. Hence, γ depends on the *density* of the data set and is usually set to the average Euclidean distance between all pairs of data points. On the other hand, α gives a certain possibility to points in different classes to be *closer*, i.e. to be more similar, than those in the same class. This procedure allows a better determination of the relevant features and will definitely improve visualization.

4   Proposed Approach

In this section, we propose a learning framework to address multiclass problems. We propose the combination of manifold learning as a feature reduction step, that increasing scalability, also promotes visualization and performance potentialities.

We start by using manifold learning to construct a reduced representation of the input space. As detailed in Section 3, we use a nonlinear embedding of data in a low dimensional space constructed with the supervised version of Isomap (Isometric Mapping) [16], enhancing geometric characteristics and preserving the geodesic distance within the manifold. Therefore, we use the multiclass training labels in the datasets to provide a better construction of features. We further apply the dissimilarity measure (3) to enhance the baseline Isomap Euclidean distance using label information, with α taking the value of 0.65 and γ the average Euclidean distance between all pairs of training data points.

When a reduced space is reached, our aim is to learn a linear-kernel structured multicass SVM [5] that can be applied in unseen examples. For testing, however, Isomap does not provide an explicit mapping of documents. Therefore we can not generate the test set directly, since we would need to use the labels. Hence, we use a generalized regression neural network (GRNN) [17] with a 0.95 spread to learn the mapping and apply it to each test document, before the SVM prediction phase, as can be gleaned from Figure 1 that summarizes the proposed approach. In the training phase the supervised Isomap procedure, that runs on features and label training instances, is captured by the GRNN using only the features. Furthermore, the reduced featured space ($\mathbb{R}^d$) is the place for the SVM multiclass modeling. When a new testing instance is to be classified, the GRNN maps it from $\mathbb{R}^D$ to $\mathbb{R}^d$ and the SVM multiclass linear-kernel model predicts the class.

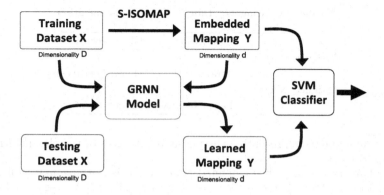

Fig. 1. Proposed approach: SIsomap+SVM

5   Experimental Setup

This section presents the conducted experiments and obtained results. First datasets and performance criteria are defined, then experiments and results are presented and analyzed.

5.1   Datasets

We have used two different multiclass datasets: a text classification toy example[1] and ISOLET, an isolated letter speech recognition problem form UCI[2].

The toy example consists of 7 classes and 2300 examples, divided in 300 training examples and 2,000 testing examples.

ISOLET task is to predict which letter-name was spoken, resulting in a 26-class problem. To generate this dataset, 150 subjects spoke the name of each letter of the alphabet twice. Hence, there are 52 training examples from each speaker. The 617 features are described in [18] and include spectral coefficients, contour features, sonorant features, pre-sonorant features, and post-sonorant features. The examples are split into 6,238 training examples and 1,559 testing examples, both with balanced class cardinality.

5.2   Performance Metrics

In multiclass problems the common performance metric is the global error, given by the percentage of wrongly classified testing instances, regardless of the incorrectly classified category or magnitude of the error.

However, the independent performance of each class is also very important. Therefore, in addition to the global error measure, we also present the error rate and F_1 performances per class. To evaluate each class performance, we first define

---

[1] http://www.cs.cornell.edu/People/tj/svm_light/svm_multiclass.html
[2] http://archive.ics.uci.edu/ml/datasets/ISOLET

Table 1. Contingency table for binary classification

	Class Positive	Class Negative
Assigned Positive	a	b
	(True Positives)	(False Positives)
Assigned Negative	c	d
	(False Negatives)	(True Negatives)

a contingency matrix representing the possible outcomes of the classification, as shown in Table 1.

Several measures have been defined based on this contingency table, such as, error rate ($\frac{b+c}{a+b+c+d}$), recall ($R = \frac{a}{a+b}$), and precision ($P = \frac{a}{a+c}$), as well as combined measures, such as, the van Rijsbergen F_β measure [19], which combines recall and precision in a single score, $F_\beta = \frac{(\beta^2+1)P \times R}{\beta^2 P + R}$. The F1 measure was chosen since it permits the identification of misclassifications even when a class has few positive examples, detecting deceiving low error rates situations.

5.3   Results and Analysis

Table 2 presents the comparison of global error measures between the baseline multiclass SVM and the proposed approach for both datasets. The feature reduction was from 47 to 10 features for Toy dataset and from 617 to 200 for ISOLET dataset. In the case of ISOLET, to speed the training procedure, of the 6,238 training examples, only 2,500 were used, maintaining balanced class cardinality.

The overall trend is that the proposed manifold multiclass SVM approach surpasses the baseline setting by around 10% and 7% for the Toy dataset and ISOLET dataset respectively (see Table 2).

Figure 2 represents the error rate and F_1 measures for each individual class of the two datasets. The error rates are seamlessly low, while the F_1 performances are more diverse. Nevertheless, the averaged values for error rates for the proposed approach improve the baseline measures: from 13.90% to 10.67% for the Toy dataset and from 1.51% to 0.96% for the ISOLET dataset. Regarding F_1 performance the values vary between the different classes, but the tendency is similar, i.e. the averaged performance values also present an improvement when using the proposed approach: from 11.49% to 34.46% for the Toy dataset and from 79.15% to 87.44% for the ISOLET dataset.

The most impressive result is achieved in visualization properties of the proposed method. As can be gleaned from Figs. 3 and 4, in the initial representation the first ten classes of ISOLET (letters a to j) are not distinguishable, while in

Table 2. Global multiclass error

	SVM	Proposed approach
Toy	48.65%	38.55%
ISOLET	19.63%	12.44%

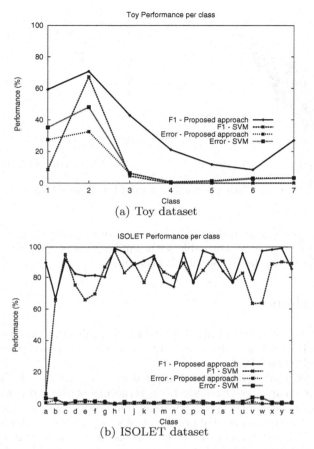

(a) Toy dataset

(b) ISOLET dataset

Fig. 2. Performance measures per class for: (a) Toy dataset; (b) ISOLET dataset

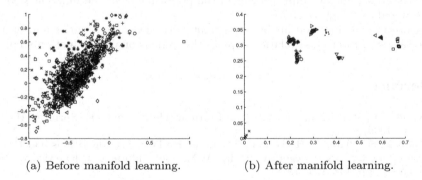

(a) Before manifold learning.   (b) After manifold learning.

Fig. 3. Training examples of ISOLET 10 first letters

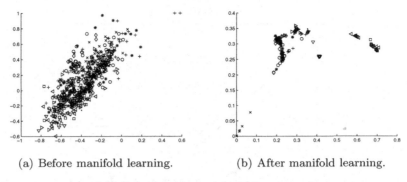

(a) Before manifold learning.   (b) After manifold learning.

Fig. 4. Testing examples of ISOLET 10 first letters

the manifold representation the ten letters are almost perfectly distinct. Illustration of the 26 classes was attainable with similar results, but the outcome was not graphically clear due to plotting limitations.

6   Conclusions

In this paper we proposed a framework to tackle multiclass problems. We use a combination of a nonlinear dimensionality reduction preprocessing method and structured multiclass SVMs.

We concluded that manifold learning, namely the supervised ISOMAP technique, efficiently captures the underlying structure of the data, preserving the distances among data points in the original dimensional space. One of the main achievements was the impressive graphical class separation that was possible in the manifold. This result can prove to be very useful to transmit information and confidence to a human user. Moreover, the use of structured multiclass SVMs permitted a significant improvement in the performance of the final classifier in the new reduced feature space.

Future work is foreseen in the refinement of the learning abilities and on the exploitation of inter-class relationships in the dimensionality reduction step.

References

1. Vapnik, V.: The Nature of Statistical Learning Theory, 2nd edn. Springer, Heidelberg (1999)
2. Dumais, S., Platt, J., Heckerman, D.: Inductive Learning Algorithms and Representations for Text categorisation. In: ACM Conf. Information Knowledge Management, pp. 148–155 (1998)
3. Crammer, K., Singer, Y.: On the Algorithmic Implementation of Multi-class kernel-based vector machines. Journal Machine Learning Research 2, 265–292 (2002)
4. Tsochantaridis, I., Hofmann, T., Joachims, T., Altun, Y.: Support vector machine learning for interdependent and structured output spaces. In: Int. Conf. Machine Learning, pp. 104–111 (2004)

5. Tsochantaridis, I., Hofmann, T., Joachims, T., Altun, Y.: Large Margin Methods for Structured and Interdependent Output Variables. Journal Machine Learning Research 6, 1453–1484 (2005)
6. Tenenbaum, J.B., de Silva, V., Langford, J.C.: A global geometric framework for nonlinear dimensionality reduction. Science 5500(290), 2319–2323 (2000)
7. Kim, H., Park, H., Zha, H.: Distance Preserving Dimension Reduction for Manifold Learning. In: Int. Conf. Data Mining, vol. II, pp. 1147–1151 (2007)
8. Navarro, D., Lee, M.D.: Spatial Visualization of Document Similarity, Defence Human Factors. Special Interest Group Meeting (2001)
9. Zhang, D., Chen, X., Lee, W.: Text Classification with Kernels on the Multinomial Manifold. In: ACM SIGIR - Special Interest Group on Information Retrieval, pp. 266–273 (2005)
10. Silva, C., Ribeiro, B.: Text Classification on Embedded Manifolds. In: Geffner, H., Prada, R., Machado Alexandre, I., David, N. (eds.) IBERAMIA 2008. LNCS (LNAI), vol. 5290, pp. 272–281. Springer, Heidelberg (2008)
11. Collins, M.: Parameter estimation for statistical parsing models: Theory and practice of distribution-free methods. In: IWPT - International Workshop on Parsing Technologies (2001)
12. Jolliffe, I.T.: Principal Component Analysis. Springer, Heidelberg (1986)
13. Comon, P.: Independent Component Analysis: a New Concept? Signal Processing 36(3), 287–314 (1994)
14. Cox, T., Cox, M.: Multidimensional Scaling. Chapman & Hall, London (1994)
15. Duraiswami, R., Raykar, V.C.: The Manifolds of Spatial Hearing. In: ICASSP 2005, vol. III, pp. 285–288 (2005)
16. Geng, X., Zhan, D., Zhou, Z.: Supervised Nonlinear Dimensionality Reduction for Visualization and Classification. IEEE Transactions Systems, Man, and Cybernetics – Part B 35(6), 1098–1107 (2005)
17. Specht, D.: A General Regression Neural Network. IEEE Transactions on Neural Networks 2(6), 568–576 (1991)
18. Fanty, M., Cole, R.: Spoken letter recognition. In: Advances in Neural Information Processing Systems, vol. 3 (1991)
19. van Rijsbergen, C.: Information Retrieval. Butterworths Ed. (1979)

A Cat-Like Robot Real-Time Learning to Run

Paweł Wawrzyński

Warsaw University of Technology, Poland

Abstract. Actor-Critics constitute an important class of reinforcement learning algorithms that can deal with continuous actions and states in an easy and natural way. In their original, sequential form, these algorithms are usually to slow to be applicable to real-life problems. However, they can be augmented by the technique of experience replay to obtain a satisfactory of learning without degrading their convergence properties. In this paper experimental results are presented that show that the combination of experience replay and Actor-Critics yields very fast learning algorithms that achieve successful policies for nontrivial control tasks in considerably short time. Namely, a policy for a model of 6-degree-of-freedom walking robot is obtained after 4 hours of the robot's time.

Keywords: reinforcement learining, actor-critics, experience replay, neural networks.

1 Introduction

Reinforcement learning (RL) addresses the problem of an agent that optimizes its reactive policy in a poorly structured and initially unknown environment [9]. Algorithms developed in this area can be viewed as computational processes that transform observations of states, actions and rewards into policy parameters. Several important RL algorithms, such as Q-Learning [10] and Actor-Critic methods [2,5,6,3], process the data sequentially. Each single observation is used for adjusting the algorithms' parameters and then becomes unavailable for further use. We shall call such methods *sequential*. They are based on a common assumption that RL applications to real-world learning control problems require large amounts of data which cannot be kept in a limited amount of memory assigned to the algorithm.

Sequential algorithms do not exploit all the information contained in the data and are known to require a large number of environment steps to obtain a satisfactory. Usually, this number is large enough to make the learning process detrimental to any real device whose control policy we would hope to optimize by means of RL. However, there are other, *non-sequential* methods that require much fewer environment steps to obtain a policy of the same quality. They achieve this at the cost of collecting data and some extensive processing thereof. This distinction between sequential and non-sequential algorithms should not be confused with the distinction between online and offline algorithms. Here, we are interested *only* in online algorithms which improve the policy as the

M. Kolehmainen et al. (Eds.): ICANNGA 2009, LNCS 5495, pp. 380–390, 2009.
© Springer-Verlag Berlin Heidelberg 2009

agent-environment interaction proceeds. Online sequential and non-sequential algorithms differ in the way of using the available computational power during the interaction.

One of the approaches to design of non-sequential algorithms consists in adding some non-sequential data processing to a sequential algorithm. The modified algorithm collects historical experiences (observations of states, actions and rewards) and applies to them the operations of the original sequential algorithm as if they have just taken place. This idea of repeating many similar operations to the same event, called *experience replay* [8,7,4], was popular a few years ago but has received little attention recently. Unfortunately, experience replay is not automatically applicable to an arbitrary RL algorithm. In particular, it cannot be directly combined with on-policy methods, i.e., those based on the assumption that the actions producing data for policy improvements are drawn from the current policy. Consequently, the same experience cannot be applied many times to adjust a continuously changing policy.

In this paper we present an experimental study on experience replay and an Actor-Critic-type learning algorithm combined in a fashion introduced in [12]. In the study we obtain a policy of an emulated cat-like robot called Half-Cheetah [11]. The robot is a kinematic string with 6-degrees of freedom. Its state space is 31 dimensional. The learning goal is to make Half-Cheetah run as fast as possible. The objective is obtained within 4 hours of Half-Cheetah time.

The paper is organized as follows. In Sec. 2 the problem of our interest is defined along with the class of algorithms that encompasses sequential Actor-Critics. Section 3 shows how to estimate improvement directions in the policy parameters' space using the data from the preceding state transition and to accelerate a sequential algorithm by combining these estimators with experience replay. The experimental study is presented in Sec. 4.

2   Problem Formulation

We will consider the standard RL setup [9]. A Markov Decision Process (MDP) defines a problem of an agent that observes its state s_t in discrete time $t = 1, 2, 3, \ldots$, performs actions a_t, receives rewards r_t and moves to other states s_{t+1}. A particular MDP is a tuple $\langle \mathcal{S}, \mathcal{A}, P_s, r \rangle$ where $\mathcal{S}$ and $\mathcal{A}$ are the state and action spaces, respectively; $\{P_s(\cdot|s, a) : s \in \mathcal{S}, a \in \mathcal{A}\}$ is a set of state transition distributions; we write $s_{t+1} \sim P_s(\cdot|s_t, a_t)$ and assume that each P_s is a density. Each state transition generates a *reward*, $r_t \in \mathfrak{R}$. Here we assume that each reward is depends deterministically on the current action and the next state, $r_t = r(a_t, s_{t+1})$.

Actions are generated according to a policy, π, which is a family of distributions parameterized by the state and a *policy vector* $\theta \in \mathfrak{R}^{n_\theta}$, namely $a_t \sim \pi(\cdot\,; s_t, \theta)$. The objective of reinforcement learning is to optimize θ to make the policy maximize future rewards. This goal may be strictly specified in various ways. We may require the policy to maximize the average reward or to maximize the sum of future discounted rewards expected in each state.

Below we analyze a large class of the existing sequential RL methods suitable for policy determination in simulations and propose a way of their acceleration based on a more extensive data processing. Our intention is to design methods that obtain a satisfactory after a much smaller amount of agent time but not necessarily after a smaller amount of computation. The control efficiency should be maximized within short period of learning to keep the controlled machine from being damaged by too many wrong actions.

Sequential Actor-Critics. Actor-Critics [2,5,6,3] constitute probably the most efficient and the most theoretically developed class of reinforcement learning algorithms. Let us analyze an example of methods that will be of interest for us: the algorithm presented in [5], quoted in the table beside, and called here the Basic Actor-Critic. The actor is represented here, as usually, by the parameterized policy π. The critic is represented by the approximator $\bar{V}(s; v)$ parameterized by the critic vector $v \in \mathfrak{R}^{n_v}$. The critic approximates the value function V^π, that is for a given policy, π, equal to the sum of future discounted rewards expected in a given state, namely

---

Algorithm 1. The Basic Actor-Critic. $\gamma \in (0, 1)$ is a discount factor, $\lambda \in [0, 1]$, $\bar{V}$ is the value function approximator (the critic) parameterized by vector v. β_t^θ and β_t^v are the step-sizes.

0: Set $y_v = 0, y_\theta = 0, t := 1$. Initialize θ and v.
1: Draw the action, $a_t \sim \pi(\cdot; s_t, \theta)$.
2: Execute a_t, evaluate the next state s_{t+1} and the reward r_t.
3: Calculate the *temporal difference* of the form
4:    $d_t(v) = r_t + \gamma \bar{V}(s_{t+1}; v) - \bar{V}(s_t; v)$.
5: Adjust the actor:
6:    $y_\theta := (\gamma\lambda)y_\theta + \beta_t^\theta \nabla_\theta \ln \pi(a_t; s_t, \theta)$
7:    $\theta := \theta + y_\theta d_t(v)$
8: Adjust the critic:
9:    $y_v := (\gamma\lambda)y_v + \beta_t^v \nabla_v \bar{V}(s_t; v)$
10:    $v := v + y_v d_t(v)$.
11: Set $t := t + 1$ and repeat from Point 1.

$V^\pi(s) = E\left(\sum_{i \geq 0} \gamma^i r_{t+i} \big| s_t = s, \pi\right)$, where $\gamma \in (0, 1)$ is a discount factor. The values β_t^θ and β_t^v are the step-sizes: they are positive reals decreasing with growing t. Also, they should satisfy the standard stochastic approximation conditions: $\sum_{t \geq 1} \beta_t = \infty, \sum_{t \geq 1} \beta_t^2 < \infty$.

Below, we provide a simplistic, appealing to intuition, analysis of this algorithm. We show it as working in the following way: It increases the probability of a given action a_t if it turns out to lead to higher rewards than expected in state s_t. If the action turns out to lead to smaller rewards, its probability is decreased. Namely, let us consider the total adjustment of the policy vector θ during the work of the algorithm. To this end, we analyze the value of y_θ by the end of the algorithm's loop. It can be seen that y_θ is then equal to

$$y_\theta = \sum_{k=0}^{t-1} (\gamma\lambda)^k \beta_{t-k}^\theta \nabla_\theta \ln \pi(a_{t-k}; s_{t-k}, \theta).$$

Therefore, the total adjustment is equal to

$$\Delta\theta = \sum_{t>0} d_t(v) \sum_{k=0}^{t-1} (\gamma\lambda)^k \beta_{t-k}^\theta \nabla_\theta \ln \pi(a_{t-k}; s_{t-k}, \theta)$$

By changing the summation order we obtain

$$\Delta\theta = \sum_{t>0} \beta_t^\theta \nabla_\theta \ln \pi(a_t; s_t, \theta) \sum_{k\geq 0} (\gamma\lambda)^k d_{t+k}(v). \tag{1}$$

We can see that an element of the above sum attributes to the state s_t the total adjustment of the policy vector that the visit in this state induces. The adjustment is equal to the product of a vector and a sum of scalars. The vector, $\nabla_\theta \ln \pi$, defines the direction in which θ must be modified to change the probability of action a_t in state s_t. The sum, $\sum_{k\geq 0}$, determines whether the action a_t leads to higher rewards than expected in state s_t (then the sum is positive and the probability of a_t is increased) or a_t leads to smaller rewards than expected in s_t (the sum is negative and the probability of a_t is decreased).

The critic training. A similar analysis reveals the compact form of the total adjustment of the critic vector. Namely, at the end of the algorithm's loop, the vector y_v is equal to

$$y_v = \sum_{k=0}^{t-1} (\gamma\lambda)^k \nabla_\theta \ln \pi(a_{t-k}; s_{t-k}, \theta).$$

Therefore, the total adjustment of the critic vector is equal to

$$\Delta v = \sum_{t>0} \beta_t^v d_t(v) \sum_{k=0}^{t-1} (\gamma\lambda)^k \nabla_v \bar{V}(s_{t-k}; v)$$

By changing the summation order we obtain

$$\Delta v = \sum_{t>0} \beta_t^v \nabla_v \bar{V}(s_t; v) \sum_{k\geq 0} (\gamma\lambda)^k d_{t+k}(v). \tag{2}$$

We can see that an element of the above sum attributes to the state s_t the total adjustment of the critic vector that the visit in this state induces. In order to understand the character of this adjustment, one may notice that the inner sum in Eq. (2) is the same as the inner sum in Eq. (1) expressing how large future turned out to be in comparison to expected in state s_t. Hence, the critic training consists in increasing $\bar{V}(s_t; v)$ when actural rewards turned to be higher than this value, and decreasing otherwise.

Generalization. In general, we will consider sequential actor-critic-type algorithms characterized by the following features:

1. Actions are generated by a stationary policy (actor) i.e., a distribution π parameterized by state s_t and the policy vector $\theta \in \Re^{n_\theta}$: $a_t \sim \pi(\cdot\,; s_t, \theta)$.
2. A visit in state s_t causes a modification of the policy vector θ by a product $\beta_t^\theta \widehat{\phi}_t$, where $\widehat{\phi}_t$ on average indicates the direction in which θ assures larger future rewards expected in state s_t whereas $(\beta_t^\theta, t = 1, 2, \ldots)$ is a vanishing sequence of step-sizes.

3. The algorithm may compute $\widehat{\phi}_t$ with the use of an auxiliary parameters $\upsilon \in \Re^{n_\upsilon}$. A visit in state s_t results in a modification of υ by a vector $\beta_t^\upsilon \widehat{\psi}_t$, where $\widehat{\psi}_t$ on average points into the direction where υ assures better quality of $\widehat{\phi}_t$ whereas $(\beta_t^\upsilon, t = 1, 2, \dots)$ is a vanishing sequence of step-sizes.

4. The vectors $\widehat{\phi}_t$ and $\widehat{\psi}_t$, different from one another, are of the same form

$$G_t(\theta, \upsilon) \sum_{k \geq 0} (\alpha\rho)^k z_{t,k}(\theta, \upsilon) \qquad (3)$$

where G_t is a vector defined by s_t and a_t, $\alpha \in [0, 1)$, $\rho \in [0, 1)$, and $z_{t,k} \in \Re$ is defined by $s_{t+k}, a_{t+k}, r_{t+k}, s_{t+k+1}$, and possibly a_{t+k+1}.

In the Basic Actor-Critic algorithm mentioned above we have

$$\widehat{\phi}_t = \nabla_\theta \ln \pi(a_t; s_t, \theta) \sum_{k \geq 0} (\gamma\lambda)^k d_{t+k}(\upsilon), \quad \widehat{\psi}_t = \nabla_\upsilon \bar{V}(s_t; \upsilon) \sum_{k \geq 0} (\gamma\lambda)^k d_{t+k}(\upsilon),$$

which means that both $\widehat{\phi}_t$ and $\widehat{\psi}_t$ are of the form (3) with $\gamma\lambda = \alpha\rho$, and

$$G_t(\theta, \upsilon) = \nabla_\theta \ln \pi(a_t; s_t, \theta), \quad z_{t,k}(\theta, \upsilon) = d_{t+k}(\upsilon)$$

for the actor, while for the critic those are equal to

$$G_t(\theta, \upsilon) = \nabla_\upsilon \bar{V}(s_t; \upsilon), \quad z_{t,k}(\theta, \upsilon) = d_{t+k}(\upsilon).$$

Important algorithms that also fit into the discussed schema are the actor-critics presented in [6,3] and OLPOMDP [1].

Let us analyze the average direction of $\widehat{\phi}_t$ and $\widehat{\psi}_t$. Namely, let ϕ be a function defined as

$$\phi(s, \theta, \upsilon) = E_{\theta, \upsilon, \beta} \left(\widehat{\phi}_t \big| s_t = s \right). \qquad (4)$$

The definition of ϕ is based on the assumption that θ, υ, and the step-sizes remain constant when $\widehat{\phi}_t$ is calculated. In fact, they slightly vary and each $\widehat{\phi}_t$ is in fact a biased estimator of $\phi(s_t, \theta, \upsilon)$ for θ and υ used at time t. However, this bias is small and since the dynamics of the parameters decreases in time, the bias asymptotically vanishes. The average $\phi(s, \theta, \upsilon)$ weighed by the steady-state distribution defines the direction of the drift of the policy vector.

The drift of υ may be analyzed in a similar way. Namely, let ψ be a function defined as

$$\psi(s, \theta, \upsilon) = E_{\theta, \upsilon, \beta} \left(\widehat{\psi}_t \big| s_t = s \right). \qquad (5)$$

As above, the definition of ψ requires that θ, υ, and the step-sizes remain constant during the time when $\widehat{\psi}_t$ is computed. The drift of υ is defined by the average $\psi(s, \theta, \upsilon)$ weighed by the steady-state distribution. The usual role of the drift of the auxiliary parameter is to move it toward the point $\upsilon^*(\theta)$ such that the average $\phi(s, \theta, \upsilon^*(\theta))$ approximates either a policy gradient or a natural policy gradient. Hence, adjustments of θ ultimately lead to policy improvement.

3   Experience Replay

The main idea analyzed in this paper is to apply to the agent's experience the same processing as a sequential actor-critic-type algorithm would, yet more intensively. A generic algorithm augmented by experience replay is presented in the table below.

After each instant t, the original sequential algorithm estimates $\phi(s_t, \theta, v)$ i.e., the direction of policy improvement, and adjusts the policy vector θ along the estimate. Within each instant t, the modified algorithm repeatedly draws one of the recently visited states, s_i, estimates $\phi(s_i, \theta, v)$, and modifies the policy vector along the estimate. Essentially both algorithms achieve the same goal but (i) the modified one does it more intensively and (ii) it employs experience gathered after visiting state s_i to adjust various policies characterized by different policy vectors. The auxiliary vector v undergoes similar operations in both algorithms.

Because the policy vector is constantly changing, each time its adjustment is performed, it is computed on the basis of the new values of θ and v. The intensity of replaying, $\nu(t)$, must be bounded for the sake of correctness of the algorithm. It is also limited by the computation power available during the agent–environment interaction. $\nu(t)$ should be additionally limited for small t to prevent many recalculations of few tuples in the database and to avoid overtraining.

Designing the estimators of ϕ and ψ for Steps 7 and 9 of Algorithm 3 we have to guarantee that their variance is bounded and their

Algorithm 2. Actor-Critic with Experience Replay. Estimators mentioned in Steps 6 and 7 are based on the data in a database.

0: $t := 1$. Initialize θ and v.
1: Draw and execute an action, $a_t \sim \pi(\cdot \, ; s_t, \theta)$.
2: Register the tuple $\langle s_t, a_t, \theta, r_t, s_{t+1} \rangle$ in the database.
3: Make sure only N most recent tuples remain in the database.
4: Repeat $\nu(t)$ times:
5:    Draw $i \in \{t - N + 1, t - N + 2, \ldots, t\}$.
6:    Adjust θ along an estimator of $\phi(s_i, \theta, v)$:
7:    $\theta := \theta + \beta_t^\theta \widehat{\phi}_i^r(\theta, v)$.
8:    Adjust v along an estimator of $\psi(s_i, \theta, v)$:
9:    $v := v + \beta_t^v \widehat{\psi}_i^r(\theta, v)$.
10: Assign $t := t + 1$ and repeat from Step 1.

bias asymptotically vanishes. This is the only way for the algorithm to preserve the limit properties of the original sequential method.

Let $b > 1$, θ_{i+j} be the policy vector applied to generate a_{i+j}, and K be drawn independently from $Geom(\rho)$, the geometric distribution[1] with parameter $\rho \in [0, 1)$. Also, let χ be equal to 0 if $z_{t,k}$ is not explicitly defined by a_{t+k+1} (as in the basic AC), and to 1 otherwise (as in AC of [6]). We introduce the *randomized-truncated estimators* $\widehat{\phi}_i^r(\theta, v)$ and $\widehat{\psi}_i^r(\theta, v)$ of the same generic form

$$\sum_{k=0}^{K} G_i(\theta, v) \alpha^k z_{i,k}(\theta, v) \min \left\{ \prod_{j=0}^{k+\chi} \frac{\pi(a_{i+j}; s_{i+j}, \theta)}{\pi(a_{i+j}; s_{i+j}, \theta_{i+j})}, b \right\}. \tag{6}$$

---

[1] That is, random variable K of values in $\{0, 1, 2, \ldots\}$ has distribution $Geom(\rho)$, iff $P(K = m) = (1 - \rho)\rho^m$ for nonnegative integer m.

where $\widehat{\phi}_i^r$ is defined by those G and z defining $\widehat{\phi}_t$, and $\widehat{\psi}_i^r$ is defined by those G and z that define $\widehat{\psi}_t$. We can see that (6) closely resembles the original form (3) of $\widehat{\phi}_t$ and $\widehat{\psi}_t$ with two important differences. First, the infinite sum is replaced by the finite one with the appropriately designed random limit. Second, truncated density ratios are introduced in order to compensate for the fact that the current policy is different than the one that generated the actions $a_t, a_{t-1}, \ldots$ contained in the database. The bias of the truncated estimator (6) is small for θ close to θ_{i+j} for all i, j, if only g is regular in a certain sense. Properties of estimator, bounded variance and asymptotic unbiaseness, (6) are analyzed in [12].

4   Experimental Study

We are interested in applications of the MDP framework to learning reactive policies of machines. In this section we analyze a challenging problem of this type, namely learning to run an emulated planar model of a large cat. The cat robot, called Half-Cheetah [11], is presented in Fig. 1. It is a planar kinematic string of 9 links, 8 joints, and 2 "paws". Because 5-th joint is fixed at 180°, and its adjacent links have the same length, joints 4-th and 6-th are always at the same position; therefore, the object does not look like a string. The angles of 4-th and 5-th joint are fixed, all the the others are controllable. Consequently, Half-Cheetah is a model of a 6-degree-of-freedom walking robot.

In all the experiments the controlled system is emulated i.e., simulated in real time. A quantum of system's real time is equal to a quantum of the corresponding computer time, which means that the computer has a lot of spare time that can be devoted to parallel computations. The setup of our experiments is designed to closely resemble a situation when control of a physical machine is to be optimized in real time by means of learning.

The torque τ_i applied at i-th joint is calculated as

$$\tau_i = T_i \min\left\{\max\{-1, \tau_i^0 + a_i^0\}, 1\right\}$$

where τ_i^0 is a "spontaneous" torque at i-th joint, a_i^0 is the output of the learning controller, and T_i expresses "strength" of i-th joint. The spontaneous torque τ_i^0 is implemented as a PD-controller with saturation. It roughly stabilizes the i-th joint at its initial angle. We follow a typical set-

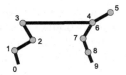

Fig. 1. The initial position of Half-Cheetah. It consists of 9 links, 8 joints among which 2 are fixed (4, 5), and 2 paws (0, 9)

ting of control system design: While it is usually relatively easy to provide a controller that stabilizes a system around a certain state (e.g. by using PD-controllers), it is much more difficult to design a controller that makes the system perform a certain nontrivial activity. In our paper, the controller is to *learn* to make Half-Cheetah run.

An interested reader may find a detailed description of Half-Cheetah in [11]. The discussion here will be limited to the main facts: Half-Cheetah is about 1 meter long, weighs 10kg, and the ,,strengths" of its joints vary from 30 to 120Nm. It is designed to be a realistic model (as far as a planar model can be realistic) of a large, yet light cat.

In order to apply the reinforcement learning to Half-Cheetah, we must define the states and the rewards the learning algorithm has access to. Both are described in detail in [11]. Here we only mention that state is 31-dimensional and the main part of reward is speed measured in meters per second. There are also other parts that play various roles in early stages of learning: (i) a penalty for an attempt to apply torque from outside of the permissible interval, (ii) a penalty for the internal force that keeps the joint angle within its bounds (if i-th joint angle is equal to either of its bounds, then i-th ,,tendon" hurts the cat), (iii) a penalty for not moving the trunk up and keeping the paws on the ground when the animal is not moving forward, (iv) penalties for touching the ground with the heel, the knee, and the head.

The learning algorithm. In order to make Half-Cheetah run, we combine the Basic Actor-Critic and the idea of experience replay. The algorithm we apply (Replaying BAC, in short) is specified below.

The policy applied to Half-Cheetah is comprised of two parts: a neural network and a normal distribution. The input of the network is the state. The output becomes a mean value of the normal distribution with covariance matrix $C = 5^2 I$. The distribution generates actions. The elements of the 6-dimensional action a are transformed into the control stimuli a^0 as $a_i^0 = a_j/30$, where the indexes $i = 1, 2, 3, 6, 7, 8$ correspond to $j = 1, 2, 3, 4, 5, 6$, respectively.

The second approximator used by the learning algorithm is the critic, $\bar{V}$, i.e. the neural approximation of the value function. Both the critic net-

---

Algorithm 3. The Basic Actor-Critic with Experience Replay (Replaying BAC).

0: $t := 1$. Initialize θ and v.
1: Draw and execute an action, $a_t \sim \pi(\,\cdot\,; s_t, \theta)$.
2: Register the tuple $\langle s_t, a_t, \theta, r_t, s_{t+1}\rangle$ in the database.
3: Make sure only N most recent tuples remain in the database.
4: Repeat $\nu(t) = \min\{c_0, c_1 t\}$ times:
5:    Draw $i \in \{t - N + 1, t - N + 2, \ldots, t\}$.
      Draw $K \sim Geom(\rho)$.
      Calculate SUM equal to
      $\sum_{k=0}^{K} \alpha^k d_{i+k} \min\left\{\prod_{j=0}^{k} \frac{\pi(a_{i+j};s_{i+j},\theta)}{\pi(a_{i+j};s_{i+j},\theta_{i+j})}, b\right\}$
      for $d_{i+k} = r_{i+k} + \gamma \bar{V}(s_{i+k+1}; v) - \bar{V}(s_{i+k}; v)$.
6:    Adjust θ along an estimator of $\phi(s_i, \theta, v)$:
7:      $\theta := \theta + \beta_t^\theta \nabla_\theta \ln \pi(a_i; s_i, \theta) SUM$.
8:    Adjust v along an estimator of $\psi(s_i, \theta, v)$:
9:      $v := v + \beta_t^v \nabla_v \bar{V}(s_i; v) SUM$.
10: Assign $t := t + 1$ and repeat from Step 1.

---

work and the actor network have the form of two layer perceptron with linear output layer. Their hidden layers consist of M^A (the actor) and M^C (the critic) sigmoidal (arctan) elements. Each neuron has a constant input (bias). The initial weights of the hidden layers are drawn randomly from the normal distribution $N(0, 1)$ and the initial weights of the output layers are set to zero.

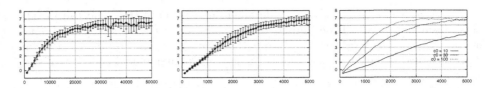

Fig. 2. Actor-Critics for Half-Cheetah: The average reward vs. trial number. *Left:* The Basic Actor-Critic. Each point averages 1000 consecutive trials. The curve averages 10 runs. The one-sigma limits are calculated to assess run-to-run variability of trial averages. *Middle:* The Basic AC with Experience Replay (Replaying BAC). Note that the number of trials in this figure is about 10 times smaller than that of the top figure for the basic method. The curve averages 5 runs and each point averages only 100 consecutive trials. *Right:* The Replaying BAC for various replaying intensity, c_0. Each curve averages 5 runs.

The parameters of the Basic Actor-Critic are as follows: (the actor) $M^A = 80$, $C = 5^2 I$, (the critic) $M^C = 160$, (step-sizes) $\beta_t^\theta \equiv \beta_t^v \equiv 5.10^{-5}$, (estimation) $\gamma = 0.99$, $\lambda = 0.9$. The resulting learning curves are depicted on the left-hand part of Fig. 2. The parameters of the replaying BAC are as follows: (the actor) $M^A = 80$, $C = 5^2 I$, (the critic) $M^C = 160$, (step-sizes) $\beta_t^\theta \equiv \beta_t^v \equiv 2.10^{-5}$, (database) $N = 3.10^4$, (estimation) $\gamma = \alpha = 0.99$, $\lambda = \rho = 0.9$, (computational effort) $c_0 = 30$, $c_1 = 0.3$. With a computer equipped with Intel Quad^{TM}Q9300, the simulations were carried on in real time of Half-Cheetah.

Experiments. Learning curves for the setting discussed above are shown in Fig. 2. A single trial lasts, on the average, for 5 sec. The left-hand part of Fig. 2 reports experiments with the Basic Actor-Critic applied to Half-Cheetah. It is seen, that the algorithm learns to control Half-Cheetah in about 3000 trials, which is about 42 hours of Half-Cheetah. The middle, and the right-hand part of Fig. 2 presents the averaged learning curve for the Replaying BAC applied to the same problem. The curve reports about 7 hours of learning. The algorithm learns to control Half-Cheetah in about 4500 trials, which is about 6 hours of Half-Cheetah time. It is interesting to observe the Half-Cheetah learned policy at various stages of learning (Fig. 3).

Fig. 3. Typical sequences of Half-Cheetah states at various stages of learning by the Basic Actor-Critic with Experience Replay (Replaying BAC).
Left: Awkward walk after 1 hour of training.
Middle: Trot after 2.5 hours of training.
Right: Nimble run after 7 hours of training.

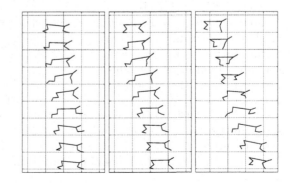

Let us now analyze whether the concepts introduced in the paper indeed improve the quality and the speed of learning. The right part of Fig. 2 demonstrates how the intensity of the computation process translates into the speed of learning. It is seen that the larger intensity, the faster convergence. Plausibly for a certain large c_0, further increase of this parameter does not yield learning speed improvement. However, it is quite time-consuming to investigate high values of c_0. In fact, for $n > 30$ the computations are too slow to take place in real time of Half-Cheetah. The computer time of a single run is then proportional to c_0 and for $c_0 = 100$ it is around 31 hours. Obviously, it is only a matter of computer power. With a fast enough computer, the processing for $c_0 = 100$ could be performed in real time of Half-Cheetah. A satisfactory could be then obtained after 3200 trails, which is about 4 hours of Half-Cheetah time.

5 Conclusions

In this paper we combined the technique of experience replay with a sequential Actor-Critic algorithm. Algorithms of this type deserve serious attention since they represent the most successful approach to applying reinforcement learning to realistic control tasks with continuous state and action spaces. As it has been verified experimentally, experience replay gives a radical learning speedup. The required number of interactions with the environment, which is critical for the applicability of reinforcement learning to real-world tasks, can be considerably reduced. For the fairly difficult Half-Cheetah task we observed a speedup factor of 10, allowing a satisfactory to be reached after as little as 4 hours of Half-Cheetah time (assuming availability of very large computation power), compared to about 42 hours required by the Basic Actor-Critic.

References

1. Bartlett, P.L., Baxter, J.: Stochastic optimization of controlled partially observable markov decision processes. In: Proc. of the 39th IEEE Conf. on Decision and Control (CDC 2000), vol. 1, pp. 124–129 (2000)
2. Barto, A.G., Sutton, R.S., Anderson, C.W.: Neuronlike adaptive elements that can learn difficult learning control problems. IEEE Trans. on SMC 13, 834–846 (1983)
3. Bhatnagar, S., Sutton, R.S., Ghavamzadeh, M., Lee, M.: Incremental natural actor-critic algorithms. In: Advances in NIPS, vol. 21 (2008)
4. Cichosz, P.: An analysis of experience replay in temporal difference learning. Cybernetics and Systems 30, 341–363 (1999)
5. Kimura, H., Kobayashi, S.: An analysis of actor/critic algorithm using eligibility traces: Reinforcement learning with imperfect value functions. In: Proc. of the 15th ICML, pp. 278–286 (1998)
6. Konda, V., Tsitsiklis, J.: Actor-critic algorithms. SIAM Journal on Control and Optimization 42(4), 1143–1166 (2003)
7. Lin, L.-J.: Reinforcement learning for robots using neural networks. Ph.D thesis, Carnegie Mellon University, Pittsburgh, PA, USA (1992)

8. Mahadevan, S., Connell, J.: Automatic programming of behavior-based robots using reinforcement learning. Artificial Intelligence 55(2-3), 311–365 (1992)
9. Sutton, R.S., Barto, A.G.: Reinforcement Learning: An Introduction. MIT Press, Cambridge (1998)
10. Watkins, C., Dayan, P.: Q-learning. Machine Learning 8, 279–292 (1992)
11. Wawrzyński, P.: Learning to control a 6-degree-of-freedom walking robot. In: Proc. of EUROCON 2007, pp. 698–705 (2007)
12. Wawrzyński, P., Pacut, A.: Truncated importance sampling for reinforcement learning with experience replay. In: Proc. CSIT Int. Multiconf., pp. 305–315 (2007)

Controlling the Experimental Three-Tank System via Support Vector Machines

Serdar Iplikci

Pamukkale University, Department of Electrical and Electronics
Engineering, Kinikli Campus, 20040, Denizli, Turkey
iplikci@pamukkale.edu.tr

Abstract. In this study, the previously proposed Support Vector Machines Based Generalized Predictive Control (SVM-Based GPC) method [1] has been applied in controlling the experimental three-tank system. The SVM regression algorithms have been successfully employed in modeling nonlinear systems due to their advantageous peculiarities such as assurance of the global minima and higher generalization capability. Thus, the fact that better modeling accuracy yields better control performance has motivated us to use an SVM model in the GPC loop [1]. In the method, the SVM model of the unknown plant is used to predict future behavior of the plant and also to extract the gradient information which is used in the Cost Function Minimization (CFM) block. The experimental results have revealed that SVM-Based GPC provides very high performance in controlling the system, i.e., the liquid level of the system can track the different types of reference inputs with very small transient- and steady-state errors even in a noisy environment when it is controlled by SVM-Based GPC.

1 Introduction

The generalized predictive control [2,3] method belongs to the class of Model-Based Predictive Control (MPC) techniques, which have been proven to be successful in controlling systems from a wide range of application area for several decades. Among other MPC techniques [4,5], which have been proposed after the first MPC technique [6] in the literature, the most widely employed one has been the GPC method [2,3]. Yet, all MPC techniques share the same idea of utilizing the model in both prediction and finding optimal control action. In the MPC techniques, the accuracy of the model of the plant plays very critical role, and consequently, many linear and nonlinear modeling techniques have been investigated in the literature. Recently, with the developments in the computational intelligence area of research, several soft computing tools namely artificial neural networks [7,8], fuzzy systems [9,10], hybrid systems [11,12] and genetic algorithms [13] have been employed to obtain the model of the unknown plant to be used in the GPC loop. Another computationally intelligent method, which can be an alternative to the soft computing tools for modeling nonlinear plants, is the so-called support vector machines [14,15,16]. The SVM algorithms rely on

M. Kolehmainen et al. (Eds.): ICANNGA 2009, LNCS 5495, pp. 391–400, 2009.
© Springer-Verlag Berlin Heidelberg 2009

the statistical learning theory and the principle of structural risk minimization, and can solve classification and regression problems without getting stuck in local minima. They achieve global minima by transforming the problem into a quadratic programming (QP) problem. Owing to such an advantageous property, the SVM algorithms have found many application areas [17,18].

The aim of this study is to show the applicability of the previously proposed SVM-Based GPC method [1] to a real system. This paper is organized as follows: in the next section, the GPC loop is investigated. Section 3 gives some brief information about the SVM regression algorithm adopted in this study. Section 4 introduces the proposed method by giving details about the extraction of the gradient information from the SVM model and the application steps of the method. Finally, the experimental results are given in Section 5.

2   Generalized Predictive Control

Consider a non-linear system, dynamics of which can be represented by the Nonlinear **A**uto**R**egressive with e**X**ogenous inputs (NARX) model,

$$y_n = f(u_n, ..., u_{n-n_u}, y_{n-1}, ..., y_{n-n_y}), \tag{1}$$

where u_n is the control input applied to the plant at the time index n, y_n is the output of the plant, and n_u and n_y stand for the number of past control inputs and the number of past plant outputs involved in the model, respectively. It is assumed that the non-linear function f is unknown. Fig. 1 illustrates the control loop of the GPC scheme, where $\hat{y}_n$ is the output of the SVM model at the time index n and $\tilde{y}$ is the reference input which is desired to be followed by the plant. One component of the GPC scheme is the SVM model of the plant that accounts for prediction of the future trajectory of the plant in response to the candidate control vector **u**. In addition, it is used to obtain necessary gradient information

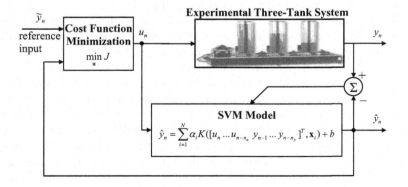

Fig. 1. The GPC architecture

that will be used in the cost function minimization (CFM) block, which is the other component of GPC. The aim of CFM is to minimize the performance index J given by (2) with respect to $\mathbf{u}$.

$$J = \sum_{j=N_1}^{N_2} (\tilde{y}_{n+j} - \hat{y}_{n+j})^2 + \sum_{j=1}^{N_u} \lambda_j (\Delta u_{n+j})^2, \tag{2}$$

where N_1 is the minimum costing horizon, N_2 is the maximum costing horizon, N_u is the control horizon, λ is the weighting factor and $\Delta u_{n+j} = u_{n+j} - u_{n+j-1}$. In the CFM algorithm, each entry of the candidate control vector $\mathbf{u} = [u_{n+1} \ u_{n+2} \ \ldots \ u_{n+N_u}]^T$, is altered within the allowable control input range by the general update rule, $\mathbf{u} \leftarrow \mathbf{u} + s\mathbf{p}$, where $\mathbf{p}$ is the search direction and s is the step-length. During each sampling period, first, the optimum $\mathbf{p}$ is determined by taking the constraints on the control signal and the output of the plant into consideration, then, the optimum s is computed, and finally, the control vector $\mathbf{u}$ is updated and its first element is applied to the plant. In order to find the optimum $\mathbf{p}$, one can employ one of the numerical optimization techniques existing in the literature [19]. Depending on the optimization technique adopted in the CFM algorithm, it may be necessary to compute the derivatives by up to second-order terms in the Taylor expansion. The first-order search algorithms, such as Gradient Descent $(\mathbf{p} = -\mathbf{g})$, need the calculation of the gradient vector,

$$\mathbf{g} = \frac{\partial J}{\partial \mathbf{u}} = \left[\frac{\partial J}{\partial u_{n+1}} \ \frac{\partial J}{\partial u_{n+2}} \ \cdots \ \frac{\partial J}{\partial u_{n+N_u}} \right]^T, \tag{3}$$

while the second-order algorithms, like Modified Newton $(\mathbf{p} = -\mathbf{H}^{-1}\mathbf{g})$, require additionally the computation of the Hessian matrix,

$$\mathbf{H} = \frac{\partial^2 J}{\partial \mathbf{u}^2} = \begin{bmatrix} \frac{\partial^2 J}{\partial u_{n+1} u_{n+1}} & \frac{\partial^2 J}{\partial u_{n+1} u_{n+2}} & \cdots & \frac{\partial^2 J}{\partial u_{n+1} u_{n+N_u}} \\ \frac{\partial^2 J}{\partial u_{n+2} u_{n+1}} & \frac{\partial^2 J}{\partial u_{n+2} u_{n+2}} & \cdots & \frac{\partial^2 J}{\partial u_{n+2} u_{n+N_u}} \\ \vdots & \vdots & \ddots & \vdots \\ \frac{\partial^2 J}{\partial u_{n+N_u} u_{n+1}} & \frac{\partial^2 J}{\partial u_{n+N_u} u_{n+2}} & \cdots & \frac{\partial^2 J}{\partial u_{n+N_u} u_{n+N_u}} \end{bmatrix}. \tag{4}$$

In the gradient vector $\mathbf{g}$, the hth element is given by (5) for $h = 1, \ldots, N_u$,

$$\frac{\partial J}{\partial u_{n+h}} = -2 \sum_{j=N_1}^{N_2} (\tilde{y}_{n+j} - \hat{y}_{n+j}) \frac{\partial \hat{y}_{n+j}}{\partial u_{n+h}} + 2 \sum_{j=1}^{N_u} \lambda_j \Delta u_{n+j} (\delta_{h,j} - \delta_{h,j-1}), \tag{5}$$

where $\delta_{i,j}$ is the Kronecker Delta function. Similarly, the mth, h^{th} element of the Hessian matrix $\mathbf{H}$, for $m = 1, \ldots, N_u$ and $h = 1, \ldots, N_u$, is given by,

$$\frac{\partial^2 J}{\partial u_{n+m} u_{n+h}} = 2 \sum_{j=N_1}^{N_2} \left(\frac{\partial \hat{y}_{n+j}}{\partial u_{n+m}} \frac{\partial \hat{y}_{n+j}}{\partial u_{n+h}} - \frac{\partial^2 \hat{y}_{n+j}}{\partial u_{n+m} u_{n+h}} (\tilde{y}_{n+j} - \hat{y}_{n+j}) \right)$$
$$+ \sum_{j=1}^{N_u} \lambda_j (\delta_{m,j} - \delta_{m,j-1})(\delta_{h,j} - \delta_{h,j-1}). \tag{6}$$

3   Modeling with ε-Support Vector Regression (ε-SVR)

Consider a data set related to the NARX model of the plant in the form below,

$$T = \left\{ u_k, ..., u_{k-n_u}, y_{k-1}, ..., y_{k-n_y}; y_k \right\}_{k=n}^{k=n+N}, \tag{7}$$

which is to be used to obtain a model of the plant dynamics. In this subsection, the ε-SVR algorithm is examined to solve the regression problem defined as follows: given a data set $T = \{\mathbf{x}_k, y_k\}_{k=1}^{k=N}$, where $\mathbf{x}_k \in X \subseteq \mathbb{R}^{n_u+n_y+1}$ is the kth input and $y_k \in Y \subseteq \mathbb{R}$ is the corresponding output value, it is desired to find a model representing the relationship between the input and output data points. This is achieved by an SVM model as $\hat{y}(\mathbf{x}) = \langle \mathbf{w}, \boldsymbol{\Phi}(\mathbf{x}) \rangle + b$, which is linear in a higher dimensional feature space $\boldsymbol{F}$, where $\mathbf{w}$ is a vector in $\boldsymbol{F}$, $\boldsymbol{\Phi}(\mathbf{x})$ is a mapping from the input space to $\boldsymbol{F}$ and b is the bias term. The ε-SVR algorithm regards this problem as an optimization problem in dual space with the model given by,

$$\hat{y}(\mathbf{x}) = \sum_{i=1}^{N} \alpha_i K(\mathbf{x}, \mathbf{x}_i) + b, \tag{8}$$

where α_i's are the coefficients of each training data and $K(\mathbf{x}_i, \mathbf{x}_j)$ is a kernel function given by $K(\mathbf{x}_i, \mathbf{x}_j) = \boldsymbol{\Phi}(\mathbf{x}_i)^T \boldsymbol{\Phi}(\mathbf{x}_j) = K_{ij}$. In the dual model (8), a training point $\mathbf{x}_i$ corresponding to a non-zero α_i is named as the *support vector*. In this work, we determine the support vectors, the coefficients and the bias term by the ε-SVR algorithm. It employs the Vapnik's ε-insensitive loss function,

$$L(\varepsilon, y, \hat{y}) = \begin{cases} 0, & y_i - \hat{y}_i \le \varepsilon \\ y_i - \hat{y}_i, & y_i - \hat{y}_i > \varepsilon \end{cases} \tag{9}$$

and formulates the primal form of the problem as follows:

$$\min_{\mathbf{w}, b, \xi, \xi^*} P_\varepsilon = \frac{1}{2} \|\mathbf{w}\|^2 + C \sum_{i=1}^{N} (\xi_i + \xi_i^*) \tag{10}$$

subject to the constraints,

$$y_i - \langle \mathbf{w}, \boldsymbol{\Phi}(\mathbf{x}_i) \rangle - b \le \varepsilon + \xi_i; \quad \langle \mathbf{w}, \boldsymbol{\Phi}(\mathbf{x}_i) \rangle + b - y_i \le \varepsilon + \xi_i^*; \quad \xi_i, \xi_i^* \ge 0, \tag{11}$$

for $i = 1, 2, \ldots, N$, where ε is the magnitude of the maximum tolerable error, ξ_i's and ξ_i^*'s are slack variables and C is a regularization parameter. Thus, the dual form of the problem becomes a quadratic programming (QP) problem as,

$$\min_{\beta, \beta^*} D_\varepsilon = \frac{1}{2} \sum_{i=1}^{N} \sum_{j=1}^{N} K_{ij}(\beta_i - \beta_i^*)(\beta_j - \beta_j^*) + \varepsilon \sum_{i=1}^{N} (\beta_i + \beta_i^*) - \sum_{i=1}^{N} y_i(\beta_i - \beta_i^*) \tag{12}$$

subject to the constraints, $0 \le \beta_i, \beta_i^* \le C$, $\sum_{i=1}^{N}(\beta_i - \beta_i^*) = 0$, $i = 1, 2, \ldots, N$. Solution of this QP problem gives the optimum values of β_i's and β_i^*'s. If α_j

is defined to be the new coefficient of $\mathbf{x}_j$ for $j = 1, 2, \ldots, N$ as $\alpha_j = \beta_j - \beta_j^*$, then we obtain an SVM model as given by (8). Moreover, if we consider only the support vectors, the model becomes,

$$\hat{y}(\mathbf{x}) = \sum_{\substack{j=1 \\ j \in SV}}^{\#SV} \alpha_j K(\mathbf{x}, \mathbf{x}_j) + b, \tag{13}$$

where $\#SV$ stands for the number of support vectors in the model [17,18].

4   The SVM-Based GPC Method

Once the SVM model is obtained, the gradient information can be extracted from the model. In what follows, the formulations for finding the gradient vector and the Hessian matrix are introduced as given in [1]. If the current state vector is formed as $\mathbf{c}_n = [u_n \ u_{n-1} \ \cdots \ u_{n-n_u} \ y_{n-1} \ y_{n-2} \ \cdots \ y_{n-n_y}]^T$, then the corresponding output of the SVM model becomes, $\hat{y}_n = \sum_{j=1}^{\#SV} \alpha_j K(\mathbf{c}_n, \mathbf{x}_j) + b$. In this study, we adopted the radial basis function (RBF), $K_{ij} = K(\mathbf{x}_i, \mathbf{x}_j) = \exp\left(-\frac{(\mathbf{x}_i - \mathbf{x}_j)^T(\mathbf{x}_i - \mathbf{x}_j)}{2\sigma^2}\right)$, as the kernel function with the width parameter σ. If d_{jn} is defined as the Euclidean distance between $\mathbf{x}_j$ and $\mathbf{c}_n$ as,

$$d_{jn} = (\mathbf{c}_n - \mathbf{x}_j)^T(\mathbf{c}_n - \mathbf{x}_j) = \sum_{i=0}^{n_u}(x_{j,i+1} - u_{n-i})^2 + \sum_{i=1}^{n_y}(x_{j,n_u+i+1} - y_{n-i})^2, \tag{14}$$

then the kernel function can be rewritten as, $K(\mathbf{c}_n, \mathbf{x}_j) = \exp\left(-\frac{d_{jn}}{2\sigma^2}\right)$ and thus the SVM model becomes, $\hat{y}_n = \sum_{j=1}^{\#SV} \alpha_j \exp\left(-\frac{d_{jn}}{2\sigma^2}\right) + b$. Now, the model can be used to predict future trajectory of the plant in response to $\mathbf{u}$ as follows:

$$\hat{y}_{n+k} = \sum_{j=1}^{\#SV} \alpha_j \exp\left(-\frac{d_{j,n+k}}{2\sigma^2}\right) + b, k = N_1, N_1 + 1, \ldots, N_2, \tag{15}$$

where

$$d_{j,n+k} = \sum_{i=1}^{min(k,n_y)}(x_{j,n_u+i+1} - \hat{y}_{n+k-i})^2 + \sum_{i=k+1}^{n_y}(x_{j,n_u+i+1} - y_{n+k-i})^2$$
$$+ \sum_{i=0}^{n_u}\begin{cases} (x_{j,i+1} - u_{n+k-i})^2, & k - N_u < i \\ (x_{j,i+1} - u_{n+N_u})^2, & k - N_u \geq i \end{cases} \tag{16}$$

Thus, the first-order derivatives can be written as,

$$\frac{\partial \hat{y}_{n+k}}{\partial u_{n+h}} = \sum_{j=1}^{\#SV} \alpha_j \frac{\partial \exp(-\frac{d_{j,n+k}}{2\sigma^2})}{\partial u_{n+h}} \tag{17}$$

where

$$\frac{\partial \exp(-\frac{d_{j,n+k}}{2\sigma^2})}{\partial u_{n+h}} = \frac{\partial \exp(-\frac{d_{j,n+k}}{2\sigma^2})}{\partial d_{j,n+k}} \frac{\partial d_{j,n+k}}{\partial u_{n+h}} = -\frac{1}{2\sigma^2} \exp(\frac{-d_{j,n+k}}{2\sigma^2}) \frac{\partial d_{j,n+k}}{\partial u_{n+h}} \quad (18)$$

and

$$\begin{aligned}
\frac{\partial d_{j,n+k}}{\partial u_{n+h}} = &-2 \sum_{i=1}^{min(k,n_y)} (x_{j,n_u+i+1} - \hat{y}_{n+k-i}) \frac{\partial \hat{y}_{n+k-i}}{\partial u_{n+h}} \delta_1(k-i-1) \\
&-2 \sum_{i=0}^{n_u} \left\{ \begin{array}{ll} (x_{j,i+1} - u_{n+k-i})\delta_{k-i,h}, & k - N_u < i \\ (x_{j,i+1} - u_{n+N_u})\delta_{N_u,h}, & k - N_u \geq i \end{array} \right.
\end{aligned} \quad (19)$$

where $\delta_1(\cdot)$ stands for the unit step function. The second-order derivatives are

$$\frac{\partial^2 \hat{y}_{n+k}}{\partial u_{n+h} \partial u_{n+m}} = \sum_{j=1}^{\#SV} \alpha_j \frac{\partial^2 \exp(-\frac{d_{j,n+k}}{2\sigma^2})}{\partial u_{n+h} \partial u_{n+m}}, \quad (20)$$

where

$$\begin{aligned}
\frac{\partial^2 \exp(-\frac{d_{j,n+k}}{2\sigma^2})}{\partial u_{n+h} \partial u_{n+m}} = &\frac{\partial \exp(-\frac{d_{j,n+k}}{2\sigma^2})}{\partial d_{j,n+k}} \frac{\partial^2 d_{j,n+k}}{\partial u_{n+h} \partial u_{n+m}} \\
&+ \frac{\partial^2 \exp(-\frac{d_{j,n+k}}{2\sigma^2})}{\partial d_{j,n+k}^2} \frac{\partial d_{j,n+k}}{\partial u_{n+h}} \frac{\partial d_{j,n+k}}{\partial u_{n+m}}
\end{aligned} \quad (21)$$

and

$$\begin{aligned}
\frac{\partial^2 d_{j,n+k}}{\partial u_{n+h} \partial u_{n+m}} = &-2 \sum_{i=1}^{min(k,n_y)} \frac{\partial \hat{y}_{n+k-i}}{\partial u_{n+h}} \frac{\partial \hat{y}_{n+k-i}}{\partial u_{n+m}} \delta_1(k-i-1) \\
&-2 \sum_{i=1}^{min(k,n_y)} (x_{j,n_u+i+1} - \hat{y}_{n+k-i}) \frac{\partial^2 \hat{y}_{n+k-i}}{\partial u_{n+h} \partial u_{n+m}} \delta_1(k-i-1).
\end{aligned} \quad (22)$$

Now, the SVM-Based GPC algorithm can be itemized as follows:

- In order to gather training data, first, the parameters u_{min}, u_{max}, τ_{min} and τ_{max} are determined, and then the plant is run for a certain period, during which the input signal is composed of a series of pulses of random magnitudes within $[u_{min}, u_{max}]$ and random durations within $[\tau_{min}, \tau_{max}]$.
- Proper values of n_u and n_y are determined, and a set of training data is formed. Then, all input/output variables are normalized to [0,1].
- SVM model is obtained with ε-SVR by using N data pairs randomly selected out of the gathered data set, remaining of which are spared for validation.
- The gradient vector and the Hessian matrix are calculated based on the SVM model by using the formulations.
- The GPC algorithm is applied: at each iteration, the candidate control vector $\mathbf{u}$ is formed by taking its previous elements, and then based on the future predictions of the SVM model in response to the control vector, $\mathbf{u}$ is updated so as to minimize J. Finally, the first element of $\mathbf{u}$ is applied to the plant.

5   Application to the Experimental Three-Tank System

The three-tank liquid system is a well-known three-dimensional nonlinear system in the literature. In this study we have tested our proposed controller on a real experimental system by assuming that its underlying dynamics is not known. Yet, it will be useful to give its dynamics which is expressed by a set of differential equations,

$$\dot{h}_1(t) = \frac{1}{A}[q_1(t) - Q_{13}(t)], \quad \dot{h}_2(t) = \frac{1}{A}[q_2(t) + Q_{32}(t) - Q_{20}(t)],$$

$$\dot{h}_3(t) = \frac{1}{A}[Q_{13}(t) - Q_{32}(t)], \tag{23}$$

where

$$Q_{13}(t) = az_{13}S_n \mathrm{sgn}(h_1(t) - h_3(t))\sqrt{2g|h_1(t) - h_3(t)|}, \quad Q_{20}(t) = az_{20}S_n\sqrt{2gh_2(t)}$$

$$Q_{32}(t) = az_{32}S_n \mathrm{sgn}(h_3(t) - h_2(t))\sqrt{2g|h_3(t) - h_2(t)|}$$

The explanations and the values regarding to the variables and the constants in the equations are given in Table 1 [20]. In this work, the aim is to control the

Table 1. The system parameters

Parameter	Value/Description
$h_i(t)$: liquid level of the tank$i$ [m]	output
$q_i(t)$: supplying flow rate of the pump$i$ [m^3/sec]	input
az_{13}: outflow coefficient between tank1 and tank3	0.5
az_{32}: outflow coefficient between tank3 and tank2	1.0
az_{10}: outflow coefficient from tank1 to reservoir	0.0
az_{20}: outflow coefficient from tank2 to reservoir	1.0
az_{30}: outflow coefficient from tank3 to reservoir	0.0
A: cross section of the cylinders [m^2]	0.0154
S_n: section of connection pipe n [m^2]	5×10^{-5}
g: gravitation coefficient [m/sec^2]	9.81

liquid level of the tank1 as the output ($y(t) = h_1(t)$) by manipulating the flow rate of the pump1 as the input ($u(t) = q_1(t)$) of the system. In the experiments, magnitude of the control signal is altered between $u_{min}=0$ m^3/sec and $u_{max}=10^{-4}$ m^3/sec, duration of the control signal is kept constant as $\tau_{min}=\tau_{max}=1.0$ sec and sampling period is 1.0 sec. The parameters in the performance index are selected as $N_1=1$, $N_2=30$, $N_u=2$ and $\lambda_j=2.0$ for $j = 1, \ldots, N_u$. We have adopted the Modified Newton method for finding **p** and the Golden-Section method for finding s. The data gathered from the system can be seen from Fig. 2. In order to eliminate the high-frequency noise components from the data, a low-pass filter,

$$H(z) = \frac{0.02483z^{-1} + 0.02224z^{-2}}{1 - 1.6720z^{-1} + 0.7190z^{-2}}. \tag{24}$$

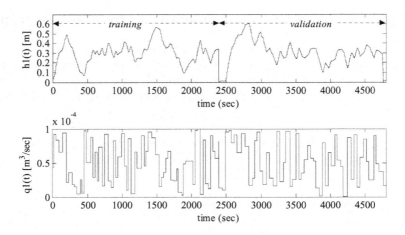

Fig. 2. Collected data

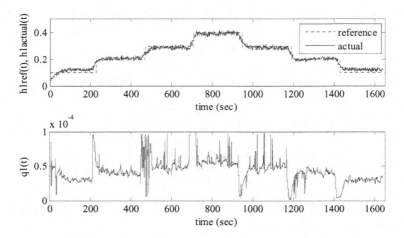

Fig. 3. Experimental results for staircase reference

is employed. The half of the 4800 data points have been used to obtain the SVM model and the remaining ones have been spared for validation of the model. The NARX parameters are selected as $n_u=5$ and $n_y=5$, while the SVM parameters are $\sigma=2$, $\varepsilon = 0.002$ and $C=1000$. Thus, the SVM model of the three-tank system is obtained with 81 support vectors. The experimental results for the staircase reference input are given in Fig. 3, while Fig. 4 illustrates the results for the sinusoidal reference input as $\tilde{y}(t) = 0.25 + 0.15\sin 0.004\pi t$.

As can be seen from the figures, liquid level of the tank1 can follow the reference trajectories with small transient- and steady-state tracking errors. It should be noted that the control of the system has been maintained when we

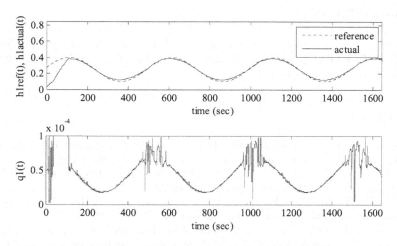

Fig. 4. Experimental results for sinusoidal reference

continue the experiments in the long run. To sum up, the results have shown that the SVM-Based GPC method provides an acceptable performance for control of the investigated system.

6   Conclusions

In this work, liquid level control of an experimental three-tank system has been carried out by the SVM-Based GPC method. First, the SVM model of the unknown system has been obtained by the ε-SVR algorithm and then it has been utilized in the GPC loop. Since the accuracy of the model play crucial role in the GPC loop, we have adopted the SVM structure due to its higher generalization potential and assurance of global minima. The experimental results have shown that the output of the three-tank system, when controlled by SVM-Based GPC, can track the staircase and sinusoidal reference trajectories with very small transient- and steady-state errors, which has proven the applicability of the proposed controller to real time systems. On the other hand, the choice of the SVM and kernel parameters is still an open problem in the machine learning area of research. Moreover, training time of the ε-SVR algorithm grows up with the number of training patterns. Yet, this drawback can be overcome by online training methods or faster training algorithms. In fact, the major advantage of the method is that it can exploit the developments in the SVM regression algorithms. In conclusion, it can be stated that the previously proposed SVM-Based GPC method has been successfully applied to the control of an experimental system and that it can be an alternative to other model-based GPC techniques.

Acknowledgments

This work is supported by The Scientific and Technological Research Council of Turkey (TÜBITAK) – National Young Researchers Career Development Programme, Project No.: 106E125.

References

1. Iplikci, S.: Support vector machines based generalized predictive control. International Journal of Robust and Nonlinear Control 16, 843–862 (2006)
2. Clarke, D.W., Mohtadi, C., Tuffs, P.C.: Generalized predictive control - part 1: the basic algorithm. Automatica 23, 137–148 (1987)
3. Clarke, D.W., Mohtadi, C., Tuffs, P.C.: Generalized predictive control - part 2: extensions and interpretations. Automatica 23, 149–163 (1987)
4. De Keyser, R.M.C., Van Cauwenberghe, A.R.: Extended prediction self-adaptive control. In: Proceedings of the 7th IFAC Symposium on Identification and System Parameter Estimation, pp. 1255–1260 (1985)
5. Soeterboek, R.: Predictive Control: A Unified Approach. Prentice Hall, Englewood Cliffs (1992)
6. Richalet, J.A., Rault, A., Testud, J.L., Papon, J.: Model predictive heuristic control: applications to an industrial process. Automatica 14, 413–428 (1978)
7. Soloway, D., Haley, P.J.: Neural generalized predictive control: a Newton-Raphson algorithm. In: Proc. of the IEEE Int'l Sym. on Int. Cont., MI, pp. 277–282 (1996)
8. Hagan, M.T., Demuth, H.B., De Jesus, O.: An introduction to the use of neural networks in control systems. Int'l J. of Rob. and Nonl. Control 12, 959–985 (2002)
9. Huang, Y.L., Lou, H.H., Gong, J.P., Edgar, T.F.: Fuzzy model predictive control. IEEE Transactions on Fuzzy Systems 8, 665–678 (2000)
10. Li, N., Li, S.Y., Xi, Y.G.: Multi-model predictive control based on the Takagi-Sugeno fuzzy models: a case study. Information Sciences 165, 247–263 (2004)
11. Hu, J.Q., Rose, E.: Generalized predictive control using a neuro-fuzzy model. International Journal of Systems Science 30, 117–122 (1999)
12. Sarimveis, H., Bafas, G.: Fuzzy model predictive control of non-linear processes using genetic algorithms. Fuzzy Sets and Systems 139, 59–80 (2003)
13. Martinez, M., Senent, J.S., Blasco, X.: Generalized predictive control using genetic algorithms (GAGPC). Eng. App. of Artificial Intelligence 11, 355–367 (1998)
14. Vapnik, V.: The Nature of Statistical Learning Theory. Springer, Heidelberg (1995)
15. Vapnik, V.: Statistical Learning Theory. John Wiley, New York (1998)
16. Vapnik, V.: The Support Vector Method of Function Estimation. In: Nonlinear Modeling Advanced Black Box Techniques. Kluwer Academic Publishers, Boston (1998)
17. Cristianini, N., Taylor, J.S.: An Introduction to Support Vector Machines and other kernel-based learning methods. Cambridge University Press, New York (2000)
18. Schölkopf, B., Burges, C.J.C., Smola, A.J.: Advances in kernel methods: Support Vector Learning. The MIT Press, Cambridge (1999)
19. Bertsekas, D.P.: Nonlinear Programming. Athena Scientific, Belmont (1999)
20. DTS200 - Laboratory Setup Three Tank System, Amira GmbH, Duisburg (2000)

Feature-Based Clustering for Electricity Use Time Series Data

Teemu Räsänen and Mikko Kolehmainen

Research Group of Environmental Informatics
Department of Environmental Sciences
University of Kuopio
P.O. Box 1627
FIN-70211 Kuopio, Finland
{Teemu.Rasanen,Mikko.Kolehmainen}@uku.fi

Abstract. Time series clustering has been shown effective in providing useful information in various applications. This paper presents an efficient computational method for time series clustering and its application focusing creation of more accurate electricity use load curves for small customers. Presented approach was based on extraction of statistical features and their use in feature-based clustering of customer specific hourly measured electricity use data. The feature-based clustering was able to cluster time series using just a set of derive statistical features. The main advantages of this method were; ability to reduce the dimensionality of original time series, it is less sensitive to missing values and it can handle different lengths of time series. The performance of the approach was evaluated using real hourly measured data for 1035 customers during 84 days testing time period. After all, clustering resulted into more accurate load curves for this set of customers than present load curves used earlier. This kind of approach helps energy companies to take advantage of new hourly information for example in electricity distribution network planning, load management, customer service and billing.

Keywords: time series clustering, feature-based clustering, feature extraction, electricity use data, load curves, electricity distribution.

1 Introduction

Data mining of multivariate time series is a well known research area where feature extraction as a data reduction technique, plays an important role in pattern recognition and data analysis [1]. There are several real world situations where large amount of data has to be reduced, variables of multivariate time-series are not timely synchronized to each other or there is lot of missing values in data. These are the main reasons for use of a feature-based method for clustering of time series [2]. Besides these problems, clustering of original time series data is more computationally demanding than feature-based approaches. Moreover, recent studies have proven that pattern recognition methodologies, such as the k-means, self-organized maps, fuzzy

M. Kolehmainen et al. (Eds.): ICANNGA 2009, LNCS 5495, pp. 401–412, 2009.
© Springer-Verlag Berlin Heidelberg 2009

k-means and hierarchical methods, can be applied for the study of the customer electricity behavior, where problems mentioned above are typically occurring [3][4].

The analysis of customer loads and load estimation is a traditional area of electricity distribution technology because electricity distribution utilities need accurate load data for pricing and tariff planning, distribution network planning and operation, power production planning, load management, customer service and billing and also for providing information to customers and public authorities [5]. Recently, there has been major technological progress in small customer energy use metering and consequently hourly measured information is available in near future for great majority of customers. Furthermore, current European Union legislation has brought new requirements to energy distributors and retail energy sales companies ordering them to provide information about consumer's energy consumption in more detailed level [6].

The electricity load curve describes the amount of electrical energy customer uses over the course of time and it is used to plan how much electricity retailer or distribution Company will need to make electricity available at any given time. Furthermore, end-use load curves (i.e. load profiles) show how the load of a particular customer varies throughout the day and week and gives understanding of peak demand [7]. The most important load information is how a customer or a group of customers uses electricity at different hours of the day, different days of the week and seasons of the year and what their share of the utility's total load is and how loads of different customers aggregate in different locations of a distribution network [5]. The factors affecting to the customer or customer groups load are 1) customer consumption behavior and residence characteristics, 2) time of day, week or year and 3) local climate factors like temperature, humidity or solar radiation [8][5].

Typically the energy companies have classified customers into groups concerning their characteristics and annual demand for electricity. Based on this classification each customer has load curve estimate which is used for billing and distribution management. However, it is typical that changes in customers life and electricity use doesn't mediate to the energy company and the needed load curve update cannot be done. Another problem is that given load curve can be wrong in the first place because customer has similar characteristics but electricity consumption behavior is different than proposed typical customer group. As a result of these problems, demand side management and distribution planning deals with misinformation causing extra costs.

The purpose of this study was to develop efficient computational approach to handle complex and large time series datasets in the context of electricity load research. Moreover, in the presented application, the main aim was to utilize large amounts of hourly measured electricity use data in order to validate and improve customer specific load curves. In this paper, we compared given load curves to real measured electricity use and investigated how well they are correlating. Furthermore, we present here computationally efficient data-based approach to create more accurate up-to-date customer specific load curves using real measurement information. Proposed methods were tested using hourly measured electricity use data from 1035 customers locating Northern-Savo, Finland. The returns showed that original load curves were not very accurate and they can be improved using data based clustering.

2 Materials and Methods

2.1 Data Used

In this study, we used data describing 1035 small customer's hourly measured electricity use (kWh) during the winter 2007. The data contained 84 days (2016 hours) starting the first of January 2007. The customers were located in Pohjois-Savo region, which is an area in eastern Finland. The region has two major cities called Kuopio and Iisalmi but major part of customer where located outside of cities in the sparsely populated area.

The energy company classifies the customer's according their characteristics when customer joins to company's distribution network. Each customer is attached to specific load curve which is used as a base of billing and distribution planning. These 1035 customers were divided to use 18 different load curves describing best each customer's electricity use and behavior. For example, house (detached house, terraced house, etc.) and heating type (use of electric heating) or type of activity of the residence (spare time cottage, agriculture residence, etc.) had been used as a classifying characters.

Used data set contained hourly energy use time series for 1035 customers and we had also original load curves for each customer. With this data we solved how customer's real electricity use corresponds to the original load curve. Furthermore, data where used to create new load curves based on each customer electricity use behavior and characteristic based clustering.

2.2 Feature Extraction

Transforming the raw time-series data into the set of features is called feature extraction. Despite of the length of the time series and missing values, a finite set of statistical measures can be used to capture the global nature of the time series [2]. Furthermore, feature extraction is used to compress large data sets by the means of dimensionality reduction. In this way, computational efficiency can be increased and use of more sophisticated algorithms is possible. Nevertheless, the majority of feature extraction methods are generic in nature, the extracted features are usually application dependent. Thus one set of features that work well on one application might not be relevant to another [9].

In this study, features were extracted from the raw hourly measured electricity use data using window of one week i.e. 168 data rows (hours). We extracted 7 features from each customer's data. The features extracted were; mean, standard deviation, skewness, kurtosis, chaos, energy and periodicity.

Mean and standard deviation (Eq. 1) are simple but useful features. Skewness (Eq. 2) is the degree of symmetry in the distribution of energy consumption data and kurtosis (Eq. 3) measures how much a distribution is peaked at the center of a distribution [10].

Many real world systems may contain chaotic behavior and especially nonlinear dynamical systems often exhibit chaos, which is characterized by sensitive dependence on initial values, or more precisely by a positive Lyapunov Exponent (LE). LE, as a measure of the divergence of nearby trajectories has been used to qualifying chaos by giving a quantitative value. It is common to just refer to the largest one, i.e. to the Maximal Lyapunov exponent (MLE), because it determines the predictability of a dynamical system. A positive MLE is usually taken as an indication that the system is chaotic. The maximal Lyapunov exponent (λ) can be defined using Eq. 4 [11].

$$\sigma = \sqrt{\frac{1}{N}\sum_{i=1}^{k}(c_i - m)^2 \cdot n_i} \tag{1}$$

$$Skew = \frac{1}{N\sigma^3}\sum_{i=1}^{k}(c_i - m)^3 \cdot n_i \tag{2}$$

$$Kurt = \frac{1}{N\sigma^4}\sum_{i=1}^{k}(c_i - m)^4 \cdot n_i \tag{3}$$

$$\lambda = \lim_{t\to\infty}\frac{1}{t}\ln\frac{|\delta Z(t)|}{|\delta Z_0|} \tag{4}$$

$$Energy = \frac{\sum_{i=1}^{w}|x_i|^2}{|w|} \tag{5}$$

The periodicity is important for determining the seasonality and examining the cyclic pattern of the time series [2]. In this case, length of occurring period was solved using Discrete Power Spectrum (periodogram), which describes the distribution of the signal strength into different frequency values. The spectrum generally enlightens the nature of the data [12]. The most powerful frequency value was transformed into hour form and it was taken into feature data set and used as a periodicity feature. Additionally, to capture data periodicity, the energy feature was calculated which is the sum of the squared discrete FFT component magnitudes of the signal. This sum was divided by the window length for normalization. Energy feature was calculated using Eq. 5, where $x1, x2, ...$ are the FFT components of the window [13].

Finally, after the feature extraction, data set contained 84 variables (7 features for each week) and 1035 rows (amount of customers) which was used in creation of new load curves using K-means clustering method.

2.3 K-Means Clustering

The number of clusters in the case specific application may not be known a priori. However, in the K-means algorithm the number of clusters has to be predefined. Therefore, it is common that the algorithm is applied with different number of clusters and then the best solution among them is selected using a validity index like the Davies-Bouldin (DB) Index [14]. It is calculated as follows,

$$DB = \frac{1}{N} \sum_{i=1}^{N} \max_{j, j \neq i} \frac{S_i + S_j}{d_{ij}} \tag{6}$$

where N is the number of clusters. The within (S_i) and between (d_{ij}) cluster distances are calculated using the cluster centroids as follows:

$$S_i = \frac{1}{|C_i|} \sum_{x \in C_i} \|x - m_j\| \tag{7}$$

$$d_{ij} = \|m_i - m_j\| \tag{8}$$

where m_i is the centre of cluster C_i, with $|C_i|$ the number of points belonging to cluster C_i. The objective is to find the set of clusters that minimizes the Eq. 8.

The Davies-Bouldin index was used to solve optimal number of clusters. The DB index varies slightly between calculations because initial starting point is set randomly. In this case, indexes were calculated 20 times and mean value of the index using different numbers of clusters was used when the optimal number of clusters was selected. After that K-means algorithm was used to cluster feature data set in order to create reasonable number of comparing groups.

The K-means algorithm was applied to the clustering of the feature vectors which were created using raw time series data. The K-means is a well-known non-hierarchical cluster algorithm [15]. The basic version begins by choosing number of clusters and randomly picking K cluster centers. After that each point is assigned to the cluster whose mean is closest in a Euclidean distances sense. Finally, the mean vectors of the points assigned to each cluster are computed, and those are used as new centers in an iterative approach until convergence criterion is met.

2.4 Estimating Goodness of Clustering

The difference between customer's electricity use and clustered load curve was calculated using Index-of-Agreement (IA). It is a dimensionless measure, limited to the range 0...1, giving a relative size of the difference [16]. It is easily understandable and ideal for making cross-comparisons between time series or models. The values range from 0 to 1, with a value of 1 indicating perfect fit between the observed and predicted data. IA is calculated as follows:

$$IA = 1 - \frac{\sum_{i=1}^{n}(P_i - O_i)^2}{\sum_{i=1}^{n}(|P_i'| + |O_i'|)^2} \tag{9}$$

$$P_i' = P_i - \overline{O} \tag{10}$$

$$O_i' = O_i - \overline{O} \tag{11}$$

In this equation, n is the number of observations, O_i is the observed variable at time i, P_i is the predicted variable at time i, $\bar{O}$ is the mean value of the observed variable over n observations.

3   Results

The aim of the study was to (1) evaluate correspondence of customers measured real electricity use and load curve set by the energy company and (2) use feature-based clustering in order to create new more accurate load curves. First, the mean of index-of-agreement was calculated between present load curves and electricity use of customers belonging to each curve. The values of index-of-agreement, standard deviation and number of customers using each load curve are illustrated in Table 1.

Table 1. The results of comparison between measured electricity use and original load curves

Load Curve	Mean IA	Mean Std	Number of customers
LC1	0.31	0.08	426
LC2	0.33	0.10	189
LC3	0.35	0.08	76
LC4	0.51	0.04	6
LC5	0.16	0.00	1
LC6	0.30	0.06	48
LC7	0.35	0.00	1
LC8	0.26	0.08	3
LC9	0.35	0.00	1
LC10	0.22	0.08	2
LC11	0.28	0.07	6
LC12	0.30	0.11	15
LC13	0.27	0.15	6
LC14	0.35	0.17	8
LC15	0.38	0.05	7
LC16	0.07	0.00	1
LC17	0.39	0.07	15
LC18	0.16	0.17	4
Unknown			220
Mean	**0.30**	**0.07**	**1035**

Next, the feature data set was created and raw time series data (1035 rows by 2016 columns matrix) were transformed into more compact format (1035 rows by 84 columns matrix). Before clustering of feature data set the Davies-Bouldin index was used to solve optimal number of clusters. In this case, there were two clear options according DB-index and 16 clusters were selected because final comparison results were better than using 32 clusters. The values of DB-index for different clusters amounts are illustrated in Figure 1.

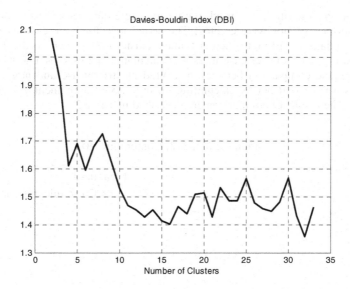

Fig. 1. The values of Davies-Bouldin index for different number of clusters. The optimum number of clusters is where index is lowest (in this case options were 32 clusters or 16 clusters).

Table 2. The results of comparison between measured electricity use and new data-based load curves

New Load Curve	Mean IA	Mean Std	Number of customers
NLC1	0.50	0.12	21
NLC2	0.68	0.10	166
NLC3	0.68	0.13	25
NLC4	0.46	0.12	20
NLC5	0.77	0.09	209
NLC6	0.83	0.09	181
NLC7	0.63	0.21	11
NLC8	0.57	0.14	25
NLC9	0.62	0.17	17
NLC10	0.69	0.14	24
NLC11	0.70	0.16	16
NLC12	0.66	0.10	91
NLC13	0.73	0.11	76
NLC14	0.55	0.20	100
NLC15	0.68	0.13	26
NLC16	0.65	0.14	27
Mean	**0.65**	**0.13**	**1035**

The K-means algorithm was used to cluster feature data set and as a result of that 16 new customer groups were created. The actual new load curves were calculated from original data according to each customers cluster id. In other words, the mean of electricity use of each cluster customers was calculated and used as a new load curve. Furthermore, the comparison between measured electricity use and new data-based load curves was carried out by calculating mean index-of-agreement values for each load curve. The results of comparison are described in Table 2.

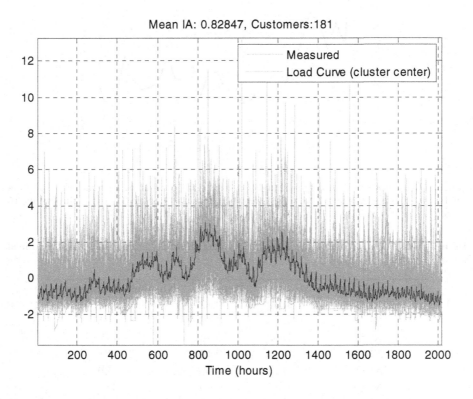

Fig. 2. An example of one of the created data-based load curves (black) and variation of customer electricity use (grey), containing all 181 customers electricity use figures during 84 days test period. Values were variance-scaled.

An example of new load curve and variation of customer electricity use is illustrated in Figure 2. In this example, mean index-of-agreement was 0.83 and 181 customers were classified to use this load curve. Additionally, the IA values were calculated for all customers in both two comparison cases; using original load curve and new data-based load curves. Results of this is shown in Figure 3, where comparison between each customer electricity use and original load curve (dash dot line) or new data based load curve (solid line) are presented using histogram.

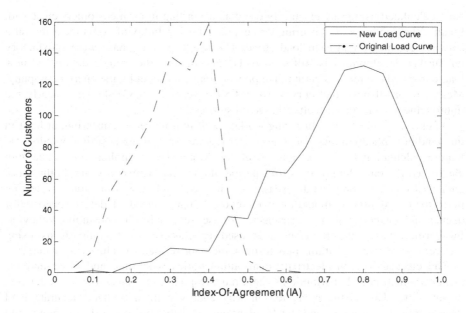

Fig. 3. The histogram of Index-Of-Agreement values for all customers in comparison between each customer electricity use and original load curve (dash dot line) or new data based load curve (solid line)

4 Discussion

The objective of this study was to apply feature-based clustering into creation of more accurate load curves for energy companies and present efficient way to take advantage of new hourly measured electricity use data. Moreover, the aim was evaluate correlation between customer real electricity use and original load curves set by energy companies.

The results of this comparison showed clearly that these present load curves do not correspond properly customer real electricity use. Mean index-of agreement for all load curves was only 0.30 (highest was 0.51) meaning almost that random values would result the same accuracy. This was actually an expected result because the energy company has already notified that there might be some problems concerning present load curves caused by changes of customer characteristics and behavior.

New hourly measured data gives whole a new view point of creation more accurate load curves. The size of the available data is huge and that is the reason why efficient computational methods are needed. Grouping customers using time series data can be done using feature-based clustering. In this case, set of statistical features were calculated and used in clustering with k-means algorithm. Features were calculated using time window of one week (168 hours). Optimal number of clusters was solved using Davies-Bouldin index resulting 16 clusters. The new data-based load curves were calculated using raw electricity use data according to each customers cluster id. Finally, the goodness of clustering was evaluated using Index-of-Agreement measure between customers and each load curve. The results in this phase were satisfactory but further development is still needed. Despite of that, the created data-based new load curves where more accurate than present load curves set by the energy company. Mean IA for all new load curves was 0.65 (highest was 0.83) showing clearly the improvement achieved by feature-based clustering.

After all, feature-based clustering worked well in time-series clustering, at least in this kind of application, but selecting proper features and setting best time-window for feature calculation has to be done carefully. Moreover, in applications concerning electricity use, data should cover time period of one year or several years. In this case, data was covering only winter season of the year and it was suitable for testing performance of used computational methods. Longer period of data is needed for deeper understanding of customer electricity use, seasonality or consuming behavior. Furthermore, only the K-means was used in clustering but comparisons using different clustering algorithms may result some improvements in clustering accuracy.

The electric load in electricity distribution varies with time and place and the power production and distribution system must respond to the customers load demand at any time. This is the mean reason why energy companies need accurate load information for pricing and tariff planning, distribution network planning and operation, power production planning, load management, customer service and billing and also providing information to customers and public authorities [5]. The methods presented in this paper, feature extraction and feature-based clustering, are suitable for creation of more accurate load curves using new hourly measured electricity use data concerning small customers. It is obvious that large number of customers and amount of raw data raises new challenges but in opposite to that, it gives great opportunity to use thousands of customers as a base of a load curve creation.

In this study we presented approach capable to cluster large and complex time-series data using feature-based clustering. The features were selected so that main characteristics, like periodicity, average, standard deviation and chaotic behavior, of electricity use data were captured. Furthermore, performance of approach was tested using real world data for over one thousand electricity customers.

5 Conclusions

This paper presents an efficient computational method for time series clustering and application concerning creation of electricity use load curves for small customers. Presented approach was based on extraction of statistical features from time series and their use in feature-based clustering of hourly measured electricity use data. The performance of approach was evaluated using data of 1035 real customers.

The feature-based clustering was able to cluster time series using just a set of derive statistical features. There were three advantages of the approach; (1) its ability to reduce the dimensionality of original time series, (2) it is less sensitive to missing values and (3) it can handle different lengths of time series. In addition, the presented approach resulted into more accurate load curves for this set of customers than present load curves set by the energy company.

Acknowledgements

This study was part of ENETE project and scientific collaboration between Research Group of Environmental Informatics (University of Kuopio), Savon Voima Oy and Enfo Ltd. in order to develop electricity distribution information systems and intelligent services for customers. We would like to thank Mr. Eero Sinkko, Mr. Ari Salovaara, Mr. Matti Huovinen, Mr. Ilkka Holmavuo and Mr. Sami Viiliäinen from Savon Voima Oy and also Mr. Harri Smolander and Mr. Jouko Kaihua from Enfo Ltd. for providing experimental data, important technical information and guidance during the research project.

References

1. Olier, I., Vellido, A.: Advances in Clustering and Visualization of Time Series Using GMT Through Time. Neural Networks 21, 904–913 (2008)
2. Wang, X., Smith, K., Hyndman, R.: Characteristic-Based Clustering for Time Series Data. Data Mining and Knowledge Discovery 13, 335–364 (2006)
3. Tsekouras, G.J., Kotoulas, P.B., Tsirekis, C.D., Dialynas, E.N., Hatziargyriou, N.D.: A Pattern Regocnition Methodology for Evaluation of Load Profiles and Typical Days of Large Electricity Customers. Elect. Power Syst. Res (2008), doi:10.1016/j.epsr.2008.01.010
4. Chicco, G., Napoli, R., Piglione, F.: Comparisons Among Clustering Techniques for Electricity Customer Classification. IEEE Transactions on Power Systems 21, 933–940 (2006)
5. Seppälä, A.: Load Research and Load Estimation in Electricity Distribution. VTT Publications 289. Technical Research Centre of Finland
6. The European Parliament and The Council of the European Union. Directive 2006/32/EC of the European Parliament and of the Council on Energy End-Use Efficiency and Energy Service and Repealing Council Directive 93/76/EEC (2006)
7. Bartels, R., Fiebig, D.G.: Metering and Modeling Residential End-Use Electricity Load Curves. Journal of Forecasting 15, 415–426 (1996)
8. Elkarmi, F.: Load Research as a Tool in Electric Power System Planning, Operation, and Control - The Case of Jordan. Energy Policy 36, 1757–1763 (2008)
9. Liao, W.: Clustering of Time Series Data - A Survey. Pattern Recognition 38, 1857–1874 (2005)
10. Baek, J., Geehyuk, L., Wonbae, P., Byoung-Ju, Y.: Accelerometer Signal Processing for User Activity Detection. In: Negoita, M.G., Howlett, R.J., Jain, L.C. (eds.) KES 2004. LNCS (LNAI), vol. 3215, pp. 610–617. Springer, Heidelberg (2004)
11. Sprott, J.C.: Chaos and Time-Series Analysis. Oxford University Press, Oxford (2003)
12. Masters, T.: Neural, Novel & Hybrid Algorithms for Time Series Prediction. John Wiley & Sons Inc., New York (1995)

13. Ravi, N., Dandekar, N., Mysore, P., Littman, M.L.: Activity Recognition from Accelometer Data. In: The Twentieth National Conference on Artificial Intelligence AAAI 2005. American Association for Artificial Intelligence, Stanford (2005)
14. Davies, D., Bouldin, D.: A Cluster Separation Measure. IEEE Transactions on Pattern Analysis and Machine Intelligence 2, 224–227 (1979)
15. MacQueen, J.: Some Methods for Classification and Analysis of Multivariate Observations. In: The Fifth Berkeley Symposium on Mathematical Statistics and Probability, vol. 1, pp. 281–297. University of California Press, Berkeley (1967)
16. Willmot, C.: Some Comments on the Evaluation of Model Perfomance. Bulletin of American Meteorological Society 63, 1309–1313 (1982)

The Effect of Different Forms of Synaptic Plasticity on Pattern Recognition in the Cerebellar Cortex

Giseli de Sousa, Rod Adams, Neil Davey, Reinoud Maex, and Volker Steuber

Science and Technology Research Institute
University of Hertfordshire
Hatfield, Herts, UK
{G.Sousa,R.G.Adams,N.Davey,R.Maex1,V.Steuber}@herts.ac.uk

Abstract. Many cerebellar learning theories assume that long-term depression (LTD) of synapses between parallel fibres (PFs) and Purkinje cells (PCs) provides the basis for pattern recognition in the cerebellum. Previous work has suggested that PCs can use a novel neural code based on the duration of silent periods. These simulations have used a simplified learning rule, where the synaptic conductance was halved each time a pattern was learned. However, experimental studies in cerebellar slices show that the synaptic conductance saturates and is rarely reduced to less than 50% of its baseline value. Moreover, the previous simulations did not include plasticity of the synapses between inhibitory interneurons and PCs. Here we study the effect of LTD saturation and inhibitory synaptic plasticity on pattern recognition in a complex PC model. We find that the PC model is very sensitive to the value at which LTD saturates, but is unaffected by inhibitory synaptic plasticity.

Keywords: Associative memory, Long-term depression, Purkinje cell, Cerebellum.

1 Introduction

The cerebellum is a part of the brain involved in a multitude of tasks, including motor control, and its functioning is responsible for the smoothness and precision of movements. These skills are improved by a process called motor learning, which is often assumed to be implemented by a form of synaptic plasticity known as long-term depression (LTD). LTD is a long-lasting decrease in synaptic strength due to a loss of AMPA receptors in the postsynaptic membrane[1]. In the cerebellum, LTD has been shown to occur at the synapses between Purkinje cells (PCs) and their excitatory inputs: climbing fibres (CFs) and parallel fibres (PFs). More specifically, cerebellar LTD is an associative process in which the strength of a PF synapse onto a PC is depressed when the CF and PF are activated at the same time.

M. Kolehmainen et al. (Eds.): ICANNGA 2009, LNCS 5495, pp. 413–422, 2009.
© Springer-Verlag Berlin Heidelberg 2009

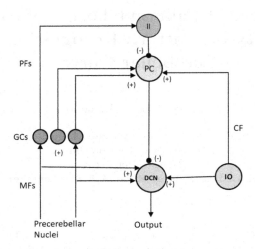

Fig. 1. Schematic diagram of the cerebellar circuitry. Purkinje cells (PCs) receive excitatory inputs (+) from 150,000 parallel fibres (PFs) and a single climbing fibre (CF), and inhibitory inputs (-) from inhibitory interneurons (II), and in turn inhibit the deep cerebellar nuclei (DCN). Also shown are: mossy fibres (MFs), granule cells (GCs) and the inferior olive (IO).

Classical cerebellar learning theories suggest that a PC can learn to discriminate between different activity patterns presented by its thousands of afferent PFs, due to LTD of the PF synapses [2]. It is assumed that as a result of LTD, the PC firing rate will be reduced when a learned pattern is presented again, and the PC will exert less inhibition on the deep cerebellar nuclei (Fig. 1). As a consequence, the cerebellar output should be increased, which could implement motor learning[1,3].

Recent work on cerebellar pattern recognition has demonstrated that this view is too simple. A combined theoretical and experimental study suggested that PCs can use a novel neural code based on the duration of their silent periods, where shorter pauses are produced in response to learned patterns [4] (Fig. 2A). This form of neural coding diverges from the classical view that uses the number or timing of individual spikes to distinguish between novel and learned patterns. In the computer simulations and experiments, the pause was compared with other spike response features like the number of spikes in a fixed time window after pattern presentation and the latency of the first spike in the response, and it was shown that the length of the pause was the best criterion for cerebellar PCs to identify learned patterns (Fig. 2B).

The previous simulations (see Methods) applied a simplified learning rule, where the AMPA receptor conductance was decreased by 50% each time a pattern was learned. After having stored a number of PF patterns, this could result in very small AMPA receptor conductances. However, experiments with LTD induction in cerebellar slices hardly ever result in mean AMPA receptor conductances of less than 50% of the pre-induction baseline [5,6]. We have therefore

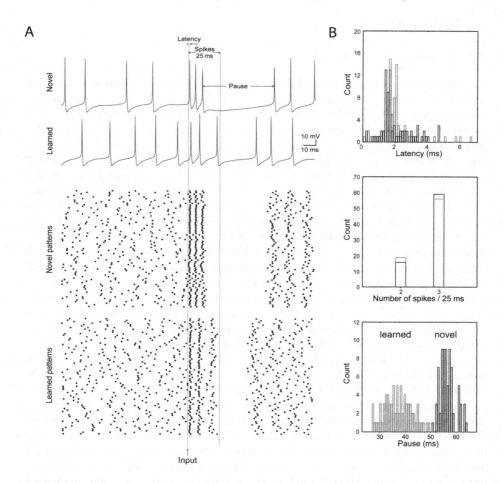

Fig. 2. Responses of a model Purkinje cell to novel and learned patterns of PF input. (A) Upper: The pause evoked by a novel pattern is longer than that for a learned pattern. Lower: Raster plot showing the responses to 75 learned and 75 novel patterns. (B) Response distribution for three different spike features. Upper: Latency of first spike after pattern presentation. Middle: Number of spikes in the first 25ms. Lower: Length of pause (modified with permission from [4]).

investigated a different learning rule with AMPA receptor conductances that saturate at varying values and have studied the effect of this learning rule in pattern recognition simulations.

Another contribution of the present work is to study the effect of LTD at the inhibitory synapses made by interneurons onto PCs (Fig. 1). It has recently been described that this inhibitory synaptic plasticity results in a mean depression of inhibitory inputs down to 75% of their original values [5]. We have run computer simulations to investigate the effect of different amounts of inhibitory synaptic plasticity on pattern recognition.

2   Methods

2.1   Purkinje Cell Model

The simulations were performed using the GENESIS neural simulator [7], with additional routines implemented in C++ and MATLAB. We simulated a multi-compartmental PC model with active dendrites and soma, as described in detail in references [8,9]. The model morphology was based on a reconstruction of a guinea-pig Purkinje cell [10]. Ten different types of voltage-dependent channels were modelled using Hodgkin-Huxley-like equations. The soma compartment had a fast and persistent Na^+ conductance, a delayed rectifier, a transient A-type K^+ conductance, a non-inactivating M-type K^+ conductance, an anomalous rectifier and a low-threshold T-type Ca^{2+} conductance. The dendritic compartments contained a Purkinje-cell specific high-threshold P-type and a low-threshold T-type Ca^{2+} conductance, two different types of Ca^{2+}-activated K^+ (KCa) conductances and an M-type K^+ conductance. Each cell was originally modelled with 147,400 dendritic spines, which were activated randomly by a sequence of PF inputs at an average frequency of 0.28 Hz. The background excitation was balanced by tonic inhibition, which made the model fire simple spikes at an average frequency of 48 Hz. Due to the large number of dendritic spines, which made the simulations computationally expensive, a simplified version of the model was constructed by decreasing the number of spines to 1% of the original number. To compensate for this reduction, the rate of PF excitation was increased to an average frequency of 28 Hz. As this simplified model gave identical results as the full model, it was used in the simulations presented here.

To study the effect of plasticity at inhibitory synapses, the model was provided with feed-forward inhibitory input by activating a variable number of inhibitory synapses onto the soma and main dendrite. The inhibitory input followed the synchronous activation of excitatory PFs synapses with a delay of 1.4 ms. Inhibition/excitation ratios were measured as ratios of the mean inhibitory postsynaptic current (IPSC) peak to the mean excitatory postsynaptic current (EPSC) peak when the model was voltage clamped to -40 mV.

2.2   Pattern Recognition

The pattern recognition simulations were performed in two steps. First, a number of random binary input patterns were generated, initially 200, and half of these patterns were learned by a corresponding artificial neural network (ANN). The ANN used was a modified version of an associative net with feed-forward connections between its inputs and output [11] and was trained by applying a modified version of the LTD learning rule [12](see below). The simulations of the ANN consisted of two phases: learning and recall.

In the learning phase, the weights of all synapses that received a positive input during the presentation of a pattern were set to a constant value. This LTD saturation value was kept constant and unaffected by further pattern presentations, different from the learning rule that had been used in the previous

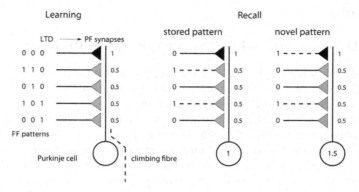

Fig. 3. Simplified schematic of the ANN model. Left side: during learning, three example PF patterns are stored by changing the synaptic weights that are associated with active input lines from their initial value of 1 to an LTD saturation value of 0.5 (this value is varied between different simulations). Right side: during recall, the responses to a stored and a novel pattern are calculated as dot product of input vector and weight vector, resulting in values of 1 and 1.5, respectively (note the difference to the original diagram in [12]).

simulations [4]. During the recall phase, the response of the ANN was given by the sum of the weights of all synapses that were associated with active inputs, which resulted in responses of the ANN to stored patterns that were lower than those to novel patterns (Fig. 3).

In the second phase of the pattern recognition simulations, the vector of synaptic weights was transferred from the ANN onto AMPA receptor conductances in the multi-compartmental PC model. This represents learning the PF patterns by depressing the corresponding AMPA receptor conductances during LTD induction. To test the recall of learned patterns, the PC model was then presented with a corresponding pattern of synchronous AMPA receptor activation at the PF synapses.

The discrimination between novel and learned pattern in the two models was evaluated by calculating a signal-to-noise ratio [13,14]:

$$s/n = \frac{(\mu_s - \mu_n)^2}{0.5(\sigma_s^2 + \sigma_n^2)} \tag{1}$$

where μ_s and μ_n represent the mean values and σ_s^2 and σ_n^2 represent the variances of the responses to stored and novel patterns, respectively. In the PC model, three different features of the spike response were tested as criteria to distinguish stored from novel patterns: the latency of the first spike fired after pattern presentation, the number of spikes in a 25ms time window after pattern presentation, and the duration of a silent period that followed the pattern presentation (see response distributions for these three different metrics in Fig. 2B). In all cases studied, the pause duration was the best criterion, and only pause based signal-to-noise ratios are presented here.

3   Results

3.1   LTD Saturation and the Number of Active PF Inputs

We initially investigated the effect of varying two parameters that were expected to affect the pattern recognition performance: the value at which LTD saturated and the number of active PFs for each pattern.

To study the effect of LTD saturation, we varied the LTD saturation value over a range from zero to 0.8, while keeping the same numbers of active PFs (1000) and PF patterns (100 novel and 100 stored) as in previous work [4]. We found that the ANN was insensitive to the amount of LTD induced (Fig. 4A). In contrast, the pattern recognition capacity based on the duration of silent periods in the PC model improved when the LTD saturation value decreased, with an optimal performance when the synaptic weights of active PFs were set to zero (Fig. 4B). The relative sensitivities of the ANN and the PC model to the amount of LTD induced are compared in Fig. 4C. While the ANN was unaffected by varying the amount of LTD, increasing the LTD saturation value to 0.8 in the PC model reduced the signal-to-noise ratio down to $0.4 \pm 0.4\%$ (n = 10) of the optimal value obtained by switching off the synapses completely. For LTD saturation values below 0.5, the PC model performed as well as or better than the previous model with a non-saturating learning rule [4].

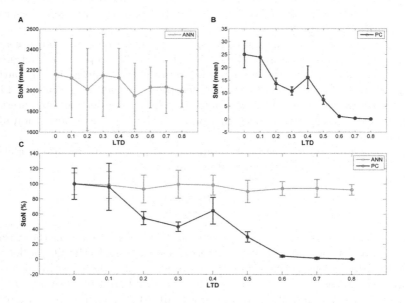

Fig. 4. Pattern recognition performance of the two models for a range of LTD values. The performance was evaluated by calculating s/n ratios for the ANN (A) and the PC model (B). The relative decreases in s/n ratio are compared in (C), showing that the PC model is more sensitive to LTD saturation than the ANN. Error bars indicate standard deviation (SD).

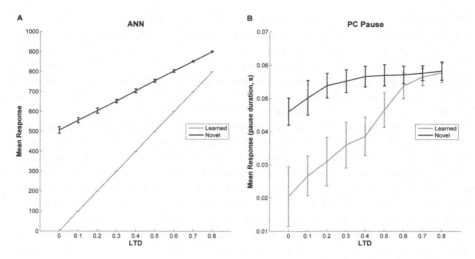

Fig. 5. Relationship between the LTD saturation value and the mean responses to stored and novel patterns in the ANN and the PC model. Although the difference between the mean responses to stored and novel patterns decreases with increasing LTD saturation values in both cases, in the ANN the variance of responses to novel patterns also decreases. This results in s/n ratios in the ANN that are independent of the LTD saturation value. Same simulation parameters as in Fig. 4. Error bars indicate SD.

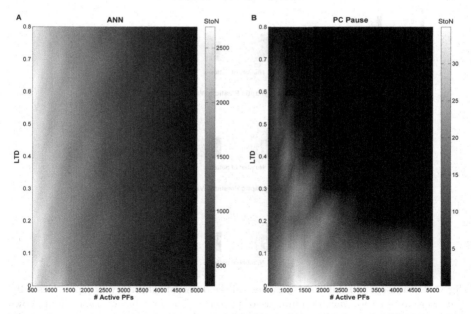

Fig. 6. Pattern recognition performance of the ANN (A) and PC model (B). The colour represents the resulting s/n ratio for each combination of a number of active PFs for each pattern (indicated on the x-axis) and an LTD saturation value (y-axis).

The reason for the difference in sensitivity of the ANN and the PC model to varying amounts of LTD became apparent when the mean responses of the two models to stored and novel patterns were plotted against the LTD saturation value (Fig. 5). In the PC model, increasing LTD saturation values reduced the difference in pause duration between stored and novel patterns, with standard deviations that were affected to a much lesser extent (Fig. 5B). This led to the drastic reduction in s/n ratio for weak LTD shown in Figure 4. In the ANN, the difference between the mean responses to stored and novel patterns was affected much less by the LTD saturation value, while the standard deviation of responses to novel patterns decreased with increasing LTD saturation values (Fig. 5A). Based on Equation (1), the constant signal-to-noise ratio of the ANN in the presence of varying amounts of LTD can be explained by a linear relationship between the squared difference of the mean responses to stored and novel patterns $(\mu_s - \mu_n)^2$ and the variance of the responses to novel patterns σ_n^2.

In a second set of simulations, we measured the effect of varying the number of active PFs in each pattern for a range of LTD values. As expected, the performance of the ANN deteriorated for larger numbers of activated PFs, while being independent of the amount of LTD induced over the whole range of numbers of active PFs tested (500-5000, Fig. 6A). In contrast, the PC model showed the

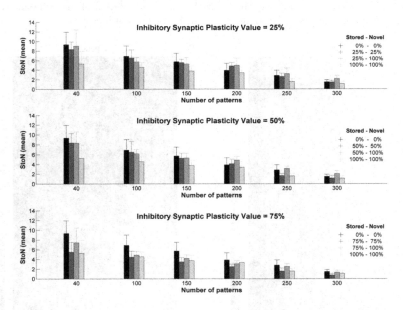

Fig. 7. Depression at inhibitory synapses. Three different inhibitory synaptic plasticity rules were applied for varying numbers of patterns. The first bar of each graph shows the s/n ratio when no inhibition is applied for both stored and novel patterns, resulting in the best pattern recognition performance. The other bars represent cases with inhibition present, with from left to right: plasticity for both stored and novel patterns, plasticity for stored patterns only and no plasticity for either type of patterns, using the original inhibitory conductances. Error bars indicate SD.

best pattern recognition capacity for a range between 1000 and 2000 active PFs and performed consistently worse for higher LTD saturation values (Fig. 6B).

3.2 Inhibitory Synaptic Plasticity

To investigate the effect of plasticity at the synapses between inhibitory interneurons and PCs, we initially used an inhibition/excitation ratio of one (see Methods), which is in the range of experimentally observed data from cerebellar slices [5]. We then introduced LTD at the inhibitory synapses and evaluated the pattern recognition performance of the PC for different numbers of patterns. The effect of inhibitory LTD was examined by depressing the inhibitory conductance to values between 25% and 75% of their pre-depression baseline. We used four different simulation setups (Fig. 7): no inhibition, plasticity at inhibitory synapse for stored and novel PF patterns, plasticity for stored patterns only and no plasticity for both patterns, that is, maintaining the baseline amplitude value for the original inhibition/excitation ratio [5].

We found that the pattern recognition performance of the PC model was unaffected by the presence of inhibitory LTD, even in the extreme case where the inhibitory plasticity was restricted to learned PFs patterns.

4   Conclusion

Previous computer simulations and experiments in cerebellar slices and awake behaving rats suggested that the cerebellum can use a novel neural code that is based on the duration of silent periods in neuronal activity [4]. These simulations used a complex multi-compartmental model of a cerebellar Purkinje cell that had been tuned to replicate a wide range of behaviours in vitro and in vivo [8,9], but they applied a simplified LTD learning rule, which involved dividing the synaptic weights of active PF inputs by two every time a PF pattern was learned. This could result in very small synaptic weights and does not fit experimental data on LTD induction in cerebellar slices, where the mean AMPA receptor conductances saturate and are hardly ever depressed to less than 50% of their pre-depression baseline values [5,6].Moreover, the previous simulations did not include the plasticity at synapses between inhibitory interneurons and PCs that has recently been characterised [5].

We have studied the effect of inhibitory synaptic plasticity and saturating LTD in the complex PC model. We found that the ability of the PC model to discriminate between learned and novel PF input patterns was unaffected by the presence of inhibitory plasticity for a wide range of parameter values.

However, the pattern recognition performance of the PC model was very sensitive to the value at which LTD saturated. In contrast to a corresponding ANN, which was unaffected by the amount of LTD induced, the performance of the PC model was improved by lower LTD saturation values. The best performance resulted from LTD saturation values of zero, which corresponds to silencing the PF synapses completely. Interestingly, large numbers of silent PF synapses have been observed by monitoring microscopically identified PF-PC connections in cerebellar slices [15]. Our simulation results indicate that the discrepancy between the existence of these silent synapses and the apparent saturation of LTD in induction experiments needs to be resolved to understand the connection between LTD and cerebellar learning.

Acknowledgment

This work was supported by a BBSRC-ANR Systems Biology Fellowship to V.S.

References

1. Ito, M.: Cerebellar long-term depression: Characterization, signal transduction, and functional roles. Physiol. Rev. 81(3), 1143–1195 (2001)
2. Marr, D.: A theory of cerebellar cortex. Journal of Physiology (London) 202, 437–470 (1969)
3. Ito, M.: The cerebellum and neural control. Raven Press, New York (1984)
4. Steuber, V., Mittmann, W., Hoebeek, F.E., Silver, R.A., De Zeeuw, C.I., Häusser, M., De Schutter, E.: Cerebellar LTD and pattern recognition by purkinje cells. Neuron 54(1), 121–136 (2007)
5. Mittmann, W., Häusser, M.: Linking synaptic plasticity and spike output at excitatory and inhibitory synapses onto cerebellar purkinje cells. J. Neurosci. 27(21), 5559–5570 (2007)
6. Wang, S.S.H., Denk, W., Häusser, M.: Coincidence detection in single dendritic spines mediated by calcium release. Nat. Neurosci. 3(12), 1266–1273 (2000)
7. Bower, J., Beeman, D.: The book of GENESIS: Exploring realistic neural models with the general neural simulation system (2003)
8. De Schutter, E., Bower, J.: An active membrane model of the cerebellar purkinje cell. I. simulation of current clamps in slice. Journal of Neurophysiology 71(1), 375–400 (1994)
9. De Schutter, E., Bower, J.: An active membrane model of the cerebellar purkinje cell: II. simulation of synaptic responses. Journal of Neurophysiology 71(1), 401–419 (1994)
10. Rapp, M., Segev, I., Yarom, Y.: Physiology, morphology and detailed passive models of guinea-pig cerebellar purkinje cells. Journal of Physiology (London) 474(1), 101–118 (1994)
11. Willshaw, D., Buneman, O., Longuet-Higgins, H.: Non-holographic associative memory. Nature 222, 960–962 (1969)
12. Steuber, V., De Schutter, E.: Long-term depression and recognition of parallel fibre patterns in a multi-compartmental model of a cerebellar purkinje cell. Neurocomputing 38, 383–388 (2001)
13. Dayan, P., Willshaw, D.J.: Optimising synaptic learning rules in linear associative memories. Biol. Cybern. 65(4), 253–265 (1991)
14. Graham, B.P.: Pattern recognition in a compartmental model of a ca1 pyramidal neuron. Network: Computation in Neural Systems 12(4), 473–492 (2001)
15. Isope, P., Barbour, B.: Properties of unitary granule cell-purkinje cell synapses in adult rat cerebellar slices. Journal of Neuroscience 22(22), 9668–9678 (2002)

Fuzzy Inference Systems for Efficient Non-invasive On-Line Two-Phase Flow Regime Identification

Tatiana Tambouratzis[1,2] and Imre Pázsit[2]

[1] Department of Industrial Management & Technology, University of Piraeus,
107 Deligiorgi St, Piraeus 185 34, Greece
[2] Department of Nuclear Engineering, Chalmers University of Technology,
SE-412 Göteborg, Sweden

The identification of two-phase flow regimes that occur in heated pipes is of paramount importance for monitoring nuclear installations such as boiling water reactors. A Sugeno-type fuzzy inference system is put forward for non-invasive, on-line flow regime identification. The proposed system is particularly efficient in that it employs a single directly computable input, four outputs calculated via subtractive clustering - each corresponding to one flow regime –, and four fuzzy inference rules. Despite its simplicity, the system accomplishes accurate identification of the flow regime of sequences of images from neutron radiography videos.

1 Introduction

Knowledge and control of the prevailing flow regime in a two-phase flow is crucial in energy producing installations, especially in BWRs. The ability of thermal hydraulic codes to predict flow regimes and parameters needs validating against measurements under well defined circumstances. Non-invasive identification methods are particularly attractive for such purposes as they avoid the need for the instrumentation to be immersed in the flow, a configuration that may not only be difficult to install but – once fitted - may disturb the flow. Statistical analysis of radiation attenuation measurements - e.g. X-rays [1] and gamma-rays [2-4] - has also been employed for investigating the two-phase flow in coolant pipes.

An interesting alternative is dynamic neutron radiography [5], which yields a two-dimensional projection of the flow structure topology in a time-resolved manner. The luminosity distribution of images has been exploited for two-phase flow identification via a number of computational intelligence approaches including

- artificial neural networks (ANNs) in [6-8],
- fuzzy inference systems (FIS's) in [9],
- neuro-fuzzy methodologies in [10].

Further to the non-invasive nature of the image capture process, these approaches promote on-line operation whereby the evolution of two-phase flow can be followed.

M. Kolehmainen et al. (Eds.): ICANNGA 2009, LNCS 5495, pp. 423–429, 2009.
© Springer-Verlag Berlin Heidelberg 2009

Along the same principles, a novel Sugeno-type FIS [11] is put forward here for non-invasive, on-line two-phase flow regime identification. The proposed system is particularly efficient; it employs (a) a single directly computable input per frame, namely the mean image intensity, (b) four outputs - each corresponding to one flow regime (bubbly, slug, churn and annular) – which are calculated via subtractive clustering [12] for the automatic generation of image-intensity flow regime-related clusters, and (c) four fuzzy inference rules. Despite its simplicity, the system accomplishes accurate identification of the flow regime of sequences of images from neutron radiography videos.

This paper is organized as follows: section 2 introduces two-phase flow; section 3 presents data pre-processing; section 4 describes the structure of the FIS employed for the flow regime identification task and details the results of five-fold cross-validation; finally, section 5 concludes the paper.

2   Two-Phase Flow Regime

As the coolant flows upwards within a heated pipe, the exchange of heat results in part of the coolant changing into its gas phase whereby two-phase flow of the coolant occurs. Depending on the proportion of liquid and gas phase in the coolant pipe and the topology of the vapour-liquid interfacial surface, a number of flow regime patterns are observed, e.g. liquid-only, bubbly, slug, churn, annular and mist, with the proportion of gas phase of the coolant progressively increasing from liquid-only to mist flow. Fig. 1 illustrates the Govier & Aziz flow-regime map [13] of vertical gas/liquid velocity at standard temperature and pressure, thus demonstrating the relationship that exists between the two phases of the coolant.

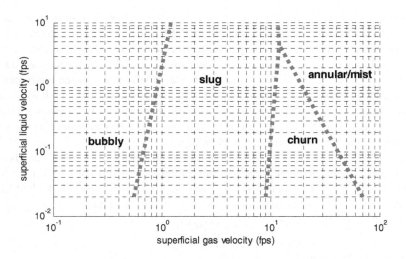

Fig. 1. Govier & Aziz flow-regime map

3 Data

3.1 Input

The dynamic neutron radiography video recorded at the Kyoto University Research Reactor Institute (KURRI) and capturing the two-phase flow in metal water loops has been employed. This video was collected during a gradual increase of the coolant temperature in the loop, whereby the four principal flow regimes (bubbly, slug, churn and annular) were recorded. The video was divided – through expert judgement - into four segments, each corresponding to one flow regime. Every video segment comprises 249 frames of size 154 x 412 pixels each; due to the image size and capture conditions, the frames are not of especially high quality. Furthermore, some frames from one flow regime may look more like those from another flow regime, a fact that was not taken into account during video segmentation.

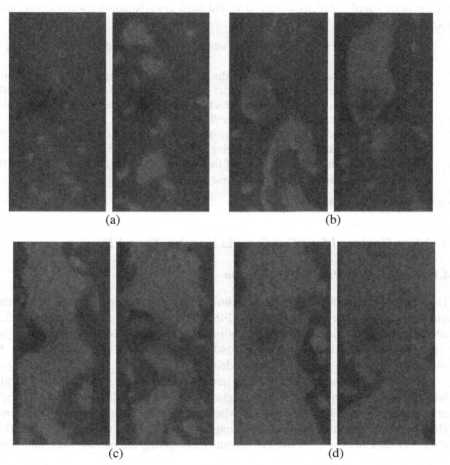

Fig. 2. KURRI video frames for progressively increasing coolant temperatures; bubbly (a), slug (b), churn (c) and annular (d) flow regimes

As can be seen in Fig. 2, the bubbly flow frames appear quite dark (liquid phase) with sporadic lighter (gas phase) small circular bubbles. The frames corresponding to slug flow are also dark but are interrupted by lighter medium-sized irregularly-shaped blobs. The churn flow frames appear roughly equally dark and light due to the lighter continuous snake-like areas of gas appearing within the dark area. Finally, the frames corresponding to annular flow appear quite light due to a unified lighter column appearing in practically the entire frame and leaving some dark areas along the two vertical edges of the frames (steam taking up most of the frame and pushing any coolant that remains in its liquid phase at the edges of the coolant pipe). In all cases, the lighter parts of the frame – or the darker parts in mist and annular flow - move upwards at subsequent frames while also gradually changing shape.

3.2 Pre-processing

Of the three simple, directly computable and most discriminating statistical operators used in [8], namely:

(A) the mean intensity of the image, which gives a measure of bubble quantity in the frame,

(B) the mean value (computed over the 412 rows of the image) of the maximum number of neighbouring pixels per row that are of intensity higher than the mean intensity of the image, which gives an indication of the bubble/blob sizes within each frame across the flow, and

(C) the mean value (computed over the 154 columns of the image) of the maximum number of neighbouring pixels per column that are of intensity higher than the mean intensity of the image, which gives an indication of the bubble/blob sizes within each frame along the flow,

only operator (A) has been utilized here.

4 Flow Regime Identification via Sugeno-Type FIS's

4.1 Training/Test Pattern Presentation

The Fuzzy Logic Toolbox [14] has been utilized for performing subtractive clustering and implementing the FIS; Sugeno-type FIS's have been used as they constitute ideal interpolating tools that can both coordinate different decision systems that are oriented towards different operating conditions and model nonlinear systems by interpolating between multiple linear models.

The FIS involves a single numerical input variable (mean image intensity) and a single categorical antecedent (flow regime). Subtractive clustering automatically generates four image-intensity clusters, where each cluster identifies one flow regime. The cluster centres (corresponding to image intensity values around 0.17, 0.3, 0.51 and 0.82) have been used as the FIS outputs representing the bubbly, slug, churn and annular flow, respectively[1]. The number of fuzzy rules equals the number of input clusters. The final output is determined by applying weights of 1 to all rule outputs. Owing to the fact that the FIS final output is continuous, the three mean values of pairs of ordered cluster centres have been used as thresholds (e.g. approximate values below 0.235 correspond to a bubbly flow regime decision, between 0.235 and 0.405 to a slug flow regime decision, between 0.405 and 0.665 to a churn flow regime decision and above 0.665 to an annular flow regime decision).

Five-fold cross-validation (CV) has been implemented for training and testing the FIS; both random and piece-wise CV have been investigated. In random CV, the frames from each video segment (flow regime) have been randomly partitioned into training and test sets such that each training set contains 199 or 200 frames of the four video segments and the corresponding test set contains the remaining 50 or 49 frames of the four video segments, respectively. More specifically, frames 1, 6, 11,..., 241 and 246 constitute the test patterns from each flow regime for the first fold, with the remaining patterns constituting the training patterns; frames 2, 7, 12,..., 242 and 247 constitute the test patterns from each flow regime for the second fold, with the remaining patterns constituting the training patterns, and so on for the third, fourth and fifth folds.

Table 1. Frames of the four video segments employed for FIS training and testing during piece-wise cross-validation

Folds	Training patterns	Test patterns
1	50-249	1-49
2	1-50 & 100-249	51-99
3	1-100 & 150-249	101-149
4	1-150 & 200-249	151-199
5	1-200	201-249

Piece-wise CV aims at simulating real situations, where data is not only limited, but - furthermore - may become available in batches and at different times. The operating conditions and data capture process may vary in such situations, whereby the identification task becomes more challenging. Table 1 illustrates the frames used in the five folds of the piece-wise CV procedure; it can be seen that the patterns of the test set are derived from consecutive frames from each flow regime, while those of the training set are derived from (partly) consecutive frames. Owing to the gradual increase of the coolant temperature in the loop during video capture (section 3.1), it is expected that the test patterns of fold 1 will tend to be underestimated, whereas those of fold 5 will tend to be overestimated.

Some extrapolation has been found necessary in all folds for both random and piece-wise CV; it is common in real situations for the training patterns not to fully capture the extremes of the phenomenon under observation.

---

[1] Although other values have been found to produce more accurate results (by up to 2%), the cluster centres have been preferred as they provide a systematic way of setting up the FIS's.

4.2 FIS Performance

Table 2 demonstrates FIS performance in terms of the mean, maximum and standard deviation of the absolute difference between predicted and actual flow regime value as well as in terms of flow-regime identification accuracy (%); the latter is further broken down into overestimations and underestimations, overall as well as per flow regime. The pooled results over the five folds of random and piece-wise CV are presented here.

Both CV procedures are satisfactory in terms of accuracy and the nature of misclassifications, which involve exclusively highly similar inputs from neighbouring flow regimes[2]. The overall accuracy of the proposed FIS is higher than that of the existing fuzzy and neuro-fuzzy approaches but slightly lower than the ANN approach of [8]; however, the latter is significantly more complex in terms of construction and computational complexity.

As expected, random CV produces more accurate results than piece-wise CV. This superiority is further expressed in the lower proportion of misclassifications, which concern only the first three flow regimes for random CV (10.04% slug-as-bubbly, 8.43% slug-to-churn and 9.64% churn-as-slug misclassifications)[3] but extend to all flow regimes for piece-wise CV (1.61% bubbly-as-slug, 11.24% slug-as-bubbly, 8.83% slug-as-churn, 10.04% churn-as-slug, 0.4% churn-as-annular and 0.4% annular-as-churn misclassifications).

Table 2. FIS accuracy over the five folds of the cross-validation process

	random CV	piece-wise CV
Mean deviation	0.0313	0.0327
maximum deviation	0.0358	0.0364
std deviation	0.2068	0.2048
accuracy (%) of which	92.97	91.57
overestimations	4.92	5.52
underestimations	2.11	2.91

5 Conclusions

A novel Sugeno-type fuzzy inference system has been put forward for the non-invasive identification of two-phase flow in boiling water reactors. The proposed system is particularly efficient in that it employs (a) a single directly computable input per frame, namely the mean image intensity, (b) four outputs calculated via subtractive clustering - each corresponding to one flow regime (bubbly, slug, churn and annular) –, and (c) four fuzzy inference rules. Despite its simplicity, the system has been found capable of accomplishing successful as well as consistent identification of the flow regime of sequences of images from neutron radiography

---

[2] A non-negligible overlap between mean intensity values is observed for frames from neighbouring flow regimes.

[3] Thus causing no misclassifications of annular flow regime which is the most crucial to identify.

videos. Further to its non-invasive nature, the proposed approach promotes on-line operation and monitoring; both image pre-processing and flow identification can be performed in real-time for each successive frame.

The proposed FIS method has been found superior to the previous on-line approaches in terms of efficiency, and comparable in terms of accuracy.

Acknowledgement

This work was financially supported by the Swedish Radiation Safety Authority, contract No. SSM 2008/3214.

References

1. Vince, M.A., Lahey, R.T.: On the development of an objective flow regime indicator. International Journal on Multiphase Flow 8, 93–124 (1982)
2. Chan, A.M.C., Banerjee, S.: Design aspects of gamma densitometers for void fraction measurements in small scale two-phase flow. Nuclear Instruments & Methods in Physics Research 190, 135–148 (1981)
3. Chan, A.M.C., Bzovey, D.: Measurement of mass flux in high temperature pressure steam-water two-phase flow using a combination of Pitot tubes and a gamma densitometer. Nuclear Engineering and Design 122, 95–104 (1990)
4. Kok, H.V., van der Hagen, T.H.J.J., Mudde, R.F.: Subchannel void-fraction measurements in a 6 x 6 rod bundle using a simple gamma-transmission method. International Journal on Multiphase Flow 27, 147–170 (2001)
5. Mishima, K., Hibiki, T., Saito, Y., Nakamura, H., Matsubayashi, M.: The review of the application of neutron radiography to thermal hydraulic research. Nuclear Instruments & Methods A 424, 66–72 (1999)
6. Mi, Y., Ishii, M., Tsoukalas, L.H.: Vertical two-phase flow identification with advanced instrumentation and neural networks. Nuclear Engineering and Design 184, 409–420 (1998)
7. Sunde, C., Avdic, S., Pázsit, I.: Classification of two-phase flow regimes via image analysis and a neuro-wavelet approach. Progress in Nuclear Energy 46, 348–355 (2005)
8. Tambouratzis, T., Pázsit, I.: Non-Invasive On-Line Two-Phase Flow Regime Identification. In: Proceedings of the 8th International FLINS Conference, Madrid, Spain, September 21-24, pp. 453–458 (2008)
9. Le Corre, J.M., Aldorwish, Y., Kim, S., Ishi, M.: Two-phase flow pattern identification using a fuzzy methodology. In: Proceedings of the 1999 International Conference on Information Intelligence and Systems, Washington D.C., November 1-3, pp. 155–161 (1999)
10. Tsoukalas, L.H., Ishi, M., Mi, Y.: A neurofuzzy methodology for impedance-based multiphase flow identification. Engineering Applications and Artificial Intelligence 10, 545–555 (1997)
11. Sugeno, M.: Fuzzy measures and fuzzy integrals: a survey. In: Fuzzy Automata and Decision Processes, pp. 89–102. North-Holland, Amsterdam (1977)
12. Chiu, S.: Fuzzy Model Identification Based on Cluster Estimation. Journal of Intelligent & Fuzzy Systems 2, 267–278 (1994)
13. Govier, G.W., Awiz, K.: The Flow of Complex Mixtures in Pipe. Van Nostrand Reinhold, N.Y. (1972)
14. The MathWorks, R2007b MatLab & Simulink, Fuzzy Logic Toolbox (September 2007)

Machine Tuning of Stable Analytical Fuzzy Predictive Controllers

Piotr M. Marusak

Institute of Control and Computation Engineering, Warsaw University of Technology,
ul. Nowowiejska 15/19, 00–665 Warszawa, Poland
P.Marusak@ia.pw.edu.pl

Abstract. Analytical fuzzy predictive controllers are composed of a few local controllers grouped in the fuzzy Takagi–Sugeno model. Usually, they are designed using the PDC method. Stability of the resulting system (controller + control plant) may be checked using variants of Lyapunov stability criterion. One of the first such variants was the criterion developed by Tanaka and Sugeno. It is rather conservative criterion because, in its basic form, it does not take into consideration the shape of the membership functions. However, this drawback can be exploited in the proposed approach. After finding the Lyapunov matrix for the system with the analytical fuzzy predictive controller, using e.g. LMIs, it is possible to change the membership functions of the controller without sacrificing stability. It is done using the heuristic method. Thus, practically any shape of membership functions may be assumed.

Keywords: fuzzy control, fuzzy systems, predictive control, nonlinear control.

1 Introduction

The model predictive control (MPC) algorithms are widely used in practice [2,8,12,17] because they can be successfully applied to control plants with difficult dynamics (inverse response, time delays) and to MIMO processes. Their success is caused by the way they are designed. During this process all information about control system operation and on conditions in which it operates can be taken into consideration.

Standard formulations of the MPC algorithms are based on linear control plant models. However, operation of the control system of a nonlinear control plant may be usually improved using the nonlinear controller. Algorithms using the approach based on Takagi–Sugeno (TS) fuzzy models [14] are relatively easy to synthesize. The idea of the fuzzy MPC algorithms is to design a few analytical MPC controllers for a few operating points and then compose them into one TS–type fuzzy controller. Once the controller is designed stability of the resulting control system may be checked using some criteria, usually based on the Lyapunov approach [3,5,6,7,13,15,16].

One of the first criteria was presented in [16]. Its basic version is, however, conservative. It does not take into consideration the membership functions used

M. Kolehmainen et al. (Eds.): ICANNGA 2009, LNCS 5495, pp. 430–439, 2009.
© Springer-Verlag Berlin Heidelberg 2009

in the controller. It is the drawback of the criterion which is creatively exploited in the paper. The idea is to first apply this criterion to the designed control system. (It can be done relatively easy using the LMI solvers, e.g. the Matlab LMI toolbox.) If the Lyapunov matrix is found, stability of the system is proven. It is in fact proven for a class of controllers (with different premises). Thus, the membership functions can be tuned and, at the same time, stability maintained. In order to tune the membership functions an optimization problem is solved. Due to the complicated form of the objective function to be minimized a heuristic tuning strategy is proposed.

In the next section fuzzy analytical MPC algorithms are described. In Sect. 3 stability criterion for fuzzy control systems is reminded. Section 4 contains description of the algorithm designed for tuning the stable fuzzy controllers. Example results obtained using the proposed tuning mechanism illustrating its usefulness are presented in Sect. 5. The paper is summarized in Sect. 6.

2 Analytical Fuzzy Predictive Controllers

In the MPC algorithms future control values are derived using a control plant model to predict future behavior of the control system. Then control signal is generated in such a way that some criteria should be maintained. Typically, minimization of the following performance function is demanded:

$$J = \sum_{i=1}^{p} \left(\overline{y}_k - y_{k+i|k}\right)^2 + \sum_{i=0}^{s-1} \lambda \left(\Delta u_{k+i|k}\right)^2 \ , \tag{1}$$

where $\overline{y}_k$ is a set–point value, $y_{k+i|k}$ is an output value for the $(k+i)^{\text{th}}$ sampling instant predicted at the kth sampling instant using control plant model, $\Delta u_{k+i|k}$ are future changes in the manipulated variables, $\lambda \geq 0$ is a weighting coefficient, p and s denote prediction and control horizons, respectively. The unconstrained optimization problem with the performance function (1) is a quadratic problem provided a linear model is used for prediction. Moreover, this problem has an analytical solution.

2.1 Standard Analytical Predictive Control Algorithms

Standard predictive control algorithms are based on linear process models, see e.g. [2,4,8,12,17], therefore the vector of predicted output values $\boldsymbol{y}$ can be described by the following formula:

$$\boldsymbol{y} = \widetilde{\boldsymbol{y}} + \boldsymbol{A} \cdot \Delta \boldsymbol{u} \ , \tag{2}$$

where $\boldsymbol{y} = \left[y_{k+1|k}, \dots, y_{k+p|k}\right]^T$, $\Delta \boldsymbol{u} = \left[\Delta u_{k|k}, \dots, \Delta u_{k+s-1|k}\right]^T$, $\widetilde{\boldsymbol{y}} = \left[\widetilde{y}_{k+1|k}, \dots, \widetilde{y}_{k+p|k}\right]^T$ is called a free response of the plant. It is because it contains future output values calculated assuming that the control signal does not change in the prediction horizon, i.e. describes influence of the manipulated variable values from the past and its form depends on the model used for prediction. Differences between MPC algorithms are caused by this fact.

A is a matrix composed of the control plant step response coefficients, called the dynamic matrix (it can be shown that it is the same in different types of MPC algorithms)

$$
A = \begin{bmatrix}
a_1 & 0 & \dots & 0 & 0 \\
a_2 & a_1 & \dots & 0 & 0 \\
\vdots & \vdots & \ddots & \vdots & \vdots \\
a_p & a_{p-1} & \dots & a_{p-s+2} & a_{p-s+1}
\end{bmatrix} .
\tag{3}
$$

The prediction (2) is used to formulate the performance function (1) which can be rewritten in the following form:

$$
J = (\overline{y} - y)^T \cdot (\overline{y} - y) + \lambda \cdot \Delta u^T \cdot \Delta u ,
\tag{4}
$$

where $\overline{y} = [\overline{y}_k, \dots, \overline{y}_k]$ is a vector of length p. The solution to the minimization problem with the performance function (4), in the case without constraints, can be expressed as:

$$
\Delta u = \left(A^T \cdot A + \lambda \cdot I \right)^{-1} \cdot A^T \cdot (\overline{y} - \widetilde{y}) .
\tag{5}
$$

From the vector Δu, the $\Delta u_{k|k}$ element is applied in the control system at each iteration. Therefore, a control law of a controller can be obtained.

The DMC algorithm is based on step responses [2,4,8,17]:

$$
\hat{y}_k = \sum_{n=1}^{p_d-1} a_n \cdot \Delta u_{k-n} + a_{p_d} \cdot u_{k-p_d} ,
\tag{6}
$$

where $\hat{y}_k$ is the output of the control plant model at the kth sampling instant, Δu_k is a change of the manipulated variable at the kth sampling instant, a_n ($n = 1, \dots, p_d$) are step response coefficients of the control plant, p_d is equal to the number of sampling instants after which the coefficients of the step response can be assumed as settled, u_{k-p_d} is a value of the manipulated variable at the $(k - p_d)^{\text{th}}$ sampling instant. The control law of the DMC controller is thus as follows, see e.g. [17]:

$$
u_k = u_{k-1} - r_0 \cdot e_k + \sum_{i=1}^{p_d-1} r_i \cdot \Delta u_{k-i} ,
\tag{7}
$$

where $e_k = \overline{y}_k - y_k$ is a value of the control error at the kth sampling instant, $r_0 = \sum_{j=1}^{p} K_{1j}$, $[r_1, \dots, r_{p_d}] = K_1 \cdot \widetilde{A}$, $K_1 = [K_{11}, \dots, K_{1p}]$ is the first row of the matrix $K = \left(A^T \cdot A + \lambda \cdot I \right)^{-1} \cdot A^T$,

$$
\widetilde{A} = \begin{bmatrix}
a_2 - a_1 & a_3 - a_2 & \dots & a_{p_d-1} - a_{p_d-2} & a_{p_d} - a_{p_d-1} \\
a_3 - a_1 & a_4 - a_2 & \dots & a_{p_d} - a_{p_d-2} & a_{p_d} - a_{p_d-1} \\
\vdots & \vdots & \ddots & \vdots & \vdots \\
a_{p+1} - a_1 & a_{p+2} - a_2 & \dots & a_{p_d} - a_{p_d-2} & a_{p_d} - a_{p_d-1}
\end{bmatrix} .
$$

Many MPC algorithms are based on a difference equation:

$$\hat{y}_k = \sum_{i=1}^{n_B} b_i \cdot y_{k-i} + \sum_{i=1}^{m_C} c_i \cdot u_{k-i},$$ (8)

where b_i, c_i are parameters of the model. In such a case the following control law is obtained:

$$u_k = u_{k-1} + r_e \cdot e_k + \sum_{i=1}^{m_C-1} r_u^i \cdot \Delta u_{k-i} + \sum_{i=1}^{n_B+1} r_y^i \cdot y_{k-i+1},$$ (9)

where r_e, r_u^i, r_y^i – coefficients obtained after the transformation.

2.2   Fuzzy Analytical Predictive Control Algorithms

Fuzzy analytical algorithms discussed in the paper are the combination of standard analytical MPC controllers. They are sets of the following rules [11,17] (for the case of the Fuzzy DMC controller):

Rule j: (10)

if y_k is B_1^j and ... and y_{k-n+1} is B_n^j and u_k is C_1^j and ... and u_{k-m+1} is C_m^j

then $u_k^j = u_{k-1} - r_0^j \cdot e_k + \sum_{i=1}^{p_d-1} r_i^j \cdot \Delta u_{k-i}$,

where y_k is an output variable value at the kth sampling instant, u_k is a manipulated variable value at the kth sampling instant, $B_1^j, \ldots, B_n^j, C_1^j, \ldots, C_m^j$ are fuzzy sets, r_i^j are the coefficients of the jth local controller, $j = 1, \ldots, l$, l is number of rules.

The output value of the FDMC controller is calculated using the following formula:

$$u_k = u_{k-1} - \tilde{r}_0 \cdot e_k + \sum_{i=1}^{p_d-1} \tilde{r}_i \cdot \Delta u_{k-i} ,$$ (11)

where $\tilde{r}_i = \sum_{j=1}^{l} \tilde{w}_j \cdot r_i^j$, $\tilde{w}_j$ are normalized weights calculated using standard fuzzy reasoning, see e.g. [14].

In the case of the MPC controller based on a difference equation only the consequences change in (10) and the output value can be calculated using the formula:

$$u_k = u_{k-1} + \tilde{r}_e \cdot e_k + \sum_{i=1}^{m_C-1} \tilde{r}_u^i \cdot \Delta u_{k-i} + \sum_{i=1}^{n_B+1} \tilde{r}_y^i \cdot y_{k-i+1} .$$ (12)

The design process of the fuzzy analytical predictive controller can be very simple:

1. A few linear models of the control plant are collected.
2. Each linear model is then used to calculate coefficients of an analytical MPC controller (e.g. (7) or (9)). These controllers will be the local controllers of the fuzzy analytical MPC controller.
3. The premise part of the fuzzy analytical MPC controller (10) is designed using expert knowledge, simulation experiments or machine tuning.

3   Stability Criterion

This section contains reminder of the stability criterion. It was developed for the systems described by the following Takagi–Sugeno fuzzy models:

Rule i: if x_k is F_1^i and ... and x_{k-n+1} is F_n^i, then

$$x_{k+1}^i = g_1^i \cdot x_k + \ldots + g_n^i \cdot x_{k-n+1}, \tag{13}$$

where $x_k, \ldots, x_{k-n+1}$ are state variables, $F_1^i, \ldots, F_n^i$ are fuzzy sets, $i = 1, \ldots, l$, l is the number of rules.

The i–th local model (13) can be thus written as:

$$\mathbf{x}_{k+1}^i = \mathbf{G}_i \cdot \mathbf{x}_k, \tag{14}$$

where

$$\mathbf{G}_i = \begin{bmatrix} g_1^i & \cdots & g_{n-1}^i & g_n^i \\ 1 & \ldots & 0 & 0 \\ 0 & \ldots & 0 & 0 \\ \vdots & \ddots & \vdots & \vdots \\ 0 & \ldots & 0 & 0 \\ 0 & \ldots & 1 & 0 \end{bmatrix}, \quad \mathbf{x}_k = \begin{bmatrix} x_k \\ x_{k-1} \\ x_{k-2} \\ \vdots \\ x_{k-n+2} \\ x_{k-n+1} \end{bmatrix}.$$

The output of the TS fuzzy model is given by the following formula:

$$\mathbf{x}_{k+1} = \frac{\sum_{i=1}^{l} w_i \cdot \mathbf{G}_i \cdot \mathbf{x}_k}{\sum_{i=1}^{l} w_i}, \tag{15}$$

where $w_i = \prod_{j=1}^{n} F_j^i \left(x(k-j+1) \right)$ is the i–th rule activation level (firing strength).

Tanaka and Sugeno formulated the following sufficient stability condition, arising from the direct Lyapunov method [16].

Theorem: The fuzzy system (14) is asymptotically stable if there is a positive definite matrix $\mathbf{P}$ such that for all matrices $\mathbf{G}_i$ the following inequalities are fulfilled:

$$\mathbf{G}_i^T \cdot \mathbf{P} \cdot \mathbf{G}_i - \mathbf{P} < 0, i = 1, \ldots, l. \tag{16}$$

$\square$

To find the matrix $\mathbf{P}$ one can solve the set of linear matrix inequalities (16) [15,18], e.g. using the Matlab LMI toolbox. The Tanaka–Sugeno criterion can be adopted for systems with delays and predictive controllers, see e.g. [10,11]. The drawback of the Tanaka–Sugeno criterion is its conservativeness. In its basic form, it does not take into consideration shape of the membership functions. However, this drawback is exploited by the proposed approach and in some sense becomes an advantage. It is because the membership functions may be changed (also their character) without sacrificing the stability which, once proven, will be maintained as long as the consequences do not change.

4   Algorithm for Machine Tuning of Fuzzy Controllers

The performance function of the control system to be minimized during the process of controller tuning can be of any standard form, e.g. the quadratic function can be used:

$$\min_{v} \left\{ \sum_{j=1}^{n} \sum_{i=1}^{t_l} \left(\overline{y}_{k+i}^{j} - y_{k+i}^{j} \right)^2 \right\} , \tag{17}$$

where v is a vector composed of the parameters of membership functions, t_l is a simulation horizon, n is the number of 'scenarios' taken into consideration during the learning process. It is advisable to prepare a few such 'scenarios' for different conditions of the control system operation (different set–point values, different disturbances). Thanks to such an approach the system may be tuned as well as possible to the conditions it will probably, according to knowledge of a designer, operate in in the future.

The other form of the performance function, also often used, is the sum of the absolute values of the control error:

$$\min_{v} \left\{ \sum_{j=1}^{n} \sum_{i=1}^{t_l} \left| \overline{y}_{k+i}^{j} - y_{k+i}^{j} \right| \right\} . \tag{18}$$

It is possible, using the method of tuning described in this section, to minimize also other typically used performance criteria (like e.g. overshoot or control time) or even combinations of them.

Due to the complicated nature of the problem and fact that the reasonable performance functions, e.g. those relied on future values of control error, depend on the parameters of membership functions in a complicated way, the following machine tuning algorithm is proposed:

1. Calculate the value $J = J_b$ of the performance function for current values of parameters of the membership functions.
2. Choose the first parameter from the vector v.
3. Using the current minimum and maximum values for the chosen parameter, generate grid of points.

4. Calculate the values of the minimized function in the grid points.
5. Pick up the point with the smallest value of the performance function.
6. If the best point (for which the value of the function is minimal) is on the boundary of the assumed range of values of the chosen parameter, then extended the range of values of the chosen parameter (if it is still possible) and pass to the step no. 4. Otherwise, pass to the next step.
7. If necessary, near the best point generate the next, more dense grid between points from the previous grid that are neighbors of the best point (to find a better solution) and pass to the step no. 4. Otherwise, pass to the next step.
8. Calculate the new value J_n of the performance function for the new value of the changed membership function parameter and compare it with J. If $J_n < J$ change the value of the parameter and assume $J := J_n$ else do not change the value of the parameter and pass to the next step.
9. Choose the next variable from the vector v. If there is no next element of the vector v then check if the value of the performance index J improved sufficiently comparing to the value from the last test J_b ($|J_b - J| > \varepsilon$, where ε is the parameter of the algorithm; during the tests it was assumed that $\varepsilon = 10^{-8}$). If yes then $J_b := J$ and return to step no. 2. If no, then stop.

In step no. 4 calculation of the values of the minimized function means the necessity to simulate operation of the control system and to derive the value of the chosen performance function, e.g. (17) or (18). It is thus advisable to choose not too dense grid of points.

Step no. 7 may be repeated a few times if needed, but during the tests it was not necessary. It should be noticed that the exact search is not crucial for the algorithm because in the next algorithm iteration the 'candidate' for the optimal point will probably change anyway.

For the search in the chosen direction the heuristic method was developed. It is because the standard golden section search technique found out to fail because of the character of the minimized function.

The first point of the algorithm resembles choice of improvement direction in the well known Gauss–Seidel optimization algorithm (directions chosen parallel to the axes). During the experiment the proposed technique found out to be sufficient. However, in the case of problems with convergence one may use an improved method of direction generation, e.g. with rotation of the axes as in the Rosenbrock method.

5   Simulation Experiments

The control plant under consideration is the ethylene distillation column. It is a highly nonlinear plant with a large time delay. The discrete–time TS plant model for sampling time $T_p = 40$min is as follows:

Rule 1: if u_{k-2} is P_1, then

$$y^1_{k+1} = 0.7659 \cdot y_k - 520.2638 \cdot u_{k-2} + 2220.9067; \quad (19)$$

Rule 2: if u_{k-2} is P_2, then

$$y^2_{k+1} = 0.7659 \cdot y_k - 253.5771 \cdot u_{k-2} + 1102.4471;$$

Rule 3: if u_{k-2} is P_3, then

$$y^3_{k+1} = 0.7659 \cdot y_k - 125.1030 \cdot u_{k-2} + 563.8767;$$

with membership functions shown in Fig. 1. More information about the control plant and profits obtained after application of a fuzzy predictive controllers in a control system of the distillation column one may find in [9].

To the control plant, the analytical fuzzy DMC controller was designed and the tuning coefficient $\lambda = 8e + 6$ was assumed. Then, utilizing procedure presented in [11], a Lyapunov matrix of the control system with this controller was found using Matlab LMI toolbox. In the next step the membership functions were tuned using the procedure described in Sect. 4.

At the beginning of experiments trapezoid membership functions were used, as in the model of the control plant. In such a case, formulation of the proper

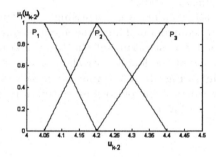

Fig. 1. Membership functions of the control plant model

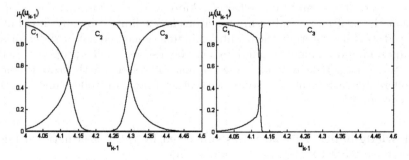

Fig. 2. Membership functions of the controller a) initial ones b) after tuning

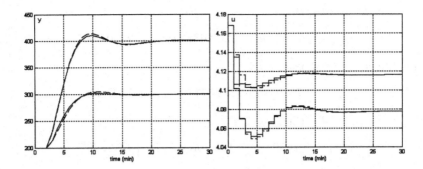

Fig. 3. Responses of the control systems to changes of the set–point value from $y_0 = 200$ ppm: before tuning – dashed line, after tuning – solid line; left – output variable, right – manipulated variable

conditions for parameters of the membership functions to guarantee their reasonable values is not a trivial task. In order to simplify the procedure it is thus advisable to use such functions that the mentioned problem will not occur. It was done during the experiments as the change of shape of the membership functions does not influence stability of the system. Therefore, membership functions shown in Fig. 2a were initially chosen.

The result obtained after tuning was surprising – the tuning algorithm eliminated the middle local controller. Thus, the membership functions depicted in Fig. 2b were obtained and structure of the fuzzy controller was simplified. Despite elimination of one of the local controllers, responses of the control system changed only slightly (Fig. 3). The tunned controller generated a little bit smaller overshoot. The value of the performance function (as in (18)) improved from $J_0 = 1394.4$ to $J_f = 1337.4$.

6   Summary

The algorithm of machine tuning of the fuzzy predictive controllers was proposed in the paper. It is simple but effective. Moreover, it is formulated in such a way that the choice of membership functions is not limited by anything but the employed stability criterion; the membership functions need not be differentiable, i.e. trapezoid and triangular functions may be also used. The proposed method found out to be satisfactory for design of control systems with fuzzy predictive controllers, but also control systems with other fuzzy controllers may be tuned using this method.

Acknowledgment. This work was supported by the Polish national budget funds for science 2007–2009 as a research project.

References

1. Blevins, T.L., McMillan, G.K., Wojsznis, W.K., Brown, M.W.: Advanced Control Unleashed. ISA (2003)
2. Camacho, E.F., Bordons, C.: Model Predictive Control. Springer, Heidelberg (1999)
3. Chadli, M., Maquin, D., Ragot, J.: Relaxed stability conditions for Takagi–Sugeno fuzzy models. In: Proc. Systems, Man an Cybernetics – 2000, USA, pp. 3514–3519 (2000)
4. Cutler, C.R., Ramaker, B.L.: Dynamic Matrix Control – a computer control algorithm. In: Proc. Joint Automatic Control Conference, San Francisco, CA, USA (1979)
5. Guerra, T.M., Vermeiren, L.: LMI–based relaxed non–quadratic stabilization conditions for nonlinear systems in Takagi–Sugeno's form. Automatica 40, 823–829 (2004)
6. Kruszewski, A., Guerra, T.M.: New Approaches for the Stabilization of Discrete Takagi-Sugeno Fuzzy Models. In: Proc. IEEE CDC/ECC, Seville, Spain (2005)
7. Lam, H.K., Leung, F.H.F.: Stability analysis of discrete–time fuzzy–model–based control systems with time delay: Time delay–independent approach. Fuzzy Sets and Systems 159, 990–1000 (2008)
8. Maciejowski, J.M.: Predictive control with constraints. Prentice Hall, Harlow (2002)
9. Marusak, P., Tatjewski, P.: Fuzzy Dynamic Matrix Control algorithms for nonlinear plants. In: Proc. 6th International Conference on Methods and Models in Automation and Robotics, Miedzyzdroje, Poland, pp. 749–754 (2000)
10. Marusak, P., Tatjewski, P.: Stability analysis of nonlinear control systems with fuzzy DMC controllers. In: Proc. IFAC Workshop on Advanced Fuzzy and Neural Control AFNC 2001, Valencia, Spain, pp. 21–26 (2001)
11. Marusak, P., Tatjewski, P.: Stability analysis of nonlinear control systems with unconstrained fuzzy predictive controllers. Archives of Control Sciences 12, 267–288 (2002)
12. Rossiter, J.A.: Model-Based Predictive Control. CRC Press, Boca Raton (2003)
13. Sonbol, A., Fadali, M.S.: Stability analysis of discrete TSK type II/III systems. IEEE Trans. Fuzzy Systems 14, 640–653 (2006)
14. Takagi, T., Sugeno, M.: Fuzzy identification of systems and its application to modeling and control. IEEE Trans. Systems, Man and Cybernetics 15, 116–132 (1985)
15. Tanaka, K., Ikeda, T., Wang, H.O.: Fuzzy regulators and fuzzy observers: relaxed stability conditions and LMI–based designs. IEEE Trans. Fuzzy Systems 6, 250–265 (1998)
16. Tanaka, K., Sugeno, M.: Stability analysis and design of fuzzy control systems. Fuzzy Sets and Systems 45, 135–156 (1992)
17. Tatjewski, P.: Advanced Control of Industrial Processes; Structures and Algorithms. Springer, London (2007)
18. Zhao, J., Wertz, V., Gorez, R.: Design a stabilizing fuzzy and/or non–fuzzy state–feedback controller using LMI method. In: Proc. 3rd European Control Conference, Italy, pp. 1201–1206 (1995)

Crisp Classifiers vs. Fuzzy Classifiers: A Statistical Study

J.L. Jara and Rodrigo Acevedo-Crespo

Universidad de Santiago de Chile (Usach); Depto. de Ingeniería Informática,
Avda. Ecuador 3659, 9170124, Santiago, Chile
jljara@informatica.usach.cl, rodrigo.acevedo@gmail.com

Abstract. A study is made of whether there is a significant statistical difference in performance between crisp and fuzzy rule-based classification. To do that, 12 datasets were chosen from the UCI repository that are widely used in the literature, and use was made of four different algorithms for rule induction —two crisp and two fuzzy— to classify them. Then a non-parametric statistical test was used for measuring the significance of the results, which indicated that both paradigms —crisp and fuzzy classification— are not different in the statistical meaning.

1 Introduction

We often hear about the use of fuzzy logic in applications that require managing uncertainty. Its main feature is that it allows linguistic labels that humans use to communicate —such as tall, short, not so cold, etc.— to be represented through simple mathematical functions. These kinds of linguistic labels are not handled naturally in crisp logic because a threshold is needed for making the rules. For example, if we want to classify people as tall or short in crisp logic, we have to fix a threshold height so that people shorter than this height will always be considered short and people taller than this height will always be considered tall. But it does not matter how close to the threshold a person is, if he/she is below the height chosen, he/she will be considered short. Hence a person who is 1 mm shorter than the threshold height will be considered short, which makes little sense in the way humans classify objects. In fuzzy classification, such a person might be considered —for example— 90% tall and 10% short. Generally, objects belong to some degree to all the classes when the fuzzy approach is used.

The contribution of fuzzy logic and fuzzy classification in solving real-world problems is undeniable. In the field of automatic control, fuzzy logic seems to offer useful simplifications of complex realities, making them more manageable [1]. Many successful applications of fuzzy classification can be found in remote sensing [2,3,4], where every pixel in a digital image representing a landscape must be classified —e.g., into different types of vegetation. This image is normally obtained from a satellite and therefore each pixel might represent several kilometers of land. A crisp classifier would assign only one class to each pixel, which does not feel natural. In this sense, fuzzy classification would represent more naturally mixtures and transition zones [2].

M. Kolehmainen et al. (Eds.): ICANNGA 2009, LNCS 5495, pp. 440–447, 2009.
© Springer-Verlag Berlin Heidelberg 2009

It is only natural to wonder whether fuzzy logic is able to improve crisp classification. Nowadays the statistical evaluation of classifiers has become an important task in machine learning research. Despite this fact, the "correct way" to carry out this task is not completely clear and, to our knowledge, there is no study that confirms or rejects whether fuzzy logic yields an improvement in classification. Nevertheless, there seems to be a common consent that fuzzy logic cannot drastically improve the performance of crisp classification. In this work we try to test this hypothesis in a more rigorous way.

The document is organised as follows: section 2 briefly describes the four learning algorithms used in the study. Section 3 explains the efficiency measure and the nonparametric statistical test utilized, as well as the experiments conducted. Section 4 presents the results in terms of both the efficiency obtained by the classifiers and the statistical comparison. Some related work is shown in section 5. Finally, section 6 contains the conclusions of this work and some ideas for future work.

2   Fuzzy and Crisp Algorithms

In this work two paradigms for classification are studied: fuzzy rules and crisp rules. We choose two representative algorithms of each paradigm in order to compare them. In general, fuzzy classification is understood as those learning algorithms that after the training process have learned a set of fuzzy rules — i.e., rules that use fuzzy logic. Crisp classifiers are those that learn rules which do not use fuzzy logic.

The two crisp algorithms selected for this study are Ripper and C4.5. Ripper [5] is an improvement of the IREP algorithm that efficiently learns propositional decision rules. It uses the minimum description length principle in the pruning mechanism to optimize the set of rules obtained. C4.5 [6] is a machine learning algorithm that builds decision trees. It uses an application of entropy for measuring how much information can be obtained with a particular split of the hypothesis space by one or more attributes to guide the tree building process.

The two fuzzy approaches selected for this work are based on evolutionary algorithms and clustering. For the former, we use 2Slave [7], which transforms the classification problem into an optimization problem which is then solved using a genetic algorithm. Thus, a population of rules mutate and reproduce, and the one that covers more examples from the training dataset is selected in each generation. The examples covered by this rule are then removed from the dataset and the process is repeated until a set of rules that covers the whole training dataset is obtained.

Clustering is an unsupervised learning technique that separates data into an arbitrary or user-defined number of clusters. The most popular clustering algorithm is the K-means Algorithm, which splits data into k clusters. In the crisp version, each data point belongs to exactly one of the clusters identified. In the fuzzy version each data point belongs to all clusters with a different membership degree. For testing this approach we use the algorithm proposed by Klawonn and Krusse [8], which can find membership functions for each dimension of any cluster. Considering that each cluster corresponds to a class, the conjunction of these membership functions constitutes a fuzzy rule that can be used for fuzzy classification.

3   Materials and Methods

In the majority of machine learning studies, two or more algorithms are compared by applying them to a particular problem and assessing the performance they achieve in the task. Although this is a valid approach when trying to solve a particular problem, these results cannot be extrapolated to another problem or domain. To be able to draw more general conclusions about two algorithms it is necessary to establish a difference in their performance which is statistically significant. Many studies report performance differences, but they do not normally apply a suitable hypothesis test to validate these conclusions.

In this work we explain and apply a procedure to establish whether the classification quality obtained by crisp and fuzzy rule inducers, two algorithms of each type, is significantly different, with statistical support. The elements involved in this procedure are detailed in the following sections.

3.1   Hypothesis Test

We use the Friedman test, which is a non-parametric version of the ANOVA test [9]. The Friedman test seems to be the best choice when evaluating more than two classifiers on different classification tasks (independent datasets), because there is no assumption about data or error distribution, as there exists with parametric tests. Moreover, good statistical tests designed for comparing two classifiers —e.g. Wilcoxon's signed rank test— might result in an increase of the family-wise error when used on more than two classifiers by evaluating repeatedly different combinations of two of them.

The Friedman test reports whether the difference in the average ranking of two or more algorithms is statistically significant, where the null hypothesis in that all algorithms have the same performance. Let r_i^j be the ranking of the jth of k algorithms on the ith of N datasets. The Friedman test compares the average ranks of the algorithms, where each average rank R_j is given by

$$R_j = \frac{1}{N} \sum_j r_i^j \qquad j = 1, \ldots, k .$$

(1)

Under the null hypothesis, the Friedman statistic has distribution χ_F^2 with $k - 1$ degrees of freedom (Eq. 2) when comparing four or more classifiers on ten or more datasets.

$$\chi_F^2 = \frac{12N}{k(k+1)} \left[\sum_j R_j^2 - \frac{k(k+1)^2}{4} \right] .$$

(2)

Friedman statistic has been criticised as being too conservative, so we actually use the improvement proposed by Iman and Davenport [10]. This statistic (Eq. 3) is distributed according to an F distribution with $k-1$ and $(k-1)(N-1)$ degrees of freedom. We will accept the null hypothesis with critical values $p < 0.05$ with a confidence of 95%, as this is the most frequent value found in the literature.

$$F_F = \frac{(N-1)\chi_F^2}{N(k-1) - \chi_F^2}.$$

(3)

It must be noted that the Friedman test reports only whether there is a significant difference between all algorithms, but it does not tell which algorithm is better than the others. To do that, *post hoc* tests such as critical distance over the average ranks should be applied. If we want all classifiers to be compared with each other, then the Nemenyi test can sort the algorithms. If we want a control classifier to be compared to the others, the Bonferroni-Dunn test can be used [11].

To apply the analysis described above we first have to build a ranking of the algorithms studied according to their performance in different classification tasks. For this we need a way to measure the performance of each algorithm and several —at least ten— datasets to test them.

3.2   Datasets

We chose 12 UCI repository datasets for training and testing every algorithm. Table 1 shows the selected datasets as well as some of their main features: number of instances, number of attributes and number of classes. Instances containing missing values were filtered out from each dataset before splitting it into training and test datasets.

Table 1. UCI datasets selected for the experiments and their main features

Dataset Name	Number of instances	Number of attributes	Number of classes
Lung Cancer	32	56	3
Musk	476	168	2
Sonar	208	60	2
Breast Cancer (Prognostic)	198	34	2
SPECTF Heart	267	44	2
Wine	178	13	3
Breast Cancer (Diagnostic)	569	32	2
Glass	214	10	6
New Thyroid	215	5	3
Iris	150	4	3
Ecoli	336	8	8
Credit Approval	690	14	2

3.3   Efficiency Measure

Efficiency here has nothing to do with the computational complexity, but is a descriptive measure of how well the classification process was performed by a particular algorithm. There are several efficiency measures, but many of them are designed for measuring the efficiency of binary classification (two class datasets). Since we use some datasets with more than two classes, such measures are not suitable. We have chosen the Cohen's Kappa statistic [12] —just *kappa* in what follows— because it measures the agreement of the class distribution over the instances, and not only the percentage of correct classified instances [13].

3.4   Experimental Design

The experiments in this work follow the 5x2 cross validation setting proposed in [14]. In this scheme each dataset is randomly split into two sets: one is used for training and the other for testing. Both sets contain approximately half of the data and keep the original class proportion. This process is repeated five times —each one is a fold— generating five pairs of training and test data.

In each fold, we register the performance of a learning algorithm on the test data —trained with the corresponding training data— as a confusion matrix. When the five folds are completed, a global confusion matrix for the algorithm is obtained by summing the individual matrices. With this global matrix we compute the kappa value for the algorithm in the task. Figure 1 shows this general experimental design.

With the kappa values collected for each algorithm, we build a ranking of the different approaches for that classification task. Repeating this procedure for each dataset in Table 1 (classification tasks), we are able to calculate an average ranking for each algorithm. Then we used the Friedman and Iman-Davenport statistics on these averages to determine whether the differences in performance are statistically significant or not.

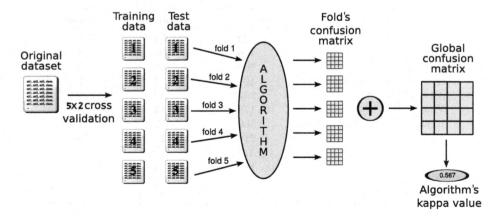

Fig. 1. 5x2 cross validation scheme for our experiments

For the clustering approach, it was necessary to apply a gain ratio filter on the original datasets, because —unlike the other three algorithms studied— this approach cannot discriminate irrelevant attributes during training.

4 Results

Table 2 shows the kappa values obtained by each algorithm for each dataset. In each case the parameters of the algorithms were individually tuned to obtain optimal results. With these efficiency measures, we can rank the algorithms for each classification tasks, as shown in Table 3.

Table 2. Performance of each algorithm for each dataset in Table 1

Dataset	Ripper	C4.5	2Slave	Clustering
Lung Cancer	0.315	0.423	0.263	0.306
Musk	0.441	0.532	0.515	0.755
Sonar	0.453	0.435	0.454	0.462
Breast Cancer (Prognostic)	0.320	0.271	0.043	0.235
SPECTF Heart	0.240	0.267	0.112	0.519
Wine	0.841	0.872	0.966	0.863
Breast Cancer (Diagnostic)	0.870	0.865	0.859	0.744
Glass	0.474	0.466	0.512	0.000
New Thyroid	0.847	0.819	0.806	0.447
Iris	0.864	0.888	0.936	0.931
Ecoli	0.726	0.722	0.733	0.106
Credit Approval	0.703	0.726	0.422	0.000

In this case the number of algorithms is $k = 4$ and the number of datasets is $N = 12$, so the Friedman statistic is calculated as follows:

$$\chi_F^2 = \frac{12 \cdot 12}{4 \cdot (4+1)} \left[2.417^2 + 2.250^2 + 2.500^2 + 2.833^2 + -\frac{4 \cdot (4+1)^2}{4} \right] = 1.3 \,. \tag{4}$$

Considering the Iman-Davenport modification, the statistic takes the value:

$$F_F = \frac{(12-1) \cdot 1.3}{12 \cdot (4-1) - 1.3} = 0.412 \,. \tag{5}$$

Using the F distribution tables, we determined that the value in Eq. 5 yields a p-value $p = 0.75$. Therefore, the null hypothesis is accepted and we can conclude that there is no statistically significant difference between the crisp and fuzzy classifiers studied.

Table 3. Ranking of algorithms. Each row shows the relative ranking of the different approaches for the given classification task.

Dataset	Ripper	C4.5	2Slave	Clustering
Lung Cancer	2	1	4	3
Musk	4	2	3	1
Sonar	3	4	2	1
Breast Cancer (Prognostic)	1	2	4	3
SPECTF Heart	3	2	4	1
Wine	4	2	1	3
Breast Cancer (Diagnostic)	1	2	3	4
Glass	2	3	1	4
New Thyroid	1	2	3	4
Iris	4	3	1	2
Ecoli	2	3	1	4
Credit Approval	2	1	3	4
Avarage rank (R_j)	2.417	2.250	2.500	2.833

5   Related Work

A theoretical and empirical study of the current practice for comparing two or more learning algorithms on multiple data sets is presented in [11], where several recommendations are made to conduct this kind of comparison which are tested experimentally with a set of (crisp) C4.5 classifiers. We have followed these recommendations in the work presented here, particularly the use of the Friedman test.

In [14], the benefit of using non-parametric tests for comparing (crisp) evolutionary-based machine learning algorithms is confirmed. This piece of work also suggests the use of the Friedman and Iman-Davenport tests as well as measuring the performance of the approaches in term of kappa. Their guidelines are closely followed in this paper.

6   Conclusions and Future Work

In this work we raise the question of whether there really exists a difference between rule-based crisp and fuzzy classifiers. Although there seems to be common agreement that the crisp and fuzzy approaches are not different for classification, we could not find a statistical demonstration of that. We have developed a valid scheme for a statistical study using 12 UCI repository datasets and a non-parametric test for comparing two crisp and two fuzzy rule learning algorithms. The results obtained confirm that the difference between both paradigms is not statistically significant.

Current work has focused on studying rule-based learning algorithms. Future work should aim at incorporating other approaches for generating crisp and fuzzy classifiers, such as probabilistic models or support vector machine, that are popular in the pattern recognition community.

It would also be interesting to look for characteristics in the datasets that might link the performance of a paradigm to the features or the nature of the datasets. This could be particularly interesting for fuzzy classifiers, since they simulate linguistic labels that are often associated with the problem's nature.

References

1. Wang, L.X.: Adaptive Fuzzy Systems and Control: Design and Stability Analysis. Prentice Hall, Englewood Cliffs (1994)
2. del Amo, A., Montero, J., Biging, G., Cutello, V.: Fuzzy classification systems. European Journal of Operational Research 156(2), 495–507 (2004)
3. Mather, E.A.: Fuzzy vs. crisp land cover classification of satellite imagery for the identification of savanna plant communities of the Oak Openings region of NW Ohio and SE Michigan. Master's thesis, The University of Toledo (2006)
4. Nakashima, T., Schaefer, G., Yokota, Y., Ishibuchi, H.: A weighted fuzzy classifier and its application to image processing tasks. Fuzzy Sets and Systems 158(3), 284–294 (2007)
5. Cohen, W.W.: Fast effective rule induction. In: Prieditis, A., Russell, S. (eds.) Proceedings of the Twelfth International Conference on Machine Learning (ICML 1995), Tahoe City, California, USA, pp. 115–123. Morgan Kaufmann, San Francisco (1995)
6. Quinlan, J.R.: C4.5: Programs for Machine Learning. Series in Machine Learning. Morgan Kaufmann, San Francisco (1993)
7. González, A., Pérez, R.: SLAVE: A genetic learning system based on an iterative approach. IEEE Transactions on Fuzzy Systems 7(2), 176–191 (1999)
8. Klawonn, F., Kruse, R.: Derivation of fuzzy classification rules from multidimensional data. In: Lasker, G., Liu, X. (eds.) Advances in Intelligent Data Analysis. The International Institute for Advanced Studies in Systems Research and Cybernetics, Windsor, Ontario, Canada, pp. 90–94 (1995)
9. Sheskin, D.J.: Handbook of Parametric and Nonparametric Statistical Procedures, 2nd edn. Chapman & Hall/CRC (2000)
10. Iman, R., Davenport, J.: Approximations of the critical region of the Friedman statistic. Communications in Statistics A 9(6), 571–595 (1980)
11. Demšar, J.: Statistical comparisons of classifiers over multiple data sets. The Journal of Machine Learning Research 7, 1–30 (2006)
12. Cohen, J.: A coefficient of agreement for nominal scales. Educational and Psychological Measurement 20(1), 37–46 (1960)
13. Dietterich, T.G.: Approximate statistical test for comparing supervised classification learning algorithms. Neural Computation 20, 1895–1923 (1998)
14. García, S., Fernández, A., Luengo, J., Herrera, F.: A study of statistical techniques and performance measures for genetics-based machine learning. Soft Computing 13(10), 959–977 (2009)

Efficient Model Predictive Control Algorithm with Fuzzy Approximations of Nonlinear Models

Piotr M. Marusak

Institute of Control and Computation Engineering, Warsaw University of Technology,
ul. Nowowiejska 15/19, 00–665 Warszawa, Poland
P.Marusak@ia.pw.edu.pl

Abstract. Model predictive control (MPC) algorithms are widely used in practical applications. They are usually formulated as optimization problems. If the model used for prediction is linear (or linearized on–line) then the optimization problem is standard, quadratic one. Otherwise, it is a nonlinear, in general, non–convex optimization problem. In the latter case the numerical problems may occur and time needed to calculate the control signals cannot be determined. Therefore approaches based on linear or linearized models are preferred in practical applications. In the paper a new algorithm is proposed, with prediction which employs heuristic fuzzy modeling. The algorithm is formulated as quadratic optimization problem but offers performance very close to that of MPC algorithm with nonlinear optimization. The efficiency of the proposed algorithm is demonstrated in the control system of the nonlinear control plant with inverse response – a chemical CSTR reactor.

Keywords: fuzzy control, fuzzy systems, predictive control, nonlinear control.

1 Introduction

The MPC algorithms offer a few advantages resulting from the way they are formulated. Therefore they can be successfully used in control systems of processes with difficult dynamics, constraints and for MIMO processes, see e.g. [3,7,11,13]. It is because during generation of the control signals a model of the control plant is used to predict behavior of the control system.

Standard formulations of the MPC algorithms are based on linear control plant models. However, operation of the control system of a nonlinear control plant may be usually improved using the MPC algorithm based on a nonlinear model. It is especially important if the MPC algorithm should operate for different operating points as in control system structures with steady–state set–point optimization, see e.g. [2,8,13].

If a nonlinear process model is used for prediction then the optimization problem solved at each iteration by the algorithm may be in general non–convex, nonlinear optimization problem – hard to solve and with unpredictable time needed to find the solution. Therefore, in practice, usually on–line linearization at each

M. Kolehmainen et al. (Eds.): ICANNGA 2009, LNCS 5495, pp. 448–457, 2009.
© Springer-Verlag Berlin Heidelberg 2009

algorithm iteration is used [9,13]. The approach proposed in the paper is in fact a hybrid one. It consists in using two models in the algorithm: a nonlinear model and its, easy to obtain, fuzzy approximation. Thanks to such an approach the proposed algorithm is formulated as the standard quadratic programming problem (as in the algorithm based on a linear process model) but it offers almost the same performance as the algorithm with nonlinear optimization.

The next section contains description of MPC algorithms. In Sect. 3 the proposed approach based on fuzzy and nonlinear models is proposed. Example results illustrating efficacy of the proposed approach are presented in Sect. 4. The paper is summarized in the last section.

2   Model Predictive Control Algorithms

The Model Predictive Control (MPC) algorithms derive future values of manipulated variables predicting future behavior of the control plant many sampling instants ahead. The values of manipulated variables are calculated in such a way that the prediction fulfills assumed criteria. Usually, the minimization of a performance function is demanded subject to the constraints put on values of manipulated and output variables [3,7,11,13]:

$$
\min_{\Delta u} \left\{ J_{\mathrm{MPC}} = \sum_{j=1}^{n_y} \sum_{i=1}^{p} \kappa_j \left(\overline{y}_k^j - y_{k+j|k}^j \right)^2 + \sum_{m=1}^{n_u} \sum_{i=0}^{s-1} \lambda_m \left(\Delta u_{k+i|k}^m \right)^2 \right\} \tag{1}
$$

subject to:

$$
\Delta u_{\min} \leq \Delta u \leq \Delta u_{\max} \ , \tag{2}
$$

$$
u_{\min} \leq u \leq u_{\max} \ , \tag{3}
$$

$$
y_{\min} \leq y \leq y_{\max} \ , \tag{4}
$$

where $\overline{y}_k^j$ is a set–point value for the jth output, $y_{k+j|k}^j$ is a value of the jth output for the $(k+i)^{\mathrm{th}}$ sampling instant predicted at the kth sampling instant using a control plant model, $\Delta u_{k+i|k}^m$ are future changes in manipulated variables, $\kappa_j \geq 0$ and $\lambda_m \geq 0$ are weighting coefficients for the predicted control errors of the jth output and for the changes of the mth manipulated variable, respectively; p and s denote prediction and control horizons, respectively; n_y, n_u denote number of output and manipulated variables, respectively;

$$
y = \left[y^1, \ldots, y^{n_y} \right]^T , \ y^j = \left[y_{k+1|k}^j, \ldots, y_{k+p|k}^j \right] \ , \tag{5}
$$

$$
\Delta u = \left[\Delta u^1, \ldots, \Delta u^{n_u} \right]^T , \ \Delta u^m = \left[\Delta u_{k+1|k}^m, \ldots, \Delta u_{k+s-1|k}^m \right] \ , \tag{6}
$$

$$
u = \left[u^1, \ldots, u^{n_u} \right]^T , \ u^m = \left[u_{k+1|k}^m, \ldots, u_{k+s-1|k}^m \right] \ , \tag{7}
$$

$\Delta u_{\min}$, $\Delta u_{\max}$, $u_{\min}$, $u_{\max}$, $y_{\min}$, $y_{\max}$ are vectors of lower and upper bounds of changes and values of the control signals and of the values of output variables, respectively. As a solution to the optimization problem (1–4) the optimal vector of changes in the manipulated variables is obtained. From this vector, the $\Delta u_{k|k}^m$ elements are applied in the control system and the algorithm passes to the next iteration.

The way the predicted values of output variables $y_{k+j|k}^j$ are derived depends on the dynamic control plant model the predictive algorithm is based on. If it is a nonlinear process model then the optimization problem (1–4) is, in general, non–convex nonlinear optimization problem, examples of such algorithms are described e.g. in [1,5,6]. In such an algorithm, different kinds of process models can be used but they are exploited in similar way, therefore an algorithm of this kind will be later referred to, in general, as Nonlinear MPC (NMPC).

2.1 MPC Algorithms Based on Linear Models

If the linear model is used for prediction then the optimization problem (1–4) is a standard quadratic programming problem [3,7,11,13]. It is because the vector of predicted output values y, after application of the superposition principle, can be described by the following formula:

$$y = \tilde{y} + A \cdot \Delta u \ , \tag{8}$$

$$\tilde{y} = \left[\tilde{y}^1, \ldots, \tilde{y}^{n_y} \right]^T , \ \tilde{y}^j = \left[\tilde{y}_{k+1|k}^j, \ldots, \tilde{y}_{k+p|k}^j \right] \ , \tag{9}$$

where $\tilde{y}$ is called a free response of the plant. It is because it contains future values of output variables calculated assuming that the control signal does not change in the prediction horizon, i.e. describes influence of the values of manipulated variables from the past. Form of the free response depends on the model used for prediction;

$$A = \begin{bmatrix} A^{11} & A^{12} & \ldots & A^{1n_u} \\ A^{21} & A^{22} & \ldots & A^{2n_u} \\ \vdots & \vdots & \ddots & \vdots \\ A^{n_y1} & A^{n_y2} & \ldots & A^{n_yn_u} \end{bmatrix}, \ A^{jm} = \begin{bmatrix} a_1^{j,m} & 0 & \ldots & 0 & 0 \\ a_2^{j,m} & a_1^{j,m} & \ldots & 0 & 0 \\ \vdots & \vdots & \ddots & \vdots & \vdots \\ a_p^{j,m} & a_{p-1}^{j,m} & \ldots & a_{p-s+2}^{j,m} & a_{p-s+1}^{j,m} \end{bmatrix} \tag{10}$$

is a matrix, called the dynamic matrix, composed of the control plant step response coefficients $a_i^{j,m}$ describing influence of the mth input on the jth output. It can be shown that the dynamic matrix has the same form in different types of MPC algorithms (using different types of linear models) [13].

Let us introduce the following vectors:

$$\overline{y} = [\overline{y}^1, \ldots, \overline{y}^{n_y}]^T , \ \overline{y}^j = \left[\overline{y}_k^j, \ldots, \overline{y}_k^j \right] \ , \tag{11}$$

where $\overline{y}^j$ are vectors of length p.

The performance function from (1) rewritten in the matrix–vector form is then as follows:

$$J_{MPC} = (\overline{\boldsymbol{y}} - \boldsymbol{y})^T \cdot \boldsymbol{K} \cdot (\overline{\boldsymbol{y}} - \boldsymbol{y}) + \Delta \boldsymbol{u}^T \cdot \boldsymbol{\Lambda} \cdot \Delta \boldsymbol{u} \ , \tag{12}$$

where $\boldsymbol{K} = \begin{bmatrix} \boldsymbol{K}_1, \ldots, \boldsymbol{K}_{n_y} \end{bmatrix} \cdot \boldsymbol{I}$; $\boldsymbol{K}_i = \begin{bmatrix} \kappa_i, \ldots, \kappa_i \end{bmatrix}$ have p elements, $\boldsymbol{\Lambda} = \begin{bmatrix} \boldsymbol{\Lambda}_1, \ldots, \boldsymbol{\Lambda}_{n_u} \end{bmatrix} \cdot \boldsymbol{I}$; $\boldsymbol{\Lambda}_i = \begin{bmatrix} \lambda_i, \ldots, \lambda_i \end{bmatrix}$ have s elements.

After applying the prediction (8) the performance function (12) can be transformed to:

$$J_{LMPC} = (\overline{\boldsymbol{y}} - \widetilde{\boldsymbol{y}} - \boldsymbol{A} \cdot \Delta \boldsymbol{u})^T \cdot \boldsymbol{K} \cdot (\overline{\boldsymbol{y}} - \widetilde{\boldsymbol{y}} - \boldsymbol{A} \cdot \Delta \boldsymbol{u}) + \Delta \boldsymbol{u}^T \cdot \boldsymbol{\Lambda} \cdot \Delta \boldsymbol{u} \ . \tag{13}$$

The performance function (13) depends quadratically on decision variables $\Delta \boldsymbol{u}$ and after using prediction (8) in the constraints (3) all constraints depend linearly on decision variables. Thus, one obtains a standard linear–quadratic optimization problem which is easy to solve by means of standard numerical routines.

3   MPC Algorithm Based on Fuzzy and Nonlinear Models

Application of the MPC algorithm based on the linear model (LMPC) to a nonlinear process may result in unsatisfactory control performance, especially if operation in different operating points is demanded. Therefore, the following fuzzy MPC (FMPC) algorithm being a combination of the LMPC and NMPC algorithms is proposed. Two models are used in this algorithm. The original, nonlinear one is used to calculate the free response, whereas the fuzzy one, being a set of a few step responses, is used to calculate the dynamic matrix, updated at each algorithm iteration. The proposed FMPC algorithm will be described now.

Let us suppose that we have the nonlinear process model (it can be practically any type of model usable in the NMPC algorithm):

$$\widehat{\boldsymbol{y}}_{k+1|k} = \boldsymbol{f}(\boldsymbol{y}_k, \boldsymbol{y}_{k-1}, \ldots, \boldsymbol{y}_{k-n_a}, \boldsymbol{u}_{k-1}, \boldsymbol{u}_{k-2}, \ldots, \boldsymbol{u}_{k-n_b}) \ , \tag{14}$$

where $\boldsymbol{y}_{k-i} = \begin{bmatrix} y_{k-i}^1, \ldots, y_{k-i}^{n_y} \end{bmatrix}^T$ is the vector of measured values of output variables at the $(k - i)^{\text{th}}$ sampling instant, $\boldsymbol{u}_{k-i} = \begin{bmatrix} u_{k-i}^1, \ldots, u_{k-i}^{n_u} \end{bmatrix}^T$ is the vector of values of manipulated variables at the $(k - i)^{\text{th}}$ sampling instant; let us also denote outputs of the model at the $(k + i)^{\text{th}}$ sampling instant as $\widehat{\boldsymbol{y}}_{k+i|k} = \begin{bmatrix} \widehat{y}_{k+i|k}^1, \ldots, \widehat{y}_{k+i|k}^{n_y} \end{bmatrix}$, n_a, n_b determine, how long the history of signals used by the model is.

The model (14) is then employed to obtain the free response, for the whole prediction horizon, in an iterative way, i.e.:

– First, the process model is used to obtain $\widehat{\boldsymbol{y}}_{k+1|k}$ (formula (14)).
– Then the values $\widehat{\boldsymbol{y}}_{k+1|k}$ are used, as the output values for the $(k + 1)^{\text{st}}$ sampling instant, to obtain output values $\widehat{\boldsymbol{y}}_{k+2|k}$, for the next sampling instant.

Moreover, the assumption that control signal does not change (the free response is calculated) is utilized:

$$\widehat{\boldsymbol{y}}_{k+2|k} = \boldsymbol{f}\left(\widehat{\boldsymbol{y}}_{k+1|k}, \boldsymbol{y}_k, \ldots, \boldsymbol{y}_{k-n_a+1}, \boldsymbol{u}_{k-1}, \boldsymbol{u}_{k-1}, \ldots, \boldsymbol{u}_{k-n_b+1}\right) ; \qquad (15)$$

– Thus, in general, in the ith iteration, using the values $\widehat{\boldsymbol{y}}_{k+1|k}, \ldots, \widehat{\boldsymbol{y}}_{k+i-1|k}$ and assuming that the control signal does not change, one obtains:

$$\widehat{\boldsymbol{y}}_{k+i|k} = \boldsymbol{f}\left(\widehat{\boldsymbol{y}}_{k+i-1|k}, \widehat{\boldsymbol{y}}_{k+i-2|k}, \ldots, \boldsymbol{y}_{k-n_a+i-1}, \boldsymbol{u}_{k-1}, \boldsymbol{u}_{k-1}, \ldots, \boldsymbol{u}_{k-n_b+i-1}\right) . \qquad (16)$$

– Then the free response is calculated taking into consideration the estimated disturbances (containing also influence of modeling errors). Thus, the final formula describing the elements of the free response is as follows:

$$\widetilde{\boldsymbol{y}}_{k+i|k} = \widehat{\boldsymbol{y}}_{k+i|k} + \boldsymbol{d}_k , \qquad (17)$$

where $\widetilde{\boldsymbol{y}}_{k+i|k} = \left[\widetilde{y}_{k+i|k}^1, \ldots, \widetilde{y}_{k+i|k}^{n_y}\right]$ and $\boldsymbol{d}_k$ is the DMC–type disturbance model, i.e. it is assumed that it is the same for all instants in the prediction horizon and

$$\boldsymbol{d}_k = \boldsymbol{y}_k - \widehat{\boldsymbol{y}}_{k|k-1} . \qquad (18)$$

After calculating the free response in the way described above, the dynamic matrix, needed to predict the influence of the future control changes (generated by the algorithm) is derived using an easy to obtain Takagi–Sugeno fuzzy model. The fuzzy model has local models in the form of step responses. Therefore, its design process is very simple. It is sufficient to collect a few sets of step responses (around a few operating points). Then, using expert knowledge, the premises can be formulated and, subsequently, they can be tuned using, e.g. fuzzy neural network. The fuzzy model is composed of the following rules:

Rule f: $\qquad\qquad\qquad\qquad\qquad\qquad\qquad\qquad\qquad\qquad\qquad\qquad$ (19)

$$\text{if } y_k^{j_y} \text{ is } B_1^{f,j_y} \text{ and } \ldots \text{ and } y_{k-n+1}^{j_y} \text{ is } B_n^{f,j_y} \text{ and}$$

$$u_k^{j_u} \text{ is } C_1^{f,j_u} \text{ and } \ldots \text{ and } u_{k-m+1}^{j_u} \text{ is } C_m^{f,j_u}$$

$$\text{then } \hat{y}_k^{j,f} = \sum_{m=1}^{n_u} \sum_{n=1}^{p_d-1} a_n^{j,m,f} \cdot \Delta u_{k-n}^m + a_{p_d}^{j,m,f} \cdot u_{k-p_d}^m ,$$

where $y_k^{j_y}$ is the $j_y{}^{\text{th}}$ output variable value at the kth sampling instant, $u_k^{j_u}$ is the $j_u{}^{\text{th}}$ manipulated variable value at the kth sampling instant, $B_1^{f,j_y}, \ldots, B_n^{f,j_y}$, $C_1^{f,j_u}, \ldots, C_m^{f,j_u}$ are fuzzy sets, $a_n^{j,m,f}$ are the coefficients of step responses in the fth local model, $j_y = 1, \ldots, n_y$, $j_u = 1, \ldots, n_u$, $f = 1, \ldots, l$, l is number of rules.

The output value of the fuzzy model (19) is calculated at each iteration using the following formula:

$$\hat{y}_k^j = \sum_{m=1}^{n_u} \sum_{n=1}^{p_d-1} \widetilde{a}_n^{j,m} \cdot \Delta u_{k-n}^m + \widetilde{a}_{p_d}^{j,m} \cdot u_{k-p_d}^m , \qquad (20)$$

where $\tilde{a}_n^{j,m} = \sum_{f=1}^{l} \tilde{w}_f \cdot a_n^{j,m,f}$, $\tilde{w}_f$ are the normalized weights calculated using fuzzy reasoning, see e.g. [10,12].

The model (20) may be interpreted as the step response describing behavior of the control plant near the current operating point. This model may be used to obtain the dynamic matrix the same way as in the LMPC algorithm, i.e. at each sampling instant of the FMPC algorithm a new dynamic matrix is generated:

$$
\boldsymbol{A}_k = \begin{bmatrix} \boldsymbol{A}_k^{11} & \boldsymbol{A}_k^{12} & \dots & \boldsymbol{A}_k^{1n_u} \\ \boldsymbol{A}_k^{21} & \boldsymbol{A}_k^{22} & \dots & \boldsymbol{A}_k^{2n_u} \\ \vdots & \vdots & \ddots & \vdots \\ \boldsymbol{A}_k^{n_y 1} & \boldsymbol{A}_k^{n_y 2} & \dots & \boldsymbol{A}_k^{n_y n_u} \end{bmatrix} , \; \boldsymbol{A}_k^{jm} = \begin{bmatrix} \tilde{a}_1^{j,m} & 0 & \dots & 0 & 0 \\ \tilde{a}_2^{j,m} & \tilde{a}_1^{j,m} & \dots & 0 & 0 \\ \vdots & \vdots & \ddots & \vdots & \vdots \\ \tilde{a}_p^{j,m} & \tilde{a}_{p-1}^{j,m} & \dots & \tilde{a}_{p-s+2}^{j,m} & \tilde{a}_{p-s+1}^{j,m} \end{bmatrix} .
$$

$$(21)$$

Then, the free response (17) and the dynamic matrix (21) are used to obtain the prediction:

$$y = \tilde{y} + \boldsymbol{A}_k \cdot \Delta \boldsymbol{u} \; . \tag{22}$$

Summing up, at each iteration of the FDMC algorithm, the following actions are done:

1. The nonlinear control plant model is used to generate free response of the control plant (17).
2. A linear model, for current sampling instant, is derived using current values of process variables, the TS fuzzy model (19) and fuzzy reasoning.
3. The obtained step response coefficients are used to generate the dynamic matrix.
4. The free response and the dynamic matrix are used to formulate the quadratic optimization problem in which the following performance function is used:

$$
J_{FMPC} = (\overline{\boldsymbol{y}} - \tilde{\boldsymbol{y}} - \boldsymbol{A}_k \cdot \Delta \boldsymbol{u})^T \cdot \boldsymbol{K} \cdot (\overline{\boldsymbol{y}} - \tilde{\boldsymbol{y}} - \boldsymbol{A}_k \cdot \Delta \boldsymbol{u}) + \Delta \boldsymbol{u}^T \cdot \boldsymbol{\Lambda} \cdot \Delta \boldsymbol{u} \; . \tag{23}
$$

5. The optimization problem is solved and, using the obtained solution, control signals are generated. Then the controller passes to the next iteration.

The approach described above is an approximate one, however, as it will be demonstrated using an example control system, it gives results comparable to those obtained with the NMPC algorithm. At the same time, it usually gives better results than the standard LMPC algorithm (based on a linear model). The proposed algorithm may be used not only as the stand–alone algorithm but also in control systems with NMPC algorithms, to improve their numerical properties. It is because it can be employed to generate the starting control trajectory for a nonlinear optimization routine. If this routine is able to improve the initial trajectory in the presumed time (during one sampling instant) then the newly derived control action can be applied to the process. Otherwise, the control signal generated by the FMPC algorithm can be used (it offers performance very close to the optimal one, anyway).

4   Simulation Experiments

4.1   Control Plant

The control plant under consideration is an isothermal CSTR in which a van de Vusse reaction carries out (Fig. 1a) [4]. Steady–state characteristics of the control plant are shown in Fig. 1b.

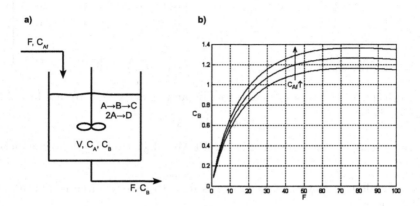

Fig. 1. Isothermal CSTR with van de Vusse reaction a) diagram of the control plant; b) steady–state characteristics

The process model of the reactor contains two composition balance equations

$$\frac{dC_A}{dt} = -k_1 \cdot C_A - k_3 \cdot C_A^2 + \frac{F}{V}(C_{Af} - C_A) \ , \\ \frac{dC_B}{dt} = k_1 \cdot C_A - k_2 \cdot C_B - \frac{F}{V}C_B \ , \tag{24}$$

where C_A, C_B are the concentrations of components A and B, respectively, F is the inlet flow rate (equal to the outlet flow rate), V is the volume in which the reaction takes place (it is assumed constant and $V = 1$ l), C_{Af} is the concentration of component A in the inlet flow stream (it is assumed that $C_{Af} = 10$ mol/l). The values of parameters are: $k_1 = 50$ 1/h, $k_2 = 100$ 1/h, $k_3 = 10$ 1/(h $\cdot$ mol).

The output variable is the concentration C_B of substance B, the manipulated variable is the inlet flow rate F of the raw substance, C_{Af} concentration is the disturbance variable.

4.2   Experiments

For the considered control plant three MPC algorithms were designed: an NMPC one (with nonlinear optimization), an LMPC one (with a linear model) and an FMPC one (proposed in the paper, exploiting a fuzzy model). The sampling period was assumed equal to $T_s = 3.6$ s; tuning parameters of all three algorithms were as follows: prediction horizon $p = 70$, control horizon $s = 35$, weighting coefficient $\lambda = 0.001$. The fuzzy model used in the FMPC algorithm is composed of step responses taken in environs of the following operating points:

P1) $C_{B0} = 0.91$ mol/l, $C_{A0} = 2.18$ mol/l, $F = 20$ l/h;
P2) $C_{B0} = 1.12$ mol/l, $C_{A0} = 3$ mol/l, $F = 34.3$ l/h;
P3) $C_{B0} = 1.22$ mol/l, $C_{A0} = 3.66$ mol/l, $F = 50$ l/h.

The model used for derivation of the dynamic matrix is thus composed of three rules. The membership functions, assumed after analysis of the steady–state characteristics of the control plant, are shown in Fig. 2.

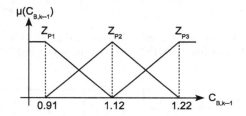

Fig. 2. Membership functions of the fuzzy model used in the FDMC controller

During the experiments operation of control systems with NMPC, LMPC and FMPC algorithms was compared. The responses obtained after the change of set–point value to $\overline{C}_B = 1.02$ are shown in Fig. 3. The responses obtained in the control system with FMPC algorithm (solid lines in Fig. 3) are very close to that obtained in the NMPC algorithm with full nonlinear optimization (dashed lines in Fig. 3). In both cases there is practically no overshoot. However, the FDMC algorithm is based on the reliable quadratic programming routine. Both algorithms outperform the standard LMPC algorithm (dotted lines in Fig. 3). In the latter case there is significant overshoot and control time is longer than in the case of NMPC and FMPC algorithms.

There was also made an experiment with disturbance change by 10% from $C_{Af0} = 10$ mol/l to $C_{Af1} = 11$ mol/l (Fig. 4). The disturbance changed at the 6th minute of simulation. The responses obtained in control systems with NMPC and FMPC algorithms are almost the same, the output variable does not decrease below the set–point value. In the case of control system with LMPC algorithm the model inaccuracy results in more aggressive control action than in the previous cases. Ranges of values achieved by both: manipulated and output variables are wider than in the case of NMPC and FMPC algorithms.

5 Summary

The numerically efficient FDMC algorithm was proposed in the paper. It uses the nonlinear model to derive the free response of the control plant and the

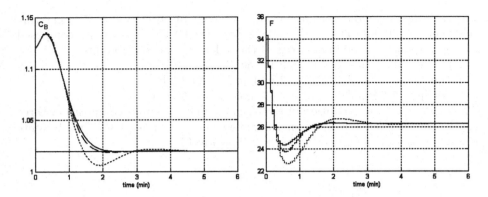

Fig. 3. Responses of the control systems to the change of the set–point value to $\overline{C}_B = 1.02$; NMPC – dashed lines, LMPC – dotted lines and FMPC – solid lines

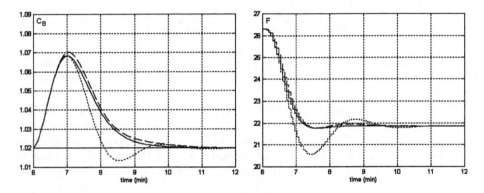

Fig. 4. Responses of the control systems to the change of the disturbance by 10% to $C_{Af1} = 11$ mol/l ; NMPC – dashed lines, LMPC – dotted lines and FMPC – solid lines

approximate, easy to obtain, fuzzy model to calculate the influence of future control action. Thanks to such an approach the algorithm offers control performance very close to that offered by the algorithm with nonlinear optimization. The FDMC algorithm, however, is formulated as the quadratic optimization problem. Thus, time needed for calculation of the control signals can be foreseen which is an important feature in control systems with determined sampling period. The FDMC algorithm can be used either as the stand–alone algorithm or in the control systems with NMPC algorithms, to improve numerical properties of the latter ones.

Acknowledgment. This work was supported by the Polish national budget funds for science 2007–2009 as a research project.

References

1. Babuska, R., te Braake, H.A.B., van Can, H.J.L., Krijgsman, A.J., Verbruggen, H.B.: Comparison of intelligent control schemes for real–time pressure control. Control Engineering Practice 4, 1585–1592 (1996)
2. Blevins, T.L., McMillan, G.K., Wojsznis, W.K., Brown, M.W.: Advanced Control Unleashed. ISA (2003)
3. Camacho, E.F., Bordons, C.: Model Predictive Control. Springer, Heidelberg (1999)
4. Doyle, F., Ogunnaike, B.A., Pearson, R.K.: Nonlinear model–based control using second–order Volterra models. Automatica 31, 697–714 (1995)
5. Fink, A., Fischer, M., Nelles, O., Isermann, R.: Supervision of nonlinear adaptive controllers based on fuzzy models. Control Engineering Practice 8, 1093–1105 (2000)
6. Foss, B.A., Johansen, T.A., Sorensen, A.V.: Nonlinear predictive control using local models–applied to a batch fermentation process. Control Engineering Practice 3, 389–396 (1995)
7. Maciejowski, J.M.: Predictive control with constraints. Prentice Hall, Harlow (2002)
8. Marusak, P.: Efficient fuzzy predictive economic set–point optimizer. In: Rutkowski, L., Tadeusiewicz, R., Zadeh, L.A., Zurada, J.M. (eds.) ICAISC 2008. LNCS (LNAI), vol. 5097, pp. 273–284. Springer, Heidelberg (2008)
9. Morari, M., Lee, J.H.: Model predictive control: past, present and future. Computers and Chemical Engineering 23, 667–682 (1999)
10. Piegat, A.: Fuzzy Modeling and Control. Physica–Verlag, Berlin (2001)
11. Rossiter, J.A.: Model-Based Predictive Control. CRC Press, Boca Raton (2003)
12. Takagi, T., Sugeno, M.: Fuzzy identification of systems and its application to modeling and control. IEEE Trans. Systems, Man and Cybernetics 15, 116–132 (1985)
13. Tatjewski, P.: Advanced Control of Industrial Processes; Structures and Algorithms. Springer, London (2007)

Dynamic Classifier Systems and Their Applications to Random Forest Ensembles*

David Štefka and Martin Holeňa

Institute of Computer Science, Academy of Sciences of the Czech Republic, v.v.i.
Pod Vodárenskou věží 2, 182 07 Prague 8, Czech Republic,
{stefka,martin}@cs.cas.cz

Abstract. Classifier combining is a popular method for improving quality of classification – instead of using one classifier, several classifiers are organized into a classifier system and their results are aggregated into a final prediction. However, most of the commonly used aggregation methods are static, i.e., they do not adapt to the currently classified pattern. In this paper, we provide a general framework for dynamic classifier systems, which use dynamic confidence measures to adapt to a particular pattern. Our experiments with random forests on 5 artificial and 11 real-world benchmark datasets show that dynamic classifier systems can significantly outperform both confidence-free and static classifier systems.

Keywords: classifier combining, dynamic classifier aggregation, random forests, classification.

1 Introduction

Classification is a process of dividing objects (called *patterns*) into disjoint sets called *classes* [1]. One comonly used technique for improving classification quality is *classifier combining* [2] – instead of using just one classifier, a team of classifiers is created and trained; each classifier in the team predicts independently, and the classifier outputs are aggregated into a final prediction. It can be shown that such a team of classifiers can perform better than any of the individual classifiers.

A common drawback of classifier aggregation methods is that they are static, i.e., they are not adapted to the particular pattern submitted for classification. However, if we use the concept of dynamic classification confidence (i.e., the extent to which we can "trust" the output of a particular classifier for the currently classified pattern), the aggregation algorithms can take into account the fact that "this classifier is/is not good *for this particular pattern*".

There has already been some research done in the field of dynamic classifier aggregation [3,4,5,6,7,8]. It is although common that the concept of dynamic

---

* The research presented in this paper was partially supported by the Program "Information Society" under project 1ET100300517 (D. Štefka) and by the grant No. 201/08/0802 of the Grant Agency of the Czech Republic and by the Institutional Research Plan AV0Z10300504 (M. Holeňa).

M. Kolehmainen et al. (Eds.): ICANNGA 2009, LNCS 5495, pp. 458–468, 2009.
© Springer-Verlag Berlin Heidelberg 2009

classification confidence is tightly bound with the aggregation method or with the particular classifier type used. However, the way a classifier classifies a pattern, the way we measure confidence of a classifier, and the way we aggregate a team of classifiers are independent on each other.

The goal of this paper is to provide a general framework of dynamic classifier systems based on three independent aspects – the classifiers in the team, the measures of confidence of the individual classifiers, and the aggregation strategy. The confidence measures and the aggregation strategy will give us three important classes of classifier systems – confidence-free (i.e., systems that do not utilize classification confidence at all), static (i.e., systems that use only "global" confidence of a classifier), and dynamic (i.e., systems that adapt to the particular pattern submitted for classification). We will then compare performance of these classifier systems on several benchmark datasets – for this purpose, we have chosen random forest [9] classifier systems.

The paper is structured as follows. Section 2 provides theoretical description of a framework for dynamic classifier systems, including formalism of classification (Sec. 2.1), classification confidence (Sec. 2.2), classifier teams (Sec 2.3) and classifier systems (Sec. 2.4). Section 3 describes random forests and Section 4 contains results of our experiments with random forests on several benchmark datasets. Section 5 then concludes the paper.

2   Dynamic Classifier Systems

2.1   Classification

Throughout the rest of the paper, we use the following notation. Let $\mathcal{X} \subseteq \mathbb{R}^n$ be a n-dimensional *feature space*, an element $\boldsymbol{x} \in \mathcal{X}$ of this space is called a *pattern*, and let $C_1, \ldots, C_N \subseteq \mathcal{X}$, $N \geq 2$, be disjoint sets called *classes*. The index of the class a pattern $\boldsymbol{x}$ belongs to will be denoted as $c(\boldsymbol{x})$ (i.e., $c(\boldsymbol{x}) = i$ iff $\boldsymbol{x} \in C_i$). Let $[0, 1]$ denote the unit interval. The goal of classification is to determine to which class a given pattern belongs, i.e., to predict $c(\boldsymbol{x})$ for unknown patterns.

Definition 1. *We call a* classifier *every mapping* $\phi : \mathcal{X} \to [0, 1]^N$, *where* $\phi(\boldsymbol{x}) = (\mu_1(\boldsymbol{x}), \ldots, \mu_N(\boldsymbol{x}))$ *are* degrees of classification *(d.o.c.) to each class.*

The d.o.c. to class C_j expresses the extent to which the pattern belongs to class C_j (if $\mu_i(\boldsymbol{x}) > \mu_j(\boldsymbol{x})$, it means that the pattern $\boldsymbol{x}$ belongs to class C_i rather than to C_j). Depending on the classifier type, it can be modelled by probability, fuzzy membership, etc.

Remark 1. This definition is of course not the only way how a classifier can be defined, but in the theory of classifier combining, this one is used most often [2].

Definition 2. *Classifier* ϕ *is called* crisp, *iff* $\forall \boldsymbol{x} \in \mathcal{X} \; \exists i$, *such that:*

$$\mu_i(\boldsymbol{x}) = 1, \; and \; \forall j \neq i \; \mu_j(\boldsymbol{x}) = 0, \text{where } \phi(\boldsymbol{x}) = (\mu_1(\boldsymbol{x}), \ldots, \mu_N(\boldsymbol{x})).$$

Definition 3. *Let* ϕ *be a classifier,* $\boldsymbol{x} \in \mathcal{X}$, $\phi(\boldsymbol{x}) = (\mu_1(\boldsymbol{x}), \ldots, \mu_N(\boldsymbol{x}))$. *Crisp output of* ϕ *on* $\boldsymbol{x}$ *is defined as* $\phi_{cr}(\boldsymbol{x}) = \arg\max_{i=1,\ldots,N} \mu_i(\boldsymbol{x})$.

2.2   Classification Confidence

Classification confidence expresses the degree of trust we can give to a classifier ϕ when classifying a pattern $\boldsymbol{x}$. It is modelled by a mapping κ_ϕ.

Definition 4. *Let ϕ be a classifier. We call a* confidence measure *of classifier ϕ every mapping $\kappa_\phi : \mathcal{X} \to [0,1]$. Let $\boldsymbol{x} \in \mathcal{X}$. $\kappa_\phi(\boldsymbol{x})$ is called* classification confidence *of ϕ on $\boldsymbol{x}$.*

The higher the confidence, the higher the probability of correct classification. $\kappa_\phi(\boldsymbol{x}) = 0$ means that the classification may not be correct, while $\kappa_\phi(\boldsymbol{x}) = 1$ means the classification is probably correct. However, κ_ϕ does not need to be modelled by a probability measure.

A confidence measure can be either *static*, i.e., it is a constant of the classifier, or *dynamic*, i.e., it adjusts itself to the currently classified pattern.

Definition 5. *Let ϕ be a classifier and κ_ϕ its confidence measure. We call κ_ϕ* static, *iff it is constant in $\boldsymbol{x}$, we call κ_ϕ* dynamic *otherwise.*

Remark 2. Since static confidence measures are constant, independent on the currently classified pattern, we will omit the pattern $\boldsymbol{x}$ in the notation, i.e., we will denote them just κ_ϕ.

Remark 3. In the rest of the paper, we will use the indicator operator I, defined as $I(\text{true}) = 1$, $I(\text{false}) = 0$.

Static (global) confidence measures. After the classifier has been trained, we can use a validation set (i.e., a set of patterns the classifier has not been trained on; we could also use training patterns, but in that case, the results would be biased) to assess its predictive power as a whole (from a global view). These methods include accuracy, precision, sensitivity, resemblance, etc. [1,10], and we can use these measures as static confidence measures. In this paper, we will use the Global Accuracy measure.

Global Accuracy (GA) of a classifier ϕ is defined as the proportion of correctly classified patterns from the validation set:

$$\kappa_\phi^{(GA)} = \frac{\sum_{\boldsymbol{y}\in\mathcal{M}} I(\phi_{cr}(\boldsymbol{y}) \stackrel{?}{=} c(\boldsymbol{y}))}{|\mathcal{M}|}, \tag{1}$$

where $\mathcal{M}$ is the validation set of ϕ and $\phi_{cr}(\boldsymbol{y})$ is the crisp output of ϕ on $\boldsymbol{y}$.

Dynamic (local) confidence measures. An easy way how a dynamic confidence measure can be defined is to compute some property on patterns neighboring $\boldsymbol{x}$. Let $N(\boldsymbol{x})$ denote a set of neighboring patterns from the validation set. In this paper, we define $N(\boldsymbol{x})$ as the set of k patterns nearest to $\boldsymbol{x}$ under Euclidean metric. Now we will define two dynamic confidence measures which use $N(\boldsymbol{x})$:

Euclidean Local Accuracy (ELA), used in [5], measures the local accuracy of ϕ in $N(x)$:

$$\kappa_\phi^{(ELA)}(x) = \frac{\sum_{y \in N(x)} I(\phi_{cr}(y) \overset{?}{=} c(y))}{|N(x)|}, \tag{2}$$

where $\phi_{cr}(y)$ is the crisp output of ϕ on y.

Euclidean Local Match (ELM), based on the ideas from [6], measures the proportion of patterns in $N(x)$ from the same class as ϕ is predicting for x:

$$\kappa_\phi^{(ELM)}(x) = \frac{\sum_{y \in N(x)} I(\phi_{cr}(x) \overset{?}{=} c(y))}{|N(x)|}, \tag{3}$$

where $\phi_{cr}(x)$ is the crisp output of ϕ on x. The difference between 2 and 3 is that in the latter case, there is $\phi_{cr}(x)$ instead of $\phi_{cr}(y)$ in the indicator.

Remark 4. The dynamic confidence measures defined in this section have one drawback – they need to compute neighboring patterns of x, which can be time-consuming, and sensitive to the similarity measure used. There are also dynamic confidence measures, which compute the classification confidence directly from the degrees of classification [7,8]. However, our preliminary experiments with such measures show that they give very poor results.

2.3   Classifier Teams

In classifier combining, instead of using just one classifier, a team of classifiers is created, and the team is then aggregated into one final classifier. If we want to utilize classification confidence in the aggregation process, each classifier must have its own confidence measure defined.

Definition 6. *Let* $r \in \mathbb{N}$, $r \geq 2$. *Classifier team is a tuple* $(\mathcal{T}, \mathcal{K})$, *where* $\mathcal{T} = (\phi_1, \ldots, \phi_r)$ *is a set of classifiers, and* $\mathcal{K} = (\kappa_{\phi_1}, \ldots, \kappa_{\phi_r})$ *is a set of corresponding confidence measures.*

If a classifier team consists only of classifiers of the same type, which differ only in their parameters, dimensionality, or training sets, the team is usually called an *ensemble of classifiers*. The restriction to classifiers of the same type is not essential, but it ensures that the outputs of the classifiers are consistent. Well-known methods for ensemble creation are *bagging* [11], *boosting* [12], *random forests* [9], or *error correction codes* [2].

If a pattern is submitted for classification, the team of classifiers gives us information of two kinds – outputs of the individual classifiers (a *decision profile*), and values of classification confidence of the classifiers (a *confidence vector*).

Definition 7. *Let* $(\mathcal{T} = (\phi_1, \ldots, \phi_r), \mathcal{K} = (\kappa_{\phi_1}, \ldots, \kappa_{\phi_r}))$ *be a classifier team, and let* $x \in \mathcal{X}$. *Then we define* decision profile $\mathcal{T}(x) \in [0,1]^{r \times N}$ *and* confidence vector $\mathcal{K}(x) \in [0,1]^r$ *as*

$$
\mathcal{T}(x) = \begin{pmatrix} \phi_1(x) \\ \phi_2(x) \\ \vdots \\ \phi_r(x) \end{pmatrix} = \begin{pmatrix} \mu_{1,1}(x) & \mu_{1,2}(x) & \cdots & \mu_{1,N}(x) \\ \mu_{2,1}(x) & \mu_{2,2}(x) & \cdots & \mu_{2,N}(x) \\ & & \ddots & \\ \mu_{r,1}(x) & \mu_{r,2}(x) & \cdots & \mu_{r,N}(x) \end{pmatrix}, \; \mathcal{K}(x) = \begin{pmatrix} \kappa_{\phi_1}(x) \\ \kappa_{\phi_2}(x) \\ \vdots \\ \kappa_{\phi_r}(x) \end{pmatrix} \quad (4)
$$

2.4 Classifier Systems

After the pattern x has been classified by all the classifiers in the team, and the confidences have been computed, these outputs have to be aggregated using a *team aggregator*, which takes the decision profile as its first argument, the confidence vector as its second argument, and returns the aggregated degrees of classification to all the classes.

Definition 8. *Let* $r, N \in \mathbb{N}$, $r, N \geq 2$. *A* team aggregator *of dimension* (r, N) *is any mapping* $\mathcal{A} : [0,1]^{r,N} \times [0,1]^r \to [0,1]^N$.

A classifier team with an aggregator will be called a *classifier system*. Such system can be also viewed as a single classifier.

Definition 9. *Let* $(\mathcal{T}, \mathcal{K})$ *be a classifier team, and let* $\mathcal{A}$ *be a team aggregator of dimension* (r, N), *where* r *is the number of classifiers in the team, and* N *is the number of classes. The triple* $\mathcal{S} = (\mathcal{T}, \mathcal{K}, \mathcal{A})$ *is called a* classifier system. *We define an* induced classifier *of* $\mathcal{S}$ *as a classifier* Φ, *defined as*

$$
\Phi(x) = \mathcal{A}(\mathcal{T}(x), \mathcal{K}(x)).
$$

Depending on the way how a classifier system utilizes the classification confidence, we can distinguish several kinds of classifier systems.

Definition 10. *Let* $(\mathcal{T}, \mathcal{K})$ *be a classifier team.* $(\mathcal{T}, \mathcal{K})$ *is called* static, *iff* $\forall \kappa \in \mathcal{K} : \kappa$ *is a static confidence measure.* $(\mathcal{T}, \mathcal{K})$ *is called* dynamic, *iff* $\exists \kappa \in \mathcal{K} : \kappa$ *is a dynamic confidence measure.*

Definition 11. *Let* $\mathcal{A}$ *be a team aggregator of dimension* (r, N). *We call* $\mathcal{A}$ confidence-free, *iff it is constant in the second argument.*

Definition 12. *Let* $\mathcal{S} = (\mathcal{T}, \mathcal{K}, \mathcal{A})$ *be a classifier system. We call* $\mathcal{S}$ confidence-free, *iff* $\mathcal{A}$ *is confidence-free. We call* $\mathcal{S}$ static, *iff* $(\mathcal{T}, \mathcal{K})$ *is static, and* $\mathcal{A}$ *is not confidence-free. We call* $\mathcal{S}$ dynamic, *iff* $(\mathcal{T}, \mathcal{K})$ *is dynamic, and* $\mathcal{A}$ *is not confidence-free.*

Confidence-free systems do not utilize the classification confidence at all. Static systems utilize classification confidence, but only as a global property (constant for all patterns). Dynamic systems utilize classification confidence in a dynamic

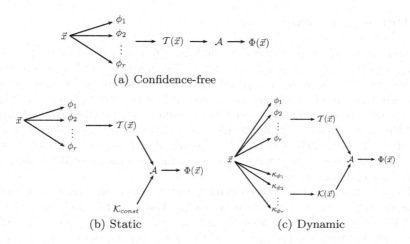

(a) Confidence-free

(b) Static   (c) Dynamic

Fig. 1. Schematic comparison of confidence-free, static, and dynamic classifier systems

way, i.e. the aggregation is adapted to the particular pattern submitted for classification. The different approaches are shown in Fig. 1.

Many methods for aggregating a team of classifiers into one final classifier have been proposed in the literature. These methods comprise simple arithmetic rules (voting, sum, product, etc.), fuzzy integral, Dempster-Shafer fusion, second-level classifiers, decision templates, and many others [2,13].

In this paper, we want to compare confidence free, static, and dynamic classifier systems. For that purpose, we define three simple aggregation algorithms, each representing one of the approaches. We will use the notation from Def. 7 and Def. 9. Let $\Phi(\boldsymbol{x}) = \mathcal{A}(\mathcal{T}(\boldsymbol{x}), \mathcal{K}(\boldsymbol{x})) = (\mu_1(\boldsymbol{x}), \ldots, \mu_N(\boldsymbol{x}))$.

Mean value aggregation (MV) is the most common (confidence-free) aggregation technique. Its aggregator is defined as

$$\mu_j(\boldsymbol{x}) = \frac{\sum_{i=1,\ldots,r} \mu_{i,j}(\boldsymbol{x})}{r}, \; j = 1, \ldots, N. \qquad (5)$$

If the classifiers in the team are crisp, MV coincides with simple voting.

Static weighted mean aggregation (SWM) computes aggregated d.o.c. as weighted mean of d.o.c. given by the individual classifiers, where the weights are static classification confidences:

$$\mu_j(\boldsymbol{x}) = \frac{\sum_{i=1,\ldots,r} \kappa_{\phi_i} \mu_{i,j}(\boldsymbol{x})}{\sum_{i=1,\ldots,r} \kappa_{\phi_i}}, \; j = 1, \ldots, N. \qquad (6)$$

Dynamic weighted mean aggregation (DWM) has the same aggregator as SWM, but the weights are dynamic classification confidences:

$$\mu_j(\boldsymbol{x}) = \frac{\sum_{i=1,\ldots,r} \kappa_{\phi_i}(\boldsymbol{x}) \mu_{i,j}(\boldsymbol{x})}{\sum_{i=1,\ldots,r} \kappa_{\phi_i}(\boldsymbol{x})}, \; j = 1, \ldots, N. \qquad (7)$$

3   Random Forests

Random forests (RF), introduced by Breiman [9], is a popular method for combining decision trees. RF is an ensemble of decision trees, which are created by bagging [11], i.e., each tree is trained on a random sample of training patterns of the same size as the training set, drawn uniformly with replacement. The individual decision trees in the ensemble are usually unpruned CART trees, with one difference – when growing the tree, not all the features of the patterns are used in each node. Only s randomly chosen features out of n possible features are examined. These two different sources of randomness make RF a successfull method for ensemble creation.

RFs are usually used as crisp classifiers, and therefore the most popular way to aggregate the individual trees is simple voting. However, for classifier combining, non-crisp outputs give us more information, and so we define non-crisp RF as follows. During training, the growing of the tree (splitting of the current node) is stopped if there are less than t training patterns in the node, and a leaf is made from the node. If a pattern x is classified by such leaf, its degrees of classification to each class are defined as the ratio of number of training patterns from that class to the number of all training patterns in the leaf.

The three confidence measures described in Sec. 2.2 are *model-indifferent*, i.e., they could be used for any classifier. However, for RF, *model-specific* confidence measures can be defined. Robnik-Šikonja [3] and Tsymbal et al. [4] use the set of k neighbors of x under so-called *Random Forest Similarity Measure* (RFSM) (we will denote the set as $RF(x)$) to compute the average value of the *margin* [3,4,9] on $RF(x)$. The margin is defined as:

$$
mg(\phi(\boldsymbol{y})) = \begin{cases} \mu_{c(\boldsymbol{y})}(\boldsymbol{y}) - \max\limits_{\substack{i=1,\ldots,N \\ i \neq c(\boldsymbol{y})}} \mu_i(\boldsymbol{y}) & \text{if } \phi_{cr}(\boldsymbol{y}) = c(\boldsymbol{y}), \\ 0 & \text{otherwise.} \end{cases} \tag{8}
$$

where $\phi(\boldsymbol{y}) = (\mu_1(\boldsymbol{y}), \ldots, \mu_N(\boldsymbol{y}))$, and $\phi_{cr}(\boldsymbol{y})$ is the crisp output of ϕ on $\boldsymbol{y}$. The RFSM is based on the idea that for a decision tree, patterns which are classified by the same leaf of the tree are similar. Given two patterns, their RFSM is defined as the number of trees in the forest in which the patterns are classified by the same leaf. For details, we refer to [3,4,9].

Random Forest Average Margin (RFAM) confidence measure is defined as the average value of the margin, computed on $RF(x)$:

$$
\kappa_\phi^{(RFAM)}(\boldsymbol{x}) = \frac{\sum_{\boldsymbol{y} \in RF(\boldsymbol{x})} mg(\phi(\boldsymbol{y}))}{|RF(\boldsymbol{x})|}. \tag{9}
$$

4   Experiments

To compare confidence-free, static, and dynamic classifier systems with different confidence measures, we implemented the random forest method and the aggregation methods described in the preceding sections (i.e., MV, SWM using GA confidence measure, and DWM using ELA, ELM, and RFAM confidence measures) in Java programming language and measured the error rate of the classifier systems. We also measured performance of the so-called *non-combined* (NC) classifier, which is simply a randomly chosen classifier from the ensemble, i.e., a single random tree. The NC classifier serves only as a reference classifier, since it does not make much sense to use it alone.

We tested the methods on 5 artificial and 11 real-world datasets from the Elena database [14] and from the UCI repository [15]. We used 10-fold cross-validation to measure the method performance (8 folds for training, 1 fold for validation, 1 fold for testing set). The mean value and standard deviation of the classifier error rate were measured. The size of the ensemble (number of trees) was set to $r = 20$, the number of features to explore in each node varied between $s = 2$ and $s = 5$ (depending on the dimensionality of the particular dataset), the maximal size of a leaf was set to $t = 10$. The values of the parameters were set based on some preliminary testing, no optimization or fine-tuning was done. All the methods used the same values of the parameters.

A key parameter of the dynamic aggregation methods is the number of neighbors k in the dynamic confidence measures. If k is small, the estimates of local classification properties are not accurate. On the other hand, as k is increasing, so is the computational complexity, and moreover, the values of the dynamic confidences tend to one global value. For the datasets with small number of patterns (Ecoli, Wine), we used $k = 10$ a priori. For the larger datasets, we set apart 10% of the data to select the best value of $k \in \{10, 20, 50\}$. The data which were used to find out the best k were not used in the rest of the experiment.

The results of the testing are shown in Table 1. We also measured statistical significance of the results (at 5% confidence level by the analysis of variance using Tukey-Kramer method [by the 'multcomp' function from the Matlab statistics toolbox]).

The results on artificial datasets show that on 4 out of 5 datasets, DWM-ELM obtains the best results. On 3 datasets, the result is a significant improvement to MV or SWM. On the real-world datasets, DWM-ELM is the best 5 times, SWM 3 times, DWM-RFAM 2 times, and DWM-ELA once. All significant improvements to MV or SWM are obtained by the DWM-ELM method only.

These results indicate that dynamic classifier systems can significantly outperform both confidence-free and static classifier systems, if the confidence measure is chosen appropriately. ELA and RFAM usually give worse results than ELM, but for some datasets, they are more feasible than ELM – this suggests that the performance of the dynamic classifier system is highly influenced by the confidence measure, and that the confidence measure should be chosen with respect to the particular dataset.

The experimental results in this paper are valid for random forests only. However, our experiments with quadratic discriminant classifiers show similar results [16].

Table 1. Mean value ± standard deviation of the classifier error rates from 10-fold crossvalidation. The best method for each dataset is displayed in boldface. Statistically significant (at 5% level) improvements to NC, MV, SWM-GA, DWM-ELA, and DWM-RFAM are marked by footnote signs (no significant improvement to DWM-ELM occurred).

Artificial datasets	Non-Combined NC	Conf.-free MV	Static κ SWM	Dynamic κ DWM	
Clouds	13.6 ± 1.5	11.9 ± 1.3	GA 11.8 ± 1.1	ELA	11.9 ± 1.9
				ELM	**11.1 ± 1.8** [1]
				RFAM	11.7 ± 1.0
Concentric	6.7 ± 2.0	2.7 ± 1.1 [1]	GA 3.0 ± 1.2 [1]	ELA	2.6 ± 0.6 [1]
				ELM	**1.9 ± 0.8** [13]
				RFAM	2.5 ± 1.2 [1]
Gauss_3D	27.8 ± 2.2	23.7 ± 1.6 [1]	GA 23.2 ± 1.3 [1]	ELA	23.7 ± 2.2 [1]
				ELM	**21.3 ± 1.5** [124]
				RFAM	22.8 ± 1.7
Gauss_8D	24.5 ± 1.9	**14.0 ± 2.0** [1]	GA 14.6 ± 2.1 [1]	ELA	14.1 ± 1.5 [1]
				ELM	16.1 ± 1.4 [1]
				RFAM	14.1 ± 2.0 [1]
Waveform	26.8 ± 1.3	17.5 ± 1.4 [1]	GA 17.8 ± 1.8 [1]	ELA	17.7 ± 1.7 [1]
				ELM	**15.2 ± 1.8** [12345]
				RFAM	17.8 ± 1.6 [1]
Real-world datasets	Non-Combined NC	Conf.-free MV	Static κ SWM	Dynamic κ DWM	
Balance	22.6 ± 4.3	16.4 ± 2.8 [1]	GA 14.6 ± 4.2 [1]	ELA	14.0 ± 5.1 [1]
				ELM	**10.7 ± 3.4** [123]
				RFAM	13.1 ± 4.5 [1]
Breast	7.5 ± 3.9	3.6 ± 2.4 [1]	GA **2.9 ± 1.8** [1]	ELA	3.9 ± 1.6 [1]
				ELM	3.7 ± 2.6 [1]
				RFAM	3.1 ± 2.4 [1]
Ecoli	25.9 ± 6.3	20.5 ± 7.9	GA **18.8 ± 4.4**	ELA	19.9 ± 6.0
				ELM	20.3 ± 7.3
				RFAM	19.8 ± 7.4
Phoneme	17.2 ± 2.3	12.8 ± 1.8 [1]	GA 13.1 ± 1.2 [1]	ELA	**12.4 ± 1.4** [1]
				ELM	13.2 ± 1.6 [1]
				RFAM	13.1 ± 0.6 [1]
Pima	27.0 ± 5.2	27.8 ± 5.3	GA **24.6 ± 6.4**	ELA	25.8 ± 5.6
				ELM	25.5 ± 4.8
				RFAM	25.6 ± 4.8
Poker	49.9 ± 2.8	46.6 ± 1.9 [1]	GA 46.2 ± 1.1 [1]	ELA	45.4 ± 1.6 [1]
				ELM	**43.8 ± 2.1** [12]
				RFAM	46.1 ± 2.0 [1]
Satimage	16.2 ± 1.3	14.9 ± 0.9	GA 14.8 ± 1.5	ELA	14.6 ± 1.2
				ELM	**14.5 ± 1.2**
				RFAM	14.9 ± 1.8
Texture	15.2 ± 2.6	2.4 ± 0.5 [1]	GA 2.4 ± 0.9 [1]	ELA	1.9 ± 0.4 [1]
				ELM	**0.8 ± 0.3** [123]
				RFAM	2.1 ± 0.6 [1]
Vowel	41.0 ± 5.4	12.2 ± 4.6 [1]	GA 14.4 ± 4.1 [1]	ELA	11.9 ± 2.9 [1]
				ELM	13.0 ± 2.8 [1]
				RFAM	**11.6 ± 3.4** [1]
Wine	9.1 ± 7.5	3.1 ± 5.1	GA 3.3 ± 2.9	ELA	3.2 ± 3.7
				ELM	4.3 ± 4.9
				RFAM	**2.1 ± 3.7** [1]
Yeast	53.6 ± 3.7	44.5 ± 2.7 [1]	GA 44.3 ± 4.2 [1]	ELA	40.9 ± 3.3 [1]
				ELM	**39.4 ± 3.7** [123]
				RFAM	41.1 ± 3.5 [1]

[1] Significant improvement to NC
[2] Significant improvement to MV
[3] Significant improvement to SWM-GA
[4] Significant improvement to DWM-ELA
[5] Significant improvement to DWM-RFAM

5   Summary and Future Work

In this paper, we have introduced a general framework for dynamic classifier combining, built on three main elements – the individual classifiers, their confidence measures, and the aggregator of the system. We have defined one static and three dynamic confidence measures which can be used in the framework, but other confidence measures can be used as well.

In our experiments with random forests on 5 artificial and 11 real-world benchmark datasets, we have shown that for several datasets, dynamic classifier systems can significantly outperform both confidence-free and static classifier systems. Furthermore, the results of the experiments suggest that the ELM dynamic confidence measure is more suitable for random forests than the RFAM dynamic confidence measure used in [3,4].

In our future work, we would like to study dynamic classifier systems of quadratic discriminant classifiers, k-NN classifiers, and support vector machines. We would like to develop model-specific dynamic confidence measures for these classifier types and to study dynamic classifier systems with state-of-the-art aggregators, for example a fuzzy t-conorm integral [17].

References

1. Duda, R.O., Hart, P.E., Stork, D.G.: Pattern Classification, 2nd edn. Wiley Interscience, Hoboken (2000)
2. Kuncheva, L.I.: Combining Pattern Classifiers: Methods and Algorithms. Wiley-Interscience, Hoboken (2004)
3. Robnik-Šikonja, M.: Improving random forests. In: Boulicaut, J.-F., Esposito, F., Giannotti, F., Pedreschi, D. (eds.) ECML 2004. LNCS, vol. 3201, pp. 359–370. Springer, Heidelberg (2004)
4. Tsymbal, A., Pechenizkiy, M., Cunningham, P.: Dynamic integration with random forests. In: Fürnkranz, J., Scheffer, T., Spiliopoulou, M. (eds.) ECML 2006. LNCS, vol. 4212, pp. 801–808. Springer, Heidelberg (2006)
5. Woods, K., Philip Kegelmeyer Jr., W., Bowyer, K.: Combination of multiple classifiers using local accuracy estimates. IEEE Trans. Pattern Anal. Mach. Intell. 19(4), 405–410 (1997)
6. Delany, S.J., Cunningham, P., Doyle, D., Zamolotskikh, A.: Generating estimates of classification confidence for a case-based spam filter. In: Muñoz-Ávila, H., Ricci, F. (eds.) ICCBR 2005. LNCS, vol. 3620, pp. 177–190. Springer, Heidelberg (2005)
7. Wilson, D.R., Martinez, T.R.: Combining cross-validation and confidence to measure fitness. In: Proceedings of the International Joint Conference on Neural Networks (IJCNN 1999), paper 163 (1999)
8. Avnimelech, R., Intrator, N.: Boosted mixture of experts: An ensemble learning scheme. Neural Computation 11(2), 483–497 (1999)
9. Breiman, L.: Random forests. Machine Learning 45(1), 5–32 (2001)
10. Hand, D.J.: Construction and Assessment of Classification Rules. Wiley, Chichester (1997)
11. Breiman, L.: Bagging predictors. Machine Learning 24(2), 123–140 (1996)
12. Freund, Y., Schapire, R.E.: Experiments with a new boosting algorithm. In: International Conference on Machine Learning, pp. 148–156 (1996)

13. Kuncheva, L.I., Bezdek, J.C., Duin, R.P.W.: Decision templates for multiple classifier fusion: an experimental comparison. Pattern Recognition 34(2), 299–314 (2001)
14. UCL MLG: Elena database (1995),
    http://www.dice.ucl.ac.be/mlg/?page=Elena
15. Newman, D.J., Hettich, S., Blake, C.L., Merz, C.J.: UCI repository of machine learning databases (1998),
    http://www.ics.uci.edu/~mlearn/MLRepository.html
16. Štefka, D., Holeňa, M.: Classifier aggregation using local classification confidence. In: Filipe, J., Fred, A., Sharp, B. (eds.) Proceedings of the First International Conference on Agents and Artificial Intelligence, ICAART 2009, Porto, Portugal, pp. 173–178 (2009)
17. Štefka, D., Holeňa, M.: The use of fuzzy t-conorm integral for combining classifiers. In: Mellouli, K. (ed.) ECSQARU 2007. LNCS, vol. 4724, pp. 755–766. Springer, Heidelberg (2007)

A Fuzzy Shape Descriptor
and Inference by Fuzzy Relaxation
with Application to Description of Bones
Contours at Hand Radiographs

Marzena Bielecka[1], Marek Skomorowski[2], and Bartosz Zieliński[3]

[1] Department of Geoinformatics and Applied Computer Science,
Faculty of Geology, Geophysics and Environmental Protection
AGH University of Science and Technology,
Al. Mickiewicza 30, 30-059 Kraków, Poland
[2] Institute of Computer Science, Jagiellonian University,
Grota-Roweckiego 26, 30-348 Kraków, Poland
[3] Faculty of Physics, Astronomy and Applied Computer Science,
Jagiellonian University,
Reymonta 2, 30-072 Kraków, Poland
bielecka@agh.edu.pl, skomorowski@ii.uj.edu.pl,
bartosz.zielinski@uj.edu.pl

Abstract. Generalization of string languages describing shapes in order
to apply them to analyze a contour of bones in hand radiographs is proposed in this paper. An algorithm to construct a fuzzy shape descriptor
is introduced. Next, basing on the fuzzy descriptor, a univocal description by fuzzy interference is realized. In prospects this method will be
used to erosion detection of hand bones visible in hand radiographs.

Keywords: contour, fuzzy shape descriptor, fuzzy parsing, hand
radiographs.

1 Introduction

Computer visual methods are used very often in medicine - see [17] in the context
of rheumatology. Therefore, fully automatic analyze of such pictures is needed.
Syntactic methods are ones of the most popular ([5], [18]). They are applied
both to contour analyze, taking advantage of string languages - see [11], [13],
[14], [15], as well to analysis of complex bones spatial relations including mutual
covering, for instance palm bones, using graph grammars - [1], [12], [16] can be
put as examples of pioneer papers concerning this topics.

Analysis of palm X-ray pictures is one of the streams in the medical pattern
analysis. It is extremely important to diagnose pathological rheumatoid changes
in early stage of a disease. This requires that differences of a rank about one
millimeter between contours of pathologically changed bones and proper ones
need to be noticed. It is difficult, because usually for majority of patients there

M. Kolehmainen et al. (Eds.): ICANNGA 2009, LNCS 5495, pp. 469–478, 2009.
© Springer-Verlag Berlin Heidelberg 2009

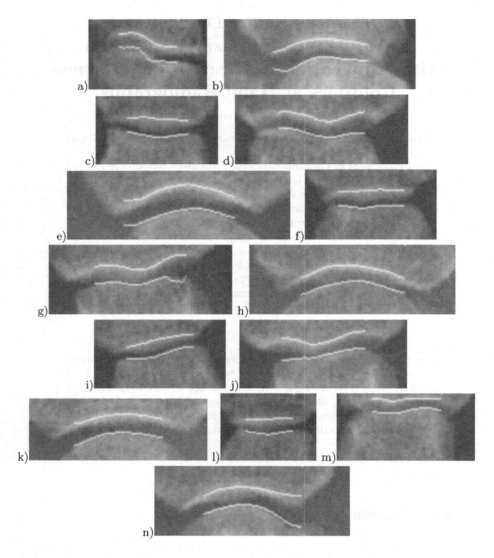

Fig. 1. Obtained contours for thumb (a, b), index finger (c, d, e), middle finger (f, g, h), ring finger (i, j, k) and small finger (l, m, n)

are no X-ray pictures of their hands before pathological changes. Moreover, the majority of experts claim that any illness progress can be noticed only when the time period between two following X-ray examinations is larger than six months, since changes proceed very slowly and cannot be diagnosed within a shorter period. To economize on time and energy spent on that kind of analysis, to make an X-ray examination more frequent and more precise, the process should be automatized.

In this paper a generalization of string languages describing shapes in order to apply them to analyze a contour of bones in hand radiographs is proposed. Fuzzy sets are used for aided of contour descriptor generalization. First, the so called alternative fuzzy descriptor is created. Then, an unambiguous fuzzy descriptor is obtained in the fuzzy inference process. The inference process is based on the parallel parsing idea presented in [4], [19], [20], [21], [22] in the context of probabilistic-aided graph languages and in [2] in the context of fuzzy-aided graph languages. The presented method is based on string shape languages [8] which has been so far applied in engineering, in particular in robotics and manufacturing - [6], [7], [9], [10], where patterns are not distorted and shapes and sizes are normalized. Introduction of fuzzy methods is needed in medical applications because transition from norm to pathology in biological systems is fluid and parameters, both in norm and in a given pathology, take values from an interval. Therefore, fuzzyness is always at least implicite introduced in medical applications - see [12].

This paper is a continuation of studies described in [3] and [23] where automatic preprocessing, including binarization, skeletization and contourization was done - see Fig.1.

2   String Shape Language

Let us recall a formalism presented in [7], [8], [9]. The basic unit of the analyzed pattern is a *primitive*. Jakubowski introduced sixteen primitives, constituting the set PRIM, being line segments or quarters of a circle - see Fig.2. This means that a primitive is a curve defined by an analytic formula describing corresponding line or curve segment. The beginning and the end of the primitive p is denoted by $hd(p)$ and $tl(p)$ respectively.

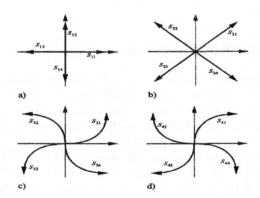

Fig. 2. The shape language primitives

Definition 21. *Let primitives* $u, v \in$ PRIM *do not lie in a single straight line. We say that* u *is joined to* v *if* $hd(u) = tl(v)$. *We denote it by* $u \odot v$. *A sequence of primitives such that* $p_i \odot p_{i+1}$ *is denoted by* $p_1 \odot ... \odot p_m$.

Definition 22. *A contour* k *composed of the primitives* $p_1, p_2, ..., p_m$ *can be expressed by* $p_1 \odot p_2 \odot ... \odot p_m = k$ *if* k *is a curve without multiple points. For such a contour the following definitions are introduced:*

- $tl(k) = tl(p_1); hd(k) = hd(p_m)$.
- $First(k) := p_1, Last(k) := p_m$.
- *A contour* k *is called a junction of contours* k', k'' *and is denoted by* $k' \odot k''$ *if* $Last(k') = First(k'')$.
- *A contour* $p_i \odot p_{i+1} \odot p_n, i \geq 1, m \geq n$ *is called a subcontour of* k.
- *If* $hd(p_m) = tl(p_1)$ *then a contour is called closed.*

Consider the contour $k = p_1 \odot ... \odot p_m$ such that each p_t is a primitive described as $s_{i_t p_t}$, $1 \leq t \leq m$.

Definition 23. *The string* $\bar{s} := s_{i_1 j_1} s_{i_2 j_2} ... s_{i_m j_m}$ *is termed the characterological description of* k *and is denoted by* $Des(k)$.

Let us specify a few features in a contour defined as above.

Definition 24. *A contour* $p_1 \odot p_2 \odot ... \odot p_n$ *is called an (l)-sinquad if all primitives* p_i, $1 \leq i \leq n$ *belong to the l-th quadrant of the Cartesian plane.*

Definition 25. *A contour* $k = k' \odot k''$ *creates a so-called (i,j)-biquad if both* k' *and* k'' *are sinquads. First(k'') is called a switch of the biquad and Last(k') is called a precursor. If* k' *or* k'' *is an axial primitive (see Fig.2a) then* k *is called an improper biquad.*

Definitions of other complex syntactic features - grooves and cascades - can be found in [7], [8] and [9].

3   Fuzzy Descriptor Creating

As it has already been mentioned, in medical practise parameters describing the investigated object, for instance a bone, take values from an interval. Therefore, it is insufficient to define single primitives as in robotic and manufacturing applications. Thus, let us introduce a fuzzy mechanism for aided the shape language.

Every one from sixteen defined primitives will be regarded as a fuzzy set. The line or curvilinear segment differ slightly from a given primitive has a value of membership function μ describing this primitive equal to 1 and its value of membership function describing other primitives is equal to 0. If the segment differs significantly from each primitive, values of its all sixteen membership function is less then 1 and is greater than 0 for this primitives to which it is most similar. The example of membership functions describing dependence on its value from the angle between a given primitive and the consider contour segment is shown

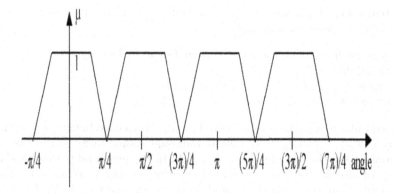

Fig. 3. The membership functions for primitives s_{11}, s_{12}, s_{13} and s_{14}

in Fig.3 for primitives s_{11}, s_{12}, s_{13} and s_{14}. It should be stressed that it is only the first component of a primitive fuzziness. The second one, ν, is connected with accuracy of a contour fragment fitting to line segment or an circle quarter. The all possibilities should be tested and all segments of which the membership function value is greater than zero are taken into consideration in an alternative fuzzy contour descriptor. The single segment of the alternative fuzzy contour descriptor is denoted as [...] and consists of primitives descriptors s_{ij}, $i, j \in \{1, 2, 3, 4\}$ and values of membership functions of the considered contour segment being a given primitive - see Examples. Let us notice that the alternative fuzzy contour descriptor segments can be nested, for instance [[...][...]] - see Example 2.

The algorithm of the alternative fuzzy contour descriptor creating has the following form.

ALGORITHM 1

1. Consider the subsequent fragment of the given contour.
2. Fit line segments and quarters of circles to the considered fragment with an assumed accuracy. If there are a few possibilities of such fitting, take all of them into consideration. For each fitting calculate fuzziness $\nu \in (0, 1]$ of the fitting using the given procedure computing ν.
3. Assign to each obtained line and curve segment values of sixteen membership functions μ comparing them with all primitives.
4. Create the next segment of the alternative fuzzy contour descriptor writing these triplets (s_{ij}, μ_k, ν_k) into square brackets for which $\mu_k > 0$. Take into consideration the alternative nesting implied by the fact that the same segment can be described both as a single primitive and a string of primitives - see Fig.6.
5. Repeat steps 1-4 until the whole contour is described.

4 Fuzzy Parsing

Let us recall definition of T-norm, widely used in fuzzy inference.

Definition 41. *A function $T : [0,1] \times [0,1] \rightarrow [0,1]$ is called a T-norm if the following conditions are satisfied:*

- *monotonicity: if $a \leq b$ and $c \leq d$ then $T(a,c) \leq T(b,d)$,*
- *commutativity: $T(a,b) = T(b,a)$,*
- *associativity: $T(T(a,b),c) = T(a,T(b,c))$,*
- *boundary conditions: $T(a,0) = 0$, $\quad T(a,1) = a$.*

Given an unknown pattern represented by an alternative fuzzy contour descriptor, say R, the problem is to obtain an outcome unique fuzzy contour descriptor r, obtained from the complex fuzzy string R. In the proposed parallel and cut-off strategy of alternative fuzzy contour descriptor parsing a number of simultaneously derived simple strings is equal to a certain number *limit*. In this case, derived simple strings spread through the search tree, but only the best, that is with maximum measure value, *limit* strings are expanded.

ALGORITHM 2

1. Choose two first elements of the alternative fuzzy contour descriptor and compose all possible substrings. Each element is of the form (s_{ij}, μ_n, ν_n), $i, j \in \{1, 2, 3, 4\}$.
2. For each substring obtained in the point 1, calculate values of their membership functions using a given T-norm using formula $T(T(\mu_1, \mu_2), T(\nu_1, \nu_2))$ where μ_1, ν_1 are membership functions for the first element and μ_2, ν_2 for the second element.
3. For further derivation choose a number (*limit*) of derived substrings with the biggest values of the calculated membership function.
4. To the chosen substrings join elements from the next element of the alternative fuzzy contour descriptor and calculate values of their membership functions using a given T-norm using formula $T(T(\mu_m, \mu_m), T(\nu_l, \nu_l))$ where μ_m, ν_m are membership functions for the derived substring and μ_l, ν_l for the next element. If the alternative fuzzy contour descriptor is finished then stop the parsing algorithm.
5. Repeat the points 3 and 4 until a complete unique fuzzy contour descriptor is obtained.
6. As a result take from the derived unique fuzzy contour descriptors the one having the maximal value of the T-norm T. Take the mean value of its membership functions as a measure of the obtained description credibility.

Example 1

Let as assume that the obtained alternative fuzzy contour descriptor is of the following form:

$R = [(s_{21}, 0.7, 0.9), (s_{41}, 0.8, 0.8)][(s_{14}, 0.9, 0.3), (s_{24}0.7, 0.8), (s_{34}, 0.6, 0.7)]$
$[(s_{14}, 0.7, 0.4), (s_{24}, 0.6, 0.9)](s_{33}, 0.9, 0.6)$.

Let, furthermore, the T-norm be given as an arithmetic product, i.e. $T(a,b) = a \cdot b$ and let *limit* $= 2$. The ALGORITHM 2 is executed as follows.

Step 1

$(s_{21}, 0.7, 0.9)(s_{14}, 0.9, 0.3)$ $T(T(0.7, 0.9), T(0.9), 0.3) = T(0.63, 0.27) = 0.1701$

$(s_{21}, 0.7, 0.9)(s_{24}, 0.7, 0.8)$ $T(T(0.7, 0.7), T(0.9, 0.8)) = T(0.49, 0.72) = 0.3528$

$(s_{21}, 0.7, 0.9)(s_{34}, 0.6, 0.7)$ $T(T(0.7, 0.6), T(0.9, 0.7)) = T(0.63, 0.42) = 0.2646$

$(s_{41}, 0.8, 0.8)(s_{14}, 0.8, 0.3)$ $T(T(0.8, 0.9), T(0.8, 0.3)) = T(0.64, 0.27) = 0.1728$

$(s_{41}, 0.8, 0.8)(s_{24}, 0.7, 0.8)$ $T(T(0.8, 0.7), T(0.8, 0.8)) = T(0.56, 0.64) = 0.2688$

$(s_{41}, 0.8, 0.8)(s_{34}, 0.6, 0.7)$ $T(T(0.8, 0.6), T(0.8, 0.7)) = T(0.64, 0.42) = 0.2588$

The strings $s_{21}s_{24}$ and $s_{41}s_{24}$ are taken for the second step of the parsing.

Step 2

$$(s_{21}, s_{24}, 0.49, 0.72)(s_{14}, 0.7, 0.4)   T(T(0.49, 0.7), T(0.72, 0.4)) =$$

$$= T(0.343, 0.288) \approx 0.0988.$$

Calculating in the same way we obtain

$(s_{21}, s_{24}, 0.49, 0.72)(s_{24}, 0.6, 0.9)$ $T(0.294, 0.648) \approx 0.1905$

$(s_{41}, s_{24}, 0.56, 0.64)(s_{14}, 0.7, 0.4)$ $T(0.392, 0.256) \approx 0.1004$

$(s_{41}, s_{24}, 0.56, 0.64)(s_{24}, 0.6, 0.9)$ $T(0.336, 0.576) \approx 0.1935$

The strings $s_{21}s_{24}s_{24}$, and $s_{41}s_{24}s_{24}$ are taken for the third step of the parsing.

Step 3

$(s_{21}s_{24}s_{24}, 0.294, 0.648)(s_{33}, 0.9, 0.6)$ $T(0.2646, 0.3888) \approx 0.1029$

$(s_{41}s_{24}s_{24}, 0.336, 0.576)(s_{33}, 0.9, 0.6)$ $T(0.3024, 0.3456) \approx 0.1045.$

The descriptor $s_{41}s_{24}s_{24}s_{33}$ is chosen. STOP.

To sum up $(s_{41}, 0.8, 0.8)(s_{24}, 0.7, 0.8)(s_{24}, 0.6, 0.9)(s_{33}, 0.9, 0.6)$ has been chosen as the most credible simple fuzzy descriptor of the shape encoded by the above fuzzy complex descriptor R. The mean value of its membership functions $\frac{0.8+0.8+0.7+0.8+0.6+0.9+0.9+0.6}{8} = 0.7625$ is a measure of the obtained description credibility.

Example 2

Let us consider a bone contour with a central erosion - see Fig.4.

If only two first cases are taken into consideration then the alternative fuzzy contour descriptor corresponding to this contour fragment would be the following form:

$$[(s_{44}, \mu_1, \nu_1)(s_{11}, \mu_2), \nu_2](s_{34}, \mu_3, \nu_3)(s_{31}, \mu_4, \nu_4)[(s_{41}\mu_5, \nu_5)(s_{11}, \mu_6, \nu_6)].$$

If all three cases are taken into consideration then the alternative fuzzy contour descriptor corresponding to this contour fragment would be far more complicated - notice the nesting of square brackets corresponding to the nesting of alternatives in the description:

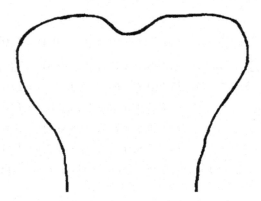

Fig. 4. The bone contour with central erosion

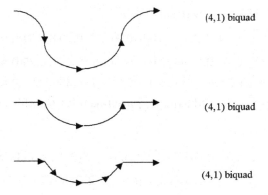

(4,1) biquad

(4,1) biquad

(4,1) biquad

Fig. 5. The examples of central erosion possible structures description

$[(s_{44}, \mu_1, \nu_1)(s_{11}, \mu_2), \nu_2]$
$[(s_{34}, \mu_3, \nu_3)[(s_{24}, \mu_4, \nu_4)[(s_{24}, \mu_5, \nu_5)(s_{11}, \mu_6, \nu_6)(s_{34}, \mu_7, \nu_7)]]]$
$[(s_{31}, \mu_8, \nu_8)[[(s_{21}, \mu_9, \nu_9)(s_{11}, \mu_{10}, \nu_{10})(s_{31}, \mu_{11}, \nu_{11})](s_{21}, \mu_{12}, \nu_{12})]]$
$[(s_{41}\mu_{13}, \nu_{13})(s_{11}, \mu_{14}, \nu_{14})].$

In both the cases the fuzzy inference algorithm (`ALGORITHM 2`) will create a unique fuzzy contour descriptor. However in all the cases the central erosion is a $(4, 1)$-biquad. Thus after creating a unique fuzzy contour descriptor it is sufficient to find a $(4, 1)$-biquad - this problem is already solved (see [7]). The derived description will allow to diagnose how the illness is advanced.

5   Concluding Remarks

The computer interpretation of palm X-ray images is extremely difficult because of a complex palm structure. The results described by other authors seemed to be very partial. It also seems that results described in [3] and [23] significantly completes the mentioned ones. Summing up, the following have been done effectively: preprocessing of the whole image, the image skeletization in the phalangeal region, segmentation in the phalangeal region, joints localization in the phalangeal region, contours description in the localized joints, computing widths of the localized joints. All the mentioned items are done automatically. The obtained contours and width description of the joints can be a basis for diagnosis of a certain diseases. In this paper the theoretical foundations and algorithmic aspect of the next stage of the problem study is presented. The proposal of the contour three-step description is introduced. During the first step the contour fragments are described by estimation using line segments or arcs. Then the obtained encoded curvilinear fragments are compared with primitives regarded as fuzzy sets. In such a way the alternative fuzzy contour description is obtained. Then, using parallel fuzzy inference, a unique fuzzy contour descriptor is created. It will be used as the starting point for inference basing on shape languages which will be applied to the detection of bone erosions. This is planned as the next stage of our studies.

References

1. Beier, D., Lehmann, T.M., Seidl, T., Thies, C.: Segmentation of medical images combining local, regional, global, and hierarchical distances into a bottom-up region merging scheme. In: Proceedings of the SPIE, vol. 5747, pp. 546–555 (2005)
2. Bielecka, M., Skomorowski, M., Bielecki, A.: Fuzzy syntactic approach to pattern recognition and scene analysis. In: Proceedings of the 4th International Conference on Informatics in Control, Automatics and Robotics ICINCO 2007, ICSO - Intelligent Control Systems and Optimization, Robotics and Automation, vol. 1, pp. 29–35 (2007)
3. Bielecki, A., Korkosz, M., Zieliński, B.: Hand radiographs preprocessing, image representation in the finger regions and joint space width measure for image interpretation. Pattern Recognition 41, 3786–3798 (2008)
4. Flasiński, M., Skomorowski, M.: Parsing of random graph languages for automated inspection in statistical-based quality assurance system. Machine Graphics and Vision 7, 565–623 (1998)
5. Fu, K.S.: Syntactic Pattern Recognition and Applications. Springer, Berlin (1977)
6. Jakubowski, R.: Syntactic characterization of machine parts shapes. Cybernetics and Systems 13, 1–24 (1982)
7. Jakubowski, R.: Extraction of shape features for syntactic recognition of mechanical parts. IEEE Trans. Systems, Man and Cybernetics 15, 642–651 (1985)
8. Jakubowski, R.: A structural representation of shape and its features. Information Sciences 39, 129–151 (1986)
9. Jakubowski, R., Bielecki, A., Chmielnicki, W.: Data structure for storing drawing being then analysed for purposes of CAD. Archiwa Informatyki Teoretycznej i Stosowanej 1, 51–70 (1993)
10. Jakubowski, R., Flasiński, M.: Towards a generalized sweeping model for designing with extraction and recognition of 3D solids. Journal of Design and Manufacturing 2, 239–258 (1992)

11. Ogiela, M.: Geometric transformations in the shape analysis of selected biomedical structures. Image Processing and Communication 4, 27–36 (1998)
12. Ogiela, L., Ogiela, M., Tadeusiewicz, R.: Image languages in intelligent radiological palm diagnostics. Pattern Recognition 39, 2157–2165 (2006)
13. Ogiela, M., Tadeusiewicz, R.: Recognition of some types X-ray images for medical diagnostics. Biocybernetics and Biomedical Engineering 17, 69–91 (1997)
14. Ogiela, M., Tadeusiewicz, R.: Syntactic analysis and languages of shape feature description in computer aided diagnosis and recognition of cancerous and inflammatory lesions of organs in selected X-ray images. Journal of Digital Imaging 12, 24–27 (1999)
15. Ogiela, M., Tadeusiewicz, R.: Artificial intelligence methods in shape feature analysis of selected organs in medical images. Image Processing and Communication 19, 3–11 (2000)
16. Ogiela, M., Tadeusiewicz, R.: Picture languages in automatic radiological palm interpretation. International Journal of Applied Mathematics and Computer Science 15, 305–312 (2005)
17. Peterfy, C., Klippel, J., Dieppe, P.: Imaging Techniques In Rheumatology. Mosby International, London (1998)
18. Schalkoff, R.: Pattern Recognition. In: Statistical, Structural and Neural Approach. Wiley, New York (1992)
19. Skomorowski, M.: Image labelling by random graph parsing for syntactic scene description. Foundadtion of Computing and Decision Sciences 23, 161–178 (1998)
20. Skomorowski, M.: Parsing of random graphs for scene analysis. Machine Graphics and Vision 7, 313–323 (1998)
21. Skomorowski, M.: Use of random graph parsing for scene labelling by probabilistic relaxation. Pattern Recognition Letters 20, 949–956 (1999)
22. Skomorowski, M.: Syntactic recognition of syntactic patterns by means of random graph parsing. Pattern Recognition Letters 28, 572–581 (2006)
23. Zieliński, B.: Fully-automated algorithm dedicated to computing metacarpophalangeal and interphalangeal joint cavity widths. Schedae Informaticae 16, 47–67 (2007)

Hough and Fuzzy Hough Transform in Music Tunes Recognition Systems

Maciej Hrebień and Józef Korbicz

Institute of Control and Computation Engineering
University of Zielona Góra, ul. Podgórna 50, 65-246 Zielona Góra
{m.hrebien,j.korbicz}@issi.uz.zgora.pl
http://www.issi.uz.zgora.pl

Abstract. This paper presents a method of music tunes recognition based on the Hough transform as well as its fuzzy version. One can also find here experimental results showing the effectiveness of the presented solutions. Perspectives of further work and quality improvements are also stated as a base for subsequent research.

1 Introduction

In last decade we can observe a very dynamic grow in the number of cell phones per person, especially in the so-called developing countries - new members of the European Union. For example, in Poland there are nearly 36758 thousands cell phones per 38157 thousands citizens and this number is still increasing [5]. The cell phone, besides the main function, which is voice communication at a distance, is starting to be a sort of mini-centre of information. Many modern cell phones have the ability of accessing the internet, exchanging data, many of them heave photo-camera, MP3 decoder or a mini-dictaphone onboard. The computational power of cell phones per Watt is increasing what gives the opportunity to threat them as a personal mini-computers with many useful features. This features can be connected with work (business notes, meeting remainders), fun (games, photo editing) or more dedicated solutions like e.g. telemedicine.

Most people like music and many of them listen to it in a car, home or even during work. How many times one was wondering what is the title of the song one is currently hearing and who is the performer, is hard to count. As most of people have cell phones it would be interesting to have an operator supporting system that would be able to recognize the played song. Such a system should include facts that a given recording is short (nobody wants to hold a cell phone more than a few seconds in front of a speaker), contains an additive noise generated by surrounding (car engine, detuned radio, wind, voices of passers-by, etc.) and its quality is limited to the capabilities of the microphone acquiring the audio signal and losses in the lossy GSM compression.

According to authors' knowledge there are currently two commercial music recognition systems in the western EU countries, that is Shazam in United Kingdom and Musiwave in Spain [13]. Because this systems are fully commercial it

M. Kolehmainen et al. (Eds.): ICANNGA 2009, LNCS 5495, pp. 479–488, 2009.
© Springer-Verlag Berlin Heidelberg 2009

is very hard to obtain the full information of how they work, what is their *real* effectiveness and computation complexity. Nevertheless some researches were conducted in this area and can be found in literature [4,6].

In this paper a method of music tunes recognition based on the Hough transform as well as its fuzzy version is presented. One can also find here experimental results collected during research and creation of a GSM based telephony simulator. Perspectives of further work and quality improvements are also stated as a base for subsequent researches.

2 System Assumptions

The main task of the system is to recognize and identify a music tune sample with the use of a cell phone. A client calls a dedicated phone number and puts its phone near sound source for a period of a few seconds. What the client expects as a result is a short message containing the information about the tune like e.g.: who is the performer, what is the title of the song and album it comes from, when it was recorded/produced, etc. Namely, it is a query-by-example problem. Unfortunately, the recognition task performed at the operators side is not very easy because of cell phones' hardware restrictions and the GSM lossy compression.

A human voice contains sound waves with base frequencies varying from about 300 to 3500 Hz [12]. A human ear, on the other hand, plays the role of a biological amplifier and is the most sensitive to frequencies characteristic to human voices generated by vibrating vocal cords [15]. This facts are used with success in the cell phone telephony. Thus, a given voice signal is passed through a band-pass

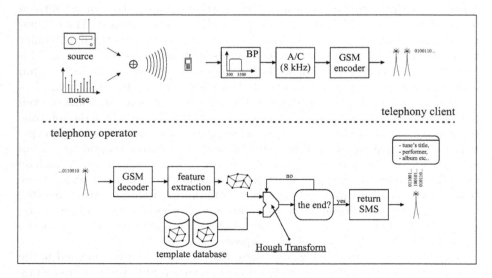

Fig. 1. Schematic of the music tunes recognition system

filter with the boundary frequencies characteristic to human voice. The signal is then sampled with $7 \div 16$ bits precision at 8 kHz frequency to satisfy the Nyquist's condition and lossy compressed by the GSM encoder [18].

The system schematic is given in Fig. 1.

3   Features Extraction

The presented solution is based on short-time frequency spectrum analysis. According to authors' observations the frequency spectrum, next to i.e. autocorrelation, is barely sensitive to the impact of any external noise, is unique for every music tune and valuably represents its nature and dynamics.

A given tune sample is at first decompressed by the GSM decoder and then analyzed frame by frame with the overlaying and frequency leak stopping technique. For each frame a set of features is calculated. In our approach the S features are these frequencies for those its amplitude estimated using the Fourier transform creates a local maxima in a given milliseconds of the analyzed tune:

$$S = \Big\{ \{(t_1, f_{1,1}), (t_1, f_{1,2}), \; \ldots, \; (t_1, f_{1,n})\}, \tag{1}$$
$$\{(t_2, f_{2,1}), (t_2, f_{2,2}), \; \ldots, \; (t_2, f_{2,n})\},$$
$$\ldots,$$
$$\{(t_k, f_{k,1}), (t_k, f_{k,2}), \; \ldots, \; (t_k, f_{k,n})\} \Big\},$$

where:

$$f_{t,1\ldots n} = \arg\max_n \Big(peaks(A(m)) \Big), \tag{2}$$

$$A(m) = |F(m)|, \tag{3}$$

$$F(m) = \sum_{n=0}^{N-1} x(n)e^{-j2\pi nm/M}, \tag{4}$$

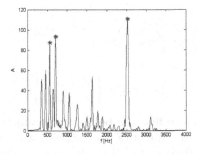

Fig. 2. Example of detected frame features (marked with "*") for $n = 3$

$$m = 1 \ldots M, \tag{5}$$

n is the number of features per frame, t is the time period and F is the frequency spectrum. In situations where the number of features in a frame is less than n the smaller group of features is also included in the S set. All empty frames (silence) are omitted.

A graphical example of feature detection is given in Fig. 2.

4 Crisp Matching

The Hough transform, originally designed for line detection [7], was many times used in literature to find a specific shape [1,9,10,11] or a group of features in a larger set [8,14]. The transform is known for its quality in the presence of noise, the ability to adopt it to the detection of a given shape [2] and the accumulator based voting.

The accumulator in our approach is two-dimensional. The first dimension, which is the time, is obvious because a given tune sample is a part of a much longer music tune and the shift in time have to be detected to find the best match. The second dimension, which is the frequency, is added because a research shown that the same music tune can be played by a musician higher or lower in the sense of sound timbre what can be observed mainly in classical music. What was additionally assumed is that the tempo has to be the same to talk about tunes equality. Thus, any tempo differences (i.e. an intentionally slowed down by a disco DJ tune) are not included in the accumulator's dimensionality but are possible at the cost of calculation speed.

The task of music tune recognition is a problem of finding the shift parameters $(\Delta t, \Delta f)$ which added to the features of a tune sample send by a client will give the higher level of fitness to the features of a template taken from an early prepared database. Since the features of a tune create an irregular shape the Hough transform have to be defined in an algorithmic form presented in Fig. 3.

For each pair of features: $(t_i^A, f_i^A) \in S^A$ and $(t_j^B, f_j^B) \in S^B$ the accumulator A collects votes in those cells that the best shows the difference between considered features. Thus, a given cell of the accumulator collects:

$$A(\Delta t, \Delta f) = \sum_{i=1}^{m} \sum_{j=1}^{n} \left[\delta_2 \big(t_j^B - t_i^A, \Delta t \big) \delta_2 \big(f_j^B - f_i^A, \Delta f \big) \right], \tag{6}$$

where:

$$\delta_2(x_1, x_2) = \begin{cases} x_1 = x_2, 1 \\ x_1 \neq x_2, 0 \end{cases} \tag{7}$$

what gives the possibility to find the best shift parameters $(\Delta t, \Delta f)$ in the accumulator by taking the argument of the maximal value of A (Fig. 4a).

As a measure of the final match (Fig. 4b) sum of Euclidian distances from features of the send by a client sample to their nearest features from a template is calculated:

$$\forall_i \forall_j A(i,j) \leftarrow 0$$

$$FOR \; \{t_i^A, f_i^A\} \in S^A, \; i = 1 \ldots m$$
$$\quad FOR \; \{t_j^B, f_j^B\} \in S^B, \; j = 1 \ldots n$$
$$\quad \{$$
$$\qquad \Delta t_{ij} = t_j^B - t_i^A$$
$$\qquad \Delta f_{ij} = f_j^B - f_i^A$$

$$\qquad \Delta t_{ij}, \Delta f_{ij} \leftarrow align(\Delta t_{ij}, \Delta f_{ij})$$
$$\qquad A(\Delta t_{ij}, \Delta f_{ij}) \leftarrow A(\Delta t_{ij}, \Delta f_{ij}) + 1$$
$$\quad \}$$

$$\Delta t, \Delta f \leftarrow \arg\max(A)$$

Fig. 3. Algorithmic form of the crisp Hough transform

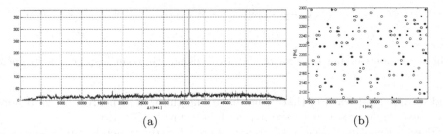

(a)   (b)

Fig. 4. Exemplary: (a) accumulator's slice for $\Delta f = 0$ Hz and low noise sample, (b) fragment of the best match with template's features marked with "×" and tested sample's features with "○" (experiment for SNR = 0 dB)

$$E = \sum_{i=1}^{m} \min \left(\| x_{i1} \|, \| x_{i2} \|, \ldots, \| x_{ij} \| \right), \qquad (8)$$

where:

$$x_{ij} = \left[(t_i^A + \Delta t) - t_j^B, (f_i^A + \Delta f) - f_j^B \right], \quad j = 1 \ldots n. \qquad (9)$$

The E measure can be considered as a match error, so the smaller the value of E the better fitness to the template.

5   Fuzzy Matching

The analysis of the features layout shows that they have a tendency to group into local clusters (Fig. 5). This is caused by the fact that a note or singer's voice sustains through several dozen of milliseconds on a given frequency while performing a given fragment of a song. Thus, one can suppose that all isolated

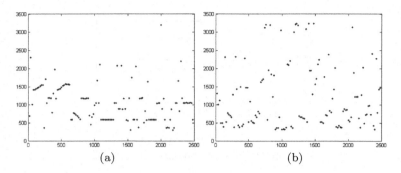

Fig. 5. Exemplary features layout: (a) thick and (b) thin clusters

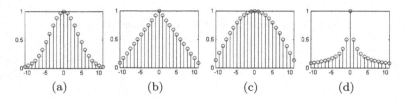

Fig. 6. Membership functions for $R = 11$: (a) gaussian, (b) triangle, (c) cosine, (d) peak

and standing out features that do not have a neighbor in its nearest neighborhood are noisy while processing a sample with very low SNR ratio.

The fuzzification process [17] in our approach assigns each feature a w weight which value depends on the membership function used and the number of neighbors in the nearest neighborhood described by the R radius. The clusters of features create some kind of "anchor" (the higher number of neighbors the higher the features weights and the more important the whole cluster) and all isolated features, which probably are noisy, get relatively small weights, thus they have smaller impact on the information collected in the accumulator.

In our approach we have tested four membership functions which are illustrated in Fig. 6:

$$f_{gaussian}(x) = \exp\left(-\frac{x^2}{2\sigma^2}\right), \quad \sigma = \frac{1}{3}, \tag{10}$$

$$f_{triangle}(x) = -x\,\text{sign}(x) + 1, \tag{11}$$

$$f_{cosine}(x) = \cos(x\frac{\pi}{2}), \tag{12}$$

$$f_{peak}(n) = \begin{cases} if \ n = 0, \ then \ 1 \\ if \ n \neq 0, \ then \ \frac{1}{|n|+1}, \end{cases} \tag{13}$$

where values close to zero at the x-axis have to be interpreted as a high number of neighboring features in the local neighborhood described by the R radius.

The algorithm that implements the fuzzy Hough transform is very similar to the one given in Fig. 3 and the main difference is the way of giving votes in the appropriate accumulator's cells. The $A(\Delta t_{ij}, \Delta f_{ij}) \leftarrow A(\Delta t_{ij}, \Delta f_{ij}) + 1$ fragment is exchanged with the $A(\Delta t_{ij}, \Delta f_{ij}) \leftarrow A(\Delta t_{ij}, \Delta f_{ij}) + \mathrm{MIN}(w_i^A, w_j^B)$ one, where w_i^A and w_j^B are the features weights. The MIN function is dictated by the assumption that a smaller weight has smaller impact on the accumulator's content. Thus the accumulator collects weights in its cells rather than crisp votes:

$$A(\Delta t, \Delta f) = \sum_{i=1}^{m} \sum_{j=1}^{n} \mathrm{MIN}(w_i^A, w_j^B) \left[\delta_2 \left(t_j^B - t_i^A, \Delta t \right) \delta_2 \left(f_j^B - f_i^A, \Delta f \right) \right]. \quad (14)$$

with the δ_2 defined as above.

6   Experimental Results

To test the quality of the presented solutions several experiments were performed in the presence of additive white noise. The first one shows the influence of tune sample length on the match quality. In this case tests were performed for 5, 10 and 15 seconds samples with one feature per frame (Fig. 7a). The second one shows the influence of the number of features used per frame with the length of a send sample equal to 5 seconds (Fig. 7b). The third experiment shows the influence of the GSM 6.10 compression with comparison to the same experiment performed but with the compression switched off (Fig. 7c). The fourth experiment shows which membership function should be used and what local neighbor radius should be considered to get the best matching results on the average (Fig. 8a). The impact of the fuzzy mechanism in the sense of accumulator's PSNR ratio defined as:

$$\mathrm{PSNR} = 20 \log_{10} \left(\frac{\mathrm{MAX}(A)}{\sqrt{\frac{1}{TF} \sum_{t=1}^{T} \sum_{f=1}^{F} A(t,f)^2}} \right), \quad (15)$$

where T and F are the number of time and frequency bins of the A accumulator respectively, was also examined (Fig. 8b).

The database consisted of 1000 randomly selected music tunes of different genres. The test set consisted of 100 randomly selected tune samples with noise levels raging from -15 to 15 dB with 3 dB interval for crisp and with 1 dB interval for fuzzy tests. All experiments were simulated with software, no telephony hardware was used.

What we can observe is that the presented method tolerates frequency restrictions, losses in the GSM compression (8.72% on average) and noise well. Additionally, experimental results illustrating the influence of the recording time on the match quality are very similar to the ones showing the influence of the number of features used. This gives the conclusion, that if a short sample is recorded by a client a higher number of features per frame is desired for better

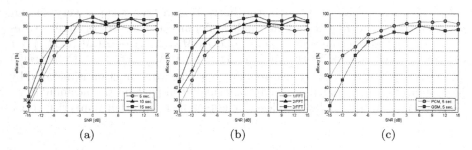

Fig. 7. Experimental results showing the influence of: (a) recording time, (b) number of features used, (c) GSM compression, on the match efficacy

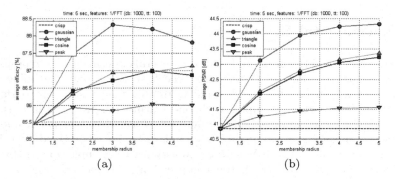

Fig. 8. Experimental results showing: (a) the quality of the recognition averaged through all tests (from -15 to 15 dB with 1 dB interval), (b) corresponding accumulator's average PSNR ratio

efficacy and vice versa – if the recording is long the smaller number of features per frame can be used to speed up the search.

The experimental results show that the fuzzy mechanism increases the match efficacy and improves sharpness of the information collected in the accumulator in the sense of the PSNR ratio (Fig. 9). This can be easily observed for very noisy samples where the PSNR ratio is about 5 dB higher than for crisp results and the match efficacy is almost two times higher in this case (see Fig. 9a and the −15 dB test). The highest match ratio and the sharpest accumulator was observed for the gaussian membership function with $R = 3$ (Fig. 8). The average efficacy through all experiments is about 2.9% higher and the PSNR ratio is about 3.1 dB higher for the Hough transform with the fuzzy mechanism than for crisp ones.

The performed experiments show that the Hough transform adopted for sound features localization can be effectively used in music tunes recognition systems. The presented solution needs less then a quarter of second on today's machines (Athlon 64 3500+ 2.8 GHz, Pentium 4 2.2 GHz) per sample–template comparison

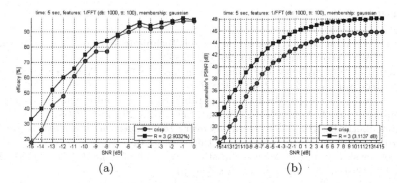

Fig. 9. Gaussian membership and crisp results comparison: (a) efficacy, (b) accumulator's PSNR ratio

depending on the hardware, the number of features and the length of sample used. For an average 3 minute music tune the method needs about 50 MB of memory for the accumulator with 7.8125 Hz and 20 milliseconds cell granularity which is not much for todays computer systems. The computational complexity of the fuzzy version of the Hough transform is very similar to the crisp one. The only difference is that the w weights have to be calculated before the main algorithm starts and the voting process is done using floating point arithmetic rather then integer one. The size of the accumulator is the same in both versions because 32-bit cell representation was used in all experiments.

7   Conclusions

In conclusion, performed experiments in the Hough transform as well as its fuzzy version adaptation to the problem of music tunes recognition are promising. This gives the base for further research in still open problems like e.g. global speed up. The database search can be accelerated by a classifier that would be able to decide what kind of music a sample represents [3] or a classifier that will create a probability vector that will show what subset of the database should be checked at first [16]. Performance and quality tests on much bigger database for mixed as well as one genre music tunes should also be considered in the future works. A real hardware (not only a software simulator) and a real user and environment (with natural noise, equalization, reverbation and audio saturated compression factors) would also be a great opportunity to test the proposed solution.

Acknowledgment

This work has been supported by the Ministry of Science and Higher Education of the Republic of Poland under the project no. N N519 4065 34 and the decision no. 9001/B/T02/2008/34.

References

1. Antiquzzamann, M.: Coarse-to-Fine Search Technique do Detect Circles in Images. Int. J. Adv. Manuf. Technol. 15, 96–102 (1999)
2. Ballard, D.: Generalizing the Hough Transform to Detect Arbitrary Shapes. Pattern Recognition 13(2), 111–122 (1981)
3. Cendrowska, D.: Strict Maximum Separability of Two Finite Sets: an Algorithmic Approach. Int. J. Appl. Math. Comput. Sci. 15(2), 295–304 (2005)
4. Dannenberg, R., Hu, N.: Pattern discovery techniques for music audio. In: Proc. 3rd Int. Conf. on Music Information Retrieval (2002)
5. Dmochowska, H. (ed.): Concise statistical yearbook of Poland. Statistical Publishing Establishment (2007), www.stat.gov.pl
6. Haitsma, J., Kalker, A.: A highly robust audio fingerprinting system. In: Proc. 3rd Int. Conf. on Music Information Retrieval, Paris (2002)
7. Hough, P.: Method and Means for Recognizing Complex Patterns, U.S. Patent No. 3,069,654 (1962)
8. Hrebień, M., Korbicz, J.: Human Identification Based on Fingerprint Local Features. In: Rutkowski, L., Tadeusiewicz, R., Zadeh, L.A., Żurada, J.M. (eds.) ICAISC 2006. LNCS, vol. 4029, pp. 796–803. Springer, Heidelberg (2006)
9. Hrebień, M., Steć, P., Nieczkowski, T., Obuchowicz, A.: Segmentation of Breast Cancer Fine Needle Biopsy Cytological Images. Int. J. Appl. Math. Comput. Sci. 18(2), 159–170 (2007)
10. Nair, P., Saunders, A.: Hough Transform Based Ellipse Detection Algorithm. Pattern Recognition Letters 17, 777–784 (1996)
11. Olson, C.: A General Method for Geometric Feature Matching and Model Extraction. Int. J. Comput. Vision 45(1), 39–54 (2001)
12. Ozimek, E.: Sound and its perception. In: Physical and psychoacoustic aspects. Polish Scientific Publishers, PWN (2002) (in Polish)
13. Prado, B.: Finding Structure in Audio for Music Information Retrieval. IEEE Signal Processing Magazine 23(3), 126–132 (2006)
14. Ratha, N., Karu, K., Chen, S., Jain, A.: A Real-time Matching System for Large Fingerprint Databases. IEEE Trans. Pattern Analysis and Machine Intelligence 28(8), 799–813 (1996)
15. Tadeusiewicz, R.: Speech signal. Transport and Communication Publishers, Warsaw (1988) (in Polish)
16. Toth, L., Kocsor, A., Csirik, J.: On Naive Bayes in Speech Recognition. Int. J. Appl. Math. Comput. Sci 15(2), 287–294 (2005)
17. Zadeh, L.: From Computing with Numbers to Computing with Words – From Manipulation of Measurements to Manipulation of Perceptions. Int. J. Appl. Math. Comput. Sci. 12(3), 307–324 (2002)
18. Zieliński, T.: Digital signal processing. In: From theory to practise. Transport and Communication Publishers, Warsaw (2005) (in Polish)

Multiple Order Gradient Feature for Macro-Invertebrate Identification Using Support Vector Machines

Ville Tirronen[1], Andrea Caponio[1,2], Tomi Haanpää[1], and Kristian Meissner[3]

[1] Department of Mathematical Information Technology,
University of Jyväskylä
{ville.tirronen,andrea.caponio,tomi.v.haanpaa}@jyu.fi
http://www.mit.jyu.fi
[2] Department of Electrotechnics and Electronics,
Technical University of Bari
caponio@deemail.poliba.it
http://www-dee.poliba.it/DEE/DEE.html
[3] Finnish Environmental Institute,
Jyväskylä Unit
kristian.meissner@ymparisto.fi
http://www.environment.fi/syke

Abstract. This paper investigates the feasibility of automated benthic macro-invertebrate taxon identification based on support vector machines and a novel gradient based feature. Biomonitoring can efficiently pinpoint subtle environmental changes and is therefore globally widely used in ecological status assessment. However, all biomonitoring is cost-intensive due to the expert work needed to identify organisms. To relieve this problem an automated image recognition system for benthic macro-invertebrate taxonomical analysis is proposed in this work. Using a novel approach, we present high accuracy classification results, suggesting that automated taxa recognition for benthic macro-invertebrates is viable. Our study indicates that automated image recognition techniques can match human taxonomic identification accuracy and greatly reduce the costs of future taxonomic analysis.

Keywords: Benthic macro-invertebrate, Biomonitoring, Classification, Machine vision, SVM, Multiple order gradient histograms.

1 Introduction

Aquatic ecosystems are facing a growing number of anthropogenic pressures operating at several time and spatial scales (e.g. global warming, eutrophication). Well planned biomonitoring is often essential to detect the cause-effect structure between subtle anthropogenic pressures and their ecosystem consequences. The growing global need to implement more biomonitoring is apparent but but due mainly to the cost-intensive human expert taxonomic identification of samples, that need cannot currently be adequately met.

M. Kolehmainen et al. (Eds.): ICANNGA 2009, LNCS 5495, pp. 489–497, 2009.
© Springer-Verlag Berlin Heidelberg 2009

Automated image recognition techniques can match human taxa identification accuracy at greatly reduced costs [1]. Despite their obvious potential, the development of automated taxa identification techniques has long been hampered by the reluctance of taxonomic experts to embrace alternative methods of taxa identification [2].

In the following, we test the applicability of pattern recognition methods on the problem of identifying a limited variety of benthic macro-invertebrate typically keyed in the biomonitoring of northern streams. To do so, we introduce a MOGH feature which allows us to avoid object segmentation in the classification process. The classification itself is done by means of *Support Vector Machines*. SVMs have been used on a variety of classification problems, i.e. to correctly identify shades in urban areas in [3] and for voice-based gender identification in [4]. A multi class SVM classifier for the classification of eight different kinds of alcohol is discussed in [5].

In Section 2 we introduce biomonitoring. Section 3 illustrates some basics on the binary and multi class SVM classifier. Section 4 explains how the feature vectors have been obtained. Section 5 depicts the experimental setup and results. In Section 6 we present our conclusions.

2   Biomonitoring

In the past, freshwater monitoring has mainly focused on chemical parameters. With the introduction of new environmental legislation (EU Water Framework Directive and the US Clean Water Act), the need for biomonitoring has significantly increased during the past two decades. Freshwater aquatic biomonitoring encompasses a number of organisms such as e.g. fish, macrophytes, planktic algae, and benthic macro-invertebrates. Where chemical samples provide the researcher with a short snap-shot of the current situation, biological organisms integrate the prevailing environmental conditions over the entire span of their life cycle.

Research on automated recognition of aquatic organisms has mainly concentrated on marine plankton [6], whereas automated benthic macro-invertebrate identification has received very little attention (but see [7]). Benthic macro-invertebrates are well suited for biomonitoring due to the intermediate length of their life cycles (typically ranging from 1-2 years). In addition, there is an abundance of studies documenting the unique responses of benthic macro-invertebrate communities to a large variety of specific environmental pressures.

Well planned biomonitoring is often essential to detect the cause-effect structure between subtle anthropogenic pressures and their ecosystem consequences. The growing global need to implement more biomonitoring is apparent but due mainly to the cost-intensive human expert taxonomic identification of samples, that need cannot curreenly be adequately met.

3   Support Vector Machines

To automatically recognize benthic macro-invertebrate, an automatic classification system based on Support Vector Machines theory was used. SVMs were chosen for this task due to their excellent real-world performance(see [8]).

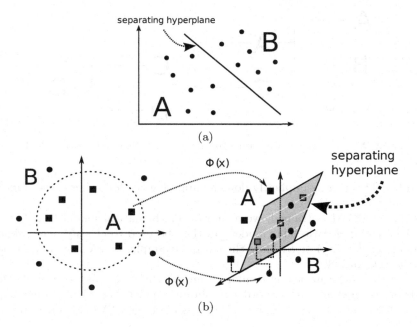

Fig. 1. An example of a linearly separable (a) and linearly non-separable (b) training set. In case (b), the domain space is mapped into the feature space in order to make it linearly separable.

A Support Vector Machine (SVM) is a binary classifier able to divide a set of data into different groups. In order for the SVM to successfully work, a training procedure must be completed. A multi class SVM can be made up of several binary class SVMs, so that its training can be reduced to the training of different binary class SVMs. In the following, we describe both a simple binary SVM and the strategy used to build the multi-class SVM.

SVMs were originally developed for pattern classification problems [9]. Geometrically an SVM creates an optimal separating hyperplane for the training data, and this plane is then used to classify all other data. For the generic SVM, let A and B be two different classes, let $X := \{x_1, x_2, \ldots, x_k\}$ be a training data set made up of k different training vectors $x_i \in \mathbb{R}^n$ each of which is associated with a given binary code $y_i \in \{-1, 1\}$ so that if $y_i = -1 \Rightarrow x_i \in A$ and if $y_i = 1 \Rightarrow x_i \in B$. The training set X can be either linearly separable or non-separable. The training of a SVM can be seen as an optimization problem in which a particular dot product function, called Kernel function, is involved (see [8])

When the training set is linearly separable, it is possible to find an optimal separating hyperplane already present in the domain space $\mathbb{R}^n$; this situation is shown in figure 1(a). If the training set is linearly non-separable, we face the situation shown in figure 1(b); in this case we need to map the domain space $\mathbb{R}^n$ into a feature space $\mathbb{R}^N$, with $N > n$, in which the training set is linearly

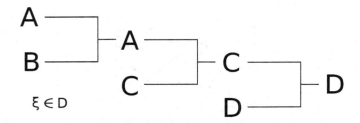

Fig. 2. An example of a tennis tournament strategy multi class SVM

separable. This can be implicitly done by applying the following *kernel* function $\Phi : \mathbb{R}^n \to \mathbb{R}^N$.

When the SVM is trained, we acquire a mathematical model that best fits the given training data. This model can then be used to classify new unknown data among classes A or B. More theory and examples on SVM can be found in [9], [10], [11] and [8].

When the data can belong to more than two different classes, several binary SVMs are trained and used in order to obtain a multi class SVM. In particular, for ζ different classes, the number of all possible binary SVMs is given by:

$$\zeta \cdot (\zeta - 1) / 2 \tag{1}$$

In this study an elimination strategy is used to build the multi class SVM classifier. Figure 2 shows an example of this: considering an element ξ belonging to class D, at first classes A and B compete with each other; the winner of these (A in the figure) competes with class C and, finally, the winning class (C) competes with class D; in the end ξ is correctly classified as a member of class D.

4   Object Identification and Global Features

Classic models of object identification rely on segmenting the object from the background, then applying feature extraction and classification methods; the main problem is that segmentation is a difficult task in the presence of cluttered backgrounds or complex objects [12].

Modern approaches to this problem often avoid the segmentation phase entirely and use a more holistic approach. Instead of finding an exact location, contour and properties of an object, the problem is seen on more general terms: "Assuming that this region contains an object of this set, what class does it belong to?"

A modern example of such thinking is *Local Binary Patterns* (LBP, [13]). Although LBP is a simple texton histogram method, it has been successfully applied to very difficult tasks such as facial expression recognition [14] and face recognition [15]. Neither of these approaches rely on segmenting facial features; instead a texture descriptor is calculated for whole facial regions and compared. Regardless of this simplification the method has proven quite effective.

Other popular methods include such examples as *SIFT* [16] which relies on gradient based local features extracted around key-points, *Bag-of-Features* (summarized in [17]) which creates 'code-books' of textons found in images, and [18] where a system randomly sampling sub-images from the original is used. All of these methods search for salient points in the image, either by factual search, or by relying on random discovery. Such points are durable when exposed to various transformations and contain *local features* that can, hopefully, discriminate the object.

As noted in [19], eliminating segmentation is not the only advantage of such modern methods. Features taken from key-points have reached a level where they are truly efficient and can be applied to sampling; in this work we use a *multiple order gradient histogram (MOGH)* as a feature. Features based on image gradients are often used in computer vision in various forms, such as weighted orientation histograms of sift and gradients estimated steerable filters. The novelty of our work is that we propose to use gradients of several orders to capture not only edges, but also other texton features such as spots or ripples. The dense feature approach was selected in order to complement the keypoint methodology in [7]. Such a feature is promising due to the fact that, unlike LBP or other dense textural features, it has a smoother response over different scales and reacts to both edge-style and textural style features in the image.

To calculate gradients of the image let us denote the standard 3×3 sobel mask by S_x when applied to calculate a horizontal gradient and by S_y when applied to calculate a vertical one. Likewise G denotes a gaussian smoothing mask with a standard deviation of 1.5 sampled to a mask of size 5×5. This mask is used for low-pass filtering of the source image. Convolution is noted by $*$. See [20] for complete definitions and practical theory of these constructs.

To calculate the MOGH feature, the source image is first decomposed into feature images $F_{u,v} = G_s^{* \max(u,v)} * S_x^{*u} * S_y^{*v}$, where I^{*n} is defined as

$$I^{*n} = \overbrace{I * I * I * \ldots * I}^{n\text{-times}}.$$

The source image is decomposed into 5 first feature images $F_{u,v}$, covering all unique gradient combinations up to second order, disregarding $F_{0,0}$. So as not to focus on extraneous details, the scale is reduced as the order is increased. Thus, the feature images represent all combinations of gradients up to second order with the filters pictured in figure 3. For each of the feature images, an 8-bin histogram is calculated and the concatenation of these histograms is

Fig. 3. Graphical representation of feature kernels

denoted as $\mathcal{M}$, or the MOGH-feature. Since this feature does not contain clear representation of image intensity, the final feature vector is augmented to form $\mathcal{F} = \{\mu, \sigma, \mathcal{M}_1, \mathcal{M}_2, \ldots, \mathcal{M}_{40}\}$, where μ denotes the image's average and σ is the standard deviation.

5   Experimental Setup and Results

The test data used consists of several pictures representing eight different taxonomical groups of benthic macro-invertebrates: *Baetis rhodani, Diura nanseni, Heptagenia sulphurea, Hydropsyche pellucidulla, Hydropsyche siltalai, Isoperla* sp., *Rhyacophila nubila* and *Taeniopteryx nebulosa*. Members belonging to the same taxonomical group were imaged through a flatbed scanner and the pictures so obtained were normalized; eventually, each sample in each scan was saved as an individual image; the final data set is made up of 1529 different pictures, unequally distributed among all the taxonomical groups.

In figure 4 and 5 an example of the data available is shown: the specimens are semi-rigid so that the actual shape can change from one sample to the other (i. e. figures 5(a) and 5(c)). Furthermore different samples may partially overlap or be mutilated (i. e. figures 4(b), 4(c)). These features in the data make the problem of automatic identification particularly challenging.

Each macro-invertebrate species is depicted in more than one scan. Since the authors cannot be confident that the image processing does not affect the features used for the classification process (i.e. the luminance level of the background can be different for each scan), all the samples were divided into two different sets

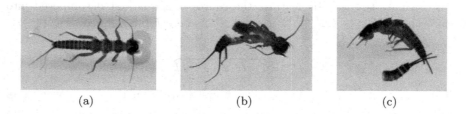

(a)   (b)   (c)

Fig. 4. Three samples of the Diura Nanseni group

(a)   (b)   (c)

Fig. 5. Three samples of the Heptageria Sulphurea group

Table 1. Classification scores for the training data in percentage. See text for class labels.

	Class A	Class B	Class C	Class D	Class E	Class F	Class G	Class H
Class A	**96.43**	0.00	0.00	1.79	1.79	0.00	0.00	0.00
Class B	0.00	**97.06**	0.00	0.00	0.00	0.00	2.94	0.00
Class C	2.38	0.00	**80.95**	0.00	9.52	0.00	7.14	0.00
Class D	2.96	0.00	0.00	**91.11**	3.70	0.00	2.22	0.00
Class E	5,11	0.73	1.46	0.73	**80.29**	1.46	6.57	3.65
Class F	3.17	0.00	0.79	0.00	3.97	**82.54**	0.00	9.53
Class G	0.00	0.00	2.44	0.00	0.00	0.00	**97.56**	0.00
Class H	0.00	0.00	0.00	0.00	5.00	0.00	0.00	**95.00**

Table 2. Classification scores for the validation data in percentage. See text for class labels.

	Class A	Class B	Class C	Class D	Class E	Class F	Class G	Class H
Class A	**77.59**	0.00	0.00	0.00	8.62	8.62	0.00	5.17
Class B	0.00	**95.08**	0.00	1.64	0.00	0.00	1.64	1.64
Class C	2.27	0.00	**70.45**	0.00	14.77	1.14	1.14	10.23
Class D	10.13	1.35	1.35	**77.70**	4.05	0.67	4.73	0.00
Class E	8.21	0.00	8.21	0.75	**69.40**	1.49	10.45	1.49
Class F	5.41	0.00	1.08	0.00	2.16	**78.92**	0.54	11.89
Class G	2.38	0.00	2.38	2.38	7.14	2.38	**80.95**	2.38
Class H	3.49	0.00	18.60	0.00	3.49	10.46	4.65	**59.30**

and then used for the training and the validation of the SVMs. Both sets were arranged so that they are composed of different scans, i.e. in the training set we will find all pictures of the Baetis rhodani taken from the first and second scans, while pictures coming from the third and fourth scans will appear only in the validation set.

The SVMs use a soft margin constraint of 7.2 and Gaussian Radial Basis function Kernel, with $\sigma = 7.6$. The Bioinformatic Toolbox of Matlab was used.

According to equation 1, for 8 different classes, we set up 28 binary SVMs, all of which were trained with 41 samples. Normalization procedure was applied separately to each binary classifier providing much more reliable binary SVMs. The multi class classifier was realized using the binary classifiers in an elimination strategy. In order to check the performance of the system, the multi class SVM was used to classify all samples from both the training and validation set. In the following, classes A, B, C, D, E, F, G and H correspond respectively to the taxonomic groups *Baetis rhodani*, *Diura nanseni*, *Heptagenia sulphurea*, *Hydropsyche pellucidulla*, *Hydropsyche siltalai*, *Isoperla* sp., *Rhyacophila nubila* and *Taeniopteryx nebulosa*. Table 1 and 2 show the performance of the multi class SVM for the training and the validation data respectively. The intersection of rows and columns indicates the amount of classified items, thus the first cell in table 1 indicates that *Baetis rhodani* was classified (correctly) as *Baetis rhodani* in 96.43% of the cases. Please note that tables 1 and 2 cannot be symmetric, since the SVMs offer a different sensitivity for different classes and because the amount of data available varies for each species.

The performance on the training data is better than that on the validation data, as shown in table 1 and 2. However, the multi class SVM correctly classified 88.17% of the training data and 75.31% of the validation data: dealing with eight different classes, this system offers performances comparable to the classifier described in [7], which had to deal with only four different species.

6   Conclusions

We have introduced a novel method for automatic taxa identification of benthic macro-invertebrates. With the ever growing demand for increased biomonitoring, automated taxa identification methods will become inevitable for meeting with budgetary constraints. We introduced a MOGH feature and showed its applicability to taxa identification in conjunction with the use of support vector machines.

The results indicate that this method is extremely successful. Given the fragile nature of the benthic samples and a resulting high degree of distortion in the data sets, our recall rate of 80% can be viewed as highly successful. We feel that recall rates are likely to increase with more sophisticated computer vision processing and less destructive field sampling methods.

We acknowledge that this study only handles a small subset of species. Increasing the number of species in future systems will lead to more challenging classification problems. These results suggest, however, that automated macro invertebrate classification is indeed possible.

References

1. Culverhouse, P.F., Williams, R., Reguera, B., Herry, V., Gonzalez-Gil, S.: Do experts make mistakes? a comparison of human and machine identification of dinoflagellates. Marine Ecology Progress Series 247, 17–25 (2003)
2. Gaston, K.J., O'Neill, M.A.: Automated species identifiction: why not? Philosophy Trans. R. Soc. Lond. B 359, 655–667 (2004)
3. Pajares, G., Cruz, J.M., Belmonte, M.: Support vector machines for shade identification in urban areas. In: ISCGAV 2004: Proceedings of the 4th WSEAS International Conference on Signal Processing, Computational Geometry & Artificial Vision, Stevens Point, Wisconsin, USA, pp. 1–4. World Scientific and Engineering Academy and Society, WSEAS (2004)
4. Kye-Hwan, L., Sang-Ick, K., Deok-Hwan, K., Joon-Hyuk, C.: A Support Vector Machine-Based Gender Identification Using Speech Signal. IEICE Trans. Commun. E91-B(10), 3326–3329 (2008)
5. Acevedo, F.J., Maldonado, S., Dominguez, E., Narvaez, A., Lopez, F.: Probabilistic support vector machines for multi-class alcohol identification. Sensors and actuators. B, Chemical 122(1), 227–235 (2007)

6. Benfield, M.C., Grosjean, P., Culverhouse, P., Irigoien, X., Sieracki, M.E., Lopez-Urrutia, A., Dam, H.G., Hu, Q., Davis, C.S., Hansen, A., Pilskaln, C.H., Riseman, E., Schultz, H., Utgoff, P.E., Gorsky, G.: Rapid research on automated plankton identification. Oceanography 20, 12–25 (2007)

7. Larios, N., Deng, H., Zhang, W., Sarpola, M., Yuen, J., Paasch, R., Moldenke, A., Lytle, D.A., Correa, S.R., Mortensen, E.N., Shapiro, L.G., Dietterich, T.G.: Automated insect identification through concatenated histograms of local appearance features: feature vector generation and region detection for deformable objects. Mach. Vision Appl. 19(2), 105–123 (2008)

8. Cortes, C., Vapnik, V.: Support vector networks. In: Machine Learning, pp. 273–297 (1995)

9. Cortes, C., Vapnik, V.: Support-vector networks. Machine Learning (20), 273–297 (1995)

10. Burges, C.J.C.: A tutorial on support vector machines for pattern recognition. Data Mining and Knowledge Discovery 2(2), 121–167 (1998)

11. Smola, A.J., Schölkopf, B.: A tutorial on support vector regression*. Journal of Statistics and Computing (14), 199–222 (2004)

12. Pal, N.R., Pal, S.K.: A review on image segmentation techniques. Pattern Recognition 26(9), 1277–1294 (1993)

13. Ojala, T., Pietikäinen, M., Mäenpää, T.: Multiresolution gray-scale and rotation invariant texture classification with local binary patterns. IEEE Transactions on Pattern Analysis and Machine Intelligence, 971–987 (2002)

14. Feng, X., Pietikainen, M., Hadid, A.: Facial expression recognition with local binary patterns and linear programming. Pattern Recognition And Image Analysis C/C of Raspoznavaniye Obrazov I Analiz Izobrazhenii 15(2), 546 (2005)

15. Ahonen, T., Hadid, A., Pietikäinen, M.: Face description with local binary patterns: Application to face recognition. IEEE Transactions on Pattern Analysis and Machine Intelligence, 2037–2041 (2006)

16. Lowe, D.G.: Object recognition from local scale-invariant features. In: International Conference on Computer Vision, Kerkyra, Greece, vol. 2, pp. 1150–1157 (1999)

17. Nowak, E., Jurie, F., Triggs, B.: Sampling strategies for bag-of-features image classification. In: Leonardis, A., Bischof, H., Pinz, A. (eds.) ECCV 2006. LNCS, vol. 3954, pp. 490–503. Springer, Heidelberg (2006)

18. Maree, R., Geurts, P., Piater, J., Wehenkel, L.: Random subwindows for robust image classification. In: IEEE Computer Society Conference on Computer Vision and Pattern Recognition, 2005. CVPR 2005, vol. 1 (2005)

19. Dalal, N., Triggs, B.: Histograms of oriented gradients for human detection. In: Proc. CVPR, vol. 1, pp. 886–893 (2005)

20. Gonzalez, R.C., Woods, R.E.: Digital Image Processing. Prentice Hall, Englewood Cliffs (2007)

Bayesian Dimension Reduction Models for Microarray Data

Albert D. Shieh

Department of Statistics, Harvard University,
Cambridge, MA 02138, USA
shieh@fas.harvard.edu

Abstract. High dimensionality, missing values, noise, and outliers are standard problems in gene expression data and are usually dealt with separately. In this paper, we propose an ideal point model that performs feature extraction, imputes missing values, and is robust to noise and outliers in a unified and unsupervised framework. We use the simplifying assumption that genes are either expressed or not expressed in order to obtain a parsimonious model. We present a fast Bayesian method for estimating the large number of parameters in the ideal point model. We apply the ideal point model to a leukemia data set, where it outperforms independent component analysis (ICA), a state of the art unsupervised feature extraction method.

1 Introduction

DNA microarrays can simultaneously measure the expression levels of thousands of genes over different experiments and have been applied with great success to problems such as cancer diagnosis [6]. However, the analysis of gene expression data is complicated by several problems. First, measurements of gene expression levels are highly contaminated with noise and outliers. As gene expression data grows larger and more complex, robust data analysis methods must be used. Second, there are often a significant number of missing gene expression levels. Most data analysis methods require complete data, so the missing values must be imputed. Finally, the number of genes is always several orders of magnitude larger than the number of experiments. The high dimensionality of microarray data makes the direct application of most data analysis methods impossible, so a dimensionality reduction step is needed.

The issues of noise and outliers, missing data, and high dimensionality are usually handled separately. Imputation is used for missing data, gene selection or feature extraction is used for high dimensionality, and robust data analysis methods are used to mitigate the effect of noise and outliers [3]. However, the lack of integration amongst all of these steps can be problematic since different methods at each step may make different assumptions. In this paper, we apply an ideal point model to microarray data that is robust to noise and outliers, imputes missing data, and performs feature extraction in a unified and unsupervised framework. We use the simplifying assumption that gene expression levels are

M. Kolehmainen et al. (Eds.): ICANNGA 2009, LNCS 5495, pp. 498–506, 2009.
© Springer-Verlag Berlin Heidelberg 2009

binary, that is, genes are either expressed or not expressed. Our approach is inspired by the recent success of biclustering algorithms [14] that also discretize microarray data. Although converting continuous data into binary data results in a loss of information, we gain a parsimonious model.

Ideal point models are a class of latent variable models originally developed in psychometrics [10] that estimate continuous latent variables called ideal points for individuals given their binary responses to a set of items. In gene expression data, the individuals are experiments, the items are genes, and the responses are expression or no expression. Rasch models, a closely related class of latent variable models, were applied to gene expression data with clustering in [12]. However, the Rasch model in [12] assumes a prohibitively simple structure for the data, does not incorporate missing data, and cannot be applied to a large number of genes. Our ideal point model can be seen as an extension of the Rasch model in [12] along all of these directions.

The rest of the paper is organized as follows. In section 2, we present the ideal point model motivated by a utility maximization approach. In section 3, we present a fast Bayesian estimation method for the ideal point model. In section 4, we apply the ideal point model to a leukemia data set and compare it to other unsupervised feature extraction methods. In section 5, we give our concluding remarks.

2   Ideal Point Model

Gene expression data on m genes for n experiments can be summarized by an $n \times m$ matrix $\mathbf{Y} = (y_{ij})$, where y_{ij} denotes the expression level of gene j in experiment i. For now, we will only consider values that are not missing. First, we need to convert the continuous values to binary values. We use a simple global threshold for gene expression from [14] based on a fold change, which was shown in [13] to be biologically reproducible. We set

$$y_{ij} = \begin{cases} 1 & \text{if } y_{ij} \geq a + (b-a)/2 \\ 0 & \text{otherwise} \end{cases} \tag{1}$$

where a and b denote the minimum and maximum values in $\mathbf{Y}$ respectively. Note that $y_{ij} = 1$ represents expression and $y_{ij} = 0$ represents no expression. Using binary values inherently gives us robustness to noise and outliers.

Now, we derive an ideal point model for a d-dimensional latent space using utility maximization as in [9]. For each gene j, experiment i chooses between alternatives of expression $\boldsymbol{\zeta}_j$ and no expression $\boldsymbol{\psi}_j$ in the latent space. We assume that the experiments have quadratic utility functions with stochastic errors over the latent space

$$U_i(\boldsymbol{\zeta}_j) = -\|\mathbf{x}_i - \boldsymbol{\zeta}_j\|^2 + \eta_{ij} \tag{2}$$

$$U_i(\boldsymbol{\psi}_j) = -\|\mathbf{x}_i - \boldsymbol{\psi}_j\|^2 + \nu_{ij} \tag{3}$$

where $\mathbf{x}_i$ is the ideal point of experiment i, $\|\cdot\|$ denotes the Euclidean norm, and η_{ij} and ν_{ij} are errors. By utility maximization, experiment i chooses $\boldsymbol{\zeta}_j$ over $\boldsymbol{\psi}_j$ such that $y_{ij} = 1$ if $U_i(\boldsymbol{\zeta}_j) > U_i(\boldsymbol{\psi}_j)$. Note that the ideal point of an experiment is the location in the latent space that maximizes its utility.

Let y_{ij}^* denote the utility difference

$$
\begin{aligned}
y_{ij}^* &= U_i(\boldsymbol{\zeta}_j) - U_i(\boldsymbol{\psi}_j) \\
&= 2\mathbf{x}_i(\boldsymbol{\zeta}_j - \boldsymbol{\psi}_j) \\
&= \boldsymbol{\beta}_j^T \mathbf{x}_i - \alpha_j + \epsilon_{ij}
\end{aligned}
\tag{4}
$$

where $\varepsilon_{ij} = \eta_{ij} - \nu_{ij}$ such that $y_{ij}^* \geq 0$ if $y_{ij} = 1$. If we assume that the error ε_{ij} is independent over both experiments and genes and is normally distributed

$$
\varepsilon_{ij} \sim \mathcal{N}(0, \sigma_j^2),
\tag{5}
$$

then

$$
\begin{aligned}
P(y_{ij} = 1) &= P(y_{ij}^* \geq 0) \\
&= P(\varepsilon_{ij} > \boldsymbol{\beta}_j^T \mathbf{x}_i - \alpha_j) \\
&= \boldsymbol{\Phi}(\boldsymbol{\beta}_j^T \mathbf{x}_i - \alpha_j)
\end{aligned}
\tag{6}
$$

where $\boldsymbol{\beta}_j = 2(\boldsymbol{\zeta}_j - \boldsymbol{\psi}_j)/\sigma_j$, $\alpha_j = (\boldsymbol{\zeta}_j^T \boldsymbol{\zeta}_j - \boldsymbol{\psi}_j^T \boldsymbol{\psi}_j)/\sigma_j$, and $\boldsymbol{\Phi}(\cdot)$ denotes the normal distribution function. It is common to set $\sigma_j = 1$ for simplicity. The coefficient vector $\boldsymbol{\beta}_j$ captures the differential expression of gene j, while the intercept α_j captures the baseline expression of gene j.

The likelihood function is

$$
L(\mathbf{X}, \mathbf{B}, \boldsymbol{\alpha} | \mathbf{Y}) = \prod_{i=1}^{n} \prod_{j=1}^{m} \boldsymbol{\Phi}(\mathbf{x}_i^T \boldsymbol{\beta}_j - \alpha_j)^{y_{ij}} (1 - \boldsymbol{\Phi}(\mathbf{x}_i^T \boldsymbol{\beta}_j - \alpha_j))^{1-y_{ij}}
\tag{7}
$$

where $\mathbf{X}$ is an $n \times d$ matrix with ith row $\mathbf{x}_i^T$, $\mathbf{B}$ is an $m \times d$ matrix with jth row $\boldsymbol{\beta}_j^T$, and $\boldsymbol{\alpha} = (\alpha_1, \ldots, \alpha_m)^T$. The ideal point model is a hierarchical probit model where the covariates are latent variables. However, the ideal point model is not identified since the ideal points can be shifted and offset by the coefficient vector and intercept. Therefore, we must place restrictions on the ideal points. In one dimension, we can simply constrain the ideal points to have mean zero and variance one. However, in multiple dimensions, we must fix the ideal points of $d + 1$ experiments. Although fixing ideal points seems difficult, it has little effect on the results in practice since the scale will automatically adjust [9]. We recommend fixing ideal points on the unit hypercube.

3   Bayesian Estimation

There are $nd + m(d + 1)$ parameters in the ideal point model. Even for a small gene expression data set with $n = 20$ experiments and $m = 1000$ genes, a one dimensional ideal point model contains 2020 parameters. Maximimum likelihood estimation of the parameters is computationally intractable since the likelihood function becomes too complex. Therefore, we use Bayesian estimation to sample the parameters from the posterior distribution and avoid exploring the likelihood function. In order to derive the posterior distribution, we must first specify prior distributions for the parameters. We use non-informative multivariate normal distributions, which yield conjugate conditional posterior distrbiutions.

For $i = 1, \ldots, m$, we use

$$\mathbf{x}_i \sim \mathcal{N}(\mathbf{v}_0, \mathbf{V}_0) \tag{8}$$

with mean vector $\mathbf{v}_0 = \mathbf{0}$ and covariance matrix $\mathbf{V}_0 = \mathbf{I}_d$ where $\mathbf{I}_d$ denotes the identity matrix of size d.

For $j = 1, \ldots, n$, we use

$$(\boldsymbol{\beta}_j, \alpha_j)^T \sim \mathcal{N}(\mathbf{t}_0, \mathbf{T}_0) \tag{9}$$

with mean vector $\mathbf{t}_0 = \mathbf{0}$ and covariance matrix $\mathbf{T}_0 = \kappa \cdot \mathbf{I}_{d+1}$. We set $\kappa = 25$.

We want to sample from the posterior distribution $p(\mathbf{X}, \mathbf{B}, \boldsymbol{\alpha} | \mathbf{Y})$. We use a Gibbs sampling algorithm to sample from the conditional posterior distributions $p(\mathbf{B}, \boldsymbol{\alpha} | \mathbf{Y}, \mathbf{X})$ and $p(\mathbf{X} | \mathbf{Y}, \mathbf{B}, \boldsymbol{\alpha})$ with an additional data augmentation step [1] for utility differences. The Gibbs sampling steps can be derived by treating the parameters as missing data. If we know $\boldsymbol{\beta}_j$ and α_j, then we can impute $\mathbf{x}_i$ by the regression of $y_{ij}^* + \alpha_j$ on $\boldsymbol{\beta}_j$ using the m expression values of experiment i. If we know $\mathbf{x}_i$, then we can impute $\boldsymbol{\beta}_j$ and α_j using the expression values of the n experiments on gene j. If we know $\mathbf{x}_i$, $\boldsymbol{\beta}_j$, and α_j, then we can impute y_{ij}^* by drawing errors from a normal distribution subject to the constraints of the binary expression levels.

At iteration t, the Gibbs sampling steps are:

1. Data augmentation

    For $i = 1, \ldots, n$ and $j = 1, \ldots, m$, we sample $y_{ij}^{*(t)}$ from the truncated normal distributions

$$y_{ij}^{*(t)} \sim \begin{cases} \mathcal{N}(\mu, 1)I(y_{ij}^{*(t)} \geq 0) & \text{if } y_{ij} = 1 \\ \mathcal{N}(\mu, 1)I(y_{ij}^{*(t)} < 0) & \text{if } y_{ij} = 0 \\ \mathcal{N}(\mu, 1) & \text{if } y_{ij} \text{ is missing} \end{cases} \tag{10}$$

    with means

$$\mu = \mathbf{x}_i^{T(t-1)} \boldsymbol{\beta}_j^{(t-1)} - \alpha_j^{(t-1)}$$

    where I denotes the indicator function evaluating to one if the argument is true and zero otherwise.

2. Impute coefficients

    For $j = 1, \ldots, m$, we sample $\boldsymbol{\beta}_j^{(t)}$ and $\alpha_j^{(t)}$ from the multivariate normal distribution

$$(\boldsymbol{\beta}_j, \alpha_j)^{T(t)} \sim \mathcal{N}(\boldsymbol{\mu}, \boldsymbol{\Sigma}) \tag{11}$$

with mean vector

$$\boldsymbol{\mu} = \left(\mathbf{X}^{*T}\mathbf{X}^* + \mathbf{T}_0^{-1}\right)^{-1}\left(\mathbf{X}^{*T}\mathbf{y}_j^{*(t)} + \mathbf{T}_0^{-1}\mathbf{t}_0\right)$$

and covariance matrix

$$\boldsymbol{\Sigma} = \left(\mathbf{X}^{*T}\mathbf{X}^* + \mathbf{T}_0^{-1}\right)^{-1}$$

where $\mathbf{X}^*$ is an $n \times (d+1)$ matrix with ith row $(\mathbf{x}_i^{T(t-1)}, -1)$.

3. Impute ideal points

For $i = 1, \ldots, n$, we sample $\mathbf{x}_i^{(t)}$ from the multivariate normal distribution

$$\mathbf{x}_i^{(t)} \sim \mathcal{N}(\boldsymbol{\mu}, \boldsymbol{\Sigma}) \qquad (12)$$

with mean vector

$$\boldsymbol{\mu} = \left(\mathbf{B}^T\mathbf{B} + \mathbf{V}_0^{-1}\right)^{-1}\left(\mathbf{B}^T\mathbf{w}_j + \mathbf{V}_0^{-1}\mathbf{v}_0\right)$$

and covariance matrix

$$\boldsymbol{\Sigma} = \left(\mathbf{B}^T\mathbf{B} + \mathbf{V}_0^{-1}\right)^{-1}$$

where $w_{ij} = y_{ij}^{*(t)} + \alpha_j^{(t)}$ and $\mathbf{B}$ is an $m \times d$ matrix with jth row $\boldsymbol{\beta}_j^{T(t)}$.

At the end of iteration t, we have a Markov chain of parameters $\boldsymbol{\theta}^{(1)}, \ldots, \boldsymbol{\theta}^{(t)}$ where $\boldsymbol{\theta}^{(t)} = (\mathbf{X}^{(t)}, \mathbf{B}^{(t)}, \boldsymbol{\alpha}^{(t)})$. We initialize $\mathbf{X}^{(1)}$ using an eigendecomposition of $\mathbf{Y}$ and $\mathbf{B}^{(1)}$ and $\boldsymbol{\alpha}^{(1)}$ using a probit regression of $\mathbf{Y}$ on $\mathbf{X}^{(1)}$ as in [9]. We recommend running 10,000 iterations, throwing away the first 5,000 iterations as burn-in, and thinning the chain by keeping every 10th iteration to produce 500 samples from the posterior distribution. The small sample sizes of gene expression data allow for good mixing and fast convergence of the chian. The mean ideal points are used as the new features.

4   Leukemia Data Set

We tested the leukemia data set from [6], which contains 3572 genes and 72 experiments, of which 47 experiments were for acute lymphoblastic leukemia (ALL), with 38 experiments on B-cells (ALL-B) and 9 experiments on T-cells (ALL-T), and 25 experiments were for acute myeloid leukemia (AML). As in [3], we preprocessed the data using thresholding, filtering, logarithm transformation, and row standardization. It is common to perform gene selection before feature extraction in order to eliminate non-informative genes [4]. We selected the top 20 genes using the original signal-to-noise ratio in [6]. A heatmap of the top 20 genes using hierarchical clustering is shown in Figure 1. The gradient from orange to yellow represents expression to no expression. The heatmap easily reveals two classes corresponding to the ALL/AML distinction, but a third class corresponding to the ALL-B/ALL-T distinction is not clear.

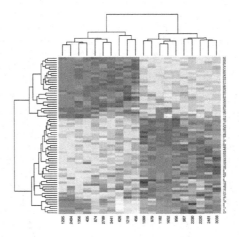

Fig. 1. Heatmap of the top 20 genes with experiments in rows and genes in columns

We consider the simplest problem of extracting one feature with the goal of summarizing class distinctions. We applied the ideal point model and compared it to independent component analysis (ICA), which is considered to be one of the best unsupervised feature extraction methods for gene expression data [11]. ICA is also robust to noise and outliers, but does not incorporate missing data. We used the R package MCMCpack to implement the ideal point model and the R package fastICA to perform ICA using the popular FastICA algorithm [8] with default parameters. Generating a chain of 10,000 samples for the ideal point model took between 3 and 5 seconds on a standard personal computer. We verified convergence by generating 5 separate chains with overdispersed initial values for the parameters and computing potential scale reduction factors [5]. Examining trace plots of the parameters revealed that most of the ideal points converged quickly within a few thousand iterations.

The features found using the ideal point model and ICA are shown in Figure 2 and have been standardized to the same scale for comparison. The ideal point model feature captures the ALL/AML distinction better than the ICA feature. All of the ALL and AML experiments are linearly separable in the ideal point model feature, while one AML experiment is mixed in with the ALL experiments in the ICA feature. More importantly, the ideal point model feature is able to capture the ALL-B/ALL-T distinction, where the ICA feature largely fails. The ALL-T experiments are grouped closely together in the ideal point model feature, while they are scattered amongst the ALL-B experiments in the ICA feature. Therefore, we expect the ideal point model feature to distinguish classes better than the ICA feature from visual examination.

We tested the ideal point model and ICA features on a classification problem with the three classes ALL-B, ALL-T, and AML. We applied a support vector machine (SVM) classifier, which is known to perform well on gene expression data [4], in a one-against-all method with a radial basis function kernel and a

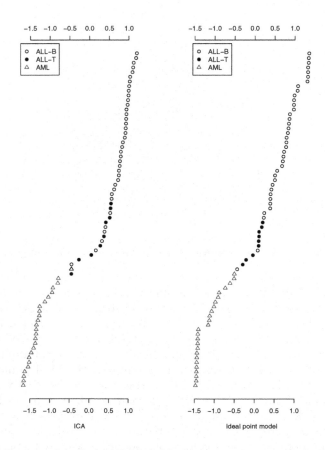

Fig. 2. Features found using ICA (left) and the ideal point model (right)

grid search for parameter selection using the LIBSVM library. We estimated the classification accuracy using a 5-fold cross validation as recommended by [7] for gene expression data. The ideal point model feature achieved 90.3% accuracy, while the ICA feature only achieved 86.1% accuracy. Therefore, the ideal point model appears to outperform ICA at separating classes with one feature.

We tested the robustness of the ideal point model and ICA features to the presence of missing values and additional noise. We introduced missing values at random for 5% of the data by sampling row and column indices and added normally distributed noise to each value from $\mathcal{N}(0, 0.01)$. Since ICA cannot handle missing data, we imputed the missing values for the ICA feature using the popular KNNImpute algorithm [15]. On the same classification problem with the degraded data, the ideal point model feature maintained 90.3% accuracy, while the ICA feature fell to 84.7% accuracy. Therefore, the ideal point model appears to be more robust than ICA to missing values and noise.

5   Conclusion

In this paper, we proposed an ideal point model to perform unsupervised feature extraction of gene expression data that is robust to noise and outliers and easily handles missing data. We used the simplifying assumption that genes are either expressed or not expressed in order to obtain robustness and a parsimonious model. Although the ideal point model contains a large number of parameters, we presented a fast Bayesian estimation method based on Gibbs sampling and data augmentation. When applied to a leukemia data set, the ideal point model outperformed ICA at classification with one feature, demonstrating promise as an unsupervised feature extraction method. Moreover, the performance of the ideal point model was not sensitive to missing values and noise.

The performance of the ideal point model is striking considering that it relies upon a reduction of gene expression levels to binary states. However, the binary representation will fail if the gene expression mechanism is more complicated and genes can be both underexpressed and overexpressed. Therefore, an important extension of the ideal point model would be to incorporate multiple expression states along the lines of [2]. The expression states could be determined using sample quantiles or more complicated methods such as clustering with finite mixture models. Incorporating more expression states will capture more of the continuous nature of the gene expression, but will also reduce robustness to noise and outliers.

Acknowledgments. The author would like to thank the anonymous reviewers, whose comments helped improve the manuscript.

References

1. Albert, J.H.: Bayesian estimation of normal ogive item response curves using Gibbs sampling. Journal of Educational Statistics 17, 251–269 (1992)
2. Albert, J.H., Chib, S.: Bayesian analysis of binary and polychotomous response data. Journal of the American Statistical Association 88, 669–679 (1993)
3. Dudoit, S., Fridlyand, J., Speed, T.P.: Comparison of discrimination methods for classification of tumors using gene expression data. Journal of the American Statistical Association 97, 77–87 (2002)
4. Furey, T.S., Cristianini, N., Duffy, N., Bednarski, D.W., Schummer, M., Haussler, D.: Support vector machine classification and validation of cancer tissue samples using microarray expression data. Bioinformatics 16, 906–914 (2000)
5. Gelman, A., Rubin, D.B.: Inference from iterative simulation using multiple sequences. Statistical Science 7, 457–511 (1992)
6. Golub, T.R., Slonim, D.K., Tamayo, P., Huard, C., Gaasenbeek, M., Mesirov, J.P., Coller, H., Loh, M.L., Downing, J.R., Caligiuri, M.A., Bloomfield, C.D., Lander, E.S.: Molecular classification of cancer: class discovery and class prediction by gene expression monitoring. Science 286, 531–537 (1999)
7. Hastie, T., Tibshirani, R., Friedman, J.: The elements of statistical learning. Springer, New York (2001)

8. Hyvarinen, A.: Fast and robust fixed-point algorithms for independent component analysis. IEEE Transactions on Neural Networks 10, 626–634 (1999)
9. Jackman, S.: Multidimensional analysis of roll call data via Bayesian simulation: identification, estimation, inference and model checking. Political Analysis 9, 227–241 (2001)
10. Johnson, V.E., Albert, J.H.: Ordinal data modeling. Springer, New York (1999)
11. Lee, S., Batzoglou, S.: Application of independent component analysis to microarrays. Genome Biology 4, 11 (2003)
12. Li, H., Hong, F.: Cluster-Rasch models for microarray gene expression data. Genome Biology 2, 8 (2001)
13. MAQC Consortium: The MicroArray Quality Control (MAQC) project shows inter- and intraplatform reproducibility of gene expression measurements. Nature Biotechnology 24, 1151–1161 (2006)
14. Prelic, A., Bleuler, S., Zimmermann, P., Wille, A., Buhlmann, P., Gruissem, W., Hennig, L., Thiele, L., Zitzler, E.: A systematic comparison and evaluation of biclustering methods for gene expression data. Bioinformatics 22, 1122–1129 (2006)
15. Troyanskaya, O., Cantor, M., Sherlock, G., Brown, P., Hastie, T., Tibshirani, R., Botstein, D., Altman, R.B.: Missing value estimation methods for DNA microarrays. Bioinformatics 17, 520–525 (2001)

Gene Selection for Cancer Classification through Ensemble of Methods

Artur Wilinski[1], Stanislaw Osowski[2,3], and Krzysztof Siwek[3]

[1] Warsaw University of Life Sciences,
[2] Military University of Technology,
[3] Warsaw University of Technology,
00-661 Warsaw, Poland
{awilinsk,sto,ksiwek}@iem.pw.edu.pl

Abstract. The paper develops the methods of selection of the most important gene sequence on the basis of the gene expression microarray, corresponding to different types of cancer. Special two stage strategy of selection has been proposed. In the first stage we apply few different methods of assessment of the importance of genes. Each method stresses different aspects of the problem. In the second stage the selected genes are compared and the genes chosen most frequently by all methods of selection are treated as the most important and representative for the particular type of problem. The results of selection are analyzed using PCA and the selected genes form the input to the SVM classifier recognizing the classes of cancer. The numerical experiments confirm the efficiency of the proposed approach.

Keywords: gene expression array, feature selection, SVM classification.

1 Introduction

The microarray experiments produce data for tens of thousands of genes simultaneously. Comparing gene expression profiles of this scale at small number of experiments (typically few hundreds) present a formidable challenge in a pattern recognition. On the other side the problem of selection of the most important genes, responsible for the development of the particular type of cancer is of great importance in diagnosis and prognosis of the illness [2].

There are many different methods used for important genes recognition on the basis of DNA microarray. To the most important belong the measures combined with the correlation analysis, clusterization of data, SOM application, Bayesian formulation, application of linear kernel Support Vector Machines, etc. [2],[3],[4],[5],[6], [7],[10],[11]. However, in spite of many existing techniques the gene selection is still an open problem, not solved in a satisfactory way. In this paper we will present the two stage procedure of the most important gene selection. In the first stage we apply few different methods of gene ranking. Each method stresses different aspect of the problem. In the second stage the selected genes are compared and the genes chosen most frequently by all methods of selection are treated as the most important and representative for the particular type of problem.

M. Kolehmainen et al. (Eds.): ICANNGA 2009, LNCS 5495, pp. 507–516, 2009.
© Springer-Verlag Berlin Heidelberg 2009

2   The Theoretical Basis of Gene Ranking Method

In the first step of our approach we apply different methods of gene selection: the mean and variance measures applied to the clusters of data belonging to the same class, the correlation analysis of the data of gene expression array, Wilcoxon statistics, three different forms of Kolmogorov-Smirnov statistical tests, and ranking of genes using linear Support Vector Machine, applied in either single-input or multi-input mode. Each method of gene selection is run thousands of times using random set containing 90% of the available data. The genes selected repeatedly in all run are treated as the most important. The set of 100 best genes chosen in this way are chosen as the representative for the each applied method. This quantity was assumed as the representative number for the purpose of gene selection in our approach. Then in the second step we select the genes commonly chosen by all selection methods. These genes are treated as the most representative for the particular data. The short description of the selection methods used in the first stage of selection follows.

2.1   The Statistical Measures Based on Clusterization

In this method the gene belonging to each class is associated with one cluster. For good candidate gene the variance of the representatives belonging to one class should be as small as possible, and at the same time the positions of mean of two different classes should be separated as much as possible. We have combined the variance and mean together to form the single Fisher quality measure [2],[9] defining the discrimination coefficient of the feature g in the form

$$S_{AB}(g) = \frac{|c_A(g) - c_B(g)|}{\sigma_A(g) + \sigma_B(g)} \tag{1}$$

In this definition c_A and c_B are the mean values of the gene g in the class A and B, respectively. The variables σ_A and σ_B represent the standard deviations determined for both classes. The large value of $S_{AB}(g)$ indicates good separation ability of the gene g for these two classes. We will refer to the results of this method by CSD.

2.2   The Correlation of the Gene with Class

The discriminative power of the gene g for the recognition of the particular class among K classes can be measured by the correlation of this gene with the class [9]. The discriminative power of g is defined in the form

$$S(g) = \frac{\sum_{k=1}^{K} P_k (m_{ck} - m_c)^2}{\text{var}(g) \sum_{k=1}^{K} P_k (1 - P_k)} \tag{2}$$

where $m_c = E\{g\}$ is the unconditional mean of the gene g for the whole set of data, $m_{ck} = E\{g|k\}$ is the conditional mean of the gene characterizing only samples belonging to kth class, var(g) is the variance and P_k is the probability of kth class. Calculating this measure for all genes we can arrange them according to their discriminative value. We will call this method of gene ranking as COR.

2.3 The Wilcoxon Statistics

This method of gene selection will use the Wilcoxon-Mann-Whitney (WMW) test [1]. This test assesses the similarity of the statistical distribution of two vectors. The WMW test checks whether two samples are drawn from the same population. The test is performed by ranking the combined data set, dividing the ranks into two sets according to the group membership of the original observations, and calculating a two sample z-statistics using the pooled variance estimate. Take the paired observations, calculate the differences, and rank them from smallest to largest by the absolute value. Add all the ranks associated with the positive and with the negative differences. Finally, the P-value associated with this statistics is found. The higher its value, the more similar are the two populations. Small value of P means large differences between the populations. The ranking of genes is created on the basis of the numerical values of P. We will call this method as WIL.

2.4 The Kolmogorov-Smirnov Tests

In Kolmogorov-Smirnov (KS) test the gene g is treated as the statistical variable. The genes representing the patients belonging to the same class have similar distribution. Our aim is to compare the statistical distribution of the particular gene g corresponding to two classes $\mathbf{x}_A$ and $\mathbf{x}_B$. The KS test is performed for these two vectors to determine if they are drawn from the same population. It is a non-parametric and distribution free test. The Matlab function *kstest2* [12] implementing this test delivers the maximum distance between the cumulative distribution functions of the data belonging to two compared classes. This distance forms the basis of the measure of difference between the distribution of both populations. Denoting the distribution of both populations as $F(\mathbf{x}_A)$ and $F(\mathbf{x}_B)$. we have defined three different discriminative measures.

- Kolmogorov-Smirnov measure (KS)

$$S_{AB}(g) = \sup|F(\mathbf{x}_A) - F(\mathbf{x}_B)| \tag{3}$$

- Additive Kolmogorov-Smirnov measure (AKS)

$$S_{AB}(g) = sum|F(\mathbf{x}_A) - F(\mathbf{x}_B)| \tag{4}$$

- Scaled Kolmogorov-Smirnov measure (SKS)

$$S_{AB}(g) = a(g) \cdot \sup|F(\mathbf{x}_A) - F(\mathbf{x}_B)| \tag{5}$$

where the scaling coefficient $\alpha(g)$ is defined as follows

$$\alpha(g) = \frac{|mean(\mathbf{x}_A) - mean(\mathbf{x}_B)|}{std(\mathbf{x}_A) + std(\mathbf{x}_B)} \tag{6}$$

High values of these measures indicate that two distributions are different (don't belong to the same population of samples).

2.5 The Linear Multi-input SVM Method

This method assesses the gene discriminative ability by considering all of them together. This is quite important since the cooperation of genes with the other may change significantly the importance of the genes. Our approach follows the idea of RFE-SVM [3]. The method is based on the idea, that the absolute values of the weights of a trained linear classifier produce a gene ranking. The gene associated with the larger weight is more important than that associated with the small one. The linear kernel SVM is used as the classifier, because this kernel does not deform the original impact of the gene on the result of the classification.

The features connected with the output of SVM through the weights of highest absolute values are regarded as the most important for recognition of these classes. All values of weights have been arranged in decreasing order and only the most important have been selected for each pair of classes. The results of this ranking will be referred by MSVM.

2.6 The Linear Single-Input SVM

The last method considered for gene selection is the application of the single-input Support Vector Machine [4],[8],[10]. The first step of this method is training the SVM network by using only one gene at a time. We train as many networks as is the number of genes. The predictive power of the single gene for a classification task is characterized by the value of the error function of the class recognition obtained by a one-dimensional linear SVM trained to classify learning samples on the basis of only one gene of interest. The smaller this error the better is the quality of the feature. Training many SVM networks by applying one feature at a time, selected in turn from the feature set, allows to create the ranking of the genes. The results of this ranking will be referred by 1SVM.

3   The Results of Numerical Experiments

The numerical experiments have been performed for the gene expression microarray data representing two problems: prostate tumour (PT) and neuroma (NR). We used the freely accessible data set concerning PT and NR [13].

3.1   The Data Base

Table 1 depicts the most important details of the data of both problems. The PT data contained 102 rows of 10509 genes (columns). The first class corresponds to the patients suffering from B-cell lymphoma and the second class represents the reference data corresponding to healthy patients. The NR data contained 47 rows of 2308 genes. The first 29 rows represent the patients suffering from Ewing Sarcoma and the rest 18 represent the malignant cancer of neuroblastoma [13]. The considered problems differ significantly with respect to the type of data.

Table 1. The most important data characterizing prostate tumour and neuroma

Name of problem	Number of genes	Number of representatives of class 1	Number of representatives of class 2
Prostate tumour (PT)	10509	52	50
Neuroma (NR)	2308	29	18

The first step in pre-processing was the normalization of the data corresponding to each gene. Standard statistical normalization has been applied. For each column (gene g) we performed the transformation $g := (g - m(g))/\sigma(g)$, where $m(g)$, $\sigma(g)$ is respectively the mean and standard deviation of the data corresponding to the gene g. These normalization was performed for PT and NR independently.

3.2 The Results of Gene Selection

The next step of experiment is the application of each of 8 mentioned selection methods of gene assessment for PT and NR independently. 10000 runs of each algorithm by using 90% of randomly selected data have been performed and only the first 100 genes have been selected in each run. This number of genes was sufficient to get the idea of the most representative genes in each case. On the basis of this we have selected the genes which appeared most frequently in these runs. Using these histograms we have selected the genes appearing with different ranges of repeatability in all runs (100%, >90%, >80%, >60%, etc.).

Table 2. The number of genes for PT and NR selected by individual ranking method according to their repeatability in 10000 runs of experiment

Ranking method	Prostate tumour (PT)				Neuroma (NR)			
	>60%	>80%	>90%	100%	>60%	>80%	>90%	100%
CSD	76	62	51	32	76	60	58	48
COR	73	60	50	31	77	58	52	37
WIL	72	55	47	32	75	66	60	37
KS	78	55	44	25	74	54	43	36
AKS	71	53	45	32	76	64	60	38
SKS	83	59	48	31	71	49	37	19
MSVM	76	54	45	31	75	59	51	36
1SVM	75	58	47	30	72	55	47	29

Table 2 presents the numerical results of such ranking for PT and NR. The numbers in the table point the quantity of genes of the defined repeatability.

In the next step we have analysed the number of the same genes selected simultaneously by different methods of ranking. Selecting the same gene as the most important by different strategy confirms its objective significance and representativeness for the particular cancer problem. This time only small portion of genes have been selected commonly. Once again we have selected the genes in different categories:

100% of repeatability, >90%, >80% of repeatability, etc. Table 3 presents the results of such selection for PT and NR.

Table 3. The final aggregated results of gene selection by all methods for PT and NR

Range of repeatability	Number of selected genes for PT	Number of selected genes for NR
100%	7	8
>90%	23	13
>80%	42	28
>70%	49	39
>60%	57	51
>50%	69	67

As a result of all experiments we have selected very small number of genes associated with each cancer problem: 7 genes for prostate tumour and 8 genes for neuroma. In further analysis we will treat them as the most important for cancer recognition.

To assess the correctness of this choice we have graphically illustrated the distribution of data at the existence of 100 randomly selected genes and at the specific number of the most important genes. Fig. 1 presents this comparison for the prostate tumour and Fig. 2 for neuroma problem in a graphical form.

It is evident in both cases that the selected genes form specific patterns well differentiating both classes of data. There is a visible border of the specific pattern of gene value distribution for both classes and this border is in a good agreement with the classes border. In the case of randomly selected gene representation this border was practically invisible.

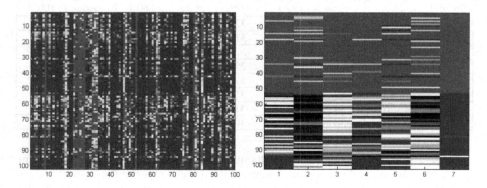

Fig. 1. The distribution of PT data at all (left) and 7 selected (right) genes representation. The horizontal axis represents genes and the vertical one – the patients: the first 52 patients from the top belong to the first class.

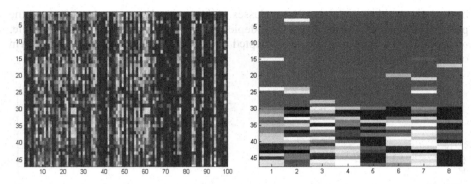

Fig. 2. The distribution of NR data at all (left) and 8 selected (right) genes representation. The horizontal axis represents genes and the vertical one – the patients: the first 29 patients from the top belong to the first class.

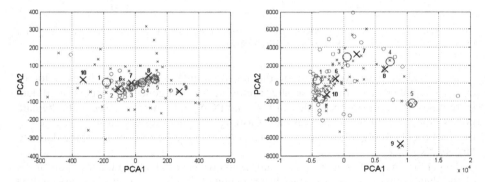

Fig. 3. The PCA distribution of PT at representation of the data by the best 100 genes (left) and after removing 7 genes of 100% repeatability (right)

The other aspect of assessment of the most important genes selection is the illustration of the distribution of data belonging to two classes using PCA analysis. PCA is a technique of mapping the N-dimensional data to K-dimensional space using the linear transformation $\mathbf{y}=\mathbf{Wx}$, where $\mathbf{x}$ represents the original N-dimensional vector, $\mathbf{y}$ – the K-dimensional vector and $\mathbf{W}$ is the transformation matrix [8]. We have mapped the multidimensional data represented by the most important 100 genes and the data deprived of the genes of 100% repeatability (7 genes for PT and 8 genes of NR) to 2-dimensional space ($K=2$) and presented the mapped data $\mathbf{y}$ on these two most important principal components PCA_1 and PCA_2. To keep equal dimension of the original data $\mathbf{x}$ under transformation we have substituted the removed data by the genes of the worse repeatability.

Fig. 3 and 4 depict the results for PT and NR. The left figure presents the distribution of data containing all 100 important genes and right figure represents the situation after removing from the set the most important genes of 100% repeatability. The sign × represents the first class and the circle – the second class. At the same time we have made the clusterization of both classes of data using K-means algorithm and the proper positions of cluster centers are depicted using large symbols of x and circle, respectively. In clustering we have adapted strategy that 10 data points should form one cluster (on average).

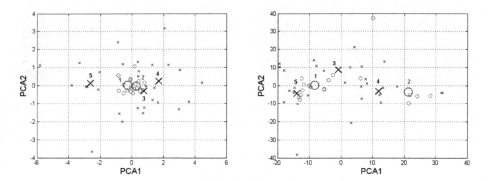

Fig. 4. The PCA distribution of NR at representation of the data by the best 100 genes (left) and after removing 8 genes of 100% repeatability (right)

It is evident that elimination of the most important genes introduces a lot of changes in data distribution. First, after removing the best genes the dispersion of data increases a lot (compare the scale of PCA_1 and PCA_2 in both cases). Secondly the data points belonging to the same class are forming compact clusters close to each other for the set containing all best genes. After removing the best genes the same data are largely dispersed.

To get the quantitative measure of both results we have calculated the mean distances of the data points to their winning cluster centers as well as the appropriate standard deviations of them. The results corresponding to PT and NR are presented in Table 4.

It is evident that after removing the best genes (7 in PT and 8 in NR) the quality measures of the clusters have dramatically deteriorated: the mean distances of all points to their winning clusters have been increased more than 15 times for PT and more than 5 times for SR. Similar rates of deterioration has been observed for standard deviations. This confirms that the best genes are very good representatives of the appropriate classes of cancer.

Table 4. The mean distances and standard deviations of the distances of data points to their winning cluster centers for the best 100 genes and after removing the best genes of 100% repeatability for PT and NR problems

	PT		NR	
	All 100 best genes	After removing 7 best genes	All 100 best genes	After removing 8 best genes
Mean of distances	0.15e3	0.23e4	1.37	7.33
Std of distances	0.08e3	0.07e4	0.86	4.37

3.3   The Results of SVM Classification

In the last step of checking the quality of the selected genes we have applied them in recognition of two different classes of data on the basis of the selected genes. As the classifier we have applied Support Vector Machine of Gaussian kernel. To get the reliable results of classification we have repeated the experiments 10000 times using each time 2/3 of the randomly selected data for learning and the other 1/3 of the data for testing the trained classifier. Each time different set of learning and testing data has been used in experiment. The average results of testing have been taken into account at the assessment of classification results. In the first set of experiments have used all 100 genes found as the best. Then we have removed the set of the best genes corresponding to their repeatability range extending from 90% to 100%, then 80%-90%, 70%-80% and from 60% to 70%. After removing the appropriate set of genes in each trial their place has been taken by the other set of the less important genes, in order to keep the number of genes equal in all experiments. In this way we can observe how the best selected genes influence the statistical accuracy of classification. Table 5 presents the results of such experiments.

It is evident that the reduction of the best genes (the range of repeatability from 90% to 100%) significantly decreases the efficiency of classifier system. The error of recognition has increased from 7.6% to 10.8% for PT and from 0.06% to 0.79% for NR. To check the general philosophy of splitting the population of genes into the best and worst sets according to their recognition ability we have also performed the same experiments of classification using each time 100 least significant genes. In the case of PT we have got the relative error equal 47.3% and for NR this error was equal 39.2%. These results confirm the effectiveness of our method of gene ranking.

It is also interesting to compare our results of classification to the other reports on the similar data basis. Our result concerning prostate tumour error (7.6% of average error) belongs to the best. The PT average error reported in the publications change from 5-13% using SVM-RCE [11], 5-7% using SVM-RFE [3] and 6-10% using PDA-RFE [11] depending on the number of genes taken into considerations.

Table 5. The statistical results of testing the SVM classifier at different composition of genes for PT and NR problems

Reduced genes	PT		NR	
	Number of reduced genes	Average testing error [%]	Number of reduced genes	Average testing error
No reduction	0	7.6±0.16	0	0.06±0.03
60%-70%	7	8.1±0.16	12	0.10±0.04
70%-80%	8	8.5±0.03	11	0.13±0.05
80%-90%	19	9.3±0.02	15	0.16±0.03
90%-100%	23	10.8±0.11	13	0.79±0.06

4   Conclusions

The paper has proposed the ensemble of methods for selection of the most important genes representing the microarray gene expression data for prostate tumour and neuroma. We have applied different ranking methods relying on various basis: the correlation parameters, the clusterization measures, the statistical hypothesis and linear Support Vector Machine. The results of ranking have been compared to each other and the genes commonly selected by all methods have been treated as the most important and representative for the considered cancer problems.

Application of the selected genes in recognition of the classes of cancer using SVM classifier has also confirmed the correctness of the proposed methodology of gene selection. The developed gene selection tools might be used in future research for predicting the potential danger of tumour on the basis of gene expression microarray analysis.

References

1. Conover, W.J.: Practical Nonparametric Statistics. Wiley, New York (1980)
2. Golub, T., et al.: Molecular classification of cancer: class discovery and class prediction by gene expression monitoring. Science 286, 531–537 (1999)
3. Guyon, I., Weston, A.J.,, Barnhill, S., Vapnik, V.: Gene selection for cancer classification using SVM. Machine Learning 46, 389–422 (2002)
4. Guyon, I., Elisseeff, A.: An introduction to variable and feature selection. Journal of Machine Learning Research 3, 1158–1182 (2003)
5. Huang, T.M., Kecman, V.: Gene extraction for cancer diagnosis by support vector machines - An improvement. Artificial Intelligence in Medicine 35, 185–194 (2005)
6. Huang, X., Pan, W.: Linear regression and two-class classification with gene expression data. Bioinformatics 19, 2072–2078 (2003)
7. Jirapech-Umpai, T., Aitken, S.: Feature selection and classification for microarray data analysis. BMC Bioinformatics 149, 1–11 (2005)
8. Schölkopf, B., Smola, A.: Learning with Kernels. MIT Press, Cambridge (2002)
9. Schurmann, J.: Pattern Classification, a Unified View of Statistical and Neural Approaches. Wiley, New York (1996)
10. Vert, J.P.: Kernel methods in genomics and computational biology. In: Camps-Vals, G., et al. (eds.) Kernel Methods in Bioengineering, pp. 42–64. Idea Group, London (2007)
11. Yusef, M., Jung, S., Showe, L., Showe, M.: Recursive cluster elimination for classification and feature selection from gene expression. BMC Bioinformatics 8(144), 1–12 (2007)
12. Matlab user manual – Statistics toolbox, MathWorks, Natick (1999)
13. Genes database:
    http://discover1.mc.vanderbilt.edu/discover/public/mcsvm

Rules versus Hierarchy: An Application of Fuzzy Set Theory to the Assessment of Spatial Grouping Techniques

Martin Watts

Centre of Full Employment & Equity, The University of Newcastle, NSW, Australia

Abstract. The geography which underpins the collection of Australian economic and social data is based on administrative areas, rather than having behavioural significance. Within most EU countries Coombes' rules-based grouping algorithm [1] which uses commuting flow data has been employed to construct Travel to Work Areas, but other approaches including Intramax (a hierarchical technique) have also been utilised. Recent developments in fuzzy set theory have enabled the comparison of the local accuracy of the solutions associated with different grouping methods. This paper will utilise both the Intramax technique and the modified version of Coombes' updated algorithm [2] to compare the properties of the solutions associated with grouping the Australian Statistical Local Areas using Journey to Work data from the 2006 Census.

Keywords: Closure, interaction, fuzzy sets, MAUP.

1 Introduction

The geography which underpins the collection of most economic and social data is typically based on administrative areas, rather than having any behavioural significance. Most spatially oriented research has been reluctant to acknowledge that the interpretation of these data can be compromised by the Modifiable Areal Unit Problem (MAUP), which has long been recognised by geographers. [3] notes that 'the areal units (zonal objects) used in many geographical studies are arbitrary, modifiable, and subject to the whims and fancies of whoever is doing, or did, the aggregating' so that the ensuing analysis is problematic. In addition, there has been a reluctance to adopt spatial econometric techniques, despite the likely spatial interdependence of contiguous areas.

Using measures of closure and interaction based on commuting flow patterns, [1] developed a rules based algorithm to identify UK Travel to Work Areas (TTWAs) based on UK Census data. TTWAs are defined as geographical areas within which a high percentage of commuting by residents occurs. It is the site for the interplay between labour supply and demand and, in principle, should be the appropriate area over which labour market statistics can be defined [4]. So each TTWA is considered to be largely self-contained (closed) from the rest of the economy, even though some commuting flows do cross boundaries. The algorithm has been adopted with some

M. Kolehmainen et al. (Eds.): ICANNGA 2009, LNCS 5495, pp. 517–526, 2009.
© Springer-Verlag Berlin Heidelberg 2009

amendments in many international studies. Also a generalisation of the algorithm, known as the European Regionalisation Algorithm, was recommended by Eurostat [5] as the standard approach for grouping areas in European countries [6]. [7] used this algorithm to analyse Australian Bureau of Statistics Journey to Work data from the 2001 Census defined across Statistical Local Areas (SLAs).

[5] quoted in [4], argues that the principles for defining these spatial clusters should be in declining order of priority, (1) autonomy: maximised self-containment; (2) homogeneity: minimised geographical area; (3) coherence: recognisible boundaries; and (4) conformity: alignment with administrative boundaries. However large groups of areas with high levels of self-containment may be inconsistent with homogeneity, because they contain pockets of both high and low unemployment. Also, within the Australian context, if the physical areas of groups of SLAs are minimised, then the final number of groups will be maximised, which imposes significant sampling costs for the monthly Labour Force Survey on the cash-strapped Australian Bureau of Statistics (ABS).

[10] has applied fuzzy set theory to the Coombes solution as a means of establishing its local properties. This entails comparing the values of a membership function (MF) across the different groups for each area.

An alternative technique Intramax, which is hierarchical, was developed by [8] and employs a simple decision rule to group areas. The key question is when to stop the grouping process. Intramax has also been utilised in the international literature, including [9] for Australia.

The absence of a formal mathematical approach to solve for the 'optimal' grouping reflects in part the idiosyncrasy of the problem, with different grouping techniques imposing different minimum criteria on the solution, including containment and levels of employment of residents. A global search procedure across all possible solutions is not feasible because of their huge number, since the following are undefined a priori a) the number of groups; and b) the size composition of the groups. and c) the allocation of areas between the groups.

Earlier papers [7,9,11] found that the Australian Standard Geographical Classification which is used as the basis for data collection does not appear to accord with economic behaviour. This geography is being revised by the ABS for the 2011 Census, using the principles of the Coombes algorithm. The creation of a more meaningful geography should enable a more sophisticated analysis of regional economic performance.

This paper will utilise the Coombes' updated algorithm [2] and the Intramax technique to generate two exhaustive and mutually exclusive groupings of Australian SLAs using Travel to Work (TTW) data from the 2006 Census. We then use the principles of FST, developed by [10] to explore the local properties of the two solutions. However the specification of the membership Function used by [10] imparts a bias to the calculations. Consequently we also compute an unbiased MF measure.

We find that, using the biased measure reveals that both solutions yield very few misallocated areas, which suggests that both approaches to grouping have strong local optimisation properties. On the other hand, the amended MF yields a large number of misallocated areas for both techniques. However the modified Coombes algorithm has markedly more groups of areas than the Intramax technique, which implies that one or both approaches has weak global properties. Hence an alternative grouping procedure should be considered, such as the use of Genetic Algorithms [12,13] or an algorithm employing the underlying principles of fuzzy set theory [14].

2 Rules Based versus Hierarchical Techniques

2.1 Coombes Algorithm

The Coombes algorithm [1] was significantly simplified for the analysis of the 2001 UK Census TTW data and here is used in the analysis of Australian commuting data defined across a much broader land mass and hence with some much lower densities of local employment and populations of employed residents. Assume that the areas have a fixed ordering according to a simple criterion such as their official ABS codes, or by actual area or the number of employed residents, etc.

First, the number of employed residents in the kth group of SLAs, R_{g_k} can be written

$$R_{g_k} = \sum T_{g_k \cdot} \quad k = 1,2,...,G$$

where $T = [T_{ij}]$ denotes the s*s TTW matrix. $T_{g_k \cdot}$ is the number of employees residing in SLA group k, and there are G groups.

The minimum closure (self containment) of grouping k can be written

$$C_{g_k} = \sum_{i,j \varepsilon g_k} T_{ij} / \max(T_{g_k \cdot}, T_{\cdot g_k})$$

where T_{ij} is the number of employees who live in SLA i and work in SLA j and $T_{\cdot g_k}$ denote total local employment in the kth group of SLAs.

[2] impose a spline function which represents a tradeoff between the number of resident employees and minimum closure. The minimum number of resident employees was set at 1,000 (R_1) with a corresponding minimum required rate of closure of 0.75, whereas 10,000 (R_2) resident employees or more necessitated a rate of closure of at least 0.70. Following [2], the spline function took the form shown below for group k, with all groups being required to assume a spline value of at least Y. [1]

$$\min(C_{g_k}, 0.75) * \min(b_1 R_{g_k}, a_2 + b_2 R_{g_k}, a_3)$$

Initially the SLA with the lowest spline value, say m is tentatively combined with say SLA n with which it has the highest rate of interaction defined as

$$I_{mn} = \max_k \{ T_{m,g_k}{}^2 / [T_{m \cdot} * T_{\cdot g_k}] + T_{g_k,m}{}^2 / [T_{g_k \cdot} * T_{\cdot m}] \}.$$

---

[1] The parameters of the spline function which are determined by the minimum closure and employment levels take the following values: 7.467 (b_1); 7407 (a_2); 0.0593 (b_2); 5600 (Y) with a_3 set at 0.8*R_2. The specification of the spline does not strictly accord with a linear tradeoff between the specified levels of resident employees and closure.

where g_k is initially a single SLA. The grouping is only consummated if the spline value for the combined SLAs exceeds the value of the spline function for SLA group n taken by itself.

The grouping of SLAs can be represented by rows of a grouping matrix A, which commences as an s*s identity matrix (A_0). One row corresponds to each grouping with unit values of the entries corresponding to the SLAs making up the particular group. The ordering of the groups, and hence rows of A, are determined by the ordering of the lowest SLA for each group. Thus the group represented in row 1, includes SLA1, while SLA2 belongs to group 1 or group 2. If m>n and combination takes place, then the grouping matrix after one iteration, A_1 can be written as

$$A_1(n,:) = A_0(m,:) + A_0(n,:)$$
$$where \quad A_0(m,m) = 1$$
$$A_1(m,:) = [0]_s.$$

where $[0]_s$ is an s dimensional row vector of zeros and A(m,:) denotes the mth row of A. So when an SLA and a group are combined, the consolidated group is represented by the nth row of A with n assumed to denote the lowest group member and unit values correspond to each group member. Then, after one iteration, the JTW matrix can be written as:

$$T_1 = A_1 T A_1'.$$

which combines the mth and n^{th} rows and columns, respectively of T.[2]

If the spline value falls, SLA m is consigned to a *reserve* list, and is not reconsidered until later. The SLA with the next lowest spline value (assumed below Y) is now considered and the same steps are enacted. If this SLA is combined with another according to the principles outlined above, the spline functions are re-ranked to identify the lowest value.

At some point, a group of 2 or more SLAs (say group m) with a spline value less than Y is selected, say at iteration t-1. It is dismembered and the individual SLAs are placed on a separate *active* list, so, at iteration t, the corresponding row of A is deleted. The SLA for possible grouping is the one from this *active* list which has the maximum flow into one of the extant (possibly singleton) groups, g_i, ie SLA k such that

$$T_{kg_n} = \max_{j,i}(T_{jg_i}) \quad j \in g_m.$$

Again SLA k is tentatively combined with group yielding the maximum interaction and is officially added to the group if the value of the group's spline function increases. If the latter does not occur, SLA k is also consigned to the *reserve* list and the next SLA from the dismembered group (active list) is chosen according to its flow to the extant groups and the same process is followed. Once a member of the dismembered group is combined with another group, the remainder of the

---

[2] In general, the grouped TTW matrix can be written as: $T_t = A_t T A_t'$, where A_t denotes the grouping matrix after iteration t.

dismembered group (ie. except those already placed on the *reserve* list) is retained as a group, because the dismembered group can no longer be reformed.[3]

Spline values can now be recalculated for the official groups with the reserve list SLAs only being reconsidered when all groups satisfy the spline constraint. Then the maximum flow requirement (with the extant groups) is imposed to identify an SLA from the *reserve* list, followed by the maximum interaction requirement, with the spline constraint being weakened, so that the group with the additional member must continue to satisfy the minimum spline constraint. Once all members of the *reserve* list are combined with groups, the algorithm ceases, because all groups still satisfy the spline constraint. Some SLAs from the reserve list may not be assigned, in which case the algorithm ceases (see below).

The final version of the matrix, A_F represents a unique representation of the groupings, and avoids degeneracy [15]. A_F can be converted into a unique grouping vector, f where $f = \sum_i i A_F(i,:)$ where f={f(i)} i=1,2,...s.

2.2 Intramax Technique

Hierarchical methods group the areas based on a criterion which is gradually lowered, until all areas satisfy the criterion [6]. Contiguity requirements can be imposed which make the grouping sub-optimal, but reduce the computational demands of the procedure, by reducing the number of permutations [16]. The hierarchical model gradually raises the internal flows of the consolidated areas and can impose the number of required regions, or minimum statistical requirements.

Under the Intramax method the TTW matrix is considered to be a contingency table. The objective function is specified in terms of the differences between the 'observed and the expected probabilities that are associated with these marginal totals' [8]. Considering the initial TTW table defined across all areas, then the expected value of each i,j commuting flow $\overline{T}_{ij}$ is derived as the product of the relevant row and column sums divided by total interaction (commuting flows), N*.

$$\overline{T}_{ij} = (T_{i.})T_{.j} / N *.$$

where the expected flows satisfy the column and row constraints, so that $T_{i.} = \sum_j \overline{T}_{ij}$ and $T_{.i} = \sum_j \overline{T}_{ji}$.

The null hypothesis for independence between the row and column marginal totals of a contingency table is $T_{ij} = \overline{T}_{ji}$ and the difference between the two terms measures the extent to which the 'observed flow exceeds (or falls below) the flow that would have been expected simply on the basis of the size of the row and column marginal totals' [8].

Treating i and j as singleton groups initially, the objective function to be maximised in this hierarchical clustering algorithm is

---

[3] This procedure prevents the program getting into an endless loop with the dismembered group being continually reformed with all its member SLAs.

$$Int = \max_{i,j}(T_{g_i g_j} / \overline{T}_{g_i g_j}) + (T_{g_i g_j} / \overline{T}_{g_i g_j}), \; i \neq j.$$

In the absence of a stopping rule, consolidation of areas would continue until there was one large grouping. There is no agreed stopping rule. [17] argues that a large increase in the within group (intrazonal) flow would not indicate 'a merger of two rather homogenous zones', so that the consolidation should cease in the previous iteration. Since it is impossible to make rigorous comparisons of the outcomes from the two grouping approaches, we choose to stop the consolidation process when total within group flows represented 75% of total commuting flows.

Thus neither grouping technique can be viewed as an optimisation procedure, per se, since both stopping rules are based on an inequality being satisfied. The local accuracy of these two grouping methods can be established by reference to Fuzzy Set Theory to which we now turn.

2.3 Fuzzy Set Theory (FST)

Fuzzy set theory has been recently applied to TTWAs [10]. The underlying principle is that an element (area) can simultaneously belong to different groups of areas to different degrees, whereas in classical theory an element can only belong to one group [10]. However, since we are seeking to define an exhaustive, mutually exclusive set of relatively areal groupings, we do not apply FST to identify a group of overlapping regions, but rather to assist in examining the rigor of grouping approaches.

The FST approach enables the identification of potential misallocations of areas across the groups by the measurement of a membership function, so that each area can be (partially) assigned to a (series of fuzzy) region(s) group(s). Commuting flow data measure both the number of residents of each area who are employed and also the level of employment in each area. These two features are combined together to construct a fuzzy group.

We can define a membership function for area i with respect to fuzzy residential grouping m, as

$$M'_{im} = \sum_{j \varepsilon g_m} T_{ji} / T_{.i} \, .$$

where area i belongs to group m on the basis of a grouping algorithm [10]. On the other hand, the membership function with respect to fuzzy local employment is:

$$M''_{im} = \sum_{j \varepsilon g_m} T_{ij} / T_{i.} \, .$$

Each expression lies between 0 and 1. The membership function with respect to a fuzzy group, m, M_{im}, can be written as the average of these two expressions, so that and $\sum_m M_{im} = 1$.

$$M_{im} = (M'_{im} + M''_{im}) / 2.$$

The membership function for area i can be defined for any fuzzy grouping but it would be expected to assume its highest value for the grouping to which it has been designated through the algorithm, [10].

If the constituent areas of a region have (say) low (residential) membership values for that region, then it would be expected that (residential) self-containment for the region would be low and hence the region would not be well defined [10]. Let $M = [M_{im}]$ denote the s*G matrix of membership values. Then if area i belongs to group m, M_{im} should assume the greatest magnitude in the ith row of M, ie $[M_{i.}]$.

This comparison of row entries is biased, however, because included in the calculation of both components of M_{im} is necessarily, T_{ii}, which typically dominates the corresponding row and column of the TTW matrix. On the other hand, for say a singleton group k, the numerators of the two components of M_{ik}, are just T_{ik} and T_{ki}, respectively.

There are two ways of addressing this issue. The first would be set the diagonal elements of [T] to zero in the calculation of the MF, but this would i) be inconsistent with the Intramax approach which is based on the consideration of gross Intrazonal flows which include intra-area flows; and ii) would undermine the use of the principles of FST to develop a new grouping algorithm, (see, for example, [14]) because each area would be combined with at least one other should any off-diagonal flows be positive. Alternatively all the numerators in the ith row should include T_{ii}, so that the elements of $[M_{i.}]$ represent the membership function corresponding to area i belonging to each of the G groups in turn.[4] In the calculations below, we adopt this strategy.

3 Data

JTW data based on 1411 areas was provided by the ABS. Small remote areas with local employment or employed residents of less than 50 were removed. This left 1365 areas. The ABS randomises low flows between SLAs to counter concerns about confidentiality. This can assume importance, given the form of the Intramax objective function, since one or both ratios can be large, despite a small numerator. Consequently all flows of 3 or less were set to zero.

4 Results and Discussion

Table 1 summarises the results for the two grouping approaches, based on both the biased and unbiased versions of the MF. First, the Coombes algorithm failed to absorb 4 SLAs into the groups. None of these areas satisfied the spline constraint in its own right, but all lay within the range of 1,000 to 10,000 resident employees. This result points to a difficulty in choosing appropriate parameters for the spline function.

Second, the Coombes algorithm which yielded 263 groups (plus the 4 exiled groups) and the Intramax algorithm, with a much lower number of groups (143), both

---

[4] The denominator, say for the first term of the MF of area i becomes $T_{.i} + (G-1)T_{ii}$ to retain the condition that the summation remains unity. This means that that the FST principles would marginally change as the number of groups declined, [14].

Table 1. Summary Statistics for Local Labour Markets

Algorithm				Self-Containment							
				Residential				Employment			
	Group	MAB	MAU	Max	Min	Mean	SD	Max	Min	Mean	SD
Coombes	263	3	225	1.00	0.708	0.905	0.071	0.998	0.718	0.911	0.056
Intramax	143	1	145	1.00	0.373	0.906	0.153	1.000	0.275	0.912	0.137

Source: ABS 2006 Census Journey to Work data and author's calculations.
Notes: MAB, MAU denotes the number of areas which were misallocated, using the biased
          and unbiased, measures based on FST, respectively.
          The groups recorded for the Coombes algorithm, exclude the 4 exiled SLAs.

had relatively high average closure rates by residence and employment (0.905, 0.911 for Coombes and 0.906, 0.912 for Intramax, respectively). The Coombes algorithm yielded 101 singleton groups, as compared to 27 under the Intramax technique. 43 of the Intramax groups did not satisfy the Coombes spline constraint.

Third, the number of areas which are misallocated according to the biased FST membership function is very low for both techniques, putting aside the 4 exiled SLAs under the Coombes algorithm. On the other hand, using the corrected specification of the MF reveals major errors in the grouping techniques with 225 of the 1361 areas being misallocated using the Coombes algorithm, and 145 of the 1365 areas under the Intramax technique. It should be noted that this measure identifies the number of separate relocations of one area at a time which enhance overall closure. This suggests that neither algorithm is efficient in achieving a local optimum solution, as defined by grouping based on high rates of closure, which implies that the algorithms lack robust global properties. Any comparison of the results needs to be undertaken very cautiously, however, because the criteria being employed in the operation of the two algorithms differ, as well as differing from the closure characteristics of the MF.

The operation of Coombes algorithm relies on the tradeoff between minimum closure and the residential employment level, but the consolidation of groups is based on an interaction function, which roughly speaking measures the sum of the product of two closure terms. The setting of the 4 key parameters assumes considerable importance, but, while it may be argued that the minimum level of residential employment can be set a priori, reflecting considerations of sample size and/or the number of groups, the rates of closure and the 'maximum' residential employment cannot be so easily determined to achieve a particular number of groups or minimum rate of closure. Relatedly, the impact of a given set of parameter values on the number and properties of the resulting groups is hard to anticipate. There are a huge number of permutations based on varying combinations of the magnitudes of the 4 parameters. It is evident that the complete dismemberment of proto-groups, which do not satisfy the spline constraint has a major impact on the properties of the final solution, with a significantly larger number of singleton groups than under the Intramax algorithm. Also the mechanics of the algorithm are complex with both active and reserve lists. While it may be possible for small remote areas to constitute singleton groups, a large number of singleton groups is undesirable.

On the other hand, the Intramax algorithm does not allow for any dismemberment of proto-groups. Also the interaction function differs from that used in the Coombes algorithm.[5] The low minimum rate of closure underlines the point however, that the grouping criterion under Intramax is quite different than those criteria characterising the Coombes algorithm. The number of groups in the final solution is more readily controlled under the Intramax technique, since the stopping rule can easily be adjusted to achieve a given form of final solution, which could be based on a minimum closure requirement, rather than the aggregate Intrazone flow. There may be limits as to the desirable rates of closure across groups. Otherwise convergence may be achieved with a singleton group, unless there is genuine geographical separation of the labour markets.

5 Conclusion

This paper has showed that neither the modified Coombes nor the Intramax algorithms appear to exhibit robust local properties, with respect to a criterion of closure, as defined by the revised MF specification. A major disadvantage of the rules-based Coombes algorithm is the difficulty in setting the underlying parameter values and also the proliferation of singleton groups, due, in part, to the dismemberment of groups when they do not satisfy the spline constraint. Dismemberment may detract from the optimal grouping process because it ensures that a group of areas which fails the spline constraint cannot be subsequently regrouped together. On the other hand, no dismembering occurs under a the hierarchical algorithm, but there is none of the uncertainty associated with setting parameter magnitudes. It can be argued, however, that, while the spline function explicitly incorporates closure, by group, this is not found in the Intramax algorithm, so that a harsh judgment of the Intramax procedure on the basis of the revised Membership Function is unwarranted.

A number of alternative approaches can be considered. The adoption of an Evolutionary Algorithm with a well specified fitness function [13,15] is worthy of consideration. A new approach which appears to have some potential to address the shortcomings of the Coombes and Intramax algorithms is to design an algorithm with an objective function based on the application of the MF principle [14]. Marginal dismemberment can occur and the revised form of the MF ensures that the algorithm does not cycle. The simplicity of the iterations which are of a similar form to those of the Intramax procedure is also an advantage of such an approach. Thus a stopping rule can be readily imposed.

In conversation with the author, an ABS official indicated that in the order of 70 Functional Economic Areas (TTWAs) would be an appropriate basis for a new Australian geographical classification. This would tend to negate any behavioural

---

[5] The respective first terms of the interaction expressions are $T_{g_m g_n}{}^2 / [T_{g_m .} * T_{.g_n}]$ and $T_{g_m g_n} / [T_{g_m .} * T_{.g_n}]$, for the Coombes and Intramax algorithms, where the common N*, which does not impact on which consolidation of groups is chosen has been removed from the second expression.

basis for the new classification, but would imply that adoption of one of the approaches outlined above, which can impose some control on the number of groups, would be appropriate.

References

1. Coombes, M.G., Green, A.E., Openshaw, S.: An Efficient Algorithm to Generate Official Statistical Reporting Areas. Journal of the Operational Research Society 37(10), 943–953 (1986)
2. Bond, S., Coombes, M.G.: 2001-based Travel to Work Areas Methodology, Mimeo (2001)
3. Openshaw, S.: The modifiable areal unit problem. Concepts and Techniques in Modern Geography 38(41) (1984)
4. Coombes, M.G.: Travel to Work Areas and the 2001 Census, Report to the Office of National Statistics, Centre for Urban and Regional Development Studies (March 2002)
5. Eurostat: Study on Employment Zones, Eurostat (E/LOC/20), Luxembourg (1992)
6. Coombes, M.G.: Defining Locality Boundaries with Synthetic Data. Environment and Planning A 32, 1499–1518 (2000)
7. Watts, M.J., Baum, S., Mitchell, W.F., Bill, A.: Identifying local labour markets and their spatial properties. Presented to the ARCRNSISS Annual Conference, Melbourne (May 2006)
8. Masser, I., Brown, P.J.B.: Hierarchical aggregation procedures for interaction data. Environment and Planning A 7, 509–523 (1975)
9. Mitchell, W.F., Watts, M.J.: Identifying functional regions in Australia using hierarchical aggregation techniques. Mimeo, Centre of Full Employment and Equity, The University of Newcastle, NSW (2006)
10. Feng, Z.: Fuzziness of Travel to Work Areas. Regional Studies 43(5), 707–720 (2009)
11. Watts, M.J.: Local Labour Markets in New South Wales: Fact or Fiction? In: Carlson, E. (ed.) A Future that Works - economics, employment and the environment, Proceedings of the 6th Path to Full Employment Conference (December 2004)
12. Florez-Revuelta, F., Casado-Diaz, J.M., Martinez-Bernabeu, L.: An Evolutionary Approach to the Delineation of Functional Areas Based on Travel-to-work Flows. International Journal of Automation and Computing 05(1), 10–21 (2008)
13. Florez-Revuelta, F., Casado-Diaz, J.M., Martinez-Bernabeu, L., Gomez-Hernandez, R.: A Memetic Algorithm for the Delineation of Local Labour Markets. In: Rudolph, G., Jansen, T., Lucas, S., Poloni, C., Beume, N. (eds.) PPSN 2008. LNCS, vol. 5199, pp. 1011–1020. Springer, Heidelberg (2008)
14. Watts, M.J.: The Application of Fuzzy Set Principles to the Design of a Spatial Grouping Algorithm. Mimeo, Centre of Full Employment and Equity, The University of Newcastle, NSW (2009)
15. Tucker, A., Crampton, J., Swift, S.: RGFGA: An Efficient Representation and Crossover for Grouping Genetic Algorithms. Evolutionary Computation 13(4), 477–499 (2005)
16. Masser, I., Scheurwater, J.: Functional regionalisation of spatial interaction data: an evaluation of some suggested strategies. Environment and Planning A 12, 1357–1382 (1980)
17. Goetgeluk, R.: Dynamic clusters in migration patterns: Intramax-analyses of inter-municipal migration flows between 1990 and 2004. In: ENHR Conference on Housing in an expanding Europe: theory, policy, participation and implementation, Slovenia (July 2006)

A Novel Signal-Based Approach to Anomaly Detection in IDS Systems

Łukasz Saganowski[1,2], Michał Choraś[1,2], Rafał Renk[1,3], and
Witold Hołubowicz[1,3]

[1] ITTI Ltd., Poznań
rafal.renk@itti.com.pl,
michal.choras@itti.com.pl
[2] Institute of Telecommunications
University of Technology & Life Sciences, Bydgoszcz
luksag@utp.edu.pl
[3] Adam Mickiewicz University, Poznań
holubowicz@amu.edu.pl

Abstract. In this paper we present our original methodology, in which
Matching Pursuit is used for networks anomaly and intrusion detection.
The architecture of anomaly-based IDS based on signal processing is presented. We propose to use mean projection of the reconstructed network
signal to determine if the examined trace is normal or attacked. Experimental results confirm the efficiency of our method in worm detection
scenario. The practical usability of the proposed approach in the intrusion detection tolerance system ($IDTS$) in the INTERSECTION project
is presented.

1 Introduction

Intrusion Detection Systems (IDS) are based on mathematical models, algorithms and architectural solutions proposed for correctly detecting inappropri-
ate, incorrect or anomalous activity within a networked systems [1].

Intrusion Detection Systems can be classified as belonging to two main groups
depending on the detection technique employed:

1. anomaly detection
2. signature-based detection.

Anomaly detection techniques rely on the existence of a reliable characterization
of what is normal and what is not, in a particular networking scenario. More
precisely, anomaly detection techniques base their evaluations on a model of
what is normal, and classify as anomalous all the events that fall outside such a
model [2].

In this paper our original methodology for networks anomaly and intrusion
detection based on Matching Pursuit is presented. In Section 2 general overview
of the proposed architecture and decision block details are shown. Moreover,
the motivation for signal processing methodologies used in intrusion detection

M. Kolehmainen et al. (Eds.): ICANNGA 2009, LNCS 5495, pp. 527–536, 2009.
© Springer-Verlag Berlin Heidelberg 2009

is given. In section 3 Matching Pursuit algorithm and base function of the proposed dictionary design is shown. Experimental results and conclusion are given thereafter.

The major contribution of this paper, is the intrusion/anomaly detection algorithm based on the Matching Pursuit. As to our best knowledge, we have not met any other IDS system based on matching pursuit.

Even though our Matching-Pursuit anomaly detection application is not working in a real time now, it is used in the off-network layer of INTERSECTION Intrusion Detection Tolerance System ($IDTS$).

2   Signal Processing Based Network Anomaly Detection

By profiling the properties of normal network traffic and modeling intrusions or unwanted traffic as anomalies, it is possible to detect the occurrence of such events within reasonable time so to activate reaction and response procedures. Determining the normal behavior model, however, is a difficult task due to the presence of different trends in data, which might be influenced by the time of day, the day of week and seasonal variations.

In our approach we store "normal" traces in a reference database. Normal traces represent traffic from days, we are sure no attacks occurred. These reference traces are compared to current, examined traces. Current traces may be either sniffed from traffic or for experimental purposes may represent old attacks (so that the ground truth is known).

The general overview of our intrusion detection system is presented in Figure 1. The overview of a decision block is explained in Figure 2.

Signal processing techniques have found application in Network Intrusion Detection Systems because of their ability to detect novel intrusions and attacks, which cannot be achieved by signature-based approaches. It has been shown that network traffic presents several relevant statistical properties when analyzed at

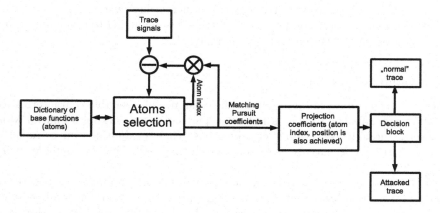

Fig. 1. IDS system block diagram

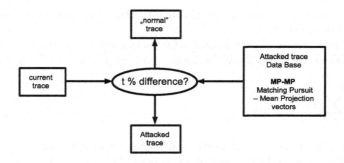

Fig. 2. IDS decision block diagram

different levels (e.g. self-similarity, long range dependence, entropy variations, etc.) [3]. Approaches based on signal processing and on statistical analysis can be powerful in decomposing the signals related to network traffic, giving the ability to distinguish between trends, noise, and actual anomalous events. Wavelet-based approaches, maximum entropy estimation, principal component analysis techniques, and spectral analysis, are examples in this regard which have been investigated in the recent years by the research community [4]-[8].

A powerful analysis, synthesis, and detection tool in this field is represented by the wavelets. Indeed, time- and scale-localization abilities of the wavelet transform, make it ideally suited to detect irregular traffic patterns in traffic traces. Recently many wavelet-based methods for detection of attacks have been tested and documented. Some are based on the continuous wavelet transform analysis, most of them however refer to the discrete wavelet transformation and the multiresolution analysis [3].

However, Discrete Wavelet Transform provides a large amount of coefficients which not necessarily reflect required features of the network signals.

Therefore, in this paper we propose another signal processing and decomposition method for anomaly/intrusion detection in networked systems. We developed original Anomaly Detection Type IDS algorithm based on Matching Pursuit.

3   Anomaly Detection Based on Matching Pursuit

Matching Pursuit signal decomposition was proposed by Mallat and Zhang [9].

Matching Pursuit is a greedy algorithm that decomposes any signal into a linear expansion of waveforms which are taken from an overcomplete dictionary D. The dictionary D is an overcomplete set of base functions called also atoms.

$$D = \{\alpha_\gamma : \gamma \in \Gamma\} \tag{1}$$

where every atom α_γ from dictionary has norm equal to 1:

$$\|\alpha_\gamma\| = 1 \tag{2}$$

Γ represents set of indexes for atom transformation parameters such as translation, rotation and scaling.

Signal s has various representations for dictionary D. Signal can be approximated by set of atoms α_k from dictionary and projection coefficients c_k:

$$s = \sum_{n=0}^{|D|-1} c_k \alpha_k \tag{3}$$

To achieve best sparse decomposition of signal s (min) we have to find vector c_k with minimal norm but sufficient for proper signal reconstruction. Matching Pursuit is a greedy algorithm that iteratively approximates signal to achieve good sparse signal decomposition. Matching Pursuit finds set of atoms α_{γ_k} such that projection of coefficients is maximal. At first step, residual R is equal to the entire signal $R_0 = s$.

$$R_0 = \langle \alpha_{\gamma_0}, R_0 \rangle \, \alpha_{\gamma_0} + R_1 \tag{4}$$

If we want to minimize energy of residual R_1 we have to maximize the projection $|\langle \alpha_{\gamma_0}, R_0 \rangle|$. At next step we must apply the same procedure to R_1.

$$R_1 = \langle \alpha_{\gamma_1}, R_1 \rangle \, \alpha_{\gamma_1} + R_2 \tag{5}$$

Residual of signal at step n can be written as follows:

$$R^n s = R^{n-1} s - \langle R^{n-1} s | \alpha_{\gamma_k} \rangle \, \alpha_{\gamma_k} \tag{6}$$

Signal s is decomposed by set of atoms:

$$s = \sum_{n=0}^{N-1} \langle \alpha_{\gamma_k} | R^n s \rangle \alpha_{\gamma_k} + R^n s \tag{7}$$

Algorithm stops when residual $R^n s$ of signal is lower then acceptable limit.

In basic Matching Pursuit algorithm atoms are selected in every step from entire dictionary which has flat structure. In this case algorithm causes significant processor burden. In our coder dictionary with internal structure was used.

Dictionary is built from:

— Atoms,
— Centered atoms,

Centered atoms groups such atoms from D that are as more correlated as possible to each other. To calculate measure of correlation between atoms function $o(a, b)$ can be used [2] .

$$o(a, b) = \sqrt{1 - \left(\frac{|\langle a, b\rangle|}{\|a\|_2 \|b\|_2}\right)^2} \tag{8}$$

The quality of centered atom can be estimated according to (9):

$$O_{k,l} = \frac{1}{|LP_{k,l}|} \sum_{i \in LP_{k,l}} o\left(A_{c(i)}, W_{c(k,l)}\right) \tag{9}$$

$LP_{k,l}$ is a list of atoms grouped by centered atom. $O_{k,l}$ is mean of local distances from centered atom $W_{c(k,l)}$ to the atoms $A_{c(i)}$ which are strongly correlated with $A_{c(i)}$.

Centroid $W_{c(k,l)}$ represents atoms $A_{c(i)}$ which belongs to the set $i \in LP_{k,l}$. List of atoms $LP_{k,l}$ should be selected according to the Equation 10:

$$\max_{i \in LP_{k,l}} o\left(A_{c(i)}, W_{c(k,l)}\right) \leq \min_{t \in D \backslash LP_{k,l}} o\left(A_{c(t)}, W_{c(k,l)}\right) \tag{10}$$

In the proposed IDS solution $1D$ real Gabor base function (Equation 11) was used to build dictionary [10]-[12].

$$\alpha_{u,s,\xi,\phi}(t) = c_{u,s,\xi,\phi} \alpha\left(\frac{t-u}{s}\right) \cos(2\pi\xi(t-u) + \phi) \tag{11}$$

where:

$$\alpha(t) = \frac{1}{\sqrt{s}} e^{-\pi t^2} \tag{12}$$

$c_{u,s,\xi,\phi}$ - is a normalizing constant used to achieve atom unit energy,

In order to create overcomplete set of $1D$ base functions dictionary D was built by varying subsequent atom parameters: Frequency ξ and phase ϕ, Position u, Scale s.

Base functions dictionary D was created with using 10 different scales (dyadic scales) and 50 different frequencies.

In Figure 3 example atoms from dictionary D are presented.

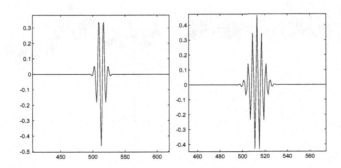

Fig. 3. Example dictionary atoms

4   Experiments and Results

In our experiments we decided to detect worm attacks. We tested our algorithms on normal and attacked traces to evaluate if our method is capable of detecting known worms.

Similarly to the work by Dainotti et al. [13] we tested the efficiency of our algorithms on Slammer and Witty worms. Slammer worm spread in 2003, while Witty spread in March 2004.

In our experiments we use TCP and UDP packets of Slammer and Witty made available by the WIDE-MAWI and CAIDA projects [15][16].

In this paper we will show our algorithm tested on attacked and normal traces.

The attacked traces represent traffic (TCP and UDP packets) from March 20th (Witty) (Figure 4) and March 25th (Slammer) (Figure 5).

The normal traces represent traffic from March 6th and March 13th (Figures 6 7).

The calculated values of Matching Pursuit Mean Projection for our test traces (normal and attacked) are presented in Tables 1-2.

In tables 1 Matching Pursuit Mean Projection values for TCP packets are presented. In tables 2 Matching Pursuit Mean Projection values for UDP packets are given, respectively.

Table 1. Mean Projection values calculated for test TCP traces

TCP Trace	MP
25.03.2004 (Slammer)	620
20.03.2004 (Witty)	667
6.03.2004	453
13.03.2004	373

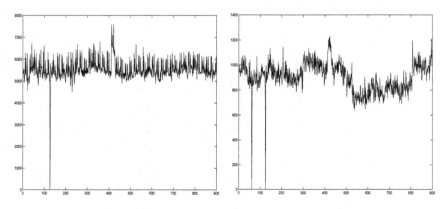

Fig. 4. Traces attacked by Witty worm from March 20th - TCP (left) and UDP (right)

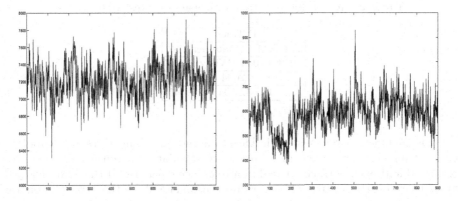

Fig. 5. Traces attacked by Slammer worm from March 25th - TCP (left) and UDP (right)

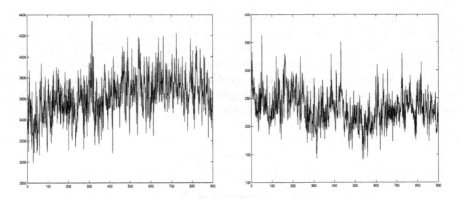

Fig. 6. Normal trace from March 6th - TCP (left) and UDP (right)

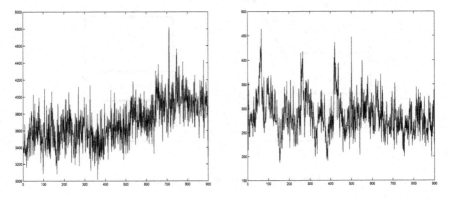

Fig. 7. Normal trace from March 13th - TCP (left) and UDP (right)

Table 2. Mean Projection values calculated for test UDP traces

UDP Trace	MP
25.03.2004 (Slammer)	82
20.03.2004 (Witty)	127
6.03.2004	32
13.03.2004	40

Decision block of our system is based on the Matching Pursuit Mean Projection values. As presented in Figure 1 we calculate difference $Diff$ between examined and normal traces stored in a reference database. If the value $Diff$ is larger than a certain threshold t our application signalizes the attack/anomaly.

In the experiments shown here, in the case of Worm attacks, our application was set to $t = 30\%$, which means that if Matching Pursuit Mean Projection differs more than 30% from the reference normal traces the attack should be detected.

As presented in Tables 1-2 mean projection values differ significantly and our IDS application successfully detects Witty and Slammer worms. In our experiments we can report 100% worm detection for TCP and UDP packets with no false alarms. However, so far we tested our method on a limited number of traces. We decided to use known and benchmark traces and worms first. Now we extensively test our method with a larger number of real-networks anonimized traces as well as with the generated traffic traces.

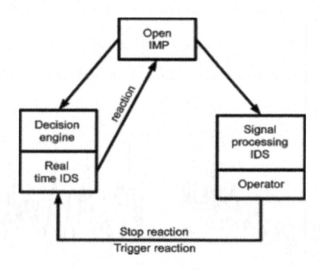

Fig. 8. IDS decision block diagram

5   Conclusion

In the article our developments in feature extraction for Intrusion Detection systems are presented. We showed that Matching Pursuit may be considered as very promising methodology which can be used in networks security framework. Upon experiments we may conclude that Matching Pursuit Mean Projection differs significantly for normal and attacked traces. Therefore our system successfully detects Slammer and Witty worms.

The major contributions of this paper is a novel algorithm for detecting anomalies based on signal decomposition. In the classification/decision module we proposed to use developed matching pursuit features such as mean projection. We tested and evaluated the presented features and showed that experimental results proved the effectiveness of our method.

The proposed Matching Pursuit signal based algorithm applied for anomaly detection IDS will be used as detection/decision module in the INTERSECTION Project security-resiliency framework for heterogeneous networks.

Signal-based anomaly detection type ADS/IDS will be used as the secondary detection/decision module to support real-time IDS. Such approach is proposed for off-network layer of the INTERSECTION framework.

The operator will have a chance to observe the results of signal-based IDS in a near real-time in order to trigger or stop the reaction of real-time IDS. Such approach will both increase the security (less detected anomalies/attacks) and increase the tolerance (less false positives). The overview of the Matching Pursuit IDS role in the INTERSECTION architecture is given in Figure 8.

Acknowledgement

The research leading to these results has received funding from the European Community's Seventh Framework Programme (FP7/2007-2013) under grant agreement no. 216585 (INTERSECTION Project).

References

1. Esposito, M., Mazzariello, C., Oliviero, F., Romano, S.P., Sansone, C.: Real Time Detection of Novel Attacks by Means of Data Mining Techniques. ICEIS (3), 120–127 (2005)
2. Esposito, M., Mazzariello, C., Oliviero, F., Romano, S.P., Sansone, C.: Evaluating Pattern Recognition Techniques in Intrusion Detection Systems. In: PRIS 2005, pp. 144–153 (2005)
3. FP7 INTERSECTION (INfrastructure for heTErogeneous, Reislient, Secure, Complex, Tightly Inter-Operating Networks) Project – Description of Work
4. Cheng, C.-M., Kung, H.T., Tan, K.-S.: Use of spectral analysis in defense against DoS attacks. In: IEEE GLOBECOM 2002, pp. 2143–2148 (2002)
5. Barford, P., Kline, J., Plonka, D., Ron, A.: A signal analysis of network traffic anomalies. In: ACM SIGCOMM InternetMeasurement Workshop 2002 (2002)

6. Huang, P., Feldmann, A., Willinger, W.: A non-intrusive, wavelet-based approach to detecting network performance problems. In: ACM SIGCOMM Internet Measurement Workshop (November 2001)
7. Li, L., Lee, G.: DDos attack detection and wavelets. In: IEEE ICCCN 2003, October 2003, pp. 421–427 (2003)
8. Dainotti, A., Pescape, A., Ventre, G.: Wavelet-based Detection of DoS Attacks. In: 2006 IEEE GLOBECOM, San Francisco, CA, USA (November 2006)
9. Mallat, S., Zhang: Matching Pursuit with time-frequency dictionaries. IEEE Transactions on Signal Processing 41(12), 3397–3415 (1993)
10. Troop, J.A.: Greed is Good: Algorithmic Results for Sparse Approximation. IEEE Transactions on Information Theory 50(10) (October 2004)
11. Gribonval, R.: Fast Matching Pursuit with a Multiscale Dictionary of Gaussian Chirps. IEEE Transactions on Signal Processing 49(5) (2001)
12. Jost, P., Vandergheynst, P., Frossard, P.: Tree-Based Pursuit: Algorithm and Properties. Swiss Federal Institute of Technology Lausanne (EPFL),Signal Processing Institute Technical Report,TR-ITS-2005.013 (May 17, 2005)
13. Dainotti, A., Pescape, A., Ventre, G.: Worm Traffic Analysis and Characterization. In: Proceedings of ICC, pp. 1435–1442. IEEE CS Press, Los Alamitos (2007)
14. Renk, R., Saganowski, Ł., Hołubowicz, W., Choraś, M.: Intrusion Detection System Based on Matching Pursuit. In: Proc. Intelligent Networks and Intelligent Systems, ICINIS 2008, pp. 213–216. IEEE CS Press, Los Alamitos (2008)
15. WIDE Project: MAWI Working Group Traffic Archive, tracer.csl.sony.co.jp/mawi/
16. The CAIDA Dataset on the Witty Worm - March 19-24, Colleen Shanon and David Moore (2004), http://www.caida.org/passive/witty

Extracting Discriminative Features Using Non-negative Matrix Factorization in Financial Distress Data

Bernardete Ribeiro[1], Catarina Silva[1], Armando Vieira[2], and João Neves[3]

[1] CISUC, Department of Informatics Engineering, University of Coimbra, Portugal
bribeiro@dei.uc.pt
[2] Physics Department, Polytechnic Institute of Porto, Portugal
asv@isep.pt
[3] ISEG - School of Economics and Management, Tech University of Lisbon, Portugal
jcneves@iseg.utl.pt

Abstract. In the recent financial crisis the incidence of important cases of bankruptcy led to a growing interest in corporate bankruptcy prediction models. In addition to building appropriate financial distress prediction models, it is also of extreme importance to devise dimensionality reduction methods able to extract the most discriminative features. Here we show that Non-Negative Matrix Factorization (NMF) is a powerful technique for successful extraction of features in this financial setting. NMF is a technique that decomposes financial multivariate data into a few basis functions and encodings using non-negative constraints. We propose an approach that first performs proper initialization of NMF taking into account original data using K-means clustering. Second, builds a bankruptcy prediction model using the discriminative financial ratios extracted by NMF decomposition. Model predictive accuracies evaluated in real database of French companies with statuses belonging to two classes (healthy and distressed) are illustrated showing the effectiveness of our approach.

1 Introduction

The Non-Negative Matrix Factorization (NMF) [1] is an algorithm able to learn a parts-based representation by imposing non-negativity constraints that allow only non-subtractive combinations. Similarly to principal component analysis (PCA) [2] that is based on finding a new representation (eigenspace) of the original data, NMF is also a projection method, since the original data is projected onto the new space. In contrast to PCA, the projected coefficients that are obtained using the NMF method are only positive. Furthermore, some of the basis components for PCA are distorted versions of the original data. The NMF basis are radically different: it is possible to extract localized features that correspond better with intuitive notions of the parts of the original data [1].

Although the concept is not new, since it has been investigated in linear algebra [3] where it was called positive matrix factorization (PMF), the last

M. Kolehmainen et al. (Eds.): ICANNGA 2009, LNCS 5495, pp. 537–547, 2009.
© Springer-Verlag Berlin Heidelberg 2009

ten years have witnessed a large amount of research on NMF since the seminal
work [1,4] was presented. Moreover, several variants of NMF have been proposed
by researchers. Hoyer [5] proposed a method of non-negative sparse coding which
minimizes a new cost function containing a positive regularization parameter.

NMF has also been applied to many areas such as bioinformatics [6,7,8] and
molecular pattern discovery [9], chemometrics [10], physics [11], multimedia data
[12], text mining [13,14], pattern recognition [1], document clustering [15], etc.

In the financial area, Non-negative Matrix Factorization was applied [16] to
the problem of identifying underlying trends in stock market data and it was
demonstrated how to impose appropriate sparsity and smoothness constraints
on the components of the decomposition. Also, in [17] parameterization of the
CreditRisk model for estimating credit portfolio risk is proposed. Therein, a
number of (non-negative) factor loadings, calculated by means of a non-negative
factorization of a positive semi-definite matrix, are used for model estimation.
The numerical optimization of the algorithm is also given.

In our work we apply NMF combined with KNN (K-Nearest Neighbor), FLD
(Fisher Linear Discriminant) Analysis and SVM (Support Vector Machines) to
a large financial database of French companies. This database is very detailed
containing information on a wide set of financial ratios spanning over a period
of five years. It contains three thousands distressed companies and about sixty
thousand healthy ones. In order to make predictions more accurate we tested
the models with data from three previous years priori to failure. The NMF
decomposition of the financial data is very useful to identify local components
of the data set. The representation of the financial statuses of the firms is a
linear combination of the basis functions weighted by the encoding factors. A
novel approach of NMF initialization is investigated using a K-means clustering
which minimizes the sum of squared distances between each data point and its
own cluster center and performs better than randomly.

The paper is organized as follows. In Section 2 a comment is given to dimensionality reduction methods. In Section 3 a brief review of Non-Negative Matrix
Factorization is presented. The proposed approach for classification using NMF
is discussed in Section 4. The financial data base is described in Section 5. In
Section 6 results are presented and discussed. Finally in Section 7 we conclude
the paper along with further lines of future research.

2   Dimensionality Reduction

Dimension reduction is desirable in many real world problems. On the other
hand, we might expect that a great many number of features will result in
more information and potentially higher accuracy. However, a raising important
paradox is that the more features, the more difficult is the process of information extraction. Consequently, the task of training the classifier is harder. This
problem is usually addressed as the curse of dimensionality which means that
the number of samples required per variable increases exponentially with the
number of variables.

The financial data used in this work (and described in Section 5) are of high-dimensionality, with redundant and possibly correlated features which might obscure the application of learning algorithms. To avoid the curse of dimensionality, the best practice is either to perform feature selection or to handle dimension reduction techniques.

Feature selection methods abound and are quite desirable to select those features that are more discriminant and can represent all the data [18]. Usually they are based on a score which is calculated for all features individually and the features with best scores are selected. However, interactions and correlations between features are omitted during selection. Dimension reduction is an alternative way, and unlike feature selection, projects the whole data into a lower dimensional space and new dimensions (components) are constructed using statistical information contained in the data. Although dimension reduction is pointed out by its lack of semantics, the new components often reflect the intrinsic structure of the data.

Principal Component Analysis (PCA) is a well known method used to reduce the feature space multidimensionality [2]. PCA computes the eigenvectors and the corresponding eigenvalues through an orthonormal transformation. By projecting onto the first few principal directions of the data, new features are obtained that are linear combinations of the original features. Principal Components are thus a set of variables that define a projection encapsulating the maximum amount of variation in a dataset and is orthogonal (and therefore uncorrelated) to the previous principle component of the same dataset.

Another method is Non-Negative Matrix Factorization, which became very popular recently due to its simplicity and impressive results in computer vision and pattern recognition.

3   Background on Non-negative Matrix Factorization

Mathematically, Non-negative Matrix Factorization (NMF) can be described as follows: given an $n \times m$ matrix V composed of non-negative elements $V_{ij} \geq 0$, the task is to factorize V into a non-negative matrix W of size $n \times r$ and another non-negative matrix H of size $r \times m$ such that $V \approx WH$ where r is a pre-specified positive integer that should satisfy the principle $r < nm/(n+m)$.

The constrained minimization problem can be put as minimizing the difference between V and WH by [19]:

$$\min_{W,H} f(W,H) \equiv \frac{1}{2} \sum_{i}^{n} \sum_{j}^{m} (V_{ij} - (WH)_{ij})^2 \tag{1}$$

$$\text{subject to } W_{ia} \geq 0, H_{bj} \geq 0 \,\forall i, j \tag{2}$$

If each column of V represents an object, NMF approximates it by a linear combination of r basis columns in W. The most popular approach [1] to solve this problem seeks to iteratively update the factorization based on a given objective function. This approach is similar to that used in Expectation-Maximization (EM) algorithms and is known as the multiplicative algorithm given below:

NMF Algorithm [1]

---

input: $V \in \mathbb{R}^{n \times m}$ and $r = rank$

Step 1. Randomize W and H with positive numbers in $[0, 1]$.
   Select a cost function to be minimized
Step 2. With W fixed, update H, then update W for the updated H.
   Iterate until the process converges.
Return: $W \in \mathbb{R}^{n \times r}$ and $H \in \mathbb{R}^{r \times m}$.

In the above algorithm, the cost function is either $C_1(V, WH) = ||V - WH||_F^2$ (where $|| \cdot ||_F$ is the Frobenius norm) or the generalized Kullback-Leibler (KL) divergence $C_2(V, WH) = \sum_{i,j} (V_{ij} \log V_{ij}/(WH)_{ij} - V_{ij} + (WH)_{ij})$. When cost function C_1 is used, the formulæ for updating of H and W are:

$$W_{ia} := W_{ia} \frac{(VH^T)_{ia}}{(WHH^T)_{ia}}, \tag{3}$$

$$H_{bj} := H_{bj} \frac{(W^T V)_{bj}}{(W^T WH)_{bj}} \tag{4}$$

whereas if cost function C_2 is used, the updating formulæ for H and W are:

$$W_{ia} := W_{ia} \sum_j \frac{V_{ij}}{(WH)_{ij}} H_{bj}, \tag{5}$$

$$W_{ia} := \frac{W_{ia}}{\sum_j W_{ja}}, \tag{6}$$

$$H_{bj} := H_{bj} \sum_i W_{ia} \frac{V_{ij}}{(WH)_{ij}}. \tag{7}$$

The above equations are obtained by minimization using non-linear programming methods such as the gradient descent. Since factors W and H are non-convex only local minimum is guaranteed to be obtained [4]. By replacing the Frobenius norm by KL divergence, NMF is equivalent to Probabilistic Latent Semantic Analysis [20]. This is intuitively reasonable since both techniques involve minimizing the distance between the model and the training data. A convergence proof of multiplicative update algorithms for nonnegative matrix factorization is given in [21]. Several bound-constrained optimization techniques have been used [5,22] to solve the problem. A simple implementation in MatLab is provided in [19] where a systematical experimentation of CPU times is exploited in image benchmark data.

4 Proposed Approach

We illustrate in Figure 1 our proposed approach while the algorithm below summarizes the main steps of the procedure. The application of the simple

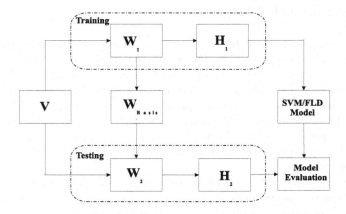

Fig. 1. NMF - FLD/SVM Approach

multiplicative NMF algorithm to the training data is straightforward. In order to speed up convergence to a desired solution initializing or seeding the algorithm is rather important. Initialization of W proceeds thus in either two ways: i) randomly or ii) using the K-means clustering algorithm in the original data. In the latter case, we choose the number of clusters K to be the rank of the factorization matrix. Once the number of clusters has been determined, the K-Means algorithm needs a centroid initialization for each cluster. One method of initializing these centroids is to randomly choose each centroid to be a different column vector from the data set. Cluster centers returned by K-means will constitute the initial basis functions, i.e. the factor W.

NMF - SVM/FLD Approach

1) Split financial data set V into $VTrain$ and $VTest$
2) Initialization of Matrices $WTrain$ and $HTrain$
2.1) If Init = 0 Perform Random initialization
2.2) else Initialize w/ K-Means clustering on matrix $VTrain$ with $K = Rank$
3) Perform NMF VTrain Decomposition: $VTrain = WTrain \times HTrain$
4) Take $WBasis = WTrain$
5) Perform NMF VTest Decomposition: $VTest = WBasis \times HTest$ starting with positive random encoding vector $HTest$
6) Classifier Design: KNN, FLD, SVM models using $HTrain$
7) Classifier Evaluation: using $HTest$

Once the basis functions and the weight encodings for the training set are obtained, the test set was projected to the models (KNN, FLD and SVM) and the encodings of the new samples were calculated. In other words, to represent a new test vector using a predefined set of basis functions, the same algorithm is iterated without modifying the matrix W. Thus, fixing W and starting with positive random encoding vector, a representation of a new data vector is obtained.

5 Experimental Setup

We used a sample obtained from Diane, a database containing financial statements of French companies. The initial sample consisted of financial ratios of about 60 000 industrial French companies, for the years of 2002 to 2006, with at least 10 employees. From these companies, about 3000 were declared bankrupted in 2007 or presented a restructuring plan to the court for approval by the creditors. The database includes information about 30 financial ratios (see Table 1) which allow the description of firms in terms of the financial strength, liquidity, solvability, productivity of labor and capital, margins, net profitability and return on investment.

Upon appropriate treatment of the database to eliminate firms with missing values, a final set of 600 default examples was obtained. In order to obtain a balanced dataset we randomly selected 600 non-default examples resulting in a set of 1200 examples. To accommodate historical information yearly variations of important financial ratios reflecting the balance sheet were then evaluated. Thus we included information from the past 3 years preceding the default. Therefore, the number of inputs was increased from 30 to 90 ratios.

Table 1. DIANE DATA BASE

	FINANCIAL RATIOS	
DIANA Data Base	1. Number of employees	2. Financial Debt/Capital Employed %
	3. Capital Employed / Fixed Assets	4. Depreciation of Tangible Assets (%)
	5. Working capital / current assets	6. Current ratio
	7. Liquidity ratio	8. Stock Turnover days
	9. Collection period	10. Credit Period
	11. Turnover per Employee (thousands euros)	12. Interest / Turnover
	13. Debt Period days	14. Financial Debt / Equity (%)
	15. Financial Debt / Cashflow	16. Cashflow / Turnover (%)
	17. Working Capital / Turnover (days)	18. Net Current Assets/Turnover (days)
	19. Working Capital Needs / Turnover (%)	20. Export (%)
	21. Value added per employee	22. Total Assets / Turnover
	23. Operating Profit Margin (%)	24. Net Profit Margin (%)
	25. Added Value Margin (%)	26. Part of Employees (%)
	27. Return on Capital Employed (%)	28. Return on Total Assets (%)
	29. EBIT Margin (%)	30. EBITDA Margin (%)

6 Results and Discussion

In this section we present and discuss the results from the proposed approach. First, we compare two ways of initialization of matrix W in the NMF algorithm. Second, we compare three algorithms (KNN, FLD and SVM) performance on the selective set of features extracted.

Experiments were carried out for various factorization ranks (rank $= 1 \cdots 81$). The highest rank corresponding to almost all features in the data set (see

Table 2. NMF with Random Initialization of W

	KNN		FLD		SVM	
Rank	Recall	Precision	Recall	Precision	Recall	Precision
4	82.46 ± 2.61	85.23 ± 2.09	85.10 ± 2.87	84.33 ± 2.22	81.60 ± 3.50	84.47 ± 2.70
9	83.43 ± 3.33	88.34 ± 1.83	87.39 ± 2.28	87.83 ± 2.27	85.12 ± 2.75	87.18 ± 2.01
16	84.86 ± 3.44	89.93 ± 2.78	90.19 ± 2.07	89.77 ± 2.80	88.74 ± 2.44	89.31 ± 2.40
25	85.16 ± 3.06	91.10 ± 2.52	90.11 ± 2.10	92.10 ± 2.46	89.99 ± 2.14	91.27 ± 2.25
36	83.12 ± 3.64	91.07 ± 2.26	90.92 ± 2.01	92.00 ± 2.46	90.35 ± 2.44	92.16 ± 1.91
49	81.67 ± 2.74	91.38 ± 1.76	91.28 ± 1.96	92.32 ± 2.03	90.34 ± 2.17	92.56 ± 2.00
64	80.70 ± 3.51	91.18 ± 1.98	91.04 ± 2.11	92.36 ± 1.63	90.67 ± 1.89	92.54 ± 1.75
81	78.32 ± 3.80	92.75 ± 2.10	91.21 ± 2.22	92.14 ± 2.77	90.31 ± 2.08	92.40 ± 2.13

Table 3. NMF with K-Means Clustering Initialization

	KNN		FLD		SVM	
Rank	Recall	Precision	Recall	Precision	Recall	Precision
4	81.81 ± 3.12	85.13 ± 2.74	85.24 ± 3.63	83.34 ± 2.08	81.23 ± 3.66	83.84 ± 2.50
9	84.70 ± 3.07	88.45 ± 2.72	88.54 ± 2.88	88.09 ± 1.84	86.31 ± 2.62	86.71 ± 2.83
16	85.96 ± 2.76	90.06 ± 1.77	91.02 ± 2.42	90.27 ± 1.91	89.72 ± 2.50	89.71 ± 1.62
25	85.58 ± 2.93	90.82 ± 2.45	91.09 ± 1.87	91.73 ± 1.98	90.83 ± 2.17	90.76 ± 1.95
36	84.11 ± 3.59	91.33 ± 2.36	91.37 ± 1.82	91.97 ± 2.27	90.46 ± 2.00	92.12 ± 2.14
49	82.91 ± 2.70	91.57 ± 2.17	91.95 ± 1.76	92.32 ± 1.90	90.78 ± 1.76	92.85 ± 1.81
64	81.23 ± 3.51	92.09 ± 2.02	91.55 ± 2.30	92.07 ± 2.11	90.81 ± 2.35	92.67 ± 2.06
81	79.18 ± 3.74	93.19 ± 1.92	91.72 ± 2.48	92.18 ± 1.78	90.45 ± 2.29	92.87 ± 1.82

Section 5). In statistical hypothesis testing a "Type I hit" occurs when a distressed firm is correctly classified and a "Type II hit" is the correct classification of a viable firm. A "Type I error" (or false positive rate) is the misclassification of a healthy firm as distressed. Conversely, a "Type II error" (or false negative rate) is one in which a distressed firm is misclassified by the predictor as viable. An "overall hit" refers to the total correct classifications for the set, regardless of type. Moreover, performance measures such as Recall and Precision defined, respectively, as $tp/(tp+fn)$ and $tp/(tp+fp)$[1] were evaluated. We also illustrate the results with $F1$-score which quantifies the tradeoff between Recall and Precision and is fairly indicative of the performance of the overall algorithm. All the results represent mean values obtained by 10-fold validation in financial data. For completion standard deviations are also indicated.

As said above, the NMF encodings of the training set were used to compute the discriminant functions for the KNN, FLD and SVM models. Likewise, for testing we used the test encodings obtained by the algorithm defined above. The initial condition of the non-convex optimization method concerning the NMF decomposition determines both its rate of convergence and the 'quality' of the local minimum found. The only free parameter is the rank r. The tolerance for the NMF factorization in the multiplicative update algorithm was setup to

---

[1] tp, fp and fn denote, respectively, true positive, false positive and false negative.

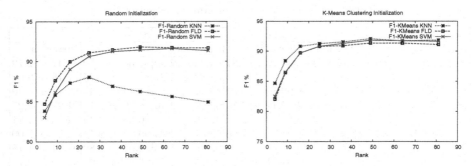

Fig. 2. F1-score for both types of initialization of W

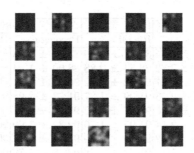

Fig. 3. Visualization of Matrix W for rank $r = 25$

$1.0E - 5$. Tables 2 and 3 report the results obtained with random initialization and using K-means clustering, respectively. In the latter case, K is fed to the clustering algorithm as the factorization rank of the initial data matrix.

In Figure 2 F1-score is evaluated and depicted for both kinds of initialization of W. We observe slightly better results with initialization by K-means as compared to random. In particular, the performance of the KNN classifier improves by around 5% with K-Means initialization of W while SVM is slightly better by around 1%.

Figure 3 depicts the resultant NMF localized features which are compatible with the intuitive idea of combining parts to form a whole. They are structured according to very discriminative financial ratios in the basis matrix W. Moreover, the intrinsic structure found by the method on the financial data is embedded in the basis functions W.

Figure 4 illustrates Type I and Type II Errors obtained for all the factorization ranks tested. We observe that for low ranks of the NMF decomposition FLD was found to perform slightly better than SVM, while for higher values of the rank SVM is preferable. The reason may be related to the capability of the Fisher Linear discriminant analysis to find the best set of vectors that minimize the intra cluster variability while maximizing the inter cluster distances particularly if the problem dimension is not too high (case of lower ranks).

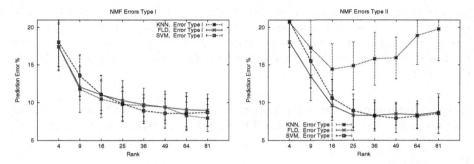

Fig. 4. Type I and Type II Errors

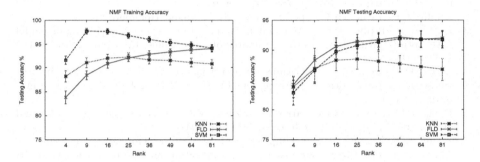

Fig. 5. Training and Testing Accuracy

In Figure 5 training and testing accuracy is depicted for the whole range of ranks in all tested algorithms. By maximizing class separability defined as the ratio of the between-class scatter matrix to the within-class scatter matrix FLD discriminates better. In case of the SVM, it is observed model overfitting for low ranks. Among the models tested, KNN presents worse results.

The ability of NMF to project financial data onto a subspace spanned by the most discriminative features is shown by the good results yielded. Although, there is clearly an optimum rank of 25 found experimentally, it is desirable that the optimal rank can be obtained automatically.

7   Conclusion and Future Work

Nonnegative matrix factorization (NMF) is a feature extraction method that has the property of intuitive parts-based representation of the original features. This unique ability makes NMF a potentially promising method for financial analysis. We developed a combined approach based on non-negative matrix decomposition (NMF) and learning classifiers, namely, KNN, FLD and SVM which was successfully tested in a financial setting. The first part of NMF was able to learn local features of the firms in distress or bankrupt and proper encoding factors useful for bankruptcy prediction. The model presents characteristics of robustness to be used in a real environment such as Diane data base of French companies. Further work along these lines will include a model to optimize the factorization rank as well as a method to incorporate prior knowledge of bankruptcy risk in the model.

Acknowledgments. Financial support from "Fundação da Ciência e Tecnologia" under the project PTDC/GES/70168/2006 is gratefully acknowledged.

References

1. Lee, D.D., Seung, H.S.: Learning the parts of objects by non-negative matrix factorization. Nature 401(6755), 788–791 (1999)
2. Jolliffe, I.T.: Principal Component Analysis. Springer, Heidelberg (2002)
3. Paatero, P., Tapper, U.: Positive matrix factorization: A non-negative factor model with optimal utilization of error estimates of data values. Environmetrics 5, 111–126 (1994)
4. Lee, D.D., Seung, H.S.: Algorithms for non-negative matrix factorization. In: Advances in Neural Information Processing 13 (Proc. NIPS 2000). MIT Press, Cambridge (2001)
5. Hoyer, P.O.: Non-negative matrix factorization with sparseness constraints. Journal of Machine Learning Research 5, 1457–1469 (2004)
6. Zhang, Z.Y., Zhang, X.S.: Two improvements of NMF used for tumor clustering. In: The First International Symposium on Optimization and Systems Biology (OSB 2007), pp. 242–249 (2007)
7. Carmona-Saez, P., Pascual-Marqui, R.D., Tirado, F., Carazo, J., Pascual-Montano, A.: Biclustering of gene expression data by non-smooth non-negative matrix factorization. BMC Bioinformatics (2006)
8. Fogel, P., Young, S., Hawkins, D.M., Ledirac, N.: Inferential, robust non-negative matrix factorization analysis of microarray data. BMC Bioinformatics 23(1), 44–49 (2007)
9. Brunet, J.P., Tamayo, P., Golub, T.R., Mesirov, J.P.: Metagenes and molecular pattern discovery using matrix factorization. In: National Academy of Science, vol. 101 (2004)
10. Guimet, F., Boqué, R., Ferré, J.: Application of non-negative matrix factorization combined with fisher's linear discriminant analysis for classification of olive oil excitationemission fluorescence spectra. Chemometrics and Intelligent Laboratory Systems 81(2006), 94–106 (2006)
11. Stefan, W., James, C., Anne, D.: Motivating non-negative matrix factorization. In: Eighth SIAM Conference on Applied Linear Algebra, Williamsburg, VA (2003)
12. Cooper, M., Foote, J.: Summarizing video using non-negative similarity matrix factorization. In: IEEE Workshop on Multimedia Signal Processing, pp. 25–28 (2002)
13. Pauca, V., Shahmazand, F., Berry, M.W., Plemmons, R.: Text mining using non-negative matrix factorization. In: SIAM International Conference on Data Mining, pp. 452–456 (2004)
14. Berry, M., Browne, M., Langville, A., Pauca, V., Plemmons, R.: Algorithms and applications for approximate nonnegative matrix factorization. Computational Statistics & Data Analysis 52(1), 155–173 (2007)

15. Xu, W., Liu, X., Gong, Y.: Document clustering based on non-negative matrix factorization. In: SIGIR 2003: Proceedings of the 26th annual International ACM SIGIR Conference on Research and Development in Information Retrieval, pp. 267–273. ACM, New York (2003)
16. Drakakis, K., Rickard, S., de Frein, R., Cichocki, A.: Analysis of financial data using non-negative matrix factorization. International Mathematical Forum 3, 1853–1870 (2008)
17. Vandendorpe, A., Ho, N.D., Vanduffel, S., Dooren, P.V.: On the parameterization of the CreditRisk+ model for estimating credit portfolio risk. Mathematics and Economics 42, 736–745 (2008)
18. Guyon, I., Gunn, S., Nikravesh, M., Zadeh, L.: Feature Extraction: Foundations And Applications. In: Studies in Fuzziness and Soft Computing. Physica Verlag, Heidelberg (2006)
19. Lin, C.-J.: Projected gradient methods for nonnegative matrix factorization. Neural Computation 19(10), 2756–2779 (2007)
20. Hofmann, T.: Probabilistic latent semantic indexing. In: Twenty-Second Annual International SIGIR Conference on Research and Development in Information Retrieval (1999)
21. Lin, C.J.: On the convergence of multiplicative update algorithms for nonnegative matrix factorization. IEEE Transactions on Neural Networks 6(18), 1589–1596 (2007)
22. Chu, M., Plemmons, R.J.: Nonnegative matrix factorization and applications. Image 34, 1–5 (2005)

Evolutionary Regression Modeling with Active Learning: An Application to Rainfall Runoff Modeling

Ivo Couckuyt[1,*], Dirk Gorissen[1], Hamed Rouhani[2],
Eric Laermans[1], and Tom Dhaene[1]

[1] Ghent University - IBBT, Department of Information Technology (INTEC), Gaston Crommenlaan 8, Bus 201, 9050 Ghent, Belgium
[2] Katholieke Universiteit Leuven, Faculty of Bioscience Engineering, Department of Land Management and Economics, Celestijnenlaan 200 E, 3001 Leuven, Belgium

Abstract. Many complex, real world phenomena are difficult to study directly using controlled experiments. Instead, the use of computer simulations has become commonplace as a feasible alternative. However, due to the computational cost of these high fidelity simulations, the use of neural networks, kernel methods, and other surrogate modeling techniques has become indispensable. Surrogate models are compact and cheap to evaluate, and have proven very useful for tasks such as optimization, design space exploration, visualization, prototyping, and sensitivity analysis. Consequently, there is great interest in techniques that facilitate the construction of such regression models, while minimizing the computational cost and maximizing model accuracy. The model calibration problem in rainfall runoff modeling is an important problem from hydrology that can benefit from advances in surrogate modeling and machine learning in general. This paper presents a novel, fully automated approach to tackling this problem. Drawing upon advances in machine learning, hyperparameter optimization, model type selection, and sample selection (active learning) are all handled automatically. Increasing the utility of such methods for the domain expert.

1 Introduction

For many problems from science and engineering it is impractical to perform experiments on the physical world directly (e.g. airfoil design, earthquake propagation). Instead, complex, physics-based simulation codes are used to run experiments on computer hardware. This allows scientists more flexibility to study phenomena under controlled conditions. However computer experiments still require a substantial investment of computation time. This is especially evident for routine tasks such as prototyping, high dimensional visualization, optimization, sensitivity analysis and design space exploration [1].

As a result researchers have turned to various approximation methods that mimic the behavior of the simulation model as closely as possible while being computationally cheap(er) to evaluate. Different types of approximation methods exist, each with

---

* Ivo Couckuyt is funded by the Institute for the Promotion of Innovation through Science and Technology in Flanders (IWT-Vlaanderen).

M. Kolehmainen et al. (Eds.): ICANNGA 2009, LNCS 5495, pp. 548–558, 2009.
© Springer-Verlag Berlin Heidelberg 2009

their relative strengths. This work concentrates on the use of data-driven, global approximations using compact surrogate models (also known as emulators, metamodels or response surface models (RSM)) in the context of computer experiments. Examples of metamodels include: Artificial Neural Networks (ANN), rational functions, Gaussian Process (GP) models, Radial Basis Function (RBF) models, and Support Vector Machines (SVM).

It is important that we stress the difference between local and global surrogate models as the two are often confused. Local surrogates are by far the most popular and involve building small, relatively low fidelity surrogates for use in optimization. Local surrogates are used as rough approximators of the (costly) optimization surface and guide the optimization algorithm towards good extrema while minimizing the number of simulations. Once the optimum is found the surrogate is discarded. Many advanced methods for constructing and managing these local surrogates have been designed (e.g., [2]).

In contrast, with global surrogate modeling the surrogate model *itself* is the goal. The objective is to construct a high fidelity approximation model that is as accurate as possible over the *complete* design space of interest using as few simulation points as possible. Once constructed, the global surrogate model (also referred to as a replacement metamodel[1]) is reused in other stages of the computational science and engineering pipeline. So optimization is not the goal, but rather a useful post-processing step.

However, constructing accurate surrogate models as efficiently as possible is an entire research domain in itself. In order to come to an acceptable approximation, numerous problems and design choices need to be overcome: what data collection strategy to use (active learning), what model type is most applicable (model selection), how should model parameters be tuned (hyperparameter optimization), how to optimize the accuracy vs. computational cost trade-off, etc. This work draws upon advances in these domains, integrating them in a coherent platform in order to better tackle the model calibration problem in hydrology.

2   Surrogate Modeling

As stated in the introduction, the principal reason driving the use of surrogate models is that the simulator is too time consuming to run for a large number of simulations. One model evaluation may take many minutes, hours, days or even weeks [1]. A simpler approximation of the simulator is needed to make optimization, design space exploration, etc. feasible. A second reason is when simulating large scale systems [3].

There are many methods involved and various choices to be made when generating surrogate models. Consequently, practical implementation leaves many options open to the designer: different model types, different experimental designs, different model selection criteria, different active learning strategies, etc. However, in practice it turns out that the designer rarely tries out more than one subset of options. All too often, surrogate model construction is done in a one-shot manner. Iterative and adaptive methods, on the other hand, have the potential of producing a much more accurate surrogate at a considerably lower cost (less data points). E.g., by applying iterative sample selection

---

[1] The terms surrogate model and metamodel are used interchangeably.

(also known as active learning and adaptive sampling) an accurate surrogate model can be constructed while minimizing the computational cost. See [4] for a good discussion on this issue. For the application in this paper we will utilize a fully featured toolbox for adaptive surrogate model generation , the SUMO toolbox [5].

3   Application

A task which is often central to hydrological modeling is the identification of suitable parameters for a given set of modeling objectives, catchment characteristics and data. However, this identification process is difficult because conceptual rainfall runoff models generally have a large number of parameters and the accuracy of their calculations depends on how the relevant parameters are defined. Additionally, because of their conceptual nature, these parameters cannot be measured directly and are therefore estimated on the basis of a calibration process, i.e., minimizing an objective function (OF).

We illustrate the strength of global surrogate modeling in improving the process of estimating the right parameters of a rainfall runoff model. The SWAT (Soil Water Assessment Tool) is an operational model that was developed to assist water resource managers in assessing water supplies and non-point source pollution at river basin scale. The model is able to assess the impact of changes in climate, landuse and management, and to simulate the transport and fate of chemicals and water quality loadings. The model is designed so that use can be made of readily available inputs. Upland components include hydrology, weather, erosion/sedimentation, soil temperature, plant growth, nutrients, pesticides, and land and water management. Stream processes include channel flood routing, channel sediment routing, nutrient and pesticide routing and transformation. The ponds and reservoirs component contains water balance, routing, sediment settling, and simplified nutrient and pesticide transformation routines. Water diversions into, out of, or within the basin can be simulated to represent irrigation and other withdrawals from the system. However, one should be aware that every process in the model is a simplification of reality.

In SWAT, a watershed is divided into multiple subwatersheds, which are then further subdivided into *hydrologic response units* (HRUs) that consist of homogeneous landuse, management, and soil characteristics. The HRUs represent percentages of the subwatershed area and are not identified spatially. The model operates in a continuous mode and has been widely used to estimate catchment runoff, nutrient and sediment loads. The SWAT model development, operation, limitations, and assumptions are extensively discussed by [6]. One of the practical problems in applying the SWAT is determining proper values for the more than 30 parameters that control the fidelity of its prediction. While many parameters can be estimated empirically a direct expensive optimization procedure is still routinely used to determine optimal settings [7], requiring many expensive simulations.

We propose to take global surrogate modeling methods routinely used in Electro-Magnetics (EM) and engineering design, and apply them to the setting of rainfall runoff modeling. Through the use of sequential modeling and active learning methods, a replacement metamodel can be generated that captures the relationship between the different SWAT parameters and provides insight in their influence on the prediction quality of the SWAT. While at the same time minimizing the number of computationally expensive simulations. Optimization can still be performed as a postprocessing step.

4   Related Work

A few studies have been reported in recent years in the field of water resources related to surrogate modeling. Savic et al. [8] applied 2 data-driven models (genetic programming and ANN) to flow prediction, results show that both are able to match up against conceptual models. Khu et al. (2003) [9] reduced the number of simulation runs required by Monte Carlo (MC). This was achieved by using an ANN and hybrid GA to respectively approximate and explore the shape of the objective function. This significantly reduces the computational effort involved in investigating hydrological model parameter uncertainty. Later on, an evolutionary-based metamodel calibration methodology was developed using a coupled genetic algorithm-RBF ANN [10]. Regis and Shoemaker (2004) [11] proposed an approach for costly black box optimization that uses space-filling experimental designs and k-nearest neighbor local function approximations to improve the performance of an EA in twelve-dimensional groundwater bioremediation problem. Broad et al., (2006) [12] evaluated six local search algorithms for purpose of improving the performance of ANN surrogate model-based optimization of water distribution systems. The results show a significant improvement in the value of the objective function by using a local search as a complementary stage of surrogate model-based optimization of water distribution systems. Kamali et al. (2007) [13] evaluate the performance of the design and analysis of computer experiments (DACE) surrogate function along with Latin Hypercube Sampling (LHS) and MC Sampling for hydrological model calibration. The results indicate that DACE along with LHS reduced the computational cost of calibration process. Recent research by Garote et al. [14] advocate the use of Bayesian networks to learn the behaviour of a rainfall runoff model.

5   Experimental Setup

5.1   SWAT

The SWAT requires spatial information about topography, river/stream reaches, landuse, soil and climate to accurately simulate the streamflow. The study basin is that of the Grote Nete (383 km²), located in the north-eastern part of Belgium. A detailed description of the study basin is given in [15]. Daily observations of precipitation, air temperature, evaporation, and daily streamflow data were obtained from the Royal Meteorological Institute and the Flemish Administration for Land and Water, Belgium. The soil map was available at a scale of 1:25.000; the soil physical data was derived from the Aardewerk-SIBIS Soil Information System and land use was derived from the multi-temporal LANDSAT 5 TM image of 18 July 1997.

The climatic inputs in SWAT include daily precipitation measured in 5 stations scattered in and outside the study area, and the potential evapotranspiration and min/max temperature collected in a station at the northern boundary of the catchment. Details of input data are given in [15]. The catchment was subdivided in 8 subcatchments and 65

HRUs. The flow separation program of [16] was used in this research as to determine the relative contribution of surface runoff and groundwater to total streamflow. The latter were created based on the various combinations of land use and soil types present in the catchment. Climate data were assigned to each HRU using the centroid method. The daily streamflows in the Varendonk outlet station were used for model calibration and verification.

Parameter sensitivity analysis was applied to identify the parameters of the SWAT model that contribute most to the variability of component flows. It is important to have an understanding of catchment characteristics and the hydrological processes involved before 'blindly' applying surrogate modeling to the available data. Based on a critical analysis of the SWAT modules to the hydrology of the study area, the parameters to calibrate were reduced to 18. Although this number of parameters is considerably smaller, to further reduce the number of parameters in the surrogate process, a sensitivity analysis was conducted to determine the most sensitive parameters of the hydrological module simulating streamflow. This analysis (through Latin Hypercube and One-factor-at-a-time) yielded the 4 most sensitive parameters.

The first parameter is p, the percentage by which CN_2 (the SCS curve number) is changed from the initial values. Thus, p, a parameter in the approximation model, is converted to CN_2, the actual parameter of the SWAT, using the following formula: $CN_2 = initial\,CN_2 + \frac{initial\,CN_2 \cdot p}{100}$. Secondly, $RCHRG\_DP$ stands for the deep aquifer percolation ratio and is a measure for the transfer between the shallow and deep aquifer system. Thirdly, $REVAPMN$ is the amount of water (mm) that must be present in the shallow aquifer store before water can move to the unsaturated zone. Finally, $ESCO$ is the soil evaporation compensation coefficient. The domains of the 4 parameters are [-40,40] (ensuring absolute bounds of [35 90] for CN_2), [0 3], [0 1] and [0 1] respectively. When the SWAT model is run it generates a time series of predicted flow during the period 1998-2002. This time series is then separated into 3 components useful for runoff prediction: $low\,flow$ (values ≤ 2), $high\,flow$ (values ≥ 5), and $total\,flow$ (all values). On each of these components the Mean Square Error (MSE) is then calculated with the true observations during that period, and that is the final output of the simulation code. Separating the total flow in more fine-grained components allows the SWAT to be calibrated for different types of flows. Thus, in sum, the SWAT simulator has 4 inputs (CN_2, $RCHRG\_DP$, $REVAPMN$, $ESCO$), and 3 outputs ($MSE_{low}, MSE_{high}, MSE_{total}$).

5.2 SUMO Toolbox

The active learning settings were set as follows: an initial optimized Latin hypercube design of size 50 is used augmented with the corner points. Modeling is allowed to commence once at least 20 of the initial samples are available. Each iteration a maximum of 50 new samples (over all outputs) are selected using the *gradient* adaptive sampling algorithm up to a maximum of 500. A full discussion of the algorithm is out of scope for this paper, details can be found in [17].

There are many surrogate modeling methods available to fit the data and many options implemented in the SUMO Toolbox. However, from the application it is not immediately clear which surrogate model type or hyperparameter optimization algorithm should be used (ANN, SVM, RBF models, ...). For this reason we shall use an automatic surrogate model type selection algorithm. The algorithm utilizes a genetic algorithm (using the island model) to simultaneously select the model type and model parameters (hyperparameter optimization). The surrogate model types included in the evolution are: single layer feed forward ANNs (using [18]), Kriging models (using [19]), rational functions and LS-SVMs (using [20]). Together with hybrid models (ensembles, that arise as a result of a crossover between two models of different type) this means that 5 model types will compete to fit the data. The population size for each model type is 10 and the maximum number of generations between each sampling iteration is 15. The final population of the previous model type selection run is used as the initial population for the next run. An extinction prevention algorithm is used to ensure no model type goes completely extinct. A full description of the algorithm, model types, and genetic operators is out of scope for this paper. Such settings can be found in [5]. Given the correlation between the outputs, they are not modeled separately (by separate models) but together in a single model with multiple outputs.

Note that this approach relieves the domain expert from technical choices related to the model generation. Besides a few high level options (which model types are of interest) and termination criteria (time limit, sample budget) no further input is required. The hyperparameter optimization, model selection, and sample selection are performed fully automatically, allowing the domain expert to concentrate on the application and not have to deal with modeling technicalities.

In order to drive the hyperparameter optimization a max-min validation set of 20% is used. Since not all data points are available at once but are chosen incrementally, the validation set grows as more data arrives. Validation points are not selected randomly but by maximizing the minimum distance between them, thus ensuring a good coverage of the domain. Note, though, that models are always trained on all the data, it is only when the error is calculated that they are temporarily re-trained on 80% of the available data. The error function that is minimized is the Average Relative Error (ARE):

$$ARE(y, \bar{y}) = \frac{1}{n} \sum_{i=1}^{n} \frac{|y_i - \bar{y}_i|}{|y_i|},$$

where y_i, $\bar{y}_i$ are the true and predicted response values respectively. Since we are dealing with multiple outputs per model, a weighted sum over the ARE values for each output is taken. Since we wish to treat all outputs equally, all weights were set to 1^2.

The SUMO Toolbox was configured to use the remote Sun Grid Engine (SGE) sample evaluation backend. This means that the toolbox will run simulations in parallel by transparently submitting them to a remote cluster. The cluster in question is the CalcUA cluster which consists of 256 nodes. Thus the SUMO Toolbox (v5.1) is running on a local machine, while the SWAT simulations are scheduled on the cluster. The number of data points selected each iteration is chosen dynamically (but never exceeding the user defined limit of 50) based on the average time needed for modeling, the average duration of a single simulation, and the number of compute nodes available at that point in time. The average time for one simulation is quite short, 4-10 minutes depending on

---

2 Alternatively, a multiobjective approach as discussed in [21] could also have been used.

cluster availability. Finally it should be noted that all the algorithms described here are available for download at http://www.sumo.intec.ugent.be.

6 Results

Figure 1 shows the evolution of the population as the modeling progresses. Some interesting dynamics can be observed. As soon as migration between the different sub-populations is allowed to take place, Kriging models quickly take over the population resulting in very smooth approximation surfaces. As the number of datapoints increases, the quality of the rational functions increases and they overtake Kriging as the most popular model type. However, the problem with the rational functions is that they are very prone to producing asymptotes in their response due to the increasing existence of poles. The implementation in the toolbox is best suited to low dimensional cases with sufficient data per dimension, in other cases the orders of the polynomials involved grow too quickly, increasing the risk of overfitting. Therefore, it is no surprise that they are finally overtaken by ANN models that, thanks to the pruning functions implemented as part of the mutation and crossover operators, are able to produce smoother responses.

Of course nothing prevents this process from recurring. The fact that the optimal solution changes with time is not necessarily a bad thing and should actually be expected since the hyperparameter optimization landscape is dynamic (due to the active learning). Note that it is the extinction prevention (EP) algorithm that makes these oscillations possible (it ensures a model type never goes completely extinct but that at least 2 individuals of each type are preserved). Without EP these dynamics are impossible and everything depends on the initial conditions. As a result the danger of converging to a poor local optimum (poorly fitted regression model) is significantly larger. Ideally these tests should be repeated many times to conclusively confirm the final outcome. However, naturally, in situations where simulations are costly such repetitions are impossible. In addition, previous work on a wide variety of benchmark problems has shown that the algorithms used here are robust across many repetitions and always perform better or equal than a set of single model type runs.

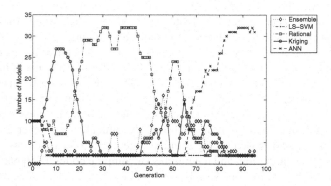

Fig. 1. Population evolution

Table 1. Errors of final ANN model (4-14-3 network)

$\|X\|$	*Output*	ARE_{TR}	ARE_V	CV	minimum x^*	$f(x^*)$
	MSE_{low}	0.08320	0.1084	0.1036	(-39.9939, 0.6907, 0.9999, 0.6549)	0.8311
500	MSE_{high}	0.02491	0.03570	0.02760	(-37.0391, 0.0000, 0.0000, 1.0000)	15.6302
	MSE_{total}	0.02336	0.03809	0.02790	(-39.6159, 0.3025, 0.9998, 0.6660)	10.3791

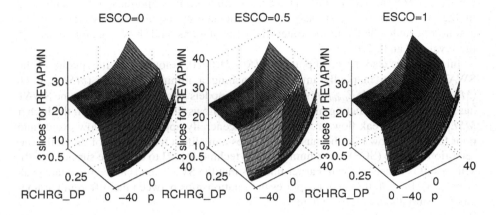

Fig. 2. Final ANN model for MSE_{total}

Table 1 shows the final average relative errors (ARE) for each of the outputs on the training (TR) and validation (V) data. In addition a 10-fold cross validation error (CV) was calculated as well. $|X|$ is the number of samples used to train the ANN model while x^* and $f(x^*)$ denote the minimum and corresponding function value of the ANN model respectively. For the MSE_{high} and MSE_{total} an error of less than 5% (acceptable for the application) is easily reached. The MSE_{low} output appears more difficult, reaching only a final ARE of 10%. Thus future runs should take this into account, placing more emphasis on the first output instead of treating all outputs equally. On the other hand, this can also be an indication that the hydrologic model parameters selected are not good enough to capture the trend to simulate base flow. Therefore incorporating more parameters like available water capacity of soils (SOL_AWC) will improve not only low flow simulated values but also high flow simulated values. This is the topic of a follow-up publication.

For space considerations plots of the MSE_{low} and MSE_{high} outputs are omitted in this paper. Figure 2 shows the plot of the final best model (a $4 - 14 - 3$ ANN) for MSE_{total}. In the figure $REVAPMN$ and $ESCO$ have been clamped at 3 values: 0, 1 and 3 for $REVAPMN$ and 0, 0.5 and 1 for $ESCO$. The remaining 2 parameters, p and $RCHRG\_DP$, are shown along the x-axis and y-axis respectively.

From the figure it is immediately clear that the 3rd and 4th parameters have virtually no influence on the quality of the SWAT prediction: the three slices of each subplot coincide and the three subplots for each output show little or no differences. This was confirmed by using the model browser of the SUMO Toolbox to browse through each of the 4 dimensions. This is an unexpected result, a further study of the study basin and HRU settings is underway to shed more light on this issue. Though, a preliminary explanation can be given as follows. The 3rd variable, *REVAPMN*, affects when and to what degree subsurface flow occurs, and therefore indirectly govern the contribution of subsurface flow to the total stream flow of the watershed of interest. These two parameters (*ESCO* and *REVAPMN*) have more influence in evapotranspiration simulated by the model. Since we just analyze flow simulated by the model, these values cause a non-noticeable change in the water yield calculations, and therefore adjustments to these values can be left out.

Interesting is also the breakpoint *RCHRG_DP* = 0, below which the quality of the SWAT prediction markedly improves, reaching a minimum of 0.8 (MSE_{low}), 16 (MSE_{high}), and 10 (MSE_{total}) respectively. The models also clearly show that the SWAT has more trouble predicting high flows than low flows (as can be seen from the higher MSE_{total} value). Peak flow predictions were generally appreciable for low events and poor for higher flow rates because SWAT uses a modified formulation of the Soil Conservation Service (SCS) curve number (CN) technique [22] to calculate surface runoff. This result is consistent with earlier findings that the SWAT tends to overestimate peak flows [23]. In sum, the model captures the relationships between the different parameters in a smooth and intuitive manner.

7   Conclusion

Global surrogate modeling is a powerful approach to facilitate the analysis of computational expensive simulation codes. However, all too often the designer only tries out a small set of techniques for a particular application.

In this paper the computationally expensive problem of parameter setting in rainfall runoff modeling was investigated (calibrating the SWAT). Therefore a replacement metamodel is generated through the use of sequential modeling and active learning methods requiring little or no knowledge about surrogate modeling. The model type and complexity was determined automatically, data points were selected iteratively, and simulations were transparently scheduled on a shared cluster. The final surrogate model produced by the SUMO Toolbox provided insight into the relationship between the different parameters (including identification of the optima) and can be used to improve the prediction quality in other settings (e.g., as part of a wider Geographic Information System (GIS) tool).

Future work will consist of increasing the number of input parameters, generalizing to other study basins, and further investigating the correlation between the different outputs. In addition the effect of applying a multiobjective optimization algorithm (such as NSGA-II) to drive the hyperparameter selection will also be explored to further improve the model quality.

References

1. Wang, G.G., Shan, S.: Review of metamodeling techniques in support of engineering design optimization. Journal of Mechanical Design 129(4), 370–380 (2007)
2. Ong, Y.S., Nair, P., Lum, K.: Max-min surrogate-assisted evolutionary algorithm for robust design. IEEE Transactions on Evolutionary Computation 10(4), 392–404 (2006)
3. Barton, R.R.: Design of experiments for fitting subsystem metamodels. In: WSC 1997: Proceedings of the 29th conference on Winter simulation, pp. 303–310. ACM Press, New York (1997)
4. Lin, Y.: An Efficient Robust Concept Exploration Method and Sequential Exploratory Experimental Design. Ph.D thesis, Georgia Institute of Technology (2004)
5. Gorissen, D., De Tommasi, L., Croon, J., Dhaene, T.: Automatic model type selection with heterogeneous evolution: An application to rf circuit block modeling. In: Proceedings of the IEEE Congress on Evolutionary Computation, WCCI 2008, Hong Kong (2008)
6. Arnold, J., Srinivasan, R., Muttiah, R., Williams, J.: Large area hydrologic modeling and assessment - part 1: Model development. Journal of the American Water Resources Association 34, 73–89 (1998)
7. Bekele, E., Nicklow, J.: Multi-objective automatic calibration of SWAT using NSGA-II. Journal of Hydrology 341, 165–176 (2007)
8. Savic, D.A., Walters, G.A., Davidson, J.W.: A genetic programming approach to rainfall-runoff modelling. Water Resources Management 13(3), 219–231 (1999)
9. Khu, S., Werner, M.G.F.: Reduction of monte carlo simulation runs for uncertainty estimation in hydrological modelling. Hydrology and Earth System Sciences 7(5), 680–692 (2003)
10. Khu, S., Savic, D., Liu, Y., Madsen, H.: A fast evolutionary-based meta-modelling approach for the calibration of a rainfall-runoff model. In: Proceedings of the First Biennial Meeting of the International Environmental Modelling and Software Society, Lugano (2004)
11. Regis, R., Shoemaker, C.: Local function approximation in evolutionary algorithms for the optimization of costly functions. IEEE Trans. Evolutionary Computation 8(5), 490–505 (2004)
12. Broad, D., Dandy, G., Maier, H., Nixon, J.: Improving metamodel-based optimization of water distribution systems with local search. In: IEEE Congress on Evolutionary Computation, pp. 710–717 (2006)
13. Kamali, M., Ponnambalam, K., Soulis, E.: Computationally efficient calibration of watclass hydrologic models using surrogate optimization. Hydrology and Earth System Sciences Discussions 4, 2307–2321 (2007)
14. Garrote, L., Molina, M., Mediero, L.: Learning Bayesian Networks from Deterministic Rainfall-Runoff Models and Monte Carlo Simulation. In: Practical Hydroinformatics, vol. 68, pp. 375–388. Springer, Heidelberg (2008)
15. Rouhani, H., Willems, P., Wyseure, G., Feyen, J.: Parameter estimation in semi-distributed hydrological catchment modelling using a multi-criteria objective function. Hydrological Processes 21(22), 2998–3008 (2007)
16. Arnold, J.G., Allen, P.M.: Automated Methods for Estimating Baseflow and Ground Water Recharge From Streamflow Records. Journal of the American Water Resources Association 35, 411–424 (1999)
17. Crombecq, K.: A gradient based approach to adaptive metamodeling. Technical report, University of Antwerp (2007)
18. Nørgaard, M., Ravn, O., Hansen, L., Poulsen, N.: The NNSYSID toolbox. In: IEEE International Symposium on Computer-Aided Control Sysstems Design (CACSD), Dearborn, Michigan, USA, pp. 374–379 (1996)

19. Lophaven, S.N., Nielsen, H.B., Søndergaard, J.: Aspects of the matlab toolbox DACE. Technical report, Informatics and Mathematical Modelling, Technical University of Denmark, DTU, Richard Petersens Plads, Building 321, DK-2800 Kgs. Lyngby (2002)
20. Suykens, J., Gestel, T.V., Brabanter, J.D., Moor, B.D., Vandewalle, J.: Least Squares Support Vector Machines. World Scientific Publishing Co., Pte, Ltd., Singapore (2002)
21. Gorissen, D., Couckuyt, I., Dhaene, T.: Multiobjective global surrogate modeling. Technical Report TR-08-08, University of Antwerp, Middelheimlaan 1, 2020 Antwerp, Belgium (2008)
22. Wischmeier, W., Smith, D.: Predicting Rainfall Erosion Losses. A Guide to Conservation Planning. Agriculture Handbook No. 537. U.S. Department of Agriculture, USDA, Washington (1978)
23. Rouhani, H., Gorissen, D., Willems, P., Feyen, J.: Improved rainfall-runoff modeling combining a semi-distributed model with artificial neural networks. In: The 4th International SWAT Conference, Delft, The Netherlands (2007)

Gene Trajectory Clustering for Learning the Stock Market Sectors

Darie Moldovan and Gheorghe Cosmin Silaghi

Babeş Bolyai University
Business Information Systems Dept.
Str. Theodor Mihali 58-60, 400591, Cluj-Napoca, Romania
{Darie.Moldovan,Gheorghe.Silaghi}@econ.ubbcluj.ro

Abstract. Hybrid Gene Trajectory Clustering (GTC) algorithm [1,2] proves to be a good candidate to cluster multi-dimensional noisy time series. In this paper we apply the hybrid GTC to learn the structure of the stock market and to infer interesting relationships out of closing prices data. We conclude that hybrid GTC can successfully identify homogeneous and stable stock clusters and these clusters can further help the investors.

1 Introduction

A stock market index is a method for measuring a section of the market. It gather together various single stocks originating from the same economic activity or geographical location. In the last few decades, building indexes and raising their performance expectations was a strong preoccupation for every fund manager [3]. Indexes can also help in portfolio management (performance estimation and structuring) or in forecasting the price evolution of a single stock. Stock exchange indexes represent an expert-based clustering method of the entire stock market. Whether a particular stock belongs or not to a particular index is a decision taken according with the interest of the potential investors. In this paper we will focus our attention to business sector indexes.

Many financial news often refer globally to an entire business sector and are closely accompanied with the performance change of the related index. But, do those news affect or concern all the individual stocks gathered by the index? Or if a rumor concerning a particular stock come out, how strong will affect it the performance of the index, or which are the other stocks that will be affected in a similar way?

In this paper we intend to learn a clustering of the stock exchange market in the sense of identifying those shares that move together on various time frames. Comparing the inferred clustering with the one supplied by the expert-created sectoral indexes can show us how firm or weak are the financial decisions based on the technical analysis pursued on those indexes. To perform the clustering, we employ the Gene Trajectory Clustering method [1,2]. This method succeeds

M. Kolehmainen et al. (Eds.): ICANNGA 2009, LNCS 5495, pp. 559–568, 2009.
© Springer-Verlag Berlin Heidelberg 2009

to cluster gene values represented by their evolution in time, even if the space dimensionality is high and there are a lot of irregularities and noise in the time series. Out of the clustering, we obtain the individual stocks grouped according with the evolution of their closing prices time series and an aggregated trajectory for each cluster. We perform the clustering procedure on the 65 stocks aggregated in the Dow Jones Composite Average index and compare it with the clustering produced by the index components: the industrial average, the transportation average and the utilities average.

The paper develops as follows. In section 2 we brief the Gene Trajectory Clustering method. Section 3 presents the stock exchange time series under study, the required preprocessing and the experiment setup. Section 4 presents the experiments and the results. Section 5 brief some related works and section 6 concludes the paper.

2   Gene Trajectory Clustering Methodology

In this section we describe the Hybrid Gene Trajectory Clustering method [1,2].

There are various alternatives for performing a clustering task, like non-model-based methods (K-Means clustering, hierarchical clustering, tree-based algorithms) or model-based ones: autoregressive models, B-splines or Multiple Linear Regression models (MLR). Gene Trajectory Clustering (GTC) is based on MLRs to account for temporal information of the in the clustering process.

The trajectory clustering problem [4] refers to clustering the set of measurements Y which are measured as a function of time. Typically, we have data for M individuals and the response variable y might be multi-dimensional. We want to cluster the M individuals in K clusters. Each individual Y_i is assigned to a cluster k with the prior probability w_k, $\sum_k^K w_k = 1$. The clustering model is a mixture of K MLRs (one regression for each cluster), each regression representing a single trajectory cluster given by [2]:

$$Y_i = S(\mu_k + \gamma_i) + \epsilon_i, \quad \gamma_i \approx N(0, \Gamma), \quad \epsilon_i \approx N(0, \sigma^2 I) \tag{1}$$

$Y_i = [y_{i,1}, y_{i,2}, ..., y_{i,l}]^t$ is the ith trajectory of length l, S is the $l \times (p+1)$-basis matrix, p is the regression order, μ_k is the $p+1$ vector of regression coefficients and γ_i and ϵ_i are uncorrelated Gaussian noises for the regression coefficients and the trajectory. Each trajectory Y_i has a cluster membership vector $z_i = (z_{i,1}, z_{i,2}, ..., z_{i,K})$ with $z_{i,k} = 1$ if the ith trajectory belongs to cluster k and 0 otherwise. The standard method for mixture model learning assume that z_i is missing and apply some Expectation Maximization algorithm [5] to estimate the regression coefficients for maximizing the complete data log likelihood. We refer the reader to [2] for a complete specification of the log likelihood iterative maximization problem.

Because the search space is multi-modal and EM is a local optimizer, the initial values of the iterative searching algorithm determines the local optima

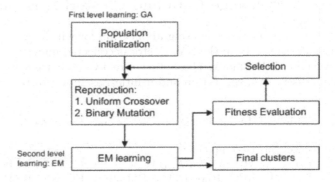

Fig. 1. The hybrid Gene Trajectory Clustering algorithm [2]

where EM will stop. Thus, hybrid Gene Trajectory Clustering [1,2] employs a genetic algorithm to select the optimal subset of data as the initial cluster centers. Figure 1 presents the steps of GTC for obtaining the final clusters.

In the genetic algorithm applied at the first level each solution is encoded as an M-bit string with the ith bit corresponding to the ith individual. A "1" in the ith bit position indicates that the ith individual is selected and "0" otherwise. For selecting K optimal "genes" Y_i as the initial cluster centers, all feasible solutions must have K numbers of 1 and $M - K$ numbers of 0. Uniform crossover selects two random parents with the crossover probability and swap the bits between the parents at the same position with 50% probability. Mutation operator simply inverts each bit of the offspring at the mutation probability and a repair operator is applied to obtain only feasible solutions. After the offspring solutions are created, they are translated as initial conditions for the mixture models and the EM algorithm is applied. The maximum log likelihood values represent the fitnesses of the corresponding offsprings. For offspring selection, GTC employs the elitist scheme $(\mu + \lambda)$ for faster convergence. This search process stops until some stopping criteria are reached. According to [1,2], this hybrid GA-based approach is less sussceptible to getting trapped in local optima, because it uses only genetic operators to create new offsprings and relies only on population information.

The hybrid Gene Trajectory Clustering is implemented in the Gene Network Explorer[1] (GNetXP) at the KEDRI at the Auckland University of Technology, New Zealand. GNetXP offers the possibility to adapt the execution of the clustering algorithm to the problem specification. One can set the number of clusters, the number of coefficients to control the smoothness of the curves, the stopping threshold for the GA, the number of parents and offspring and the number of generations.

---

[1] http://www.aut.ac.nz/research/research_institutes/kedri/research_centres/centre_for_bioinformatics/gene_network.htm

3   The Stock Exchange Data and Clustering Experiment

In this section we present the clustering application for which we applied GTC.

We selected 65 stocks[2] from the NYSE, stocks collected under the Dow Jones Composite Average (DJA) index. These stocks are divided in three sector-specific sub-indexes: industry, transportation and utilities, as shown in table 1.

Table 1. DJA sub-indexes (as at the end of year 2007)

DJA sub-index	Size	Cluster components
industry	30	AA, AIG, AXP, BA, BAC, C, CAT, CVX, DD, DIS, GE, GM, HD, HPQ, IBM, INTC, JNJ, JPM, KO, MCD, MMM, MRK, MSFT, PFE, PG, T, UTX, VZ, WMT, XOM
transportation	20	AMR, ALEX, BNI, CAL, CHRW, CNW, CSX, EXPD, FDX, GMT, JBHT, JBLU, LSTR, LUV, NSC, OSG, R, UNP, UPS, YRCW
utilities	15	AEP, AES, CNP, D, DUK, ED, EIX, EXC, FE, FPL, NI, PCG, PEG, SO, WMB

For each stock we considered daily adjusted closing prices p_i between years 2000 and 2007. To obtain a homogeneous image of the stocks' evolution during time, we computed the daily logarithmic returns $log(p_i/p_{i-1})$, as advised by the financial investments literature [6]. To be able to project the data as trajectories, we scaled it, considering a start point of 100 points for every stock at the beginning of each year and next, applied the logarithmic returns on this 100 points basis. Figure 2 depicts several stocks trajectories for one year.

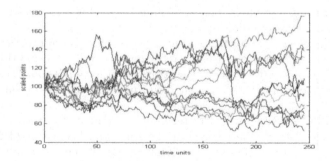

Fig. 2. Stock scaled points trajectories to be clustered

We want to perform a clustering of the stock trajectories for each year, in order to see which are the stocks that move together on a medium term and if

---

[2] JBLU entered the DJA transportation subindex during 2002. We used this stock in our experiments only from 2003.

there is any sectoral or other business-related relationship that can be learned from the stock movements.

We compared the standard EM algorithm for generating the mixture models, the hybrid genetic algorithm based GTC described in section 2 with the expert-based division of the DJA stocks in the three categories of the index: industry, transportation and utilities. Because the hybrid GTC is based on randomization, each experiment was run 10 times and we report the results averaged over the runs. The maximum log likelihood value represents the fitness of each clustering experiment.

4 Results

First, we want to show that hybrid GTC performs better than the standard EM algorithm. In figure 3 we plot the performance of the hybrid GTC against the standard EM. We considered the clustering tasks with 3 to 8 clusters for all stocks of the DJA during year 2000. Similar results were obtained for the remaining years.

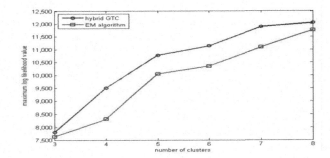

Fig. 3. Hybrid GTC against the standard EM algorithm for clusterization

We can note that the hybrid GTC algorithm leads toward higher values of the maximum likelihood measure. More, the outcomes of the hybrid GTC algorithm are more stable and homogeneous in sense of the cluster components. For example, running 30 times the hybrid GTC algorithm for learning three clusters, we obtain the same structural division of the stocks, with very few examples which will be discussed later on the paper. But, when running the standard EM algorithm, the stocks distribution on the clusters changes from run to run. This is a clear sign that the hybrid GTC succeeds to avoid local optima and there is some intrinsic linkage in the data under analysis that force the stocks to distribute well into clusters. This outcome remains valid even when we run the clusterization algorithms for more than 3 clusters and for any of the considered years.

Out of figure 3 we note that the slope of the curve for the relevant hybrid GTC clustering decreases when the number of clusters exceeds 5. Figure 4 depicts the outcomes of the hybrid GTC algorithm for 3 to 8 clusters. Each subfigure plots

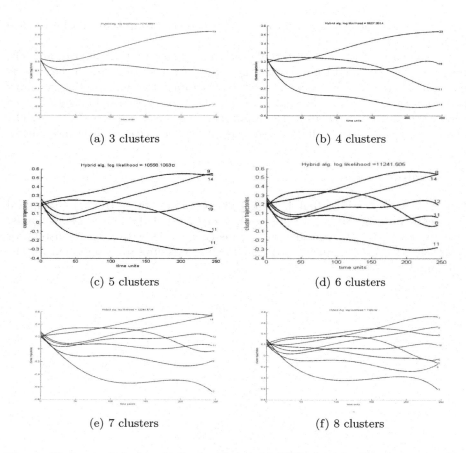

(a) 3 clusters   (b) 4 clusters

(c) 5 clusters   (d) 6 clusters

(e) 7 clusters   (f) 8 clusters

Fig. 4. Cluster trajectories obtained with hybrid GTC algorithm for various number of clusters

the cluster trajectories. Further, we note that for the experiments with more that 5 clusters, there are some clusters with very few individuals and the trajectories of those clusters are similar with the ones of the remaining clusters. Thus, we conclude that clustering experiments up generating up to 5 clusters are relevant.

In table 2 we show the clusters obtained when searching for a 5-bins clusterization. In table 3 we show how well maps the 5-bins clusterization on the sectoral DJA subindexes. We note that clusters 1 and 3 map well to the industry sector, cluster 5 goes to utilities while cluster 2 is a mixture of transportation and utilities and cluster 4 a mixture of industry and transportation. Also, we can note that the transportation sector is highly heterogeneous and utilities is the most homogeneous. As clusters 2 and 5 reports the highest performances (see figure 4c), we can learn that year 2000 was a good one for the utilities sector. The only stocks that exhibited another behavior are ED (Consolidated Edison) and D (Dominion Resources), both on the electricity business.

Table 2. Clusters' components for the 5-bins clustering task

Cluster no.	No. of stocks	Cluster components
1	11	AA, CAT, CSX, DD, EIX, HD, MCD, MSFT, NSC, PG, WMT
2	14	AES, C, CHRW, CNP, D, DUK, EXC, GMT, LSTR, LUV, OSG, PCG, PFE, WMB
3	11	BAC, CNW, DIS, GM, HPQ, IBM, INTC, JBHT, JPM, R, VZ
4	19	AMR, AXP, BNI, CAL, CVX, ED, EXPD, FDX, GE, JNJ, KO, MMM, MRK, T, UNP, UPS, UTX, XOM, YRCW
5	9	AEP, AIG, ALEX, BA, FE, FPL, NI, PEG, SO

Table 3. Maping of the 5-bins clusters to the sectoral DJA subindexes

DJA subindex	Cluster 1	Cluster 2	Cluster 3	Cluster 4	Cluster 5
Industry	8	2	8	10	2
Transportation	2	5	3	8	1
Utilities	1	7	0	1	6

To prove how strong is the cohesion between the individuals classified in the same cluster, for each cluster we computed the pairwise distance between the individuals, using the Euclidean distance [7]. Based on the pairwise distances, we depicted a dendrogram containing all stocks (figure 5) and 5 dendrograms for each cluster in the above-presented clusterization (figure 6). From figure 5 we can clearly note that the following stocks go together in clusters and these cluster kernels are learned by the GTC clusterization: {BNI, UNP, MMM, UPS, KO}, {AA, DD, CAT, CSX, MCD, WMT}, {D, EXC, DUK} and {AEP, FE, PEG, AIG, ALEX}. From figure 6 we can infer which are the stocks that are less related with the cluster, situated at the boundary of the cluster and prone to be classified in another class by another clusterization method: stock PG in cluster 1, stocks CNP and OSG in cluster 2, stocks HPQ and INTC in cluster 3; clusters 4 and 5 are almost homogeneous, the inside distances between all stocks being almost the same. We can note a strong adhesion inside clusters 4 and 5 and the cluster 2 is the one with the individuals highly spread out on the space.

We applied the procedure presented above on each year from 2000 and 2007. Table 4 shows the synthesized results, indicated the optimal number of clusters and the corresponding log likelihood value. We can note that in general, we obtain more clusters than the number of sub-indexes, which clearly show and sustain the conclusions we draw out from the cluster analysis for year 2000: the DJA sub-indexes contains heterogeneous stocks.

We note that the only year that plays distinct is year 2003, which is characterized by a strong rebound of the stock market after the dot com crisis.

Regarding the distribution of the stocks on the sectors, we note that the utilities sub-index is the most homogeneous one, and usually the stocks in the electricity moved together. IT-industry stocks (IBM, HPQ, VZ, INTC) with

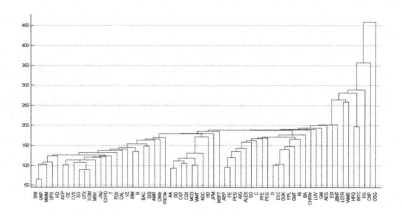

Fig. 5. Dendrogram with all stocks showing the adhesion between all individuals in the population

Table 4. Number of clusters obtained for each year using the hybrid GTC algorithm

Year	No. of clusters	Log likelihood
2000	5	10779.22
2001	5	14632.35
2002	5	12763.99
2003	3	19978.61
2004	5	15974.88
2005	4	14582.66
2006	5	12092.22
2007	5	11087.16

the exception of MSFT usually go together on the same cluster and MSFT, being very volatile represents the exception. Also, the financial sector (AIG, C, JPM, BAC, GE) is heterogeneous, being impossible to get these stocks connected on the same cluster on the long term.

5   Related Work

Hybrid Gene Trajectory Clustering was developed by [1] with the intention to cluster multi-dimensional gene data. Although the method is highly computational intensive, it was successfully applied on bio-informatics [2].

In the computational financial analysis, learning the stock markets sector is one challenge. Growing Neural Gas was applied to produce a tree linking together the FTSE stocks that exhibit a similar behavior [8]. Gavrilov et al. [7] tries to

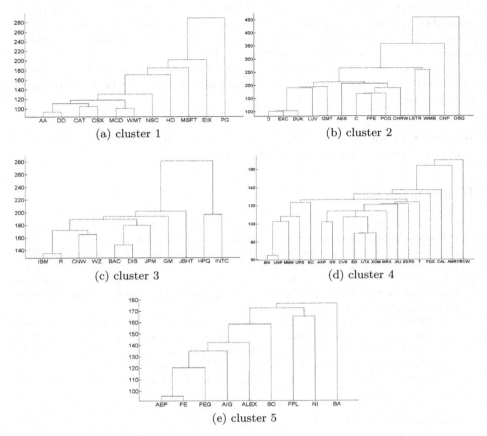

Fig. 6. Dendrograms for the clusters, showing the adhesion between the individuals inside cluster

identify which distance measure is the best to characterize the similarity between stock movements.

More, each stock market has its own indexes which are said to characterize the market. But, investing on global indexes without a further in-depth analysis might be hazardous because the index structure is usually composed of many stocks that exhibit contracting behaviors. Thus, investment houses tries to build proprietary indexes and to increase their performance. These indexes perform a clustering of the stocks, but this clustering is subjective and the investor has to trust or not the owner of the index.

6   Conclusion

In this paper we applied the hybrid gene trajectory clustering algorithm [1,2] to characterize the division of the NYSE stocks in the Dow Jones Composite

Average index. For each stock in the DJA index, we took daily closing prices from 2000 to 2007 and on each year we applied the clusterizaton method. Our target was to identify those stocks that go together on the medium term and to see if there is any sectoral linkage between them.

We concluded that the hybrid GTC algorithm succeeds to perform a stable division of the stocks and interesting financial relations can be learned out of the clustering.

Acknowledgements. This work is supported by the Romanian Authority for Scientific Research under project IDEL573.

References

1. Chan, Z.S.H., Kasabov, N.K.: Gene trajectory clustering with a hybrid genetic algorithm and expectation maximization method. In: Proceedings of 2004 IEEE International Joint Conference on Neural Networks, 2004, vol. 3, pp. 1669–1674. IEEE Computer Society, Los Alamitos (2004)
2. Chan, Z.S.H., Kasabov, N.K., Collins, L.: A hybrid genetic algorithm and expectation maximization method for global gene trajectory clustering. J. Bioinformatics and Computational Biology 3(5), 1227–1242 (2005)
3. Schoenfeld, S.A.: Active Index Investing: Maximizing Portfolio Performance and Minimizing Risk Through Global Index Strategies, 1st edn. Wiley, Chichester (2004)
4. Gaffney, S., Smyth, P.: Trajectory clustering with mixtures of regression models. Technical report, No. 99-15, University of California, Irvine (1999)
5. McLachlan, G., Krishnan, T.: The EM Algorithm and Extensions. John Wiley and Sons, Chichester (1997)
6. Elton, E.J., Gruber, M.J., Brown, S.J., Goetzmann, W.N.: Modern Portfolio Theory and Investment Analysis. Wiley, Chichester (2006)
7. Gavrilov, M., Anguelov, D., Indyk, P., Matwani, R.: Mining the stock market (extended abstract): which measure is best? In: KDD 2000: Proceedings of the sixth ACM SIGKDD international conference on knowledge discovery and data mining, pp. 487–496. ACM, New York (2000)
8. Doherty, K.A., Adams, R.G., Davey, N., Pensuwon, W.: Hierarchical topological clustering learns stock market sectors. In: ICSC Congress on Computational Intelligence Methods and Applications, 2005. IEEE Computer Society, Los Alamitos (2005)

Accurate Prediction of Financial Distress of Companies with Machine Learning Algorithms

Armando S. Vieira[1], João Duarte[1], Bernardete Ribeiro[2], and João C. Neves[3]

[1] ISEP, Rua de S. Tomé, 4200 Porto, Portugal
[2] Depart. of Informatics Engineering, University of Coimbra, P-3030-290 Coimbra, Portugal
[3] ISEG - School of Economics, Rua Miguel Lupi 20, 1249-078 Lisboa, Portugal
asv@isep.ipp.pt

Abstract. Prediction of financial distress of companies is analyzed with several machine learning approaches. We used Diane, a large database containing financial records from small and medium size French companies, from the year of 2002 up to 2007. It is shown that inclusion of historical data, up to 3 years priori to the analysis, increases the prediction accuracy and that Support Vector Machines are the most accurate predictor.

1 Introduction

One of the most important threats for business is the credit risk associated with counterparts. The rate of bankruptcies have increased in recent years and its becoming harder to estimate as companies become more complex and develop sophisticated schemes to hide their real situation. Due to the recent financial crisis and regulatory concerns, credit risk assessment is a very active area both for academic and business community. The ability to discriminate between faithful customers from potential bad ones is thus crucial for commercial banks and retailers.

The problem of bankruptcy prediction can be addressed as follows: given a set of financial ratios describing the situation of a company over a given period, predict the probability that this company may become bankrupt in a near future, normally during the following year.

Prediction of financial distress of companies with financial ratios has been addressed by several models. Despite all its limitations, Linear Discriminant Analysis is still largely used as a standard tool for bankruptcy prediction [1,2]. In particular, versions of the Logistic model [3] are widely used by credit ranking agencies.

In previous works we have shown that some recent machine learning approaches, like Genetic Algorithms and Support Vector Machines, are able to achieve superior accuracy in early detection of bankruptcy [4, 5]. For a review on the application of machine learning algorithms to financial distress prediction of companies see [6-8].

For this study we use a large database of French companies. This new database is very detailed containing information on a wide set of financial ratios spanning over a period of several years. It contains up to three thousands distressed companies and about sixty thousands healthy ones. We used this dataset to compare the efficiency of the Logistic model with other machine learning algorithms, namely: Support Vector Machines, Neural Networks and AddaboostM1.

M. Kolehmainen et al. (Eds.): ICANNGA 2009, LNCS 5495, pp. 569–576, 2009.
© Springer-Verlag Berlin Heidelberg 2009

In order to make predictions more accurate we tested the models with data from three previous years priori to failure. It is shown that inclusion of these historical records can boost precision and robustness of the classifiers, particularly in early detection.

This paper is organized as follows: Section 2 describes the dataset, Section 3 presents the models used, Section 4 contains the results obtained and in Section 5 the conclusions are presented.

2   The Dataset

We used a sample obtained from Diane, a database containing financial statements of French companies. The initial sample consisted of financial ratios of about 60 000 industrial French companies, for the years of 2002 to 2006, with at least 10 employees. From these companies, about 3000 were declared bankrupted in 2007 or presented a restructuring plan ("Plan de Redressement") to the court for approval by the creditors. We decided not to distinguish these two categories as both signals companies in financial distress.

The dataset includes information about 30 financial ratios defined by COFACE of the companies covering a wide range of industrial sectors.

2.1   Preprocessing and Feature Selection

Our database contains many cases with missing values, especially for defaults companies. For this reason we sorted the default cases by the number of missing values and selected the examples with 10 missing values at most. A final set of 600 default examples was obtained. In order to obtain a balanced dataset we selected randomly 600 non-default examples resulting in a set of 1200 examples.

The remaining missing data was treated as follows. For the ratios of the years 2003 and 2006 each missing value was replaced by the value of the closest available year; for 2004 and 2005, if values of the next and previous years were available, each missing value was replaced by their mean, otherwise it was replaced by the remaining value. In some cases there was no data available for a ratio in any of the years. In this very few cases the missing data was replaced by the median value of the ratio in each year. Finally, all ratios were logarithmized and then standardized to zero mean and unity variance.

The 30 financial ratios produced by COFACE are described in Table 1. These ratios allow a very comprehensive financial analysis of the firms including the financial strength, liquidity, solvability, productivity of labor and capital, margins, net profitability and return on investment. Although, in the context of linear models, some of these variables have small discriminatory capabilities for default prediction, the non-linear approaches here used may extract relevant information contained in these ratios to improve the classification accuracy without compromising generalization.

Due to the large number of attributes available, we used several ranking algorithms to select the most relevant. We used the following methods: SVM Attribute evaluation, Chisquared, Consistency Subset, GainRatio. Before running these algorithms we evaluated the correlation matrix and found that the majority of attributes are only slightly correlated. For instance, Principal Component Analysis will choose 23 components out of 30.

In SVMA [13] attributes are ranked by the square of the weight assigned by the SVM. Attribute selection is handled by ranking attributes for each class separately using a one-vs-all method and then "dealing" from the top of each pile to give a final ranking.

In Chisquared algorithm the worth of an attribute in calculated by computing the chi-squared statistic with respect to the class. Consistency Subset [9] evaluates the worth of a subset of attributes by the level of consistency in the class values when the training instances are projected onto the subset of attributes. GainRatio evaluates the worth of an attribute A by measuring the information gain ratio with respect to the class.

2.2 Historical Data

A company is a dynamic entity, subjected to fluctuation of the market, economy cycles and unavoidable contingencies related to its business activity. Therefore, yearly variations of important financial ratios reflecting the balance sheet, sometimes quite relevant, are common particularly for small companies. Yearly variations of over 50% in some ratios are not atypical.

In order to accommodate these fluctuations, we decided to use an extended record from years preceding the default. However care must be taken in choosing the relevant information. If many years are used, we increase the complexity of the problem and may obscure the present situation of the company by averaging over a remote past. On the other hand if few years are used we may not properly characterize the company background. In this study we considered data from 3 years priori to the bankruptcy event.

This adds complexity to the analysis as the number of inputs is increased three fold - from 30 to 90 ratios. Furthermore, we found that more relevant than the ratios themselves, are the variations that occurs over the period range of the analysis.

We consider the following parameters: ratios of the current year, R_0^i, ratios from previous year, R_1^i, and fluctuations of the ratios over the period considered, R_2^i. They are defined as:

$$R_0^i = R^i$$
$$R_1^i = R^{i-1}$$
$$R_2^i = \frac{\max(R^i) - \min(R^i)}{\overline{R}}$$

where $\overline{R}$ is the ratio average over the period considered. The variables selected by the feature ranking algorithms are presented in Table 1. Note that many of selected attributes are the ratios variations over the period, in this case three years. Features selected by different algorithms differ considerably meaning that important correlations can exist in the ratios.

Table 1. Ratios used with historical data set. Selection procedure: [SVMA - Top 20 (**I**), Top 15 (**II**), Top 10 (**III**), Top 5 (**IV**)], [Cfs Subset Evaluation - Greedy (**V**), Genetic (**VI**)], [Chisquared: Top 20 (**VII**)], [Consistency Subset: Greedy (**VIII**), Genetic (**IX**)] and [GainRatio: Top 20 (**X**)]. The labels means: 0 – current year, 1 – previous year, 2 – variation over three previous years.

#	Designation	I	II	III	IV	V	VI	VII	VIII	IX	X
1	Number of employees	0	0	0	0	0	0		0	0	
2	Financial Debt / Capital Employed %										
3	Capital Employed / Fixed Assets	2	2	2							
4	Depreciation of Tangible Assets (%)	0	0		0	0					
5	Working capital / current assets						0		2	1,2	
6	Current ratio	0					2		0		
7	Liquidity ratio	1				0,2	0	0	0,2	0,2	
8	Stock Turnover days						2			0	
9	Collection period										
10	Credit Period						2				
11	Turnover per Employee	0,2	0,2	0,2	0,2	0,2	0,2	2	2		2
12	Interest / Turnover	0	0							0	
13	Debt Period (days)	2	2	2						2	
14	Financial Debt / Equity (%)					0	0	0			0
15	Financial Debt / Cashflow	0	0	0		0,2	0,2	0,2			0,2
16	Cashflow / Turnover (%)	0,1	0,1	0	0	0,2	0,2	0,2	2		0,2
17	Working Capital / Turnover (days)	0,1	0,1	1			2			0	
18	Net Current Assets/Turnover (days)	0	0								

#		C1	C2	C3	C4	C5	C6	C7	C8	C9	C10
19	Working Capital Needs / Turnover (%)	2									
20	Export (%)					2			1		
21	Value added per employee					2	0,2		1	0	
22	Total Assets / Turnover					2	2		2		
23	Operating Profit Margin (%)						0,2	0,2		0,2	0,2
24	Net Profit Margin (%)						0,2	0,2			0,2
25	Added Value Margin (%)								0		
26	Part of Employees (%)					0	0,1	0		1	0
27	Return on Capital Employed (%)	2				0	0,2	0,2		1,2	0,1,2
28	Return on Total Assets (%)	0	0	0	0	0,2	0	0,2	0		0,2
29	EBIT Margin (%)							0,2		2	0,2
30	EBITDA Margin (%)	0					0	0,2			0,2

3 Models Used and Results

We analyze the data with four machine learning algorithms: Logistic, Neural Networks with a Multilayer Perceptron (MLP), Support Vector Machines (SVM) [10, 11] and AdaBoost M1.

For MLP, we used a neural network trained with backpropagation and one hidden layer with a number of neurons defined by: $(number\_ratios + 1) / 2$. The learning rate was set to 0.3 and the momentum to 0.2.

For the C-SVM algorithm we used the LibSVM [12] library with a radial basis function as kernel with the cost parameter $C = 1$ and a shrinking heuristic.

For AdaBoost M1 algorithm we used a Decision Stump as weak learner and set the number of iterations to 100. No resampling was used.

First we compare the efficiency of the classifiers using data from a single year. Table 2 presents the results obtained when all ratios are used. We used 10-fold cross validation in all classifiers.

Support Vector Machines achieved the highest accuracy, 92.42% and the lowest error types. For 2005, two years before bankruptcy, the Adaboost retrieved the best results. It is remarkable such a high accuracy taking into account the fluctuations on ratios occurring from one year to the other. The highest error is type II, as expected. Neural Networks (MLP) were the worst classifier due to the large dimensionality of the training data exposing it to the corresponding risk of overfitting.

Table 2. Accuracy and error types I and II for different models in 2007 using data from the previous year (2006) and from 2005. All 30 variables were used.

	Classifier	Accuracy	Type I	Type II
2006	Logistic	91.25	6.33	11.17
	MLP	91.17	6.33	11.33
	C-SVM	**92.42**	**5.16**	**10.00**
	AdaboostM1	89.75	8.16	12.33
	Classifier	**Accuracy**	**Type I**	**Type II**
2005	Logistic	79.92	**19.50**	20.67
	MLP	75.83	24.50	23.83
	C-SVM	**80.00**	21.17	**18.83**
	AdaboostM1	78.17	20.50	23.17

Most default prediction models use a small set of financial ratios, between 5 and 10, usually from a single year, to quantifying the profitability, cashflow and liabilities of the company. Since we have a large pool of data, we tested the models with several sets of attributes. The first, containing the top 5 attributes, selected by SVMA, the second the top 10, the third the top 15, the fourth the top 20 and finally the fifth with all the 90 attributes.

The results are presented in Table 3. The best accuracy (94%) was obtained again with SVM using 20 variables, which means a reduction of about 30% in type I error and 20% in type II error. This improvement is justified by the fact that more data is used. With the 5 top ratios we achieved a performance similar to the previous dataset with all ratios included.

The Adaboost algorithm is the least sensitive to overfiting and therefore is relatively immune to the curse of dimensionality when using the full set of attributes. In practice it is unwise to use the full set of attributes and the accuracy has to be sacrificed to simplicity. For the top 20 ratios, selected by SVMA, the Logistic model achieved again an accuracy very close to SVM.

The importance of using a large set of ratios is clearly evident on prediction two years before bankruptcy (Table 4). In this case, the accuracy of SVM increased substantially from 76.42% to 81.42%.

Table 3. Accuracy in predicting failures during 2007 using data from 2006, 2005 and 2004

	# variables				
Classifier	**5**	**10**	**15**	**20**	**ALL**
Logistic	91.17	**93.33**	**93.42**	93.58	92.25
MLP	91.33	93.25	93.08	93.17	92.50
C-SVM	**91.67**	**93.33**	**93.42**	**94.00**	**93.17**
AdaboostM1	90.50	**93.33**	93.00	92.25	91.58

Table 4. Accuracy in predicting failures during 2007 using data from 2005, 2004 and 2003

Classifier	# variables				
	5	10	15	20	ALL
Logistic	75.83	77.58	79.25	79.33	79.00
MLP	75.83	75.00	76.17	75.33	77.42
C-SVM	76.42	76.58	78.83	78.41	**81.42**
AdaboostM1	**77.08**	**77.83**	**79.92**	**80.33**	81.33

4 Conclusions

In this work it is shown that bankruptcy of small and medium size companies can be accurately predicted if a detailed training dataset is available. Of all the models tested Support Vector Machines achieved the best performance, but all approaches show comparable results.

We show that inclusion of information from previous years before default, especially fluctuations of relevant ratios, like debt to cash-flow, is crucial to achieve a good precision. Furthermore, we proved that the use of larger sets of inputs in the classifier can reduced both error types by up to 30%.

In future work we will consider inclusion of more years and the use of more efficient feature selection algorithms.

Acknowledgements. The authors like to thank "Fundação da Ciência e Tecnologia" for financial support under the project PTDC/GES/70168/2006.

References

1. Altman, E.I.: Financial Ratios, Discriminant Analysis and the Prediction of Corporate Bankruptcy. Journal of Finance 23, 589–609 (1968)
2. Eisenbeis, R.A.: Pitfalls in the Application of Discriminant Analysis in Business, Finance and Economics. Journal of Finance 32(3), 875–900 (1977)
3. Martin, D.: Early Warning of Bank Failure: A Logit Regression Approach. Journal of Banking and Finance 1, 249–276 (1977)
4. Vieira, A.S., Neves, J.C.: Improving Bankruptcy Prediction with Hidden Layer Learning Vector Quantization. Eur. Account. Rev. 15(2), 253–271 (2006)
5. Vieira, A.S., Ribeiro, B., Mukkamala, S., Neves, J.C., Sung, A.H.: On the Performance of Learning Machines for Bankruptcy Detection. In: IEEE International Conference on Computational Cybernetics, Vienna, Austria, August 30 – September 1, pp. 223–227 (2004)
6. Atiya, F.: Bankruptcy prediction for credit risk using neural networks: A survey and new results. IEEE Trans. Neural Net. 4, 12–16 (2001)
7. Udo, G.: Neural Network Performance on the Bankruptcy Classification Problem. Computers and Industrial Engineering 25, 377–380 (1993)
8. Grice, J.S., Dugan, M.T.: The limitations of bankruptcy prediction models: Some cautions for the researcher. Rev. of Quant. Finance and Account. 17(2), 151 (2001)

9. Liu, H., Setiono, R.: A probabilistic approach to feature selection - A filter solution. In: 13th International Conference on Machine Learning, pp. 319–327 (1996)
10. Vapnik, V.: The Nature of Statistical Learning Theory. Springer, Heidelberg (1995)
11. Cortes, C., Vapnik, V.: Support vector networks. Machine Learning 20, 273–297 (1995)
12. Chang, C.-C., Lin, C.-J.: LIBSVM a library for support vector machines, Technical Report, Department of Computer Science and Information Engineering, National Taiwan University, Taipei, Taiwan (2000)
13. Guyon, I., Weston, J., Barnhill, S., Vapnik, V.: Gene selection for cancer classification using support vector machines. Machine Learning 46, 389–422 (2002)
14. Guyon, I., Gunn, S., Nikravesh, M., Zadeh, L.: Feature Extraction Foundations and Applications. In: Studies in Fuzziness and Soft Computing. Springer, Heidelberg (2006)

Approximation Scheduling Algorithms for Solving Multi-objects Movement Synchronization Problem

Zbigniew Tarapata

Military University of Technology, Cybernetics Faculty,
Gen. S. Kaliskiego Str. 2, 00-908 Warsaw
zbigniew.tarapata@wat.edu.pl

Abstract. The paper presents some models and algorithms for the nonlinear optimization problem of multi-objects movement scheduling to synchronize their movement as well as properties of presented algorithms. Similarities and differences between defined problems and the classical tasks scheduling problem on parallel processors are shown. Two algorithms for synchronous movement scheduling are proposed and their properties are considered. One of the algorithm is based on dynamic programming approach and the second one uses some approximation techniques. Theoretical and experimental analysis of complexity and effectiveness of the algorithms as well as their practical usefulness are discussed.

1 Introduction

Scheduling of object movement is an essential element of numerous systems: for routing in computer networks, for movement planning of mobile robots [3], for tasks processed inside distributed or parallel computing system [2], [6], etc. A special type of movement is such that objects must be moved simultaneously [11] and a special type of system with this requirement is a system for movement planning and simulation of military objects (units) in combat simulators [7]. Movement scheduling has an influence on accuracy, adequateness, effectiveness and other characteristics of these systems. Afterwards, the problem is to model and optimize such movements of detachments as to achieve intended goals of commands (such as: achievement of destinations on restricted time, avoiding losses during redeployment etc.) [9]. One of the techniques of providing the simulated opponent is to use a computer system that generates and controls multiple simulation entities using software. Such a system is known as a semi-automated force (SAF or SAFOR) or a computer generated force (CGF) [8]. Regardless of the kind of military actions, military objects are moved according to a group pattern. From the point of view of mission realization, preservation of group pattern during military actions is very important. For example, each object being moved in a group (e.g. during attack, during redeployment) must keep specific distances between each other inside the group [7]. Therefore, the paper presents a few problems of movement scheduling for many objects to synchronize their movement and algorithms for solving them with theoretical and experimental analysis.

M. Kolehmainen et al. (Eds.): ICANNGA 2009, LNCS 5495, pp. 577–589, 2009.
© Springer-Verlag Berlin Heidelberg 2009

2 Scheduling Models of Synchronous Movement

2.1 Notations and Definitions

Let us assume that we have a directed graph G that defines structure of the terrain (divided into squares, hexagons, etc.) [8], [9], $G = \langle V_G, A_G \rangle$, $V = |V_G|$, V_G – set of graph nodes (as centre of terrain squares, crossroads), A_G – set of graph arcs, $A_G \subset V_G \times V_G$, $A = |A_G|$. On each arc we have a defined value $d_{n,n'}$ of function d, which describes the terrain distance between the graph nodes n and n'. K objects (columns, trucks, tasks) move from source nodes vector $s = (s_1, s_2, ..., s_K)$ to destination nodes vector $t = (t_1, t_2, ..., t_K)$ of G. For further discussion we accepted the following notations:

$$I_k(s_k, t_k) = I_k = \left(i^0(k) = s_k, \, i^1(k), \, \ldots, \, i^r(k), \, \ldots, \, i^{R_k}(k) = t_k \right) \tag{1}$$

$$T_k(I_k) = T_k = \left(\tau^0(k), \tau^1(k), ..., \tau^r(k), ..., \tau^{R_k}(k) = \tau(I_k) \right) \tag{2}$$

$$V_k(I_k) = V_k = \left(v_{i^0(k), i^1(k)}, v_{i^1(k), i^2(k)}, \, \ldots, \, v_{i^{R_k-1}(k), i^{R_k}(k)} \right) \tag{3}$$

where I_k - vector of nodes describing the path for the k-th object, $\underset{m \in \{1,...,R_k\}}{\forall} \left(i^{m-1}(k), i^m(k) \right) \in A_G$; $i^r(k)$ - the r-th node on the path for the k-th object; s_k, t_k – source and destination nodes for the k-th object; T_k - vector of time instances of achieving the nodes belonging to the path for the k-th object; $\tau^r(k)$ - time instance of achieving node $i^r(k)$ by the head of the k-th object, $\underset{k=1,K}{\forall} \underset{r=0,R_k-1}{\forall} \tau^{r+1}(k) \geq \tau^r(k) \geq 0$ and $\underset{k=1,K}{\forall} \tau^0(k) = 0$; $\tau^{R_k}(k) = \tau(I_k)$ - time of achieving destination node by the k-th object; V_k - vector of velocities $v_{i^r(k), i^{r+1}(k)}$ of the k-th object on the arc $\left(i^r(k), i^{r+1}(k) \right)$ of its path; R_k - number of arcs belonging to the path of the k-th object. For the set $\Pi(s,t)$ describing the set of vectors $I(s,t)$ of paths from $s = (s_1, s_2, ..., s_K)$ to $t = (t_1, t_2, ..., t_K)$ we have defined time τ^* as the earliest time of achieving the destination node by the most delayed object:

$$\tau^* = \min_{I(s,t)=(I_1, I_2, ..., I_K) \in \Pi(s,t)} \max_{k \in \{1,...,K\}} \tau(I_k) \tag{4}$$

Let k^* denotes index of object for which the moment of achieving destination node for its path is the latest among paths for other objects, i.e. $k = k^* \Leftrightarrow \tau^{R_{k^*}}(k^*) = \max_{k \in \{1,...,K\}} \tau^{R_k}(k)$. Let $IP_k = \{i_1(k), i_2(k), ..., i_p(k), ..., i_{P_k}(k)\}$ denotes a set of nodes (checkpoints) at which we must align the head of the k-th object in relation to the heads of other objects, where $i_p(k)$ - the p-th element of IP_k satisfying: $\underset{p=1,P_k}{\forall} \underset{r \in \{1,...,R_k\}}{\exists} i_p(k) = i^r(k)$ and $r_p(k) = r \in \{1,...,R_k\} \Leftrightarrow i_p(k) = i^r(k)$. The form of IP_k and $r_p(k)$ indicate that the path for the k-th object must cross by nodes belonging to IP_k.

Let, by analogy $TP_k = \left(\tau_1(k),\ \tau_2(k),...,\ \tau_p(k),...,\ \tau_{P_k}(k) \right)$ denotes ordered set of time instances of achievement particular alignment nodes from set IP_k by the k-th object head, $\tau_p(k)$ denotes moment of achieving the p-th alignment node by the k-th object,

$$\tau_p(k) = \tau^0(k) + \sum_{r \in \{0,...,r_p(k)-1\}} \frac{d_{i^r(k),i^{r+1}(k)}}{v_{i^r(k),i^{r+1}(k)}} \tag{5}$$

Additionally, we made the assumption that $P_1 = P_2 = ... = P_K = N$, i.e. for all objects exist the same number of alignment points (nodes). Let us define for each $p=1,..,N$ the following characteristics:

$$\tau_p^{max} = \max_{k \in \{1,...,K\}} \tau_p(k) \ , \qquad \tau_p^{avg} = \frac{1}{K}\sum_{k=1}^{K} \tau_p(k) \tag{6), (7}$$

2.2 Formulation of Optimization Problems for Movement Synchronization

One of the formulations of optimization problem for movement synchronization of K objects can be defined as follows: for fixed paths I_k of each k-th object to determine such $v_{i^r(k),i^{r+1}(k)}$, $r = \overline{0,R_k - 1}$, $k = \overline{1,K}$ that

$$\sum_{p=1}^{N} \sum_{k=1}^{K} \left(\tau_p^{max} - \tau_p(k) \right) \rightarrow \min \tag{8}$$

with constraints:

$$v_{i^r(k),i^{r+1}(k)} \leq v^{max}(k), \qquad r = \overline{0,R_k - 1}, \ k = \overline{1,K} \tag{9}$$

$$v_{i^r(k),i^{r+1}(k)} > 0, \qquad r = \overline{0,R_k - 1}, \ k = \overline{1,K} \tag{10}$$

where $v^{max}(k)$ describes maximal velocity of the k-th object resulting from its technical properties. Taking into consideration (5) and (6) we can write (8) as follows:

$$\sum_{p=1}^{N} \sum_{k=1}^{K} \left(\max_{j \in \{1,...,K\}} \left(\tau^0(j) + \sum_{\substack{r \in \{0,...,R_j - 1\} \\ r \leq r_p(j)}} \frac{d_{i^r(j),i^{r+1}(j)}}{v_{i^r(j),i^{r+1}(j)}} \right) - \left(\tau^0(k) + \sum_{\substack{r \in \{0,...,R_k - 1\} \\ r \leq r_p(k)}} \frac{d_{i^r(k),i^{r+1}(k)}}{v_{i^r(k),i^{r+1}(k)}} \right) \right) \rightarrow \min \tag{11}$$

Path I_k for the k-th object may be disjoint or not and must cross at fixed alignment points or we have to dynamically determine these points (e.g. during movement simulation/realization). In the first case we have NP-hard optimization problem and we can solve it using an approximation algorithm for finding disjoint paths [10]. In the second case we can use a two-stage approach: (*) finding the best paths for K objects iteratively using methods for finding the m-th (1st, 2nd, 3rd, etc.) best path for each of the K objects [4] and visiting specified nodes [5]; (**) synchronizing movement of K objects by solving problem (8)-(10) and using algorithms described in section 3 [13]. We can consider one of extensions of problem (8)-(10): adding constraint as follows

$$\tau^0(k) + \sum_{r \in \{0,...,R_k - 1\}} \frac{d_{i^r(k),i^{r+1}(k)}}{v_{i^r(k),i^{r+1}(k)}} \leq T^{max}, \quad k = \overline{1,K} \tag{12}$$

we will find such a movement schedule that achieving the moment of destination node by the latest object is no greater than $T^{\max} \geq \tau^*$.

To solve the problem (8)-(10) with the additional constraint (12), in generality, we define this problem in its changed form: for fixed paths I_k of each k-th object to determine such $x_{k,p}$, $k=1,...,K$, $p=1,...,N$ that:

$$\sum_{p=1}^{N}\sum_{k=1}^{K}\left(\max_{j\in\{1,...,K\}}\left(\tau_p(j)+\sum_{i=1}^{p}x_{j,i}\right)-\left(\tau_p(k)+\sum_{i=1}^{p}x_{k,i}\right)\right)\rightarrow\min \tag{13}$$

with constraints:

$$\sum_{p=1}^{N}x_{k,p}\leq FT(k), \qquad k=1,...,K \tag{14}$$

$$x_{k,p}\geq 0, \quad k=1,...,K, \quad p=1,...,N \tag{15}$$

where $x_{k,p}$ describes time instance which is added to $\tau_p(k)$ for the k-th object in its p-th alignment point (node). It can be observed that $\tau_p(k)+\sum_{i=1}^{p}x_{k,i}=\tau'_p(k)$. Therefore, if we denote $\Delta\tau_p'^{\max}(k)=\tau_p'^{\max}-\tau'_p(k)$, where $\tau_p'^{\max}$ is defined like in (6), then function (13) has an equivalent form of $\sum_{p=1}^{N}\sum_{k=1}^{K}\Delta\tau_p'^{\max}(k)\rightarrow\min$ and we obtain (8). Free time $FT(k)$ for the k-th object we define as: $FT(k)=T^{\max}-\tau^{R_k}(k)$.

We can observe that problem (8)-(10) is similar to a problem of task scheduling on parallel processors [2], [6]. There are the following similarities: (a) scheduling the problem before critical lines to minimize the sum of maximal delays in alignment points (nodes); the p-th critical line is created by nodes $i_p(1), i_p(2),...,i_p(K)$; (b) we have parts of the path (arcs) as tasks; (c) we have moved objects as processors (K); (d) tasks are indivisible and dependent (the dependence is defined by each of the arc $\forall_{m\in\{1,...,R_k\}}\left(i^{m-1}(k), i^m(k)\right)\in A_G$ belonging to the path for each of the object). Differences: (a) tasks (arcs of the path) are assigned to processors (objects) (we have no influence on this assignment) and we decide only on the delays of operation of processors (to increase realization time of tasks).

3   Scheduling Algorithms for Movement Synchronization

Two movement scheduling algorithms are presented: the first one ($A.1$) for solving the problem (8)-(10) and the second one ($A.2$) for solving the problem (13)-(15). Let us denote by $\tau'_p(k)$ modified (by algorithms) moment of achieving the p-th alignment point by the k-th object and $\Delta\tau'_p(k)=\tau'_p(k)-\tau_p(k)$.

3.1 Dynamic Programming Algorithm

Algorithm A.1

For each $p \in \{1, \ldots, N\}$ recurrently compute the modified moments of achieving alignment nodes for K objects:

$$\tau_p'(k) = \max_{j \in \{1,\ldots,K\}} \left(\Delta \tau_{p-1}'(j) + \tau_p'(j) \right), \quad \text{for } 1 \le k \le K \tag{16}$$

and in addition $\tau_0'(k) = \tau_0(k) = \tau^0(k)$, $1 \le k \le K$.

It is important that $A.1$ algorithm solves the problem (8)-(10) optimally (properties of the algorithm presented in [13]). Notice that $\underset{k \in \{1,\ldots,K\}}{\forall} \Delta \tau_p'(k) \ge 0$. It results from (16)

and from the assumption that $\underset{k=1,K}{\forall} \underset{r=0,R_k-1}{\forall} \tau^{r+1}(k) \ge \tau^r(k) \ge 0$. Having

$\underset{p \in \{1,\ldots,N\}}{\forall} \underset{k \in \{1,\ldots,K\}}{\forall} \tau_p'(k)$ and $\Delta \tau_p'(k)$ we can easily compute:

$\underset{k \in \{1,\ldots,K\}}{\forall} \underset{r \in \{0,\ldots,R_k\}}{\forall} \tau^{'r}(k) := \tau^r(k) + \Delta \tau_{q(r)}'(k)$, $q(r) = \max\left\{ p \in \{1,\ldots,N\} : r_p(k) \le r \right\}$ and

$\underset{k \in \{1,\ldots,K\}}{\forall} \underset{r \in \{0,\ldots,R_k\}}{\forall} v_{i^r(k),i^{r+1}(k)}' := \dfrac{d_{i^r(k),i^{r+1}(k)}}{\tau^{'r+1}(k) - \tau^{'r}(k)}$. The complexity of the algorithm $A.1$ is equal

to $\Theta(K^2 N)$ but we can obtain complexity $\Theta(KN)$ because for each $p \in \{1,\ldots,N\}$ $\tau_p'(1) = \tau_p'(2) = \ldots = \tau_p'(K)$.

3.2 Cost-Profit Approximation Algorithm

We can present the heuristic (greedy) algorithm $A.2$ which solves the problem (13)-(15) (it is equivalent to the problem (8)-(10) with (12)). We define the notations used inside the algorithm: $card(x)$ – strength of set x; $a_p(k)$ - time instance which is

added to $\tau_p'(k)$, $P_s^+(k) = \left\{ p \in \{s,\ldots,N\} : \Delta \tau_p^{max}(k) > 0 \right\}$,

$P_s^{\ge}(k) = \left\{ p \in \{s,\ldots,N\} : \Delta \tau_p^{max}(k) - a_p(k) \ge 0 \right\}$, $P_s^<(k) = \left\{ p \in \{s,\ldots,N\} : \Delta \tau_p^{max}(k) - a_p(k) < 0 \right\}$.

Functions $Z(\cdot)$ and $L(\cdot)$ describe „profit" and "cost" of decreasing $\Delta \tau_p^{max}(k)$ with value $a_{s_k}(k)$, $s_k \in P_{s_k}^+(k)$:

$$Z(a_{s_k}(k)) = a_{s_k}(k) \cdot card\left(P_{s_k}^{\ge}(k) \right) + \sum_{p \in P_{s_k}^<(k)} \Delta \tau_p^{max}(k) \tag{17}$$

$$L(a_{s_k}(k)) = (K-1) \cdot \sum_{p \in P_{s_k}^<(k)} \left| \Delta \tau_p^{max}(k) - a_{s_k}(k) \right| \tag{18}$$

Value $x_{k,p} := x_{k,p} + a_p(k)$ (in step 10) is equal to the sum of $a_p(k)$ values that are determined for all iterations of $A.2$ and for every k and p. The idea of algorithm $A.2$ consists of decreasing the value of $OBJ = \sum_{p=1}^{N} \sum_{k=1}^{K} \Delta \tau_p^{'max}(k)$ by decreasing the value of $\Delta \tau_p^{'max}(k)$ for any k and p.

Algorithm A.2

Given sets: I_k, T_k, IP_k, TP_k for each $k=1,\dots,K$ and values *ObjOrder*, *Strategy*;

Initialize: $\underset{k\in\{1,\dots,K\}}{\forall}\ \underset{p\in\{1,\dots,N\}}{\forall} a_p(k):=0$; $\underset{k\in\{1,\dots,K\}}{\forall}\ \underset{p\in\{1,\dots,N\}}{\forall} x_{k,p}:=0$; counter:=N;

$\underset{k\in\{1,\dots,K\}}{\forall} FT(k):=\tau^{R_{k^*}}(k^*)-\tau^{R_k}(k)$; $\underset{k\in\{1,\dots,K\}}{\forall}\ \underset{p\in\{1,\dots,N\}}{\forall} \Delta\tau_p^{'\max}(k):=\tau_p^{\max}-\tau_p(k)$;

1. WHILE $\left(\underset{k\in\{1,\dots,K\}}{\exists} FT(k)>0\right)\wedge(\text{counter}>0)$ DO

2.   counter:=0;

3.   To determine *KO* vector using *ObjOrder*;

3a.   FOR $k=KO[1],\dots,KO[K]$ DO

4.   IF $FT(k)>0$ THEN

5.   Use current *Strategy* to find s_k and $a_{s_k}(k)$;

6.   IF $a_{s_k}(k)>0$ THEN

7. $\underset{p\in\{s_k,\dots,N\}}{\forall}\Delta\tau_p^{'\max}(k):=\Delta\tau_p^{'\max}(k)-a_{s_k}(k)$;

8. $\underset{p\in P_k^<(k)}{\forall}\ \underset{j\in\{1,\dots,K\}}{\forall}\Delta\tau_p^{'\max}(j):=\Delta\tau_p^{'\max}(j)+\left|\Delta\tau_p^{'\max}(k)\right|$;

9. $FT(k):=FT(k)-a_{s_k}(k)$;

10. $x_{k,s_k}:=x_{k,s_k}+a_{s_k}(k)$;

11.   counter:=counter+1; $a_{s_k}(k):=0$;

12.   END IF;

13.   END IF;

14.   END WHILE;

15. END WHILE.

To set an examination order vector *KO* of K objects in *A.2* algorithm we use some object order *ObjOrder*$\in\{0,\dots,3\}$ strategy (step 3 of the algorithm): *ObjOrder*=0 – set elements of *KO* iteratively, from $k=1$ to $k=K$; *ObjOrder*=1 – set elements of *KO* randomly, with uniform distribution on the set $\{1,\dots,K\}$; *ObjOrder*=2 – set elements of *KO* iteratively, starting from such a k which corresponds to the first greatest, second greatest, ..., the K-th greatest values of vector *FT*; *ObjOrder*=3 – set elements of *KO* iteratively, starting from such a k which corresponds to the first smallest, second smallest,..., the K-th smallest values of vector *FT*.

To find values of $s_k \in P_1^+(k)$ and $a_{s_k}(k)\in\left(0,\ \min\{\Delta\tau_{s_k}^{'\max}(k),FT(k)\}\right]$ we use some *Strategy*$\in\{0,\dots,4\}$ (step 5 of the algorithm): *Strategy*=0 – find such a value s_k and maximal value $a_{s_k}(k)$ for which condition $Z(a_{s_k}(k))>L(a_{s_k}(k))$ is satisfied; *Strategy*=1 – find such a value s_k and value $a_{s_k}(k)$ for which value $Z(a_{s_k}(k))-L(a_{s_k}(k))$ is maximal and positive; *Strategy*=2 – find N times such a value s_k and randomly $a_{s_k}(k)$ for which value $Z(a_{s_k}(k))-L(a_{s_k}(k))$ is maximal and positive; *Strategy*=3 – find N times randomly such values s_k and $a_{s_k}(k)$ for which value

$Z(a_{s_k}(k)) - L(a_{s_k}(k))$ is maximal and positive;   *Strategy*=4 – like for *Strategy*=3 but we draw values s_k and $a_{s_k}(k)$ only one time.

For example, when *ObjOrder*=0 and *Strategy*=0, the *OBJ* will be decreased when we select such a maximum value of $a_{s_k}(k) \in \left(0, \; \min\{\Delta\tau_{s_k}^{'max}(k), FT(k)\}\right]$ for any $s_k \in P_1^+(k)$ that $Z(a_{s_k}(k)) > L(a_{s_k}(k))$. Let us take into account the third row of Table 2 (for *k*=2). It is profitable to set a_1=2=min{max{2,4,3,1}, 2, 2} because when we decrease values of $\Delta\tau_p^{max}(2)$ for $p \in P_1^\geq(2) = \{1, 2, 3\}$ then our "profit" (decreasing the value of *OBJ*) is equal:

$$Z(a_1(2)) = a_1(2) \cdot card\left(P_1^\geq(2)\right) + \sum_{p \in P_1^<(2)} \Delta\tau_p^{max}(2) = 2 \cdot 3 + 1 = 7.$$ "Cost"   is   equal

$$L(a_1(2)) = (3-1) \cdot \sum_{p \in P_1^<(2)} \left|\Delta\tau_p^{max}(2) - a_1(2)\right| = 2 \cdot 1 \text{ (increasing value of } OBJ\text{)}. \text{ Afterwards, in}$$

the steps 7÷9 we decrease the value of $\Delta\tau_p^{'max}(k)$ and *FT*(k) with $a_{s_k}(k)$ for all $p \geq s_k$. In the case of $\Delta\tau_p^{max}(k) - a_{s_k}(k) < 0$ in step 7, then we must increase this value like in step 8. The algorithm tries to decrease the value of *OBJ* until the free time *FT*(k) for all *k* will be equal to zero or when $a_{s_k}(k) > 0$ (for which condition $Z(a_{s_k}(k)) > L(a_{s_k}(k))$ is satisfied)   does   not   exist   for   any   *k*   and   *p*   (variable   counter=0). Let $\Delta\tau^{min}(k) = \min_{p \in \{1,..,N\}} \{\tau_p^{max} - \tau_p(k)\}$, if $\tau_p^{max} - \tau_p(k) > 0$ and $\Delta\tau^{min}(k) = 1$ -   otherwise. Iteration number L_{WHILE} of the   WHILE   loop   can   be   estimated   as   follows:

$L_{WHILE} < \max_{k \in \{1,...,K\}} \left\lceil \dfrac{FT(k)}{\Delta\tau^{min}(k)} \right\rceil$. It is easy to observe that the complexity of separate steps of the algorithm is as follows: step 5 – $O(N^2)$, step 7 – $O(N)$, step 8 – $O(KN)$, steps 9÷11 – $O(1)$. Steps 4÷14 are realized in the FOR loop *K* times, hence complexity of the *A*.2 algorithm is equal $O\left(L_{WHILE}\left(K^2N + KN^2\right)\right)$.

It is possible to improve value of objective function (8) and computational time in *A*.2 algorithm using some preprocessing step (algorithm *A*.2.0). In *A*.2.0 algorithm we try to decrease value of objective function (8) by decreasing $\underset{p \in \{1,...,N\}}{\forall} \Delta\tau_p^{'max}(k)$ values (for each *k*-th object), to obtain all nonnegative values of $\Delta\tau_p^{'max}(k)$ (like in *A*.2 algorithm). Notice that method of value $a_{s_k}(k)$ selection in the 4 step of the algorithm guarantees, that value of cost function $L(a_{s_k}(k)) = 0$ (see (18)) because of $P_s^<(k) = \varnothing$. After running *A*.2.0 algorithm, we start *A*.2 algorithm taking into initialization   step   values $\underset{k \in \{1,...,K\}}{\forall} \underset{p \in \{1,...,N\}}{\forall} x_{k,p}$, $\underset{k \in \{1,...,K\}}{\forall} FT(k)$ and $\underset{k \in \{1,...,K\}}{\forall} \underset{p \in \{1,...,N\}}{\forall} \Delta\tau_p^{'max}(k)$ obtained from *A*.2.0 algorithm. Computational complexity of *A*.2.0 algorithm can be estimated as follows: external loop FOR realizes *K* times, number of iteration L_{WHILE} of WHILE loop for fixed *k* is bounded by value L_{WHILE} (like in *A*.2 algorithm), step 4 has complexity $O(N)$, and steps 6÷8 – $O(N)$. Hence, total complexity of the *A*.2.0 algorithm is equal $O\left(KL_{WHILE}N\right)$.

Algorithm A.2.0

Given sets: I_k, T_k, IP_k, TP_k for each $k=1,\ldots,K$;

Initialize: $\displaystyle\mathop{\forall}_{k\in\{1,\ldots,K\}} \mathop{\forall}_{p\in\{1,\ldots,N\}} a_p(k):=0$; $\displaystyle\mathop{\forall}_{k\in\{1,\ldots,K\}} \mathop{\forall}_{p\in\{1,\ldots,N\}} x_{k,p}:=0$; $Exit:=false$;

$\displaystyle\mathop{\forall}_{k\in\{1,\ldots,K\}} FT(k):=\tau^{R_{k^*}}(k^*)-\tau^{R_k}(k)$; $\displaystyle\mathop{\forall}_{k\in\{1,\ldots,K\}} \mathop{\forall}_{p\in\{1,\ldots,N\}} \Delta\tau_p^{'\max}(k):=\tau_p^{\max}-\tau_p(k)$;

1. FOR $k=1,\ldots,K$ DO
2.   WHILE $Exit=false$ DO
3.   IF FT(k)>0
4.   Find such a minimal value $s_k \in P_1^+(k)$ and maximal value $a_{s_k}(k)\in\left(0,\ \min\left\{\mathop{\max}_{p\in\{s_k,\ldots,N\}}\Delta\tau_p^{'\max}(k),FT(k)\right\}\right]$, for which condition $\displaystyle\mathop{\forall}_{p\in\{s_k,\ldots,N\}}\Delta\tau_p^{'\max}(k)-a_{s_k}(k)\ge0$ is satisfied;
5.   IF $a_{s_k}(k)>0$ THEN
6. $\displaystyle\mathop{\forall}_{p\in\{s_k,\ldots,N\}}\Delta\tau_p^{'\max}(k):=\Delta\tau_p^{'\max}(k)-a_{s_k}(k)$;
7. $FT(k):=FT(k)-a_{s_k}(k)$;
8. $x_{k,s_k}:=x_{k,s_k}+a_{s_k}(k)$;
9.   ELSE
10. $Exit=true$;
11.   END IF;
12.   ELSE
13. $Exit=true$;
14.   END IF;
15.   END WHILE;
16. END FOR;

4 An Experimental Analysis of Algorithms

Presented in Fig.1 are examples of using A.1 and A.2 algorithms. It can be observed (Table 1) that the value of the criterion function (8) before using A.2 algorithm is equal to 20 (sum of values in the table excluding the last column) and after using the A.2 algorithm (Table 4) is equal to 14. Table 2 presents initial values of functions $\Delta\tau_p^{'\max}(k)$ and $FT(k)$ before running algorithm A.2 (it has been assumed $T^{\max}=\tau^*$). Table 3 contains final values of these functions, after running the A.2 algorithm. Values of $x_{k,p}$ determined by algorithm are equal zero excluding two values: $x_{3,4}=1$,

$x_{2,1}=2$. Taking into account values of $x_{k,p}$ and formula $\displaystyle\tau_p'(k)=\tau_p(k)+\sum_{i=1}^{p}x_{k,i}$ we can obtain modified moments of achieving alignment nodes by all objects (Table 4). Taking into account the explanation presented in the section 3.1, values of $\tau_p'(k)$ and geometric distances $d_{i^r(k),i^{r+1}(k)}$ between nodes $i^r(k),i^{r+1}(k)$ we can calculate modified velocities $v_{i^r(k),i^{r+1}(k)}'$ as follows: $\displaystyle\mathop{\forall}_{k\in\{1,\ldots,K\}}\mathop{\forall}_{r\in\{0,\ldots,R_k-1\}} v_{i^r(k),i^{r+1}(k)}':=\frac{d_{i^r(k),i^{r+1}(k)}}{\tau^{'r+1}(k)-\tau^{'r}(k)}$.

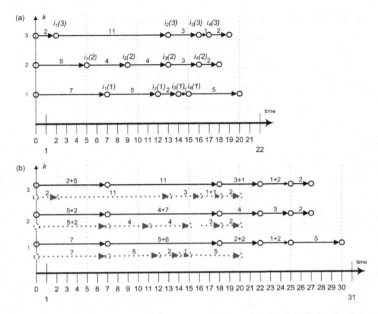

Fig. 1. (a) Vector of paths for $K=3$ objects with achieved times of each $N=4$ alignment nodes for each object; (b) Results of realization of $A.1$ (regular line) and $A.2$ (dashed line) algorithms

Table 1. Values of functions $\tau_p(k)$ and $\tau^{R_k}(k)$ for example from Fig.1a

		p			
k	1	2	3	4	$\tau^{R_k}(k)$
3	2	13	16	17	19
2	5	9	13	16	18
1	7	12	14	15	20

Table 2. Initial values of functions $\Delta\tau_p^{'\max}(k)$ and $FT(k)$ (before running algorithm $A.2$)

		p			
k	1	2	3	4	$FT(k)$
3	5	0	0	0	1
2	2	4	3	1	2
1	0	1	2	2	0

Table 3. Final values of functions $\Delta\tau_p^{'\max}(k)$ and $FT(k)$ (after running algorithm $A.2$)

		p			
k	1	2	3	4	$FT(k)$
3	5	0	0	0	0
2	0	2	1	0	0
1	0	1	2	3	0

Table 4. Modified moments $\tau'_p(k)$ of achieving checkpoints by all objects

			p	
k	1	2	3	4
3	2	13	16	17+1
2	5+2	9+2	13+2	16+2
1	7	12	14	15

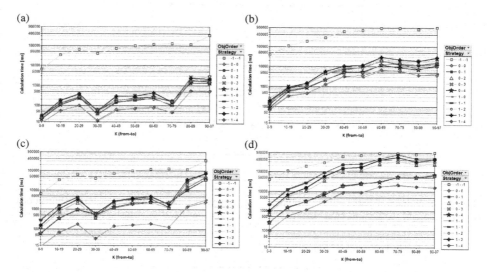

Fig. 2. Average computation time in logarithmic scale [msec] for *A.2* algorithm, with (*Preprocessing*=true, using *A.2.0* algorithm) or without preprocessing (*Preprocessing*=false), using different pairs *ObjOrder-Strategy*. *ObjOrder*=-1 and *Strategy*=-1 deal with solving nonlinear optimization problem (13)-(15) using GAMS/*CONOPT* solver: (a) *Preprocessing*=true, *ObjOrder*∈{0,1}, *Strategy*∈{0,...,4}, *N*∈{1,...,50}; (b) *Preprocessing*=true, *ObjOrder*∈{0,1}, *Strategy*∈{0,...,4}, *N*∈{51,...,100}; (c) *Preprocessing*=false, *ObjOrder*∈{0,1}, *Strategy*∈{0,...,4}, *N*∈{1,...,50}; (d) *Preprocessing*=false, *ObjOrder*∈{0,1}, *Strategy*∈{0,...,4}, *N*∈{51,...,100}.

At the Fig.2 average computation time (computer with Intel Pentium IV 3GHz processor) in logarithmic scale [msec] for *A.2* algorithm, with preprocessing (*A.2.0* before *A.2* algorithm) and without it using different pairs *ObjOrder-Strategy* (*ObjOrder*∈{0,1} because for *ObjOrder*∈{2,3} similar results have been obtained) is presented. The size of the problem (13)-(15) has been set as follows: values of $K∈\{1,...,100\}$ and values of $N∈\{1,...,100\}$ (values of K are divided into group with range 10, values of N are grouped into two sets: for $1 \le N \le 50$ and for $50 < N \le 100$). Over 200 000 randomly generated input data for the problem (13)-(15) has been examined. To compare obtained results from *A.2* algorithm, the problem (13)-(15) has been also solved using GAMS/*CONOPT* solver (*ObjOrder–Strategy*=-1--1).

It can be observed (comparing Fig.2a and Fig.2c or Fig.2b and Fig.2d) that using preprocessing step (running algorithm *A.2.0* before *A.2*) we can accelerate

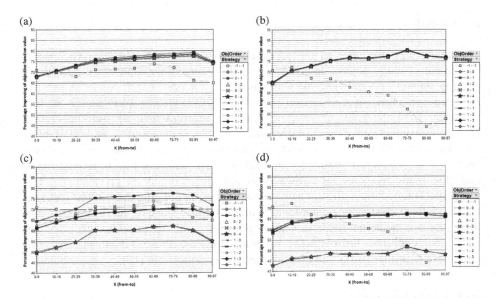

Fig. 3. Average percentage improvement of objective function (13) value for *A.2* algorithm, with (*Preprocessing*=true, using *A.2.0* algorithm) or without preprocessing (*Preprocessing*=false), using different pairs *ObjOrder-Strategy*. *ObjOrder*=-1 and *Strategy*=-1 deal with solving nonlinear optimization problem (13)-(15) using GAMS/*CONOPT* solver: (a) *Preprocessing*=true, *ObjOrder*∈{0,1}, *Strategy*∈{0,...,4}, *N*∈{1,...,50}; (b) *Preprocessing*=true, *ObjOrder*∈{0,1}, *Strategy*∈{0,...,4}, *N*∈{51,...,100}; (c) *Preprocessing*=false, *ObjOrder*∈{0,1}, *Strategy*∈{0,...,4}, *N*∈{1,...,50}; (d) *Preprocessing*=false, *ObjOrder*∈{0,1}, *Strategy*∈{0,...,4}, *N*∈{51,...,100}.

computation time between a few and twenty times faster than without preprocessing step. For all pairs *ObjOrder–Strategy* we have obtained faster computation time than for using GAMS/*CONOPT* solver. We have obtained the best computation time for *ObjOrder–Strategy*: 0–0, 1–0 (also for 2–0 and 3–0).

At the Fig.3 average percentage improvement of objective function (13) value for *A.2* algorithm, with or without preprocessing (*A.2.0* algorithm) using different pairs *ObjOrder-Strategy* is presented (*ObjOrder*∈{0,1} because for *ObjOrder*∈{2,3} similar results have been obtained). Value *PI* of percentage improvement of objective function value is calculated as follows: $PI = \dfrac{OBJ_0 - OBJ_1}{OBJ_0} \cdot 100\%$, where OBJ_0, OBJ_1 – values of objective function (13) before and after running *A.2* algorithm, respectively. It can be observed that for *K*>20 almost for all pairs *ObjOrder-Strategy* in *A.2* algorithm percentage improvement of objective function value is better than for using GAMS/*CONOPT* solver. This difference is growing when value of *K* is growing. We have obtained the best results using preprocessing step (Fig.3a and Fig.3b) and following pairs of *ObjOrder–Strategy*: 0–1, 1–1 (also for 2–1 and 3–1). Percentage improvement of objective function (13) value for the best pairs of *ObjOrder–Strategy* is equal 65% to 80%.

5 Summary

The approaches presented in the paper give possibilities to schedule synchronous movement of many objects and they are used in some simulation-based operational training support system [1] on the planning stage of action [12]. It can be shown that they are very fast (in comparison with GAMS/CONOPT solver) and it is very important from the point of view of simulator reaction time on user interaction. During movement simulation (movement schedule realization) it is important to movement control and reaction to deviations from determined schedule [11]. These problems are essentials especially in CGF or SAF systems [8]. It is possible to consider many problems for synchronous movement based on given approaches: multicriteria scheduling, disjoint path scheduling, using not only the shortest paths but also the k-th shortest paths (faster to compute) [4]. Since some of the algorithms being discussed are heuristic ($A.2$) it seems to be essential to provide necessary and sufficient conditions for obtaining optimal solutions. Presented suggestions may contribute to further works and they are partially considered in [13].

Acknowledgements

This work was partially supported by MUT grant PBW GD-604 and by project number MNiSW OR00005506 titled "Simulation of military actions in heterogeneous environment of multiresolution and distributed simulation".

References

1. Antkiewicz, R., Najgebauer, A., Tarapata, Z., Rulka, J., Kulas, W., Pierzchala, D., Wantoch-Rekowski, R.: The Automation of Combat Decision Processes in the Simulation Based Operational Training Support System. In: Proceedings of the IEEE Symposium on Computational Intelligence for Security and Defense Applications (CISDA 2007), Honolulu, Hawaii, USA, 01-05.04 (2007) ISBN 1-4244-0698-6
2. Blazewicz, J., Ecker, K.H., Pesch, E., Schmidt, G., Weglarz, J.: Scheduling Computer and Manufacturing Processes. Springer, Heidelberg (2001)
3. Buchli, J. (ed.): Mobile Robotics Moving Intelligence. Pro Literatur Verlag, Germany (2006)
4. Eppstein, D.: Finding the K shortest Paths. SIAM J.Computing 28(2), 652–673 (1999)
5. Ibaraki, T.: Algorithms for obtaining shortest paths visiting specified nodes. SIAM Review 15(2), Part 1, 309–317 (1973)
6. Leung, J.Y.-T. (ed.): Handbook of Scheduling: Algorithms, Models and Performance Analysis. Chapman & Hall/CRC, Boca Raton (2004)
7. Logan, B.: Route planning with ordered constraints. In: Proceedings of the 16th Workshop of the UK Planning and Scheduling Special Interest Group, Durham, UK, pp. 133–144 (December 1997)
8. Petty, M.D.: Computer generated forces in Distributed Interactive Simulation. In: Proceedings of the Conference on Distributed Interactive Simulation Systems for Simulation and Training in the Aerospace Environment, Orlando, USA, April 19-20, pp. 251–280 (1995)

9. Rajput, S., Karr, C.: Unit Route Planning, Technical Report IST-TR-94-42, Institute for Simulation and Training, Orlando, USA (1994)
10. Schrijver, A., Seymour, P.: Disjoint paths in a planar graph – a general theorem. SIAM Journal of Discrete Mathematics 5, 112–116 (1992)
11. Tarapata, Z.: Automatization of decision processes in conflict situations: modelling, simulation and optimization. In: Arreguin, J.M.R. (ed.) Automation and Robotics, pp. 297–328. I-Tech Education and Publishing, Vienna (2008)
12. Tarapata, Z.: Modeling, simulation and optimization of selected decision processes in conflict situations- a case study. Polish Journal of Environmental Studies 17(3B), 467–474 (2008)
13. Tarapata, Z.: Selected scheduling problems for synchronization of multi-objects movement. Bulletin of Military University of Technology, 4 (652) LVII, 25–37 (2008)

Automatic Segmentation of Bone Tissue in X-Ray Hand Images

Ayhan Yuksel and Tamer Olmez

Department of Electronics and Communication Engineering, Istanbul Technical University,
34469 Istanbul, Turkey
{yukselay,olmezt}@itu.edu.tr

Abstract. Automatic segmentation of X-ray hand images is an important process. In studies such as skeletal bone age assessment, bone densitometry and analyzing of bone fractures, it is a necessary extremely difficult and complicated task. In this study, hand X-ray images were segmented by using C-means classifier. Extraction of bone tissue was realized in three steps: i) preprocessing, ii) feature extraction and iii) automatic segmentation. In preprocessing scheme, inhomogeneous intensity distribution is eliminated and some structural pre-information about hand was obtained in order to use in feature extraction block. In feature extraction process, edges between soft and bone tissues were extracted by proposed enhancement process. In automatic segmentation process, the image was segmented using C-mean classifier by taking care of local information. In the study, hand images of ten different people were segmented with high performances above 95%.

Keywords: Hand radiographs, image segmentation, classifiers.

1 Introduction

X-ray imaging is a highly available and low cost imaging method [1]. Although it is a popular imaging system, computer assisted analyzing of X-ray images have some difficulties which originate from nature of X-rays, X-ray machine, X-ray film or scanning methods. According to Ogiela et al [2], some specific problems in the analysis of 2D hand X-ray images are listed: i) Some of the details become blurred because of the overlapping bones. As a result, some of the bones may disappear. Therefore, the segmentation algorithm must estimate the form of the bone and its relation with the other elements on the image. ii) There can be some extra bones or bone decrements which are not described as a priori by anatomical maps in some patients. iii) Fractures and displacements caused by injuries or some pathological conditions can be displayed on the image. In addition to these problems, X-ray beams do not touch to the examined material at equal strength due to the X-ray beam source location. As a result, non uniform intensity distribution occurs on the resulted X-ray image.

X-ray bone images are used in the areas such as bone age assessment [3-7], bone mass assessment [8,9] and examination of bone fractures [10]. Extraction of bone

M. Kolehmainen et al. (Eds.): ICANNGA 2009, LNCS 5495, pp. 590–599, 2009.
© Springer-Verlag Berlin Heidelberg 2009

tissue from other tissues and background (segmentation) is one of the main steps in such applications. Segmentation of medical images is a challenging problem and a necessary first step in many image analysis and classification processes [10]. There are three approaches in the segmentation step: Manual segmentation, semi-automatic segmentation and fully automatic segmentation. In manual segmentation, selection of different anatomical structures is realized by an operator. Hence, they lead to operator dependent and subjective results [11]. In semi-automatic segmentation, some parameters and initial conditions of the method are set in supervision of an expert user. Semi-automatic segmentation and manual methods generally have better performances due to expert knowledge. However, non-automatic segmentation methods are difficult and time consuming [12].

In fully automatic segmentation systems, the image is segmented by developed algorithms in a computer and there is no need for expert knowledge. In the literature, there are many studies for segmenting the X-ray images. However, a fully automatic segmentation method that segments all X-ray images has not been met yet. Most of the studies were applied to limited number of images, not to general [4,6,11] Also, a number of automatic segmentation methods are not applicable for all bones in the hand X-ray image [3,5,8,9,13]. In addition, some studies needs user intervention in some part of the algorithm [13,14]. An automatic segmentation is the first step in automatic bone age assessment studies [4-7]. Likewise, an automatic segmentation step is a need in bone density assessment [8,9] and bone fracture analysis [10].

In this study, an automatic segmentation method for X-ray hand images was presented. Here, X-ray images were segmented using C-means algorithm by incorporating some structural pre-information.

2   Method

In this study, a new method for segmentation of bones in X-Ray hand images was proposed. Segmentation procedure was studied in three steps. In the first step, the intensity non-uniformity of X-ray image was removed. Then, some structural information about image was obtained in order to use for improving feature extraction block.

In the second step, the features which emphasize the bone boundaries were analyzed. Here, two feature images were generated. The difference operation between the two feature images emphasizes the bone edges. Also a feature improvement method was applied in this step.

In the third step, an unsupervised neural network was used for segmenting bone tissues in the decision process. In this step, feature image was given to neural network by parceling small windows in order to take care of local feature variations. Segmented image contains only bone edges. Finally, by using morphological image processing techniques bone tissue was extracted.

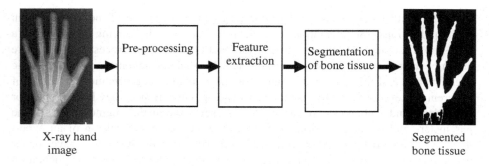

X-ray hand
image

Segmented
bone tissue

Fig. 1. Main steps of automatic hand bone segmentation

2.1 Pre-processing

Pre-processing step consists of removing non-uniform intensity distribution, creating a hand mask and determining the locations of the fingertips and the wrist region of hand.

X-ray images have non-uniform intensity distribution because X-ray beams do not reach to the examined material at equal strength due to the X-ray beam source location. In order to eliminate this non-uniformity, a model for generation and distribution of X-ray beams was created. Then, a correction strategy which determines parameters of the X-ray source and corrects intensity distribution of the image retrospectively was developed. This method had been proposed and applied in [15]. Result of intensity distribution procedure is given in Fig. 2.

Employing some pre-information about hand image to be segmented provides better segmentation results with less error. In order to do this, location of hand was determined. Note that, intensity correction procedure makes hand extracting method easier with a simple automatic thresholding method [15].

Other pre-information about hand is the location information of the fingertips and the wrist region. Previously created hand mask was used for this step. Fingertips were

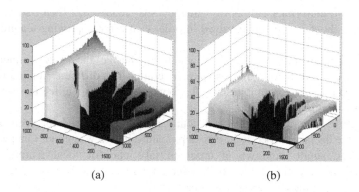

(a)   (b)

Fig. 2. Result of intensity distribution procedure. 3-D representation of the original image (a) and the corrected image (b).

found by analyzing binary hand mask image. The task was realized by looking for peaks in the boundary of hand mask. Wrist region was determined by creating a hand silhouette from X-ray image. To do this, X-ray image is masked by the hand mask and extracted hand region is filtered with a CWT filter with a scaling factor of 0.5. Obtained hand silhouette gives its maximum gray level value at the position where wrist bones stands because their thicknesses cause brighter regions in X-ray images.

2.2 Feature Extraction

An X-ray image consists of three different regions; background, soft tissue and bone tissue. Also, two types of boundaries can be thought; background- soft tissue boundary and soft tissue-bone tissue boundary. The boundaries between two regions form edges which give higher variance values than other areas.

The goal of the feature extraction step is to expose the features which emphasize the bone edges. Generating a variance map is an easy method for focusing on sudden feature differences or edges. However, when the variance of the image is calculated directly, other soft tissue-background edges are revealed near edges of bone tissue. Therefore, two variance maps are calculated: i) variance map for emphasizing bone tissue -soft tissue edges, ii) variance map for emphasizing soft tissue- background edges.

For the first variance map, an averaged square image is created by the equation (1). In this equation, energy value of a pixel is calculated by averaging square values of pixels in N-neighborhood window, where N was selected 1 in the study. This operation makes edges between bone tissue and soft tissue more strong than other edges.

$$E_{(x,y)} = \frac{1}{(2N+1)^2} \sum_{j=-N}^{N} \sum_{i=-N}^{N} I_{(x+i,y+j)}^2 \qquad (1)$$

Value of a pixel in variance map is calculated by equation (2). Here, $\mu_{E_{(x,y)}}$ is average gray level value of pixels which are inside the N- neighborhood of pixel at position (x,y). Similar to energy map, neighborhood degree N was selected 1 for this study.

$$\sigma^2 E_{(x,y)} = \frac{1}{(2N+1)^2} \sum_{j=-N}^{N} \sum_{i=-N}^{N} (E_{(x+i,y+j)} - \mu_{E_{(x,y)}})^2 \qquad (2)$$

Created variance map has high values at bone edges and the boundaries between soft tissue-background. In order to eliminate the unwanted boundaries between soft tissue-background and obtain a result image which have only bone edges, another feature image which is dominated by soft tissue-background borders is created. For this purpose, a logarithmic image was created and variance map of the image was obtained. Boundary of soft tissue-background is dominant in created logarithmic variance image. Calculations of logarithmic image and variance map of this image are given in equations 3 and 4.

$$L_{(x,y)} = \ln\left(\frac{1}{(2N+1)^2} \sum_{j=-N}^{N} \sum_{i=-N}^{N} I_{(x+i,y+j)}\right) \tag{3}$$

$$\sigma^2 L_{(x,y)} = \frac{1}{(2N+1)^2} \sum_{j=-N}^{N} \sum_{i=-N}^{N} (L_{(x+i,y+j)} - \mu_{L_{(x,y)}})^2 \tag{4}$$

Where $\mu_{L_{(x,y)}}$ is average gray level value of pixels which are inside the N-neighborhood of pixel at position (x,y). N is the neighborhood degree which was selected 1 for the study.

Initially, two images, created by equations 2 and 4, were normalized in order to have the same average gray level value. Next, difference image, ϕ was obtained by subtracting $\mu_{L_{(x,y)}}$ feature from $\mu_{E_{(x,y)}}$ feature as given in equation (5). This procedure makes bone edges dominant while eliminating soft tissue-background borders in the image. Variance features of energy image, logarithmic image and ϕ difference image are depicted in Fig. 3.

$$\phi = \sigma^2 E - \sigma^2 L \tag{5}$$

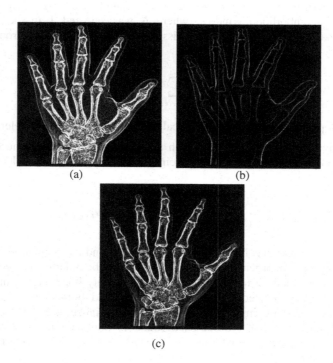

(a)   (b)

(c)

Fig. 3. Obtained feature images: (a) Variance map of energy image, (b) variance map of logarithmic image, (c) difference image

2.2.1 Feature Improving

Difference of variance maps, ϕ, contains bone edges with higher values and gives good results when used as a feature for segmentation of bones but segmentation with this feature generally fails at some specific regions where variance values are not enough high like fingertips and wrist region. Improving the features at those specific locations will increase the segmentation performance.

For improving the features, ϕ was multiplied with a mask which has bell shaped curves with peaks at the pre-determined locations of fingertips and wrist region. 3D representation of multiplication mask is seen in Figure 4-a. Here, there are five curves at fingertips and one curve at the wrist region. Widths and peak values of each curve were predetermined by taking anatomical features into consideration. Improvement of ϕ is also depicted in Figure 4-b with 3-D coordinates where $-z$ coordinate stands for the feature value at related position.

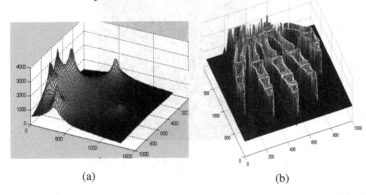

(a)   (b)

Fig. 4. Improving of features: (a) 3-D representation of multiplication mask (b) feature values in 3-D after improvement step

2.3 Segmentation

A segmentation method that analyses feature image with sub windows was used in the study. Sub windows were then segmented by neural network and local segmentation results from all sub windows were obtained. Then a method was developed for combining local segmentation methods and generating a segmentation result for whole image. Finally, segmentation result was processed by a series of structural image processing methods in order to get final segmentation result. Segmentation process can be examined by three steps: i) segmentation in sub windows, ii) obtaining global segmentation result from local segmentation method, iii) converting segmentation result into final segmented bone tissue by structural image processing methods.

Because the feature values of the bone tissue in the feature image varies among whole hand region, a segmentation method which analyses all of input image at a time will fail. So, there is a need for a method which is able to observe local feature variations in the image. For this purpose, segmentation process was realized by examining the feature image inside small sized, randomly located sub windows. Square sub window size was set at about width of a finger. Sub window locations were selected randomly during segmentation process. This approach will generate more reliable results because it provides a lot of decisions for any pixel with a lot of point of views. Sub windows are shown in Figure 5. In order to show more comprehensible image, only fifty sub windows are shown.

Sub windows were segmented by using C-means classifier with two nodes. The nodes were labeled as edge (high valued) and non-edge (low valued) according to their position in one dimensional feature space. Distribution of feature values inside the sub window determines node position. Nodes of classifier are attracted to class centers automatically.

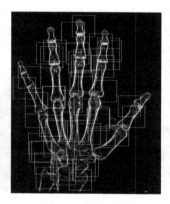

Fig. 5. Randomly selected fifty sub windows

After the training step of the classifier has been completed for the selected sub window, segmentation process starts. In segmentation process, each pixel in sub window is given to classifier as a one dimensional feature vector. Label of this pixel was defined by the label of nearest (according to Euclidian distance in feature space) node of the classifier.

Every pixel in the hand will be covered by a sub window due to the high number (1000) of sub windows. Since there should be many sub windows which cover the same pixel, different labels may be obtained for that pixel at different sub windows. Therefore, In order to store the segmentation results, two label counters was assigned for each pixel in the image. When it is covered by a sub window, only one of the counters of a pixel is incremented by one according to the result of the neural network. This process is iterated until determined number of sub window is selected and segmented. After sub window segmentation step has been completed, global segmentation result is generated by examining label counters for each pixel in the image. Label of any pixel counter with highest value will be assigned to it.

In the previous step, only bone edges were found. However, the extraction of the whole bone tissue is needed. To do this, bone tissue was obtained by using bone edge image found in previous step. By applying a simple region growing algorithm, it is possible to extract all bones.

Firstly, a dilation operation is applied in order to remove all small spaces. For filling interiors of bone edges and obtaining final segmented bone tissue, a region growing algorithm which grows outside the bone tissue is applied. This process finds outside of the bone tissue. Inverted image of region growing result will give the final bone tissue image.

3   Computer Simulation Results

In order to examine the validity of the proposed method, X-Ray hand images of ten different people were segmented and quantitative performance values of segmentation process were investigated by applying a manually created test set which was created by selecting total 300 points from inside and outside of the bone tissue. Then, segmentation performance for each image was evaluated by the ratio of true decision over test size. Hand X-Ray images used in the study were obtained from X-Ray database of Image Processing and Informatics Laboratory of South California University [16].

Proposed segmentation result was coded in Matlab ® Release 14 and executed with a 1.73 GHz Intel Centrino Notebook. Segmentation of an image with sizes about 1000x1200 pixels takes about three minutes (Preprocessing 1 minute, feature extraction 0.5 minute and segmentation 1.5 minute, approximately).

In Table 1, Performance results for segmentation of ten images are given. Original and segmentation results for two of the images are shown in Figure 6.

Table 1. Segmentation performance for selected images

Image Name	Performance(%)
3229	99.7
3842	100
4335	100
4482	98
4494	96
4524	99.3
5104	99.7
5209	100
5268	99
5365	99.7

Images names are the original image names. More information about images can be obtained from the database [16].

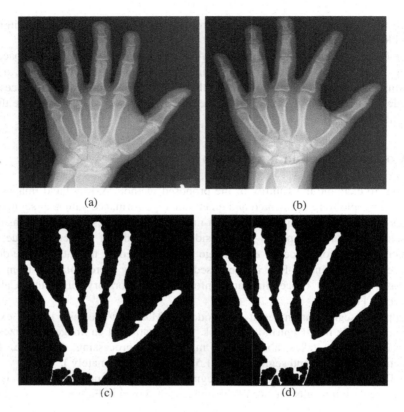

Fig. 6. Original X-Ray hand images (c,d) Segmentation results of (a) and (b) respectively

3   Discussion

In this work, a method for segmenting of bone tissue from X-Ray hand images was developed. Features of bone tissue vary over whole hand image. Because of this, methods that segment the bone tissue by looking general image often fail. Better and more reliable results were obtained with segmentation of the image by searching local feature variations and generating global segmentation results by interpreting local segmentation results as a kind of statistical data.

In the study, using some structural information about hand improved the segmentation performance. It was observed that, without adding structural information, segmentation process had given poor performance value. So, using structural information of hand with extracted features will give better results.

Segmentation of bones automatically is an extremely challenging task. Steps of segmentation process takes nearly three minutes for an image with sizes 1000x1200. The method gave similar results for different people's images which is the evidence for generality of method.

In the future work, segmentation of bones from different organs will be studied. Also, in this work, bones are segmented as a whole tissue. A method for segmenting of the image bone by bone will be searched.

References

1. Felisberto, M.K., Lopes, H.S., Centeno, T.M., Arruda, L.V.R.: Object detection and recognition system for weld bead extraction from digital radiographs. Computer Vision and Image Understanding 102, 238–249 (2006)
2. Ogiela, M.R., Tadeusiewicz, R., Ogiela, L.: Graph image language techniques supporting radiological, hand image interpretations. Computer Vision and Image Understanding 103, 112–120 (2006)
3. Garcia, R.L., Fernandez, M.M., Arribas, J.I., Lopez, C.A.: A fully automatic algorithm for contour detection of bones in hand radiographs using active contours. In: International Conference on Image Processing, pp. 421–424 (2003)
4. Han, C.C., Lee, C.H., Peng, W.L.: Hand radiograph image segmentation using a coarse-to-fine strategy. Pattern Recognition 40, 2994–3004 (2007)
5. Zhang, A., Gertych, A., Liu, B.J.: Automatic bone age assessment for young children from newborn to 7-year-old using carpal bones. Computerized Medical Imaging and Graphics 31, 299–310 (2007)
6. Mahmoodi, S., Sharif, B.S., Chester, E.G., Owen, J.P., Lee, R.: Skeletal growth estimation using radiographic image processing and analysis. IEEE Transactions on Information Technology in Biomedicine 4, 292–297 (2000)
7. Pietka, E., Kurkowska, S.P., Gertych, A., Cao, F.: Integration of computer assisted bone age assessment with clinical PACS. Computerized Medical Imaging and Graphics 27, 217–228 (2003)
8. Sotoca, J.M., Inesta, J.M., Belmonte, M.A.: Hand bone segmentation in radioabsorptiometry images for computerised bone mass assessment. Computerized Medical Imaging and Graphics 27, 459–467 (2003)
9. Haidekker, M.A., Stevens, H.Y., Frangos, J.A.: Computerised methods for X-ray-based small bone densitometry. Computer Methods and Programs in Biomedicine 73, 35–42 (2004)
10. Jiang, Y., Babyn, P.: X-ray bone fracture segmentation by incorporating global shape model priors into geodesic active contours. In: Computer Assisted Radiology and Surgery, pp. 219–224 (2004)
11. Sharif, B.S., Chester, E.G., Owen, J.P., Lee, E.J.: Bone edge detection in hand radiographic images. In: Engineering Advances: New Opportunities for Biomedical Engineer, pp. 514–515 (1994)
12. Kurnaz, M.N.: Artımsal Yapay Sinir Ağları Kullanılarak Ultrasonik Görüntülerin Bölütlenmesi (in Turkish). Ph.D Thesis, İstanbul Technical University (2006)
13. Bocchi, L., Ferrara, F., Nicoletti, I., Valli, G.: An artificial neural network architecture for skeletal age assessment. In: International Conference on Image Processing, vol. 1, pp. 1077–1080 (2003)
14. Behiels, G., Maes, F., Vandermeulen, D., Suetens, P.: Evaluation of image features and search strategies for segmentation of bone structures in radiographs using Active Shape Models. Medical Image Analysis 6, 47–62 (2002)
15. Yuksel, A., Dokur, Z., Korurek, M., Olmez, T.: Modeling of inhomogeneous intensity distribution of X-ray source in radiographic images. In: 23rd International Symposium on Computer and Information Sciences, 2008. ISCIS 2008, pp. 1–5 (2008)
16. Image Processing and Informatics Laboratory, http://www.ipilab.org/

Automatic Morphing of Face Images

Vittorio Zanella[1], Geovany Ramirez[2], Héctor Vargas[1], and Lorna V. Rosas[1]

[1] Universidad Popular Autónoma del Estado de Puebla
21 sur 1103 Col. Santiago Puebla México
{vittorio.zanella,hectorsimon.vargas,
lornaveronica.rosas}@upaep.mx
www.upaep.mx
[2] The University of Texas at El Paso
500 West University Avenue El Paso, TX 79968-0518
garamirez@miners.utep.edu
www.cs.utep.edu

Abstract. Image metamorphosis, commonly known as morphing, is a powerful tool for visual effects that consists of the fluid transformation of one digital image into another. There are many techniques for image metamorphosis, but in all of them is a person who supplies the correspondence between the features in the source image and target image. In this paper we use a method to find the faces in the image and the Active Shape Models to find the features in the face images and perform the metamorphosis of face images in frontal view automatically.

Keywords: Automatic Image Metamorphosis, Active Shape Models, Facial Features.

1 Introduction

Image metamorphosis is a powerful tool for visual effects that consists of the fluid transformation of one digital image into another. This process, commonly known as *morphing* [1], has received much attention in recent years. This technique is used for visual effects in films and television [2,3], and it is also used for recognition of faces and objects [4].

Image metamorphosis is performed by coupling image warping with color interpolation. Image warping applies 2D geometric transformations to images to retain geometric alignment between their features, while color interpolation blends their colors.

The quality of a morphing sequence depends on the solution of three problems: feature specification, warp generation and transition control. Feature specification is performed by a person who chooses the correspondence between pairs of feature primitives. In actual morphing algorithms, meshes [3, 5, 6], line segments [7, 8, 9], or points [10, 11, 12] are used to determine feature positions in the images. Each primitive specifies an image feature, or landmark. Feature correspondence is then used to compute mapping functions that define the spatial relationship between all points in both images. These mapping functions are known as warp functions and are used to interpolate the positions of the features across the morph sequence. Once both images have been warped into alignment for intermediate feature positions, ordinary color

M. Kolehmainen et al. (Eds.): ICANNGA 2009, LNCS 5495, pp. 600–608, 2009.
© Springer-Verlag Berlin Heidelberg 2009

interpolation (cross-dissolve) is performed to generate image morphing. Transition control determines the rate of warping and color blending across the morph sequence.

Feature specification is the most tedious aspect of morphing, since it requires a person to determine the landmarks in the images. A way to determine the landmarks automatically, without the participation of a human, would be desirable. In this work, we use a method for to find a face in an image and after that to use the Active Shape Models for to find the facial features and in this form perform the spatial relationship between all points in both images, without the intervention of a human expert. We initially chose work with images of faces in frontal view with uniform illumination and without glasses to simplify the problem.

2 Active Shape Models

The Active Shape Models (ASM) was originally proposed by Cootes [13]. The ASM are statistical models which iteratively move toward structures in images similar to those on which they were trained. The aim is to build a model that describes shapes and typical variations of an object. To make the model able of capturing typical variations we collect different images of that object, and aim that the object is appearing in different ways reflecting its possible variations. This set of images is named the training set. In this work, we used the Stegmann's trained set that comprises 37 different frontal human image faces, all without glasses and with a neutral expression [14]. To collect information about the shape variations needed to build the model, we represent each shape with a set of landmarks points. In this work we use a model of 58 points that represents the eyes, eyebrows, nose, mouth and jaw, see Figure 1.

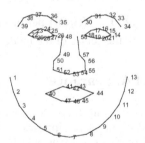

Fig. 1. The face model

Each image in the training set is labeled by a set of points; each labeled point represents a particular part of the face or its boundary. Each point will thus have a certain distribution in the image space

3 Point Distribution Models

The Point Distribution Models are generated from examples of shapes, where each shape is represented by a set of labeled points. A given point corresponds to a particular location on each shape or object to be modeled [15]. The examples shapes are all aligned into a standard co-ordinate frame, and a principal component analysis is applied to the co-ordinates of the points. This produces the mean position for each of the points and description of the main ways in which the points tend to move together. The model can be used to generate new shapes using the equation

$$\mathbf{x} = \bar{\mathbf{x}} + \mathbf{Pb} \tag{1}$$

Where $\mathbf{x} = (x_0, y_0, \ldots, x_{n-1}, y_{n-1})^T$, (x_k, y_k) is the kth model point. $\bar{\mathbf{x}}$ represents the mean shape, $\mathbf{P}$ is a $2n$ x t matrix of t unit eigenvectors and $\mathbf{b} = (b_1, \ldots, b_t)^T$ is a set of shape parameters b_i.

If the shape parameters b_i are chosen such that the square of the Mahalanobis distance D_m^2 is limited, then the shape generated by (1) will be similar to those given in the training set.

$$D_m^2 = \sum_{k=1}^{t} \left(\frac{b_k^2}{\lambda_k} \right) \le D_{max}^2 \tag{2}$$

λ_k is the variance of parameter b_k in the original training set and $D_{max}^2 = 3.0$.

By choosing a set of shape parameters $\mathbf{b}$ for a Point Distribution Model, we define the shape of a model object in an object centred co-ordinate frame. We can then create an instance, $\mathbf{X}$, of the model in the image frame by defining the position, orientation and scale:

$$\mathbf{X} = M(s, \theta)[\mathbf{x}] + \mathbf{X}_c \tag{3}$$

Where $\mathbf{X}_c = (X_c, Y_c, \ldots, X_c, Y_c)^T$, $M(s, \theta)[]$ performs a rotation by θ and scaling by s, and (X_c, Y_c) is the position of the centre of the model in the image frame.

4 Modelling Grey Level Appearance

We wish to use our models for locating facial features in new face images. For this purpose, not only shape, but also grey-level appearance is important. We account for this by examining the statistics of the grey levels in regions around each of the labelled model points [16]. Since a given point corresponds to a particular part of the object, the grey-level patterns about that point in images of different examples will often be similar. We need to associate an orientation with each point of our shape model in order to align the region correctly, in this case normal to the boundary. For every point i in each image j, we can extract a profile g_{ij}', of length n_p pixels, centred at the point. We choose to sample the derivative of the grey levels along the profile in the image and normalise. If the profile runs from p_{start} to p_{end} and is of length n_p pixels, the kth element of the derivative profile is

$$g'_{jk} = I_j(\mathbf{y}_{k+1}) - I_j(\mathbf{y}_{k-1}) \tag{4}$$

where $\mathbf{y_k}$ is the kth point along the profile:

$$\mathbf{y}_k = \mathbf{p}_{start} + \frac{k-1}{n_p - 1}(\mathbf{p}_{end} - \mathbf{p}_{start}) \tag{5}$$

and $I_j(\mathbf{y}_k)$ is the grey level in image j at that point.
We then normalise this profile,

$$\mathbf{g}_{ij} = \frac{\mathbf{g}_{ij}'}{\displaystyle\sum_{k=1}^{n_p} |g_{ijk}'|} \tag{6}$$

The normalised derivative profile tends to be more invariant to changes in the image caused by variations in lighting than a simple grey-level profile,

$$\overline{\mathbf{g}}_i = \frac{1}{N_s} \sum_{j=1}^{N_s} \mathbf{g}_{ij} \tag{7}$$

We can then calculate an n_p x n_p covariance matrix, $\mathbf{Sg}_i$, giving us a statistical description of the expected profiles about the point. Having generated a flexible model and a description of the grey levels about each model point we would like to find new examples of the modelled face in images.

5   Calculating a Suggested Movement for Each Model Point

Given an initial estimate of the positions of a set of model points which we are attempting to fit to a face image we need to estimate a set of adjustments which will move each point toward a better position. At a particular model point we extract a derivative profile $\mathbf{g}$, from the current image of some length l ($l > n_p$), centered at the point and aligned normal to the boundary. We then run the profile model along this sampled profile and find the point at which the model best matches. Given a sampled derivative profile the fit of the model at a point d pixels along it is calculated as follows

$$f_{prof}(d) = (\mathbf{h}(d) - \overline{\mathbf{g}})^T \mathbf{S}_{\mathbf{g}}^{-1}(\mathbf{h}(d) - \overline{\mathbf{g}}) \tag{8}$$

Where $\mathbf{h}(d)$ is a sub-interval of g of length np pixels centred at d, normalised using (6). This is the Mahalanobis distance of the sample from the mean grey model, the value of f_{prof} decreases as the fit improves. The point of best fit is thus the point at which $f_{prof}(d)$ is minimum [16].

6   Estimates the Initial Position of Model Points

The initial position of the model points is estimated using the method for faces detection showed in [17]. This method is based in two stages, each one based in one heuristic. The first stage is based in the next heuristic.

In a face with uniform lighting, the average intensity of the eyes is lower than the intensity of the part of the nose that is between the eyes.

In this stage we find regions that correspond to the eyes regions of possible faces in the image. To apply the heuristic to an image of a possible face first it is necessary to resize the image to 9x9 pixels. The region of the eyes in a 9x9 image corresponds to rows 2, 3 and 4. Then the eyes region has a size of 9x3 (See Fig. 2). The equation to compute the *ENdif* (Eyes-Nose differential) from the eyes region, which we will call *I* is:

$$ENdif = I_{5,2} - \left(I_{2,2} + I_{3,2} + I_{7,2} + I_{8,2}\right)/4 \tag{9}$$

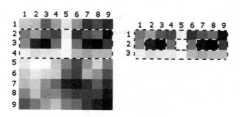

Fig. 2. Face region of size 9x9 pixels; and the eyes region

It is necessary to perform the search on the whole image to locate all the possible regions that can contain a face. The size of the regions depends on the size of the face to look for. It is necessary to determine the value of the *ENdif* of each region using (9). The regions that have an *ENdif* > *thENdif* (Eye-Nose differential acceptation threshold, calculated from one set of faces) are the regions that will be selected for the second stage. The regions selected by the first stage correspond to the eyes region of the possible faces; these regions are extended to the size of the face region. To determine if the possible face is really a face we apply a second discriminator, which is based on a heuristic:

The histograms of the image in grayscale of a face with uniform lighting always have a specific shape

We know that a face has eyes, nose, mouth, eyebrows and it is covered by skin, and some elements of the face are darker than others. The relationship between the elements of the face results in a histogram with a specific shape it is showed in the Figure 3.

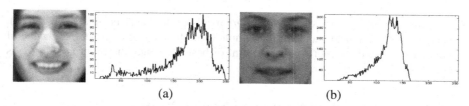

(a)   (b)

Fig. 3. Examples of the histograms from two face images. The shapes of histograms (a) and (b) are similar, one is wider than the other one and the values are different, this due to the lighting conditions and the different sizes in the images.

The discriminator uses a curve model to compare it with the histogram of the possible face. Due to the variability of the size and lighting conditions of the possible faces, their histograms can vary in magnitude and in number of elements. So it is necessary to normalize the histogram of the possible face and after that to compare with a Pearson IV curve obtained from our training set of faces as show in the figure 4.

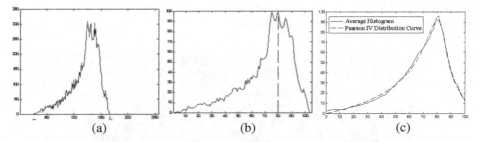

(a)   (b)   (c)

Fig. 4. (a) Original histogram of the face image in Fig. 3b. (b) The new histogram normalized (c) Pearson IV distribution curve and the average normalized histogram from our set of faces.

In this stage we reduced the number of regions of possible faces found in the first stage. Finally we use a mask of seven segments (Figure 5) for finding the face in the remaining regions of the second stage using the horizontal edge image. This discriminator first segments the image using the Sobel method for edge-finding in horizontal direction; after that, dilation is performed to enhance the edges. The resulting image will contain only the eyes, the eyebrows, the mouth and the nose; we use a the mask to evaluate the presence or absence of these elements, it is showed in the figure 5. One example of the whole process is presented in the Figure 6, for details see [17].

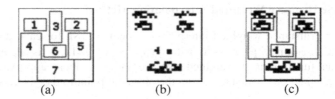

(a)   (b)   (c)

Fig. 5. (a) The mask of seven segments, (b) The horizontal edge image, (c) The mask used like discriminator

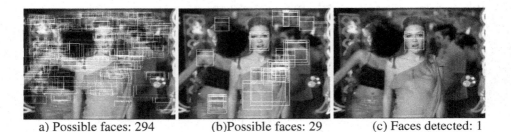

a) Possible faces: 294   (b)Possible faces: 29   (c) Faces detected: 1

Fig. 6. Process for to detect a face. (a) After the first stage. (b) After the second stage. (c) Final result.

7   Results

In the Figure 7 is showed the process to find the face feature in the face image. In (a) is showed the result of the face detection algorithm and in the Figures (b) to (f) is showed the result of the active shape models process.

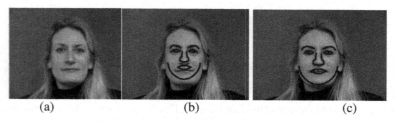

(a)   (b)   (c)

Fig. 7. (a) Position of the face in the image, (b) The initial position of the model, (c) Model after 10 iterations

8   Warp Generation

Once the model has been adjusted to the images, the next step is to perform image deformation, or warping, by mapping each feature in the source image to its corresponding feature in the target image. In this work we use the inverse distance weighted interpolation method.

8.1   Inverse Distance Weighted Interpolation Method

In the inverse distance weighted interpolation method [18], for each data point $\mathbf{p}_i$, a local approximation $f_i(\mathbf{p}):\Re^2 \to \Re$ with $f_i(\mathbf{p}_i) = y_i$, $i=1,..,n$ is determined. The interpolation function is a weighted average of these local approximations, with weights dependent on the distance from the observed point to the given points,

$$f(\mathbf{p}) = \sum_{i=1}^{n} w_i(\mathbf{p}) f_i(\mathbf{p})$$

(10)

Where $f_i(\mathbf{p}_i) = y_i$, $i=1,..,n$. $w_i:\Re^2 \to \Re$ is the weight function:

$$w_i(\mathbf{p}) = \frac{\sigma_i(\mathbf{p})}{\sum_{j=1}^{n} \sigma_j(\mathbf{p})}$$

(11)

with

$$\sigma_i(\mathbf{p}) = \frac{1}{\left\| \mathbf{p} - \mathbf{p}_i \right\|^{\mu}}$$

(12)

The exponent μ controls the smoothness of the interpolation.

9 Transition Control

To obtain the transition between the source image and the target image we use linear interpolation of their attributes. If I_S and I_T are the source and target images we generate the sequence of images I_λ, $\lambda \in [0,1]$, such that

$$I_\lambda = (1 - \lambda) \cdot I_S + \lambda \cdot I_T \tag{13}$$

This method is called cross-dissolve.

In the Figure 8 is showed the model adjusted to four different face images using the method. And in Figure 9 is showed the morphing process of the images in Figure 8.

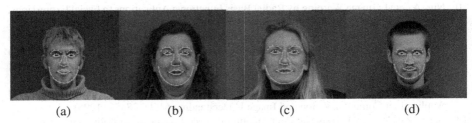

(a)   (b)   (c)   (d)

Fig. 8. Model adjusted to different face images

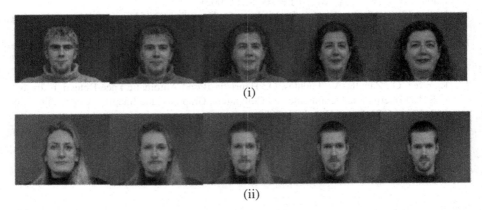

(i)

(ii)

Fig. 9. (i) Morphing between the Figures 8a and 8b,   (ii) Morphing between the Fig. 8c and 8d

References

1. Wolberg, G.: Image Morphing: a Survey. The Visual Computer 14, 360–372 (1998)
2. Litwinowicz, P., Williams, L.: Animating Images with Drawings. In: Proceedings of the SIGGRAPH Annual Conference on Computer Graphics, pp. 409–412 (1994)
3. Wolberg, G.: Digital Image Warping. IEEE Computer Society Press, Los Alamitos (1990)
4. Bichsel: Automatic Interpolation and Recognition of Face Images by Morphing. In: The 2nd International Conference on Automatic Face and Gesture Recognition, pp. 128–135. IEEE Computer Society Press, Los Alamitos (1996)

5. Aaron, W., et al.: Multiresolution Mesh Morphing. In: Proceedings of the SIGGRAPH Annual Conference on Computer Graphics, August 1999, pp. 343–350 (1999)
6. Wolberg, G.: Recent Avances in Image Morphing. In: Computer Graphics International, Pohang, Korea (June 1996)
7. Beier, T., Shawn, N.: Feature-Based Image Metamorphosis. In: Proceedings of the SIGGRAPH AnnualConference on Computer Graphics, July 1992, vol. 26(2), pp. 35–42 (1992)
8. Lee, S., Chwa, K., Shin, S.: Image Metamorphosis Using Snakes and Free-Form Deformations. In: Proceedings of the SIGGRAPH AnnualConference on Computer Graphics (August 1995)
9. Lee, S., et al.: Image Metamorphosis with Scattered Feature Constraints. IEEE Transactions on Visualization and Computer Graphics 2(4) (1996)
10. Nur, A., et al.: Image Warping by Radial Basis Functions: Aplications to Facial Expressions. CVGIP: Graph Models Image Processing 56(2), 161–172 (1994)
11. Lee, S., et al.: Image Morphing Using Deformable Surfaces. In: Proceedings of the Computer Animation Conference, pp. 31–39. IEEE Computer Society, Los Alamitos (1994)
12. Lee, S., et al.: Image Morphing Using Deformation Techniques. J. Visualization Comp. Anim. No. 7(1), 3–23 (1996)
13. Cootes, T., Taylor, C., Cooper, D., Graham, J.: Actives Shape Models- Their Training and Application. Computer Vision and Image Understanding 61(1), 38–59 (1995)
14. Stegmann, M.B.: Analysis and Segmentation of Face Images using Point Annotations and Linear Subspace Techniques, Informatics and Mathematical Modelling, Technical University of Denmark, IMM-REP-2002-xx (2002)
15. Cootes, T., Hill, A., Taylor, C., Haslam, J.: Use of active shape models for locating structures in medical images. Image and Vision Computing 12(6), 355–366 (1994)
16. Cootes, T., Taylor, C., Lanitis, A., Cooper, D., Graham, J.: Building and Using Flexible Models Incorporating Grey-Level Information. In: Proc. 4th. ICCV, pp. 242–246. IEEE Computer Society Press, Los Alamitos (1993)
17. Ramirez, G., Zanella, V., Fuentes, O.: Heuristic-Based Automatic Face Detection. In: 6th IASTED International Conference on Computers, Graphics, and Imaging, CGIM 2003, pp. 267–272. ACTA Press, Honolulu (2003)
18. Ruprecht, D., Muller, H.: Image warping with scattered data interpolation. IEEE Computer Graphics and Applications 15(2), 37–43 (1995)

A Comparison Study of Strategies for Combining Classifiers from Distributed Data Sources

Ireneusz Czarnowski and Piotr Jędrzejowicz

Department of Information Systems, Gdynia Maritime University
Morska 83, 81-225 Gdynia, Poland
irek@am.gdynia.pl, pj@am.gdynia.pl

Abstract. Distributed data mining (DDM) is an important research area. The task of distributed data mining is to extract and integrate knowledge from different sources. Solving such tasks requires a special approach and tools, different from those applied to learning from data located in a single database. One of the approaches suitable for the DDM is to select relevant local patterns from the distributed databases. Such patterns often called prototypes, are subsequently merged to create a compact representation of the distributed data repositories. Next, the global classifier, called combiner, can be learned from such a compact representation. The paper proposes and reviews several strategies for constructing combiner classifiers to be used in solving the DDM tasks. Suggested strategies are evaluated experimentally. The evaluation process is based on several well-known benchmark data sets.

1 Introduction

One of basic approaches to the distributed data mining (DDM) is applying the technique known as the meta-learning. This technique assumes combining the global classifier from independent local classifier, where each one classifier is learned from the separated data set [11].

The other approach to learning from distributed data sets is based on moving all of the data to a central site, merging the data and building a single global model. However, moving all data into a centralized location may not be feasible due to, for example, the restricted communication bandwidth among sites or high expenses involved. Selecting out of the distributed databases only the relevant data can eliminate or reduce the above restrictions and speed up the global knowledge extraction process. Selection of the relevant data is the process often referred to as the data reduction with an objective to find patterns, also called prototypes or references vectors, or regularities within certain attributes (see, for example [10]). Thus, the goal of data reduction approaches is to reduce the number of instances in each of the distributed data subsets, without loss of the extractable information, to enable either pooling the data together and using some mono-database mining tools or effectively applying meta-learning techniques.

M. Kolehmainen et al. (Eds.): ICANNGA 2009, LNCS 5495, pp. 609–618, 2009.
© Springer-Verlag Berlin Heidelberg 2009

The process of learning from the distributed data can be further complicated by differences of structures and features within the considered distributed data sets. In case of the heterogeneous datasets each site can store values of the different sets of features with possibly some common features among the sites. Extracting knowledge from the distributed and heterogeneous data sets is a challenge for the distributed learning. Differences between feature sets can be dealt with through the data reduction process. The aim of the distributed data reduction is to select prototypes at each local site. In such case prototypes are selected by simultaneously reducing data set in the two dimensions through selecting reference instances and removing irrelevant attributes. If thus reduced data sets are still heterogeneous then some special techniques for combining the global classifier would be required.

The main contribution of the paper is investigating several strategies for constructing combiner classifiers intended to be used in solving the DDM tasks, assuming that the data reduction process carried out independently at separate sites has produced a set of the heterogeneous datasets. Several strategies for learning combiner classifiers in such a situation are reviewed and experimentally evaluated. Among the compared strategies are the extended bagging, boosting and majority voting approaches.

The paper is organized as follows. Section 2 contains overview of several selected techniques for combining global classifiers. Section 3 includes general assumptions for the discussed approach, gives a brief description of the classic combiner strategies and proposes how to extend them to be used in case of the DDM with a heterogeneous datasets. Section 4 contains results of the computational experiment carried out with a view to validate the discussed strategies. Finally, the last section contains conclusions and suggestions for future research.

2   Related Work

The main aim of combining classifiers is to improve predictive performance and overcome some disadvantages of the base classification algorithm. Combining multiple classifiers can be also useful when it is impossible or impractical to mine the whole dataset. In such case the dataset can be partitioned and classifiers can be learned from the separated data sets. The final prediction model can be constructed through combining thus obtained local classifiers. Combining classifiers (prediction models) is a fairly general method for distributed learning, where data are physically distributed. In such case different models of data from different sites are combined.

Among widely used approaches developed for aggregating multiple models are statistical methods, like bootstrap, bagging, stacking and boosting [4], [12]. In general, the bootstrap and bagging approaches generate multiple models from different data sets and then average the output of the models. The stacking approach combines the outputs of multiple models that are learned from independent data sets. The boosting strategy creates multiple learners that are based on the weighted training set. The approach assumes that a weak classifier can be learned from each of the considered weighted training sets and then weak classifiers are combined into a final classifier. Others methods for constructing combiner classifiers include voting pools of classifiers [3], committees of classifiers and their combinations [14]. In particular, the committees can consist of models of the same types or different types. Committees may also differ in instance distribution and subsets of attributes used for learning [14]. Boosting and bootstrap techniques can also be used to construct committees of classifiers. The above approaches to aggregating multiple models have been initially introduced with a view of increasing the accuracy of data mining algorithms. Subsequently their applications have been also extended to combining models in distributed data mining (see, for example [9] and [14]).

The related notion of the so called meta-learning provides mechanisms for combining and integrating a number of classifiers learned at separated locations finally producing a meta-classifier. Meta-learning also improves predictive performance by combining different learning models, thus sharing a lot of similarities with the above approaches. However all above approaches to combining classifiers are mostly suitable for mining homogenous distributed data. Several systems, like for example JAM, PADMA, Kensington, Papyrus and BODHI ([2],[6],[7],[11],[13]), have been developed for the distributed data mining from the homogenous and the heterogenous data sets. Although several examples of such systems have been recently described effective methods for learning from heterogenous data sets, including these obtained by data reduction, are still an active field of research (see, for example [7]).

3   Strategies for Combining Classifiers

3.1   Distributed Data Reduction

The classic problem of learning from data is, as a rule, based on the assumption that a dataset D (with N examples) is centralized. Each example consists of a set of attribute values. The set of attributes (or features) A, common to all examples, has the total number of attributes equal to n . However, in the distributed learning a dataset D is distributed among data sources $D_1, \ldots, D_K$, with $N_1, \ldots, N_K$ examples, which are stored in separate sites, where K is a number of sites and where the following properties hold: $\sum_{i=1}^{K} N_i = N$ and where all attributes are presented at each location (i.e. $\forall_{ij} A_i = A_j$, where $i, j = 1, \ldots, K$).

When data reduction is carried out subsets $D_1, \ldots, D_K$ are replaced by the reduced data sets $S_1, \ldots, S_K$ of local patterns. When data reduction is carried out in both dimensions, i.e. by example (or instance) selection and attribute (or feature) selection simultaneously, reduced data sets are likely to become heterogenous. In such case $A_1, \ldots, A_K$ are sets of attributes which values are stored at sites $1, \ldots, K$ respectively. However, it is possible that some attributes can be shared across more then one reduced data set S_i, where $i = 1, \ldots, K$.

Ideally, from the point of view of the DDM, data reduction process should result in obtaining, at a single central site, a set of the compact datasets that retain extractable features and knowledge from each of the distributed data sources. Such set of the compact datasets can be used to induce a global classifier possibly applying the preferred combiner strategy. In the other words, results from the local level mining, i.e. local models, are merged and a global classifier, called also a meta-classifier, can be combined.

3.2 Combiner Strategies Studied

To induce the combiner (or meta-) classifier under the assumption that at the global level a set of the compact datasets is available, one can follow several strategies. In the reported research the following strategies to induce meta-classifiers have been considered: the bagging combiner strategy, the AdaBoost combiner strategy, the simple voting strategy and the hybrid feature voting strategy.

The bagging combiner strategy is based on the bagging approach as described in [12]. The bootstrap sample is obtained by uniformly sampling instances from the given reduced dataset S_i with replacement. From each sample set a classifier is induced. This procedure is carried out independently for each reduced data set $S_1, \ldots, S_K$, and next the global classifier is generated by aggregating classifiers representing each sample set.

Formally, let K be the number of the distributed data sets, T - the number of the bootstrap samples. Then the combiner classifier is generated by aggregating the $T + K$ classifiers. The final classification of a new object x is obtained by the uniform voting scheme on h_{ij}, where h_{ij} is a classifier learned on j trial and where $j = 1, \ldots, T$, $i = 1, \ldots, K$. This means that a new object x is assigned to the most often predicted class. Thus the voting decision for a new object x is computed as:

$$h(x) = \arg\max_{c \in C} \sum_{j=1}^{T} \sum_{i=1}^{K} (h_{ij}(x) = c), \tag{1}$$

where c is the class label, C is a set of decision classes and $h_{ij}(x)$ is the class label predicted for object x by ij-th classifier, respectively. The detailed pseudo-code of the above algorithm is shown below as the Algorithm 1.

Algorithm 1. Bagging combiner strategy

Input: S_i, with N_i examples and correct labels, where $i = 1, \ldots, K$, and integer T specifying the number of bootstrap iterations
Output: h classifier
1. For each separated reduced data set S_i repeat points 2-4
2. For $j = 1, \ldots, T$ repeat points 3-4
3. Take sample S_i' from S_i
4. Generate a classifier h_{ij} using S_i' as the training set
5. Run x on the input h_{ij}
6. The classification of vector x is built by voting scheme based on condition (1)

The AdaBoost combiner strategy is based on the AdaBoost approach [12]. In this case each classifier's vote is a function of its accuracy. Weak classifiers are constructed based on the reduced datasets to produce the set of heterogeneous weak classifiers. The boosted classifier h is combined from all weak classifiers h_{ij}, where h_{ij} is the weak classifier induced on the j-th boosting trial ($j = 1, \ldots, T$) and based on the i-th reduced data set ($i = 1, \ldots, K$). In particular, the combination decision is taken through summing up votes of the weak classifiers h_{ij}. The vote of each h_{ij} is worth $\log(1/\beta_{ij})$ units, where $\beta_{ij} = \varepsilon_{ij}/(1 - \varepsilon_{ij})$ is a correct classification factor calculated from the error ε_{ij} of the weak classifier h_{ij}. Thus the voting decision for a new object x is computed as:

$$h(x) = \arg\max_{c \in C} \sum_{j=1}^{T} \sum_{i=1}^{K} ((\log \frac{1}{\beta_{ij}}) \cdot h_{ij}(x) = c) \tag{2}$$

and a detailed pseudo-code of the above strategy is shown as the Algorithm 2.

Algorithm 2. *AdaBoost combiner strategy*

Input: S_i, with N_i examples and correct labels, where $i = 1, \ldots, K$, and integer T specifying the number of boosting trials
Output: h classifier
1. For each separated reduced data set S_i repeat points 2-9
2. Initialize the distribution $w_1(k) = \frac{1}{N_i}$ ($k = 1, \ldots, N_i$) for each example from S_i
3. For $j = 1, \ldots, T$ repeat points 4-9
4. Select a training data set S'_i from S_i based on the current distribution
5. Generate a classifier h_{ij} using S'_i as the training set
6. Calculate the error of the classifier h_{ij}
7. If $\varepsilon_{ij} > \frac{1}{2}$ then goto 10.
8. Set β_{ij}
9. Update the distribution $w_j(k)$ and normalize $w_j(k)$
10. Run x on the input h_{ij}
11. The classification of vector x is built by voting scheme based on condition (2)

A simple voting strategy proposed by the authors aims at obtaining the global classifier from the global set of prototypes. The global set of prototypes is created by integrating local level solutions. Since local level solutions represent heterogeneous sets of prototypes it has been decided to use the unanimous voting mechanism to support integration process. Hence, only features that were selected by the data reduction algorithms from all distributed sites are retained and the global classifier is formed. Formally, each local solution has been induced from the set of prototypes, which, at each local level, are homogenous and based on the locally selected common subset of attributes $A_i \supseteq A$, where $i = 1, \ldots, K$. Then, at the global level all prototypes from different sites are further reduced in the feature dimension to obtain prototypes with features belonging to the subset of attributes A', where $A' = \bigcup_{i=1}^{K} A_i$. Thus the decision for new object x is computed by a classifier induced from the global set of prototypes $S' = \bigcup_{i=1}^{K} S_i$, where each example is a vector of attribute values with attributes belonging to the set of attributes A'. The detailed pseudo-code of the above strategy is shown as the Algorithm 3.

Algorithm 3. *Voting strategy*

Input: A_i, S_i and where $i = 1, \ldots, K$
Output: h classifier
1. Create a global set of attributes $A' = \bigcup_{i=1}^{K} A_i$
2. Based on A' update each S_i by deleting values of attributes not in A'
3. Create the global set of prototypes $S' = \bigcup_{i=1}^{K} S_i$, where each example is described by the set of attributes A'
4. Generate a classifier h using S' as the training set
5. Run x on the input h

Finally, the hybrid feature voting strategy, proposed by the authors, derives the global classifier from the global set of prototypes. After integrating local level solutions features are selected independently by two feature selection techniques i.e. by the forward and backward sequential selection (FSS and BSS) [8]. In the process of feature evaluation the ten cross validation approach has been used. The learning and the validation sets have been obtained by randomly splitting the global set of prototypes. In each of the 10-C-V runs the feature selection process is carried out. The final set of features is obtained through the unanimous voting mechanism. Thus, only features that have been selected in each out of all of the 10-C-V runs are retained and the global classifier is induced. The detailed pseudo-code of the above strategy is shown as the Algorithm 4.

Algorithm 4. *Hybrid feature voting strategy*

Input: A_i, S_i and where $i = 1, \ldots, K$
Output: h classifier
1. Create a global set of attributes $A' = \bigcup_{i=1}^{K} A_i$
2. Based on A' update each S_i by deleting values of attributes not in A'
3. Create the global set of prototypes $S' = \bigcup_{i=1}^{K} S_i$, where each example is described by the set of attributes A'
4. For ten cross validation iteration repeat points 5-6
5. Divide S' into the learning and the validating set
6. Run wrapper approaches and create appropriate features sets A_i^{FSS} and A_i^{BSS}, where i is a iteration number of cross validation fold
7. Create a final set of attributes $A'' = \bigcup_{i=1}^{10} (A_i^{FSS} \cup A_i^{BSS})$
8. Generate a classifier h using S' as the training set and where prototypes are described by the set of attributes A''
9. Run x on the input h

Further strategies can be obtained by modification or extension of the above described ones.

4   Computational Experiment Results

The aim of the experimental study was to evaluate to what extent the above described combiner strategies could contribute towards increasing classification accuracy of the global classifier induced from the set of prototypes selected at each of the autonomous distributed sites. Classification accuracies of global classifiers obtained using the proposed approaches have been compared with: results obtained by pooling together all instances from the distributed databases (without data reduction) into the centralized database, and results obtained by pooling together all instances selected from distributed databases through the reduction of the example space only.

To obtain the reduced data sets, at each local sites, the agent-based population evolution algorithm has been used. Its detailed description is available in [5]. Generalization accuracy has been used as the performance criterion. The classifier used in all cases has been the C 4.5 algorithm [12].

The experiment involved three datasets - customer (24000 instances, 36 attributes, 2 classes), adult (30162, 14, 2) and waveform (30000, 21, 2). For the first two datasets the best known reported classification accuracies are respectively 75.53% and 84.46% [1], [15]. The reported computational experiment was based on the ten cross validation approach. At first, the available datasets have been randomly divided into the training and test sets in approximately 9/10 and 1/10 proportions. The second step involved the random partition of the previously generated training sets into the training subsets each representing a different dataset placed in the separate location. Next, each of the obtained datasets has been reduced. The reduced subsets have been then used to compute the global classifier using the proposed combiner strategies. The above scheme was repeated ten times, using a different dataset partition as the test set for each trial.

The experiment has been repeated four times for the four different partitions of the training set into a multi-database. The original data set was randomly partitioned into 2, 3, 4 and 5 datasets. The respective experiment results are shown in Table 1 and have been averaged over ten cross validation runs. The results cover seven independent cases. In the first case only reference instance selection at the local level has been carried out, and next, the global classifier has been computed based on the homogenous set of prototypes. In the next cases full data reduction at the local level has been carried out and the global classifier has been computed using one of the selected combiner strategies.

For combiner strategy based on bagging, the parameter T (number of bootstraps) was set to 3 and 5. Choosing these small values of T of was inspired by good results obtained by Quinlan for the C 4.5 classifier with bagging [12]. In case of the AdaBoost - based strategy the value of T (number of boosting rounds) has been arbitrary set to 10.

Generally, it should be noted that data reduction in two dimensions (selection of reference instances and features) assures better results in comparison to data reduction only in one dimension i.e. instance dimension, and that the above conclusion holds true independently from the combiner strategy used at the global

Table 1. Average classification accuracy (%) and its standard deviation obtained by the C 4.5 algorithm

Problem	number of distributed data sources			
	2	3	4	5
A: Selection of reference instances at the local level only				
customer	68.45±0.98	70.40±0.76	74.67±2.12	75.21±0.7
adult	86.20 ±0.67	87.20±0.45	86.81±0.51	87.10±0.32
waveform	75.52±0.72	77.61±0.87	78.32±0.45	80.67±0.7
B: Combiner strategy based on the Bagging approach (number of bootstraps equals 3)				
customer	69.67±1.34	72.13±1.12	77.36±0.97	76.13±2.03
adult	88.34±1.67	88.86±1.3	89.15±2.5	89.17±1.45
waveform	74.12±1.29	75.2±2.68	77.4±2.05	78.34±2.13
C: Combiner strategy based on the Bagging approach (number of bootstraps equals 5)				
customer	69.99±1.56	72.38±1.3	77.65±1.25	77.07±1.98
adult	87.78±2.7	88.34±2.1	89.67±1.98	88.57±3.12
waveform	75.23±2.14	74.67±1.67	77.87±3.23	79.23±2.54
D: Combiner strategy based on the AdaBoost approach (number of boosting repetitions equals 10)				
customer	70.05±1.43	74.78±1.01	76.23±1.23	77.41±0.74
adult	89.64±1.98	89.28±1.21	90.23±1.58	91.78±1.1
waveform	77.12±1.7	78.4±1.3	77.23±2.1	79.51±1.65
E: Combiner strategy based on the simple feature voting				
customer	69.10 ±0.63	73.43 ±0.72	75.35 ±0.53	77.20 ±0.49
adult	88.90 ±0.41	87.45 ±0.31	91.13 ±0.23	91.58 ±0.41
waveform	80.12 ±1.03	82.46 ±0.98	85.04±0.73	83.84±0.64
F: Combiner strategy based on the hybrid feature voting				
customer	71.02±1.2	74.53±0.97	76.85±1.09	78.15±0.81
adult	88.24±0.67	89.47±0.76	91.87±0.55	92.48±0.51
waveform	80.67±0.75	82.15±0.96	83.45±0.43	82.04±1.3
G: The AddBoost algorithm applied after the combiner strategy like in case F				
customer	72.13±0.32	74.84±0.7	77.21±1.01	78.32±0.94
adult	88.67±0.54	90.02±0.57	92.6±0.87	91.34±0.7
waveform	82.45±0.72	83.62±0.9	84.43±0.63	85±0.62

level. It has been also confirmed, that learning classifiers from distributed data and performing data reduction at the local level, produces reasonable to very good results in comparison with the case in which all instances from distributed datasets are pooled together.

For example, pooling all instances from distributed datasets assures classification accuracy of 73.32%(+/-1.42), 82.43%(+/-1.03) and 71.01%(+/-0.8) for customer, adult and vaweform datasets, respectively. On the other hand, the global classifier based on instance selection only assures classification accuracy of

75.21%, 87.1% and 80.67%. These results can still be considerably improved using the hybrid feature voting strategy assuring classification accuracy of 78.15%, 92.48% and 82.04%, respectively for the investigated datasets.

To reinforce our conclusions the experiment results, shown in Table 1, have been used to perform the one-way analysis of variance (ANOVA), where the following null hypothesi was formulated: Choice of the combiner strategy does not influence the classifier performance. One-way ANOVA has been performed for the investigated datasets and it was established, at the 5% significance level, that our hypothesi should be rejected in all cases.

Comparison of the computational experiment results obtained using the proposed strategies shows that the hybrid feature voting strategy at the global level and the AdaBoost - based strategy produce good classifiers and a comparable accuracy of classification. Merging the two as in case G in Table 1, that is applying the AdaBoost algorithm after the hybrid feature voting strategy should further improve the classification accuracy.

5   Conclusions

This paper investigates and compares four basic strategies and their combinations used for constructing the combiner classifier at the global level of the distributed learning. The discussed strategies have been formally defined and experimentally evaluated using several benchmark datasets. Computation experiment results confirmed that the quality of results depends on the strategy used for constructing the combiner classifier. The scope of the reported experiment does not allow to conclude that some of the investigated strategies would always outperform the others. However hybrid feature voting at the global level, possibly combined with the AdaBoost algorithm should be seriously considered as a potential winner strategy for solving the distributed data mining problems.

Future work will focus on evaluating other combiner classifier strategies in terms of classification accuracy and computation costs and on carrying more extensive experiments with a view to obtain statistically validated conclusions. Moreover, the experiment was carried using the C 4.5 learning algorithm as the classification tool. The future work is needed to extend computational experiments to other learning algorithms.

Acknowledgements. This research has been supported by the Polish Ministry of Science and Higher Education with grant for years 2008-2010.

References

1. Asuncion, A., Newman, D.J.: UCI Machine Learning Repository. University of California, School of Information and Computer Science, Irvine (2007), http://www.ics.uci.edu/~mlearn/MLRepository.html
2. Bailey, S., Grossman, R., Sivakumar, H., Turinsky, A.: Papyrus: a system for data mining over local and wide area clusters and super-clusters. In: Proc. of ACM/IEEE SC Conference (SC 1999), p. 63 (1999)

3. Battiti, R., Coalla, A.M.: Democracy in neural nets: Voting schemes for classification. Neural Network 7(4), 691–707 (1994)
4. Bauer, E., Kohavi, R.: An empirical comparison of voting classification algorithhms: Bagging, boosting, and variants. Machine Learning 36(1-2), 691–707 (1994)
5. Czarnowski, I., Jędrzejowicz, P.: Data reduction algorithm for machine learning and data mining. In: Nguyen, N.T., Borzemski, L., Grzech, A., Ali, M. (eds.) IEA/AIE 2008. LNCS (LNAI), vol. 5027, pp. 276–285. Springer, Heidelberg (2008)
6. Guo, Y., Rueger, S.M., Sutiwaraphun, J., Forbes-Millott, J.: Meta-leraning for parallel data mining. In: Proc. of the Seventh Parallel Computing Workshop, pp. 1–2 (1997)
7. Kargupta, H., Park, B.-H., Hershberger, D., Johnson, E.: Collective data mining: A new perspective toward distributed data analysis. In: Kargupta, H., Chan, P. (eds.) The Advances in Distributed Data Mining. AAAI/MIT Press (1999) (accepted)
8. Kohavi, R., John, G.H.: Wrappers for feature subset selection. Artificial Intelligence 97(1-2), 273–324 (1997)
9. Lazarevic, A., Obradovic, Z.: The distributed boosting algorithm. In: Proc. ACM-SIG KDD Internetional Conference on Knowledge Discovery and Data Mining, San Francisco, pp. 311–316 (2001)
10. Liu, H., Lu, H., Yao, J.: Identifying relevant databases for multidatabase mining. In: Proceedings of Pacific-Asia Conference on Knowledge Discovery and Data Mining, pp. 210–221 (1998)
11. Prodromidis, A., Chan, P.K., Stolfo, S.J.: Meta-learning in distributed data mining systems: issues and approaches. In: Kargupta, H., Chan, P. (eds.) Advances in Distributed and Parallel Knowledge Discovery, ch. 3. AAAI/MIT Press (2000)
12. Quinlan, J.R.: Bagging, boosting and C 4.5. In: Proceedings of the 13th National Conference on Artificial Intelligence, pp. 725–730 (1996)
13. Stolfo, S., Prodromidis, A.L., Tselepis, S., Lee, W., Fan, D.W.: JAM: Java Agents for Meta-Learning over Distributed Databases. In: Proceedings of the 3rd International Conference on Knowledge Discovery and Data Mining, Newport Beach, CA, pp. 74–81. AAAI Press, Menlo Park (1997)
14. Ting, K.M., Low, B.T.: Model combination in the multiple-data-base scenario. In: van Someren, M., Widmer, G. (eds.) ECML 1997. LNCS (LNAI), vol. 1224, pp. 250–265. Springer, Heidelberg (1997)
15. The European Network of Excellence on Intelligence Technologies for Smart Adaptive Systems (EUNITE) - EUNITE World Competition in domain of Intelligent Technologies, http://neuron.tuke.sk/competition2 (accesed September 1, 2002)
16. Zhang, X.-F., Lam, C.-M., William, K.C.: Mining local data sources for learning global cluster model via local model exchange. IEEE Intelligence Informatics Bulletin 4(2) (2004)

Visualizing Time Series State Changes with Prototype Based Clustering

Markus Pylvänen, Sami Äyrämö, and Tommi Kärkkäinen

University of Jyväskylä, Department of Mathematical Information Technology
P.O.Box 35 (Agora) FIN-40014 Jyväskylä, Finland
markus.t.pylvanen@jyu.fi, sami.ayramo@jyu.fi, tka@mit.jyu.fi

Abstract. Modern process and condition monitoring systems produce a huge amount of data which is hard to analyze manually. Previous analyzing techniques disregard time information and concentrate only for the indentification of normal and abnormal operational states. We present a new method for visualizing operational states and overall order of the transitions between them. This method is implemented to a visualization tool which helps the user to see the overall development of operational states allowing to find causes for abnormal behaviour. In the end visualization tool is tested in practice with real time series data collected from gear unit.

1 Introduction

Industrial processes and systems of a condition management produce nowadays a huge amount of time series data. The data are often monitored in industrial applications by defining separate limits for attribute values. This type of monitoring is easy to implement and understand, but it is unable to show if more than one attributes behave abnormally without breaking their limits. Altogether, a process state is characterized and controlled individually, without overall utilization of the measurements.

Clustering has been used before for finding states of industrial process and abnormal behaviour from multivariate data [18][17]. However, this method loses time information between the states. We can examine in which states observations represent, but we can not examine in which order they occurred. This information can be meaningful if we are interested in causes that expose abnormal behaviour. This can only be seen by examining states that have occurred before the abnormal states.

This paper presents a new concept for visualizing time series data with cluster prototypes where information about the transitions between states is added. Implementation of this method is presented and it is tested with time series data collected from a gear unit. For this case, we present shortly the whole knowledge mining process [2] and the role of techniques presented here on that.

M. Kolehmainen et al. (Eds.): ICANNGA 2009, LNCS 5495, pp. 619–628, 2009.
© Springer-Verlag Berlin Heidelberg 2009

2 Related Work

Methods for identifying operational states from industrial data have been presented in several publications. Wang [20] presents adaptive resonance theory (ART) which is an unsupervised learning algorithm and the Bayesian automatic classification system (AutoClass) for indentifying operational states. Also self-orgazing maps are used for identifying states in processes like Heikkinen et al. [9] have done. Alhoniemi et al. [1] used SOM for monitoring and modelling industrial processes.

Different clustering algorithms produce different results and it can be difficult to compare them [11]. Visualization of clusters offers a user-friendly method for comparing and presenting their dissimilarities. Data and cluster visualization are overlapping approaches to the analysis of large data sets because some data visualization techniques might present clusters at the same time like Self-organizing maps [14] and parallel coordinates [12]. A problem in self-orgazing maps is that the clustering algorithm is embedded to the method and that prevents selecting the best clustering algorithm for each data set.

Huang and Lin [11] have developed a visualization technique for validating clusters. They use Fastmap for visualizing high dimensional data in 2D and a k-prototypes algorithm for clustering. Also Hoffman and Grinstein [10] have presented many visualization techniques in their survey, but the problem is that they are not meant to visualize time series data, i.e. temporal information is lost when whole data is concerned.

3 The Approach

This method is originally developed for visualizing time series data collected from the sensors attached to a gear unit. The primary use of collected data is to detect faults before they cause a serious damage to the gear unit. In this case goal is to form states that present either normal or abnormal behaviour of a gear unit. This knowledge can be used afterwards for finding patterns which may precede to a malfunction and this way predict faults even sooner. For example, the system may run smoothly if two normal states take turns occasionally, but when these states take turns rapidly it could expose faults. This kind of behaviour can not be seen by examining only values of single attributes.

The used data has to be in a chronological order and complete. Observations that have missing values can be removed or they can be estimated based on other values of same attribute. In future the clustering algorithm can be replaced with a version which allows missing values.

The system where the data is collected does not have delays between depended attribute values. If two attributes are depended to each other and there is time delay between their changes blur clusters so that it is hard to find basic features for each cluster. This kind of problem might come out in industrial processes like the waste water treatment process that Sànchez et al. [17] have studied.

3.1 Clustering

A base element of the new visualization method is the gear unit state identification with data clustering. For this prototype construction, the K-means method is chosen. The K-means clustering method is an iterative process that divides a given data set into K disjoint groups [16], [7]. It is one of the most widely used clustering principles, and the best-known partitioning-based clustering method that utilize prototypes for cluster presentation. Due to its straightforward implementation, gaussian assumptions, and computational efficiency, K-means is popular principle for many problems. It also has smaller memory requirements than, for instance, hierarchical methods. The K-means algorithm converges to a partition for which the cluster prototypes minimizes the clustering error with respect to the sum of the within-cluster squared errors.

$$\min \mathcal{J}(\mathbf{c}, \{\mathbf{m}_k\}_{k=1}^K)_{\mathbf{c} \in \mathbb{N}^n, \mathbf{m}_k \in \mathbb{R}^p} = \sum_{i=1}^n \|\mathbf{x}_i - \mathbf{m}_{(\mathbf{c})_i}\|_2^2 \tag{1}$$

$$\text{subject to } (\mathbf{c})_i \in \{1, \ldots, K\} \text{ for all } i = 1, \ldots, n,$$

where $\mathbf{c}$ is a code vector, which represents the cluster assignments of the objects, and $\mathbf{m}_{(\mathbf{c})_i}$ is the mean of the cluster, where the data point $\mathbf{x}_i$ is assigned to.

A general iterative relocation algorithm for solving the problem of K-means is given by the following algorithm:

Input: The number of clusters K, $n \times p$ data set $\mathbf{X}$.
Output: Allocation of each data point to one of K clusters.
Step 1. (*Initialization*) Compute the initial K cluster centers.
Step 2. (*Recomputation*) (Re)compute memberships of the data points to the current cluster centers.
Step 3. (*Update*) Update the cluster centers for the assignments of the data points.
Step 4. (*Stopping rule*) Repeat from Step 2 until no data point changes cluster.

One should note that the K-means is very sensitive to the initial partition and towards outliers. Since this work presents initial experiments with a new clustering-based visualization method, K-means is a sufficient method.

Another typical option for the clustering step in a state identification problem are provided by the hierarchical clustering methods [6]. The problem with the hierarchical clustering is the $O(n^2)$ costs due to the use of the $n \times n$ distance matrix. Another problem are the missing data values since the gear unit data can contain them. The similarities computed in different sub-spaces are not easy to compare. Sometimes the comparison may be impossible. For instance, let us consider the distance computation for the following three 3-dimensional data vectors

$$\mathbf{x}_1 = (1 \quad 0 \quad \text{NaN})^T, \mathbf{x}_2 = (\text{NaN} \quad 1 \quad 1)^T, \mathbf{x}_3 = (1 \quad \text{NaN} \quad 1)^T.$$

Straightforward comparison of the between-object distances is difficult, since all the points lie in the different sub-spaces. Hence, the use of prototype-based methods enables us to represent the recognized states with explicit prototype vectors and, on the other hand, provides more straightforward solutions for missing data treatment.

3.2 Dimensionality Reduction

"Curse of dimensionality" is often a problem in data mining applications. Real life data have often many variables or attributes which makes visualization difficult. A human being can realize one, two or three dimensional space easily, but when there are more dimensions than this visualization is not straightforward. This is one reason why dimensionality reduction techniques are developed.

The easiest way to reduce dimensions is just to reduce variables. We can select only the most interesting variables based on domain-knowledge and visualize them. This technique inevitably losts some information and that is why we have to use advanced ones.

Principal component analysis (PCA) and multi dimensional scaling (MDS) are used often for dimension reduction [8]. Also linear discriminant analysis (LDA) [4] can be used for this. They all present original n-dimensional data where are n-axes with fewer axes so that differences in the original data are showing as much as possible.

Principal component analysis aims at finding such linear combinations of a data set that preserve the maximum amount of information assuming that the information is measured by variance. Hence, it is natural to use it for explorative data mining. A reduced dimension is obtained when the original high dimensional data is projected from the original $\Re^p$ space into the lower dimensional $\Re^q$ space ($p \gg q$) that is determined by the principal components.

4   Method Description

The overall process of this method is presented in picture 1.

At first time series data is clustered with some prototype based clustering algorithm. Clustering is used for dividing observations into classes so that observations in the same cluster are similar to each other and observations in different clusters are dissimilar [13]. All the used algorithms in this prototype version are very general and in the future these can be replaced with the application specific algorithms.

Protoype based methods produce mainly spherical clusters. This is a big difference when we compare them for example to density based clustering algorithms like DBSCAN which can produce arbitrary cluster shapes [5]. Arbitrary shape clusters are not the best presentation for operational states because obervations that belong to the same cluster can be very dissimilar. In spherical clusters most of the observations are similar to each other.

When we use prototype based clustering, each cluster has a prototype which presents for example mean or median observation of all other observations in

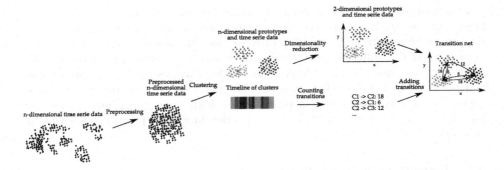

Fig. 1. Steps for visualizing time series data with prototype based clustering

one cluster. In this way we get an explicit presentation for each cluster in n-dimensional space. Also prototypes can be examined for finding the most common values of each attribute in every cluster.

In this case we selected the K-means algorithm for clustering because it is simple and scalable for data where are numerous attributes and many observations. K-means has also been used for same kind of data before [17]. However the K-means algorithm needs K-value as a parameter which depends always on the clustered data. This means that we can not know optimal K in advance and that is why we have to run the algorithm with different values and try to find best result. Another K-means problem is that it converges to a local minimum and we can not be sure if that is the best clustering result for the selected K value. We reduced the effect of this problem by running the algorithm several times with random initial partitions.

In a stable system it is possible that there are not any natural clusters and then visualizing time series can not offer any meaningful results. In systems which are unstable due changing environment or changes in system itself visualizing of time series can offer reasonable information about the process.

4.1   Cluster Prototypes

N-dimensional data is first clustered with K-means. The idea is that found clusters are representing specific states of monitored system and each prototype of cluster is a presentation of observations which belong to specific cluster. In this way we get a simple computational presentation for each cluster which can be used later in this work.

Cluster prototypes are n-dimensional like the clustered data itself. If n is greater than three visualizing prototypes is not simple. There is need to use some kind of dimension reduction technique and in this case we selected often used PCA. PCA can be used by itself for detecting anomalies like Liu et al. [15] have done. This brings additional information to our method but the main the point of PCA is in this case dimensionality reduction.

Principal components were formed from the original n-dimensional data. With these components the prototypes were represented in the same plane with original observations. From the position of these two dimensional prototypes can be seen how similar different prototypes are. If two prototypes are close to each other it means that they may have more similar features than those prototypes which are far from each other. It is also possible that the two projected prototypes are close to each other in plane but in high dimensional space they are not. This can not be seen after PCA, but different cluster prototypes can point this out.

4.2   Transitions between States

When examining only the projected states the time and order information is totally lost. This is why there has to be added some kind of information for showing the order where states occur.

Clustering gives a label to every observation. When these labels are ordered with respect to original observations order they form a timeline which present the whole measuring period. From this timeline can be count transitions between states.

Picture 2 shows a presentation of timeline where white colour presents observations with missing data and other four colours present occurrence of clusters.

Fig. 2. Example of cluster occurence in timeline

Counted transitions can be present with a line between prototypes of clusters in a two dimensional picture. In this way we get a transition net which shows roughly dissimilarities of clusters and their order of occurrence.

4.3   Implementation

The presented method was implemented with Matlab and Java programming language. Matlab was used for clustering, PCA and calculating statistical values and Java program was used for creating user interface where results were visualized. We used both open source and self-developed components for creating user the interface. The graphs were created with JFreeChart library and time line component was designed and implemented by self. The user interface for visualizing is shown in picture 3.

On the left there are cluster labels, colour of clusters, observations in clusters and their percentages. The user can change cluster names, their colour and give short descriptions.

The transition net is shown in the middle. Observations of each cluster are shown with different colour and prototypes of clusters are shown with a cross.

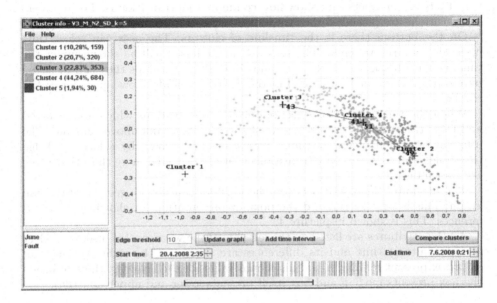

Fig. 3. Developed user interface for visualizing cluster prototypes and their transitions

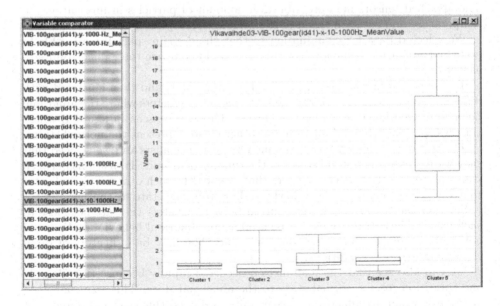

Fig. 4. Window used for comparing differences between clusters

The transitions between cluster prototypes are shown with gray lines. Over the lines there are numbers that show how common each transition is. The numbers of transitions are placed so that they show arrivals to the nearest prototype.

On the bottom there is the timeline of clusters. The user can select time intervals from the timeline. The selected time interval will be shown under the timeline with black line. If the user finds interesting intervals they can be saved by clicking "Add time interval"-button. Saved intervals show up in lower left corner.

When the time interval is selected only the observations which are occurred and their transitions in that time are showed in the graph above timeline. The user can also define how common transitions are shown by selecting "Edge threshold" value. In this way transitions that are occurred more than that value are shown only.

The absolute differences between the clusters can not be seen in transition graph. This is why we selected the boxplot presentation for showing those differences. This is shown in picture 4.

All the attributes are listed on the left side of the window. The user can select one attribute at time and its differences are shown on the right. Whiskers of each box present minimum and maximum values of each cluster, the line inside the box presents median and box itself presents lower and upper quartiles.

5   Case: Visualizing Data from Gear Unit

The presented method was tested with data collected from sensors of gear unit. The attached sensors measure vibration, amount of particles in lubrication oil, temperature and rotation speed of the gear unit.

We selected data for examination so that the domain specialist was able to point out abnormal behaviour. In this way could compare the gained results to earlier ones.

The sensor data was first normalized to range $[0, 1]$ and then clustered with different values of K. $K = 5$ was the first amount of clusters where the abnormal behaviour was shown in a separate cluster. This can be seen in figure 4, where cluster 5 presents the malfunction and other clusters present the normal states.

The rest of the clusters were examined by gear specialists and we noticed that they were presenting normal states of the running gear unit. Transitions between states occur mostly between clusters that are near to each other. This indicates that the gear changes its state slowly, sliding from one state to another.

Some of the transitions have occurred between clusters that are not close to each other. These transitions can be seen as an abnormal behaviour. We could not find explanation for these transitions but they may be indicators of rapid changes in external conditions.

We can also see that the malfunction occurred rapidly because there is only one transition from a normal state to the abnormal state. This was confirmed afterwards when manufacturer of gear units examined this gear and found that cause of the malfunction could not be seen with the attached sensors. That is why it was impossible to find causes for malfunction with presented method.

6   Conclusions

We presented a new method for visualizing multivariate time series data with the transition network. The method can be used for visualizing processes and systems that produce time dependent data from their behaviour. This data can be used for forming operational states of a process or system. When we add transitions between these states we can examine also their causalities. This can offer valuable information when we try to find causes for abnormal behaviour and after this prepare for it or even prevent it.

The method works fine if there are under ten clusters. With a large amount of clusters transition network is hard to examine because the amount of transitions also increases. This problem can be avoided if user could select only small part of transitions like the transitions between selected clusters.

In the presented method the user has to select the right number of clusters. We used iterative method for finding the correct K value but this requires too much manual work. In future the right value for K can be found by using advanced heuristics.

In the presented implementation PCA was made only once and the result was shown to the user. If there are many different anomalies in data it is possible that some of them are not shown separately because some anomalies have stronger influence on PCA than the others. One way to avoid this is that PCA is done after the user has selected a certain time interval. Also the user could select certain clusters and after that PCA is done only to observations in these clusters. In this way it would be possible to see dissimilarities between clusters.

All the algorithms behind this method can be replaced with advanced ones. For example randon initial prototypes can be replaced with robust initial prototypes [3] and clustering algorithm can be replaced with the one which accepts missing values of variables. Robust methods can also be used with PCA [19]. By using robust algorithms we can reduce the effects of outliers in data.

In this case we could not find causes for the abnormal behaviour which was our original objective, but we found a promising method which can be used for this purpose in the future.

Acknowledgements

This work supported by the Agora Center research unit at the University of Jyväskylä.

References

1. Alhoniemi, E., Hollmén, J., Simula, O., Vesanto, J.: Process monitoring and modeling using the self-organizing map. Integr. Comput.-Aided Eng. 6(1), 3–14 (1999)
2. Äyrämö, S.: Knowledge Mining using Robust Clustering, Ph.D thesis (monograph). University of Jyväskylä, Jyväskylä (2006)

3. Äyrämö, S., Kärkkäinen, T., Majava, K.: Robust refinement of initial prototypes for partitioning-based clustering algorithms. In: Recent Advances in Stochastic Modeling and Data Analysis, pp. 473–482. World Scientific, Singapore (2007)
4. Dillon, W.R., Goldstein, M.: Multivariate analysis: methods and applications. Wiley series in probability and mathematical statistics, Applied probability and statistics. Wiley, New York (1984)
5. Ester, M., Kriegel, H.-P., Sander, J., Xu, X.: A density-based algorithm for discovering clusters in large spatial databases with noise. In: Simoudis, E., Han, J., Fayyad, U. (eds.) Second International Conference on Knowledge Discovery and Data Mining, Portland, Oregon, pp. 226–231. AAAI Press, Menlo Park (1996)
6. Everitt, B.S., Landau, S., Leese, M.: Cluster analysis. Arnolds, a member of the Hodder Headline Group (2001)
7. Forgy, E.W.: Cluster analysis of multivariate data: Efficiency versus interpretability of classifications. Biometrics 21(3), 768–769 (1965) (abstract)
8. Hand, D.J., Smyth, P., Mannila, H.: Principles of data mining. MIT Press, Cambridge (2001)
9. Heikkinen, M., Kettunen, A., Niemitalo, E., Kuivalainen, R., Hiltunen, Y.: Sombased method for process state monitoring and optimization in fluidized bed energy plant. In: Duch, W., Kacprzyk, J., Oja, E., Zadrożny, S. (eds.) ICANN 2005. LNCS, vol. 3696, pp. 409–414. Springer, Heidelberg (2005)
10. Hoffman, P.E., Grinstein, G.G.: A survey of visualizations for high-dimensional data mining, pp. 47–82 (2002)
11. Huang, Z., Lin, T.: A visual method of cluster validation with fastmap. In: PADKK 2000: Proceedings of the 4th Pacific-Asia Conference on Knowledge Discovery and Data Mining, Current Issues and New Applications, London, UK, pp. 153–164. Springer, Heidelberg (2000)
12. Inselberg, A.: The plane with parallel coordinates. The Visual Computer V1(4), 69–91 (1985)
13. Jain, A.K., Dubes, R.C.: Algorithms for clustering data. Prentice-Hall, Inc., Upper Saddle River (1988)
14. Kohonen, T.: The self-organizing map. Neurocomputing 21(1–3), 1–6 (1998)
15. Liu, J., Lim, K.-W., Rajagopalan, S., Doan, X.-T.: On-line process monitoring and fault isolation using pca. In: Intelligent Control, 2005. Proceedings of the 2005 IEEE International Symposium on, Mediterrean Conference on Control and Automation, June 2005, pp. 658–661 (2005)
16. Macqueen, J.B.: Some methods for classification and analysis of multivariate observations. In: Procedings of the Fifth Berkeley Symposium on Math, Statistics, and Probability, vol. 1, pp. 281–297. University of California Press, Berkeley (1967)
17. Sànchez, M., Cortés, U., Béjar, J., De Grácia, J., Lafuente, J., Poch, M.: Concept formation in wwtp by means of classification techniques: Acompared study. Applied Intelligence 7(2), 147–165 (1997)
18. Singhal, A., Seborg, D.E.: Clustering multivariate time-series data. Journal of Chemometrics 19(8), 427–438 (2005)
19. Visuri, S., Koivunen, V., Oja, H.: Sign and rank covariance matrices. Journal of Statistical Planning and Inference 91(2), 557–575 (2000)
20. Wang, X.Z., McGreavy, C.: Data Mining and Knowledge Discovery for Process Monitoring and Control. Springer, London (1999)

Author Index